FIFTH EDITION

SOCIOLOGY

THE ESSENTIALS

Margaret L. Andersen

University of Delaware

Howard F. Taylor

Princeton University

Australia • Brazil • Canada • Mexico • Singapore • Spain • United Kingdom • United States

Sociology, The Essentials, Fifth Edition
Margeret L. Andersen and Howard F. Taylor

Acquisitions Editor: *Chris Caldeira*
Development Editor: *Kate Barnes*
Assistant Editor: *Christina Beliso*
Editorial Assistant: *Erin Parkins*
Marketing Manager: *Michelle Williams*
Marketing Assistant: *Ileana Shevlin*
Marketing Communications Manager: *Linda Yip*
Project Manager, Editorial Production: *Cheri Palmer*
Creative Director: *Rob Hugel*
Art Director: *Caryl Gorska*
Print Buyer: *Linda Hsu*
Permissions Editor: *Mardell Schultz*
Image Account Manager: *Don Schlotman*
Production Service: *Dusty Friedman, The Book Company*
Text Designer: *Yvo Riezebos*
Art Editor: *Dusty Friedman, The Book Company*
Photo Researcher: *Sarah Evertson*
Copy Editor: *Jane Loftus*
Illustrator: *Impact Publications, Graphic World, Inc.*
Cover Designer: *Riezebos Holzbaur Design Group, LLC*
Cover Image: © *Grant V. Faint / Getty Images*
Compositor: *Graphic World, Inc.*

To Richard and Pat, with love

Printed in the United States of America
1 2 3 4 5 6 7 11 10 09 08 07

Library of Congress Control Number: 2007938667

ISBN-13: 978-0-495-39093-0
ISBN-10: 0-495-39093-3

Thomson Higher Education
10 Davis Drive
Belmont, CA 94002-3098
USA

For more information about our products, contact us at:
Thomson Learning Academic Resource Center
1-800-423-0563
For permission to use material from this text or product, submit a request online at **http://www.thomsonrights.com.**
Any additional questions about permissions can be submitted by email to **thomsonrights@thomson.com.**

BRIEF CONTENTS

Deviance and Crime 144

PART THREE *Social Inequalities*

Social Class and Social Stratification 176

Chapter 8

Global Stratification 208

Education and Health Care 342

Chapter 14

Politics and the Economy 370

Chapter 15

Population, Urbanization, and the Environment 400

Chapter 16

Social Change and Social Movements 424

BOXES

Doing Sociological Research

Understanding Diversity

A Sociological Eye on the Media

MAPS

Mapping America's Diversity

Viewing Society in Global Perspective

Preface

Studying sociology opens new ways of looking at the world. It is a perspective that is grounded in careful observation of social facts, as well as analytical interpretations of how society operates. For students and faculty alike, studying sociology can be exciting, interesting, and downright fun. We have tried to capture the excitement of the sociological perspective in this book, and to do so in a way that is engaging and accessible to undergraduate readers, while also preserving the integrity of sociological research and theory. Our book introduces students to the basic concepts and theories of sociology.

With each new edition of the book, we think about the new generation of students that the book reaches. Although the basic perspective of sociology persists over time, the issues that new generations of learners face change—as do the ways that different students learn. One way you will see this reflected in the Fifth Edition is the new format that is intended to appeal to a more visually oriented student body. While we always retain a strong focus on core sociological concepts and ideas, we try to present our book in such a way that students will continue to tell us that they learn from it. As authors, one of our most rewarding moments is when a student tells us something about an insight they gained from *Sociology: The Essentials.*

Each time we revise the book, we are also surprised by how much new research in sociology has emerged, even in a short period of time. And, of course, events in the world continue to shape the examples and explanations that we provide for our readers. We do so while retaining the strong focus on sociological research and theory that this book is known for. Our experience in teaching introductory students shows us that students can appreciate the revelations of sociological research and theory if presented in a way that engages them and connects to their lives. We have kept this in mind throughout this revision and have focused on material that students can understand and apply to their own social worlds.

An Approach That Reflects the Significance of Sociology

Diversity

The study of diversity is central to our book. Unlike other introductory texts that add diversity to a preexisting approach to sociology, we see diversity as part of the texture of society. Diversity is central to how society is organized, how inequality shapes the experiences of different groups, and how diversity is shaping (and is shaped by) contemporary social changes. Our attention to diversity pervades the body of the book as well as some of the book's special features (see especially the box feature "Understanding Diversity").

We define diversity to include the differences in experience created by social factors, including race, ethnicity, class, gender, age, religion, sexual orientation, and region of residence. We do not think of diversity as just the study of victims, although systems of disadvantage are clearly part of society. We stress the positive aspects of a diverse society, as well as its problems. We are pleased that our thorough integration of diversity has led reviewers to comment that our book provides the most comprehensive coverage of diversity of any book on the market.

Current Theory and Research

According to readers and reviewers of this book, our presentation of current theory and research is one of its strongest features. We use the most current research throughout the book to show students the value of a sociological education. The box feature **Doing Sociological Research** showcases different research studies. We include this so that students will understand not just the concepts and procedures of research methods, but also how sociologists do their work and how the questions they ask are linked to the methods of inquiry that they use. Each box begins by identifying the question asked by the researcher. Then it describes the method of research, presents the findings and conclusions, and briefly discusses the implications. These boxes also show students the diverse ways that sociologists conduct research and thus feature the rich and varied content of the discipline.

We also help students understand how different theoretical frameworks in sociology interpret different topics by including tables in every chapter that concisely compare different theoretical viewpoints, showing how each illuminates certain questions and principles. We think this comparison of theories helps students understand an important point: Starting from a different set of assumptions can change how you interpret different social phenomena.

Critical Thinking and Debunking

We use the theme of *debunking* in the manner first developed by Peter Berger (1963)—to look behind the facades of everyday life, challenging the ready-made assumptions that permeate commonsense thinking. Debunking is a way for students to develop their critical thinking, and we use the debunking theme to help students understand how society is constructed and sustained. This theme is highlighted in the **Debunking Society's Myths** feature found throughout every chapter.

In this edition, we have also added a feature to help students see the relevance of sociology in their everyday lives. The box feature **See for Yourself** allows students to apply a sociological concept to an observation from their own lives, thus helping them develop their critical abilities and understand the importance of the sociological perspective.

Critical thinking is a term widely used but often vaguely defined. We use it to define the process by which students learn to apply sociological concepts to observable events in society. Throughout the book, we ask students to use sociological concepts to analyze and interpret the world they inhabit. This is reflected in the **Thinking Sociologically** feature that is also present in every chapter.

Because contemporary students are so strongly influenced by the media, we also encourage their critical thinking through the box feature **A Sociological Eye on the Media.** These boxes examine sociological research that challenges some of the ideas and images portrayed in the media. This not only improves students' critical thinking skills, but also shows them how research can debunk these ideas and images.

Social Change

The sociological perspective helps students see society as characterized both by constant change and social stability. We use the change theme in every chapter by examining social change at the conclusion of each chapter. Current events, such as the impact of Hurricane Katrina on different groups of people or the tremendous impact of technological developments, are an important part of this book. For example, the new **Seeing Society in Everyday Life** photo essay in Chapter 4 discusses the impact of new technologies on social interaction—a topic that will be of special interest to students very much affected by these changes. Also, we examine the aftermath of Hurricane Katrina in several places, including in the analysis of race, gender, and poverty.

Global Perspective

Diversity includes the increasingly global character of society. The United States is increasingly being changed by globalization. We use a global perspective to examine how global changes are affecting all parts of life within the United States, as well as other parts of the world. This means more than including cross-cultural examples. It means, for example, examining phenomena such as migration and immigration, the formation of world cities, the increasing cultural diversity found within the United States, and the impact of an international division of labor on work within the United States. This global perspective is found in the research and examples cited throughout the book, as well as in various chapters that directly focus on the influence of globalization on particular topics, such as work, culture, and crime. The map program **Viewing Society in Global Perspective** also brings a global perspective to the subject matter.

New to the Fifth Edition

We have made many changes to the fifth edition to make it stronger and more effective. Together, these changes should make the fifth edition easier for instructors to teach and more accessible and interesting for students.

Sociology: The Essentials is organized into five major parts: "Introducing the Sociological Imagination" (Chapter 1); "Society, Individuals, and Social Structure" (Chapters 2 through 6); "Social Inequalities" (Chapters 7 through 11); "Social Institutions" (Chapters 12 through 14); and "Social Change" (Chapters 15 and 16).

Part I, Introducing the Sociological Imagination, introduces students to the unique perspective of sociology, differentiating it from other ways of studying society, particularly the individualistic framework students tend to assume. Within this section, **Chapter 1, Sociological Perspectives and Sociological Research,** introduces students to the sociological perspective and the basics of sociological research methods. This chapter briefly reviews the development of sociology as a discipline, with a focus on the classical frameworks of sociological theory, as well as contemporary theories, such as feminist theory and postmodernism. This edition streamlines the discussion of individual theorists, while keeping the material on the development of social theory. We have also added material in the section on research to help students learn how to judge the reliability of material on the Internet.

In **Part II, Society, Individuals, and Social Structure,** students learn some of the core concepts of sociology. It begins with the study of culture in **Chapter 2, Culture,** and has a strong focus on popular culture and the media given students' high interest in these topics. New topics, such as youth countercultures and language changes coming as the result of text messaging, keep this chapter relevant to contemporary generations. We include here a discussion of the growing significance of blogs as a form of cultural exchange. And we illustrate the concept of culture shock by examining the experiences of residents displaced by Hurricane Katrina and relocated in very different parts of the country. **Chapter 3, Socialization and the Life Course,** includes new material on socialization theory and research, such as research on children's understanding of race. The chapter also includes discussion of aging and the life course. **Chapter 4, Social Interaction and Social Structure,** has been completely reorganized to emphasize how changes in the macrostructure of society influence the microlevel of social interaction. We do this by opening with a focus on technological changes that are now part of students' everyday lives and making the connection between changes at the societal level in the everyday realities of people's lives. The chapter thus includes new material on Internet dating, music downloading, and new research on young people's use of the Internet.

In **Chapter 5, Groups and Organizations,** we study social groups and formal organizations, using sociology to understand the complex processes of group influence, organizational dynamics, and the bureaucratization of society. The chapter includes updates of research on juries and on social networks, as well as a new report on the governmental failures in responding to the 9/11 terrorist attacks on the World Trade Center in 2001. **Chapter 6, Deviance and Crime,** includes a look at the sociological theories of deviance with attention to labeling theory; corporate crime and deviance; and the effects of race, class, and gender on arrest rates. The core material is illustrated with contemporary events, such as updates on corporate and "white collar" crime, a new box and analysis of the rampage killings at Virginia Tech University in the spring of 2007, and recent research on what has become *mass racialized incarceration* of Blacks and Hispanics in U.S. prisons.

In **Part III, Social Inequalities,** each chapter explores a particular dimension of stratification in society. Beginning with the significance of class, **Chapter 7, Social Class and Social Stratification,** provides an overview of basic concepts central to the study of social stratification, as well as current sociological perspectives on class inequality, poverty, and welfare. There is new material on status symbols, a photo essay on class inequality, current research on welfare and teen motherhood, as well as attention to the growing debt among the middle class. Updated data on poverty and income are included throughout.

Chapter 8, Global Stratification, follows, with a particular emphasis on understanding the significance of global stratification, the inequality that has developed among, as well as within, various nations. Throughout this text, we see globalization as a process that is transforming many societies, including the United States. Here we examine global events and processes, including corporate globalization, and explore their consequences worldwide. We have added new material on the link between inequality and risk factors for violence against women in different national settings.

Chapter 9, Race and Ethnicity, is a comprehensive review of the significance of race and ethnicity in society. Although these concepts are integrated throughout the book because of our focus on diversity, they also require particular focus on how race and ethnicity differentiate the experiences of diverse groups in society. Included are topics such as an update on forms and types of racism, the relevance of race in the devastation caused by Hurricane Katrina, a new discussion of the combined effects of race and class, a new look at ways to attain racial equality, and a discussion of recent court cases involving affirmative action.

Chapter 10, Gender, focuses on gender as a central concept in sociology closely linked to systems of stratification in society. The chapter links the social construction of gender to issues of homophobia, then is followed by the separate chapter on sexuality. This edition maintains a separate chapter on sexuality (**Chapter 11, Sexuality**), reflecting the increased attention in sociological research to sexuality as a component of social stratification. The new edition strengthens the discussion of the social construction of sexual identity. We have also added significant new material on the sexualization of young girls in an increasingly consumer-based culture.

Part IV, Social Institutions, includes three chapters, each focusing on basic institutions within society. Beginning with **Chapter 12, Families and Religion,** these chapters explore the basic structure of social institutions and examine how different theoretical perspectives within sociology help us interpret different dimensions of people's experiences within social institutions. In Chapter 12, there is updated material on the same-sex marriage debate and a strong discussion of religious extremism. The chapter maintains its inclusion of important topics in the study of families, such as interracial dating, gay and lesbian households, fatherhood, gender roles within families, and family violence.

Chapter 13, Education and Health Care, is on the institutions of education and health care and overviews education in the United States, including recent developments in education and inequality, plus a look at health and sickness seen globally. It includes theoretical perspectives on health care and the health care crisis in America. New to this chapter is recent research and data on inequality and standardized testing, on the "cognitive elite" and the nature versus nurture argument, and on schooling and gender. The chapter also contains an update of the health care crisis in America.

Chapter 14, Politics and the Economy, analyzes the state, power, and authority, and bureaucratic government. It also contains a detailed discussion of theories of power, in addition to coverage of the economy seen globally and characteristics of the labor force. This chapter contains new material on women and minorities in government, new material on joblessness and unemployment, and a new discussion of diversity in the overall labor market.

Part V, Social Change, includes two chapters. **Chapter 15, Population, Urbanization, and the Environment,** covers demographic processes, urbanization, and theories of population growth, as well as research on pollution and depletion of the physical environment. New data are included on racial–ethnic migration, global warming, and the role of environmental racism during the aftermath of Hurricane Katrina. The last chapter, **Chapter 16, Social Change and Social Movements,** addresses social change and social movements, with detailed coverage on theories of social change, the causes of social change, and the nature and types of social movements. New discussions include inequality, powerlessness and the individual, as well as a broader discussion of social movements.

Features and Pedagogical Aids

The special features of this book flow from its major themes: diversity, current theory and research, debunking and critical thinking, social change, and a global perspective. The features are also designed to help students develop critical thinking skills so that they can apply abstract concepts to observed experiences in their everyday life and learn how to interpret different theoretical paradigms and approaches to sociological research questions.

Critical Thinking Features

The feature **Thinking Sociologically** takes concepts from each chapter and asks students to think about these concepts in relationship to something they can easily observe in an exercise or class discussion. The feature **Debunking Society's Myths** takes certain common assumptions and shows students how the sociological perspective would inform such assumptions and beliefs.

See for Yourself

Each chapter includes a new feature, **See for Yourself,** intended to provide students with the chance to apply sociological concepts and ideas to their own observations. This feature can also be used as the basis for writing exercises, helping students improve both their analytic skills and their writing skills.

Seeing Society in Everyday Life

New to this edition, these new photo essays help students understand concepts more clearly. Essays include "Social Interaction," in Chapter 4; "Social Inequality" in Chapter 7, and "Gender, Bodies, and Beauty" in Chapter 10.

Unparalleled Integration of Web-Based Resources

Instructors will find that several technology-based teaching enhancements are integrated throughout the book, making this book the best conceived in using Internet tools for teaching and learning introductory sociology.

The Sociology: The Essentials Companion Website (**www.thomsonedu.com/sociology/andersen**) has a multitude of resources for students. Links to the student website are provided at the end of every chapter. On the companion website students will find Suggested Readings, weblinks, a MicroCase® Online feature that teaches students how to research society, learning objectives, Internet exercises, quizzes, and flash cards.

In addition, **ThomsonNOW™,** a self-study assessment web-based assessment and study tool is integrated in every chapter. (See the Supplements section later in this preface for more details on this robust study tool.)

An Extensive and Content-Rich Map Program

We use the map feature that appears throughout the book to help students visualize some of the ideas presented, as well as to learn more about regional and international diversity. One map theme is **Mapping America's Diversity,** and the other is **Viewing Society in Global Perspective.** These maps have multiple uses for instructional value, beyond instructing students about world and national geography. The maps have been designed primarily to show the differentiation by county, state, and/or country on key social facts. For example, in Chapter 3 we show the dispersion of the population under five years of age, both nationally and worldwide. Students can use this information to ask questions about how the age distribution of the population might be related to immigration, poverty, or global stratification.

We have included a critical thinking component to the maps to integrate them more effectively with the chapter material. Thus, each map includes **critical thinking questions** that ask students to interpret the map data within the context of concepts and ideas from the chapter.

High-Interest Theme Boxes

We use three high-interest themes for the box features that embellish our focus on diversity and sociological research throughout the text. **Understanding Diversity** further explores the approach to diversity taken throughout the book. In most cases, these box features provide personal narratives or other information designed to teach students about the experiences of different groups in society.

Because many are written as first-person narratives, they can invoke student empathy toward groups other than those to which they belong—something we think is critical to teaching about diversity. We hope to show students the connections between race, class, and other social groups that they otherwise find difficult to grasp.

The box feature **Doing Sociological Research** is intended to show students the diversity of research questions that form the basis of sociological knowledge, and, equally important, how the question a researcher asks influences the method used to investigate the question.

We see this as an important part of sociological research—that how one investigates a question is determined as much by the nature of the question as by allegiance to a particular research method. Some questions require a more qualitative approach, others, a more quantitative approach. In developing these box features, we ask, What is the central question sociologists are asking? How did they explore this question using sociological research methods? What did they find? and What are the implications of this research? We deliberately selected questions that show the full and diverse range of sociological theories and research methods, as well as the diversity of sociologists. Each box feature ends with **Questions to Consider** to encourage students to think further about the implications and applications of the research.

The feature **A Sociological Eye on the Media,** found in several chapters, examines some aspect of how the media influence public understanding of some of the subjects in this book. We think this is important because sociological research often debunks taken-for-granted points of view presented in the media, and we want students to be able to look at the media with a more critical eye. Because of the enormous influence of the media, we think this is increasingly important in educating students about sociology.

In-Text Learning Aids

In addition to the features just described, there is an entire set of learning aids within each chapter that promotes student mastery of the sociological concepts.

Chapter Outlines. A concise chapter outline at the beginning of each chapter provides students with an overview of the major topics to be covered.

Key Terms. Key terms and major concepts appear in bold when first introduced in the chapter. A list of the key terms is found at the end of the chapter, which makes study more effective. Definitions for the key terms may be found in the glossary.

Theory Tables. Each chapter includes a table that summarizes different theoretical perspectives by comparing and contrasting how these theories illuminate different aspects of different subjects.

Chapter Summary in Question-and-Answer Format. Questions and answers highlight the major points in each chapter and provide a quick review of major concepts and themes covered in the chapter.

A **Glossary** and complete **Bibliography** for the whole text is found at the back of the book.

Supplements

Sociology: The Essentials, Fifth Edition, is accompanied by a wide array of supplements prepared to create the best learning environment inside as well as outside the classroom for both the instructor and the student. All the continuing supplements for *Sociology: The Essentials,* Fifth Edition, have been thoroughly revised and updated, and several are new to this edition. We invite you to take full advantage of the teaching and learning tools available to you.

For the Instructor

Instructor's Resource Manual. This supplement offers the instructor brief chapter outlines, student learning objectives, key terms and people, detailed chapter lecture outlines, lecture and discussion suggestions, discussion questions for WebTutor, student activities, chapter worksheets, film suggestions/activities, and Internet exercises. Also included is a listing of resources offered with the text, as well as an assortment of useful appendices, which include activities that help integrate ASA task force recommendations in your course.

Test Bank. This test bank consists of a myriad of multiple-choice and true/false questions for each chapter of the text, all with answer explanations and page references to the text. Each multiple-choice item has the question type (factual, applied, or conceptual) indicated. Also included are short-answer and essay questions for each chapter. New to this edition, each test question will be mapped to a learning objective for the chapter. All questions are also labeled as new, modified, or pick-up so instructors know if the question is new to this edition of the test bank, modified but picked up from the previous edition of the test bank, or picked up straight from the previous edition of the test bank.

PowerLecture with JoinIn™ and ExamView® CD-ROM. This easy-to-use, one-stop digital library and presentation tool includes the following:

- Preassembled Microsoft® PowerPoint® lecture slides with graphics from the text, making it easy for you to assemble, edit, publish, and present custom lectures for your course.
- The PowerLecture CD-ROM includes video-based polling and quiz questions that can be used with the JoinIn™ on TurningPoint® personal response system.
- PowerLecture also features ExamView® testing software, which includes all the test items from the printed Test Bank in electronic format, enabling you to create customized tests of up to 250 items that can be delivered in print or online.

Classroom Activities for Introductory Sociology. Made up of contributions from Introductory Sociology instructors from around the country, this brand-new print supplement will be offered free to adopters of Andersen/Taylor's *Sociology: The Essentials,* Fifth Edition. The booklet features classroom activities, student projects, and lecture ideas to help instructors make topics fun and interesting for students. With general teaching tactics as well as topic-focused activities, it has never been easier to find a way to integrate new ideas into your classroom.

Introduction to Sociology Group Activities Workbook. This supplement is both a workbook for students and a repository of ideas for instructors. The workbook contains both in- and out-of-class group activities on activity sheets that students can complete, tear out, and turn in to the instructor. Also included are teaching ideas for video clips to anchor group discussions, maps, case studies, group quizzes, ethical debates, group questions, group project topics, and ideas for outside readings.

Extension: Wadsworth's Sociology Reader Collection. Create your own customized reader for your sociology class, drawing from dozens of classic and contemporary articles found on the exclusive Thomson Wadsworth TextChoice database. Using the TextChoice website (**www.TextChoice.com**) you can preview articles, select your content, and add your own original material. TextChoice will then produce your materials as a printed supplementary reader for your class.

Classroom Presentation Tools for the Instructor

Videos. Adopters of *Sociology: The Essentials,* Fifth Edition, have several different video options available with the text. Please consult with your Thomson Learning sales representative to determine if you are a qualified adopter for a particular video.

Wadsworth's Lecture Launchers for Introductory Sociology. An exclusive offering jointly created by Thomson Wadsworth and DALLAS TeleLearning, this video contains a collection of video highlights taken from the *Exploring Society: An Introduction to Sociology Telecourse* (formerly *The Sociological Imagination*). Each three- to six-minute video segment has been specially chosen to enhance and enliven class lectures and discussions of twenty key topics covered in the Introduction to Sociology course. Accompanying the video is a brief written description of each clip, along with suggested discussion questions to help effectively incorporate the material into the classroom. Available on VHS or DVD.

Sociology: Core Concepts Video. Another exclusive offering jointly created by Thomson Wadsworth and DALLAS TeleLearning, this video contains a collection of video highlights taken from the *Exploring Society: An Introduction to Sociology Telecourse* (formerly *The Sociological Imagination*). Each fifteen- to twenty-minute video segment will enhance student learning of the essential concepts in the introductory course and can be used to initiate class lectures, discussion, and review. The video covers topics such as the sociological imagination, stratification, race and ethnic relations, social change, and more. Available on VHS or DVD.

ABC® Video Series, Volumes I–III. This series of videos, comprising footage from ABC broadcasts, is specially selected and arranged to accompany your Introductory Sociology course. The segments may be used in conjunction with Wadsworth's Introductory Sociology texts to help provide a real-world example to illustrate course concepts or to instigate discussion.

Wadsworth Sociology Video Library. Bring sociological concepts to life with videos from Wadsworth's Sociology Video Library, which includes thought-provoking offerings from *Films for the Humanities,* as well as other excellent educational video sources. This extensive collection illustrates important sociological concepts covered in many sociology courses.

Supplements for the Student

Study Card. This handy card provides all the important sociological concepts highlighted in Andersen/Taylor, broken down by chapter. Providing a large amount of information at a glance, this study card is an invaluable tool for a quick review.

CengageNOW™ (also called ThomsonNOW™). This online tool provides students with a customized study plan based on a diagnostic "pretest" that they take after reading each chapter. The study plan provides interactive exercises, learning modules, animations, video exercises, and other resources to help students master concepts that are central to sociology. After the study plan has been reviewed, students can take a posttest to monitor their progress in mastering the chapter concepts. Instructors may bundle this product for their students with each new copy of the text for free! If your instructor did not order the free access code card to be packaged with your text—or if you have a used copy of the text—you can still obtain an access code for a nominal fee. Just visit the Thomson Wadsworth E-Commerce site at **www.ichapters.com**, where easy-to-follow instructions help you purchase your access code.

Study Guide. This student study tool contains brief chapter outlines, a chapter focus, questions to guide your reading, a list of key terms and key people with page references to the text, detailed chapter outlines, Internet & InfoTrac College Edition exercises, and practice tests consisting of multiple-choice, true/false, fill-in-the-blank, and essay questions. All questions include answer explanations and page references to the text.

Practice Tests. This collection of practice tests helps students adequately prepare for exams by presenting them with multiple-choice, true/false, short-answer, and essay questions that are similar in quality to the test bank questions. Multiple-choice and true/false answers are included, and page references are provided for all questions.

Researching Society with MicroCase® Online booklet. This new supplement contains MicroCase exercises for each chapter. Students can see the results of actual research by using the Wadsworth MicroCase® Online feature available from the Wadsworth website. This feature allows students to look at some of the results from national surveys, census data, and other data sources. They can either explore this easy-to-use feature on their own or use the examples provided. Previously in the textbook, these Microcase exercises are now in their own supplement.

Thomson Audio Study Tools. Your students will enjoy the MP3-ready Audio Lecture Overviews for each chapter and a comprehensive audio glossary of key terms for quick study and review. Whether walking to class, doing laundry, or studying at their desk, students now have the freedom to choose when, where, and how they interact with their audio-based educational media. Contact your Thomson Wadsworth sales rep for more information.

Internet-Based Supplements

InfoTrac® College Edition with InfoMarks®. Available as a free option with newly purchased texts, InfoTrac College Edition gives instructors and students four months of free access to an extensive online database of reliable, full-length articles (not just abstracts) from thousands of scholarly and popular publications going back as much as twenty-two years. Among the journals available are *American Journal of Sociology, Social Forces, Social Research,* and *Sociology.* InfoTrac College Edition now also comes with InfoMarks, a tool that allows you to save your search parameters, as well as save links to specific articles.

WebTutor™ Advantage on WebCT® and Blackboard®. This web-based software for students and instructors takes a course beyond the classroom to an anywhere, anytime environment. Students gain access to a full array of study tools, including chapter outlines, chapter-specific quizzing material, interactive games and maps, and videos. With WebTutor Advantage, instructors can provide virtual office hours, post syllabi, track student progress with the quizzing material, and even customize the content to suit their needs.

Wadsworth's Sociology Home Page at www.thomsonedu.com/sociology/. Combine this text with the exciting range of Web resources on Wadsworth's Sociology Home Page and you will have truly integrated technology into your learning system. Wadsworth's Sociology Home Page provides instructors and students with a wealth of information and resources, such as *Sociology in Action, Census 2000: A Student Guide for Sociology,* Research Online, a Sociology Timeline, a Spanish glossary of key sociological terms and concepts, and more.

Turnitin™ Online Originality Checker. This online "originality checker" is a simple solution for professors who want to put a strong deterrent against plagiarism into place and make sure their students are employing proper research techniques. Students upload their papers to their professor's personalized website and within seconds, the paper is checked against three databases—a constantly updated archive of over 4.5 billion web pages; a collection of millions of published works, including a number of Thomson Higher Education texts; and the millions of student papers already submitted to Turnitin. For each paper submitted, the professor receives a customized report that documents any text matches found in Turnitin's databases. At a glance, the professor can see if the student has used proper research and citation skills, or if he or she has simply copied the material from a source and pasted it into the paper without giving credit where credit was due. Our exclusive deal with iParadigms, the producers of Turnitin, gives instructors the ability to package Turnitin with the *Sociology: The Essentials,* Fifth Edition, Thomson textbook. Please consult with your Thomson Learning sales representative to find out more!

Companion Website for *Sociology: The Essentials,* (www.thomsonedu.com/sociology/andersen). The book's companion website includes chapter-specific resources for instructors and students. For instructors, the site offers a password-protected instructor's manual, PowerPoint presentation slides, and more. For students, there is a multitude of text-specific study aids, including the following:

- Tutorial practice quizzes that can be scored and emailed to the instructor
- Web links
- InfoTrac College Edition exercises
- Flash cards
- MicroCase Online data exercises
- Crossword puzzles
- Virtual Explorations
- And much more!

Acknowledgments

We relied on the comments of many reviewers to improve the book, and we thank them for the time they gave in developing very thoughtful commentaries on the different chapters. Thanks to Darlaine Gardetto, St. Louis Community College—Meramec; Shannon Houvouras, University of West Georgia; Christine Monnier, College of DuPage; Fernando Rivera, University of Central Florida; Beverly Stiles, Midwestern State University; and Richard Sullivan, Illinois State University.

We would also like to extend our thanks to those who reviewed the first four editions of *Sociology: The Essentials.* Brenda N. Bauch, Jefferson College; E. M. Beck, University of Georgia; Alessandro Bonanno, Sam Houston State University; G. M. Britten, Lenoir Community College (North Carolina); James E. Coverdill, University of Georgia; Susan Crafts, Niagara Community College; Jean E. Daniels, California State University; Angela D. Danzi, Farmingdale State University, Northridge; Ione Y. DeOllos, Ball State University; Marlese Durr, Wright State University; Lois Easterday, Onondaga Community College; Cynthia K. Epperson, St. Louis Community College at Meramec; Lynda Ann Ewen, Marshall University; Grant Farr, Portland State University; Irene Fiala, Kent State University—Ashtabula; James Fillman, Bucks County Community College; Lorna Forster, Clinton Community College; Patricia Gibbs, Foothill College; Bethany Gizzi, Monroe Community College; Jennifer Hamer, Wayne State University; Sara E. Hanna, Oakland Community College, Highland Lakes Campus; James R. Hunter, Indiana University—Purdue University Indianapolis; Jon Iannitti, SUNY College of Agriculture and Technology at Morrisville; Carol A. Jenkins, Glendale Community College; Diane E. Johnson, Kutztown University of Pennsylvania; Katherine Johnson, Niagara Community College; Anna Karpathakis, Kingsborough Community College, CUNY; Alice Abel Kemp, University of New Orleans; Keith Kirkpatrick, Victoria College (Texas); Tim Kubal, California State University, Fresno; Elizabeth D. Leonard, Vanguard University; James Lindberg, Montgomery College, Rockville Campus, Maryland; Martha O. Loustaunau, New Mexico State University; Brad Lyman, Baltimore City Community College; Susan A. Mann, University of New Orleans; Brian L. Maze, Franklin University; Leland C. McCormick, Minnesota State University—Mankato; Brian Moss, Oakland Community College—Waterford Campus (MI); Timothy Owens, Purdue University; Tara Perrello, Fordham University; Linda L. Petroff, Central Community College; David L. Phillips, Arkansas State University; Billie Joyce Pool, Homes Community College; Ralph Pyle, Michigan State University; David Redburn, Furman University; Lesley Williams Reid, Georgia State University; Lisa Riley, Creighton University; Michael C. Smith, Milwaukee Area Technical College; Tracey Steele, Wright State University; Melvin Thomas, North Carolina State University; Judith Warner, Texas A&M International University; Stephani Williams, Arizona State University; Sheryline A. Zebroski, St. Louis Community College, Florissant; Carl W. Zeigler, Elgin Community College; and Brenda S. Zicha, Mott Community College.

We also thank the following people, each of whom provided critical support in different, but important ways: Alison Bianchi, Cindy Gibson, Linda Keen, Nancy Quillen, and Judy Watson. We particularly thank Michelle Wilcox for her research assistance.

We are fortunate to be working with a publishing team with great enthusiasm for this project. We thank all of the people at Wadsworth who have worked with us on this and other projects, but especially we thank Chris Caldeira, Kate Barnes, Cheri Palmer, and Tali Beesley for their efforts on behalf of our book and the guidance and advice, not to mention the hard work, they have given to this project. We especially thank Dusty Friedman of The Book Company for her extraordinary attention to detail; we appreciate enormously her talent and perseverance. We thank Jane Loftus for her very careful copyediting of the manuscript and Sarah Evertson for her work on the photo research. Finally, our special thanks also go to our spouses Richard Morris Rosenfeld and Patricia Epps Taylor for their ongoing support of this project.

About the Authors

Margaret L. Andersen, raised in Oakland, California; Rome, Georgia; and Boston, is Edward F. and Elizabeth Goodman Rosenberg Professor of Sociology at the University of Delaware. She received her Ph.D. from the University of Massachusetts, Amherst, and her B.A. from Georgia State University. She is the author of ***Thinking About Women: Sociological Perspectives on Sex and Gender*** (Allyn and Bacon) and the best-selling Wadsworth text, ***Race, Class, and Gender: An Anthology*** (with Patricia Hill Collins). In 2006, she was awarded the prestigious Jessie Bernard Award by the American Sociological Association. In 2003, she was awarded the SWS Feminist Lecturer Award, given annually by SWS (Sociologists for Women in Society) to a social scientist whose work has contributed to improving the status of women in society. She currently serves as Chair of the National Advisory Board of the Center for Comparative Studies in Race and Ethnicity at Stanford University. She has served as the Interim Dean of the College of Arts and Science and Vice Provost for Academic Affairs at the University of Delaware, where she has also won the University's Excellence in Teaching Award. She was recently elected Vice President of the American Sociological Association. She lives on the Elk River in Maryland with her husband Richard Rosenfeld.

Howard F. Taylor was raised in Cleveland, Ohio. He graduated Phi Beta Kappa from Hiram College and has a Ph.D. in sociology from Yale University. He has taught at the Illinois Institute of Technology, Syracuse University, and Princeton University, where he is presently Professor of Sociology and former Director of the African American Studies Program. He has published over fifty articles in sociology, education, social psychology, and race relations. His books include ***The IQ Game*** (Rutgers University Press), a critique of hereditarian accounts of intelligence; ***Balance in Small Groups*** (Van Nostrand Reinhold), translated into Japanese; and the forthcoming ***Race and Class and the Bell Curve in America.*** He has appeared widely before college, radio, and TV audiences, including ABC's ***Nightline.*** He is past president of the Eastern Sociological Society, and a member of the American Sociological Association and the Sociological Research Association, an honorary society for distinguished research. He is a winner of the DuBois-Johnson-Frazier Award, given by the American Sociological Association for distinguished research in race and ethnic relations, and the President's Award for Distinguished Teaching at Princeton University. He lives in Pennington, New Jersey, with his wife, a corporate lawyer.

1

Chapter one

CHAPTER ONE

Chapter

Sociological Perspectives and Sociological Research

Imagine that you had been switched with another infant at birth and raised in very different conditions. How different would your life have been? What if your accidental family was very poor . . . or very rich? How might this have affected the schools you attended, the health care you received, the possibilities for your future career? What if you had been raised in a different religion—how would this have affected your beliefs, values, and attitudes? Taking a greater leap, what if you had been born another sex or a different race? What would you be like now?

We are talking about changing the basic facts of your life—your family, social class, education, religion, sex, and race. Each has major consequences for who you are and how you will fare in life. These factors play a major part in writing your life script. Social location (meaning a person's place in society) establishes the limits and possibilities of a life. Consider this:

- The pay gap between women and men, which had been declining since the 1980s, has recently *increased* between college-educated women and men (Blau and Kahn 2004; Leonhardt 2006).

continued

chapter one

CHAPTER ONE

- Black Americans who kill Whites are much more likely to face the death penalty than Blacks who kill other Black people (Paternoster et al. 2003).
- Although people think of the standard work week as "9 to 5," one-third of employed Americans work on Saturday, Sunday, or both, and one-fifth of all employed people work other than such standard hours; this is even more likely among dual-earner couples (Presser 2004).

These conclusions, drawn from current sociological research, describe some consequences of particular social locations in society. Although we may take our place in society for granted, our social location has a profound effect on our chances in life. The power of sociology is that it teaches us to see how society influences our lives and the lives of others, and it helps us explain the consequences of different social arrangements.

What Is Sociology?

Sociology is the study of human behavior in society. Sociologists are interested in the study of people and have learned a fundamental lesson: All human behavior occurs in a societal context. That context—the institutions and culture that surround us—shapes what people do and think. In this book, we will examine the dimensions of society and analyze the elements of social context that influence human behavior.

Sociology is a scientific way of thinking about society and its influence on human groups. Observation, reasoning, and logical analysis are the tools of the sociologist, coupled with knowledge of the large body of theoretical and analytical work done by previous sociologists and others. Sociology is inspired by the fascination people have for the thoughts and actions of other people, but it goes far beyond casual observations. It attempts to build on observations that are objective and accurate to create analyses that are reliable and that can be validated by others.

Every day, the media in their various forms (television, film, video, digital, and print) bombard us with social commentary. Whether it is Oprah Winfrey or Dr. Phil, media commentators provide endless opinion about the various and sometimes bizarre forms of behavior in our society. Sociology is different. Sociologists may appear in the media and often study the same subjects that the media examine, such as domestic violence or crime, but sociologists use specific research techniques and well-tested theories to explain social issues. Indeed, sociology can provide the tools for testing whether the things we hear about society are actually true. Much of what we hear in the media and elsewhere about society, delivered as it may be with perfect earnestness, is misstated and sometimes completely wrong, as you will see in some of the "Debunking Society's Myths" examples featured throughout this book.

The subject matter of sociology is everywhere. Social behavior and social change—these are the topics of sociological study. Psychologists, anthropologists, political scientists, economists, social workers, and others also study social behavior and social change. Along with sociologists, these disciplines make up what are called the social sciences.

The Sociological Perspective

Think back to the opening of this chapter where you were asked to imagine yourself having grown up under completely different circumstances. Our goal in that passage was to make you feel the stirring of the *sociological perspective*—the ability to see the societal patterns that influence individual and group life. The beginnings of the sociological perspective can be as simple as the pleasures of watching people or wondering how society influences people's lives. Indeed, many students begin their study of sociology because they are "interested in people." Sociologists convert this curiosity into the systematic study of how society influences different people's experiences within it.

Sociology is the study of human behavior, including the significance of diversity.

C. Wright Mills (1916–1962) was one of the first to write about the sociological perspective in his classic book, *The Sociological Imagination* (1959). He wrote that the task of sociology was to understand the relationship between individuals and the society in which they live. He defined the **sociological imagination** as the ability to see the societal patterns that influence the individual as well as groups of individuals. Sociology should be used, Mills argued, to reveal how the context of society shapes our lives. He thought that to understand the experience of a given person or group of people, one had to have knowledge of the social and historical context in which people lived.

Think, for example, about the time and effort that many people put into their appearance. Although, on the one hand, you can think about this merely as personal grooming or an individual attempt to "look good," there are significant social origins of this behavior. An individual may stand in front of the mirror, seemingly just one person not particularly thinking about how society is present in his or her reflection. But we likely look in the mirror to see how others see us; therefore, this is a very social act. Furthermore, if by looking we are trying to achieve a particular look, that look has likely been established by the social forces that tell us to achieve a particular ideal—ideals that are produced by industries that profit enormously from the products and services that people buy.

Some industries suggest that you should be thinner or curvier, your pants should be baggy or straight, your breasts should be minimized or maximized—either way you need more products. Maybe you should have a complete makeover! Many people then go to great lengths—sometimes extreme lengths that are hazardous to physical and mental health—to try to achieve a constantly changing beauty ideal, one that is probably not even attainable (such as flawless skin, hair never out of place, perfectly proportioned body parts).

The point is that the alleged standard is produced by social factors that extend far beyond the individual's immediate concerns with personal appearance. The ideals are produced in particular social and historical contexts. People may come up with all kinds of personal strategies for achieving a presumed ideal—they may buy more products, try to lose more weight, get a Botox treatment, or even become extremely depressed and anxious if they perceive their efforts as failing. These personal behaviors may seem to be only individual issues, but they have a basic social causes. That is, the origins of these behaviors originate beyond personal lives. The sociological imagination, which permits us to see that something as seemingly personal as how you look—and what you do to create that look—arises from a social context, not just individual behavior. Sociologists are certainly concerned about individuals, but they direct their attention to the social and historical context that shapes the experiences of individuals and groups.

A fundamental distinction within the sociological imagination is that between *troubles* and *issues*. **Troubles** are privately felt problems that spring from events or feelings in a person's life. **Issues** affect large numbers of people and have their origins in the institutional arrangements and history of a society (Mills 1959). This distinction is the crux of the difference between individual experience and **social structure,** which is defined as the organized pattern of social relationships and social institutions that together constitute society. Issues shape the context within which troubles arise. Sociologists employ the sociological perspective to understand how issues are shaped by social structures.

Personal troubles are felt by individuals who are experiencing problems; social issues arise when large numbers of people experience problems that are rooted in the social structure of society.

Mills used the example of unemployment to explain the meaning of troubles versus issues. When a person becomes unemployed, he or she has a personal trouble. In addition to financial problems, the person may feel a loss of identity, may become depressed, may lose touch with former work associates, or may have to uproot a family and move. The problem of unemployment, however, is deeper than the experience of one person. Unemployment is rooted in the structure of society; this is what interests sociologists. What societal forces cause unemployment? Who is most likely to become unemployed at different times? How does unemployment affect an entire community (for instance, when a large plant shuts down) or an entire nationwide group (such as the corporate downsizing of older workers)? Sociologists know that unemployment causes personal troubles, but understanding unemployment is more than understanding one person's experience. It requires understanding the social structural conditions that influence people's lives.

The specific task of sociology, according to Mills, is to comprehend the whole of human society—its personal and public dimensions, historical and contemporary—and its influence on the lives of human beings. Mills had an important point: People often feel

understanding diversity

Becoming a Sociologist

Individual biographies often have a great influence on the subjects sociologists choose to study. The authors of this book are no exception. Margaret Andersen, a White woman, now studies the sociology of race and women's studies. Howard Taylor, an African American man, studies race, social psychology, and especially race and intelligence testing. Here, each of them writes about the influence of their early experiences on becoming a sociologist.

Margaret Andersen

As I was growing up in the 1950s and 1960s, my family moved from California to Georgia, then to Massachusetts, and then back to Georgia. Moving as we did from urban to small-town environments and in and out of regions of the country that were very different in their racial character, I probably could not help becoming fascinated by the sociology of race. Oakland, California, where I was born, was highly diverse; my neighborhood was mostly White and Asian American. When I moved to a small town in Georgia in the 1950s, I was ten years old, but I was shocked by the racial norms I encountered. I had always loved riding in the back of the bus—our major mode of transportation in Oakland—and could not understand why this was no longer allowed. Labeled by my peers as an outsider because I was not southern, I painfully learned what it meant to feel excluded just because of "where you are from."

When I moved again to suburban Boston in the 1960s, I was defined by Bostonians as a southerner and ridiculed. Nicknamed "Dixie," I was teased for how I talked. Unlike in the South, where despite strict racial segregation, Black people were part of White people's daily lives, Black people in Boston were even less visible. In my high school of 2500 or so students, Black students were rare. To me, the school seemed not much different from the strictly segregated schools I had attended in Georgia. My family soon returned to Georgia, where I was an outsider again; when I later returned to Massachusetts for graduate school in the 1970s, I worried about how a southerner would be accepted in this "Yankee" environment. Because I had acquired a southern accent, I think many of my teachers stereotyped me and thought I was not as smart as the students from other places.

These early lessons, which I may have been unaware of at the time, must have kindled my interest in the sociology of race relations. As I explored sociology, I wondered how the concepts and theories of race relations applied to women's lives. So much of what I had experienced growing up as a woman in

that things are "beyond their control," meaning that they are being shaped by social forces larger than their own individual life. Social forces influence our lives in profound ways, even though we may not know how all the time. Consider this. Most likely you remember what you were doing on September 11, 2001, when you first heard that terrorists had flown planes into the World Trade Center in New York City. Obviously, this affected people's personal lives, but its impact—and its causes—go beyond the personal troubles it produced. The sociological perspective explains many dimensions of this event and its aftermath, including, as one example, the significance of cultural symbols that have emerged in the aftermath of 9/11. Think of the T-shirts that many people now wear honoring New York's firefighters. Of course, the social forces that influence people's lives are not always that drastic and include the ordinary events of everyday life.

It is important to note that sociology is an **empirical** discipline. This means conclusions are based on careful and systematic observations. Thus sociology is very different from ordinary common sense. For empirical observations to be useful to other observers, they must be gathered and recorded rigorously. Sociologists are also obliged to reexamine their assumptions and conclusions constantly. Only careful, unprejudiced observations add to the fund of sociological knowledge. Although the specific methods that sociologists use to examine different problems vary, as we will see, the empirical basis of sociology is what distinguishes it from mere opinion or other forms of social commentary.

this society was completely unexamined in what I studied in school. As the women's movement developed in the 1970s, I found sociology to be the framework that helped me understand the significance of gender and race in people's lives. To this day, I write and teach about race and gender, using sociology to help students understand their significance in society.

Howard Taylor

I grew up in Cleveland, Ohio, the son of African American professional parents. My mother, Murtis Taylor, was a social worker and the founder and then president of a social work agency, called the Murtis H. Taylor Multi-Service Center, in Cleveland, Ohio. She is well known for her contributions to the city of Cleveland and was an early "superwoman," working days and nights, cooking, caring for her two sons, and being active in many professional and civic activities. I think this gave me an early appreciation for the roles of women and the place of gender in society, although I surely would not have articulated it as such at the time.

My father was a businessman in a then all-Black life insurance company. He was also a "closet scientist," always doing experiments and talking about scientific studies. He encouraged my brother and me to engage in science, so we were always experimenting with scientific studies in the basement of our house. In the summers, I worked for my mother in the social service agency where she worked, as a camp counselor and in other jobs. Early on, I contemplated becoming a social worker, but I was also excited by science. As a young child, I acquired my father's love of science and my mother's interest in society. In college, the one field that would gratify both sides of me, science and social work, was sociology. I wanted to study human interaction, but I also wanted to be a scientist, so the appeal of sociology was clear.

At the same time, growing up African American meant that I faced the consequences of race everyday. It was always there, and like other young African American children, I spent much of my childhood confronting racism and prejudice. When I discovered sociology, in addition to bridging the scientific and humanistic parts of my interests, I found a field that provided a framework for studying race and ethnic relations. The merging of two ways of thinking, coupled with the analysis of race that sociology has long provided, made sociology fascinating to me.

Today, my research on race and intelligence testing and Black leadership networks seems rooted in these early experiences. I do quantitative research in sociology and see sociology as a science that reveals the workings of race, class, and gender in society.

Discovering Inconvenient Facts

In studying sociology, it is crucial to examine the most controversial topics and to do so with an open mind, even when you see the most disquieting facts. The facts we learn through sociological research can be "inconvenient" because the data can challenge familiar ways of thinking. Consider the following:

- Despite the idea that Asian Americans are a "model minority," poverty among Asian American families is higher than among White American families and has increased in recent years. Among certain Asian American groups, namely Laotians and Cambodians, poverty strikes two-thirds of all families (DeNavas-Walt et al. 2006; Lee 1994).
- Women's income (on average) has generally risen, while men's has generally fallen. Still, women with a college degree earn less on average than men with only some college experience (DeNavas-Walt et al. 2006).
- The number of women prisoners has increased at twice the rate of increase for men; most women in prison are mothers (U.S. Bureau of Justice Statistics 2005; Greenfield and Snell 2000).

These facts provide unsettling evidence of persistent problems in the United States, *problems that are embedded in society,* not just in individual behavior. Sociologists try to reveal the social factors that shape society and determine the chances of success for different groups. Some never get the chance to go to

© William Thomas Cain/Getty Images

© Lindsay Hebberd/CORBIS

Cultural practices that seem bizarre to outsiders may be taken for granted or defined as appropriate by insiders.

college; others are likely never to go to jail. These divisions persist because of people's placement within society.

Sociologists study not just the disquieting side of society. Sociologists may study questions that affect everyday life, such as how families adapt to changing societal conditions (Lempert and DeVault 2000), how children of immigrants fare (Portes and Rumbaut 2001), or how racism has changed in recent years (Zuberi 2001; Bonilla-Silva 2001). There are also many intriguing studies of unusual groups, including cyberspace users (Turkle 1995; Kendall 2002) and heavily tattooed people, known as collectors (Irwin 2001). The subject matter of sociology is vast. Some research illuminates odd corners of society; other studies address urgent problems of society that may affect the lives of millions.

Debunking in Sociology

The power of sociological thinking is that it helps us see everyday life in new ways. Sociologists question actions and ideas that are usually taken for granted. Peter Berger (1963) calls this process debunking. **Debunking** refers to looking behind the facades of everyday life—what Berger called the "unmasking tendency" of sociology (1963: 38). In other words, sociologists look at the behind-the-scenes patterns and processes that shape the behavior they observe in the social world.

Take schooling, for example: We can see how the sociological perspective debunks common assumptions about education. Most people think that education is primarily a way to learn and get ahead. Although this is true, a sociological perspective on education reveals something more. Sociologists have concluded that more than learning takes place in schools; other social processes are at work. Social cliques are formed where some students are "insiders" and others are excluded "outsiders." Young school children acquire not just formal knowledge, but also the expectations of society and people's place within it. Race and class conflicts are often played out in schools. Relative to boys, girls are often shortchanged by the school system—receiving less attention and encouragement, less interaction with teachers, less instruction in the sciences, and many other deficits disproportionately forced upon them (American Association of University Women 1998; Sadker and Sadker 1994). Poor children seldom have the same resources in schools as middle-class or elite children, and they are often assumed to be incapable of doing schoolwork and are treated accordingly. The somber reality is that schools may actually stifle the opportunities of some children rather than launch all children toward success.

Debunking is sometimes easier to do when looking at a culture or society different from one's own. Consider how behaviors that are unquestioned in one society may seem positively bizarre to an outsider. For a thousand years in China, it was usual for the elite classes to bind the feet of young girls to keep the feet from growing bigger—a practice allegedly derived from a mistress of the emperor. Bound feet were a sign of delicacy and vulnerability. A woman with large feet (defined as more than 4 inches long!) was thought to bring shame to her husband's household. The practice was supported by the belief that men were highly aroused by small feet even though men

doing sociological research

Debunking the Myths of Black Teenage Motherhood

Research Question: Sociologist Elaine Bell Kaplan knew that there was a stereotypical view of Black teen mothers, that they had grown up in fatherless households where their mothers had no moral values and no control over their children. The myth of Black teenage motherhood also depicts teen mothers as unable to control their sexuality, as having children to collect welfare checks, and as having families who condone their behavior. Is this true?

Research Method: Kaplan did extensive research in two communities in the San Francisco Bay Area—East Oakland and Richmond, both communities with a large African American population and typical of many inner-city, poor neighborhoods. Once thriving Black communities, East Oakland and Richmond are now characterized by high rates of unemployment, poverty, inadequate schools, crime, drug-related violence, and high numbers of single-parent households. Having grown up herself in Harlem, Kaplan knew that communities like those she studied have not always had these problems, nor have they condoned teen pregnancy. She spent several months in these communities, working as a volunteer in a community teen center that provided educational programs, day care, and counseling to teen parents, and "hanging out" with a core group of teen mothers. She did extensive interviews with thirty-two teen mothers, supplementing them when she could with interviews with their mothers and, sometimes, the fathers of their children.

Research Results: Kaplan found that teen mothers adopt strategies for survival that help them cope with their environment even though these same strategies do not help them overcome the problems they face. Unlike what the popular stereotype suggests, she did not find that the Black community condones teen pregnancy; quite the contrary, the teens felt embarrassed and stigmatized by being pregnant and experienced tension and conflict with their mothers, who saw their pregnancy as disrupting the hopes they had for their daughters' success. When the women had to go on welfare, they also felt embarrassed, often developing strategies to hide the fact that they were dependent on welfare to make ends meet. These conclusions run directly counter to the public image that such women do not value success and live in a culture that promotes welfare dependency.

Conclusions and Implications: Instead of simply stereotyping these teens as young and tough, Kaplan sees them as struggling to develop their own gender and sexual identity. Like other teens, they are highly vulnerable, searching for love and aspiring to create a meaningful and positive identity for themselves. But failed by the educational system and locked out of the job market, the young women's struggle to develop an identity is compounded by the disruptive social and economic conditions in which they live.

Kaplan's research is a fine example of how sociologists debunk some of the commonly shared myths that surround contemporary issues. Carefully placing her analysis in the context of the social structural changes that affect these young women's lives, Kaplan provides an excellent example of how sociological research can shed new light on some of our most pressing social problems.

Questions to Consider

1. Suppose that Kaplan had studied middle-class teen mothers. What similarities and differences would you predict in the experiences of middle-class and poor teen mothers? Does race matter? In what ways does your answer *debunk* myths about teen pregnancy?
2. Make a list of the challenges you would face were you to be a teen parent. Having done so, indicate those that would be considered personal troubles and those that are social issues. How are the two related?

Source: Kaplan, Elaine Bell. 1996. *Not Our Kind of Girl: Unraveling the Myths of Black Teenage Motherhood.* Berkeley, CA: University of California Press.

Careers in Sociology

Now that you understand a bit more what sociology is about, you may ask, What can I do with a degree in sociology? This is a question we often hear from students. There is no single job called "sociologist" like there is "engineer" or "nurse" or "teacher," but sociology prepares you well for many different kinds of jobs, whether with a bachelor's degree or a postgraduate education. The skills you acquire from your sociological education are useful for jobs in business, health care, criminal justice, government agencies, various nonprofit organizations, and other job venues.

For example, the research skills one gains through sociology can be important in analyzing business data or organizing information for a food bank or homeless shelter. Students in sociology also gain experience working with and understanding those with different cultural and social backgrounds; this is an important and valued skill that employers seek. Also, the ability to dissect the different causes of a social problem can be an asset for jobs in various social service organizations.

Some sociologists have worked in their communities to deliver more effective social services. Some are employed in business organizations and social services where they use their sociological training to address issues such as poverty, crime and delinquency, population studies, substance abuse, violence against women, family social services, immigration policy, and any number of other important issues. Sociologists also work in the offices of U.S. representatives and senators doing background research on the various issues addressed in the political process.

These are just a few examples of how sociology can prepare you for various careers. A good way to learn more about how sociology prepares you for work is to consider doing an internship while you are still in college.

For more information about careers in sociology, see the booklet, *Careers in Sociology,* available through the American Sociological Association. You can find it online at either **www.asanet.org/student/career/homepage.html or http://sociology.wadsworth.com**

Question:

What do the following people have in common?

Dan Akroyd (actor; comedian)

Debra Winger (actress)

Saul Bellow (novelist; Nobel Prize recipient)

Joe Theissman (football player)

(Former) Senator Robert Torricelli

Congresswoman Maxine Waters (from California)

Senator Barbara Mikulski (from Maryland)

Regis Philbin (TV personality)

Rev. Jesse Jackson

Ruth Westheimer (the "sex doctor")

Robin Williams (comedian; actor)

Martin Luther King, Jr.

former President Ronald Reagan

Answer:

They were all sociology majors!

Source: Compiled by Peter Dreier, Occidental College. Full list available on the home page of the American Sociological Association **(http://www.asanet.org/page.ww?section=Students&name=Famous+Sociology+Majors)**.

never actually saw the naked foot. If they had, they might have been repulsed, because a woman's actual foot was U-shaped and often rotten and covered with dead skin (Blake 1994).

Outside the social, cultural, and historical context in which it was practiced, foot binding seems bizarre, even dangerous. Feminists have pointed out that Chinese women were crippled by this practice, making them unable to move about freely and more dependent on men (Chang 1991). This is an example of outsiders debunking a practice that was taken for granted by those within the culture.

Debunking can also call into question practices in one's own culture that may normally go unexamined. Strange as the practice of Chinese foot binding may seem to you, how might someone from another culture view wearing shoes that make it difficult to walk? Or piercing one's tongue or eyebrow? These practices of contemporary U.S. culture are taken for granted by many, just as was Chinese foot binding. Until these cultural processes are debunked, seen as if for the first time, they might seem normal.

Debunking Society's Myths

Myth: Violence is at an all-time high.

Sociological research: Although rates of violence are high in the United States, at least twice before (in 1930 and 1980) the murder rate was at a comparable level. It is also true, however, that rates of violence in the United States are higher than in any other industrialized nation (Levine and Rosich 1996; Best 1999).

Key Sociological Concepts

As you build your sociological perspective, you must learn certain key concepts to begin understanding how sociologists view human behavior. *Social structure, social institutions, social change,* and *social interaction* are not the only sociological concepts, but they are fundamental to grasping the sociological perspective.

Social Structure. Earlier, we defined social structure as the organized pattern of social relationships and social institutions that together constitute society. Social structure is not a "thing," but refers to the fact that social forces, not always visible to the human eye, guide and shape human behavior. Acknowledging that social structure exists does not mean that humans have no choice in how they behave, only that those choices are largely conditioned by one's location in society.

Social Institutions. In this book, you will also learn about the significance of **social institutions,** defined as established and organized systems of social behavior with a particular and recognized purpose. The family, religion, marriage, government, and the economy are examples of major social institutions. Social institutions confront individuals at birth, and they transcend individual experience—but still influence individual behavior.

Social Change. As you can tell, sociologists are also interested in the process of **social change**—the alteration of society over time. As much as sociologists see society as producing certain outcomes, they do not see society as fixed, nor do they see humans as passive recipients of social expectations. Sociologists view society as stable but constantly changing.

Social Interaction. Sociologists see **social interaction** as behavior between two or more people that is given meaning. Through social interaction, people react and change, depending on the actions and reactions of others. Since society changes as new forms of human behavior emerge, change is always in the works. As you read this book, you will see that these key concepts—social structure, social institutions, social change, and social interaction—are central to the sociological imagination.

The Significance of Diversity

The analysis of diversity is one central theme of sociology. Differences among groups, especially differences in the treatment of groups, are significant in any society, but they are particularly compelling in a society as diverse as that in the United States.

Defining Diversity

Today, the United States includes people from all nations and races. In 1900, one in eight Americans was not White; today, racial and ethnic minority groups, including African Americans, Latinos, American Indians, and Asian Americans, represent one-quarter of Americans, and that proportion is growing (see Figure 1.1; U.S. Census Bureau 2007). These broad categories themselves are internally diverse, including for example, those with long-term roots in the United States, as well as Cuban Americans, Salvadorans, Cape Verdeans, Filipinos, and many others.

Understanding diversity is important in a society comprising so many different groups, each with unique, but interconnected, experiences.

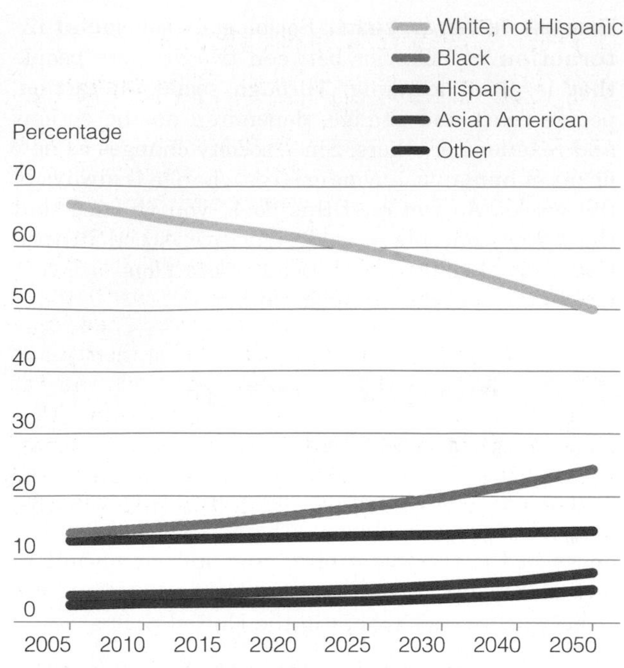

FIGURE 1.1 *Share of Minorities in the U.S. Population*

It is a social fact that by 2050 the U.S. population is projected to be half non-Hispanic Whites. How do you think this will affect your life? What will the population percentage be for different racial-ethnic groups when you are forty years old? How does this illustrate how social facts influence the lives of individual people and diverse groups in society?

Source: U.S. Census Bureau. 2004. "U.S. Interim Projections by Age, Sex, Race, and Hispanic Origin." www.census.gov/ipc/www/usinterimproj/

Perhaps the most basic lesson of sociology is that people are shaped by the social context around them. In the United States, with so much cultural diversity, people will share some experiences, but not all. Experiences not held in common can include some of the most important influences on social development, such as language, religion, and the traditions of family and community. Understanding diversity means recognizing this diversity and making it central to sociological analyses.

In this book, we use the term *diversity* to refer to the variety of group experiences that result from the social structure of society. **Diversity** is a broad concept that includes studying group differences in society's opportunities, the shaping of social institutions by different social factors, the formation of group and individual identity, and the process of social change. Diversity includes the study of different cultural orientations, although diversity is not exclusively about culture.

Understanding diversity is crucial to understanding society because fundamental patterns of social change and social structure are increasingly patterned by diverse group experiences. There are numerous sources of diversity, including race, class, gender, and others as well. Age, nationality, sexual orientation, and region of residence, among other factors, also differentiate the experience of diverse groups in the United States. And as the world is increasingly interconnected through global communication and a global economy, the study of diversity also encompasses a global perspective—that is, an understanding of the international connections existing across national borders and the impact of such connections on life throughout the world.

Society in Global Perspective

No society can be understood apart from the global context that now influences the development of all societies. The social and economic system of any one society is increasingly intertwined with those of other nations. Coupled with the increasing ease of travel and telecommunication, this means that a global perspective is necessary to understand change both in the United States and in other parts of the world.

To understand globalization, you must look beyond the boundaries of your own society to see how patterns in any given society are increasingly being shaped by the connections between societies. Comparing and contrasting societies across different cultures is valuable. It helps you see patterns in your own society that you might otherwise take for granted, and it enriches your appreciation of the diverse patterns of culture that mark human society and human history. A global perspective, however, goes beyond just comparing different cultures; it also helps you see how events in one society or community may be linked to events occurring on the other side of the globe.

For example, return to the example of unemployment that C. Wright Mills used to distinguish troubles and issues. One man may lose his job in Peoria, Illinois, and a woman in Los Angeles may employ a Latina domestic worker to take care of her child while she pursues a career. On the one hand, these are individual experiences—for all three people—but they are linked in a pattern of globalization that shapes

Globalization brings diverse cultures together, but it is also a process by which Western markets have penetrated much of the world.

AP Images/Eugene Hoshiko

"Actually, Lou, I think it was more than just my being in the right place at the right time. I think it was my being the right race, the right religion, the right sex, the right socioeconomic group, having the right accent, the right clothes, going to the right schools . . ."

the lives of all three. The Latina domestic may have a family whom she has left in a different nation so that she can afford to support them. The corporation for which the Los Angeles woman works may have invested in a new plant overseas that employs cheap labor, resulting in the unemployment of the man in Peoria. The man in Peoria may have seen immigrant workers moving into his community, and one of his children may have made a friend at school who speaks a language other than English.

Such processes are increasingly shaping many of the subjects examined in this book—work, family, education, politics, just to name a few. Without a global perspective, you would not be able to fully understand the experience of any one of the people just mentioned—much less how society is being shaped by these processes of change and global context. Throughout this book, we will use a global perspective to understand some of the developments shaping contemporary life in the United States.

SEE FOR YOURSELF

TROUBLES AND ISSUES

Write a short paragraph identifying a social issue that interests you. This paragraph should include a statement about the personal troubles that this issue generates: who is most affected by the issue you have identified and how it affects them in a personal way. Next, write a second paragraph in which you move beyond thinking about the personal troubles of this issue and identify what you think might be some of its societal origins. Then, using the example you have written about, explain what C. Wright Mills means in arguing that personal biographies are linked to the social and historical context in which they are lived.

The Development of Sociological Theory

Like the subjects it studies, sociology is itself a social product. Sociology first emerged in western Europe during the eighteenth and nineteenth centuries. In this period, the political and economic systems of Europe were rapidly changing. Monarchy, the rule of society by kings and queens, was disappearing in western Europe. These changes generated new ways of thinking. Religion as the system of authority and law was giving way to scientific authority. At the same time, capitalism grew. Contact between different societies increased, and worldwide economic markets developed. The traditional ways of the past were giving way to a new social order. The time was ripe for a new understanding.

The Influence of the Enlightenment

The **Enlightenment** in eighteenth- and nineteenth-century Europe had an enormous influence on the development of modern sociology. Also known as the Age of Reason, the Enlightenment was characterized by faith in the ability of human reason to solve society's problems. Intellectuals believed that there were natural laws and processes in society to be discovered and used for the general good. Modern science was gradually supplanting traditional and religious explanations for natural phenomena with theories confirmed by experiments.

The earliest sociologists promoted a vision of sociology grounded in careful observation. **Auguste Comte** (1798–1857), a French philosopher who coined the term *sociology,* believed that just as science had discovered the laws of nature, sociology could discover the laws of human social behavior and thus help solve society's problems. This approach is called **positivism,** a system of thought, still prominent today, in which scientific observation and description is considered the highest form of knowledge, as opposed to, say, religious dogma or poetic inspiration. The modern scientific method, which guides sociological research, grew out of positivism.

Alexis de Tocqueville (1805–1859), a French citizen, traveled to the United States as an observer beginning in 1831. Tocqueville thought that democratic values and the belief in human equality positively influenced American social institutions and transformed personal relationships. Less admiringly, he felt that in the United States the tyranny of kings had been replaced by the "tyranny of the majority." He was referring to the ability of a majority to impose its will on everyone else in a democracy. Tocqueville also felt that despite the emphasis on individualism in American culture, Americans had little independence of mind, making them self-centered and anxious about their social class position (Collins and Makowsky 1972).

As one of the earliest observers of American culture, Harriet Martineau used the powers of social observation to record and analyze the social structure of American society. Long ignored for her contributions to sociology, she is now seen as one of the founders of early sociological thought.

Another early sociologist is **Harriet Martineau** (1802–1876). Like Tocqueville, Martineau, a British citizen, embarked on a long tour of the United States in 1834. She was fascinated by the newly emerging culture in America. Her book *Society in America* (1837) is an analysis of the social customs that she observed. This important work was overlooked for many years, probably because the author was a woman. It is now recognized as a classic. Martineau also wrote the first sociological methods book, *How to Observe Manners and Morals* (1838), in which she discussed how to observe behavior when one is a participant in the situation being studied.

Classical Sociological Theory

Of all the contributors to the development of sociology, the giants of the European tradition were Emile Durkheim, Karl Marx, and Max Weber. They are classical thinkers because the analyses they offered more than 150 years ago continue to enlighten our understanding of society—not just in sociology, but in other fields as well (such as political science and history).

Emile Durkheim. During the early academic career of the Frenchman **Emile Durkheim** (1858–1917), France was in the throes of great political and religious upheaval. Anti-Semitism (hatred of Jews) was being expressed, along with ill feeling among other religions, as well. Durkheim, himself Jewish, was fascinated by how the public degradation of Jews by non-Jews seemed to calm and unify a large segment of the divided French public. Durkheim later wrote that public rituals have a special purpose in society, creating social solidarity, referring to the bonds that link the members of a group. Some of Durkheim's most significant works explore the question of what forces hold society together and make it stable.

According to Durkheim, people in society are glued together by belief systems (Durkheim 1947/1912). The rituals of religion and other institutions symbolize and reinforce the sense of belonging. Public ceremonies create a bond between people in a social unit. Durkheim thought that by publicly punishing people, such rituals sustain moral cohesion in society. Durkheim's views on this are further examined in Chapter 6, which discusses deviant behavior.

Durkheim also viewed society as an entity larger than the sum of its parts. He described this as society *sui generis* (meaning "thing in itself"), meaning that society is a subject to be studied separate from the sum of the individuals who compose it. Society is external to individuals, yet its existence is internalized in people's minds—that is, people come to believe what society expects them to believe. Durkheim conceived of society as an integrated whole—each part contributing to the overall stability of the system. His work is the basis for *functionalism,* an important theoretical perspective that we will return to later in this chapter.

One contribution from Durkheim was his conceptualization of the *social.* Durkheim created the term **social facts** to indicate those social patterns that are *external* to individuals. Things such as customs and social values exist outside individuals, whereas psychological drives and motivation exist inside people. Social facts, therefore, are not to be explained by biology or psychology, but are the proper subject of sociology; they are its reason for being.

A striking illustration of this principle was Durkheim's study of suicide (Durkheim 1951/1897). He analyzed rates of suicide in a society, as opposed to looking at individual (psychological) causes of suicide. He showed that suicide rates varied according to how clear the norms and customs of the society were, whether the norms and customs were consistent with each other and noncontradictory. Where norms were either grossly unclear or contradictory, a condition that he called *anomie* ("normlessness") existed, and the suicide rates were higher in such societies or such parts of a society. It is important to note that this condition was "external," thus outside of each individual taken singly. In this sense such a condition is truly societal.

Durkheim held that social facts, though they exist outside individuals, nonetheless pose constraints on individual behavior. Durkheim's major contribution was the discovery of the social basis of human behavior. He proposed that society could be known through the discovery and analysis of social fact. This is the central task of sociology (Bellah 1973; Coser 1977; Durkheim 1950/1938).

Emile Durkheim established the significance of society as something larger than the sum of its parts. Social facts stem from society and have profound influence on the lives of people within society.

Karl Marx. It is hard to imagine another scholar who has had as much influence on intellectual history as **Karl Marx** (1818–1883). Along with his collaborator, Friedrich Engels, Marx not only changed intellectual history but world history, too.

Marx's work was devoted to explaining how capitalism shaped society. He argued that capitalism is an economic system based on the pursuit of profit and the sanctity of private property. Marx used a class analysis to explain capitalism, describing capitalism as a system of relationships among different classes—including capitalists (also known as the bourgeois class), the proletariat (or working class), the petty bourgeoisie (small business owners and managers), and the *lumpenproletariat* (those "discarded" by the capitalist system, such as the homeless). In Marx's view, profit, the goal of capitalist endeavors, is produced through the exploitation of the working class. Workers sell their labor in exchange for wages, and capitalists make certain that wages are worth less than the goods the workers produce. The difference in value is the profit of the capitalist. In the Marxist view, the capitalist class system is inherently unfair because the entire system rests on workers getting less than they give.

Marx thought that the economic organization of society was the most important influence on what humans think and how they behave. He found that the beliefs of the common people tended to support the interests of the capitalist system, not the interests of the workers themselves. Why? Because the capitalist class controls not only the production of goods but also the production of ideas. It owns the publishing companies, endows the universities where knowledge is produced, and controls information industries.

Marx considered all of society to be shaped by economic forces. Laws, family structures, schools, and other institutions all develop, according to Marx, to suit economic needs under capitalism. Like other early sociologists, Marx took social structure as his subject rather than the actions of individuals. It was the *system* of capitalism that dictated people's behavior. Marx saw social change as arising from tensions inherent in a capitalist system—the conflict between the capitalist and working classes. Marx's ideas are often misperceived by U.S. students because communist revolutionaries throughout the world have claimed Marx as their guiding spirit. It would be naive to reject his ideas solely on political grounds. Much that Marx predicted has not occurred—for instance, he claimed that the "laws" of history made a worldwide revolution of workers inevitable, and this has not happened. Still, he left us an important body of sociological thought springing from his insight that society is systematic and structural and that class is a fundamental dimension of society that shapes social behavior.

Max Weber. **Max Weber** (1864–1920; pronounced "Vay-ber") was greatly influenced by Marx's work and built upon it. But, whereas Marx saw economics as the basic organizing element of society, Weber theorized that society had three basic dimensions: political, economic, and cultural. According to Weber, a complete sociological analysis must recognize the interplay between economic, political, and cultural institutions (Parsons 1947). Weber is credited with developing a *multidimensional* analysis of society that goes beyond Marx's more one-dimensional focus on economics.

Weber also theorized extensively about the relationship of sociology to social and political values. He did not believe there could be a value-free sociology, because values would always influence what sociologists considered worthy of study. Weber thought sociologists should acknowledge the influence of values so that ingrained beliefs might not interfere with objectivity. Weber professed that the task of sociologists is to teach students the uncomfortable truth about the world. Faculty should not use their positions to promote their political opinions, he felt; rather, they have a responsibility to examine all opinions, including unpopular ones, and use the tools of rigorous sociological inquiry to understand why people believe and behave as they do.

An important concept in Weber's sociology is *verstehen* (pronounced "ver-shtay-en"). **Verstehen,** a German word, refers to understanding social behavior from the point of view of those engaged in it.

Karl Marx analyzed capitalism as an economic system with enormous implications for how society is organized, in particular how inequality between groups stems from the economic organization of society.

Max Weber used a multidimensional approach to analyzing society, interpreting the economic, cultural, and political organization of society as together shaping social institutions and social change.

Weber believed that to understand social behavior, one had to understand the meaning that a behavior had for people. He did not believe sociologists had to be born into a group to understand it (in other words, he didn't believe "it takes one to know one"), but he did think sociologists had to develop some subjective understanding of how other people experience their world. One major contribution from Weber was the definition of *social action* as a behavior to which people give meaning (Gerth and Mills 1946; Parsons 1951b; Weber 1962), such as placing a bumper sticker on your car that states pride in U.S. military troops.

Sociology in America

American sociology was built on the earlier work of Europeans, but unique features of U.S. culture contribute to its distinctive flavor. Less theoretical and more practical than their European counterparts, early American sociologists believed that if they exposed the causes of social problems, they could alleviate some of the consequences, which are measured in human suffering.

Early sociologists in both Europe and the United States conceived of society as an organism, a system of interrelated functions and parts that work together to create the whole. This perspective is called the **organic metaphor.** Sociologists saw society as constantly evolving, like an organism. The question many early sociologists asked was to what extent humans could shape the evolution of society.

Many were influenced in this question by the work of British scholar **Charles Darwin** (1809–1882), who revolutionized biology when he identified the process termed *evolution,* a process by which new species are created through the survival of the fittest. **Social Darwinism** was the application of Darwinian thought to society. According to the social Darwinists, the "survival of the fittest" is the driving force of social evolution as well. They conceived of society as an organism that evolved from simple to complex in a process of adaptation to the environment. They theorized that society was best left alone to follow its natural evolutionary course. Because social Darwinists believed that evolution always took a course toward perfection, they advocated a *laissez-faire* (that is, "hands-off") approach to social change. Social Darwinism was thus a conservative mode of thought; it assumed that the current arrangements in society were natural and inevitable (Hofstadter 1944).

Most other early sociologists in the United States took a more reform-based approach. Nowhere was the emphasis on application more evident than at the University of Chicago, where a style of sociological thinking known as the Chicago School developed. The Chicago School is characterized by thinkers who were interested in how society shaped the mind and identity of people. We study some of these thinkers, such as George Herbert Mead and Charles Horton Cooley in Chapter 3. They thought of society as a human laboratory where they could observe and understand human behavior to be better able to address human needs, and they used the city in which they lived as a living laboratory.

Robert Park (1864–1944), from the University of Chicago, was a key founder of sociology. Originally a journalist who worked in several Midwestern cities, Park was interested in urban problems and how different racial groups interacted with each other. He was also fascinated by the sociological design of cities, noting that cities were typically sets of concentric circles. At the time, the very rich and the very poor lived in the middle, ringed by slums and low-income neighborhoods (Collins and Makowsky 1972; Coser 1977; Park and Burgess 1921). Park would still be intrigued by how boundaries are defined and maintained in urban neighborhoods. You might notice this yourself. A single street crossing might delineate a Vietnamese neighborhood from an Italian one, a White affluent neighborhood from a barrio. The social structure of cities continues to be a subject of sociological research.

Many early sociologists of the Chicago School were women whose work is only now being rediscovered. **Jane Addams** (1860–1935) was one of the most renowned sociologists of her day. She was a leader in the settlement house movement, which provided community services and did systematic research designed to improve the lives of slum dwellers, immigrants, and other dispossessed groups. Addams, the only practicing sociologist ever to win a Nobel Peace Prize (in 1931), never had a regular teaching job. Instead, she used her skills as a research sociologist to develop community projects that assisted people in need (Deegan 1988).

Although not part of the Chicago School, **W.E.B. Du Bois** (1868–1963; pronounced "due boys") was one of the most important early sociological thinkers in America. Du Bois was a prominent Black scholar, a cofounder in 1909 of the NAACP (National Association for the Advancement of Colored People), a prolific writer, and one of America's best

© CORBIS

Jane Addams, the only sociologist to win the Nobel Peace Prize, used her sociological skills to try to improve people's lives. The settlement house movement provided social services to groups in need, while also providing a social laboratory in which to observe the sociological dimensions of problems such as poverty.

Courtesy of University of Massachusetts at Amherst

An insightful observer of race and culture, W.E.B. Du Bois was one of the first sociologists to use community studies as the basis for sociological work. His work, long excluded from the "great works" of sociological theory, is now seen as a brilliant and lasting analysis of the significance of race in the United States.

minds. He received the first Ph.D. ever awarded to a Black person at Harvard University in any field, and he studied for a time in Germany, hearing several lectures by Max Weber. Du Bois was deeply troubled by the racial divisiveness in society, writing in a classic essay published in 1901 that "the problem of the twentieth century is the problem of the color line" (Du Bois 1901: 354). Like many of his women colleagues, he envisioned a community based, activist profession committed to social justice (Deegan 1988); he was a friend and collaborator with Jane Addams. He believed in the importance of a scientific approach to sociological questions, but also thought that convictions always directed one's studies.

Theoretical Frameworks: Functionalism, Conflict Theory, and Symbolic Interaction

The founders of sociology have established theoretical traditions that ask basic questions about society and inform sociological research. The idea of theory may seem dry to you because it connotes something that is only hypothetical and divorced from "real life." However, sociological theory is one of the tools that sociologists use to interpret real life. Sociologists use theory to organize their observations and apply them to the broad questions sociologists ask, such as How are individuals related to society? How is social order maintained? Why is there inequality in society? How does social change occur?

Different theoretical frameworks within sociology make different assumptions—and provide different insights—about the nature of society. In the realm of *macrosociology* are theories that strive to understand society as a whole. Durkheim, Marx, and Weber were macrosociological theorists. Theoretical frameworks that center on face-to-face social interaction are known as *microsociology.* Some of the work derived from the Chicago School—research that studies individuals and group processes in society—is microsociological. Although sociologists draw from diverse theoretical perspectives to understand society, three broad traditions form the major theoretical perspectives that they use: functionalism, conflict theory, and symbolic interaction.

Functionalism. Functionalism has its origins in the work of Durkheim, who you will recall was especially interested in how social order is possible or how society remains relatively stable. **Functionalism** interprets each part of society in terms of how it contributes to the stability of the whole. As Durkheim suggested, functionalism conceptualizes society as more than the sum of its component parts. Each part is "functional" for society—that is, contributes to the stability of the whole. The different parts are primarily the institutions of society, each of which is organized to fill different needs and each of which has particular consequences for the form and shape of society. The parts each then depend on one another.

The family as an institution, for example, serves multiple functions. At its most basic level, the family has a reproductive role. Within the family, infants receive protection and sustenance. As they grow older, they are exposed to the patterns and expectations of their culture. Across generations, the family supplies a broad unit of support and enriches individual experience with a sense of continuity with the past and future. All these aspects of family can be assessed by how they contribute to the stability and prosperity of society. The same is true for other institutions.

The functionalist framework emphasizes the consensus and order that exist in society, focusing on social stability and shared public values. From a functionalist perspective, disorganization in the system, such as deviant behavior and so forth, leads to change because societal components must adjust to achieve stability. This is a key part of functionalist theory—that when one part of society is not working (or is *dysfunctional,* as they would say), it affects all the other parts and creates social problems. Change may be for better or worse; changes for the worse stem from instability in the social system, such as a breakdown in shared values or a social institution no longer meeting people's needs (Collins 1994; Eitzen and Baca Zinn 2006; Turner 1974).

Functionalism was a dominant theoretical perspective in sociology for many years, and one of its major theorists was **Talcott Parsons** (1902–1979). In Parsons's view, all parts of a social system are interrelated with different parts of society having different basic functions. Functionalism was further developed by **Robert Merton** (1910–2003). Merton saw that social practices often have consequences for society that are not immediately apparent, nor are they necessarily the same as the stated purpose of a given practice. He suggested that human behavior has both manifest and latent functions. *Manifest functions* are the stated and intended goals of social

thinking sociologically

What are the *manifest functions* of grades in college? What are the *latent functions*?

behavior. *Latent functions* are the unintended consequences of behavior (Merton 1968). For example, reforming social welfare programs may have the manifest function of reducing federal budget expenditures, but the policy may also have the latent function of increasing crime (because people have to support themselves through illegitimate means) or increasing homelessness and street violence.

Critics of functionalism argue that its emphasis on social stability is inherently conservative and that it understates the roles of power and conflict in society. Critics also disagree with the explanation of inequality offered by functionalism—that it persists because social inequality creates a system for the fair and equitable distribution of societal resources. Functionalists would, for example, argue that it is fair and equitable that the higher social classes earn more money since they—so it is argued—are more important (functional) to society. Critics of functionalism argue that functionalism is too accepting of the status quo. Functionalists would counter this argument by saying that, regardless of the injustices that inequality produces, inequality serves a purpose in society: It provides an incentive system for people to work and promotes solidarity among groups linked by common social standing.

Conflict Theory. **Conflict theory** emphasizes the role of coercion and power, a person's or group's ability to exercise influence and control over others, in producing social order. Whereas functionalism emphasizes cohesion within society, conflict theory emphasizes strife and friction. Derived from the work of Karl Marx, conflict theory pictures society as fragmented into groups that compete for social and economic resources. Social order is maintained not by consensus, but by domination, with power in the hands of those with the greatest political, economic, and social resources. When consensus exists, according to conflict theorists, it is attributable to people being united around common interests, often in opposition to other groups (Dahrendorf 1959; Mills 1956).

According to conflict theory, inequality exists because those in control of a disproportionate share of society's resources actively defend their advantages. The masses are not bound to society by their shared values, but by coercion at the hands of the powerful. In conflict theory, the emphasis is on social control, not consensus and conformity. Groups and individuals advance their own interests, struggling over control of societal resources. Those with the most resources exercise power over others; inequality and power struggles are the result. Conflict theory gives great attention to class, race, and gender in society because these are seen as the grounds of the most pertinent and enduring struggles in society.

Whereas functionalists find some benefit to society in the unequal distribution of resources, conflict theorists see inequality as inherently unfair, persisting only because groups who are economically advantaged use their social position to their own betterment. Their dominance even extends to the point of shaping the beliefs of other members of the society by controlling public information and having major influence over institutions such as education and religion. From the conflict perspective, power struggles between conflicting groups are the source of social change. Typically, those with the greatest power are able to maintain their advantage at the expense of other groups.

Conflict theory has been criticized for neglecting the importance of shared values and public consensus in society while overemphasizing inequality. Like functionalist theory, conflict theory finds the origins of social behavior in the structure of society, but it differs from functionalism in emphasizing the importance of power.

Symbolic Interaction. The third major framework of sociological theory is **symbolic interaction theory.** Instead of thinking of society in terms of abstract institutions, symbolic interactionists consider immediate social interaction to be the place where "society" exists. Because of the human capacity for reflection, people give meaning to their behavior, and this is how they interpret the different behaviors, events, or things that are significant for sociological study.

Because of this, symbolic interaction, as its name implies, relies extensively on the symbolic meaning that people develop and rely upon in the process of social interaction. Symbolic interaction theory emphasizes face-to-face interaction and thus is a form of microsociology, whereas functionalism and conflict theory are more macrosociological.

Derived from the work of the Chicago School, symbolic interaction theory analyzes society by addressing the subjective meanings that people impose on objects, events, and behaviors. Subjective meanings are given primacy because, according to symbolic interactionists—and according to Thomas's earlier mentioned dictum—people behave based on what they *believe,* not just on what is objectively true. Thus, society is considered to be socially constructed through human interpretation (Berger and Luckmann 1967; Blumer 1969; Shibutani 1961). Symbolic interactionists see meaning as constantly modified through social interaction. People interpret

thinking sociologically

Think about the example given about smoking and using a *symbolic interaction* framework. How would you explain other risky behaviors, such as steroid use among athletes or eating disorders among young women?

one another's behavior, and it is these interpretations that form the social bond. These interpretations are called the "definition of the situation." For example, why would young people smoke cigarettes even when all objective medical evidence points to the danger of doing so? The answer is in the definition of the situation that people create. Studies find that teenagers are well informed about the risks of tobacco, but they also think that "smoking is cool," that they themselves will be safe from harm, and that smoking projects an image, a positive identity—for boys, as a "tough guy" and for girls, as fun-loving, mature, and glamorous. Smoking is also defined by young women as keeping you thin—an ideal constructed through dominant images of beauty. In other words, the symbolic meaning of smoking overrides the actual facts regarding smoking and risk (Sternja et al. 2004).

Symbolic interaction interprets social order as constantly negotiated and created through the interpretations people give to their behavior. In observing society, symbolic interactionists see not simply facts but "social constructions," the meanings attached to things, whether those are concrete symbols (like a certain way of dress or a tattoo) or nonverbal behaviors. To a symbolic interactionist, society is highly subjective—existing in the minds of people, even though its effects are very real.

Functionalism, conflict theory, and symbolic interaction theory are by no means the only theoretical frameworks in sociology. For some time, however, they have provided the most prominent general explanations of society. Each has a unique view of the social realm. None is a perfect explanation of society, yet each has something to contribute. Functionalism gives special weight to the order and cohesion that usually characterizes society. Conflict theory emphasizes the inequalities and power imbalances in society. Symbolic interaction emphasizes the meanings that humans give to their behavior. Together, these frameworks provide a rich, comprehensive perspective on society, individuals within society, and social change (see Table 1.1).

Diverse Theoretical Perspectives

Sociological theory is in transition. In addition to the three frameworks just discussed, other theories are also used within sociology. Contemporary sociological theory has been greatly influenced by the development of **feminist theory,** which analyzes the status of women and men in society with the purpose of using that knowledge to better women's lives. Feminist

table 1.1 Three Sociological Frameworks

Basic Questions:	Functionalism	Conflict Theory	Symbolic Interaction
What is the relationship of individuals to society?	Individuals occupy fixed social roles.	Individuals are subordinated to society.	Individuals and society are interdependent.
Why is there inequality?	Inequality is inevitable and functional for society.	Inequality results from a struggle over scarce resources.	Inequality is demonstrated through the importance of symbols.
How is social order possible?	Social order stems from consensus on public values.	Social order is maintained through power and coercion.	Social order is sustained through social interaction and adherence to social norms.
What is the source of social change?	Society seeks equilibrium when there is social disorganization.	Change comes through the mobilization of people struggling for resources.	Change derives from an ever-evolving set of social relationships and the creation of new meaning systems.
Major criticism:	This is a conservative view of society that underplays power differences among and between groups.	The theory understates the degree of cohesion and stability in society.	There is little analysis of inequality and it overstates the subjective basis of society.

theory has created vital new knowledge about women and transformed what is understood about men. Feminist scholarship in sociology, by focusing on the experiences of women, provides new ways of seeing the world and contributes to a more complete view of society. We examine feminist theory throughout this book in the context of particular topics, such as family (see Chapter 12) and politics (see Chapter 14).

Many contemporary theorists are increasingly influenced by postmodernism—a strand of thinking now influencing many disciplines. **Postmodernism** is based on the idea that society is not an objective thing. Instead, it is found in the words and images that people use to represent behavior and ideas. Postmodernists think that images and texts reveal the underlying ways that people think and act. Postmodernist studies typically involve detailed analyses of images, words, film, music, and other forms of popular culture. In a civilization such as that in the United States, saturated by the imagery of the mass media, postmodernist analysis illuminates much about society. Postmodernist thinkers see contemporary life as involving multiple experiences and interpretations, but they avoid categorizing human experience into broad and abstract concepts such as institutions or society (Rosenau 1992).

Whatever the theoretical framework used, theory is evaluated in terms of its ability to explain observed social facts. The sociological imagination is not a single-minded way of looking at the world. It is the ability to observe social behavior and interpret that behavior in light of societal influences.

© Nina Berman/Redux

Sociologists find social order even when it seems that there is just mass movement. One of the goals of sociological research is to systematically discover the processes involved in creating such order.

Doing Sociological Research

Sociological research is the tool sociologists use to answer questions. There are various methods that sociologists use to do research, all of which involve rigorous observation and careful analysis.

Suppose you want to know how homeless people live. What is life like for them? Do they interact with each other? How do others treat them? Do they have any sense of living in a community? Sociologist Mitch Duneier (1999) wanted to know these things so he studied a group of homeless people by living with them on their park benches and in doorways on New York City's lower East Side. He spent four years interacting with them—largely a group of African American men who sold books and magazines on the street. Duneier is White. He tells how becoming accepted into this society of African American men was itself an interesting process. Contrary to popular belief, he discovered that these men lived in a rather well-organized "mini-society," with a social structure, rules, norms, and culture.

Duneier was engaged in what is called **participant observation**—a sociological research technique in which the researcher actually becomes both participant in and observer of that which she or he studies.

There are other kinds of research that sociologists do, as well. Some approaches are more structured and focused than participant observation, such as survey research. Other methods may involve the use of official records or interviews. The different approaches used reflect the different questions asked. Other methods may require statistical analysis of a large set of information. Either way, the chosen research method must be appropriate to the sociological question being asked. (In the "Doing Sociological Research" boxes in this book, we explore different research projects that sociologists have done, showing what question they started with, how they did their research, and what they found.)

However it is done, research is an engaging and demanding process. It requires skill, careful observation, and the ability to think logically about the things that spark your sociological curiosity.

Sociology and the Scientific Method

Sociological research derives from the scientific method. The **scientific method** involves several steps in a research process, including observation, hypothesis testing, analysis of data, and generalization. Since its beginnings, sociology has attempted to adhere to the scientific method. To the degree that it has succeeded, sociology is a science; yet, there is also an art to developing sociological knowledge. Sociology aspires to be both scientific and humanistic, but sociological research varies in how strictly it adheres to the

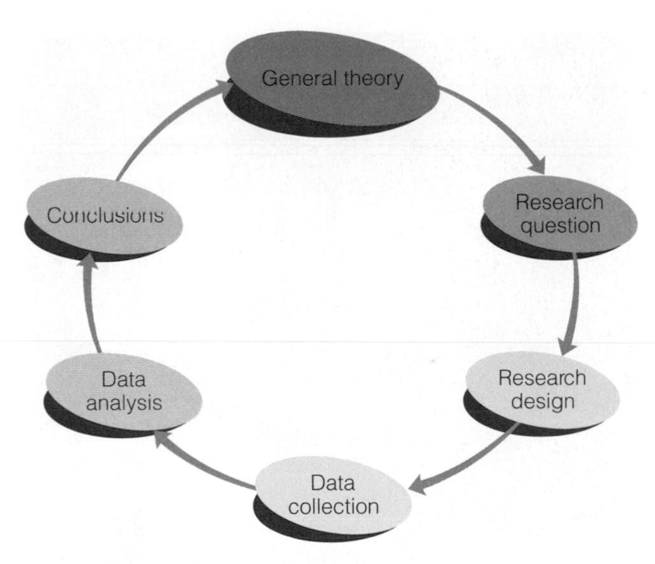

FIGURE 1.2 *The Research Process*

Research can begin by asking a research question derived from general theory or earlier studies, but it can also begin with an observation or even from the conclusion of prior research. One's research question is the basis for a research design and the subsequent collection of data. As this figure shows, the steps in the research process flow logically from what is being asked.

scientific method. Some sociologists test hypotheses (discussed later); others use more open-ended methods, such as in Duneier's study of homeless men.

Science is empirical, meaning it is based on careful and systematic observation, not just on conjecture. Although some sociological studies are highly *quantitative* and statistically sophisticated, others are *qualitatively* based—that is, based on more interpretive observations, not statistical analysis. Both quantitative and qualitative studies are empirical. Sociological studies may be based on surveys, observations, and many other forms of analysis, but they always depend on an empirical underpinning.

One wellspring of sociological insight is **deductive reasoning.** When a sociologist uses deductive reasoning, he or she creates a specific research question about a focused point that is based on a more general or universal principle (see Figure 1.2). Here is an example of deductive reasoning: One might reason that because Catholic doctrine forbids abortion, Catholics would then be less likely than other religious groups to support abortion rights. (You could test this via a survey.)

Inductive reasoning—another source of sociological insight—reverses this logic: that is, it arrives at general conclusions from specific observations. For example, if you observe that most of the demonstrators protesting abortion in front of a family planning clinic are evangelical Christians, you might infer that strongly held religious beliefs are important in determining human behavior. Again, referring to Figure 1.2, inductive reasoning would begin with one's observations. Either way—deductively or inductively—you are engaged in research.

The Research Process

When sociologists do research, they engage in a process of discovery. They organize their research questions and procedures systematically—their research site being the social world. Through research, sociologists organize their observations and interpret them.

Developing a Research Question. Sociological research is an organized practice that can be described in a series of steps (see Figure 1.2). The first step in sociological research is to develop a research question. One source of questions is past research. For any number of reasons, the sociologist might disagree with a research finding and decide to carry out further research or develop a detailed criticism of previous research. A research question can also begin from an observation that you make in everyday life, such as wondering about the lives of street people.

Developing a sociological research question typically involves reviewing the existing literature on the subject, such as past studies and research reports. Digital technology has vastly simplified the task of reviewing the literature. Researchers who once had to burrow through paper indexes and card catalogs to find material relevant to their studies can now scan much larger swaths of material in far less time, using online databases. The catalogs of most major libraries in the world are accessible on the Internet, as are specialized indexes, discussion groups, and other

The research process involves several operations that can be performed on the computer, such as entering data in numerical form and writing the research report.

research tools developed to assist sociological researchers. Increasingly, many journals that report new sociological research are now available online in full-text format. You must be careful using the Internet for research, however: How do you know when something found on the web is valid or true? Much of what is found on the web is invalid—that is, unsubstantiated by accurate research or empirical study. Pay attention, for example, to what person or group has posted the website. Is it a political organization? An organization promoting a cause? A person expressing an opinion? See the box "A Sociological Eye on the Media" for some guidelines about interpreting what you see on the web and in the media.

When you review prior research, you may wonder if the same results would be found if the study were repeated, perhaps examining a different group or studying the phenomenon at a different time. Research that is repeated exactly, but on a different group of people or in a different time or place, is called a **replication study.** Suppose earlier research found that women managers have fewer opportunities for promotion than men. You might want to know if this still holds true. You would then replicate the original study, probably using a different group of women and men managers, but asking the same questions that were asked earlier. A replication study can tell you what changes have occurred since the original study and may also refine the results of the earlier work. Research findings should be reproducible; if research is sound, other researchers who repeat a study should get the same results unless, of course, change has occurred in the interim.

© Bob Daemmrich/PhotoEdit

Some research is done by analyzing the content of various cultural artifacts. Content analysis *is one tool of sociological research.*

Creating a Research Design. A **research design** is the overall logic and strategy underlying a research project. Sociologists engaged in research may distribute questionnaires, interview people, or make direct observations in a social setting or laboratory. They might analyze cultural artifacts, such as magazines, newspapers, television shows, or other media. Some do research using historical records. Others base their work on the analysis of social policy. All these are forms of sociological observation. Research design consists of choosing the observational technique best suited to a particular research question.

Suppose you wanted to study the career goals of student athletes. In reviewing earlier studies, perhaps you found research discussing how athletics is related to academic achievement (Schacht 1996). You might also have read an article in your student newspaper reporting that the rate of graduation for women college athletes is much higher than the rate for men athletes and wondered if women athletes are better students than men athletes. In other words, are athletic participation, academic achievement, and gender interrelated and, if so, how?

Your research design would lay out a plan for investigating these questions. Which athletes would you study? How will you study them? To begin, you will need to get sound data on the graduation rates of the groups you are studying to verify that your assumption of better graduation rates among women athletes is actually true. Perhaps, you think, the differences between men and women are not so great when the men and women play the same sports. Or perhaps the differences depend on other factors, such as what kind of financial support they get or whether coaches encourage academic success. To observe the influence of coaches, you might observe interactions between coaches and student athletes, recording what coaches say about class work. As you proceed, you would probably refine your research design and even your research question. Do coaches encourage different traits in men and women athletes? To answer this question, you have to build into your research design a comparison of coaches interacting with men and with women. Perhaps you even want to compare female and male coaches and how they interact with women and men. *The details of your research design flow from the specific questions you ask.*

thinking sociologically

If you wanted to conduct research that would examine the relationship between student alcohol use and family background, what *measures* would you use to get at the two variables: alcohol use and family background? How might you *design* your study?

Quantitative versus Qualitative Research. The research design often involves deciding whether the research will be qualitative or quantitative or perhaps some combination of both. **Quantitative research** is research that uses numerical analysis. In essence this approach reduces the data into numbers, for example, the percentage of teenage mothers in California. **Qualitative research** is somewhat less structured than quantitative research, yet still focuses on a central research question. Qualitative research allows for more interpretation and nuance in what people say and do and thus can provide an in-depth look at a particular social behavior. Both forms of research are useful, and both are used extensively in sociology.

Some research designs involve the testing of hypotheses. A **hypothesis** (pronounced "hy-POTH-i-sis") is a prediction, or a hunch, a tentative assumption that one intends to test. If you have a research design that calls for the investigation of a very specific hunch, you might formulate a hypothesis. Hypotheses are often formulated as if–then statements. For example,

> *hypothesis:* If a person's parents are racially prejudiced, then that person will, on average, be more prejudiced than a person whose parents are free of prejudice.

This is merely a hypothesis or expectation, not a demonstration of fact. Having phrased a hypothesis, the sociologist must then determine if it is true or false. To test the preceding example, one might take a large sample of people and determine their prejudice level by interviews or some other mechanism. One would then determine the prejudice level of their parents. According to the hypothesis, one would expect to find more prejudiced children among prejudiced parents and more nonprejudiced children among nonprejudiced parents. If this association is found, the hypothesis is supported. If it is not found, then the hypothesis would be rejected.

Not all sociological research follows the model of hypothesis testing, but all research includes a plan for how **data** will be gathered. (Note that *data* is the plural form; one says, "data are used . . . ," not "data is used . . ..") Data can be qualitative or quantitative; either way, they are still data. Sociologists often try to convert their observations into a quantitative form (see "Statistics in Sociology" on page 25).

Sociologists frequently design research to test the influence of one variable on another. A **variable** can have more than one value or score. A variable can be relatively straightforward, such as age or income, or a variable may be more abstract, such as social class or degree of prejudice. In much sociological research, variables are analyzed to understand how they influence each other. With proper measurement techniques and a good research design, the relationships between different variables can be discerned. In the example of student athletes given above, the variables you use would likely be student graduation rates, gender, and perhaps the sport played.

An **independent variable** (see Figure 1.3) is one that the researcher wants to test as the presumed cause of something else. The **dependent variable** is one on which there is a presumed effect. That is, if X is the independent variable, then X leads to Y, the dependent variable. In the previous example of the hypothesis, the amount of prejudice of the parent is the independent variable, and the amount of prejudice of the child (offspring) is the dependent variable. In some sociological research, *intervening variables* are also studied: variables that fall between the independent and dependent variables. Sociological variables are often sensitive to a variety of influences, and trying to control for the influence of factors other than the one being studied is a challenge to the creativity of the researcher.

Variables are sometimes used to show more abstract concepts that cannot be directly measured, such as the concept of social class. In such cases the variables are **indicators**—something that points to or reflects an abstract concept. An example is shown in Map 1.1 on the United Nations' human development index. Here, the human development index is an *indicator* used to show different levels of well-being around the world. The index involves several variables, including life expectancy and educational attainment, combined to show levels of well-being.

The **validity** of a measurement is the degree to which it accurately measures or reflects a concept. To ensure the validity of their findings, researchers usually use more than one measure for a particular concept. If two or more chosen measures of a concept give similar results, it is likely that the measurements are giving an accurate—that is, valid—depiction of the concept.

Sociologists also must be concerned with the **reliability** of their research results. A measurement is reliable if repeating the measurement gives the same result. If a person is given a test daily and every day the test gives different results, then the reliability of the test is poor. One way to ensure that sociological measurements are reliable is to use measures that have proved sound in past studies. Another technique is to have a variety of people gather the data to make certain the results are not skewed by the tester's appearance, personality, and so forth. The researcher must be sensitive to all factors that affect the reliability of a study.

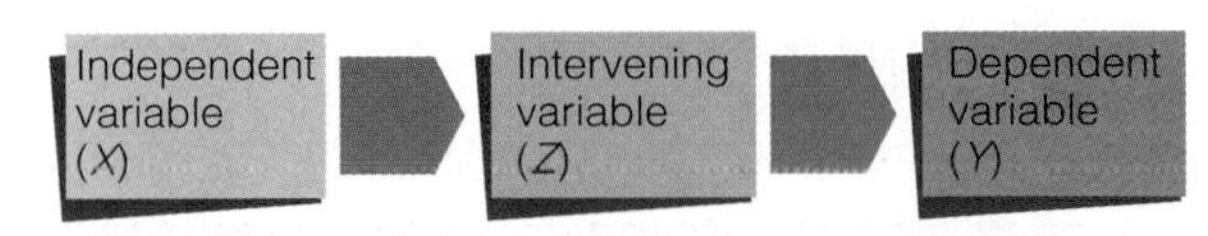

FIGURE 1.3 *The Analysis of Variables*

Sociological research seeks to find out whether some independent variable (X) affects an intervening variable (Z), which in turn affects a dependent variable (Y).

[a sociological eye on the media

Research and the Media

On any given day, if you watch the news, read a newspaper, or search the web, you are likely to learn about various new research studies purporting some new finding. How do you know if the research results reported in the media are accurate?

Most people are not likely to check the details of the study or have the research skills to verify the study's claims. But, one benefit of learning the basic concepts and tools of sociological research is to be able to critically assess the research frequently reported in the media. The following questions will help.

1. **What are the major variables in the study? Are the researchers claiming a causal connection between two or more variables?** For example, the press reported that one way parents can reduce the chances of their children becoming sexually active at an early age is to quit smoking (O'Neil 2002). The researcher who conducted this study actually claimed there was no direct link between parental smoking and teen sex, although she did find a correlation between parents' risky behaviors—smoking, heavy drinking, and not using seat belts—and children's sexual activity. She argued that parents who engage in unsafe activities provide a model for their children's own risky behavior (Wilder and Watt 2002).

   Just because there is a link, it does not mean one caused the other. Seeing parental behavior as a model for what children do is hardly the same thing as seeing parents' smoking as the cause of early sexual activity!

2. **How have researchers defined and measured the major topics of their study?** For example, if someone claims that 10 percent of all people are gay, how is "being gay" defined? Does it mean having had one such experience over one's entire lifetime or does it mean actually having a gay identity? The difference matters because one definition will likely inflate the number reported. Sometimes you must look up the original study to learn how things are defined or how they are measured. Ask yourself if the same conclusions would be reached had the researchers used different definitions and measurements.

3. **Is the research based on a random scientific sample or is it biased?** You might have to go to the original source of the study to learn this, but often the sample will be reported in the press (even if in nonscientific language). For example, a study widely reported in the media had headlines exclaiming: "Study Links Working Mothers to Slower Learning" (Lewin 2002). But, if you read even the news report closely, you will learn that this study included only White, non-Hispanic families. Black and Hispanic children were dropped from some of the published results because there were too few cases in the sample to make meaningful statistical comparisons (Brooks-Gunn et al. 2002). A later study by the same research team found that there were no significant effects of mother's employment on children's intellectual development among African American or Hispanic children (Waldfogel et al. 2002). The point is not that the study is invalid, but that its results have more limited implications than the headlines suggest.

4. **Is there false generalization in the media report?** Often a study has more limited claims in the scientific version than what is reported in the media. Using the example just given about the connection between maternal employment and children's learning, it would be a big mistake to generalize from the study's results to all children and families. Remember that some groups were not included. Even within the study, the researchers found less effect due to mothers' employment in female-headed households than among married couples. You cannot generalize the findings reported in the media to all families.

5. **If a study uses numerical data to support its claims, what do the numbers mean and how were they generated?** Numbers can also be misleading. How were the numbers or statistics in the reports you see generated? When people do statistical research, how they record and analyze data may be poorly decided. Are the numbers reported reasonable? Sometimes people can use statistics to exaggerate—or downplay—a given issue (Best 2001).

6. **Can the study be replicated?** Unless there is full disclosure of the research methodology (that is, how the study was conducted), this will not be possible. But you can ask yourself how the study was conducted, whether the procedures used were reasonable and logical, and whether the researchers made good decisions in constructing their research question and research design.

7. **Who sponsored the study and do they have a vested interest in the study's results?** Find out if a group or organization with a particular interest in the outcome sponsors the research. For example, would you give as much validity to a study of environmental pollution that was funded and secretly conducted by a chemical company as you would a study on the same topic conducted by independent scientists who openly report their research methods and results? Research sponsored by interested parties does not necessarily negate research findings, but it can raise questions about the researchers' objectivity and the standards of inquiry they used.

8. **Who benefits from the study's conclusions?** Although this question does not necessarily challenge the study's findings, it can help you think about whom the findings are likely to help.

9. **What assumptions did the researchers have to make to ask the question they did?** For example, if you started from the assumption that poverty is not the individual's fault, but is the result of how society is structured, would you study the values of the poor or perhaps the values of policy makers? When research studies explore matters where social values influence people's opinions, it is especially important to identify the assumptions made by certain questions.

10. **What are the implications of the study's claims?** Thinking through the policy implications of a given result can often help you see things in a new light, particularly given how the media tend to sensationalize much of what is reported.

    Consider the study of maternal employment and children's intellectual development examined in Question 3. If you take the media headlines at face value, you might leap to the conclusion that working mothers hurt their children's intellectual development, and you might then think it would be best if mothers quit their jobs and stayed at home. But is this a reasonable implication of this study? Does the study not have just as many implications for day-care policies as it does for encouraging stay-at-home mothers? Especially when reported research studies involve politically charged topics (such as issues of "family values"), it is important to ask questions that explore various implications of social policies.

11. **Do these questions mean you should never believe anything you hear in the media?** Of course not. Thinking critically about research does not mean being negative or cynical about everything you hear or read. The point is not to reject all media claims out of hand, but instead to be able to evaluate good versus bad research. All research has limitations. Learning the basic tools of research, even if you never conduct research yourself or pursue a career where you would use such skills, can make you a better informed citizen and prevent you from being duped by claims that are neither scientifically nor sociologically valid.

map 1.1 VIEWING SOCIETY IN GLOBAL PERSPECTIVE

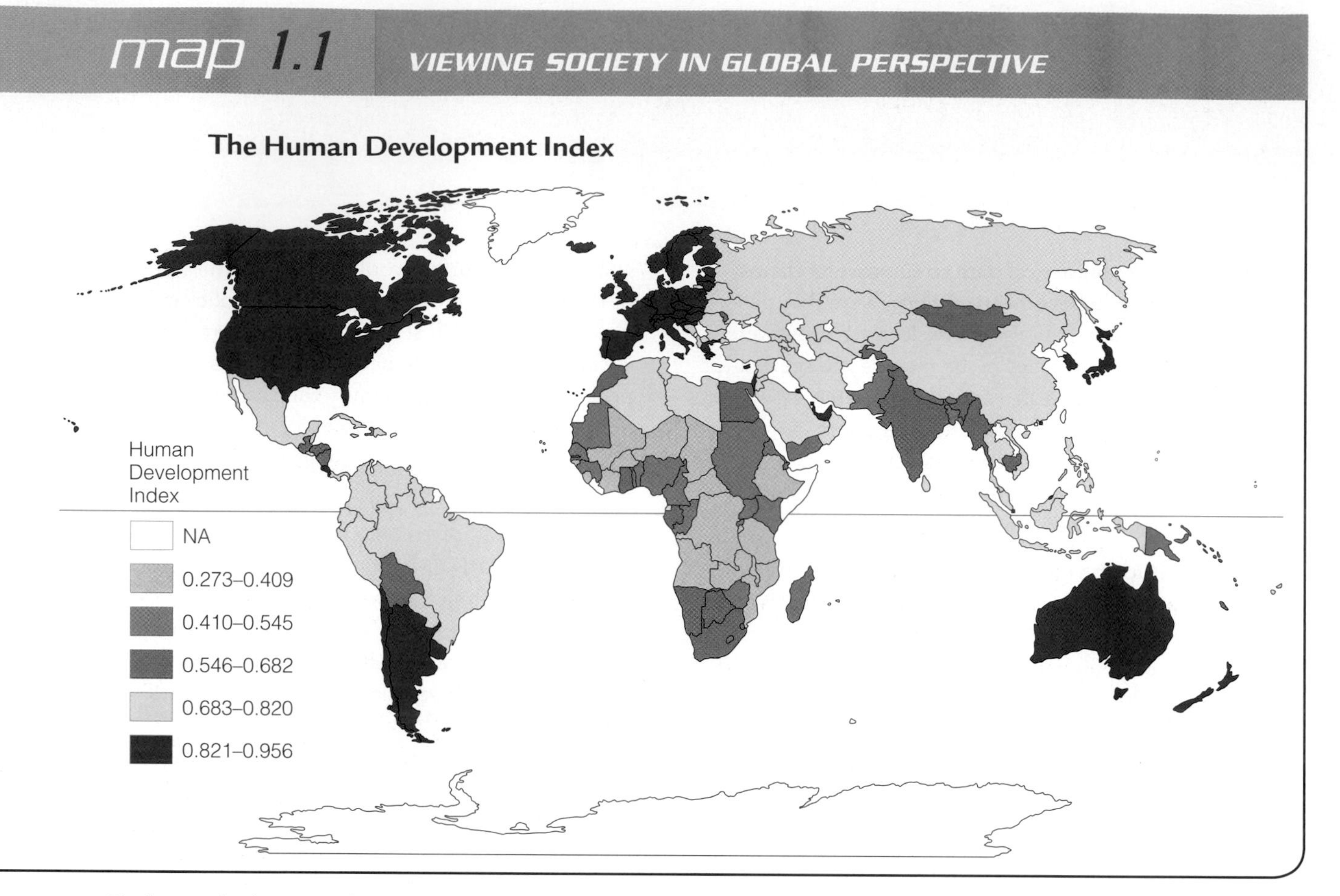

The human development index is an indicator developed by the United Nations and used to show the differing levels of well-being in nations around the world. The index is calculated using a number of measures, including life expectancy, educational attainment, and standard of living. Are these reasonable indicators of well-being? What else might you use?

Data: United Nations. 2003. Human Development Report 2003. **www.undp.org**

Gathering Data. After research design comes data collection. During this stage of the research process, the researcher interviews people, observes behaviors, or collects facts that throw light on the research question. When sociologists gather original material, the product is known as *primary data.* Examples include the answers to questionnaires or notes made while observing group behavior. Sociologists often rely on *secondary data,* namely data that have already been gathered and organized by some other party. This can include national opinion polls, census data, national crime statistics, or data from an earlier study made available by the original researcher. Secondary data may also come from official sources, such as university records, city or county records, national health statistics, or historical records.

When gathering data, often the groups that sociologists want to study are so large or so dispersed that research on the whole group is impossible. To construct a picture of the entire group, sociologists take data from a subset of the group and extrapolate to get a picture of the whole. A **sample** is any subset of a population. A **population** is a relatively large collection of people (or other units) that a researcher studies and about which generalizations are made. Suppose a sociologist wants to study the students at your school. All the students together constitute the population being studied. A survey could be done that reached every student, but conducting a detailed interview with every student would be highly impractical. If the sociologist wants the sort of information that can be gathered only during a personal interview, she would study only a portion, or sample, of all the students at your school.

How is it possible to draw accurate conclusions about a population by studying only part of it? The secret lies in making sure that the sample is *representative* of the population as a whole. The sample should have the same mix of people as the larger population and in the same proportions. If the sample is representative, then the researcher can generalize what she finds from the sample to the entire population. For example, if she interviews a sample of 100 students and finds that 10 percent of them are in favor of a tuition increase, and if the sample is representative of the population, then she can conclude that about

Statistics in Sociology

Certain fundamental statistical concepts are basic to sociological research. Although not all sociologists do quantitative research, basic statistics are important to interpreting sociological studies.

A **percentage** is the same as parts per hundred. To say that 22 percent of U.S. children are poor tells you that for every 100 children randomly selected from the whole population, approximately 22 will be poor. A **rate** is the same as parts per some number, such as per 10,000 or 100,000. The homicide rate in 2005 was 5.6, meaning that for every 100,000 in the population, approximately 5.6 were murdered (Federal Bureau of Investigation 2005). A rate is meaningless without knowing the numeric base on which it is founded—it is always the number per some other number.

A **mean** is the same as an average. Adding a list of fifteen numbers and dividing by fifteen gives the mean. The **median** is often confused with the mean but is actually quite different. The median is the midpoint in a series of values arranged in numeric order. In a list of fifteen numbers arrayed in numeric order, the eighth number (the middle number) is the median. In some cases, the median is a better measure than the mean because the mean can be skewed by extremes at either end. Another often used measure is the **mode,** which is simply the value (or score) that appears most frequently in a set of data.

Let's illustrate the difference between mean and median using national income distribution as an example. Suppose that you have a group of ten people. Two make $10,000 per year, seven make $40,000 per year, and one makes $1 million per year. If you calculate the mean, it comes to $130,000. The median, on the other hand, is $40,000—a figure that more accurately suggests the income profile of the group. That single million-a-year earner dramatically distorts, or skews, the picture of the group's income. If we want information about how the group in general lives, we are wiser to use the median income figure as a rough guide, not the mean. Note also that in this example the mode is the same as the median: $40,000.

Sociologists frequently choose to relate different variables to each other. **Correlation** is a widely used technique for analyzing the patterns of association, or correlation, between pairs of variables such as income and education. We might begin with a questionnaire that ask for annual earnings (*Y*) and level of education (*X*). Correlation analysis delivers two dimensions of information: It tells us the "direction" of the relationship between *X* and *Y* and also the strength of that relationship. The direction of a relationship is *positive* (that is, a positive correlation exists) if *X* is low when *Y* is low *and* if *X* is high when *Y* is high. But there is also a correlation if *Y* is low when *X* is high (or vice versa); this is a *negative,* or inverse, correlation. The strength of a correlation is simply how closely or tightly the variables are associated, regardless of the direction of correlation.

A correlation does not necessarily imply cause and effect. A correlation is simply an association, one whose cause must be explained by means other than simple correlation analysis. A *spurious correlation* exists when there is no meaningful causal connection between apparently associated effects.

Statistical information is notoriously easy to misinterpret, willfully or accidentally. Examples of some statistical mistakes include the following:

- **Citing a correlation as a cause.** A correlation reveals an association between things. Correlations do not necessarily indicate that one thing causes the other. Sociologists often say "Correlation is not proof of causation."
- **Overgeneralizing.** Statistical findings are limited by the extent to which the sample group actually reflects or represents the population from which the sample was obtained. Generalizing beyond the population is a misuse of statistics. Studying only men and then generalizing conclusions to both men and women would be an example of overgeneralizing.
- **Building in bias.** In a famous advertising campaign, public taste tests were offered between two soft drinks. A wily journalist verified that in at least one site, the brand sold by the sponsor of the test was a few degrees colder than its competitor when it was given to testers, which biased the results. Bias can also be built into studies by careless wording on questionnaires.
- **Faking data.** Perhaps one of the worse misuses of statistics is actually making up, or faking, data. A famous instance of this occurred in a study of separated identical twins (Burt 1966). The researcher wished to show that despite their separation, the twins remained similar in certain traits, such as measured intelligence (IQ), thus suggesting that their (identical) genes caused their striking similarity in intelligence. It was later shown that the data were fabricated (Kamin 1974; Hearnshaw 1979; Taylor 1980, 2002; Mackintosh 1995).
- **Using data selectively.** Sometimes a survey includes many questions, but the researcher reports on only a few of the answers. Doing so makes it quite easy to misstate the findings. Researchers often do not report findings that show no association between variables, but these can be just as telling as associations that do exist. For example, researchers on gender differences typically report the differences they find between men and women, but seldom publish their findings when the results for men and women are identical. This tends to exaggerate the differences between women and men and falsely confirms certain social stereotypes about gender differences.

10 percent of *all* the students at your school are in favor of a tuition increase. Note that a sample of 5 or 6 students would probably result in generalizations of poor quality, because the sample is not large enough to be representative. A *biased* (nonrepresentative) sample can lead to grossly inaccurate conclusions.

The best way to ensure a representative sample is to make certain that the sample population is selected randomly. A scientific **random sample** gives everyone in the population an equal chance of being selected. Quite often, striking and controversial research findings prove to be distorted by inadequate sampling. The man-on-the-street survey, much favored by radio and TV news reports, is the least scientific type of sample and the least representative.

Analyzing the Data. After the data have been collected, whether primary or secondary data, they must be analyzed. **Data analysis** is the process by which sociologists organize collected data to discover the patterns and uniformities that the data reveal. The analysis may be statistical or qualitative. When the data analysis is completed, conclusions and generalizations can be made. Data analysis is labor intensive, but it is also an exciting phase of research. Here is where research discoveries are made.

Reaching Conclusions and Reporting Results. The final stage in research is developing conclusions, relating findings to sociological theory and past research, and reporting the findings.

An important question researchers will ask at this stage is whether their findings can be generalized. **Generalization** is the ability to draw conclusions from specific data and to apply them to a broader population. Researchers ask, do my results apply only to those people who were studied, or do they also apply to the world beyond? Assuming that the results have wide application, the researcher can then ask if the findings refine or refute existing theories and whether the research has direct application to social issues. Using the earlier example of the relationship between parent and offspring prejudice, if you found that racially prejudiced people did tend to have racially prejudiced parents (thus supporting your hypothesis), then you might report these results in a paper or research report. You might also ask, what kinds of programs for reducing prejudice do the results of your study suggest?

The Tools of Sociological Research

There are several tools or techniques sociologists use to gather data. Among the most widely used are survey research, participant observation, controlled experiments, content analysis, historical research, and evaluation research.

The Survey: Polls, Questionnaires, and Interviews

Whether in the form of a questionnaire, interview, or telephone poll, surveys are among the most commonly used tools of sociological research. Questionnaires are typically distributed to large groups of people. The *return rate* is the percentage of questionnaires returned out of all those distributed. A low return rate introduces possible bias because the small number of responses may not be representative of the whole group.

Like questionnaires, interviews provide a structured way to ask people questions. They may be conducted face to face, by phone, even by electronic mail. Interview questions may be open-ended or closed-ended, though the open-ended form is particularly accommodating if respondents wish to elaborate.

Debunking Society's Myths

Myth: People who are just hanging out together and relaxing don't care much about social differences between them.

Sociological research: Even casual groups have organized social hierarchies. That is, they make distinctions within the group that give some people higher status than others (Anderson 1976, 1990; Whyte 1943).

As a research tool, surveys make it possible to ask specific questions about a large number of topics and then to perform sophisticated analyses to find patterns and relationships among variables. The disadvantages of surveys arise from their rigidity. Responses may not accurately capture the opinions of the respondent or fail to capture nuances in people's behavior and attitudes. Also, what people say and what they do are not always the same. Survey researchers must be persistent to get answers that are truthful, one reason for allowing respondents to be anonymous.

Participant Observation

A unique and interesting way for sociologists to collect data and study society is to actually become part of the group they are studying. This is the method of participant observation. Two roles are played at the same time: subjective participant and objective observer. Usually, the group is aware that the sociologist is studying them, but not always. Participant observation is sometimes called *field research,* a term from the field of anthropology.

Participant observation combines subjective knowledge gained through personal involvement and objective knowledge acquired by disciplined recording of what one has seen. The subjective component supplies a dimension of information that is lacking in survey data.

There are a few built-in weaknesses to participant observation as a research technique. We already mentioned that it is very time-consuming. Participant observers have to cull data from vast amounts of notes. Such studies usually focus on fairly small groups, posing problems of generalization. Observers may also lose their objectivity by becoming too much a part of what they study. These limitations aside, participant observation has been the source of some of the most arresting and valuable studies in sociology.

Controlled Experiments

Controlled experiments are highly focused ways of collecting data and are especially useful for determining a pattern of cause and effect. To conduct a controlled experiment, two groups are created, an *experimental group,* which is exposed to the factor one is examining, and the *control group,* which is not. In a controlled experiment, external influences are either eliminated or equalized, that is, held constant. This is necessary in order to establish cause and effect.

Suppose you wanted to study whether violent television programming causes aggressive behavior in children. You could conduct a controlled experiment to investigate this question. The behavior of children would be the dependent variable; the independent variable is exposure to violent programming. To investigate your question, you would expose an experimental group of children (under monitored conditions) to a movie containing lots of violence (martial arts, for example, or gunfighting). The control group would watch a movie that is free of violence. Aggressiveness in the children would be measured twice: a pretest measurement made before the movies are shown and a posttest measurement made afterward. You would take pre- and posttest measures on both the control and the experimental groups. Studies of this sort actually find that the children who watched the violent movie are indeed more violent and aggressive afterward than those who watched a movie containing no violence (Worchel et al. 2000; Goldstein 1994).

Among its advantages, a controlled experiment can establish causation, and it can zero in on a single independent variable. On the downside, controlled experiments can be artificial. They are for the most part done in a contrived laboratory setting (unless it is a field experiment), and they tend to eliminate many real-life effects. Analysis of controlled experiments includes making judgments about how much the artificial setting has affected the results.

Content Analysis

Researchers can learn a vast amount about a society by analyzing cultural artifacts such as newspapers, magazines, TV programs, or popular music. **Content analysis** is a way of measuring by examining the cultural artifacts of what people write, say, see, and hear. The researcher studies not people but the communications the people produce as a way of creating a picture of their society.

Playing a violent video game often causes the player to be somewhat more aggressive afterward.

Content analysis is frequently used to measure cultural change and to study different aspects of culture (Lamont 1992). Sociologists also use content analysis as an indirect way to determine how social groups are perceived—they might examine, for example, how Asian Americans are depicted in television dramas or how women are depicted in advertisements.

Content analysis has the advantage of being *unobtrusive.* The research has no effect on the person being studied, because the cultural artifact has already been produced. Content analysis is limited in what it can study, however, because it is based only on mass communication—either visual, oral, or written. It cannot tell us what people really think about these images or whether they affect people's behavior. Other methods of research, such as interviewing or participant observation, would be used to answer these questions.

Historical Research

Historical research examines sociological themes over time. It is commonly done in historical archives, such as official records, church records, town archives, private diaries, or oral histories. The sources of this sort of material are critical to its quality and applicability. Oral histories have been especially illuminating, most dramatically in revealing the unknown histories of groups that have been ignored or misrepresented in other historical accounts. For example, when developing an account of the spirituality of Native Americans, one would be misguided to rely solely on the records left by Christian missionaries or U.S. Army officials. These records would give a useful picture of how Whites perceived Native American religion, but they would be a very poor source for discovering how Native Americans understood their own spirituality.

Handled properly, comparative and historical research is rich with the ability to capture long-term social changes and is the perfect tool for sociologists who want to ground their studies in historical or comparative perspectives.

Evaluation Research

Evaluation research assesses the effect of policies and programs on people in society. If the research is intended to produce policy recommendations, then it is called *policy research.*

Suppose you want to know if an educational program is actually improving student performance. You could design a study that measured the academic performance of two groups of students, one that participates in the program and one that does not. If the academic performance of students in the program is better, and if the groups are alike in other ways (they are often matched to accomplish this), you would conclude that the program was effective. If you use this research to recommend social policy, you would be doing policy research.

Table 1.2 summarizes the six research techniques we have just discussed.

Research Ethics: Is Sociology Value-Free?

The topics dealt with by sociology are often controversial. People have strong opinions about social questions, and in some cases, the settings for sociological work are highly politicized. Imagine spending time in an urban precinct house to do research on police brutality or doing research on acquired immune deficiency syndrome (AIDS) and sex education in a conservative public school system. Under these conditions, can sociology be scientifically objective? How do researchers balance their own political and moral commitments against the need to be objective and open-minded? Sociological knowledge has an intimate connection to political values and social views. Often the very purpose of sociological research is to gather data as a step in creating social policy. Can sociology be value free? Should it be?

This is an important question without a simple answer. Most sociologists do not claim to be value free, but they do try to produce objective research. It must be acknowledged that researchers make choices throughout their research that can influence their

table 1.2 A Comparison of Six Research Techniques

Technique (Tool)	Qualitative Analysis or Quantitative Analysis	Advantages	Disadvantages
The survey (polls, questionnaires, interviews)	Usually quantitative	Permits the study of a large number of variables; results can be generalized to a larger population if sampling is accurate.	Difficult to focus in great depth on a few variables; difficult to measure subtle nuances in people's attitudes
Participant observation	Usually qualitative	Studies actual behavior in its home setting; affords great depth of inquiry	Is very time-consuming; it is difficult to generalize beyond the research setting.
Controlled experiment	Usually quantitative	Focuses on only two or three variables; able to study cause and effect	Difficult or impossible to measure large number of variables; may have an artificial quality
Content analysis	Can be either qualitative or quantitative	A way of measuring culture	Limited by studying only cultural products or artifacts (music, TV programs, stories, other), rather than people's actual attitudes
Historical research	Usually qualitative	Saves time and expense in data collection; takes differences over time into account	Data often reflect biases of the original researcher and reflect cultural norms that were in effect when the data were collected.
Evaluation research	Can be either qualitative or quantitative	Evaluates the actual outcomes of a program or strategy; often direct policy application	Limited in the number of variables that can be measured; maintaining objectivity is problematic if research is done or commissioned by administrators of the program being evaluated.

results. The problems sociologists choose to study, the people they decide to observe, the research design they select, and the type of media they use to distribute their research can all be influenced by the values of the researcher.

Sociological research often raises ethical questions. In fact, ethical considerations of one sort or another exist with any type of research. In a survey, the person being questioned is often not told the purpose of the survey or who is funding the study. Is it ethical to conceal this type of information? In controlled experiments, deception is often employed. Many experiments depend on respondents giving natural, spontaneous responses to staged situations. Researchers often reveal the true purpose of an experiment only after it is completed. The deception is therefore temporary. Does that lessen the potential ethical violation?

The American Sociological Association (ASA) has developed a professional code of ethics (see the ASA website for the full code of ethics). The federal government also has many regulations about the protection of human subjects. Ethical researchers adhere to these guidelines and must ensure that research subjects are not subjected to physical, mental, or legal harm. Research subjects must also be informed of the rights and responsibilities of both researcher and subject. Sociologists, like other scientists, also should not involve people in research without what is called *informed consent*—that is, getting agreement to participate from the research subjects. There can be exceptions to this, such as observing people in public places. Sociologists also take measures not to identify their subjects through the use of pseudonyms or by avoiding the use of names at all.

This chapter has covered the basics of doing sociological research and introduced you to the general contours of a sociological perspective. The remaining chapters of the book explore different dimensions of human society and the different conclusions that sociologists have drawn from the study of society. As you read, you should begin to see the fascination with which sociologists view the social world and how social change and social stability influence the people who constitute society.

1 Chapter Summary

- ***What is sociology?***

  Sociology is the study of human behavior in society. The *sociological imagination* is the ability to see societal patterns that influence individuals. Sociology is an *empirical* discipline, relying on careful observations as the basis for its knowledge.

- ***What is debunking?***

  Debunking in sociology refers to the ability to look behind things taken for granted, looking instead to the origins of social behavior.

- ***Why is diversity central to the study of sociology?***

  One of the central insights of sociology is its analysis of social diversity and inequality. Understanding *diversity* is critical to sociology because it is necessary to analyze *social institutions* and because diversity shapes most of our social and cultural institutions.

- ***When and how did sociology emerge as a field of study?***

  Sociology emerged in western Europe during the *Enlightenment* and was influenced by the values of critical reason, humanitarianism, and positivism. *Auguste Comte,* one of the earliest sociologists, emphasized sociology as a positivist discipline. *Alexis de Tocqueville* and *Harriet Martineau* developed early and insightful analyses of American culture.

- ***What are some of the basic insights of classical sociological theory?***

  Emile Durkheim is credited with conceptualizing society as a social system and with identifying *social facts* as patterns of behavior that are external to the individual. *Karl Marx* showed how capitalism shaped the development of society. *Max Weber* sought to explain society through cultural, political, and economic factors.

- ***What are the major theoretical frameworks in sociology?***

  Functionalism emphasizes the stability and integration in society. *Conflict theory* sees society as organized around the unequal distribution of resources and held together through power and coercion. *Symbolic interaction* theory emphasizes the role of individuals in giving meaning to social behavior, thereby creating society.

  Feminist theory is the analysis of women and men in society and is intended to improve women's lives. *Postmodernism* argues that constructed cultural products are the realities of our postindustrial society.

- ***Is sociology scientific?***

  Sociological research is derived from the *scientific method,* meaning that it relies on *empirical* observation and, sometimes, the testing of *hypotheses.* The research process involves several steps: developing a research question, designing the research, collecting data, analyzing data, and developing conclusions.

- ***What different tools of research do sociologists use?***

  The most common tools of sociological research are surveys and interviews, participant observation, controlled experiments, content analysis, comparative and historical research, and evaluation research. Each has its own strengths and weaknesses. You can better generalize from surveys, for example, than participant observation, but participant observation is better for capturing subtle nuances and depth in social behavior.

- ***Can sociology be value free?***

  Although no research in any field can always be value free, sociological research nonetheless strives for objectivity while recognizing that the values of the research may have some influence on the work. There are ethical considerations in doing sociological research, such as whether one should collect data without letting research subjects (people) know they are being observed.

Key Terms

conflict theory 16
content analysis 27
controlled experiment 27
correlation 25
data 21
data analysis 26
debunking 6
deductive reasoning 19
dependent variable 21
diversity 10
empirical 4
Enlightenment 11
evaluation research 28
feminist theory 17
functionalism 15
generalization 26
hypothesis 21
independent variable 21
indicator 21
inductive reasoning 19
issues 3
mean 25
median 25
mode 25
organic metaphor 14
participant observation 18
percentage 25
population 24
positivism 11
postmodernism 18
qualitative research 21
quantitative research 21
random sample 26
rate 25
reliability 21
replication study 20
research design 20
sample 24
scientific method 18
social change 9
social Darwinism 14
social facts 12
social institution 9
social interaction 9
social structure 3
sociological imagination 2
sociology 2
symbolic interaction theory 16
troubles 3
validity 21
variable 21
verstehen 13

Online Resources

Sociology: The Essentials Companion Website

www.thomsonedu.com/sociology/andersen

Visit your book companion website where you will find more resources to help you study and write your research papers. Resources include Suggested Readings, web links, and a MicroCase Online feature that teaches you how to research society. Other resources include Learning Objectives, Internet exercises, quizzing, and flash cards.

is an easy-to-use online resource that helps you study in less time to get the grade you want NOW.

www.thomsonedu.com/login

Need help studying? This site is your one-stop study shop. Take a Pre-Test and Thomson NOW will generate a Personalized Study Plan based on your test results. The Study Plan will identify the topics you need to review and direct you to online resources to help you master those topics. You can then take a Post-Test to determine the concepts you have mastered and what you still need to work on.

2

CHAPTER TWO

Culture

Defining Culture

The Elements of Culture

Cultural Diversity

Popular Culture and the Media

Theoretical Perspectives on Culture

Cultural Change

In one contemporary society known for its technological sophistication, people—especially the young—walk around with plugs in their ears. The plugs are connected to small wires that are themselves coated with a plastic film. These little plastic-covered wires are then connected to small devices made of metal, plastic, silicon, and other modern components, although most of the people who use them have no idea how they are made. When turned on, the device puts music into people's ears, or, in some cases, shows pictures and movies on a screen about the size of a postage stamp. Some of the people who use these devices wouldn't even consider walking around without them; it is as if the device shields them from some of the other elements of their culture.

The same people who carry these devices around have other habits that, when seen from the perspective of someone unfamiliar with this culture, might seem peculiar and certainly highly ritualized. Apparently when the young people in this society go away to school, most take a large number of various technological devices along with them. According to recent reports from these young people, many of them sleep with one of these devices turned on all night. It looks like a large box—some square, others relatively flat—and it projects pictures and sound when the user clicks buttons on another small device that, though detached from the bigger box, can be placed anywhere in the room. If you click the buttons on this portable device, the pictures and sound coming forth from the larger box will change possibly hundreds of times, revealing a huge assortment of images that seem to influence what people in this culture believe and, in many cases, how they behave. They say that in over 40 percent of the households in this culture, this device is turned on 24 hours a day (Gitlin 2002)!

continued

The Tchikrin people of Brazilian rain forest paint elaborate and beautiful designs on their bodies that define the relationship of people to social groups. Are there ways that cultural practices in the United States also define social relationships?

© Joan Bamburger/Anthro Photos

The young people in this culture seem to get up every day and immediately go to another device where they do things with unusual names, such as to "IM" their friends (who, by the way, may be nowhere near them). Or, they might also do something called "text messaging," where they push buttons with their thumbs on yet another small device with a tiny screen. Indeed, it seems that everything these young people do involves looking at some kind of screen—enough so that one of the authors of this book has labeled their generation "Screenagers."

Not everyone in this culture has access to all of these devices, although many want them. Indeed, having more of the devices seems to be a mark of one's social status—that is, how you are regarded in this culture. But very few people know where the devices are made, what they are made of, or how they work—even though the young often ridicule older people for not understanding how the devices work or why they are so important to them.[1]

From outside the culture, these practices seem strange, yet few within the culture think the behaviors associated with these devices are anything but perfectly ordinary. Most of the time, people do not spend much time thinking about the meaning of the behaviors associated with these devices unless, for some reason, they suddenly do not work.

You have surely guessed that the practices described here are taken from U.S. culture: iPods, MP3 players, text messaging, instant messaging, television/video viewing. These are such daily practices that they practically define modern American culture. Unless they are somehow interrupted—as we will ask you to do in the media blackout exercise described on page 52—most do not think much about their influence on society, on people's relationships, or on people's definitions of themselves.

When viewed from the outside, cultural habits that seem perfectly normal take on a strange aspect. Take an example from a different culture. The Tchikrin people—a remote culture of the central Brazilian rain forest—paint their bodies in elaborate designs. Painted bodies communicate to others about the relationship of the person to his or her body, society, and the spiritual world. The designs and colors symbolize the balance the Tchikrin think exists between biological powers and the integration of people into the social group. The Tchikrin associate hair with sexual powers, and lovers get a special thrill from using their teeth to pluck an eyebrow or eyelash from their partner's face (Turner 1969). To the Tchikrin people, these practices are no more unusual or exotic than the technological habits we practice in the United States.

To study culture, to analyze it and measure its significance in society, we must separate ourselves from judgments such as "strange" or "normal." We must see a culture as it is seen by insiders, but we cannot be completely taken in by that view. One might say that we should know the culture as insiders and understand it as outsiders.

1 This introduction is inspired by a classic article on the "Nacirema"—*American*, backwards—by Horace Miner (1956). But it is also written based on the essays students at the University of Delaware have written based on the media blackout exercise described on page 52. Students have written that, without access to their usual media devices, they felt they "had no personality!" and that the period of the blackout was "the worst forty-eight hours of my life!"

Defining Culture

Culture is the complex system of meaning and behavior that defines the way of life for a given group or society. It includes beliefs, values, knowledge, art, morals, laws, customs, habits, language, and dress, among other things. Culture includes ways of thinking as well as patterns of behavior. Observing culture involves studying what people think, how they interact, and the objects they use.

In any society, culture defines what is perceived as beautiful and ugly, right and wrong, good and bad. Culture helps hold society together, giving people a sense of belonging, instructing them on how to behave, and telling them what to think in particular situations. Culture gives meaning to society.

Culture is both material and nonmaterial. **Material culture** consists of the objects created in a given society—its buildings, art, tools, toys, print and broadcast media, and other tangible objects, such as those discussed in the chapter opener. In the popular mind, material artifacts constitute culture because they can be collected in museums or archives and analyzed for what they represent. These objects are significant because of the meaning they are given. A temple, for example, is not merely a building, nor is it only a place of worship. Its form and presentation signify the religious meaning system of the faithful.

Culture is an emergent phenomenon, transformed sometimes by innovations that change how people think about their social worlds.

Nonmaterial culture includes the norms, laws, customs, ideas, and beliefs of a group of people. Nonmaterial culture is less tangible than material culture, but it has a strong presence in social behavior. Examples of nonmaterial culture are numerous and found in the patterns of everyday life. In some cultures, people eat with silverware; in others, with chopsticks; and in some, with their fingers. Such are the practices of nonmaterial culture, but note that the eating utensils are part of material culture.

There is a link between nonmaterial and material culture. Material culture can be the manifestation of nonmaterial culture; at the same time, nonmaterial culture can be shaped by material culture (such as when widespread marketing fosters a culture of consumerism).

Characteristics of Culture

Across societies, certain features of culture are noted by sociologists. These different characteristics of culture are examined here.

1. ***Culture is shared.*** Culture would have no significance if people did not hold it in common. Culture is collectively experienced and collectively agreed upon. The shared nature of culture makes human society possible. The shared basis of culture may be difficult to see in complex societies where groups have different traditions, perspectives, and ways of thinking and behaving. In the United States, for example, different racial and ethnic groups have unique histories, languages, and beliefs—that is, different cultures. Even within these groups, there are diverse cultural traditions. Latinos, for example, comprise many groups with distinct origins and cultures. Still, there are features of Latino culture, such as the Spanish language and some values and traditions, that are shared. Latinos also share a culture that is shaped by their common experiences as minorities in the United States. Similarly, African Americans have created a rich and distinct culture that is the result of their unique experience within the United States. What identifies African American culture are the practices and traditions that have evolved from both the U.S. experience and African and Caribbean traditions. Placed in another country, such as an African nation, African Americans would likely recognize elements of their culture, but they would also feel culturally distinct as Americans.

   Within the United States, culture varies by age, race, region, gender, ethnicity, religion, class, and other social factors. A person growing up in the South is likely to develop different tastes, modes of speech, and cultural interests than a person raised in the West. Despite these differ-

© David Reed/Corbis

© Paul Chesley/Stone/Getty Images

© Lindsay Hebberd/Corbis

© Jack Kurtz/The Image Works

© Chris Ryan/Getty Images/OJO Images

Different culture traditions create rituals that mark the passage from one phase of life to another. Here, a young Jewish boy undertakes the Bar Mitzvah ceremony marking his passage to adulthood (upper left); a young Apache girl experiences four-day long puberty rites following the first onset of menstruation to celebrate her new-found womanhood (top right); a child takes part in a Navjote ceremony, the most important event in a young child's life, noting acceptance into the Parsi community in India (middle left); a young girl awaits her quinceañera *ceremony in Chiapas, Mexico (middle right); a group of young women in the United States celebrate a friend's 16th birthday (bottom left).*

ences, there is a common cultural basis to life in the United States. Certain symbols, language patterns, belief systems, and ways of thinking are distinctively American and form a common culture even though great cultural diversity exists.

2. ***Culture is learned.*** Cultural beliefs and practices are usually so well learned that they seem perfectly natural, but they are learned nonetheless. How do people come to prefer some foods to others? How is musical taste acquired? Culture may be taught through direct instruction, such as a parent teaching a child how to use silverware or teachers instructing children in songs, myths, and other traditions in school.

   Culture is also learned indirectly through observation and imitation. Think of how a person learns what it means to be a man or a woman. Although the "proper" roles for men and women may never be explicitly taught, one learns what is expected from observing others. A person becomes a member of a culture through both formal and informal transmission of culture. Until the culture is learned, the person will feel like an outsider. The process of learning culture is referred to by sociologists as *socialization* (discussed in the next chapter).

3. ***Culture is taken for granted.*** Because culture is learned, members of a given society seldom question the culture of which they are a part, unless for some reason they become outsiders or establish some critical distance from the usual cultural expectations. People engage unthinkingly in hundreds of specifically cultural practices every day; culture makes these practices seem "normal." If you suddenly stopped participating in your culture and questioned each belief and every behavior, you would soon find yourself feeling detached and perhaps a little disoriented; you would also become increasingly ineffective at functioning within your group. Little wonder that tourists stand out so much in a foreign culture. They rarely have much knowledge of the culture they are visiting and, even when they are well-informed, typically approach the society from their own cultural orientation.

   Think, for example, of how you might feel if you were a Native American student in a predominantly White classroom. Many, though not all, Native American people are raised to be quiet and not outspoken. If students in a classroom are expected to assert themselves and state what is on their minds, a Native American student may feel awkward, as will others for whom these expectations are contrary to their cultural upbringing. If the professor is not aware of these cultural differences, he or she may penalize students who are quiet, resulting perhaps in a lower grade for the student from a different cultural background. Culture binds us together, but lack of communication across cultures can have negative consequences, as this example shows.

4. ***Culture is symbolic.*** The significance of culture lies in the meaning it holds for people. **Symbols** are things or behaviors to which people give meaning; the meaning is not inherent in a symbol but is bestowed by the meaning people give it. The U.S. flag, for example, is literally a decorated piece of cloth. Its cultural significance derives not from the cloth of which it is made, but from its meaning as a symbol of freedom and democracy, as was witnessed by the widespread flying of the flag after the terrorist attacks on the United States on September 11, 2001.

   That something has symbolic meaning does not make it any less important or influential than objective facts. Symbols are powerful expressions of human culture. Think of the Confederate flag. Those who object to the Confederate flag being displayed on public buildings see it as a symbol of racism and the legacy of slavery. Those who defend it see it as representing Southern heritage, a symbol of group pride and regional loyalty. Similarly, the use of Native American mascots to name and represent sports teams is symbolic of the exploitation of Native Americans. Native American activists and their supporters see the use of Native American mascots as derogatory and extremely insulting, representing gross caricatures

Cultural values can clash when groups have strongly held, but clashing, value systems. A good example is the conflict that developed over a Supreme Court judge in Alabama placing the Ten Commandments in the lobby of the county courthouse. When he refused to remove them because they violated the constitutional separation of church and state, he was forced to resign.

understanding diversity

The Social Meaning of Language

Language reflects the assumptions of a culture. This can be seen and exemplified in several ways:

- Language affects people's perception of reality.
  Example: Researchers have found that using male pronouns, even when intended to be gender-neutral, produces male-centered imagery and ideas. Studies also find that when college students look at job descriptions written using masculine pronouns, they assume that women are not qualified for the job (Gastil 1990; Hamilton 1988; Switzer 1990).
- Language reflects the social and political status of different groups in society.
  Example: A term such as *woman doctor* suggests that men are the standard and women the exception. The term *working woman* (used to refer to women who are employed) also suggests that women who do not work for wages are not working. Ask yourself what the term *working man* connotes and how this differs from *working woman.*
- Groups may advocate changing language referring to them as a way of asserting a positive group identity.
  Example: Some advocates for the disabled challenge the term *handicapped,* arguing that it stigmatizes people who may have many abilities, even if they are physically distinctive. Also, though someone may have one disabling condition, they may otherwise be perfectly able.
- The implications of language emerge from specific historical and cultural contexts.
  Example: The naming of so-called races comes from the social and historical processes that define different groups as inferior or superior. Racial labels do not come just from physical, national, or cultural differences. The term *Caucasian,* for example, was coined in the seventeenth century when racist thinkers developed alleged scientific classification systems to rank different societal groups. Alfred Blumenbach used the label *Caucasian* to refer to people from the Caucasus of Russia whom he thought were more beautiful and intelligent than any group in the world.
- Language can distort actual group experience.
  Example: Terms used to describe different racial and ethnic groups homogenize experiences that may be unique. Thus, the terms *Hispanic* and *Latino* lump together Mexican Americans, island Puerto Ricans, U.S.-born Puerto Ricans, as well as people from Honduras, Panama, El Salvador, and other Central and South American countries. *Hispanic* and *Latino* point to the shared experience of those from Latin cultures, but like the terms *Native American* and

of Native American traditions. (Think of the Washington Redskins, the Cleveland Indians, or the Atlanta Braves' "tomahawk chop.") The protests that have developed over controversial symbols are indicative of the enormous influence of cultural symbols.

Symbolic attachments can guide human behavior. For example, people stand when the national anthem is sung and may feel emotional from displays of the cross or the Star of David. Under some conditions, people organize mass movements to protest what they see as the defamation of important symbols, such as the burning of a flag or the burning of a cross. The significance of the symbolic value of culture can hardly be overestimated. Learning a culture means not just engaging in particular behaviors but also learning their symbolic meanings within the culture.

5. ***Culture varies across time and place.*** Culture develops as humans adapt to the physical and social environment around them. Culture is not fixed from one place to another. In the United States, for example, there is a strong cultural belief in scientific solutions to human problems; consequently, many think that problems of food supply and environmental deterioration can be addressed by scientific breakthroughs, such as genetic engineering to create high-yield tomatoes or cloning cows to eliminate mad cow disease. In other cultural settings, different solutions may seem preferable. Indeed, some religions think of genetic engineering as trespassing on divine territory.

   Culture also varies over time. As people encounter new situations, the culture that emerges is a mix of the past and present. Second generation immigrants to the United States are raised in the traditions of their culture of origin, and chil-

American Indian, they obscure the experiences of unique groups, such as the Sioux, Nanticoke, Cherokee, Yavapai, or Navajo.

- Language shapes people's perceptions of groups and events in society.
  Example: Following Hurricane Katrina in New Orleans, African American people taking food from abandoned stores were described as "looting" and White people as "finding food." Also, Native American victories during the nineteenth century are typically described as "massacres"; comparable victories by White settlers are described in heroic terms (Moore 1992).
- Terms used to define different groups change over time and can originate in movements to assert a positive identity.
  Example: In the 1960s, *Black American* replaced the term *Negro* because the civil rights and Black Power movements inspired Black pride and the importance of self-naming (Smith 1992). Earlier, *Negro* and *colored* were used to define African Americans. Currently, it is popular to refer to all so-called racial groups as "people of color." This phrase was derived from the phrase "women of color," created by feminist African American, Latina,[2] Asian American, and Native American women to emphasize their common experiences. Some people find the use of "color" in this label offensive since it harkens back to the phrase "colored people," a phrase generally seen as paternalistic and racist because it was a label used by dominant groups to refer to African Americans prior to the civil rights movement. The phrase "women of color" now has a more positive meaning than the earlier term *colored women* because it is meant to recognize common experiences, not just label people because of their presumed skin color.

In this book, we have tried to be sensitive to the language used to describe different groups. We recognize that the language we use is fraught with cultural and political assumptions and that what seems acceptable now may be offensive later. Perhaps the best way to solve this problem is for different groups to learn as much as they can about one another, becoming more aware of the meaning and nuances of naming and language and more conscious of the racial assumptions embedded in the language. Greater sensitivity to the language used in describing different group experiences is an important step in promoting better intergroup relationships.

2 Latina is the feminine form in Spanish and refers to women; Latino, to men.

dren of immigrants typically grow up with both the traditional cultural expectations of their parents' homeland and the cultural expectations of a new society. Adapting to the new society can create conflict between generations, especially if the older generation is intent on passing along their cultural traditions. The children may be more influenced by their peers and may choose to dress, speak, and behave in ways that are characteristic of their new society but unacceptable to their parents (Portes and Rumbaut 2001).

To sum up, culture is concrete because we can observe the cultural objects and practices that define human experience. Culture is abstract because it is a way of thinking, feeling, believing, and behaving. Culture links the past and the present because it is the knowledge that makes us part of human groups. Culture gives shape to human experience.

Ethnocentrism and Cultural Relativism

Because culture tends to be taken for granted, it can be difficult for people within a culture to see their culture as anything but "the way things are." It can thus be difficult to view other cultures without making judgments based on one's own cultural views. **Ethnocentrism** is the habit of only seeing things from the point of view of one's own group. Judging one culture by the standards of another culture is ethnocentric. An ethnocentric perspective prevents you from understanding the world as it is experienced by others, and it can lead to narrow-minded conclusions about the worth of diverse cultures.

Any group can be ethnocentric. Also, ethnocentrism can be extreme or subtle—as in the example of social groups who think their way of life is better than that of any other group. Is there such a ranking among groups in your community? Fraternities and sororities often build group rituals around such claims, youth groups see their way of life as superior to adults, and urbanites may think their cultural habits are more sophisticated than those of groups labeled "country hicks." Ethnocentrism is a powerful force because it combines a strong sense of group solidarity with the idea of group superiority.

Ethnocentrism can build group solidarity, but it also discourages intergroup understanding. Understanding ethnocentrism is critical to understanding some of the major conflicts that are shaping current history. Consider the issue of *nationalism,* the sense of identity that arises when a group exalts its own culture over all others and organizes politically and socially around this principle. Nationalist groups (think of Hamas, the militant Palestinian group) tend to reject those who do not share their cultural experience and judge all other cultures to be inferior.

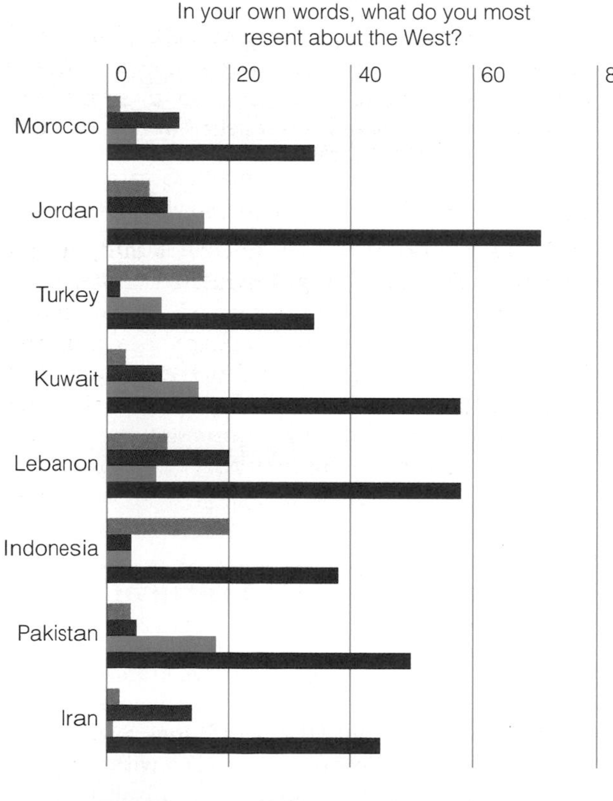

FIGURE 2.1 *Islamic Views of the West*

Data: Burkholder, Richard. 2003. "Iraq and the West: How Wide Is the Morality Gap?" *The Gallup Poll*, Princeton, NJ. **www.gallup.com**

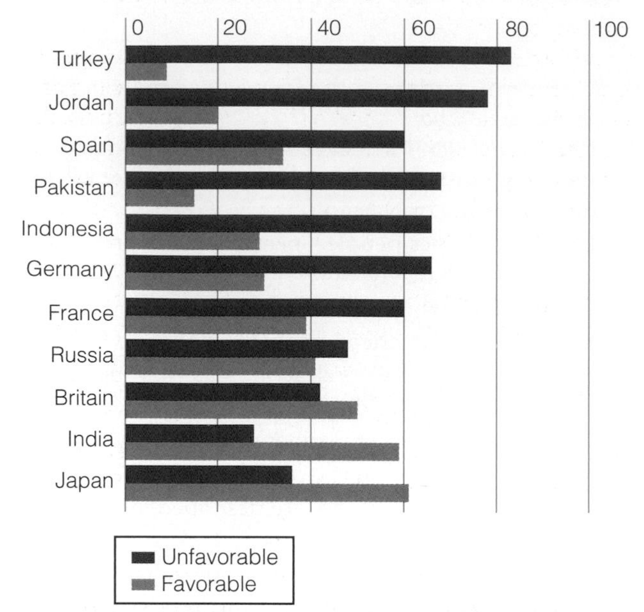

FIGURE 2.2 *Views of the United States from Abroad*

*Opinions change over time, but in every case favorable opinions have declined since 2002. The same survey indicates that countries' opinions of "Americans" are more favorable than opinions about "the United States."

Source: Pew Research Center. 2007. "Global Unease with Major World Powers." Pew Global Attitudes Project, June 22, 2007. Washington, DC: Pew Research Center. **www.pewglobal.org**

Nationalist movements tend to use extreme ethnocentrism as the basis for nation building. Taken to its limits, ethnocentrism can lead to overt political conflict, war, terrorism, even *genocide,* which is the mass killing of people based on their membership in a particular group. You might wonder how someone could believe in the righteousness of their religious faith so much that they would murder people to express that commitment. Understanding ethnocentrism does not excuse such behavior, but it can help you understand how it can occur. Think of Al Qaeda and the extremist belief that terrorism is justified as a religious *jihad* (defined as a religious struggle to defend Islamic faith). Although it would be overly simple to explain current political conflicts only in terms of ethnocentrism, as Figure 2.1 shows, cultural values in the Islamic world can clash with those of the West and are part of the complexity of U.S. relations with those cultures.

Ethnocentrism can also help you understand the view that many nations now have of the United States—a fact that those within the United States have difficulty understanding because of their ethnocentric views of their own culture. As Figure 2.2 shows, many other nations do not see U.S. culture in the positive light that U.S. citizens might expect;

this figure also shows you how such views can change over time.

Contrasting ethnocentrism is cultural relativism. **Cultural relativism** is the idea that something can be understood and judged only in relationship to the cultural context in which it appears. This does not make every cultural practice morally acceptable, but it suggests that without knowing the cultural context, it is impossible to understand why people behave as they do. For example, in the United States, burying or cremating the dead is the cultural practice. It may be difficult for someone from this culture to understand that in Tibet, with a ruggedly cold climate and the inability to dig the soil, the dead are cut into pieces and left for vultures to eat. Although this would be repulsive (and illegal) in the United States, within Tibetan culture, this practice is understandable.

Biology and Human Culture

It is cultural patterns that make humans so interesting. Is it culture that distinguishes human beings from animals? Some animal species develop what we might call culture. Chimpanzees, for example, learn behavior through observing and imitating others—a point proved by observing the different eating practices among chimpanzees in the same species but raised in different groups (Whiten et al. 1999). Others have observed elephants picking up the dead bones of other elephants and fondling them, perhaps evidence of grieving behavior (Meredith 2003). Dolphins are known to have a complex auditory language. And most people think that their pets communicate with them.

Apparently, humans are not unique in their ability to develop systems of communication. But some scientists generally conclude that animals lack the elaborate symbol-based cultures common in human societies. Perhaps, as even Charles Darwin wrote, "The difference in mind between man and the higher animals, great as it is, certainly is one of degree and not of kind" (Darwin, cited in Gould 1999).

Studying animal groups reminds us of the interplay between biology and culture. Human biology sets limits and provides certain capacities for human life and the development of culture. Similarly, the environment in which humans live establishes the possibilities and limitations for human society. Nutrition, for instance, is greatly influenced by environment, thereby affecting human body height and weight. Not everyone can drive a golf ball like Tiger Woods or lob a tennis ball like Venus and Serena Williams, but with training and conditioning, people can enhance their physical abilities. Biological limits exist, but cultural factors have an enormous influence on the development of human life.

The Elements of Culture

Culture is multifaceted, consisting both of material and nonmaterial things, some parts of culture being abstract, others more concrete. The different elements of culture include language, norms, beliefs, and values (see Table 2.1).

Language

Language is a set of symbols and rules that, put together in a meaningful way, provides a complex communication system. The formation of culture among humans is made possible by language. Learning the language of a culture is essential to becoming part of a society, and it is one of the first things children learn. Indeed, until children acquire at least a rudimentary command of language, they seem unable to acquire other social skills. Language is so important

table 2.1 Elements of Culture

	Definition	Examples
Language	A set of symbols and rules that, put together in a meaningful way, provides a complex communication system	English; Spanish; hieroglyphics
Norms	The specific cultural expectations for how to behave in a given situation	Behavior involving use of personal space; manners
Folkways	General standards of behavior adhered to by a group	Cultural forms of dress; food habits
Mores	Strict norms that control moral and ethical behavior	Religious doctrines; formal law
Values	Abstract standards in a society or group that define ideal principles	Liberty; freedom
Beliefs	Shared ideas held collectively by people within a given culture	Belief in a higher being

to human interaction that it is difficult to think of life without it; indeed, as one commentator on language has said, "Life is lived as a series of conversations" (Tannen 1990: 13).

Think about the experience of becoming part of a social group. When you enter a new society or a different social group, you have to learn its language to become a member of the group. This includes any special terms of reference used by the group. Lawyers, for example, have their own vocabulary and their own way of constructing sentences called, not always kindly, "legalese." Becoming a part of any social group—a friendship circle, fraternity or sorority, or any other group—involves learning the language they use. Those who do not share the language of a group cannot participate fully in its culture.

Debunking Society's Myths

Myth: The use of Native American names for school mascots is just for fun and is no big deal.

Sociological perspective: Language carries with it great meaning that reflects the perceived social value of diverse groups. Research finds that exposure to the trivial or degrading use of Native American images for such things as school mascots and sports teams actually lowers Native American children's sense of self-worth (Fryberg 2003).

Language is fluid and dynamic and evolves in response to social change. Think, for example, of how the introduction of computers has affected the English language. People now talk about needing "downtime" and providing "input." Only a few years ago, had you said you were going to "IM" your friends, no one would have known what you were talking about. Even now, those without exposure to advanced electronic technologies might not understand that IM refers to the instant messaging increasingly common among e-mail users. IM has also introduced its own language—BFN ("bye for now"), LOL ("laughing out loud"), and GTG ("got to go")—a new language shared among those in the IM culture. There are now even online dictionaries listing and defining such "words" (see Table 2.2).

table 2.2 The Language of the Internet

Language often changes as society and culture change. One example is the new language being created through text and instant messaging, some samples of which are provided below:

AAS	alive and smiling	LMAO	laughing my a** off
AFK	away from keyboard	MTFBWU	May the force be with you.
AML	all my love	OP	on phone
B/F	boyfriend	POS	parent over shoulder
BM&Y	between you and me	QT	cutie
BTDT	been there, done that	SLAP	sounds like a plan
CRBT	crying really big tears	SNAG	sensitive New Age guy
FYEO	for your eyes only	SSINF	so stupid it's not funny
GFI	go for it	SUP	what's up
HAGO	have a good one	TMB	text me back
INAL	I am not a lawyer.	UCMU	you crack me up
JK	just kidding	WDYK	what do you know
KIR	keeping it real	WOMBAT	waste of money, brains, and time
KOTL	kiss on the lips	WWJD	What would Jesus do?
KWIM	know what I mean?	YRYOCC	You're running your own cuckoo clock.
LD	later dude	ZZ	bored

Source: "Text Messaging Abbreviations" 2007. **www.webopedea.com**; **www.netlingo.com**

Does Language Shape Culture? Language is clearly a big part of culture. Edward Sapir (writing in the 1920s) and his student Benjamin Whorf (writing in the 1950s) thought that language was central in determining social thought. Their theory, the **Sapir–Whorf hypothesis,** asserts that language determines other aspects of culture because language provides the categories through which social reality is defined. Sapir and Whorf thought that language determines what people think because language forces people to perceive the world in certain terms (Sapir 1921; Whorf 1956).

If the Sapir–Whorf hypothesis is correct, then speakers of different languages have different perceptions of reality. Whorf used the example of the social meaning of time to illustrate cultural differences in how language shapes perceptions of reality. He noted that the Hopi Indians conceptualize time as a slowly turning cylinder, whereas English-speaking people conceive of time as running forward in one direction at a uniform pace. Linguistic constructions of time shape how the two different cultures think about time and therefore how they think about reality. In Hopi culture, events are located not in specific moments of time, but in "categories of being"—as if everything is in a state of becoming, not fixed in a particular time and place (Carroll 1956). In contrast, the English language locates things in a definite time and place, placing great importance on verb tense—things are located precisely in the past, present, or future.

AP Images/Paul Sakuma

Living in a multicultural society often juxtaposes diverse cultures, even in public places.

Recent critics do not think that language determines culture to the extent that Sapir and Whorf proposed. Language does not single-handedly dictate the perception of reality—but, no doubt, language has a strong influence on culture. Most scholars now see two-way causality between language and culture. Asking whether language determines culture or vice versa is like asking which came first, the chicken or the egg. Language and culture are inextricable, each shaping the other.

Consider again the example of time. Contemporary Americans think of the week as divided into two parts: *weekdays* and *weekends*, words that reflect how we think about time. When does a week end? Having language that defines the weekend encourages us to think about the weekend in specific ways. It is a time for rest, play, chores, and family. In this sense, language shapes how we think about the passage of time—we look forward to the weekend, we prepare ourselves for the work week—but the language itself (the very concept of the weekend) stems from patterns in the culture—specifically, the work patterns of advanced capitalism. The capitalist work ethic makes it morally offensive to merely "pass the time"; instead, time is to be managed. Concepts of time in preindustrial, agricultural societies follow a different pattern. In agricultural societies, time and calendars are based on agricultural and seasonal patterns; the year proceeds according to this rhythm, not the arbitrary units of time of weeks and months. This shows how language and culture shape each other.

Social Inequality in Language. The language of any culture reflects the nature of that society. Thus, in a society where there is group inequality, language is likely to communicate assumptions and stereotypes about different social groups. What people say—including what people are called—reinforces patterns of inequality in society (Moore 1992). We see this in what different groups in the United States are called (see also the box (on page 38–39) "Understanding Diversity: The Social Meaning of Language"). What someone is called can be significant because it imposes an identity on that person. This is why the names for various racial and ethnic groups have been so heavily debated. Thus, for years, many Native Americans objected to being called "Indian," because it was a term created about them by White conquerors. To emphasize their native roots in the Americas, the term *Native American* was adopted. Now, though many prefer to be called by their actual origin, *Native American* and *American Indian* are also used interchangeably. Likewise, Asian Americans tend to be offended by being called "Oriental"—an expression that stemmed from Western (that is, European and American) views of Asian nations.

Language reflects the social value placed on different groups and it reflects power relationships, depending on who gets to name whom. Derogatory terms such as *redneck, white trash,* and *trailer park trash* stigmatize people based on regional identity and social class. This is also why it is so demeaning when derogatory terms are used to describe racial ethnic groups. For example, throughout the period of Jim Crow segregation in the American South, Black men, regardless of their age, were routinely called "boy" by Whites. Calling a grown man a "boy" is an insult; it diminishes his status by defining him as childlike. Referring to a woman as a "girl" has the same effect. Why are young women—even well into their twenties—routinely referred to as "girls"? Just as calling a man "boy," this diminishes women's status.

Debunking Society's Myths

Myth: Bilingual education discourages immigrant children from learning English and thus blocks their assimilation into American culture and reduces their chances for a good education.

Sociological perspective: Studies of students who are fluent bilinguals show that they outperform both English only students and students with limited bilingualism, as measured by grade point average and performance on standardized tests. Moreover, preserving the use of native languages can better meet the need for skilled bilingual workers in the labor market (Portes 2002).

Note, however, that terms such as *girl* and *boy* are pejorative only in the context of dominant and subordinate group relationships. African American women, as an example, often refer to each other as "girl" in informal conversation. The term *girl* used between those of similar status is not perceived as derogatory, but when used by someone in a position of dominance, such as when a male boss calls his secretary a "girl," it is demeaning. Likewise, terms like *dyke, fag,* and *queer* are terms lesbians and gay men sometimes use without offense in referring to each other even though the same terms are offensive to lesbians and gays when used about them by others. By reclaiming these terms as positive within their own culture, lesbians and gays build cohesiveness and solidarity (Due 1995). These examples show that power relationships between groups supply the social context for the connotations of language.

In sum, language can reproduce the inequalities that exist in society. At the same time, changing the language that people use can, to some extent, alter social stereotypes and thereby change the way people think.

Norms

Social norms are another component of culture. **Norms** are the specific cultural expectations for how to behave in a given situation. Society without norms would be chaos; with norms in place, people know how to act, and social interactions are consistent, predictable, and learnable. There are norms governing every situation. Sometimes they are implicit; that is, they need not be spelled out for people to understand them. For example, when joining a line, there is an implicit norm that you should stand behind the last person, not barge in front of those ahead of you. Implicit norms may not be formal rules, but violation of these norms may nonetheless produce a harsh response. Implicit norms may be learned through specific instruction or by observation of the culture; they are part of a society's or group's customs. Norms are explicit when the rules governing behavior are written down or formally communicated. Typically, specific sanctions are imposed for violating explicit norms.

thinking sociologically

Identify a *norm* that you commonly observe. Construct an experiment in which you, perhaps with the assistance of others, violate the norm. Record how others react and note the sanctions engaged through this norm violation exercise. NOTE: Be careful not to do anything that puts you in danger or causes serious problems for others.

In the early years of sociology, **William Graham Sumner** (1906) identified two types of norms: folkways and mores. **Folkways** are the general standards of behavior adhered to by a group. You might think of folkways as the ordinary customs of different group cultures. Men wearing pants and not skirts is an example of a cultural folkway. Other examples are the ways that people greet each other, decorate their homes, and prepare their food. Folkways may be loosely defined and loosely adhered to, but they nevertheless structure group customs and implicitly govern much social behavior.

Mores (pronounced "more-ays") are strict norms that control moral and ethical behavior. Mores provide strict codes of behavior, such as the injunctions, legal and religious, against killing others and committing adultery. Mores are often upheld through rules or **laws,** the written set of guidelines that define right and wrong in society. Basically, laws are formalized mores. Violating mores can bring serious repercussions. When any social norm is violated, the violator is typically punished. **Social sanctions** are mechanisms of social control that enforce norms. The seriousness of a social sanction depends on how strictly the norm is held. The strictest norms in society are **taboos**—those behaviors that bring the most serious sanctions. Dressing in an unusual way that violates the folkways of dress may bring ridicule but is usually not seriously punished. In some cultures the rules of dress are strictly interpreted, such as the requirement by Islamic fundamentalists that women who appear in public have their bodies cloaked and faces veiled. It would be considered a taboo for a woman in this culture to appear in public without being veiled. The sanctions for doing so can be as severe as whipping, branding, banishment, even death.

Sanctions can be positive or negative, that is, based on rewards or punishment. When children learn social norms, for example, correct behavior may elicit positive sanctions—the behavior is reinforced through praise, approval, or an explicit reward. Early on, for example, parents might praise children for learning to put on their own clothes; later, children might get an allowance if they keep their rooms clean. Bad behavior earns negative sanctions, such as getting spanked or grounded. In society, negative sanctions may be mild or severe, ranging from subtle mechanisms of control, such as ridicule, to overt forms of punishment, such as imprisonment, physical coercion, or death.

One way to study social norms is to observe what happens when they are violated. Once you become aware of how social situations are controlled by norms, you can see how easy it is to disrupt situations where adherence to the norms produces social order. **Ethnomethodology** is a theoretical approach in sociology based on the idea that you can discover the normal social order through disrupting it. As a technique of study, ethnomethodologists often deliberately disrupt social norms to see how people respond and try to reinstate social order (Garfinkel 1967).

In a famous series of ethnomethodological experiments, college students were asked to pretend they were boarders in their own homes for a period of fifteen minutes to one hour. They did not tell their families what they were doing. The students were instructed to be polite, circumspect, and impersonal; to use terms of formal address; and to speak only when spoken to. After the experiment, two of the participating students reported that their families treated the experiment as a joke; another's family thought the daughter was being extra nice because she wanted something. One family believed that the student was hiding some serious problem. In all the other cases, parents reacted with shock, bewilderment, and anger. Students were accused of being mean, nasty, impolite, and inconsiderate; the parents demanded explanations for their sons' and daughters' behavior (Garfinkel 1967). Through this experiment, the student researchers were able to see that even the informal norms governing behavior in one's home are carefully structured. By violating the norms of the household, the norms were revealed.

Ethnomethodological research teaches us that society proceeds on an "as if" basis. That is, society exists because people behave as if there were no other way to do so. Usually people go along with what is expected of them. Culture is actually "enforced" through the social sanctions applied to those who violate social norms. Usually, specific sanctions are unnecessary because people have learned the normative expectations. When the norms are violated, their existence becomes apparent (see also Chapter 4).

Beliefs

As important as social norms are the beliefs of people in society. **Beliefs** are shared ideas held collectively by people within a given culture about what is true. Shared beliefs are part of what binds people together in society. Beliefs are also the basis for many norms and values of a given culture. In the United States, beliefs that are widely held and cherished are the belief in God and the belief in democracy.

Some beliefs are so strongly held that people find it difficult to cope with ideas or experiences that contradict them. Someone who devoutly believes in God may find atheism intolerable; those who believe in reincarnation may seem irrational to those who think life ends at death. Similarly, those who believe in magic may seem merely superstitious to those with a more scientific and rational view of the world.

Whatever beliefs people hold, they orient us to the world. They provide answers to otherwise imponderable questions about the meaning of life. Beliefs provide a meaning system around which culture is organized. Whether belief stems from religion, myth, folklore, or science, it shapes what people take to be possible and true. Although a given belief may be logically impossible, it nonetheless guides people through their lives.

Values

Deeply intertwined with beliefs are the values of a culture. **Values** are the abstract standards in a society or group that define ideal principles. Values define what is desirable and morally correct; thus, values determine what is considered right and wrong, beautiful and ugly, good and bad. Although values are abstract, they provide a general outline for behavior. Freedom, for example, is a value held to be important in U.S. culture, as is equality. Values are ideals, forming the abstract standards for group behavior, but they are also ideals that may not be realized in every situation.

Values can be a basis for cultural cohesion, but they can also be a source of conflict. You may remember the conflicts that developed over the case of Terri Schiavo—a woman who collapsed from a chemical imbalance brought on by an eating disorder. She was left severely brain-damaged in a "permanent vegetative state" for over fifteen years. Her husband, Michael Schiavo, wanted to remove her feeding tube, thereby ending her life, something he argued she would have wanted. But her parents objected on religious grounds and took the case through numerous court decisions over several years and ultimately to the U.S. Supreme Court. By the time Terri Schiavo died in 2005 (after her feeding tube had been removed), the public had become embroiled in a huge national debate about differing values. There were those who believed she had the right to die in peace and with dignity, those who defined her death as murder, and others (in fact, the majority of Americans) who thought that the government simply had no business interfering in the private affairs of this family. The intensity of this issue—even the language by which people described Terri Schiavo—all reflected different positions in a national conflict over competing values.

Values guide the behavior of people in society; they also shape the social norms in a given culture. An example of the impact that values have on people's behavior comes from an American Indian society known as the Kwakiutl (pronounced "kwa-kee-YOO-ta-l"), a group from the coastal region of southern Alaska, Washington State, and British Columbia. The Kwakiutl developed a practice known as *potlatch,* in which wealthy chiefs would periodically pile up their possessions and give them away to their followers and rivals (Benedict 1934; Harris 1974; Wolcott 1996). The object of potlatch was to give away or destroy more of one's goods than did one's rivals. The potlatch reflected Kwakiutl values of reciprocity, the full use of food and goods, and the social status of the wealthiest chiefs in Kwakiutl society. (By the way, chiefs did not lose their status by giving away their goods, because the goods were eventually returned in the course of other potlatches. They would even burn large piles of goods, knowing that others would soon replace their wealth through other potlatches.)

Compare this practice with the patterns of consumption in the United States. Imagine the CEOs of major corporations regularly gathering up their wealth and giving it away to their workers and rival CEOs! In the contemporary United States, *conspicuous consumption* (consuming for the sake of displaying one's wealth) celebrates values similar to those of the potlatch: High-status people demonstrate their position by accumulating more material possessions than those around them (Veblen 1899).

Together, norms, beliefs, and values guide the behavior of people in society. It is necessary to understand how they operate in a situation in order to understand why people behave as they do.

Cultural Diversity

It is rare for a society to be culturally uniform. As societies develop and become more complex, different cultural traditions appear. The more complex the society, the more likely its culture will be internally varied and diverse. The United States, for example, hosts enormous cultural diversity stemming from religious, ethnic, and racial differences, as well as regional, age, gender, and class differences. Currently, more than 12 percent of people in the United States are foreign born. In a single year, immigrants from more than 100 countries come to the United States (U.S. Census Bureau 2007). Whereas earlier immigrants were predominantly from Europe, now Latin America and Asia are the greatest sources of new immigrants. One result is a large increase in the number of U.S. residents for whom English is the second language. Cultural diversity is clearly a characteristic of contemporary American society.

The richness of American culture stems from the many traditions that different groups have brought with them to this society, as well as from the cultural

Values can be a source of cultural cohesion, but also of cultural conflict. What are some of the different values that are being debated in society?

map 2.1 MAPPING AMERICA'S DIVERSITY

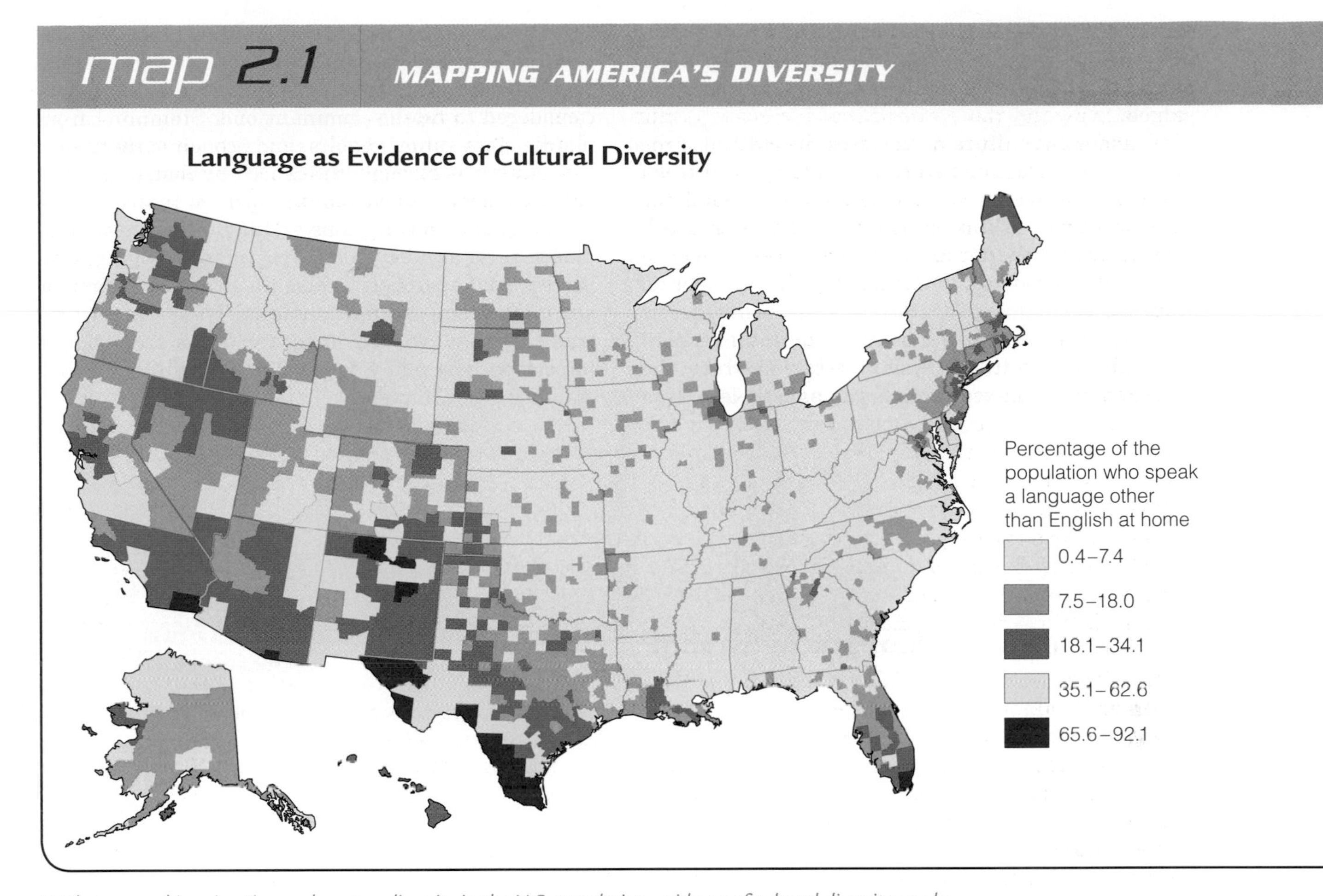

With increased immigration and greater diversity in the U.S. population, evidence of cultural diversity can be seen in many homes—language being one type of evidence. This map shows the regional differences in the percentage of the population over age 5 who speak a language other than English at home. For the United States as a whole, 17.9 percent of the population—almost one-fifth—fit into this category. Eight percent of the population say they speak English less than very well. What implications does this have for the regions most affected? How might it influence relations between different generations within households?

Source: U.S. Census Bureau. 2007. American FactFinder. www.census.gov

forms that have emerged through their experience within the United States. Jazz, for example, is one of the few musical forms indigenous to the United States. An indigenous art form refers to something that originated in a particular region or culture. However, jazz also has roots in the musical traditions of slave communities and African cultures. Since the birth of jazz, cultural greats such as Ella Fitzgerald, Count Basie, Duke Ellington, Billie Holiday, and numerous others have not only enriched the jazz tradition, but have also influenced other forms of music, including rock and roll.

Native American cultures have likewise enriched the culture of our society, as have the cultures that various immigrant groups have brought with them to the United States. With such great variety, how can the United States be called one culture? The culture of the United States, including its language, arts, food customs, religious practices, and dress, can be seen as the sum of the diverse cultures that constitute this society.

Dominant Culture

Two concepts from sociology help us understand the complexity of culture in a given society: dominant culture and subculture. The **dominant culture** is the culture of the most powerful group in a society. It is the cultural form that receives the most support from major institutions and that constitutes the major belief system. Although the dominant culture is not the only culture in a society, it is commonly believed to be "the" culture of a society despite the other cultures present. Social institutions in the society perpetuate the dominant culture and give it a degree of legitimacy that is not shared by other cultures. Quite often, the dominant culture is the standard by which other cultures in the society are judged.

A dominant culture need not be the culture of the majority of people; rather, it is simply the culture of that group in society with enough power to define the cultural framework. As an example, think of a college or university that has a strong system of fraternities and sororities. On campus, the number of students belonging to fraternities and sororities is probably a numerical minority of the total student body, but the cultural system established by the Greeks may dominate campus life nonetheless. In a society as complex as the United States, it is hard to isolate a single dominant culture, although there is a widely acknowledged "American" culture that is considered to be the dominant one. Stemming from middle-class values, habits, and economic resources, this culture is strongly influenced by instruments of culture such as television, the fashion industry, and Anglo-European traditions and includes diverse elements such as fast food, Christmas shopping, and professional sports. It is also a culture that emphasizes achievement and individual effort—a cultural tradition that we will later see has a tremendous impact on how many in the United States view inequality (see Chapter 7).

doing

doing sociological research

Tattoos: Status Risk or Status Symbol?

Research Question: Not so long ago tattoos were considered a mark of social outcasts. They were associated with gang members, sailors, and juvenile delinquents. But now tattoos are in vogue—a symbol of who's trendy and hip. How did this happen—that a once stigmatized activity associated with the working class became a statement of middle-class fashion?

Research Method: This is what sociologist Katherine Irwin wanted to know when she first noticed the increase in tattooing among the middle class. Irwin first encountered the culture of tattooing when she accompanied a friend getting a tattoo in a shop she calls Blue Mosque. She started hanging out in the shop, eventually married the shop's owner, and began a four-year study using participant observation in the shop along with interviews of people getting their first tattoos. Irwin also interviewed some of the parents of tattooees and potential tattooees.

Research Results: Irwin found that middle-class tattoo patrons were initially fearful that their desire for a tattoo would associate them with low-status groups, but they reconciled this by adopting attitudes that associated tattooing with middle-class values and norms. Thus, they defined tattooing as symbolic of independence, liberation, and freedom from social constraints. Many of the women defined tattooing as symbolizing toughness and strength—values they thought rejected more conventional ideals of femininity.

Some saw tattoos as a way of increasing their attachment to alternative social groups or to gain entrée into "fringe" social worlds. Although tattoos held different cultural meanings to different groups, people getting tattooed used various techniques (what Irwin calls "legitimation techniques") to counter the negative stereotypes associated with tattooing.

Conclusions and Implications: Irwin concludes that people try to align their behavior with legitimate cultural values and norms—even when that behavior seemingly falls outside of prevailing standards.

Questions to Consider

1. Do you think of tattoos as fashionable or deviant? What do you think influences your judgment about this, and how might your judgment be different were you in a different culture, age group, or historical moment?
2. Are there fashion adornments that you associate with different social classes? What are they, and what kinds of judgment (positive and negative) do people make about them? Where do these judgments come from and why are they associated with social class?

Source: Katherine Irwin. 2001. "Legitimating the First Tattoo: Moral Passage through Informal Interaction." *Symbolic Interaction* 24 (March): 49–73.

Subcultures

Subcultures are the cultures of groups whose values and norms of behavior differ to some degree from those of the dominant culture. Members of subcultures tend to interact frequently with one another and share a common worldview. They may be identifiable by their appearance (style of clothing or adornments) or perhaps by language, dialect, or other cultural markers. You can view subcultures along a continuum of how well they are integrated into the dominant culture. Subcultures typically share some elements of the dominant culture and coexist within it, although some subcultures may be quite separated from the dominant one. This separation occurs because they are either unwilling or unable to assimilate into the dominant culture, that is, share its values, norms, and beliefs (Dowd and Dowd 2003).

thinking sociologically

Identify a group on your campus that you would call a *subculture*. What are the distinctive *norms* of this group? Based on your observations of this group, how would you describe its relationship to the dominant *culture* on campus?

Rap and hip-hop music first emerged as a subculture where young African Americans developed their own style of dress and music to articulate their resistance to the dominant White culture. Now, rap and hip-hop have been incorporated into mainstream youth culture. Indeed, they are now global phenomena, as cultural industries have turned hip-hop and rap into a profitable commodity. Even so, rap still expresses an oppositional identity for Black and White youth and other groups who feel marginalized by the dominant culture (Rose 1994).

Some subcultures actually retreat from the dominant culture, as do the Amish, some religious cults, and some communal groups. In these cases, the subculture is actually a separate community that lives as independently from the dominant culture as possible. Other subcultures may coexist with the dominant society, and members of the subculture may participate in both the subculture and the dominant culture.

Subcultures also develop when new groups enter a society. Puerto Rican immigration to the U.S. mainland, for example, has generated distinct Puerto Rican subcultures within many urban areas. Although Puerto Ricans also partake in the dominant culture, their unique heritage is part of their subcultural experience. Parts of this culture are now entering the

AP Images/Amy Sancetta

The Amish people form a subculture in the United States, although preserving their traditional way of life can be a challenge in the context of contemporary society.

dominant culture. Salsa music, now heard on mainstream radio stations, was created in the late 1960s by Puerto Rican musicians who were expressing the contours of their working-class culture (Sanchez 1999; Boggs 1992). The themes in salsa reflect the experience of barrio people and mix the musical traditions of other Latin music, including rumba, mambo, and cha-cha. As with other subcultures, the boundaries between the dominant culture and the subculture are permeable, resulting in cultural change as new groups enter society.

Countercultures

Countercultures are subcultures created as a reaction against the values of the dominant culture. Members of the counterculture reject the dominant cultural values, often for political or moral reasons, and develop cultural practices that explicitly defy the norms and values of the dominant group. Nonconformity to the dominant culture is often the hallmark of a counterculture. Youth groups often form countercultures. Why? In part, they do so to resist the culture of older generations, thereby asserting their independence and identity. Some also argue that young people establish countercultures because they have so little power in society that they have to construct their own cultures to have some sort of status—or social standing, at least among their peers (Milner 2004). Thus, countercultures among youth, like other countercultures, usually have a unique way of dress, their own special language, perhaps even different values and rituals.

© Dennis MacDonald/PhotoEdit

The styles and practices of subcultures can be a source of innovation in society. Hip hop, for example, once a subculture associated only with Black urban youth has now influenced so-called mainstream style.

Some countercultures directly challenge the dominant society. The white supremacist movement is an example. People affiliated with this movement have an extreme worldview, one that is in direct opposition to dominant values. White supremacist groups have developed a shared worldview—one based on extreme hostility to racial minorities, gays, lesbians, and feminists. Because of their self-contained culture—one focused on hate—they can be very dangerous (Ferber 1998; Stern 1996).

Countercultures may also develop in situations where there is political repression and some groups are forced "underground." Under a dictatorship, for example, some groups may be forbidden to practice their religion or speak their own language. In Spain under the dictator Francisco Franco, people were forbidden to speak Catalan—the language of the region around Barcelona. When Franco died in 1975 and Spain became more democratic, the Catalan language flourished—both in public speaking and in the press.

The Globalization of Culture

The infusion of Western culture throughout the world seems to be accelerating as the commercialized culture of the United States is marketed worldwide. One can go to quite distant places in the world and see familiar elements of U.S. culture, whether it is McDonald's in Hong Kong, The Gap in South Africa, or Disney products in western Europe. From films to fast food, the United States dominates international mass culture—largely through the influence of capitalist markets, as conflict theorists would argue. The diffusion of a single culture throughout the world is referred to as **global culture.** Despite the enormous diversity of cultures worldwide, fashion, food, entertainment, and other cultural values are increasingly dominated by U.S. markets, thereby creating a more homogenous world culture. Global culture is increasingly marked by capitalist interests, squeezing out the more diverse folk cultures that have been common throughout the world (Rieff 1993).

Does increasing globalization of culture change traditional cultural values? Some worry that globalization imposes western values on nonwestern cultures, thus eroding long-held cultural traditions. But global economic change can also introduce more tolerant values to cultures that might have had a more narrow worldview previously. As globalization occurs, *both* economic changes *and* traditional cultural values shape the emerging national culture of different societies (Inglehart and Baker 2000).

The conflict between traditional and more commercial values is now being played out in world affairs, with some arguing that the conflicts we see in international relations are partially rooted in a struggle between the values of a consumer-based, capitalist Western culture and the traditional values of local communities. Benjamin Barber (1995) expresses this as the struggle between "McWorld" and "Jihad"—a struggle he interprets as the tension between global commerce and parochial values. As some people resist the influence of market-driven values, movements to reclaim or maintain ethnic and cultural identity can intensify. Thus, you can witness a proliferation of culturally based movements—including strong feelings of nationalism, such as among extremist groups in the Middle East.

Popular Culture and the Media

Some aspects of culture pervade the whole society, such as common language, general patterns of dress, and dominant value systems. **Popular culture** includes the beliefs, practices, and objects that are part of everyday traditions. This includes mass-produced culture, such as popular music and films, mass-marketed books and magazines, large-circulation newspapers, and other parts of the culture that are shared by the general populace (Gans 1999). Popular culture is distinct from elite culture, which is shared by only a select few but is highly valued. Unlike elite culture (sometimes referred to as "high culture"), popular culture is mass-produced and mass-consumed and has enormous significance in the formation of public attitudes and values. Popular culture is also supported by patterns of mass consumption, as the many objects associated with popular culture are promoted and sold to a consuming public.

Different groups partake of popular and elite culture in different ways. First, social class affects the ability of groups to participate in certain forms of culture, as you can see in Figure 2.3 showing which

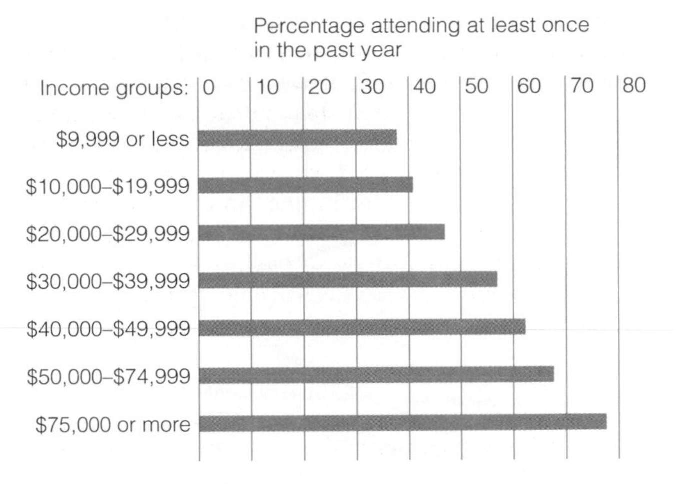

FIGURE 2.3 *Who Goes to the Movies?*

Source: U.S. Census Bureau. 2005. *Statistical Abstract of the United States 2004.* Washington, DC: U.S. Department of Commerce, p. 769.

income groups are most likely to attend movies. Some groups may derive their culture from expensive theater shows or opera performances where tickets can cost hundreds of dollars. Meanwhile, millions of "ordinary" citizens get their primary cultural experience from television, movie rentals, and increasingly, the Internet.

Familiarity with different cultural forms stems from patterns of exclusion throughout history, as well as integration into networks that provide information about the arts. As a result, African Americans are much more likely than White Americans to attend jazz concerts and listen to soul, blues, rhythm and blues, and other historically African American musical forms (DiMaggio and Ostrower 1990).

© Frank Micelotta/American Idol/Getty Images for Fox

The reach of the mass media can be staggering. In 2007, there were 74 million votes cast for the final winner, by comparison, 122 million people voted in the presidential election of 2004.

The Influence of the Mass Media

Popular culture is increasingly characterized by mass distribution (see Figure 2.4). **Mass media** are the channels of communication available to wide segments of the population—the print, film, and electronic media (radio and television), and increasingly, the Internet. The mass media have extraordinary power to shape information and public perceptions in an era when complex issues are reduced to "sound bites" and "photo opportunities." If you doubt this, try the experiment in the "See for Yourself" feature on page 53.

For most Americans, leisure time is dominated by television. Ninety-eight percent of all homes in the United States have at least one television—more than have telephone service. The average person consumes some form of media sixty-eight hours per week—more time than they likely spend in school or at work; thirty of these hours are spent watching television (U.S. Census Bureau 2007). Watching television is also the most popular leisure activity of

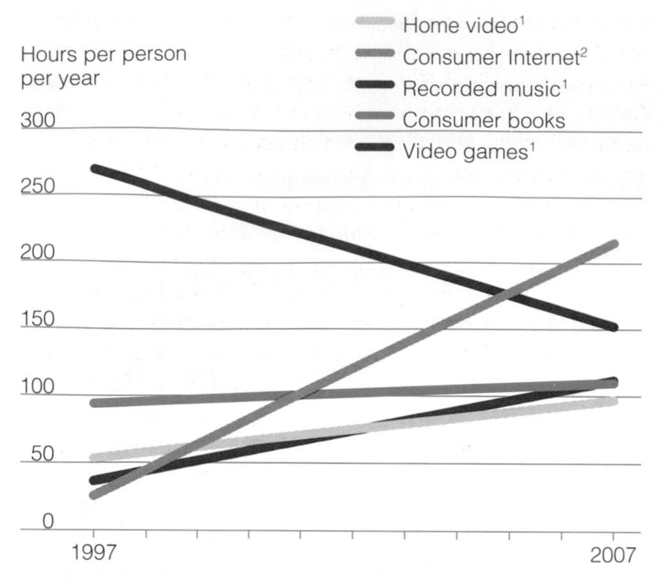

FIGURE 2.4 *The Changing Use of the Media*

Data: U.S. Census Bureau. 2007. *Statistical Abstract of the United States 2007.* Washington, DC: U.S. Government Printing Office, www.census.gov; U.S. Census Bureau. 2003. *Statistical Abstract of the United States 2002.* Washington, DC: U.S. Government Printing Office, p. 699.

[1]Does not include Internet-related use of traditional media.
[2]Examples include music downloaded from a computer, reading downloaded to e-books, listening to radio via the Internet, reading web-based newspapers, and so forth.

Americans; 26 percent say it is their favorite way to spend an evening, compared to 9 percent who would rather read and 1 percent whose favorite evening activity is listening to music (Saad 2002).

Television is a powerful transmitter of culture, but it also portrays a very homogeneous view of culture because in seeking the widest possible audience, networks and sponsors find the most common ground and take few risks. Television and other forms of mass media have enormous power to shape public opinion and behavior. If you doubt this, observe how much a part of everyday life are certain characters from television sitcoms and dramas. People at work may talk about last night's episode of a particular show or laugh about the antics of their favorite sitcom character. The media is also everywhere—present in airports, elevators, classrooms, bars and restaurants, and hospital waiting rooms. You may even be born to the sounds and images of television, since they are turned on in many hospital delivery rooms. Television is now so ever-present in our lives that 42 percent of all U.S. households are now called "constant television households"—that is, those households where television is on most of the time. Black children are more likely than White or Hispanic children to live in such households (Gitlin 2002). For many families, TV is the "babysitter."

The mass media (and television, in particular) play a huge role in shaping people's perception and awareness of social issues. For example, even though crime has actually decreased, the amount of time spent reporting crime in the media has actually increased. Sociologists have found that people's fear of crime is directly related to the time they spend watching television or listening to the radio (Angotti 1997; Chiricos et al. 1997).

thinking sociologically

Watch a particular kind of television show (situation comedy, sports broadcast, children's cartoon, or news program, for example) and make careful written notes on the depiction of different groups in this show. How often are women and men, boys and girls shown?

How are they depicted? You could also observe the portrayal of Asian Americans, Native Americans, African Americans, or Latinos. What do your observations tell you about the cultural ideals that are communicated through *popular culture*?

Although people tend to think of the news as authentic and true, news is actually manufactured in a complex social process. From a sociological perspective, it is not objective reality that determines what news is presented and how it is portrayed, but commercial interests, the values of news producers, and perceptions of what matters to the public. The media shape our definition of social problems by determining the range of opinion or information that is defined as legitimate and by deciding which experts will be called on to elaborate an issue (Gans 1979; Gitlin 2001).

Race, Gender, and Class in the Media

Many sociologists have argued that the mass media promote narrow definitions of who people are and what they can be. What is considered beauty, for example, is not universal. Ideals of beauty change as cultures change and depend upon what certain cultural institutions promote as beautiful. Aging is not beautiful, youth is. Light skin is promoted as more beautiful than dark skin, regardless of race, although being tan is seen as more beautiful than being pale. In African American women's magazines, the models typified as most beautiful are generally those with the clearly Anglo features of light skin, blue eyes, and straight or wavy hair. European facial features are also pervasive in the images of Asian and Latino women appearing in U.S. magazines. The media communicate that only certain forms of beauty are cultur-

SEE FOR YOURSELF

A MEDIA BLACKOUT EXPERIMENT

Try the following experiment:

1. Keep a written log of how much time you spend over a forty-eight-hour period engaged with some form of media (television, film, video, books, cell phone, and the like).
2. Eliminate all use of the media, except for that required for work, school, and emergencies for a forty-eight-hour period, keeping a journal as you go of what happens—what you are thinking, what others say, and how people interact with you.
3. At the conclusion of the forty-eight-hour period, write a brief paper describing what happened during the blackout. Use your paper to illustrate the concepts of *norms, values,* and *cultural hegemony.*
4. How would each of the following theoretical frameworks explain what happened during your media blackout: functionalism, conflict theory, feminist theory, symbolic interaction?
5. What does your experiment teach you about the power of the media to influence society and people within it?

ally valued. These ideals are not somehow "natural." They are constructed by those who control cultural and economic institutions (Craig 2002; Gimlin 2002).

Images of women and racial and ethnic minorities in the media are similarly limiting. Content analyses of television reveal that during prime time men are a large majority of the characters shown. There has been an increase in the extent to which women are depicted in professional jobs, but such images usually depict professional women as young, suggesting that career success comes early, especially to thin and beautiful women (Signorielli and Bacue 1999). On soap operas, women are cast either as evil or good, but also naive. In music videos, women characters appear less frequently, but typically wear sexy and skimpy clothing and are more often the object of another's gaze than their male counterparts; Black women especially are represented in sexualized ways (Emerson 2002; Signorielli et al. 1994; Collins 2004).

Even though African Americans and Hispanics watch more television than Whites do, they represent a small proportion of TV characters, generally confined to a narrow variety of character types, depicted in stereotypical ways. Latinos are often stereotyped as criminals or passionate lovers. African American men are most typically seen as athletes and sports commentators, criminals, or entertainers, and African American women in domestic or sexual roles or as sex objects. It is difficult to find a single show where Asians are the principal characters—usually they are depicted in silent roles as domestics or behind-the-scenes characters. Native Americans make occasional appearances, where they usually are depicted as mystics or warriors. Jewish women are generally invisible on popular TV programming, except when they are ridiculed in stereotypical roles (Kray 1993). Arab Americans are likewise stereotyped, depicted as terrorists, rich oil magnates, or in the case of women, as perpetually veiled and secluded (Read 2003; Mandel 2001).

Class stereotypes abound as well, with working-class men typically portrayed as being ineffectual, even buffoonish (Butsch 1992; Dines and Humez 2002). A recent study of TV talk shows also demonstrates how such shows exploit working-class people. The researcher, Laura Grindstaff, spent six months working on two popular talk shows, carefully doing participant observation and interviewing staff and guests. She found that guests had to enact stereotypes of their groups to get air time. She argues that, although these shows give ordinary people a place to air their problems and be heard, they exploit the working class, making a spectacle of their troubles (Grindstaff 2002; Press 2002).

Recently, there has been increased representation of gays and lesbians in the media, after years of being virtually invisible or only the subject of ridicule. As advertisers have sought to expand their commercial markets, they are showing more gay and lesbian characters on television. This makes gays and lesbians more visible, although critics point out that they are still cast in narrow and stereotypical terms, showing little about real life for gays and lesbians. Nonetheless, cultural visibility for any group is important because it validates people and can influence the public's acceptance of and generate support for equal rights protection (Gamson 1998).

Television is not the only form of popular culture that influences public consciousness, class, gender, and race. Music, film, books, and other industries play a significant role in molding public consciousness. What images are produced by these cultural forms? You can look for yourself. Try to buy a birthday card that contains neither an age or gender stereotype, or watch TV or a movie and see how different gender and race groups are portrayed. You will likely find that women are depicted as trying to get the attention of men; African Americans are more likely than Whites to be seen singing and dancing.

Do these images matter? Studies find that exposure to traditional sexualized imagery in music videos has a negative effect on college students' attitudes, for example, holding more adversarial attitudes about sexual relationships (Kalof 1999). Other studies find that even when viewers see media images as unrealistic, they think that others find the images important and will evaluate them accordingly; this has been found especially true for young White girls who think boys will judge them by how well they match the media ideal (Milkie 1999). Although people do not just passively internalize media images and do distinguish between fantasy and reality (Hollander 2002; Currie 1997), such images form cultural ideals that have a huge impact on people's behavior, values, and self-image.

Theoretical Perspectives on Culture

Sociologists study culture in a variety of ways, asking a variety of questions about the relationship of culture to other social institutions and the role of culture in modern life (see Table 2.3). One important question for sociologists studying the mass media is whether these images have any effect on those who see them. Do the media create popular values or reflect them? The **reflection hypothesis** contends that the mass media reflect the values of the general population (Tuchman 1979). The media try to appeal to the most broad-based audience, so they aim for the middle ground in depicting images and ideas. Maximizing popular appeal is central to television program development; media organizations spend huge amounts on market research to uncover what people think and believe and what they will like. Characters are then created with whom people will identify. Interestingly, the images in the media with which we identify are distorted versions of reality. Real people seldom live like the characters on television, although part of the appeal of these shows is how they build upon, but then mystify, the actual experiences of people.

The reflection hypothesis assumes that images and values portrayed in the media reflect the values existing in the public, but the reverse can also be true—that is, the ideals portrayed in the media also influence the values of those who see them. As an example, social scientists have studied the stereotyped images commonly found in children's programming. Among their findings, they have shown that the children who watch the most TV hold the most stereotypic gender attitudes (Signorielli 1991). Although there is not a simple and direct relationship between the content of mass media images and what people think of themselves, clearly these mass-produced images can have a significant impact on who we are and what we think.

Culture and Group Solidarity

Many sociologists have studied particular forms of culture and have provided detailed analyses of the content of cultural artifacts, such as images in certain television programs or genres of popular music. Other sociologists take a broader view by analyzing the relationship of culture to other forms of social organization. Beginning with some of the classical sociological theorists (see Chapter 1), sociologists have studied the relationship of culture to other social institutions. Max Weber looked at the impact of culture on the formation of social and economic institutions. In his classic analysis of the Protestant work ethic and capitalism, Weber argued that the Protestant faith rested on cultural beliefs that were highly compatible with the development of modern capitalism. By promoting a strong work ethic and a need to display material success as a sign of religious salvation, the Protestant work ethic indirectly but effectively promoted the interests of an emerging capitalist economy. (We revisit this issue in Chapter 14.) In other words, culture influences other social institutions.

Many sociologists have also examined how culture integrates members into society and social groups.

table 2.3 Theoretical Perspectives on Culture

According to: Functionalism Culture . . .	Conflict Theory	Symbolic Interaction	New Cultural Studies
Integrates people into groups	Serves the interests of powerful groups	Creates group identity from diverse cultural meanings	Is ephemeral, unpredictable, and constantly changing
Provides coherence and stability in society	Can be a source of political resistance	Changes as people produce new cultural meanings	Is a material manifestation of a consumer-oriented society
Creates norms and values that integrate people in society	Is increasingly connected by economic monopolies	Is socially constructed through the activities of social groups	Is best understood by analyzing its artifacts—books, films, television images

[a sociological eye on the media

Media Blackout

Suppose that you lived for a few days without use of the mass media that permeates our lives. How would this affect you? In an intriguing experiment, Charles Gallagher (a sociologist at Georgia State University) has developed a research project for students in which he asks them to stage a media blackout in their lives for just forty-eight hours. First, students write a log of the amount of time they spend engaged with the media in the week prior to the blackout. Included is time spent watching television, on the Internet, reading books and magazines, listening to music, viewing films, even using cell phones—any activity that can be construed as part of the media monopoly on people's time.

Next, students take a forty-eight hour period during which they eliminate all discretionary time with the media. (Note that you are allowed to do required work and schoolwork during the imposed blackout.) When one of the authors of this book (Andersen) had her students do this experiment, they complained—even before starting—that they wouldn't be able to do it! But they had to try. What did they find?

First, Andersen's students got a big assist—the week of the assignment came during Hurricane Isabel on the East Coast when many were without power for several days. This did not deter the students from thinking they just *had to have* their DVD players, music, TV, and cell phones! Many of the students said they could only stand being without access to the media for a few hours and couldn't go the full two days without using the media. What did they report happened to them when they were not engaged with the media to the extent that is normal in their lives?

Most reported that they felt isolated during the media blackout—not just from information, but mostly from other people. They were excluded from conversations with friends—about what happened on a given television episode or about film characters or movie stars profiled in magazines—and from playing computer games. One even wrote that without the media she felt that she had no personality! Overall, without their connection to the media, students were alienated, isolated, and detached—although most also reported that they studied more without the distraction of the media. A most interesting finding was that several reported that they were much more reflective during this time and had more meaningful conversations with friends.

Should you try this experiment, you can think about the enormous influence that the mass media have in shaping everyday life—including our self-concepts and our relationships with other people. A warning: If you try the media blackout, be sure to have some plan in place for having your family and/or friends contact you in case of an emergency!

Sources: Gitlin, Todd. 2002. *Media Unlimited: How the Torrent of Images and Sounds Overwhelms Our Lives.* New York: Metropolitan Books; Personal correspondence, Charles Gallagher, Georgia State University.

Functionalist theorists, for example, believe that norms and values create social bonds that attach people to society. Culture therefore provides coherence and stability in society. Robert Putnam examines this idea in his book *Bowling Alone* (2000), in which he argues that there has been a decline in *civic engagement*—defined as participation in voluntary organizations, religious activities, and other forms of public life—in recent years. As people become less engaged in such activities, there is a decline in the shared values and norms of the society so that social disorder results. Sociologists are debating the extent to which there has been such a decline in public life, but from a functionalist perspective, the point is that participation in a common culture is an important social bond—one that unites society (Etzioni et al. 2001).

Classical theoretical analyses of culture have placed special emphasis on nonmaterial culture—the values, norms, and belief systems of society. Sociologists who use this perspective emphasize the integrative function of culture, that is, its ability to give people a sense of belonging in an otherwise complex social system (Smelser 1992a). In the broadest sense, they see culture as a major integrative force in society, providing societies with a sense of collective identity and commonly shared worldviews.

Culture, Power, and Social Conflict

Whereas the emphasis on shared values and group solidarity drives one sociological analysis of culture, conflicting values drives another. Conflict theorists (see Chapter 1) have analyzed culture as a source of power in society. You can find numerous examples throughout human history where conflict between different cultures has actually shaped the course of world affairs. One such example comes from the Middle East and the situation for the Kurdish people. The Kurds are an ethnic group (see Chapter 9) who speak their own language and inhabit an area in the Middle East that includes parts of Iraq, Iran, Turkey, and Syria, though they mostly live in northern Iraq. Most are Sunni Muslims, and they have experienced years of political and economic repression and, under Saddam Hussein, mass murder. Attempting to eliminate Kurds altogether, Saddam Hussein ordered the execution of over 180,000 people in Kurdish villages, often through chemical and biological weapons (O'Leary 2002). This and other examples of so-called "ethnic cleansing" show how cultural conflict can be driven by intense group hatred and powerful forms of domination.

From a conflict-oriented perspective, culture is also dominated by economic interests. A few powerful groups are the major producers and distributors of culture. Now, with corporate mergers, a single corporation can control a huge share of television, radio, newspapers, music, publishing, film, and the Internet, as shown in Figure 2.5. As the production of popular culture becomes concentrated in the hands of just a few, there may be less diversity in the content.

Conflict theorists see contemporary culture as produced within institutions that are based on inequality and capitalist principles. As a result, the cultural values and products that are produced and sold promote the economic and political interests of the few—those who own or benefit from these cultural industries. This is especially evident in the study of

Viacom owns:

MTV; BET (Black Entertainment Television); Nickelodeon; Paramount Pictures; Dreamworks; CBS (split in 2006); Simon and Schuster Publishers; College TV Network; Blockbuster Video; 39 television stations; 185 local radio stations; VH1; Comedy Central; Outdoor Systems billboards; Madison Square Garden (sold in 1995 for $1 billion); TV Land; Pearson Publishing (publisher of elementary, high school, and college textbooks); Rumored to be considering purchasing facebook.com for $2 billion; and more . . .

Time Warner owns:

HBO; CNN; AOL Instant Messenger; Warner Brothers; Warner Brothers Studios; WB Television Network; Hanna-Barbera Cartoons; Castle Rock Entertainment; Warner Home Video; Turner Network Television; TBS Superstation; Netscape; Amazon.com (partial owner); Warner Brothers International Theaters (owns/operates multiplex theatres in over 12 countries); over 50 magazines (including *Time; Fortune; People; Life; Sports Illustrated; Money; In Style; Cooking Light; This Old House; Field and Stream; Popular Science; Snowboard Life; Yachting Magazine; Travel and Leisure;* and others); Atlanta Braves; and more . . .

Hearst Corporation owns:

17 magazines: including *Cosmo, Cosmo Girl, Marie Claire, O* (The Oprah Magazine), *Good Housekeeping, Redbook, Seventeen, Popular Mechanics,* and others), 12 newspapers, 27 local TV stations, several cable networks, including The History Channel, part of ESPN, Lifetime, A&E, Hearst Books, King Features TV syndicate; holds investments in Netscape, XM Satellite radio, drugstore.com, and more . . .

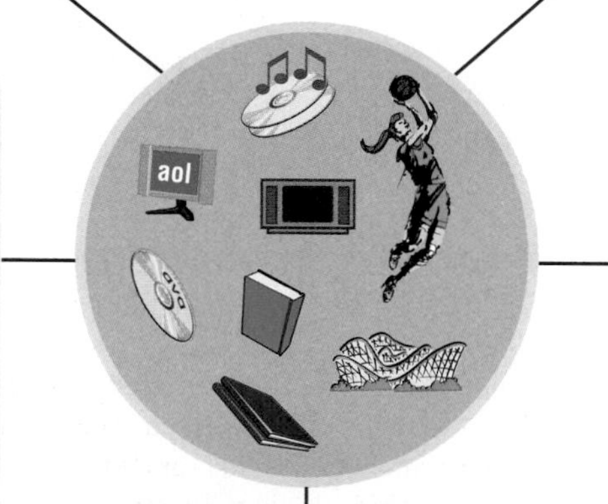

Disney owns:

Disney Pictures; Miramax Films; Touchstone Pictures; Buena Vista Home Entertainment; Pixar; ABC television network; 10 local television stations; ESPN network; ABC Family Television; Disney Channel; 68 local radio stations; 15 magazines; Buena Vista Music Group; 5 Disneylands; Disney stores; Disney Cruise Line; Disney Shopping Inc.; and more . . .

News Corporation owns:

World's largest newspaper publisher; Fox News; myspace.com (purchased in 2005 for $580 million); William Morrow and Avon Books; 35 local radio stations; 25 international newspapers; New York Post; Broadcasting rights for NFL in Asia; Harper Collins Publishing; TV Guide; 20th Century Fox Film; Los Angeles Dodgers (sold in 2004); Los Angeles Lakers; Los Angeles Kings; Staples Center (LA); and more . . .

FIGURE 2.5

Cultural Monopolies

popular culture, which is marketed to the masses by entities with a vast economic stake in distributing their products. Conflict theorists conclude that the cultural products most likely to be produced are consistent with the values, needs, and interests of the most powerful groups in society. The evening news, for example, typically is sponsored by major financial institutions, oil companies, and automobile makers. Conflict theorists then ask how this commercial sponsorship influences the content of the news. If the news were sponsored by labor unions, would conflicts between management and workers always be defined as "labor troubles," or might newscasters refer instead to "capitalist troubles"?

Conflict theorists see culture as increasingly controlled by economic monopolies. Whether it is books, music, films, news, or other cultural forms, monopolies in the communications industry (where culture is increasingly located) have a strong interest in protecting the status quo. As media conglomerates swallow up smaller companies and drive out smaller, less-efficient competitors, the control that economic monopolies have over the production and distribution of culture becomes enormous. Mega-communications companies then influence everything—from the movies and television shows you see to the books you read in school.

Sociologists refer to the concentration of cultural power as **cultural hegemony** (pronounced "heh-JEM-o-nee"), defined as the pervasive and excessive influence of one culture throughout society. Cultural hegemony creates a homogeneous mass culture. It is a means through which powerful groups gain the assent of those they rule. The concept of cultural hegemony implies that culture is highly politicized, even if it does not appear so. Through cultural hegemony, those who control cultural institutions can also control people's political awareness because they create cultural beliefs that make the rule of those in power seem inevitable and right. As a result, political resistance to the dominant culture is blunted (Gramsci 1971).

Culture can also be a source of political resistance and social change. Reclaiming an indigenous culture that had been denied or repressed is one way that groups mobilize to assert their independence. An example from within the United States is the *repatriation movement* among American Indians who have argued for the return of both cultural artifacts and human remains held in museum collections. Many American Indians believe that, despite the public good that is derived from studying such remains and objects, cultural independence and spiritual respect outweigh such scientific arguments (Thornton 2001). Other social movements, such as the gay and lesbian movement, have also used cultural performance as a means of political and social protest. Cross-dressing, drag shows, and other forms of "gender play" can be seen as cultural performances that challenge homophobia and traditional sexual and gender roles (Rupp and Taylor 2003).

A final point of focus for sociologists studying culture from a conflict perspective lies in the concept of cultural capital. **Cultural capital** refers to the cultural resources that are deemed worthy (such as knowledge of elite culture) and that give advantages to groups possessing such capital. This idea has been most developed by the French sociologist Pierre Bourdieu (1984), who sees the appropriation of culture as one way that groups maintain their social status.

Bourdieu argues that members of the dominant class have distinctive lifestyles that mark their status in society. Their ability to display this cultural lifestyle signals their importance to others; that is, they possess cultural capital. From this point of view, culture has a role in reproducing inequality among groups. Those with cultural capital use it to improve their social and economic position in society. Sociologists have found a significant relationship, for example, between cultural capital and grades in school—that is, those from the more well-to-do classes (those with more cultural capital) are able to parlay their knowledge into higher grades, thereby reproducing their social position by being more competitive in school admissions and, eventually, in the labor market (Hill 2001; Treiman 2001).

Symbolic Interaction and the Study of Culture

Especially productive when applied to the study of culture has been *symbolic interaction theory*—a perspective that analyzes behavior in terms of the meaning people give it. (See chapter 1). The concept of culture is central to this orientation. Symbolic interaction emphasizes the interpretive basis of social behavior, and culture provides the interpretive framework through which behavior is understood.

Symbolic interaction also emphasizes that culture, like all other forms of social behavior, is socially constructed. That is, culture is produced through social relationships and in social groups, such as the media organizations that produce and distribute culture. People do not just passively submit to cultural norms; rather, they actively make, interpret, and respond to the culture around them. Culture is not one-dimensional; it contains diverse elements and provides people with a wide range of choices from which to select how they will behave (Swidler 1986). Culture, in fact, represents the creative dimension of human life.

In recent years, a new interdisciplinary field known as *cultural studies* has emerged that builds on the insights of the symbolic interaction perspective in sociology. Sociologists who work in cultural studies are often critical of classical sociological approaches to studying culture, arguing that the classical approach has overemphasized nonmaterial culture—that is, ideas, beliefs, values, and norms. The new scholars of cultural studies find that material culture has increasing importance in modern society (Crane 1994; Walters 1999). This includes cultural forms that are recorded through print, film, artifacts, or the electronic media. Postmodernist theory has greatly influenced new cultural studies (see Chapter 1). *Postmodernism* is based on the idea that society is not an objective thing; rather, it is found in the words and images that people use to represent behavior and ideas. Given this orientation, postmodernism often analyzes common images and cultural products found in everyday life.

Classical theorists have tended to study the unifying features of culture; cultural studies researchers tend to see culture as more fragmented and unpredictable. To them, culture is a series of images—images that can be interpreted in multiple ways depending on the viewpoint of the observer. From the perspective of new cultural studies theorists, the ephemeral and rapidly changing quality of contemporary cultural forms is reflective of the highly technological and consumer-based culture on which the modern economy rests. Modern culture, for example, is increasingly dominated by the ever-changing, but ever-present, images that the media bombard us with in everyday life. The fascination that cultural studies theorists have for these images is partially founded in illusions that such a dynamic and rapidly changing culture produces.

Cultural Change

In one sense, culture is a conservative force in society; it tends to be based on tradition and is passed on through generations, conserving and regenerating the values and beliefs of society. Culture is also increasingly based on institutions that have an economic interest in maintaining the status quo. People are also often resistant to cultural change because familiar ways and established patterns of doing things are hard to give up. But in other ways, culture is completely taken for granted, and it may be hard to imagine a society different from that which is familiar.

Imagine, for example, the United States without fast food. Can you do so? Probably not. Fast food is so much a part of contemporary culture that it is hard to imagine life without it. Consider these facts about fast-food culture:

- The average person in the United States consumes three hamburgers and four orders of french fries per week.
- People in the United States spend more money on fast food than on movies, books, magazines, newspapers, videos, music, computers, and higher education combined.
- One in eight workers has at some point been employed by McDonald's.
- McDonald's is the largest private operator of playgrounds in the United States and the single largest purchaser of beef, pork, and potatoes.
- Ninety-six percent of American schoolchildren can identify Ronald McDonald—only exceeded by the number who can identify Santa Claus (Schlosser 2001).

Eric Schlosser, who has written about the permeation of society by fast food culture, has written that "a nation's diet can be more revealing than its art or literature" (2001: 3). He relates the growth of the fast-food industry to other fundamental changes in American society, including the vast entry of women into the paid labor market, the development of an automobile culture, the increased reliance on low-wage service jobs, the decline of family farming, and the growth of agribusiness. One result is a cultural emphasis on uniformity, not to mention increased fat and calories in people's diets.

This example shows how cultures can change over time, sometimes in ways that are hardly visible to us unless we take a longer-range view or, as sociologists would do, question that which surrounds us. Culture is a dynamic, not static, force in society, and it develops as people respond to various changes in their physical and social environments.

Culture Lag

Sometimes cultures adjust slowly to changing cultural conditions, and the result can be **culture lag** (Ogburn 1922). Some parts of culture may change more rapidly than others; thus, one aspect of culture may "lag" behind another. Rapid technological change is often attended by culture lag because some elements of the culture do not keep pace with technological innovation. In today's world, we have the technological ability to develop efficient, less-polluting rapid transit, but changing people's transportation habits is difficult.

When culture changes rapidly or someone is suddenly thrust into a new cultural situation, the result can be **culture shock,** the feeling of disorientation when one encounters a new or rapidly changed cultural situation. Even moving from one cultural environment to another within one's own society can make a person feel out of place. The greater the difference between cultural settings, the greater the culture shock. Some people displaced from New Orleans following hurricane Katrina experienced culture shock in the host communities where they relocated. Accustomed to the food, customs, and environment in their New Orleans home, many evacuees were relocated to remote, mostly white, rural parts of the country. On top of their disorientation from the trauma of the storm itself, living in these new environments could be very disorienting (Wilkerson 2005).

Sources of Cultural Change

There are several causes of cultural change, including (1) a change in the societal conditions, (2) cultural diffusion, (3) innovation, and (4) the imposition of cultural change by an outside agency. Let us examine each.

1. *Cultures change in response to changed conditions in the society.* Economic changes, population changes, and other social transformations all influence the development of culture. A change in the makeup of a society's population may be enough by itself to cause a cultural transformation. The high rate of immigration in recent years has brought many cultural changes to the United States. Some cities, such as Miami and Los Angeles, have a Latin feel because of the large Hispanic population. But even outside urban areas, cultural change from immigration is apparent. Markets selling Asian, Mexican, and Middle Eastern foods are increasingly common; school districts include students who speak a huge variety of languages; popular music bears the imprint of different world cultures. This is not the first time U.S. culture has changed because of immigration. Many national traditions stem from the patterns of immigration that marked the earlier part of the twentieth century—think of St. Patrick's Day parades, Italian markets, and Chinatowns.
2. *Cultures change through cultural diffusion.* **Cultural diffusion** is the transmission of cultural elements from one society or cultural group to another. In our world of instantaneous communication, cultural diffusion is swift and widespread. This is evident in the degree to which worldwide cultures have been Westernized. Cultural diffusion also occurs when subcultural influences enter the dominant group. Dominant cultures are regularly enriched by minority cultures. An example is the influence of Black and Latino music on other musical forms. Rap music, for example, emerged within inner-city African American neighborhoods, describing and analyzing in its own form the economic and political conditions of the urban ghetto. Now, rap music is listened to by White as well as Black audiences and is part of youth culture in general. Cultural diffusion is one thing that drives cultural evolution, especially in a society such as ours that is lush with diversity.
3. *Cultures change as the result of innovation, including inventions and technological developments.* Cultural innovations can create dramatic changes in society. Think, for example, of how the invention of trolleys, subways, and automobiles changed the character of cities. People no longer walked to work; instead, cities expanded outward to include suburbs. Further, the invention of the elevator let cities expand not just out, but up.

   Now the development of computer technology infiltrates every dimension of life It is hard to overestimate the effect of innovation on contemporary cultural change. Technological innovation is so rapid and dynamic that one generation can barely maintain competence with the hardware of

The Granger Collection, New York

© Michael Newman/PhotoEdit

Cigarette smomking historically was glamorized, but the anti-smoking movement has now defined smoking as a deviant activity.

the next. The smallest laptop or handheld computer today weighs hardly more than a few ounces, and its capabilities rival that of computers that filled entire buildings only twenty years ago. Downloading music was not even imaginable just a few years ago; now it is a common practice.

What are some of the social changes that technology change is creating? People can now work and be miles—even nations—away from their place of employment. Families can communicate from multiple sites; children can be paged; grandparents can receive live photos of a family event; criminals are tracked via cellular technology; music can be stolen without even going in to a music store. Conveniences multiply with the growth of such technology, but so do the invasions of privacy and, perhaps, identity theft. In such a rapidly changing technological world, it is hard to imagine what will be common in just a few years.

4. *Cultural change can be imposed.* Change can occur when a powerful group takes over a society and imposes a new culture. The dominating group may arise internally, as in a political revolution, or it may appear from outside, perhaps as an invasion. When an external group takes over the society of a "native," or indigenous, group—as White settlers did with Native American societies—they typically impose their own culture while prohibiting the indigenous group from expressing its original cultural ways. Manipulating the culture of a group is a way of exerting social control. Many have argued that public education in the United States, which developed during a period of mass immigration, was designed to force White, northern European, middle-class values onto a diverse immigrant population that was perceived to be potentially unruly and politically disruptive. Likewise, the schools run by the Bureau of Indian Affairs have been used to impose dominant group values on Native American children (Snipp 1996).

Resistance to political oppression often takes the form of a cultural movement that asserts or revives the culture of an oppressed group; thus, cultural expression can be a form of political protest. Identification with a common culture can be the basis for group solidarity, as found in the example of the "Black is beautiful" movement in the 1970s that encouraged Black Americans to celebrate their African heritage with Afro hairstyles, African dress, and African awareness. Cultural solidarity has also been encouraged among Latinos through La Raza Unida (meaning "the race," or "the people, united"). Cultural change can promote social change, just as social change can transform culture.

Chapter Summary

- ***What is culture?***

  Culture is the complex and elaborate system of meaning and behavior that defines the way of life for a group or society. It is shared, learned, taken for granted, symbolic, and emergent and varies from one society to another.

- ***How do sociologists define norms, beliefs, and values?***

  Norms are rules of social behavior that guide every situation and may be formal or informal. When norms are violated, *social sanctions* are applied. *Beliefs* are strongly shared ideas about the nature of social reality. *Values* are the abstract concepts in a society that define the worth of different things and ideas.

- ***What is the significance of diversity in human cultures?***

  As societies develop and become more complex, cultural diversity can appear. The United States is highly diverse culturally, with many of its traditions influenced by immigrant cultures and the cultures of African Americans, Latinos, and Native Americans. The *dominant culture* is the culture of the most powerful group in society. *Subcultures* are groups whose *values* and cultural patterns depart significantly from the *dominant culture.*

- ***What is the sociological significance of popular culture?***

  Popular culture includes the beliefs, practices, and objects of everyday traditions. Elements of popular culture, such as the *mass media,* have an enormous influence on groups' beliefs and values, including images associated with racism and sexism.

- ***What do different sociological theories reveal about culture?***

  Sociological theory provides different perspectives on the significance of culture. *Functionalist theory* emphasizes the influence of values, norms, and beliefs on the whole society. *Conflict theorists* see culture as influenced by economic interests and power relations in society. *Symbolic interactionists* emphasize that culture is socially constructed. This has influenced new cultural studies, which interpret culture as a series of images that can be analyzed from the viewpoint of different observers.

- ***How do cultures change?***
  There are several sources of cultural change, including change in societal conditions, *cultural diffusion,* innovation, and the imposition of change by dominant groups. As cultures change, *culture lag* can result, meaning that sometimes cultural adjustments are out of synchrony with each other. Persons who experience new cultural situations may experience *culture shock.*

Key Terms

beliefs 45
counterculture 50
cultural capital 57
cultural diffusion 59
cultural hegemony 57
cultural relativism 41
culture 35
culture lag 58
culture shock 58
dominant culture 47
ethnocentrism 39
ethnomethodology 45
folkways 44
global culture 50
language 41
laws 44
mass media 51
material culture 35
mores 44
nonmaterial culture 35
norms 44
popular culture 51
reflection hypothesis 54
Sapir–Whorf hypothesis 43
social sanctions 44
subculture 49
symbols 37
taboo 44
values 45

Online Resources

Sociology: The Essentials Companion Website

www.thomsonedu.com/sociology/andersen
Visit your book companion website where you will find more resources to help you study and write your research papers. Resources include Suggested Readings, web links, and a MicroCase Online feature that teaches you how to research society. Other resources include Learning Objectives, Internet exercises, quizzing, and flash cards.

is an easy-to-use online resource that helps you study in less time to get the grade you want NOW.

www.thomsonedu.com/login
Need help studying? This site is your one-stop study shop. Take a Pre-Test and Thomson NOW will generate a Personalized Study Plan based on your test results. The Study Plan will identify the topics you need to review and direct you to online resources to help you master those topics. You can then take a Post-Test to determine the concepts you have mastered and what you still need to work on.

3

Chapter three

CHAPTER THREE

Chapter

Socialization and the Life Course

[**During the summer** of 2000, scientists working on the human genome project announced that they had deciphered the human genetic code. By mapping the complex structure of DNA (deoxyribonucleic acid) on high-speed computers, scientists identified the 3.12 billion chemical base pairs in human DNA and put them in proper sequence, unlocking the genetic code of human life. Scientists likened this to assembling "the book of life," that is, having the knowledge to make and maintain human beings. The stated purpose of the human genome project is to see how genetics influences the development of disease, but it raises numerous ethical questions about human cloning and the possibility of creating human life in the laboratory. Is our genetic constitution what makes us human? Suppose you created a human being in the laboratory but left that creature without social contact. Would the "person" be human?

continued

three chapter three CHAPTER TH

Rare cases of *feral children,* who have been raised in the absence of human contact, provide some clues as to what happens during human development when a person has little or no social contact. One such case, discovered in 1970, involved a young girl given the pseudonym of Genie. When her blind mother appeared in the Los Angeles County welfare office seeking assistance for herself, case workers first thought the girl was six years old. In fact, she was thirteen, although she weighed only 59 pounds and was 4 feet, 6 inches tall. She was small and withered, unable to stand up straight, incontinent, and severely malnourished. Her eyes did not focus, she had two nearly complete sets of teeth, and a strange ring of calluses circled her buttocks. She could not talk.

As the case unfolded, it was discovered that the girl had been kept in nearly total isolation for most of her life. The first scientific report about Genie states:

> In the house Genie was confined to a small bedroom, harnessed to an infant's potty seat. Genie's father sewed the harness himself; unclad except for the harness, Genie was left to sit on that chair. Unable to move anything except her fingers and hands, feet and toes, Genie was left to sit, tied-up, hour after hour, often into the night, day after day, month after month, year after year. (Curtiss 1977: 5)

At night, she was restrained in a handmade sleeping bag that held her arms stationary and placed in a crib. If she made a sound, her father beat her. She was given no toys and was allowed to play only with two old raincoats and her father's censored version of *TV Guide.* (He had deleted everything "suggestive," such as pictures of women in bathing suits.) Genie's mother, timid and blind, was also victimized by her husband. Shortly after the mother sought help after years of abuse, the father committed suicide.

Genie was studied intensively by scientists interested in language acquisition and the social effects of extreme confinement. They hoped that her development would throw some light on the question of nature versus nurture; that is, are people the product of their genes or their social training? After intense language instruction and psychological treatment, Genie developed some verbal ability and, after a year, showed progress in her mental and physical development. Yet the years of isolation and severe abuse took their toll. Genie now lives in a home for mentally disabled adults (Rymer 1993).

This rare case of a feral child sheds some light on the consequences of life without social contact. Knowing the sequence of the human genome may raise the specter of making human beings in the laboratory, but without society, what would humans be like? Genes may confer skin and bone and brain, but only by learning the values, norms, and roles that culture bestows on people do we become the social beings that define us in the eyes of others and to ourselves. Sociologists refer to this process as *socialization*—the subject of this chapter.

The Socialization Process

Socialization is the process through which people learn the expectations of society. **Roles** are the expected behavior associated with a given status in society. When you occupy a given social role, you tend to take on the expectations of others. For example, when you enter a new group of friends, you probably observe their behavior, their language, their dress—perhaps even their opinions of others—and often modify your own behavior accordingly. Before you know it, you are a member of the group, perhaps socializing others into the same set of expectations. We explore roles further in Chapter 4, but it is important to know here that roles are learned through the socialization process.

Through socialization, people absorb their culture—customs, habits, laws, practices, and means of expression. Socialization is the basis for **identity**—how one defines oneself. Identity is both personal and

Formal and informal learning, through schools and other socialization agents, are important elements of the socialization process. Here deaf children are learning sign language.

[understanding diversity

My Childhood (Bong Hwan Kim)

Childhood is a time when children learn their gender, as well as their racial and ethnic identity. This excerpt from an interview with Bong Hwan Kim, a Korean American man, is a reflection on growing up and learning both Korean and American culture.

I came to the United States in 1962, when I was three or four years old. My father had come before us to get a Ph.D. in chemistry. He had planned to return to Korea afterward, but it was hard for him to support three children in Korea while he was studying in the United States, and he wasn't happy alone, so he brought the family over. . .

The Bergenfield, New Jersey, community where I grew up was a blue-collar town of about 40,000 people, mostly Irish and Italian Americans. I lived a schizophrenic existence. I had one life in the family, where I felt warmth, closeness, love, and protection, and another life outside—school, friends, television, the feeling that I was on my own. I accepted that my parents would not be able to help me much.

I can remember clearly my first childhood memory about difference. I had been in this country for maybe a year. It was the first day of kindergarten, and I was very excited about having lunch at school. All morning, I could think only of the lunch that was waiting for me in my desk. My mother had made kimpahp *[rice balls rolled up in dried seaweed] and wrapped it all up in aluminum foil. I was eagerly looking forward to that special treat. I could hardly wait. When the lunch bell rang, I happily took out my foil-wrapped kimpahp. But all the other kids pointed and gawked. "What is that? How could you eat that?" they shrieked. I don't remember whether I ate my lunch or not, but I told my mother I would only bring tuna or peanut butter sandwiches for lunch after that.*

I have always liked Korean food, but I had to like it secretly, at home. There are things you don't show to your non-Korean friends. At various times when I was growing up, I felt ashamed of the food in the refrigerator, but only when friends would come over and wonder what it was. They'd see a jar of garlic and say, "You don't eat that stuff, do you?" I would say, "I don't eat it, but my parents do; they do a lot of weird stuff like that."

. . . As a child you are sensitive; you don't want to be different. You want to be like the other kids. They made fun of my face. They called me "flat face." When I got older, they called me "Chink" or "Jap" or said "Remember Pearl Harbor." In all cases, it made me feel terrible. I would get angry and get into fights. Even in high school, even the guys I hung around with on a regular basis, would say, "You're just a Chink" when they got angry. Later, they would say they didn't mean it, but that was not much consolation. When you are angry, your true perceptions and emotions come out. The rest is a facade.

They used to say, "We consider you to be just like us. You don't seem Korean." That would give rise to such mixed feelings in me. I wanted to believe that I was no different from my White classmates. It was painful to be reminded that I was different, which people did when they wanted to put me in my place, as if I should be grateful to them for allowing me to be their friend.

I wanted to be as American as possible —playing football, dating cheerleaders. I drank a lot and tried to be cool. I had convinced myself that I was "American," whatever that meant, all the while knowing underneath that I'd have to reconcile myself to try to figure out where I would fit in a society that never sanctioned that identity as a public possibility. Part of growing up in America meant denying my cultural and ethnic identity . . . When I got to college, I experienced an identity crisis. . . . I decided to go to Korea, hoping to find something to make me feel more whole. Being in Korea somehow gave me a sense of freedom I had never really felt in America. It also made me love my parents even more. I could imagine where they came from and what they experienced. I began to understand and appreciate their sacrifice and love and what parental support means. Visiting Korea didn't provide answers about the meaning of life, but it gave me a sense of comfort and belonging, the feeling that there was somewhere in this world that validated that part of me that I knew was real but few others outside my immediate family ever recognized.

Source: Edited by Karin Aguilar-San Juan. Copyright © 1994. *The State of Asian America.* Boston, MA: South End Press. Reprinted with permission of South End Press.

social. To a great extent, it is bestowed by others, because we come to see ourselves as others see us. Socialization also establishes **personality,** defined as a person's relatively consistent pattern of behavior, feelings, predispositions, and beliefs.

The socialization experience differs for individuals, depending on factors such as race, gender, and class, as well as more subtle factors such as attractiveness and personality. Women and men encounter different socialization patterns as they grow up because each gender brings with it different social expectations (see Chapter 10). Likewise, growing up Jewish, Asian, Latino, or African American involves different socialization experiences, as the box Understanding Diversity: "My Childhood (Bong Hwan Kim)" shows. In the example, a Korean American man reflects on the cultural habits he learned growing up in two cultures, Korean and American. His comments reveal the strain felt when socialization involves competing expectations, a strain that can be particularly acute when a person grows up within different, even overlapping, cultures.

Through socialization, people internalize cultural expectations, then pass these expectations on to others. *Internalization* occurs when behaviors and assumptions are learned so thoroughly that people no longer question them, but simply accept them as correct. Through socialization, one internalizes the expectations of society. The lessons that are internalized can have a powerful influence on attitudes and behavior. For example, someone socialized to believe that homosexuality is morally repugnant is unlikely to be tolerant of gays and lesbians. If such a person, say a man, experiences erotic feelings about another man, he is likely to have deep inner conflicts about his identity. Similarly, someone socialized to believe that racism is morally repugnant is likely to be more accepting of different races. However, people can change the cultural expectations they learn. New experiences can undermine narrow cultural expectations. Attending college often has a liberalizing effect, supplanting old expectations with new ones generated by exposure to the diversity of college life.

The Nature–Nurture Controversy

Examining the socialization process helps reveal the degree to which our lives are *socially constructed,* meaning that the organization of society and the life outcomes of people within it are the result of social definitions and processes. Is it "nature" (what is natural) or is it "nurture" (what is social) that makes us human? This question has been the basis for debate for many years.

From a sociological perspective, what a person becomes results more from social experiences than *innate* (inborn or natural) traits, although innate traits have some influence, as we saw in the preceding chapter on culture. For example, a person may be born with a great capacity for knowledge, but without a good education, that person is unlikely to achieve his or her full potential and may not be recognized as intellectually gifted.

From a sociological perspective, nature provides a certain stage for what is possible, but society provides the full drama of what we become. The expression *tabula rasa*, for example, means humans are born as a "blank slate." But our values and social attitudes are not inborn; they emerge through the interactions we have with others and our social position in society. Such factors as your family environment, how people of your social group are treated, and the historic influences of the time all shape how we are nurtured by society—or not.

Perhaps the best way to understand the nature–nurture controversy is not that one or the other fully controls who we become, but that life involves a complex interplay, or *interaction*, between natural and social influences on human beings. The emphasis in sociology, however, is to see the social realities of our lives as far more important in shaping human experience (Ridley 2003).

Socialization as Social Control

Sociologist Peter Berger pointed out that not only do people live in society, but society also lives in people (Berger 1963). Socialization is, therefore, a mode of social control. **Social control** is the process by which groups and individuals within those groups are brought into conformity with dominant social expectations. Sometimes the individual attempts to resist this conformity (see the box on page 65). Because socialized people conform to cultural expectations, socialization gives society a certain degree of predictability. Patterns are established that become the basis for social order.

To understand how socialization is a form of social control, imagine the individual in society as surrounded by a series of concentric circles (see Figure 3.1). Each circle is a layer of social controls, ranging from the most subtle, such as the expectations of others, to the most overt, such as physical coercion and violence. Coercion and violence are usually not necessary to extract conformity, because learned beliefs and the expectations of others are enough to keep people in line. These socializing forces can be subtle, because even when a person disagrees with others, he or she can feel pressure to conform and may experience stress and discomfort in

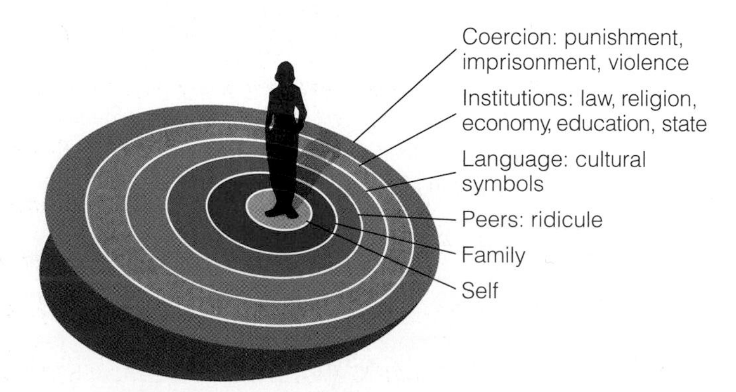

FIGURE 3.1 *Socialization as Social Control*

Though we are all individuals, the process of socialization also keeps us in line with society's expectations. This may occur subtly through peer pressure or, in some circumstances, through coercion and/or violence.

choosing not to conform. People learn through a lifetime of experience that deviating from the expectations of others invites peer pressure, ridicule, and other social judgments that remind one of what is expected.

Conformity and Individuality

Saying that people conform to social expectations does not eliminate individuality. We are all unique to some degree. Our uniqueness arises from different experiences, different patterns of socialization, the choices we make, and the imperfect ways we learn our roles; furthermore, people resist some of society's expectations. Sociologists warn against seeing human beings as totally passive creatures because people interact with their environment in creative ways (Wrong 1961). Yet, most people conform, although to differing degrees. Socialization is profoundly significant, but this does not mean that people are robots. Instead, socialization emphasizes the adaptations people make as they learn to live in society.

Some people conform too much, for which they pay a price. Socialization into men's roles can encourage aggression and a zeal for risk-taking. Men have a lower life expectancy and higher rate of accidental death than women, probably because of the risky behaviors associated with men's roles, that is, simply "being a man" (National Center for Health Statistics 2004; Kimmel and Messner 2004). Women's gender roles carry their own risks. Striving excessively to meet the beauty ideals of the dominant culture can result in feelings of low self-worth and may encourage harmful behaviors, such as smoking or restricting eating to keep one's weight down. It is not that being a man or woman is inherently bad for your health, but conforming to gender roles to an extreme can compromise your physical and mental health.

The Consequences of Socialization

Socialization is a lifelong process with consequences that affect how we behave toward others and what we think of ourselves. First, socialization *establishes self-concepts.* How we think of ourselves is the result of the socialization experiences we have over a lifetime. Socialization is also influenced by various social factors, as shown in Figure 3.2, which describes how students' self-concepts are shaped by gender.

Second, *socialization creates the capacity for role-taking,* or, put another way, for the ability to see

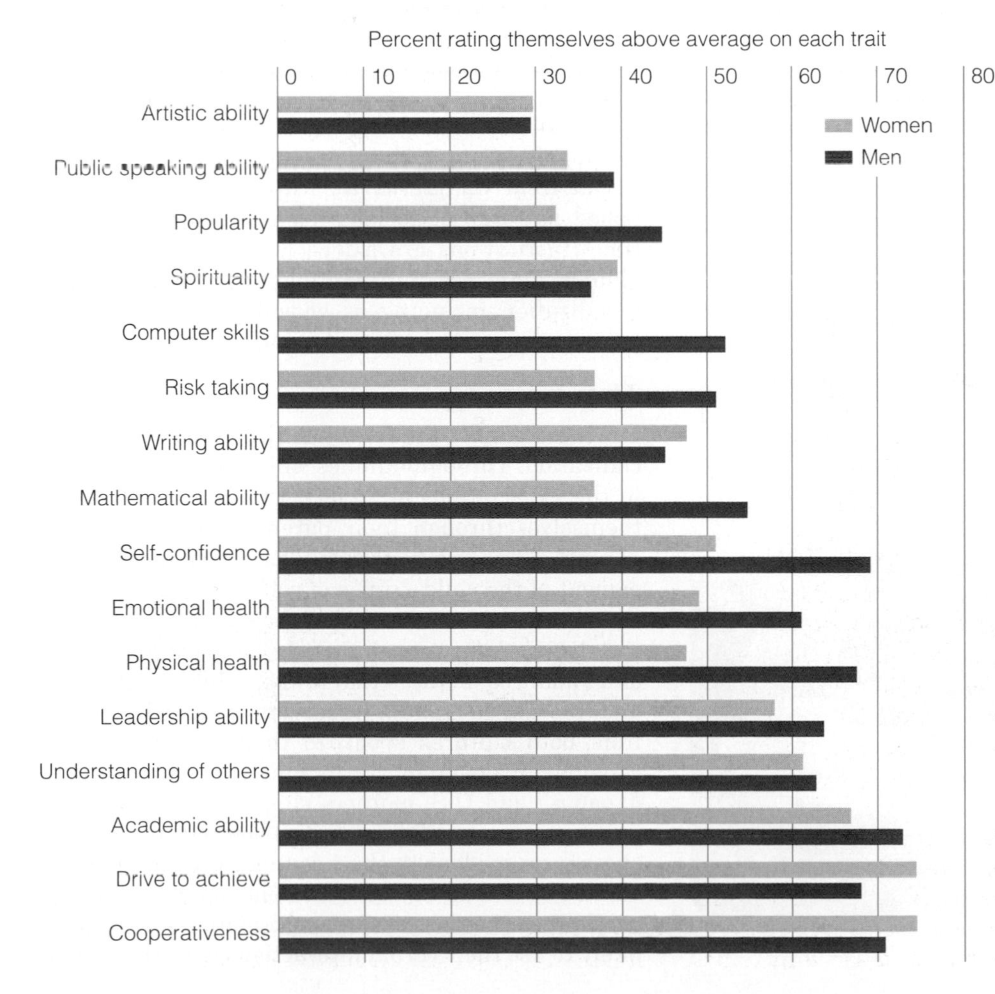

FIGURE 3.2 *Student Self-Concepts: The Difference Gender Makes**

*Based on national sample of first-year college students, Fall 2003.

Source: Sax, L.J., et al. 2003. *The American Freshman: National Norms for Fall 2003.* Higher Education Research Institute. Los Angeles, CA: University of California, Los Angeles. Used by permission.

oneself through the perspective of another. Socialization is fundamentally reflective; that is, it involves self-conscious human beings seeing and reacting to the expectations of others. The capacity for reflection and the development of identity are ongoing. As we encounter new situations in life, such as going away to college or getting a new job, we are able to see what is expected and to adapt to the situation accordingly. Of course, not all people do so successfully. This can become the basis for social deviance (explored in Chapter 6) or for many common problems in social and psychological adjustment.

Third, *socialization creates the tendency for people to act in socially acceptable ways.* Through socialization, people learn the normative expectations attached to social situations and the expectations of society in general. As a result, socialization creates some predictability in human behavior and brings some order to what might otherwise be social chaos.

Finally, *socialization makes people bearers of culture.* Socialization is the process by which people learn and internalize the attitudes, beliefs, and behaviors of their culture. At the same time, socialization is a two-way process—that is, a person is not only the recipient of culture but also is the creator of culture, passing on cultural expectations on to others. The main product of socialization, then, is society itself. By molding individuals, socializing forces perpetuate the society into which individuals are born. Beginning with the newborn infant, socialization contributes to the formation of a self; the self learns roles and rules, and thus outfitted, the self becomes a bearer of culture, passing on all that it has acquired.

© Yellow Dog Productions/Getty Images/The Image Bank

Dress at work is a form of socialization.

Agents of Socialization

Socialization agents are those who pass on social expectations. Everyone is a socializing agent, because social expectations are communicated in countless ways and in every interaction people have, whether intentionally or not. When people are simply doing what they consider "normal," they are communicating social expectations to others. When you dress a particular way, you may not feel you are telling others they must dress that way. Yet, when everyone in the same environment dresses similarly, some expectation about appropriate dress is clearly being conveyed. People feel pressure to become what society expects of them even though the pressure may be subtle and unrecognized.

thinking sociologically

Think about the first week that you were attending college. What expectations were communicated to you and by whom? Who were the most significant *socialization agents* during this period? Which expectations were communicated formally and which informally? If you were analyzing this experience sociologically, what would be some of the most important concepts to help you understand how one "becomes a college student"?

Socialization does not occur simply between individuals; it occurs in the context of social institutions. Recall from Chapter 1 that institutions are established patterns of social behavior that persist over time. Institutions are a level of society above individuals. Many social institutions shape the process of socialization, including, as we will see, the family, the media, peers, religion, sports, and schools.

The Family

For most people, the family is the first source of socialization. Through families, children are introduced to the expectations of society. Children learn to see themselves through their parents' eyes. Thus, how parents define and treat a child is crucial to the development of the child's sense of self.

An interesting example of the subtlety in familial socialization comes from a study comparing how U.S. and Japanese mothers talk to their children (Fernald and Morikawa 1993). Observers watched mothers from both cultures speak to their infant children, ranging in age from six to nineteen months. Both Japanese and U.S. mothers simplified and repeated words for their children, but strong cultural differences were evident in their context. U.S. mothers focused on naming objects for their babies: "Is that a *car*?" "Kiss the *doggy.*" Japanese mothers were more likely to use their verbal interactions with children as an opportunity to practice social routines such as "give me" and "thank you." The behavior of the Japanese mothers implied that the name of the object was less important than the polite exchange. Japanese mothers were also more likely to use sounds to represent the objects, such as "oink-oink" for a pig, or "vroom-vroom" for a vehicle. U.S. mothers were more likely to use the actual names for objects. A U.S. mother might say, "See the car? Nice color!", whereas a Japanese mother might say," Here! It's a vroom-vroom! I give it to you. Now you give this to me. Give me. Yes! Thank you." The researchers interpreted these interactions as reflecting the beliefs and practices of each culture. Japanese mothers used objects as part of a ritual of social exchange, emphasizing polite routines, whereas U.S. mothers focused on labeling things. In each case, the child receives a message about what is most significant in the culture.

What children learn in families is certainly not uniform. Even though families pass on the expectations of a given culture, within that culture families may be highly diverse, as we will see in Chapter 12. Some families may emphasize educational achievement; some may be more permissive, whereas others emphasize strict obedience and discipline. Even within families, children my experience different expectations based on gender or birth order (being first, second, or third born). Researchers have found, for example, that fathers' and mothers' support for gender equity increases when they have only daughters (Warner and Steel 1999). Living in a family experiencing the strain of social problems such as alcoholism, unemployment, domestic violence, or teen pregnancy also affects how children are socialized. The specific effects of different family structures and processes are the basis for ongoing and extensive sociological research.

Virginia Sherwood/© NBC/Courtesy: Everett Collection

Even when women are depicted in the media as "tough" and "intelligent," the tendency is to depict women as sexual objects, as shown in this image of Marcia Gay Harden and Mariska Hargitay in the television show Law and Order: Special Victims Unit.

As important as the family is in socializing the young, it is not the only socialization agent. As children grow up, they encounter other socializing influences, sometimes in ways that might contradict family expectations. Parents who want to socialize their children in less gender-stereotyped ways might be frustrated by the influence of the media, which promotes highly gender-typed toys and activities to boys and girls. These multiple influences on the socialization process create a reflection of society in us.

The Media

Increasingly, the mass media are important agents of socialization. Television alone has a huge impact on what we are socialized to believe and become. Add to that the print messages received in books, comics, newspapers, and now blogs and the Internet, plus images from film, music, video games, and radio, and you begin to see the enormous influence the media have on the values we form, our images of society, our desires for ourselves, and our relationships with others. These images are powerful throughout our lifetimes (as we have seen in Chapter 2 on culture), but many worry that their effect during childhood may be particularly deleterious.

Take the issue of violence. The high degree of violence in the media has led to the development of a rating system for televised programming. No doubt, violence is extensive in the media. Analysts estimate that by age eighteen, the average child will have witnessed at least 18,000 simulated murders on television (Wilson et al. 2002). Moreover, violence in children's programming is frequently shown as humorous with no serious consequences (National Television Violence Study 1997).

Violence is pervasive on television, but does it have any effect? This is what researchers debate. There is strong evidence of the unhealthy effects of violence, especially on children. There is evidence that violence in the media encourages both antisocial behavior and fear among children. Media violence also tends to desensitize children to the effects of violence, including engendering less sympathy for victims of violence (Cantor 2000). Children also tend to imitate the aggressive behavior they see in the media. Violence in the media is not solely to blame for violent behavior in society, however. Children, for example, do not watch television in a vacuum—they live in families where they learn different values and attitudes about violent behavior, and they observe the society around them, not just the images they see in fictional representations. Most likely, children are influenced not only by the images of televised and filmed violence, but also by the broader social context in which they live. The images of violence in the media in some ways only reflect the violence in society.

The media expose us to numerous images that shape our definitions of ourselves and the world around us. What we think of as beautiful, sexy, politically acceptable, and materially necessary is strongly influenced by the media. If every week, as you read a weekly newsmagazine, someone shows you the new car that will give you status and distinction, or if every weekend, as we watch televised sports, someone tells us that to really have fun we should drink the right beer, it is little wonder that we begin to think that our self-worth can be measured by the car we drive and that parties are perceived as better when everyone is drunk. The values represented in the media, whether they are about violence, racist and sexist stereotypes, or any number of other social images, have a great effect on what we think and who we come to be.

Support from peers and family is an important source of strong self-esteem. Organized peer groups such as the Special Olympics can also foster a desire for achievement and enhance one's sense of self-worth.

Peers

Peers are those with whom you interact on equal terms, such as friends, fellow students, and coworkers. Among peers, there are no formally defined superior and subordinate roles, although status distinctions commonly arise in peer group interactions. Without peer approval, most people find it hard to feel socially accepted.

Peers are enormously important in the socialization process. Peer cultures for young people often take the form of *cliques*—friendship circles where members identify with each other and hold a sense of common identity. You probably had cliques in your high school and may even be able to name them. Did your school have "jocks," "preps," "goths," "nerds," "druggies," and so forth? Sociologists studying cliques have found that they are formed based on a sense of exclusive membership, like the in-groups and out-groups we will examine in Chapter 5. Cliques are cohesive but also have an internal hierarchy with certain group leaders having more power and status than other members. Interaction techniques, like making fun of people, produce group boundaries, defining who's in and who's out (Adler and Adler 1998). The influence of peers is strong in childhood and adolescence, but it also persists into adulthood.

Peers are an important agent of socialization. Young girls and boys learn society's images of what they are supposed to be through the socialization process.

As agents of socialization, peers are important sources of social approval, disapproval, and support. This is one reason groups without peers of similar status are often at a disadvantage in various settings, such as women in male-dominated professions or minority students on predominantly White campuses. Being a "token" or an "only," as it has come to be called, places unique stresses on those in settings with relatively few peers from whom to draw support. This is one reason those who are minorities in a dominant group context often form same-sex or same-race groups for support, social activities, and the sharing of information about how to succeed in their environment.

Religion

Religion is another powerful agent of socialization, and religious instruction contributes greatly to the identities children construct for themselves. Children tend to develop the same religious beliefs as their parents; switching to a religious faith different from the one in which a person is raised is rare (Hadaway and Marler 1993). Even those who renounce the religion of their youth are deeply affected by the attitudes, images, and beliefs instilled by early religious training. Very often those who disavow religion return to their original faith at some point in their life, especially if they have strong ties to their family of origin and if they form families of their own (Wilson 1994).

Religious socialization shapes the beliefs that people develop. An example comes from studies of people who believe in creationism. Creationism is a set of beliefs that largely reject the theory of human biological evolution and instead argue that human beings as now exist were created by a central force or God. Those who believe this have generally been taught to believe so over a long period—that is, they have been specifically socialized to believe in the creationist view of the world's origin and to reject scientific explanations, such as the theory of evolution. Sociological research further finds that socialization into creationist beliefs is more likely to be effective among people who grow up in small-town environments where they are less exposed to other influences. Those who believe in creationism are also likely to have mothers who have filled the traditional homemaker's role (Eckberg 1992). This shows the influence that social context has on the religious socialization people experience.

Religious socialization influences a large number of beliefs that guide adults in how they organize their lives, including beliefs about moral development and behavior, the roles of men and women, and sexuality, to name a few. One's religious beliefs strongly influence belief about gender roles within the family, including men's engagement in housework and the odds that wives will be employed outside the home (Ellison and Bartkowski 2002; Becker and Hofmeister 2001; Scott 2000). Religious socialization also influences beliefs about sexuality, including the likelihood of tolerance for gay and lesbian sexuality (Reynolds 2003; Sherkat 2002). Religion can even influence child-rearing practices. Thus, sociologists have found that conservative Protestants are more likely to use strict discipline in raising children, but they are also more likely to hug and praise their children than are parents with less conservative theological views.

Sports

Most people perhaps think of sports as something that is just for fun—or perhaps to provide opportunities for college scholarships and athletic careers—but sports are also an agent of socialization. Through sports, men and women learn concepts of self that stay with them in their later lives.

Sports are also where many ideas about gender differences are formed and reinforced (Dworkin and Messner 1999; Messner 2002). For men, success or failure as an athlete can be a major part of a man's identity. Even for men who have not been athletes, knowing about and participating in sports is an important source of men's gender socialization. Men learn that being competitive in sports is considered a part of manhood. Indeed, the attitude that "sports builds character" runs deep in the culture. Sports are supposed to pass on values such as competitiveness, the work ethic, fair play, and a winning attitude. Sports are considered to be where one learns to be a man.

Debunking Society's Myths

Myth: Sports are basically played just for the fun of it.

Sociological perspective: Although sports are a form of entertainment, playing sports is also a source for socialization into roles, such as gender roles.

Michael Messner's research on men and sports reveals the extent to which sports shape masculine identity. Messner interviewed thirty former athletes: Latino, Black, and White men from poor, working-class, and middle-class backgrounds. All of them spoke of the extraordinary influence of sports on them as they grew up. Not only are sports a major source of gender socialization, but working-class, African American, and Latino men often see sports as their only possibility for a good career, even though the number of men who succeed in athletic careers is a minuscule percentage of those who hold such hopes.

Messner's research shows that, for most men, playing or watching sports is often the context for developing relationships with fathers, even when the father is absent or emotionally distant in other areas of life. Older brothers and other male relatives also socialize young men into sports. For many of the men in Messner's study, the athletic accomplishments of other family members created uncomfortable pressure to perform and compete, although on the whole, they recalled their early sporting years with positive emotions. It was through sports relationships with male peers, more than anyone else, however, that the men's identity was shaped. As boys, the men could form "safe" bonds with other men; still, through sports activity, men learned homophobic attitudes (that is, fear and hatred of homosexuals) and rarely developed intimate, emotional relationships with each other (Messner 1992, 2002).

Sports historically have been less significant in the formation of women's identity, although this has changed, largely as the result of Title IX. Title IX opened more opportunities in athletics to girls and women by legally defining the exclusion of women from school sports as sex discrimination. Women who participate in sports typically develop a strong sense of bodily competence—something usually denied to them by the prevailing, unattainable cultural images of women's bodies. Sports also give women a strong sense of self-confidence and encourage them to seek challenges, take risks, and set goals (Blinde et al. 1993, 1994).

Still, athletic prowess, highly esteemed in men, is not tied to cultural images of womanliness. Quite the contrary, women who excel at sports are often stereotyped as lesbians and may be ridiculed for not being womanly enough. These stereotypes reinforce traditional gender roles for women, as do media images of women athletes that emphasize family images and the personality of women athletes (Blinde and Taub 1992a, 1992b; Cavalier 2003). Research in the sociology of sports shows how activities as ordinary as shooting baskets on a city lot, playing on the soccer team for one's high school, or playing touch football on a Saturday afternoon can convey powerful cultural messages about our identity and our place in the world. Sports are a good example of the power of socialization in our everyday lives.

Schools

Once young people enter kindergarten (or, even earlier, day care), another process of socialization begins. At home, parents are the overwhelmingly dominant source of socialization cues. In school, teachers and other students are the source of expectations that encourage children to think and behave in particular ways. The expectations encountered in schools vary for different groups of students. These differences are shaped by a number of factors, including teachers' expectations for different groups and the resources that different parents can bring to bear on the educational process. The parents of children attending elite, private schools, for example, often have more influence on school policies and classroom activities than do parents in low-income communities. In any context, studying socialization in the schools is an excellent way to see the influence of gender, class, and race in shaping the socialization process.

Debunking Society's Myths

Myth: Schools are primarily places where young people learn skills and other knowledge.

Sociological perspective: There is a hidden curriculum in schools where students learn expectations associated with race, class, and gender relations in society as influenced by the socialization process.

For example, research finds that teachers respond differently to boys and girls in school. Boys receive more attention from teachers than do girls. Even when teachers respond negatively to boys who are misbehaving, they are calling more attention to the boys (American Association of University Women 1998; Sadker and Sadker 1994). Social class stereotypes also affect teachers' interactions with students. Teachers are likely to perceive working-class children and poor children as less bright and less motivated than middle-class children; teachers are also more likely to define working-class students as troublemakers (Bowditch 1993). These negative appraisals are *self-fulfilling prophecies*, meaning that the expectations they create often become the basis for actual behavior; thus, they affect the odds of success for children.

Boys also receive more attention in the curriculum than girls. The characters in texts are more frequently boys; the accomplishments of boys are more likely portrayed in classroom materials; and boys and men are more typically depicted as active players in history, society, and culture (American Association of University Women 1998; Sadker and Sadker 1994). This is called the *hidden curriculum* in the schools—the informal and often subtle messages about social roles that are conveyed through classroom interaction and classroom materials—roles that are clearly linked to gender, race, and class.

In schools, boys and girls are quite often segregated into different groups, with significant sociological consequences. Differences between boys and girls become exaggerated when they are defined as distinct groups (Thorne 1993). Seating boys and girls in separate groups or sorting them into separate play groups heightens gender differences and greatly increases the significance of gender in the children's interactions with each other. Equally important is that gender becomes less relevant in the interactions between boys and girls when they are grouped together in common working groups, although gender does not disappear altogether as an influence. Barrie Thorne, who has observed gender interaction in schools, concludes from her observations that gender has a "fluid" character and that gender relations between boys and girls can be improved through conscious changes that discourage gender separation.

While in school, young people acquire identities and learn patterns of behavior that are congruent with the needs of other social institutions. Sociologists using conflict theory to understand schools would say that U.S. schools reflect the needs of a capitalist society. School is typically the place where children are first exposed to a hierarchical, bureaucratic environment. Not only does school teach them the skills of reading, writing, and other subject areas, but it is also where children are trained to respect authority, be punctual, and follow rules—thereby preparing them for their future lives as workers in organizations that value these traits. Schools emphasize conformity to societal needs, although not everyone internalizes these lessons to the same degree (Bowles and Gintis 1976; Lever 1978). Research has found, for example, that working-class school children form subcultures in school that resist the dominant culture.

thinking sociologically

Visit a local day-care center, preschool, or elementary school and observe children at play. Record the activities they are involved in, and note what both girls and boys are doing. Do you observe any differences between boys' and girls' play? What do your observations tell you about *socialization* patterns for boys and girls?

Theories of Socialization

Knowing that people become socialized does not explain how it happens. Different theoretical perspectives explain socialization, including psychoanalytic theory, social learning theory, and symbolic interaction theory. Each perspective, including functionalism as well as conflict theory, carries a unique set of assumptions about socialization and its effect on the development of the self (see Table 3.1).

Psychoanalytic Theory

Psychoanalytic theory originates in the work of **Sigmund Freud** (1856–1939). Perhaps Freud's greatest contribution was the idea that the unconscious mind shapes human behavior. Freud is also known for developing the technique of *psychoanalysis* to help discover the causes of psychological problems in the recesses of troubled patients' minds.

Psychoanalytic theory depicts the human psyche in three parts: the id, the superego, and the ego. The **id** consists of deep drives and impulses. Freud was particularly absorbed by the sexual component of the id, which he considered an especially forceful denizen of the unconscious mind. The **superego** is the dimension of the self that represents the standards of society. The superego incorporates or internalizes acquired values and norms—in short, society's collective expectations. According to Freud, an ordered society requires that people repress the wild impulses generated by the id. Consequently, the id is in permanent conflict with the superego. The superego represents what Freud saw as the inherent repressiveness of society. People cope with the tension between social expectations (the superego) and their impulses (the id) by developing defense mechanisms, typically repression, avoidance, or denial (Freud 1960/1923, 1961/1930, 1965/1901). Suppose someone has a great desire for the wrong person or even for another person's property. This person might refuse to admit this (repression); or, acknowledging the impulse, might avoid the opportunity for temptation (avoidance); or might indulge in misconduct, believing it was not misconduct (denial).

The third component of the self in Freud's theory, the **ego** is the seat of reason and common sense. The ego plays a balancing act between the id and the superego, adapting the desires of the id to the social expectations of the superego. In psychoanalytic theory, the conflict between the id and the superego occurs in the subconscious mind, yet it shapes human behavior. We get a glimpse of the unconscious mind in dreams and in occasional slips of the tongue—the famous "Freudian slip" that is believed to reveal an underlying state of mind. For example, someone might intend to say, "There were six people at the party," but instead says, "There were sex people at the party," revealing that either the memory of the

table 3.1 Theories of Socialization

	Psychoanalytic Theory	Social Learning Theory	Functionalism	Conflict Theory	Symbolic Interaction Theory
How each theory views:					
Individual learning process	The unconscious mind shapes behavior.	People respond to social stimuli in their environment.	People internalize the role expectations that are present in society.	Individual and group aspirations that are shaped by the opportunities available to different groups.	Children learn through taking the role of significant others.
Formation of self	The self (ego) emerges from tension between the id and the superego.	Identity is created through the interaction of mental and social worlds.	Internalizing the values of society reinforces social consensus.	Group consciousness is formed in the context of a system of inequality.	Identity emerges as the creative self interacts with the social expectations of others.
Influence of society	Societal expectations are represented by the superego.	Young children learn the principles that shape the external world.	Society relies upon conformity to maintain stability and social equilibrium.	Social control agents exert pressure to conform.	Expectations of others form the social context for learning social roles.

party or the person he or she is speaking to is causing sexual thoughts to lurk in the unconscious mind.

Although Freud's work is largely psychological, there is an understanding of society built into psychoanalytic thought. The superego forces constant awareness of society, and the ego negotiates an uneasy balance between the superego and the id. In this balancing act of the three parts of the self, we see socialization happening, with the ego bridging the gap between the primal id and the socialized superego.

Some sociologists have criticized Freud's work for not being generalizable because he worked with only a small and unrepresentative group of clients. Still, psychoanalytic theory is an influential and popular analysis of human personality. We often speak of what motivates people, as if motives were internal, unconscious states of mind that direct human behavior.

The psychoanalytic perspective interprets human identity as relatively fixed at an early age in a process greatly influenced by one's family. Nancy Chodorow, a sociologist, uses psychoanalytic theory to explain how gendered personalities develop. She argues that infants are strongly attached to their primary caregiver—in our society, typically the mother. As they grow older, they try to separate themselves from their parents (or individuate), both physically and emotionally, becoming a freestanding individual. However, early attachments to the primary caregiver persist. At the same time, children identify with their same-sex parent, meaning that boys and girls separate themselves from their parents differently. Because girls identify with the mother, they are less able to detach themselves from their primary caregiver. This is the basis, according to Chodorow, for why—even in adulthood—women tend to have personalities based on attachment and an orientation toward others, and men, personalities based on greater detachment (Chodorow 1978).

To sum up, psychoanalytic theory interprets the development of social identity as an unconscious process, developed from dynamic tensions between strong instinctual impulses and the social standards of society. Most important, psychoanalysis sees human behavior as directed and motivated by underlying psychic forces that are largely hidden from ordinary view.

Social Learning Theory

Whereas psychoanalytic theory places great importance on the unconscious processes of the human mind, **social learning theory** considers the formation of identity to be a learned response to social stimuli (Bandura and Walters 1963). Social learning theory emphasizes the societal context of socialization. Identity is regarded not as the product of the unconscious, but as the result of modeling oneself in response to the expectations of others. According to social learning theory, behaviors and attitudes develop in response to reinforcement and encouragement from those around us. Social learning theorists acknowledge the importance of early childhood experience, but they think that the identity people acquire is based more on the behaviors and attitudes of others than on the interior landscape of the individual.

Early models of social learning theory regarded learning rather simplistically in terms of stimulus and response. People were seen as passive creatures who merely responded to stimuli in their environment. This mechanistic view of social learning was transformed by the work of the Swiss psychologist **Jean Piaget** (1896–1980), who believed that learning was crucial to socialization but that imagination also had a critical role. He argued that the human mind organizes experience into mental categories, or configurations, he called *schema,* which are modified and developed as social experiences accumulate. Schema might be compared to a person's understanding of the rules of a game. Humans do not simply respond to stimulus but actively absorb experience and figure out what they are seeing to construct a picture of the world.

Piaget proposed that children go through distinct stages of cognitive development as they learn the basic rules of reasoning. They must master the skills at each level before they go on to the next (Piaget 1926). In the initial *sensorimotor* stage, children experience the world only through their senses—touch, taste, sight, smell, and sound. Next comes the *preoperational* stage, in which children begin to use language and other symbols. Children in the preoperational stage cannot think in abstract terms, but they do gain an appreciation of meanings that go beyond their immediate senses. They also begin to see things as others might see them. Third, the *concrete operational*

Social learning theory emphasizes how people model their behaviors and attitudes on those of others.

stage is when children learn logical principles regarding the concrete world. This stage prepares them for more abstract forms of reasoning. In the *formal operational stage,* children are able to think abstractly and imagine alternatives to the reality in which they live.

Building on Piaget's model of stages of development, psychologist Lawrence Kohlberg (1969) developed a theory of what he called moral development. Kohlberg interpreted the process of developing moral reasoning as occurring in several stages grouped into three levels: the *preconventional stage,* the *conventional stage,* and the *postconventional stage.* In the first, young children judge right and wrong in simple terms of obedience and punishment, based on their own needs and feelings. In adolescence (the conventional stage), Kohlberg argued, young people develop moral judgment in terms of cultural norms, particularly social acceptance and following authority. In the final stage of moral development, the postconventional stage, people are able to consider abstract ethical questions, thereby showing maturity in their moral reasoning. In Kohlberg's original research, men, he argued, reached a higher level of moral development than women because women remained more concerned with feelings and social opinions (a lower phase), whereas men were concerned with authority.

Kohlberg's work was later criticized by psychologist Carol Gilligan. Gilligan (1982) found that women conceptualize morality in different terms than men. Instead of judging women by a standard set by men's experience, Gilligan showed that women's moral judgments were more contextual than those of men. In other words, when faced with a moral dilemma, women were more likely to consider the different relationships in the social context that would be affected by any decision instead of making moral judgments according to abstract principles. Gilligan's research made an important point—not just about the importance of including women in studies of human development, but also about being careful not to assume that social learning follows a universal course for all groups.

Functionalism and Conflict Theory

Sociologists use a variety of theoretical perspectives to understand the socialization process, including those just described. They can also draw from the major theoretical frameworks we have introduced to understand socialization. From the vantage point of functionalist theory, socialization integrates people into society because it is the mechanism through which they internalize social roles and the values of society. This reinforces social consensus because it encourages at least some degree of conformity. Thus, socialization is one way that society maintains its stability.

Conflict theorists would see this differently. Because of the emphasis in conflict theory on the role of power and coercion in society, in thinking about socialization conflict theorists would be most interested in how group identity is shaped by patterns of inequality in society. A person's or group's identity always emerges in a context, and if that context is one marked by different opportunities for different groups, then one's identity will be shaped by that fact. This may help you understand why, for example, women are more likely to choose college majors in areas of study that have traditionally been associated with women's work opportunities (that is, in the so-called helping professions and in the arts and humanities and less frequently in math and sciences). Furthermore, though social control agents pressure people to conform, people also resist oppression. Thus, the identities of people oppressed in society often include some form of resistance to oppression. This can help you understand why members of racial groups who identity with their own group, not the dominant White group, tend to have higher self-esteem (that is, a stronger valuing of self). In other words, resisting the expectations of a dominant group (such as being subservient or internalizing a feeling of inferiority) can actually heighten one's perceived self-worth (see Table 3.1).

Symbolic Interaction Theory

Recall that *symbolic interaction* theory centers on the idea that human actions are based on the meanings people attribute to behavior; these meanings emerge through social interaction (Blumer 1969). Symbolic interaction has been especially important in developing an understanding of socialization. People learn identities and values through socialization For example, learning to become a good student, then, means taking on the characteristics associated with that role. Because roles are socially defined, they are not real, like objects or things, but are real because of the meanings people give them. As did Piaget, symbolic interactionists see the human capacity for reflection and interpretation as having an important role in the socialization process.

For symbolic interactionists, meaning is constantly reconstructed as people act within their social environments. The **self** is what we imagine we are; it is not an interior bundle of drives, instincts, and motives. Because of the importance attributed to reflection in symbolic interaction theory, symbolic interactionists use the term *self,* rather than the term *personality,* to refer to a person's identity. Symbolic interaction theory emphasizes that human beings make conscious and meaningful adaptations to their social environment. From a symbolic interactionist perspective, identity is not something that is unconscious and hidden from view, but is socially bestowed and socially sustained (Berger 1963).

Two theorists have greatly influenced the development of symbolic interactionist theory in sociology. **Charles Horton Cooley** (1864–1929) and **George Herbert Mead** (1863–1931) were both sociologists at the University of Chicago in the early 1900s (see Chapter 1). Cooley and Mead saw the self developing in response to the expectations and judgments of others in their social environment.

Charles Horton Cooley postulated the **looking glass self** to explain how a person's conception of self arises through considering our relationships to others (Cooley 1902, 1967/1909). The development of the looking glass self emerges from (1) how we think we appear to others; (2) how we think others judge us; and (3) how the first two make us feel—proud, embarrassed, or other feelings. The looking glass self involves perception and effect, the perception of how others see us and the effect of others' judgment on us (see Figure 3.3).

How others see us is fundamental to the idea of the looking glass self. In seeing ourselves as others do, we respond to the expectations others have of us. This means that the formation of the self is fundamentally a social process—one based in the interaction people have with each other, as well as the human capacity for self-examination. One unique feature of human life is the ability to see ourselves through others' eyes. People can imagine themselves in relationship to others and develop a definition of themselves accordingly. From a symbolic interactionist perspective, the reflective process is key to the development of the self. If you grow up with others who think you are smart and sharp-witted, chances are you will develop this definition of yourself. If others see you as dull-witted and withdrawn, chances are good that you will see yourself this way. George Herbert Mead agreed with Cooley that children are socialized by responding to other's attitudes toward them. According to Mead, social roles are the basis of all social interaction.

Taking the role of the other is the process of putting oneself into the point of view of another. To Mead, role-taking is a source of self-awareness. As people take on new roles, their awareness of self changes. According to Mead, identity emerges from the roles one plays. He explained this process in detail by examining childhood socialization, which he saw as occurring in three stages: the imitation stage, the play stage, and the game stage (Mead 1934). In each phase of development, the child becomes more proficient at taking the role of the other. In the first stage, the **imitation stage,** children merely copy the behavior of those around them. Role-taking in this phase is nonexistent because the child simply mimics the behavior of those in the surrounding environment without much understanding of the social meaning of the behavior. Although the child in the imitation stage has little understanding of the behavior being copied, he or she is learning to become a social being. For example, think of young children who simply mimic the behavior of people around them (such as pretending to read a book, but doing so with the book upside down). In Mead's analysis, this behavior is one way that children begin to learn the expectations of others.

In the second stage, the **play stage,** children begin to take on the roles of significant people in their environment, not just imitating but incorporating their relationship to the other. Especially meaningful

FIGURE 3.3 *The Looking Glass Self*

The looking glass self refers to the process by which we see ourselves as others see us.

is when children take on the role of **significant others,** those with whom they have a close affiliation. A child pretending to be his mother may talk to himself as the mother would. The child begins to develop self-awareness, seeing himself or herself as others do.

In the third stage of socialization, the **game stage,** the child becomes capable of taking on multiple roles at the same time. These roles are organized in a complex system that gives the child a more general or comprehensive view of the self. In this stage, the child begins to comprehend the system of social relationships in which he or she is located. The child not only sees himself or herself from the perspective of a significant other, but also understands how people are related to each other and how others are related to him or her. This is the phase where children internalize (incorporate into the self) an abstract understanding of how society sees them.

Mead compared the lessons of the game stage to a baseball game. In baseball, all roles together make the game. The pitcher does not just throw the ball past the batter as if they were the only two people on the field; rather, each player has a specific role, and each role intersects with the others. The network of social roles and the division of labor in the baseball game is a social system, like the social systems children must learn as they develop a concept of themselves in society.

In the game stage, children learn more than just the roles of significant others in their environment. They also acquire a concept of the **generalized other**—the abstract composite of social roles and social expectations. In the generalized other, they have an example of community values and general social expectations that adds to their understanding of self; however, children do not all learn the same generalized other. Depending on one's social position (that is, race, class, gender, region, or religion), one learns a particular set of social and cultural expectations.

If the self is socially constructed through the expectations of others, how do people become individuals? Mead answered this by saying that the self has two dimensions: the "I" and the "me." The I is the unique part of individual personality, the active, creative, self-defining part. The me is the passive, conforming self, the part that reacts to others. In each person, there is a balance between the I and the me, similar to the tension Freud proposed between the id and the superego. Mead differed from Freud, however, in his judgment about when identity is formed. Freud felt that identity was fixed in childhood and henceforth driven by internal, not external, forces. In Mead's version, social identity is always in flux, constantly emerging (or "becoming") and dependent on social situations. Over time, identity stabilizes as one learns to respond consistently to common situations.

Social expectations associated with given roles change as people redefine situations and as social and historical conditions change; thus, the social expectations learned through the socialization process are not permanently fixed. For example, as more women enter the paid labor force and as men take on additional responsibilities in the home, the expectations associated with motherhood and fatherhood are changing. Men now experience some of the role conflicts that women have faced in balancing work and family. As the roles of mother and father are redefined, children are learning new socialization patterns; however, traditional gender expectations maintain a remarkable grip. Despite many changes in family life and organization, young girls are still socialized for motherhood, and young boys are still socialized for greater independence and autonomy.

SEE FOR YOURSELF

WRITING ABOUT SOCIETY—CHILDHOOD PLAY AND SOCIALIZATION

The purpose of this exercise is to explain how *childhood socialization* is a mechanism for passing on social *norms and values*. Begin your essay by identifying a form of play that you engaged in as a young child. What did you play? Who did you play with? Was it structured or unstructured play? What were the rules? Were they formal or informal, and who controlled whether they were observed?

1. Now think about what *norms and values* were being taught to you by way of this play. Do they still affect you today? If so, how?
2. How does your experience compare to those of students in your class who differ from you in terms of gender, race, ethnicity, regional origin, and so forth. Are there differences in learned norms and values that can be attributed to these different social characteristics?

Growing Up in a Diverse Society

Understanding the institutional context of socialization is important for understanding how socialization affects different groups in society. Socialization makes us members of our society. It instills in us the values of the culture and brings society into our self-definition, our perceptions of others, and our understanding of the world around us. Socialization is not, however, a uniform process, as the different examples developed in this chapter show. In a society as complex and diverse as the United States, no two people will have exactly the same experiences. We can find similarities between us, often across vast social and cultural differences,

© Kate Powers/Getty Images/Taxi

The family serves as a major agent of socialization, especially of the young.

but variation in social contexts creates vastly different socialization experiences.

Furthermore, current changes in the U.S. population are creating new multiracial and multicultural environments in which young people grow up. Schools, as an example, are in many places being transformed by the large number of immigrant groups entering the school system. In such places, children come into contact with other children from a variety of different groups. This creates a new context in which children form their social values and learn their social identities (see Doing Sociological Research: Children's Understanding of Race).

One task of the sociological imagination is to examine the influence of different contexts on socialization. Where you grow up; how your family is structured; what resources you have at your disposal; your racial–ethnic identity, gender, and nationality—all shape the socialization experience. Socialization experiences for all groups are shaped by many factors that intermingle to form the context for socialization.

One way that this has been demonstrated is in research by sociologist Annette Lareau. Over an extended period of time, Lareau and her research assistants carefully observed White and Black families from middle-class, working-class, and poor backgrounds. The researchers spent many hours in the homes of the families studied, including following the children and parents as they went about their daily routines. Based on these detailed observations, they observed important class differences in how families—both Black and White—socialize their children.

The middle-class children were highly programmed in their activities, their lives filled with various organized activities—music lessons, sports, school groups, and so forth. In contrast, the working-class and poor children, regardless of race, were less structured in their activities, and economic constraints were a constant theme in their daily lives. But the pace of life for working-class and poor children was slower and more relaxed. These children had more unstructured play time, whereas middle-class children's lives were a constant barrage of highly structured activities with intense time demands. Lareau argues that middle-class families engaged in *concerted cultivation* of childhood, meaning they made "deliberate and sustained effort to stimulate children's development and to cultivate their cognitive and social skills" (Lareau 2003: 238). Working-class and poor children experienced more "natural growth"—that is, childhood experiences that allow them to develop in a less-structured environment with more time for creative play.

As a result, middle-class children tend to learn a more individualized self-concept and a sense of entitlement, but the price is an overly programmed daily life. Working-class and poor children experience obvious costs in that they have more financial constraints, but even more fundamentally, Lareau argues, they are left unable to negotiate their way through various social institutions as effectively as the middle-class. In a sense, these childhood socialization patterns are also reshaping the class system that children will likely find themselves in as adults. Middle-class children are being prepared, even if inadvertently, for lives with a sense of privilege and entitlement; working-class children, for responding to the directives of others. In this way, patterns of socialization occurring because of social class origins are training children to take their place in the class system that will likely mark their adult lives. Thus, social class is an important—though often invisible—force shaping the socialization of young people.

Aging and the Life Course

Socialization begins the moment a person is born. As soon as the sex of a child is known (which now can be even before birth), parents, grandparents, brothers, and sisters greet the infant with different expectations, depending on whether it is a boy or a girl. Socialization does not come to an end as we reach adulthood; rather, it continues through our lifetime. As we enter new situations, and even as we interact in familiar ones, we learn new roles and undergo changes in identity.

doing sociological research

Children's Understanding of Race

In a racially stratified society, people learn concepts about race that shape their interactions with others. Sociologists Debra Van Ausdale and Joe Feagin wanted to know how children understand racial and ethnic concepts and how this influences their interaction with other children.

Research Question: How do children learn about race? Prior to Van Ausdale and Feagin's study, most knowledge about children's understandings of race came from experimental studies in a laboratory or from psychological tests and interviews with children. Van Ausdale and Feagin wanted to study children in a natural setting so they observed children in school, systematically observing children's interactions with one another.

Research Method: They observed three-, four-, and five-year-olds in an urban preschool. Twenty-four of the children were White; nineteen, Asian; four, Black; three, biracial; three, Middle Eastern; two, Latino; and three classified as "other." The children's racial designations were provided by their parents.

The researchers observed in the school five days a week for eleven months and saw one to three episodes involving significant racial or ethnic matters every day.

Research Results: The researchers found that young children use racial and ethnic concepts to exclude other children from play. Sometimes language is the ethnic marker; other times, skin color. The children showed an awareness of negative racial attitudes, even though they were attending a school that prided itself on limiting children's exposure to prejudice and discrimination and used a multicultural curriculum to teach students to value racial and ethnic diversity. At times, the children also used racial–ethnic understandings to include others—teaching other students about racial–ethnic identities. Race and ethnicity were also the basis for children's concepts of themselves and others. As an example, one four-year-old White child insisted that her classmate was Indian because she wore her long, dark hair in a braid. When the classmate explained that she was not Indian, the young girl remarked that maybe her mother was Indian.

Conclusions and Implications: Throughout the research, the children showed how significant racial–ethnic concepts were in their interactions with others. Race and ethnicity are powerful identifiers of self and others. Despite the importance of race in the children's interactions, Van Ausdale and Feagin also noted a strong tendency for the adults they observed to *deny* that race and ethnicity were significant to the children. The implication is that while adults tend to deny the reality of race in their everyday lives, observing the interaction of children helps to instruct adults about the relevance of race and how racial awareness develops.

Through the socialization process, young children learn the values of their culture. These values shape their relationships with other people.

Source: Van Ausdale, Debra, and Joe R. Feagin. 2000. *The First R: How Children Learn Race and Racism.* Lanham, MD: Rowman and Littlefield.

Sociologists use the term **life course** perspective to describe and analyze the connection between people's personal attributes, the roles they occupy, the life events they experience, and the social and historical aspects of these events (Stoller and Gibson 2000). The life course perspective underscores the point made by C. Wright Mills (introduced in Chapter 1) that personal biographies are linked to specific social–historical periods. Thus, different generations are strongly influenced by large-scale events (such as a war, immigration, economic prosperity, or depression, for example).

The phases of the life course are familiar: childhood, youth and adolescence, adulthood, and old age. These phases of the life course bind different generations and define some of life's most significant events, such as birth, marriage, retirement, and death.

Childhood

During childhood, socialization establishes one's initial identity and values. In this period, the family is an extremely influential source of socialization, but experiences in school, peer relationships, sports, religion, and the media also have a profound effect. Children acquire knowledge of their culture through countless subtle cues that provide them with an understanding of what it means to live in society.

Socializing cues begin as early as infancy, when parents and others begin to describe their children based on their perceptions. Frequently, these perceptions are derived from the cultural expectations parents have for children. Parents of girls may describe their babies as "sweet" and "cuddly," whereas boys are described as "strong" and "alert." Even though it is difficult to physically identify baby boys and girls, in this culture parents dress even their infants in colors and styles that typically distinguish one gender from the other.

The lessons of childhood socialization come in a myriad of ways, some more subtle than others. In an example of how gender influences childhood socialization, researchers observed mothers and fathers who were walking young children through public places. Although the parents may not have been aware of it, both mothers and fathers were more protective toward girl toddlers than boy toddlers. Parents were more likely to let boy toddlers walk alone but held girls' hands, carried them, or kept them in strollers. The children were not the only ones learning gender roles. The researchers also observed that when the child was out of the stroller, the mother was far more likely than the father to be pushing the empty stroller (Mitchell et al. 1992). In countless ways, sometimes subtle, sometimes overt, we learn society's expectations.

Much socialization in early childhood takes place through play and games. Games that encourage competition help instill the value of competitiveness throughout someone's life. Likewise, play with other children and games that are challenging give children important intellectual, social, and interpersonal skills. Extensive research has been done on how children's play and games influence their identities as boys and girls (Campenni 1999; Raaj and Rackliff 1998). Generally, the research finds that boys' play tends to be rougher, more aggressive, and involve more specific rules. Boys are also more likely to be involved in group play, and girls engage in more conversational play (Moller et al. 1992). Sociologists have concluded that the games children play significantly influence their development into adults.

Another enormous influence on childhood socialization is what children observe of the adult world. Children are keen observers, and what they perceive will influence their self-concept and how they relate to others. This is vividly illustrated by research revealing how many adult child abusers were themselves victims of child abuse (Fattah 1994). Children become socialized by observing the roles of those around them and internalizing the values, beliefs, and expectations of their culture.

Adolescence

Only recently has adolescence been thought of as a separate phase in the life cycle. Until the early twentieth century, children moved directly from childhood roles to adult roles. It was only when formal education was extended to all classes that adolescence emerged as a particular phase in life when young people are regarded as no longer children, but not yet adults. There are no clear boundaries to adolescence, although it generally lasts from junior high school until the time one takes on adult roles by getting a job, marrying, and so forth. Adolescence can include the period through high school and extend right up through college graduation.

Erik Erikson (1980), the noted psychologist, stated that the central task of adolescence is the formation of a consistent identity. Adolescents are trying to become independent of their families, but they have not yet moved into adult roles. Conflict and confusion can arise as the adolescent swings between childhood and adult maturity. Some argue that adolescence is a period of delayed maturity. Although society expects adolescents to behave like adults, they are denied many privileges associated with adult life. Until age 18, they cannot vote or marry without permission, and sexual activity is condemned. In addition, until age 21 they can not legally drink alcohol. The tensions of adolescence have been blamed for numerous social problems, such as drug and alcohol abuse, youth violence, and the school dropout rate.

The issues that young people face are a good barometer of social change across generations. Today's young people face an uncertain world where adult roles are less predictable than in the past. Marriage later in life, high divorce rates, a volatile labor market, and frequent technological change create a confusing environment for young people (Csikzentmihalyi and Schneider 2000). Studies of adolescents find that, in this context, young people understand the need for flexibility, specialization, and, likely, frequent job change. Although the media stereotype adolescents as slackers, most teens are willing to work hard, do not engage in criminal or violent activity, and have high expectations for an education that will lead to a good job. Many, however, find that their expectations are out of alignment with the opportunities that are

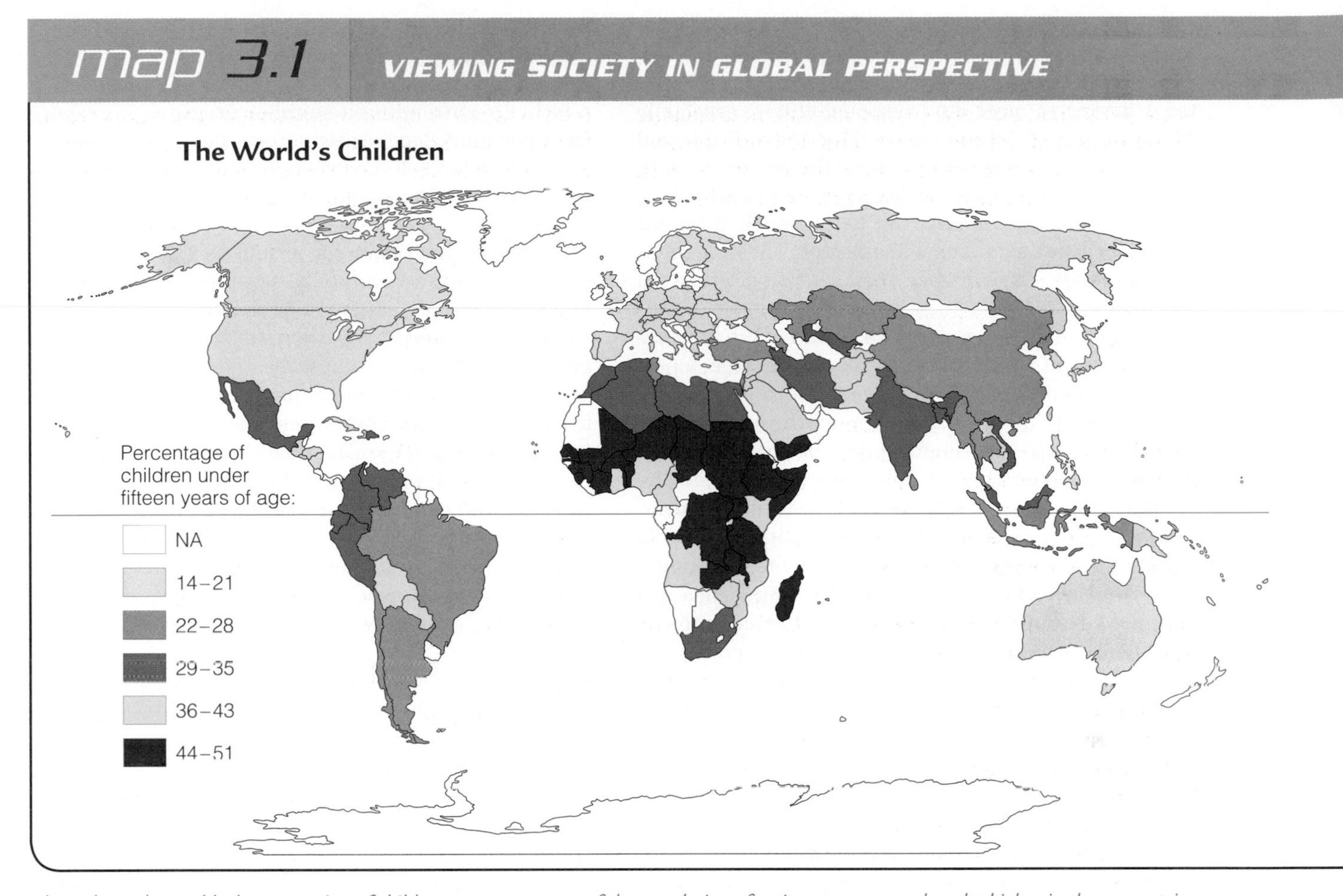

Throughout the world, the proportion of children as a percentage of the population of a given country tends to be higher in those countries that are most economically disadvantaged and most overpopulated. In such countries, children are also more likely to die young but are needed to contribute to the labor that families do. What consequences do you think the proportion of children in a given society has for the society as a whole?

Data: From the U.S. Census Bureau. 2006. *Statistical Abstract of the United States 2005.* Washington, DC: U.S. Government Printing Office.

actually available, creating a paradox of a generation that is "motivated but directionless" (Schneider and Stevenson 1999).

Patterns of adolescent socialization vary significantly by race and social class. National surveys find some intriguing class and race differences in how young people think about work and play in their lives. In general, the most economically privileged young people see their activities as more like play than work, whereas those less well off are more likely to define their activities as work. Likewise, White youth (boys especially) are more likely than other groups to see their lives as playful. The researchers interpret these findings to mean that being economically privileged allows you to think of your work as if it were play. Being in a less advantaged position, on the other hand, makes you see the world as more "worklike." This is supported by further findings that young people from less advantaged backgrounds spend more time in activities they define as purposeless (Schneider and Stevenson 1999).

Thus, differences in adolescent socialization are associated with significantly different life paths. The values we form as a young person have a decided influence on where we find ourselves later in life. However it is also true that the values formed in youth reflect the life chances the young believe are possible.

Adulthood

Socialization does not end when one becomes an adult. Building on the identity formed in childhood and adolescence, adult socialization is the process of learning new roles and expectations in adult life. More so than at earlier stages in life, **adult socialization** involves learning behaviors and attitudes appropriate to specific situations and roles.

Youths entering college, to take an example from young adulthood, are newly independent and have new responsibilities. In college, one acquires not just an education, but also a new identity. Those who enter college directly from high school may encounter conflicts with their family over their newfound status. Older students who work and attend college may experience difficulties (defined as *role conflict*; see Chapter 4) trying to meet dual responsibilities, especially if their family is not supportive. Meeting multiple and conflicting demands may require the returning student to develop different expectations about how much she can accomplish or to establish different priorities about what she will attempt. These changes reflect a new stage in her socialization (Settersten and Lovegreen 1998).

Adult life is peppered with events that may require the adult to adapt to new roles. Marriage, a new career, starting a family, entering the military, getting a divorce, or dealing with a death in the family all transform an individual's previous social identity. In today's world, these transitions through the life course are not as orderly as they were in the past. Where there was once a sequential and predictable trajectory of schooling, work, and family roles through one's twenties and thirties, that is no longer the case. Younger generations now experience diverse patterns in the sequencing of work, schooling, and family formation—even returning home—than was true in the past. These changes complicate the life course, and people have to make different adaptations to these changing roles (Cooksey and Rindfuss 2001; Rindfuss et al. 1999). Becoming an adult is also taking longer than before. People stay in school longer, are marrying later, and postponing childbearing (see Table 3.2). With these social changes, people have to be inventive in the roles they occupy because some of the old expectations no longer apply.

Another part of learning a new role is **anticipatory socialization,** the learning of expectations associated with a role a person expects to enter in the future. One might rehearse the expectations associated with being a professor by working as a teaching assistant, taking a class in preparation for becoming a father, or attending a summer program to prepare for entering college. Anticipatory socialization allows a person to foresee the expectations associated with a new role and learn what is expected in that role in advance.

In the transition from an old role to a new one, individuals often vacillate between their old and new identities as they adjust to fresh settings and expectations. An interesting example is *coming out,* the process of identifying oneself as gay or lesbian. This can be either a public coming out or a private acknowledgment of sexual orientation. The process can take years and generally means coming out to a few people, at first selective family members or friends who are likely to have the most positive reaction. Coming out is rarely a single event but occurs in stages on the way to developing a new identity—one that is not only a new sexual identity, but also a new sense of self (Due 1995).

Age and Aging

Passage through adulthood involves many transitions. In our society, one of the most difficult transitions is the passage to old age. We are taught to fear aging in this society, and many people spend a lot of time and money trying to keep looking young. Unlike many societies, ours does not revere the elderly but instead devalues them, making the aging process even more difficult.

It is easy to think that aging is just a natural fact. Despite desperate attempts to hide gray hair, eliminate wrinkles, and reduce middle-aged bulge, aging is inevitable. The skin creases and sags, the hair thins,

table 3.2 Slowing the Transition to Adulthood

	1980	2002
Percentage aged 20–21 in school	31.9%	47.4%
Median age at first marriage		
Women	22.0 yrs.	25.3 yrs.
Men	24.7 yrs.	26.9 yrs.
Fertility rate, women aged 15–19	68.4 (per 1000 women)	60.2 (per 1000 women)
Percentage aged 16–19 in labor force		
Women	52.9%	47.3%
Men	60.5%	47.5%

Source: U.S. Census Bureau. 2005. *Statistical Abstract of the United States 2004.* Washington, DC: U.S. Government Printing Office, pp. 61, 63, 138, 371; Fields, Jason, and Lynne M. Casper. 2001. *America's Families and Living Arrangements.* Washington, DC: U.S. Census Bureau.

The stresses of life that accompany age can change a person in many ways, as is evident in these "before" and "after" photographs of former President Bill Clinton.

metabolism slows, and bones become less dense and more brittle by losing bone mass. Although aging is a physical process, the social dimensions of aging are just as important, if not more so, in determining the aging process. Just think about how some people appear to age much more rapidly than others. Some sixty-year-olds look only forty, and some forty-year-olds look much older. These differences result from combinations of biological and social factors, such as eating and exercise habits, stress, smoking habits, pollution in the physical environment, and many other factors. The social dimensions of aging are what interest sociologists.

Although the physiology of aging proceeds according to biological processes, what it *means* to grow older is a social phenomenon. **Age stereotypes** are preconceived judgments about what different age groups are like. Stereotypes abound for both old and young people. Young people, especially teenagers, are stereotyped as irresponsible, addicted to loud music, lazy, sloppy, and so on; the elderly are stereotyped as forgetful, set in their ways, mentally dim, and unproductive. Though like any stereotype, these stereotypes are largely myths, they are widely believed. Age stereotypes also differ for different groups. Older women are stereotyped as having lost their sexual appeal, contrary to the stereotype of older men as handsome or "dashing" and desirable. Gender is, in fact, one of the most significant factors in age stereotypes. Women may even be viewed as becoming old sooner than men because people describe women as old a decade sooner than they describe men as old (Stoller and Gibson 2000).

Age stereotypes are also reinforced through popular culture. Advertisements depict women as needing creams and lotions to hide "the tell-tale signs of aging." Men are admonished to cover the patches of gray hair that appear or to use other products to prevent baldness. Entire industries are constructed on the fear of aging that popular culture promotes. Face-lifts, tummy tucks, and vitamin advertisements all claim to "reverse the process of aging," even though the aging process is a fact of life.

Age Prejudice and Discrimination. **Age prejudice** refers to a negative attitude about an age group that is generalized to all people in that group. Prejudice against the elderly is prominent. As an example, people may talk "baby talk" to the elderly. Doing so defines the elderly as childlike and incompetent. Prejudice relegates people to a perceived lower status in society and stems from the stereotypes associated with different age groups.

Age discrimination is the different and unequal treatment of people based solely on their age. Whereas age prejudice is an attitude, age discrimination involves actual behavior. Some forms of age discrimination are illegal. The Age Discrimination Employment Act, first passed in 1967 but amended several times since, protects people from age discrimination in employment. An employer can neither hire nor fire someone based solely on age, nor segregate or classify workers based on age. Age discrimination cases have become one of the most frequently filed cases through the Equal Employment Opportunity Commission (EEOC), the federal agency set up to monitor violations of civil rights in employment.

Ageism is a term sociologists use to describe the institutionalized practice of age prejudice and discrimination. More than a single attitude or an explicit act of discrimination, ageism is structured into the institutional fabric of society. Like racism and sexism, ageism encompasses both prejudice and discrimination, but it is also manifested in the structure of institutions. As such, it does not have to be intentional or overt to affect how age groups are treated. Ageism in society means that, regardless of laws that prohibit age discrimination, people's age is a significant predictor of their life chances. Resources are distributed in society in ways that advantage some age groups and disadvantage others; cultural belief systems devalue the elderly; society's systems of care are often inadequate to meet people's needs as they grow old—these are the manifestations of ageism, a persistent and institutionalized feature of society.

Age Stratification. Most societies produce age hierarchies—systems in which some age groups have more power and better life chances than others. **Age stratification** refers to the hierarchical ranking of different age groups in society. Age stratification exists because processes in society ensure that people of different ages differ in their access to society's rewards, power, and privileges. As we will see, in the United States and elsewhere, age is a major source of inequality (see Figure 3.4).

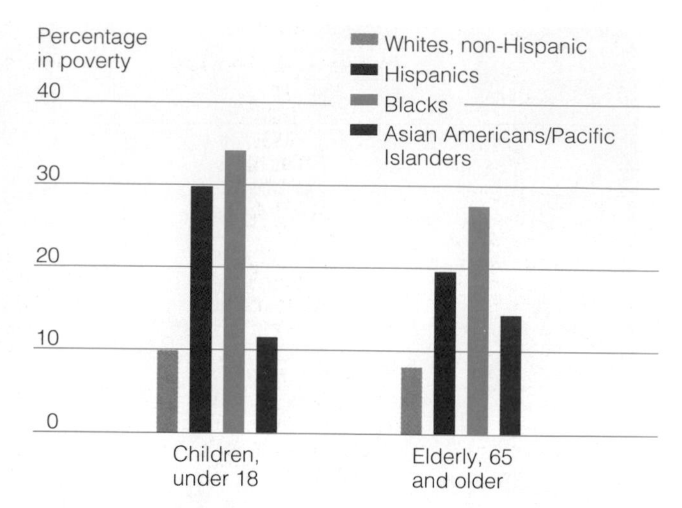

FIGURE 3.4 *Poverty by Age and Race*

Source: DeNavas-Walt, Carmen, Bernadette D. Proctor, and Cheryl Hill Lee. *Income, Poverty, and Health Insurance Coverage in the United States: 2005.* Washington, DC: U.S. Census Bureau. **www.census.gov**

Age is an *ascribed status*; that is, age is determined by when you were born. Different from other ascribed statuses, which remain relatively constant over the duration of a person's life, age changes steadily throughout your life. Still, you remain part of a particular generation—something sociologists call an **age cohort**—an aggregate group of people born during the same period (Stoller and Gibson 2000). Cohorts are not homogenous. Within a given cohort, there will be considerable diversity on many dimensions: sexual orientation, gender, race, class, nationality, ethnicity. How these cohorts are arrayed in a given society at a given time shapes the character of society and the social issues within it.

People in the same age cohort share the same historical experiences—wars, technological developments, economic fluctuations—although they might do so in different ways, depending on other life factors. Living through the Great Depression, for example, shaped an entire generation's attitudes and behaviors, as did growing up in the 1960s, as will being a member of the contemporary youth generation. Recall from Chapter 1 that C. Wright Mills saw the task of the sociological imagination as analyzing the relationship between biography and history. Understanding the experiences of different age cohorts is one way you can do this. People who live through the same historic period experience a similar impact of that period in their personal lives. The troubles and triumphs they experience and the societal issues they face are rooted in the commonality established by their age cohort. The shared historical experiences of age cohorts result in discernible generational patterns—in social attitudes and similarity of life chances (Stoller and Gibson 2000).

Different generations must grapple with and respond to different social contexts. Someone born just after World War II would, upon graduation from high school or college, enter a labor market where there was widespread availability of jobs and, for many, expanding opportunities. Now young people face a labor market where entry-level jobs in secure corporate environments are rare and where many are trapped in low-level jobs with little opportunity for advancement. Many young people worry, as a result, about whether they will be able to achieve even the same degree of economic status as their parents—the first time this has happened in U.S. history. Understanding how society shapes the experiences of different generations is what sociologists mean by say-

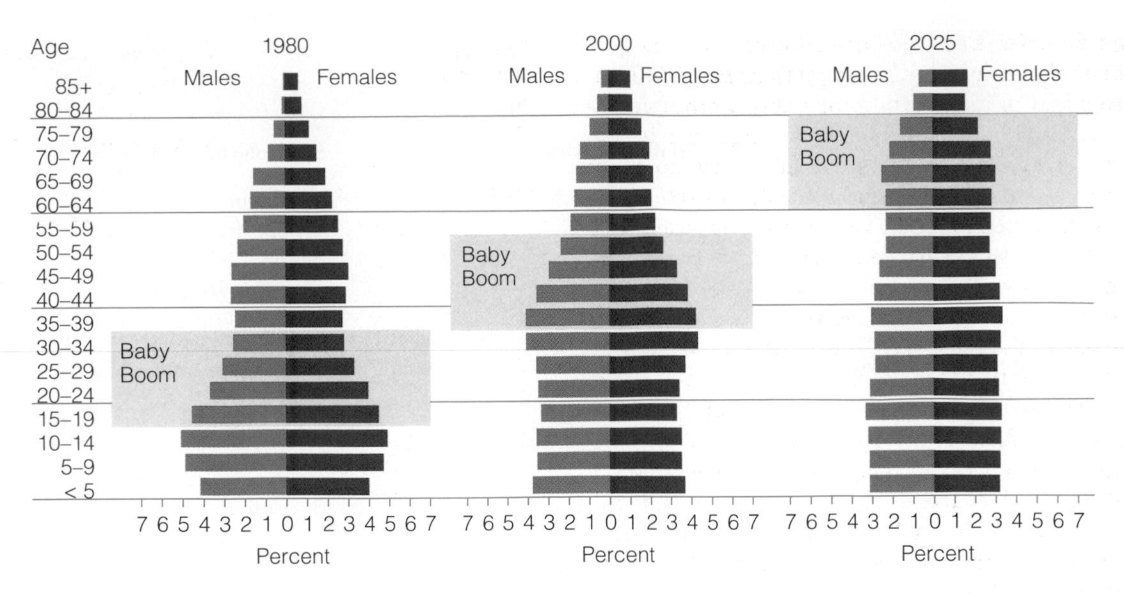

FIGURE 3.5 *An Aging Society*

Source: U.S. Census Bureau. 2005. *Statistical Abstract of the United States 2004.* Washington, DC: U.S. Department of Commerce, pp. 12–13.

ing that age is a structural feature of society (see Figure 3.5).

The age structure of society shapes people's opportunities and is the basis for cultural understandings of age itself. In the very poor regions of northeast Brazil, for example, mothers show little attachment to those who are born small and weak; if they die, there is little ceremony, and their graves remain unmarked. Because sick infants are believed to be angels who fly to heaven, mother's tears are believed to dampen their wings, risking their flight. Anthropologists interpret this as the mothers' reaction to their impoverishment: They cannot invest attention or emotion in the lives of children who are unlikely to live (Scheper-Hughes 1992; Peoples and Bailey 2003).

There is also great variation in how the old are treated. In many societies, older people are given enormous respect. There may be traditions to honor the elders, and they may be given authority over decisions in society, as they are perceived as most wise. On the other hand, among some cultures, adults who can no longer contribute to the society because of old age or illness may be perceived as extreme burdens. You might ask yourself how cultural definitions in this society affect how people grieve for different age groups. Is the death of a young person perceived as somehow more tragic than the death of a very old person? How do people react in each circumstance? Your answer will likely reveal the cultural beliefs surrounding aging in this culture.

Why does society stratify people on the basis of age? Once again we find that the three main theoretical perspectives of sociological analysis—functionalism, conflict theory, and symbolic interaction—offer different explanations (see Table 3.3). Functionalist sociologists ask whether the grouping of individuals contributes in some way to the common good of society. From this perspective, adulthood is functional to society because adults are seen as the group contributing most fully to it; the elderly are not. Functionalists argue that older people are seen as less useful and are therefore granted lower status in society. Youth are in between. The constraints and expectations placed on youth—they are prohibited from engaging in a variety of "adult" activities, expected to go to school, not expected to support themselves—are seen to free them from the cares of adulthood and give them time and opportunity to learn an occupation and prepare to contribute to society.

According to the functionalist argument, the elderly voluntarily withdraw from society by retiring and lessening their participation in social activities such as church, civic affairs, and family. **Disengagement theory,** drawn from functionalism, predicts that as people age, they gradually withdraw from participation in society and are simultaneously relieved of responsibilities. This withdrawal is functional to society because it provides for an orderly transition from one generation to the next. The young presumably infuse the roles they take over from the elderly with youthful energy and stamina. According to the functionalist argument, the diminished usefulness of the elderly justifies their depressed earning power and their relative neglect in social support networks.

table 3.3 Sociological Theories of Aging

	Functionalism	Conflict Theory	Symbolic Interaction
Age differentiation	Contributes to the common good of society because each group has varying levels of utility in society	Results from the different economic status and power of age cohorts	Occurs in most societies, but the social value placed on different age groups varies across diverse cultures
Age groups	Are valued according to their usefulness in society	Compete for resources in society, resulting in generational inequities and thus potential conflict	Are stereotyped according to the perceived value of different groups
Age stratification	Results from the functional value of different age cohorts	Intertwines with inequalities of class, race, and gender	Promotes ageism, which is institutionalized prejudice and discrimination against old people

Conflict theory focuses on the competition over scarce resources between age groups. Among the most important scarce resources are jobs. Unlike functionalist theory, conflict theory offers an explanation of why both youth and the elderly are assigned lower status in society and are most likely to be poor. Barring youth and the elderly from the labor market eliminates these groups from competition, improving the prospects for middle-aged workers. Removed from competition, both the young and the old have very little power, and like other minorities, they are denied access to the resources they need to change their situation. Conflict theory also helps explain that competition can emerge between age groups, such as deciding whether to limit Social Security payments to save for future generations.

Symbolic interaction theory analyzes the different meanings attributed to social entities. Symbolic interactionists ask what meanings become attached to different age groups and to what extent these meanings explain how society ranks such groups. Definitions of aging are socially constructed, as we saw in our discussion of age stereotypes. Moreover, in some societies, the elderly may be perceived as having higher status than in other societies. Symbolic interaction considers the role of social perception in understanding the sociology of age. Age clearly takes on significant social meaning—meaning that varies from society to society for a given age group and that varies within a society for different age groups.

Growing old in a society such as the United States with such a strong emphasis on youth means encountering social stereotypes about the old, adjusting to diminished social and financial resources, and sometimes living in the absence of social supports, even when facing some of life's most difficult transitions, such as declining health and the loss of loved ones. Still, many people experience old age as a time of great satisfaction and enjoy a sense of accomplishment connected to work, family, and friends. The degree of satisfaction during old age depends to a great extent on the social support networks established earlier in life—evidence of the continuing influence of socialization.

For many, old age is a time for new accomplishments and achievements, such as this marathon runner.

Rites of Passage

A **rite of passage** is a ceremony or ritual that marks the transition of an individual from one role to another. Rites of passage define and legitimize abrupt role changes that begin or end each stage of life. The ceremonies surrounding rites of passage are often dramatic and infused with awe and solemnity. Examples include graduation ceremonies; weddings; and religious affirmations, such as the Jewish ceremony of the bar mitzvah for boys or the bat mitzvah for girls, confirmation for Catholics, and adult baptism for many Christian denominations.

Formal promotions or entry into some new careers may also include rites of passage. Completing police academy training or being handed one's diploma are examples. Such rites usually include family and friends, who watch the ceremony with pride; people frequently keep mementos of these rites as markers of the transition through life's major stages. Bridal showers and baby showers have been analyzed as rites of passage. At a shower, the person who is being honored is about to assume a new role and identity—from young woman to wife or mother. Rites of passage entail public announcement of the new status for the benefit of both the individual and those with whom the newly anointed person will interact. In the absence of such rituals, the transformation of identity would not be formally recognized, perhaps leaving uncertainty in the youngster or the community about the individual's worthiness, preparedness, or community acceptance.

Sociologists have noted that in the United States there is no standard and formalized rite of passage marking the transition from childhood to adulthood. As a consequence, the period of adolescence is attended by ambivalence and uncertainty. As adolescents hover between adult and child status, they may not have the clear sense of identity that a rite of passage can provide. However, although there is no universal ceremony in our culture by which young people are noted as moving from child to adult, some subcultures do mark the occasion. Among the wealthy, the debutante's coming out celebration is a traditional introduction of a young

Every culture has important rites of passage that mark the transition from one phase in the life course to another. Here different cultural traditions distinguish the rites of passage associated with marriage—a traditional Nigerian wedding (upper left); a young American couple (upper right); a Shinto (Japanese) bride taking a marital pledge by drinking sake (lower left); and a newlywed orthodox Christian couple in Eritrea (lower right).

woman to adult society. Latinos may celebrate the *quinceañera* (fifteenth birthday) of young girls. A tradition of the Catholic Church, this rite recognizes the girl's coming of age, while also keeping faith with an ethnic heritage. Dressed in white, she is introduced by her parents to the larger community. Formerly associated mostly with working-class families in the barrios, the quinceañera has also become popular among affluent Mexican Americans, who may match New York debutante society by spending as much as $30,000–$50,000 on the event (McLane 1995).

Resocialization

Most transitions people experience in their lifetimes involve continuity with the former self as it undergoes gradual redefinition. Sometimes, however, adults are forced to undergo a radical shift of identity. **Resocialization** is the process by which existing social roles are radically altered or replaced (Fein 1988). Resocialization is especially likely when people enter institutional settings where the institution claims enormous control over the individual. Examples include the military, prisons, monastic orders, and some cults (see also Chapter 5 for a discussion of total institutions). When military recruits enter boot camp, they are stripped of personal belongings, their heads are shaved, and they are issued identical uniforms. Although military recruits do not discard their former identities, the changes brought about by becoming a soldier can be dramatic and are meant to make the military primary, not one's family, friends, or personal history. The military represents an extreme form of resocialization in which individuals are expected to subordinate their identity to that of the group. In such organizations, individuals are interchangeable, and group consensus (meaning, in the military, unanimous, unquestioned subordination to higher ranks) is an essential component of group cohesion and effectiveness. Military personnel are expected to act as soldiers, not as individuals. Understanding the importance of resocialization on entry to the military helps us understand such practices as "the rat line" at VMI (Virginia Military Institute) where members of the senior class taunt and harass new recruits.

Resocialization often occurs when people enter hierarchical organizations that require them to respond to authority on principle, not out of individual loyalty. The resocialization process promotes group solidarity and generates a feeling of belonging. Participants in these settings are expected to honor the symbols and objectives of the organization; disloyalty is seen as a threat to the entire group. In a convent, for example, nuns are expected to subordinate their own identity to the calling they have taken on, a calling that requires obedience both to God and to an abbess.

thinking sociologically

Find a group of adults (young or old) who have just entered a new stage of life (getting a new or first job, getting married, becoming a grandparent, retiring, entering a nursing home, and so forth), and ask them to describe this new experience. Ask questions such as what others expect of them in this new role, how these expectations are communicated to them, what changes they see in their own behavior, and what expectations they have of their new situation. What do your observations tell you about *adult socialization*?

Resocialization may involve degrading initiates physically and psychologically with the aim of breaking down or redefining their old identity. They may be given menial and humiliating tasks and be expected to act in a subservient manner. Social control in such a setting may be exerted by peer ridicule or actual punishment. Fraternities and sororities offer an interesting everyday example of this pattern of resocialization. Intense resocialization rituals, whether in jailhouses, barracks, convents, or sorority and fraternity houses, serve the same purpose: imposing some sort of ordeal to cement the seriousness and permanence of new roles and expectations.

Hazings are a good example of the rites of passage that often accompany integration into a new group. This hazing incident became national news when an annual hazing ritual at a suburban Chicago high school injured several young women and was broadcast on most major media networks.

The Process of Conversion

Resocialization also occurs during what people popularly think of as conversion. A conversion is a far-reaching transformation of identity, often related to religious or political beliefs. People usually think of conversion in the context of cults, but it happens in other settings as well.

The case of John Walker Lindh—a U.S. citizen who several years ago joined the Taliban in Afghanistan and was later charged with conspiring to kill Americans abroad and supporting terrorist organizations—is an example of an extreme conversion. Lindh was raised Catholic in an affluent family, but he converted to Islam as a teenager, changing not just his ideas, but also his dress. Neighbors described him as being transformed from "a boy who wore blue jeans and T-shirts to an imposing figure in flowing Muslim garb" (Robertson and Burke 2001). As a young man, he traveled to Yemen and Pakistan to study language and the Koran and was introduced there to the Taliban.

As when people join religious cults, this is extreme conversion, but conversion happens in less extreme situations, too. People may convert to a different religion, thereby undergoing resocialization by changing beliefs and religious practices. Or someone may become strongly influenced by the beliefs of a social movement and abruptly or gradually change beliefs—even identity—as a result.

The Brainwashing Debate

Extreme examples of resocialization are seen in the phenomenon popularly called brainwashing. In the popular view of brainwashing, converts have their previous identities totally stripped; the transformation is seen as so complete that only deprogramming can restore the former self. Potential candidates of brainwashing include people who enter religious cults, prisoners of war, and hostages. Sociologists have examined brainwashing to illustrate the process of resocialization. As the result of their research, sociologists have cautioned against using the word *brainwashing* when referring to this form of conversion. The term implies that humans are mere puppets or passive victims whose free will can be taken away during these conversions (Robbins 1988). In religious cults, however, converts do not necessarily drop their former identity.

Sociological research has found that the people most susceptible to cult influence are the most suggestible, primarily young adults who are socially isolated, drifting, and having difficulty performing in other areas (such as in their jobs or in school). Such people may choose to affiliate with cults voluntarily. Despite the widespread belief that people have to be deprogrammed to be freed from the influence of cults, many people are able to leave on their own (Robbins 1988). So-called brainwashing is simply a manifestation of the social influence people experience through interaction with others. Even in cult settings, socialization is an interactive process, not just a transfer of group expectations to passive victims.

Forcible confinement and physical torture can be instruments of extreme resocialization. Under severe captivity and deprivation, a captured person may come to identify with the captor; this is known as the **Stockholm Syndrome.** In such instances, the captured person has become dependent on the captor. On release, the captive frequently needs debriefing. Prisoners of war and hostages may not lose free will altogether, but they do lose freedom of movement and association, which makes prisoners intensely dependent on their captors and therefore vulnerable to the captor's influence. The Stockholm Syndrome can help explain why some battered women do not leave their abusers. Dependent on their abuser both financially and emotionally, battered women often develop identities that keep them attached to men who abuse them. In these cases, outsiders often think the women should leave instantly, whereas the women themselves may find leaving difficult, even in the most abusive situations.

In sum, resocialization involves establishing a radically new definition of oneself. The new identity may seem dramatically different from the former one, but the process by which it is established is much the same as the ordinary socialization process—a process that, as we have seen, is a critical part of society.

Chapter Summary

- ***What is socialization, and why is it significant for society?***

Socialization is the process by which human beings learn the social expectations of society. Socialization creates the expectations that are the basis for people's attitudes and behaviors. Through socialization, people conform to social expectations, although people still express themselves as individuals.

- ***What are the agents of socialization?***

Socialization agents are those who pass on social expectations. They include the family, the media, peers, sports, religious institutions, and schools, among others. The family is usually the first source of socialization. The media also influence people's values and behaviors. Peer groups are an important source of individual identity; without peer approval, most people find it hard to be socially accepted. Schools also pass on expectations that are influenced by gender, race, and other social characteristics of people and groups.

- ***What theoretical perspectives do sociologists use to explain socialization?***

Psychoanalytic theory sees the self as driven by unconscious drives and forces that interact with the expectations of society. *Social learning theory* sees identity as a learned response to social stimuli such as role models. *Functionalism* interprets socialization as key to social stability because socialization establishes shared roles and values. *Conflict theory* interprets socialization in the context of inequality and power relations. *Symbolic interaction theory* sees people as constructing the self as they interact with the environment and give meaning to their experience. Charles Horton Cooley described this process as the *looking glass self*. Another sociologist, George Herbert Mead, described childhood socialization as occurring in three stages: imitation, play, and games.

- ***Does socialization mean that everyone grows up the same?***

Socialization is not a uniform process. Growing up in different environments and in such a diverse society means that different people and different groups are exposed to different expectations. Factors such as family structure, social class, regional differences, and many others influence how one is socialized.

- ***Does socialization end during childhood?***

Socialization continues through a lifetime, although childhood is an especially significant time for the formation of identity. Adolescence is also a period when peer cultures have an enormous influence on the formation of people's self-concepts. *Adult socialization* involves the learning of specific expectations associated with new roles.

- ***What are the social dimensions of the aging process?***

Although aging is a physiological process, its significance stems from social meanings attached to aging. *Age prejudice* and *age discrimination* result in the devaluation of older people. *Age stratification*—referring to the inequality that occurs among different age groups—is the result.

- ***What does resocialization mean?***

Resocialization is the process by which existing social roles are radically altered or replaced. It can take place in an organization that maintains strict social control and demands that the individual conform to the needs of the group or organization.

Key Terms

adult socialization 81
age cohort 84
age discrimination 84
age prejudice 83
age stereotype 83
age stratification 84
ageism 84
anticipatory socialization 82
disengagement theory 85
ego 73
game stage 77
generalized other 77
id 73
identity 64
imitation stage 76
life course 79
looking glass self 76
peers 70
personality 65
play stage 76
psychoanalytic theory 73
resocialization 88
rite of passage 87
role 64
self 75
significant others 77
social control 66
social learning theory 74
socialization 64
socialization agents 68
Stockholm Syndrome 89
superego 73
taking the role of the other 76

Online Resources

Sociology: The Essentials Companion Website

www.thomsonedu.com/sociology/andersen

Visit your book companion website where you will find more resources to help you study and write your research papers. Resources include Suggested Readings, web links, and a MicroCase Online feature that teaches you how to research society. Other resources include Learning Objectives, Internet exercises, quizzing, and flash cards.

is an easy-to-use online resource that helps you study in less time to get the grade you want NOW.

www.thomsonedu.com/login

Need help studying? This site is your one-stop study shop. Take a Pre-Test and Thomson NOW will generate a Personalized Study Plan based on your test results. The Study Plan will identify the topics you need to review and direct you to online resources to help you master those topics. You can then take a Post-Test to determine the concepts you have mastered and what you still need to work on.

4
Chapter four
CHAPTER FOUR
Chapter

Social Interaction and Social Structure

[**Picture a college** classroom on your campus. Students sit, and some are taking notes; others, listening; a few perhaps, sleeping. The class period ends and students stand, gathering their books, backpacks, bags, and other gear. As they stand, many whip out their cell phones, place them to their ears, and quickly push buttons that connect them to a friend. As the students exit the room, many are engaged in *social interaction*—chatting with their friends: some by phone, others by text messaging, some by talking face-to-face. Few, if any, of them realize that their behavior is at that moment influenced by *society*—a society whose influence extends into their immediate social relationships, even when the contours of that society—its *social structure*—is likely invisible to them.

These same students might plug a music player into their ears as they move on to their next class, possibly tuning in to the latest sounds while tuning out the sounds of the environment around them. Some will return to their residences and perhaps text message friends, download some music, or connect with "friends" on Facebook or MySpace. Some might watch a video or podcast on a small, handheld device. Surrounding all of this behavior are social changes that are taking place in society, including changes in technology, in global communication, and in how people now interact with each other. How we make sense of these changes requires an understanding of the connection between society and social interaction. In this way, a sociological perspective can help you see the relationship between individuals and the larger society of which they are a part.

chapter four

CHAPTER FOUR

What Is Society?

In Chapter 2, we studied culture as one force that holds society together. *Culture* is the general way of life, including norms, customs, beliefs, and language. Human **society** is a system of social interaction that includes both culture and social organization. Within a society, members have a common culture, even though there may also be great diversity within it. Members of a society think of themselves as distinct from other societies, maintain ties of social interaction, and have a high degree of interdependence. The interaction they have, whether based on harmony or conflict, is one element of society. Within society, **social interaction** is behavior between two or more people that is given meaning by them. Social interaction is how people relate to each other and form a social bond.

Social interaction is the foundation of society, but society is more than a collection of individual social actions. Emile Durkheim, the classical sociological theorist, described society as *sui generis*—a Latin phrase meaning "a thing in itself, of its own particular kind." To sociologists, seeing society *sui generis* means that society is more than just the sum of its parts. Durkheim saw society as an organism, something comprising different parts that work together to create a unique whole. Just as a human body is not just a collection of organs but is alive as a whole organism, society is not only a simple collection of individuals, groups, or institutions but is a whole entity that consists of all these elements and their interrelationships.

Durkheim's point—central to sociological analysis—is that society is much more than the sum of the individuals in it. Society takes on a life of its own. It is patterned on humans and their interactions, but it is something that endures and takes on shape and structure beyond the immediacy of any given group of people. This is a basic idea that guides sociological thinking.

© Bob Daemmrich/The Image Works

The introduction of new technologies is transforming how people interact with each other.

You can think of it this way: Imagine how a photographer views a landscape. The landscape is not just the sum of its individual parts—mountains, pastures, trees, or clouds—although each part contributes to the whole. The power and beauty of the landscape is that all its parts *relate* to each other, some in harmony and some in contrast, to create a panoramic view. The photographer who tries to capture this landscape will likely use a wide-angle lens. This method of photography captures the breadth and comprehensive scope of what the photographer sees. Similarly, sociologists try to picture society as a whole, not only by seeing its individual parts but also recognizing the relatedness of these parts and their vast complexity.

Macro- and Microanalysis

Sociologists use different lenses to see the different parts of society. Some views are more macroscopic—that is, sociologists try to comprehend the whole of society, how it is organized and how it changes. This is called **macroanalysis,** a sociological approach that takes the broadest view of society by studying large patterns of social interaction that are vast, complex, and highly differentiated. You might do this by looking at a whole society or comparing different total societies to each other. For example, as we opened this chapter, you saw that large-scale changes in technology influence even the most immediate social interaction that we have with other people. Thus, whereas years ago it would not have been imaginable to create a network of friends in cyberspace, today it is a common practice, especially for young people.

Other views are more microscopic—that is, the focus is on the smallest, most immediately visible parts of social life, such as specific people interacting with each other. This is called **microanalysis.** In this approach, sociologists study patterns of social interactions that are relatively small, less complex, and less differentiated—the microlevel of society. Again, thinking of how this chapter opened, you saw how social interaction at the microlevel can be shaped by large-scale technological developments in society. A sociologist who studies social interaction on the Internet, for example, would be engaging in microanalysis but might interpret what is found in the context of macrolevel processes. Just as a photographer might use a wide-angle lens to photograph a landscape or a telephoto lens for a closer view, sociologists use both macro- and microanalyses to reveal different dimensions of society.

In this chapter, we continue our study of sociology by starting with the macrolevel of social life (by studying total social structures), then continuing through the microlevel (by studying groups and face-to-face interaction). The idea is to help you see how large-scale dimensions of society shape even the most immediate forms of social interaction.

Sociologists use the term **social organization** to describe the order established in social groups at any level. Specifically, social organization brings regularity and predictability to human behavior; social organization is present at every level of interaction, from the whole society to the smallest groups.

Social Institutions

Societies are identified by their cultural characteristics and the social institutions that compose each society. A **social institution** (or simply an institution) is an established and organized system of social behavior with a recognized purpose. The term refers to the broad systems that organize specific functions in society. Unlike individual behavior, social institutions cannot be directly observed, but their impact and structure can still be seen. For example, the family is an institution that provides for the care of the young and the transmission of culture. Religion is an institution that organizes sacred beliefs. Education is the institution through which people learn the skills needed to live in the society.

The concept of social institutions is important to sociological thinking. You can think of social institutions as the enduring consequences of social behavior, but what fascinates sociologists is how social institutions take on a life of their own. For example, you are likely born in a hospital, which itself is part of the health care institution. The simple act of birth, which you might think of as an individual experience, is shaped by the structure of this social institution. Thus, you are likely delivered by a doctor, accompanied by nurses and, perhaps, a midwife—each of whom exists in a specific social relationship to the health care institution. Moreover, the practices surrounding your birth are also shaped by this social institution. Thus, you might be initially removed from your mother and examined by a doctor, which is very different from the institutional practices in other societies. These practices reflect the institutional structure of the health care system in society. Although social institutions change over time, they also endure. It is the connection between stability and change that makes social institutions so interesting to sociologists.

The major institutions in society include the family, education, work and the economy, the political institution (or state), religion, and health care, as well as the mass media, organized sports, and the military. These are all complex structures that exist to meet certain needs that are necessary for society to exist. Functionalist theorists have identified these needs (functions) as follows (Aberle et al. 1950; Parsons 1951a; Levy 1949):

1. *The socialization of new members of the society.* This is primarily accomplished by the family, but involves others institutions as well, such as education.
2. *The production and distribution of goods and services.* The economy is generally the institution that performs this set of tasks, but this may also involve the family as an institution—especially in societies where production takes place within households.
3. *Replacement of society's members.* All societies must have a means of replacing its members who die, move or migrate away, or otherwise leave the society. Families are typically organized to do this.

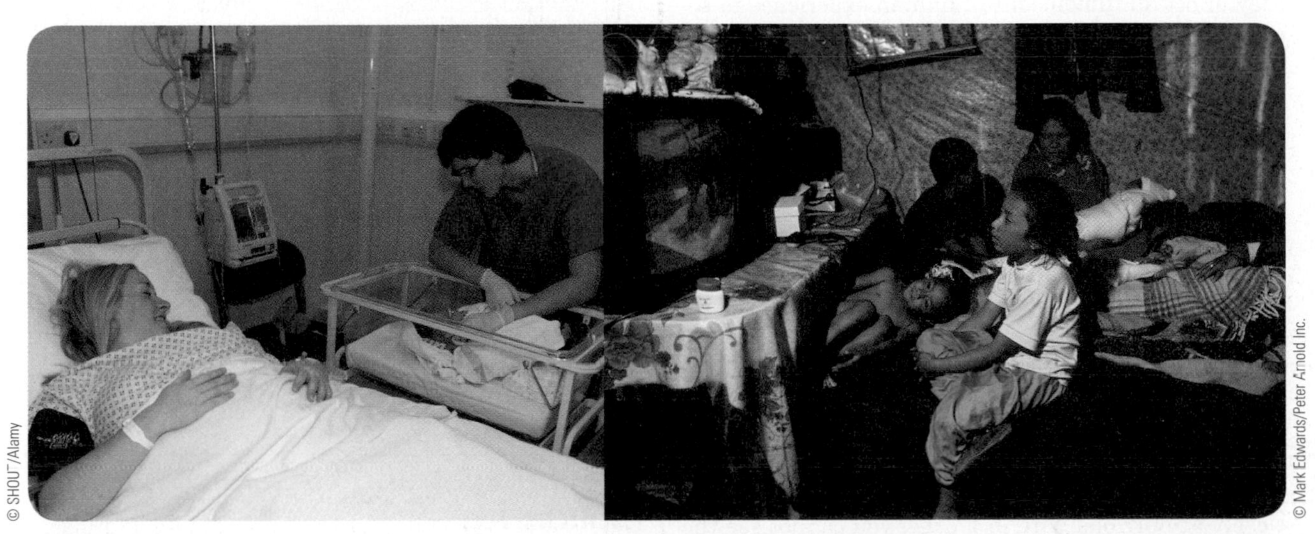

Birth, though a natural process, occurs within social institutions*—institutions that vary in different societies, depending on the* social organization *of society. Here you see the contrast in how birth is mainly defined as a medical event in the United States, contrasted with a health assistant attending a birth in rural Mexico.*

4. *The maintenance of stability and existence.* Certain institutions within a society (such as the government, the police force, and the military) contribute toward the stability and continuance of the society.
5. *Providing the members with an ultimate sense of purpose.* Societies accomplish this task by creating national anthems and encouraging patriotism in addition to providing basic values and moral codes through institutions such as the family, religion, and education.

In contrast to functional theory, conflict theory further notes that because conflict is inherent in most societies, the social institutions of society do not provide for all its members equally. Some members are provided for better than others, thus demonstrating that institutions affect people by granting more power to some social groups than to others. Using the example of the health care institution given above, some groups have considerably less power within the institution than others. Thus, nurses are generally subordinate to doctors and doctors to hospital administrators. And beyond these specific actors within the health care institutions, different social groups in society have more or less power within social institutions. Thus, racial and ethnic minorities in general have poorer access to health care than others; the poor have less access, as do those of lower social class status. (For more information, see Chapter 13 on health care.)

Social Structure

Sociologists use the term **social structure** to refer to the organized pattern of social relationships and social institutions that together compose society. Social structures are not immediately visible to the untrained observer; nevertheless, they are present, and they affect all dimensions of human experience in society. Social structural analysis is a way of looking at society in which the sociologist analyzes the patterns in social life that reflect and produce social behavior.

Social class distinctions are an example of a social structure. Class shapes the access that different groups have to the resources of society, and it shapes many interactions people have with each other. People may form cliques with those who share similar class standing, or they may identify with certain values associated with a given class. Class then forms a social structure—one that shapes and guides human behavior at all levels, no matter how overtly visible or invisible this structure is to someone at a given time.

The philosopher Marilyn Frye aptly describes the concept of social structure in her writing. Using the metaphor of a birdcage, she writes that if you look closely at only one wire in a cage, you cannot see the other wires. You might then wonder why the bird within does not fly away. Only when you step back and see the whole cage instead of a single wire do you understand why the bird does not escape. Frye writes:

> It is perfectly obvious that the bird is surrounded by a network of systematically related barriers, no one of which would be the least hindrance to its flight, but all of which, by their relations to each other, are as confining as the solid walls of a dungeon. It is now possible to grasp one reason why oppression can be hard to see and recognize: one can study the elements of an oppressive structure with great care and some good will without seeing or being able to understand that one is looking at a cage and that there are people there who are caged, whose motion and mobility are restricted, whose lives are shaped and reduced (Frye 1983: 4–5).

Just as a birdcage is a network of wires, society is a network of social structures, both micro and macro.

thinking sociologically

Using Marilyn Frye's analogy of the birdcage, think of a time when you believed your choices were constrained by *social structure*. When you applied to college, for example, could you go anywhere you wanted? What social structural conditions guided your ultimate selection of schools to attend?

What Holds Society Together?

What holds societies together? We have been asking this question throughout this chapter. This central question in sociology was first addressed by Emile Durkheim, the French sociologist writing in the late 1800s and early 1900s. He argued that people in society had a **collective consciousness,** defined as the body of beliefs common to a community or society that give people a sense of belonging and a feeling of moral obligation to its demands and values. According to Durkheim, collective consciousness gives groups social solidarity because members of a group feel they are part of one society.

Where does the collective consciousness come from? Durkheim argued that it stems from people's participation in common activities, such as work, family, education, and religion—in short, society's institutions.

Mechanical and Organic Solidarity

According to Durkheim, there are two types of social solidarity: mechanical and organic. **Mechanical solidarity** arises when individuals play similar roles within the society. Individuals in societies marked by mechanical solidarity share the same values and hold the same things sacred. This particular type of cohesiveness is weakened when a society becomes more complex. Contemporary examples of mechanical solidarity are rare because most societies of the world have been absorbed in the global trend for greater complexity and interrelatedness. Native American groups before European conquest were bound together by mechanical solidarity; indeed, many Native American groups are now trying to regain the mechanical solidarity on which their cultural heritage rests, but they are finding that the superimposition of White institutions on Native American life interferes with the adoption of traditional ways of thinking and being, which prevents mechanical solidarity from gaining its original strength.

In contrast, **organic solidarity** occurs when people play a great variety of roles, and unity is based on role differentiation, not similarity. The United States and other industrial societies are built on organic solidarity, and each is cohesive because of the differentiation within each. Roles are no longer necessarily similar, but they are necessarily interlinked—the performance of multiple roles is necessary for the execution of society's complex and integrated functions.

Durkheim described this state as the **division of labor,** defined as the relatedness of different tasks that develop in complex societies. The labor force within the contemporary U.S. economy, for example, is divided according to the kinds of work people do. Within any division of labor, tasks become distinct from one another, but they are still woven into a whole.

The division of labor is a central concept in sociology because it represents how the different pieces of society fit together. The division of labor in most contemporary societies is often marked by age, gender, race, and class. In other words, if you look at who does what in society, you will see that women and men tend to do different things; this is the gender division of labor. Similarly, old and young to some extent do different things; this is a division of labor by age. This is crosscut by the racial division of labor, the pattern whereby those in different racial–ethnic groups tend to do different work—or are often forced to do different work—in society. At the same time, the division of labor is also marked by class distinctions, with some groups providing work that is highly valued and rewarded and others doing work that is devalued and poorly rewarded. As you will see throughout this book, gender, race, and class intersect and overlap in the division of labor in society.

Durkheim's thinking about the origins of social cohesion can bring light to contemporary discussions over "family values." Some want to promote traditional family values as the moral standards of society. Is such a thing necessarily good, or even possible? The United States is an increasingly diverse society, and family life differs among different groups. It is unlikely that a single set of family values can be the basis for social solidarity.

Gemeinschaft and Gesellschaft

Different societies are held together by different forms of solidarity. Some societies are characterized by what sociologists call **gemeinschaft,** a German word that means "community"; others are characterized as **gesellschaft,** which literally means "society" (Tönnies 1963/1887). Each involves a type of solidarity or cohesiveness. Those societies that are gemeinschafts (communities) are characterized by a sense of "we" feeling, a moderate division of labor, strong personal ties, strong family relationships, and a sense of personal loyalty. The sense of solidarity between members of the *gemeinschaft* society arises from personal ties; small, relatively simple social institutions; and a collective sense of loyalty to the whole society. People tend to be well integrated into the whole, and social cohesion comes from deeply shared values and beliefs (often, sacred values). Social control need not be imposed externally because control comes from the internal sense of belonging that members share. You might think of a small community church as an example.

In contrast, in societies marked by gesellschaft, an increasing importance is placed on the secondary relationships people have—that is, less intimate and more instrumental relationships such as work roles instead of family or community roles. Gesellschaft is characterized by less prominence of personal ties, a somewhat diminished role of the nuclear family, and a lessened sense of personal loyalty to the total society. The solidarity and cohesion remain, and it can be very cohesive, but the cohesion comes from an elaborated division of labor (thus, *organic* solidarity), greater flexibility in social roles, and the instrumental ties that people have to one another.

Social solidarity under gesellschaft is weaker than in the gemeinschaft society, however. Gesellschaft is more likely than gemeinschaft to be torn by class conflict because class distinctions are less prominent, though still present, in the gemeinschaft. Racial–ethnic conflict is more likely within gesellschaft societies because the gemeinschaft tends to be ethically and racially very homogeneous; often it is characterized by only one racial or ethnic group. This means that conflict between gemeinschaft societies, such as ethnically based wars, can be very high, because both groups have a strong internal sense of group identity that may be intolerant of others (for example, Palestinians versus Israelis or Shiite versus Sunni Muslims in Iraq).

In sum, complexity and differentiation are what make the gesellschaft cohesive, whereas similarity and unity cohere the gemeinschaft society. In a single society, such as the United States, you can conceptualize the whole society as gesellschaft, with some internal groups marked by gemeinschaft. Our national motto seems to embody this idea: *E pluribus unum* (Unity within diversity), although clearly this idealistic motto has only been partly realized.

Types of Societies

In addition to comparing how different societies are bound together, sociologists are interested in how social organization evolves in different societies. Simple things such as the size of a society can also shape its social organization, as do the different roles that men and women engage in as they produce goods, care for the old and young, and pass on societal traditions. Societies also differ according to their resource base—whether they are predominantly agricultural or industrial, for example, and whether they are sparsely or densely populated.

Thousands of years ago, societies were small, sparsely populated, and technologically limited. In the competition for scarce resources, larger and more technologically advanced societies dominated smaller ones. Today, we have arrived at a global society with highly evolved degrees of social differentiation and inequality, notably along class, gender, racial, and ethnic lines (Nolan and Lenski 2005). Sociologists distinguish six types of societies based on the complexity of their social structure, the amount of overall cultural accumulation, and the level of their technology. They are *foraging, pastoral, horticultural, agricultural* (these four are called *preindustrial* societies), then *industrial* and *postindustrial* societies (see Table 4.1). Each type of society can still be found on Earth, though all but the most isolated societies are rapidly moving toward the industrial and postindustrial stages of development.

These different societies vary in the basis for their organization and the complexity of their division of labor. Some, such as foraging societies, are subsistence economies, where men and women hunt and gather food but accumulate very little. Others, such as pastoral societies and horticultural societies, develop a more elaborate division of labor as the social roles that are needed for raising livestock and farming become more numerous. With the development of agricultural societies, production becomes more large scale and strong patterns of social differentiation develop, sometimes taking the form of a caste system or even slavery.

The key driving force that distinguishes these different societies from each other is the development of technology. All societies use technology to help fill human needs, and the form of technology differs for the different types of society.

Preindustrial Societies

A **preindustrial society** is one that directly uses, modifies, and/or tills the land as a major means of survival. There are four kinds of preindustrial societies, listed here by degree of technological development: foraging (or hunting–gathering) societies, pastoral societies, horticultural societies, and agricultural societies (see Table 4.1).

In *foraging (hunting–gathering) societies,* the technology enables the hunting of animals and gathering of vegetation. The technology does not permit the refrigeration or processing of food, hence these individuals must search continuously for plants and

Different types of societies produce different kinds of social relationships. Some may involve more direct and personal relationships, whereas others produce more fragmented and impersonal relationships.

game. Because hunting and gathering are activities that require large amounts of land, most foraging societies are nomadic; that is, they constantly travel as they deplete the plant supply or follow the migrations of animals. The central institution is the family, which serves as the means of distributing food, training children, and protecting its members. There is usually role differentiation on the basis of gender, although the specific form of the gender division of labor varies in different societies. They occasionally wage war with other clans or similar societies, and spears and bows and arrows are the weapons used. Examples of foraging societies are the Aborigines of Australia and the Pygmies of Central Africa.

In *pastoral societies,* technology is based on the domestication of animals. Such societies tend to develop in desert areas that are too arid to provide rich vegetation. The pastoral society is nomadic, necessitated by the endless search for fresh grazing grounds for the herds of their domesticated animals. The animals are used as sources of hard work that enable the creation of a material surplus. Unlike a foraging society, this surplus frees some individuals from the tasks of hunting and gathering and allows them to create crafts, make pottery, cut hair, build tents, and apply tattoos. The surplus generates a more complex and differentiated social system with an elite or upper class and more role differentiation on the basis of gender. The nomadic Bedouins of Africa and the Middle East are pastoral societies.

In *horticultural societies,* hand tools are used to cultivate the land, such as the hoe and the digging stick. The individuals in horticultural societies practice ancestor worship and conceive of a deity or deities (God or gods) as a creator. This distinguishes them from foraging societies that generally employ the notion of numerous spirits to explain the unknowable. Horticultural societies recultivate the land each year and tend to establish relatively permanent settlements and villages. Role differentiation is extensive, resulting in different and interdependent occupational roles such as farmer, trader, and craftsperson. The Aztecs of Mexico and the Incas of Peru represent examples of horticultural societies.

The *agricultural society* is exemplified by the pre–Civil War American South. Such societies have a large and complex economic system that is based on large-scale farming. Such societies rely on technologies such as use of the wheel, and use of metals. Farms tend to be considerably larger than the

table 4.1 *Types of Societies*

		Economic Base	Social Organization	Examples
Preindustrial Societies	*Foraging Societies*	Economic sustenance dependent on hunting and foraging	Gender is important basis for social organization, although division of labor is not rigid; little accumulation of wealth	Pygmies of Central Africa
	Pastoral Societies	Nomadic societies, with substantial dependence on domesticated animals for economic production	Complex social system with an elite upper class and greater gender role differentiation than in foraging societies	Bedouins of Africa and Middle East
	Horticultural Societies	Society marked by relatively permanent settlement and production of domesticated crops	Accumulation of wealth and elaboration of the division of labor, with different occupational roles (farmers, traders, craftspeople, and so on)	Aztecs of Mexico; Inca empire of Peru
	Agricultural Societies	Livelihood dependent on elaborate and large-scale patterns of agriculture and increased use of technology in agricultural production	Caste system develops that differentiates the elite and agricultural laborers; may include system of slavery	American South, pre-Civil War
Industrial Societies		Economic system based on the development of elaborate machinery and a factory system; economy based on cash and wages	Highly differentiated labor force with a complex division of labor and large formal organizations	Nineteenth- and most of twentieth-century United States and Western Europe
Postindustrial Societies		Information-based societies in which technology plays a vital role in social organization	Education increasingly important to the division of labor	Contemporary United States, Japan, and others

cultivated land in horticultural societies. Large and permanent settlements characterize agricultural societies, which also exhibit dramatic social inequalities. A rigid caste system develops, separating the peasants, or slaves, from the controlling elite caste, which is then freed from manual work allowing time for art, literature, and philosophy, activities of which they can then claim the lower castes are incapable.

Industrial Societies

An *industrial society* is one that uses machines and other advanced technologies to produce and distribute goods and services. The Industrial Revolution began over 200 years ago when the steam engine was invented in England, delivering previously unattainable amounts of mechanical power for the performance of work. Steam engines powered locomotives, factories, and dynamos and transformed societies as the Industrial Revolution spread. The growth of science led to advances in farming techniques such as crop rotation, harvesting, and ginning cotton, as well as industrial-scale projects such as dams for generating hydroelectric power. Joining these advances were developments in medicine, new techniques to prolong and improve life, and the emergence of birth control to limit population growth.

Unlike agricultural societies, industrial societies rely on a highly differentiated labor force and the intensive use of capital and technology. Large formal organizations are common. The task of holding society together, falling on institutions such as religion in preindustrial societies, now falls more on the institutions that have a high division of labor, such as the economy and work, government, politics, and large bureaucracies.

Within industrial societies, the forms of gender inequality that we see in contemporary U.S. society tend to develop. With the advent of industrialization, societies move to a cash-based economy, with labor performed in factories and mills paid on a wage basis and household labor remaining unpaid. This introduced what is known as the *family–wage economy,* in which families become dependent on wages to support themselves, but work within the family (housework, child care, and other forms of household work) is unpaid and therefore increasingly devalued (Tilly and Scott 1978). In addition, even though women (and young children) worked in factories and mills from the first inception of industrialization, the family–wage economy is based on the idea that men are the primary breadwinners. A system of inequality in men's and women's wages was introduced—an economic system that today continues to produce a wage gap between men and women.

Industrial societies tend to be highly productive economically, with a large working class of industrial laborers. People become increasingly urbanized as they move from farmlands to urban centers or other areas where factories are located. Immigration is common in industrial societies, particularly because industries are forming where there is a high demand for more, cheap labor.

Industrialization has brought many benefits to U.S. society—a highly productive and efficient economic system, expansion of international markets, extraordinary availability of consumer products, and for many, a good working wage. Industrialization has, at the same time, also produced some of the most serious social problems that our nation faces: industrial pollution, an overdependence on consumer goods, wage inequality and job dislocation for millions, and problems of crime and crowding in urban areas. The portrait of population density depicted in Map 4.1 illustrates where crowding is most extreme. Understanding the process of industrialization and its accompanying process of urbanization is a major avenue for sociological research and is explored further in Chapter 15.

Postindustrial Societies

In the contemporary era, a new type of society is emerging. Whereas most twentieth-century societies can be characterized in terms of their generation of material goods, **postindustrial society** depends economically on the production and distribution of services, information, and knowledge (Bell 1973). Postindustrial societies are information-based societies in which technology plays a vital role in the social organization. The United States is fast becoming a postindustrial society, and Japan may be even further along. Many of the workers provide services such as administration, education, legal services, scientific research, and banking, or they engage in the development, management, and distribution of information, such as in the areas of computer use and design. Central to the economy of the postindustrial society are the highly advanced technologies of computers, robotics, genetic engineering, and laser technology. Multinational corporations globally link the economies of postindustrial societies.

The transition to a postindustrial society has a strong influence on the character of social institutions. Educational institutions acquire paramount importance in the postindustrial society, and science takes an especially prominent place. For some, the transition to a postindustrial society means more discretionary income for leisure activities—tourism, entertainment, and relaxation industries (spas, massage centers, and exercise) become more prominent—at least for people in certain classes. For others, the transition to postindustrialism can mean permanent joblessness or the need to hold down more than one job simply to make ends meet. Workers without highly technical skills may not fit in such a society, and millions may find themselves stuck in low-paid, unskilled work.

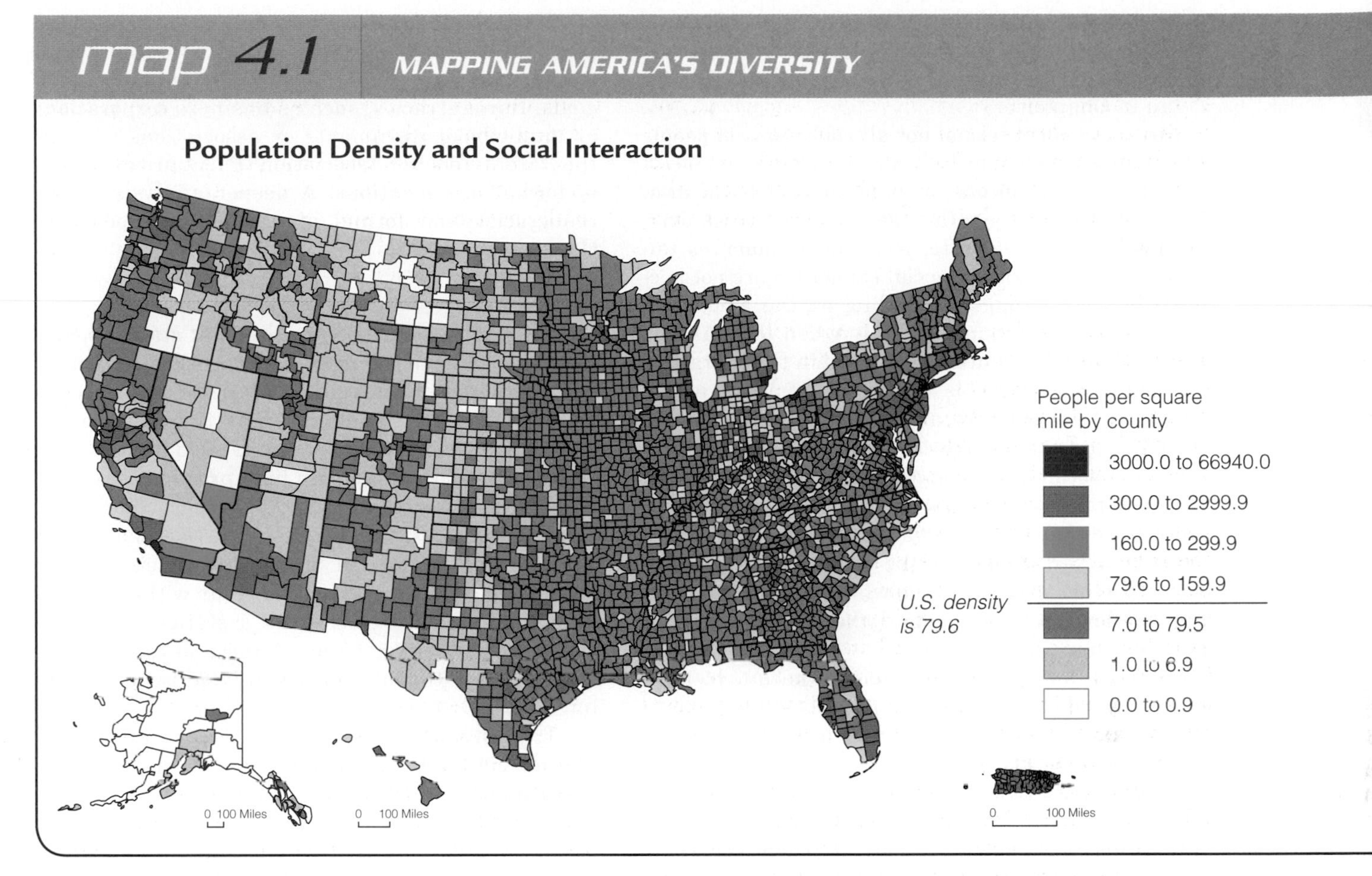

As this map shows, population density (measured as the number of people per square mile) varies enormously in different regions and areas of the country. In what ways do you think the density of a given area might affect people's social interaction?

Data from: From the U.S. Census Bureau. 2006. **http://www.census.gov/population/www/censusdata/2000maps.html**

The United States is suspended between the industrial and postindustrial phases. Manufacturing jobs are still a major segment of the labor force, although they are in decline because most workers are employed in the service sector of the economy that involves the delivery of services and information, not the actual production of material goods. Postindustrial societies are also increasingly dependent on a global economy because goods tend to be produced in economically dependent areas of the world for consumption in the wealthier nations. As a result, the world, not just individual societies, becomes characterized by poverty and inequality resulting from the social structure that postindustrialism produces.

Social Interaction and Society

You can see by now that society exists above and beyond individuals. Also, different societies are marked by different forms of *social organization*. Although societies differ, emerge, and change, they are also highly predictable. Your society shapes virtually every aspect of your life from the structure of its social institutions to the more immediate ways that you interact with people. It is to that level—the microlevel of society, that we now turn.

Groups

At the microlevel society is made up of many different social groups. At any given moment, each of us is a member of many groups simultaneously, and we are subject to their influence: family, friendship groups, athletic teams, work groups, racial and ethnic groups, and so on. Groups impinge on every aspect of our lives and are a major determinant of our attitudes and values regarding everything from personal issues such as sexual attitudes and family values to major social issues such as the death penalty and physician-assisted suicide.

To sociologists a **group** is a collection of individuals who

- interact and communicate with each other;
- share goals and norms; and,
- have a subjective awareness of themselves as "we," that is, as a distinct social unit.

To be a group, the social unit in question must possess all three of these characteristics. We will examine the nature and behavior of groups in greater detail in Chapter 5.

In sociological terms, not all collections of people are groups. People may be lumped together into *social categories* based on one or more shared characteristics, such as teenagers (an age category), truck drivers (an occupational category), and billionaires (an economic category). But social categories are not necessarily social groups, depending on the amount of "we" feeling the group has. Only when there is this sense of common identity, as defined in the characteristics of groups above, is a collection of people an actual group. For example, all people nationwide watching TV programs at 8 o'clock Wednesday evening form a distinct social unit, an *audience*. But they are not a group, because they do not interact with one another, nor do they possess an awareness of themselves as "we." However, if many of the same viewers were to come together in a TV studio where they could interact and develop a "we" feeling, then they would constitute a group.

We now know that groups do not need to be face-to-face in order to constitute a group. Online communities, for example, are people who interact with each other regularly, share a common identity, and think of themselves as being a distinct social unit. On the Internet community Facebook, for example, you may have a group of "friends"—some of whom you know personally and others of whom you only know online. But these *friends*, as they are known on Facebook, make up a social group that might interact on a regular, indeed, daily basis—possibly even across great distances.

Groups also need not be small or "close-up" and personal. *Formal organizations* are highly structured social groupings that form to pursue a set of goals. Bureaucracies, such as business corporations or municipal governments or associations such as the Parent-Teacher Association (PTA), are examples of formal organizations. A deeper analysis of bureaucracies and formal organizations appears in Chapter 5.

Status

Within groups, people occupy different statuses. **Status** is an established position in a social structure that carries with it a degree of prestige (that is, social value). A status is a rank in society. For example, the position "Vice President of the United States" is a status, one that carries very high prestige as well as a complex set of expectations. "High school teacher" is another status; it carries less prestige than "Vice President of the United States," but more prestige than, say, "cabdriver." Statuses occur within institutions. "High school teacher" is a status within the education institution. Other statuses in the same institution are "student," "principal," and "school superintendent."

Typically, a person occupies many statuses simultaneously. The combination of statuses composes a **status set,** which is the complete set of statuses occupied by a person at a given time (Merton 1968). A person may occupy different statuses in different institutions. Simultaneously, a person may be a daughter (in the family institution), bank president (in the economic institution), voter (in the political institution), church member (in the religious institution), and treasurer of the PTA (in the education

Social groups are organized around different kinds of relationships, but involve a "we" feeling.

institution). Each status may be associated with a different level of prestige.

Sometimes the multiple statuses of an individual conflict with one another. **Status inconsistency** exists where the statuses occupied by a person bring with them significantly different amounts of prestige and thus differing expectations. For example, someone trained as a lawyer, but working as a cabdriver, experiences status inconsistency. Some recent immigrants from Vietnam and Korea have experienced status inconsistency. Many refugees who had been in high-status occupations in their home country, such as teachers, doctors, and lawyers, could find work in the United States only as grocers or nail technicians—jobs of relatively lower status than the jobs they left behind. A relatively large body of research in sociology has demonstrated that status inconsistency (*in addition to* low status) can lead to stress and depression (Taylor et al. 2006; Min 1990).

Achieved statuses are those attained by virtue of individual effort. Most occupational statuses—police officer, pharmacist, or boat builder—are achieved statuses. In contrast, **ascribed statuses** are those occupied from the moment a person is born. Your biological sex is an ascribed status. Yet, even ascribed statuses are not exempt from the process of social construction. For most individuals, race is an ascribed status fixed at birth, although an individual with one light-skinned African American parent and one White parent may appear to be White and may go through life as a White person. This is called *passing*, although this term is used less often now than it was several years ago. Ascribed status may not be clearly defined, as for individuals who are *biracial* or *multiracial*. Finally, ascribed statuses can arise through means beyond an individual's control, such as severe disability or chronic illness.

Some seemingly ascribed statuses, such as gender, can become achieved statuses. Gender, typically thought of as fixed at birth, is a social construct. You can be born female or male, but becoming a woman or a man is the result of social behaviors associated with your ascribed status. In other words, gender is also achieved. People who cross-dress, have a sex change, or develop some characteristics associated with the other sex are good examples of how gender is achieved, but you do not have to see these exceptional behaviors to observe that. People "do" gender in everyday life. They put on appearances and behaviors that are associated with their presumed gender (West and Zimmerman 1987; West and Fenstermaker 1995; Andersen 2006). If you doubt this, ask yourself what you did today to "achieve" your gender status. Did you dress a certain way? Wear "manly" cologne or deodorant? Splash on a "feminine" fragrance? These behaviors—all performed at the microlevel—reflect the macrolevel of your gender status.

Debunking Society's Myths

Myth: Gender is an *ascribed status* where one's gender identity is established at birth.

Sociological perspective: Although one's biological sex identity is an ascribed status, gender is a social construct and thus is also an *achieved status*—that is, accomplished through routine, everyday behavior, including patterns of dress, speech, touch, and other social behaviors (Andersen 2006).

The line between achieved and ascribed status can be hard to draw. Social class, for example, is determined by occupation, education, and annual income—all of which are achieved statuses—yet one's job, education, and income are known to correlate strongly with the social class of one's parents. Hence, one's social class status is at least partly—though not perfectly—determined at birth. It is an achieved status that includes an inseparable component of ascribed status as well.

Although people occupy many statuses at one time, it is usually the case that one status is dominant, called the **master status,** overriding all other features of the person's identity. The master status may be imposed by others, or a person may define his or her own master status. A woman judge, for example, may carry the master status "woman" in the eyes of many. She is seen not just as a judge, but as a woman judge, thus making gender a master status (Webster and Hysom 1998). A master status can completely supplant all other statuses in someone's status set. Being in a wheelchair is another example of a master status. People may see this, at least at first, as the most important, or salient, part of identity, ignoring other statuses that define someone as a person.

thinking sociologically

Make a list of terms that describe who you are. Which of these are *ascribed statuses* and which are *achieved statuses*? What do you think your *master status* is in the eyes of others? Does one's *master status* depend on who is defining you? What does this tell you about the significance of social judgments in determining who you are?

Roles

A **role** is the behavior others expect from a person associated with a particular status. Statuses are occupied; roles are acted or "played." The status of police officer carries with it many expectations; this is the role of police officer. Police officers are expected to uphold the law, pursue suspected criminals, assist victims of crimes, and so on. Usually, people behave

© Lon C. Diehl/PhotoEdit

In role modeling, a person imitates the behavior of an admired other.

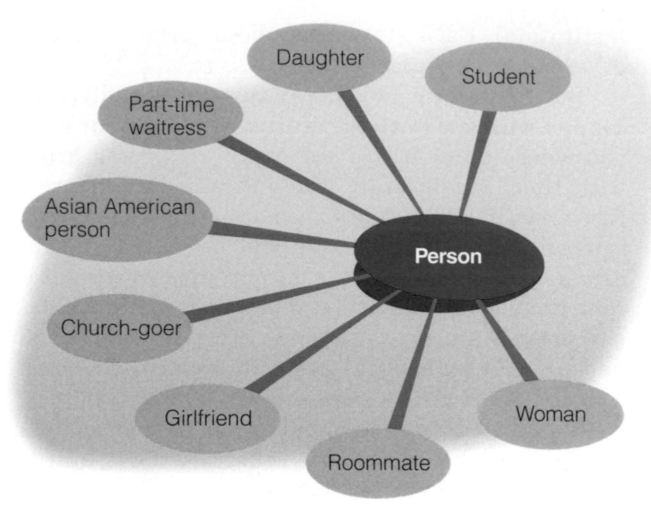

FIGURE 4.1 *Roles in a College Student's Role Set*

Identify the different roles that you occupy and draw a similar diagram of your own role set. Then identify which roles are consistent with each other and which might produce *role conflict* and *role strain*. Are there institutional reasons why these conflicts and strains occur in your case?

in their roles as others expect them to, but not always. When a police officer commits a crime, such as physically brutalizing someone, he or she has violated the role expectations. Role expectations may vary according to the role of the observer—whether the person observing the police officer is a member of a minority group, for example.

As we saw in Chapter 3, social learning theory predicts that we learn attitudes and behaviors in response to the positive reinforcement and encouragement received from those around us. This is important in the formation of our own identity in society. "I am Linda, the skater" or "I am John, the guitarist." These identities are often obtained through **role modeling,** a process by which we imitate the behavior of another person we admire who is in a particular role. A ten-year-old girl or boy who greatly admires the teenage expert skateboarder next door will attempt, through role modeling, to closely imitate the tricks that neighbor performs on the skateboard. As a result, the formation of the child's self-identity is significantly influenced.

A person may occupy several statuses and roles at one time. A person's **role set** includes all the roles occupied by the person at a given time. Thus, a person may be not only Linda the skater but also Linda the student, the daughter, and the lover. Roles may clash with each other, a situation called **role conflict,** wherein two or more roles are associated with contradictory expectations. Notice that in Figure 4.1 some of the roles diagrammed for this college student may conflict with others. Can you speculate about which might and which might not?

In U.S. society, some of the most common forms of role conflict arise from the dual responsibilities of job and family. The parental role demands extensive time and commitment and so does the job role. Time given to one role is time taken away from the other. Although the norms pertaining to working women and men are rapidly changing, it is still true that women are more often expected to uphold traditional role expectations associated with their gender role and are more likely held responsible for minding the family when job and family conflict. The sociologist Arlie Hochschild captured the predicament of today's women when she described the "second shift": a working mother spends time and energy all day on the job, only to come home to the "second shift" of family and home responsibilities. These are sometimes delegated to the a husband or boyfriend, who encounters less well-formed role expectations that he will take on those responsibilities and is therefore more likely to leave the jobs undone (Hochschild 1997).

Hochschild has found that some companies have instituted "family friendly" policies, designed to reduce the conflicts generated by the "second shift." Ironically, however, in her study she found that few workers take advantage of programs such as more flexible hours, paid maternity leave, and job sharing—except for the on-site child care that actually allowed parents to work more!

Hochschild's studies point to the conflict between two social roles: family roles and work roles. Her research is also illustrative of a different sociological concept: **role strain,** a condition wherein a single role brings conflicting expectations. Different from

© Denis Boissavy/Getty Images/Taxi

Changes in the roles for women who are mothers can create role strain.

role conflict, which involves tensions *between* two roles, role strain involves conflicts within a single role. In Hochschild's study, the work role has not only the expectations traditionally associated with work, but also the expectation that one "love" one's work and be as devoted to it as to one's family. The result is role strain. The role of student often involves role strain. For example, students are expected to be independent thinkers, yet they feel—quite correctly—that they are often required to simply repeat on an exam what a professor tells them. The tension between the two competing expectations is an example of role strain.

Everyday Social Interaction

You can also see the influence of society in everyday behavior, including such basics as how you talk, patterns of touch, and who you are attracted to. Although you might think of such things as "just coming naturally," they are deeply patterned by society. The cultural context of social interaction really matters in our understanding of what given behaviors mean. An action is defined as positive or negative by the cultural context because social behavior is that to which people give meaning. An action that is positive in one culture can be negative in another. For example, shaking the right hand in greeting is a positive action in the United States, but the same action in East India or certain Arab countries might be an insult. Social and cultural context matter. A kiss on the lips is a positive act in most cultures, yet if you were kissed on the lips by a stranger, you would probably consider it a negative act, perhaps even repulsive.

Verbal and Nonverbal Communication. We saw in the culture chapter (Chapter 2) how patterns of social interaction are embedded in the language we use, and language is deeply influenced by culture and society. Furthermore, communication is not just what you say, but also how you say it and to whom. You can see the influence of society on *how* people speak, especially in different contexts. Under some circumstances, a pause in speaking may communicate emphasis; for others, it may indicate uncertainty. Cultural differences across society make this obvious. Thus, during interactions between Japanese businessmen, long periods of silence often occur. Unlike U.S. citizens, who are experts in "small talk" and who try at all costs to avoid periods of silence in conversation, Japanese people do not need to talk all the time and regard periods of silence as desirable opportunities for collecting their thoughts (Worchel et al. 2000; Fukuda 1994). American businesspeople in their first meetings with Japanese executives often think, erroneously, that these silent interludes mean the Japanese are responding negatively to a presentation. Even though some find the Japanese mode of conversation highly uncomfortable, getting used to it is a key tool in successful negotiations. The fate of a deal may depend on a glance, an exhalation, or a smile (Mizutami 1990).

Nonverbal communication is also a form of social interaction and can be seen in various social patterns. A surprisingly large portion of our everyday communication with others is nonverbal, although we are generally only conscious of a small fraction of the nonverbal "conversations" in which we take part. Consider all the nonverbal signals exchanged in a casual chat: body position, head nods, eye contact, facial expressions, touching, and so on. Studies of nonverbal communication, like those of verbal communication, show that it is much influenced by social forces, including the relationships between diverse groups of people. The meanings of nonverbal communications depend heavily on race, ethnicity, social class, and gender, as we shall see.

For example, patterns of touch are strongly influenced by gender. Parents vary their touching behavior depending on whether the child is a boy or a girl. Boys tend to be touched more roughly; girls, more tenderly and protectively. Such patterns continue into adulthood, where women touch each other more often in everyday conversation than do men. Women are on the average more likely to touch and hug as an expression of emotional support, whereas men touch and hug more often to assert power or to express sexual interest (Worchel et al. 2000; Wood 1994). Clearly, there are also instances where women touch to express sexual interest and/or dominance, but research shows that,

© Michael Newman/PhotoEdit

Patterns of touch reflect differences in the power that is part of many social relationships

SEE FOR YOURSELF

RIDING IN ELEVATORS

1. Try a simple experiment. Ride in an elevator and closely observe the behavior of everyone in the elevator with you. Write down in a notebook such things as how far away people stand from each other. What do they look at? Do they converse with strangers or the people they are with? If so, what do they talk about?

2. Now return to the same elevator and do something that breaks the usual norms of elevator behavior. Write a second part to your essay explaining what you did. How did people react? How did you feel?

Now write a short paragraph discussing how social norms are maintained through informal norms of social control. In doing so, explain the perspective of *ethnomethodology*.

in general, for women touch is a supporting activity. For men, touch is often a dominance-asserting activity, except in athletic contexts where hugging and patting among men is a supportive activity (Worchel et al. 2000; Wood 1994; Tannen 1990).

In observing patterns of touch, you can see where social status influences the meaning of nonverbal behaviors. Professors, male or female, may pat a man or woman student on the back as a gesture of approval; students will rarely do this to a professor. Male professors touch students more often than do female professors, showing the additional effect of gender. Because such patterns of touching reflect power relationships between women and men, they can also be offensive and may even involve sexual harassment (see Chapters 10 and 14).

You can also see the social meaning of interaction by observing how people use personal space. **Proxemic communication** refers to the amount of space between interacting individuals. Although people are generally unaware of how they use personal space, usually the more friendly people feel toward each other, the closer they will stand. In casual conversation, friends stand closer to each other than strangers. People who are sexually attracted to each other stand especially close, whether the sexual attraction is gay, lesbian, or heterosexual (Taylor et al. 2005). According to anthropologist E.T. Hall (1966; Hall and Hall 1987), we all carry around us a *proxemic bubble* that represents our personal, three-dimensional space. When people we do not know enter our proxemic bubble, we feel threatened and may take evasive action. Friends stand close; enemies tend to avoid interaction and keep far apart. According to Hall's theory, we attempt to exclude from our private space those whom we do not know or do not like even though we may not be fully aware that we are doing so.

The proxemic bubbles of different groups have different sizes. Hispanic people tend to stand much closer to each other than do White, middle-class Americans; their proxemic bubble is, on average, smaller. Similarly, African Americans also tend to stand close to each other while conversing. Interaction distance is quite large between White, middle-class, British males—their average interaction distances can be as much as several feet.

Proxemic interactions also differ between men and women (Taylor et al. 2006; Romain 1999; Tannen 1990). Women of the same race and culture tend to stand closer to each other in casual conversation than do men of the same race and culture. When a Middle Eastern man (who has a relatively small proxemic bubble) engages in conversation with a White, middle-class, U.S. man (who has a larger proxemic bubble), the Middle Eastern man tends to move toward the White American, who tends to back away. You can observe the negotiations of proxemic space at cocktail parties or any other setting that involves casual social interaction.

In a society as diverse as the United States, understanding how diversity shapes social interaction is an essential part of understanding human behavior (Taylor et al. 2006; Gilbert et al. 1998). Ignorance of the meanings that gestures have in a society can get you in trouble. For example, Mexicans and Mexican Americans may display the right hand held up, palm inward, all fingers extended, as an obscene gesture meaning "screw you many times over." This provocative gesture has no meaning at all in Anglo (White) society. Likewise, people who grow up in urban environments learn to avoid eye contact on the streets. Staring at someone for only two or three seconds can be interpreted as a hostile act, if done man to man (Anderson 1999, 1990). If a woman maintains mutual eye contact with a male stranger for more than two or three seconds, she may be assumed by the man to be sexually interested in him. In contrast, during sustained conversation with acquaintances, women maintain mutual eye contact longer than do men (Romain 1999; Gilbert et al. 1998; Wood 1994).

Interpersonal Attraction

We have already asked, "What holds society together?" This was asked at the macroscopic level—that is, the level of society. But what holds relationships together—or, for that matter, makes them fall apart? You will not be surprised to learn that formation of relationships has a strong social structural component—that is, it is patterned by social forces and can to a great extent be predicted.

Humans have a powerful desire to be with other human beings; in other words, they have a strong need for *affiliation*. We tend to spend about 75 percent of our time with other people when doing all sorts of activities—eating, watching television, studying, doing hobbies, working, and so on (Cassidy and Shaver 1999). People who lack all forms of human contact are very rare in the general population, and their isolation is usually rooted in psychotic or schizophrenic disorders. Extreme social isolation at an early age causes severe disruption of mental, emotional, and language development, as we saw earlier in Chapter 3.

The affiliation tendency has been likened to *imprinting*, a phenomenon seen in newborn or newly hatched animals who attach themselves to the first living creature they encounter, even if it is of another species (Lorenz 1966). Studies of ducks and squirrels show that once the young animal attaches itself to a human experimenter, the process is irreversible. The young animal prefers the company of the human to the company of its own species. A degree of imprinting may be discernible in human infant attachment, but researchers note that in infants the process is more complex, more changeable, and more influenced by social factors (Brown 1986).

Somewhat similar to affiliation is interpersonal attraction, a nonspecific positive response toward another person. Attraction occurs in ordinary day-to-day interaction and varies from mild attraction (such as thinking your grocer is a nice person) all the way to deep feelings of love. According to one view, attractions fall on a continuum ranging from hate to strong dislike to mild dislike to mild liking to strong liking to love. Another view is that attraction and love are two different continua, able to exist separately. In this view, you can actually like someone a whole lot, but not be in love. Conversely, you can feel passionate love for someone, including strong sexual feelings and

Nina Leen/Time Life Pictures/Getty Images

Konrad Lorenz, the animal behaviorist, shows that adult ducks that have imprinted on him the moment they were hatched will follow him anywhere, as though he were their mother duck.

intense emotion, yet not really "like" the person. Have you ever been in love with someone you did not particularly like?

Can attraction be scientifically predicted? Can you identify with whom you are most likely to fall in love? The surprising answer to these questions is a loud, although somewhat qualified, yes. Most of us have been raised to believe that love is impossible to measure and certainly impossible to predict scientifically. We think of love, especially romantic love, as quick and mysterious—a lightning bolt. Couples report falling in love at first sight, thinking that they were "meant to be together" (McCollum 2002). Countless novels and stories support this view, but extensive research in sociology and social psychology suggests otherwise: In a probabilistic sense, love can be predicted beyond the level of pure chance. Let us take a look at some of these intriguing findings.

A strong determinant of your attraction to others is simply whether you live near them, work next to them, or have frequent contact with them. You are more likely to form friendships with people from your own city than with people from a thousand miles away. One classic study even showed that you are more likely to be attracted to someone on your floor, your residence hall, or your apartment building than to someone even two floors down or two streets over (Festinger et al. 1950). Such is the effect of proximity in the formation of human friendships.

Now, though the general principle still holds, many people form relationships without being in close proximity, such as in online dating. Studies of Internet dating show that, even in this cyberworld, social norms still apply. Studies of Internet dating

© Alex Mares-Manton/Asia Images/Jupiter Images

Romantic love is idealized in this society as something that "just happens," but research shows that interpersonal attraction follows rather predictable patterns.

find, for example, that unlike other dating behavior, on the Internet there is pressure to disclose more in a shorter period of time (Lawson and Leck 2006).

Our attraction to another person is also greatly affected simply by how frequently we see that person or even his or her photograph. When watching a movie, have you ever noticed that the central character seems more attractive at the end of the movie than at the beginning? This is particularly true if you already find the person very attractive when the movie begins. Have you ever noticed that the fabulous-looking person sitting next to you in class looks better every day? You may be experiencing *mere exposure effect*—the more you see someone in person or in a photograph, the more you like him or her. In studies where people are repeatedly shown photographs of the same face, the more often a person sees a particular face, the more he or she likes that person (Moreland and Beach 1992; Zajonc 1968). There are two qualifications to the effect. First, overexposure can result when a photograph is seen too often. The viewer becomes saturated and ceases to like the pictured person more with each exposure. Second, the initial response of the viewer can determine how much liking will increase. If someone starts out liking a particular person, seeing that person frequently will increase the liking for that person; however, if one starts out disliking the pictured person, the amount of dislike tends to remain about the same, regardless of how often one sees the person (Taylor et al. 2006).

We hear that "beauty is only skin deep." Apparently that is deep enough. To a surprisingly large degree, the attractions we feel toward people of either gender are based on our perception of their physical attractiveness. A vast amount of research over the years has consistently shown the importance of attractiveness in human interactions: adults react more leniently to the bad behavior of an attractive child than to the same behavior of an unattractive child (Berscheid and Reis 1998; Dion 1972). Teachers evaluate cute children of either gender as "smarter" than unattractive children with identical academic records (Worchel et al. 2000; Clifford and Walster 1973). In studies of mock jury trials, attractive defendants, male or female, receive lighter sentences on average than unattractive defendants convicted of the same crime (Gilbert et al. 1998; Sigall and Ostrove 1975).

Of course, standards of attractiveness vary between cultures and between subcultures within the same society. What is highly attractive in one culture may be repulsive in another. In the United States, there is a maxim that you can never be too thin—a major cause of eating disorders such as *anorexia* and *bulimia*, especially among White women (Hesse-Biber 2007). The maxim is oppressive for women in U.S. society, yet it is clearly highly culturally relative, even within U.S. culture. Among certain African Americans, chubbiness in women is considered attractive. Such women are called "healthy" and "phatt" (not fat), which means the same as "stacked" or curvaceous. Similar cultural norms often apply in certain U.S. Hispanic populations. The skinny woman is not considered attractive. Nonetheless, studies show that anorexia and bulimia are now increasing among women of color, showing how cultural norms can change—even though Black women, in general, are more satisfied with their body image than White women (Lovejoy 2001; Fitzgibbon and Stolley 2000).

Studies of dating patterns among college students show that the more attractive one is, the more likely one will be asked on a date. This applies to gay and lesbian dating as well as to heterosexual dating (Cohen and Tannenbaum 2001; Berscheid and Reiss 1998; Speed and Gangestad 1997). However, one very important exception can be added to this finding: Physical attractiveness predicts only the early stages of a relationship. When one measures relationships that last a while, other factors come into play, principally religion, political attitudes, social class background, educational aspirations, and race. Perceived physical attractiveness may predict who is attracted to whom initially, but other variables are better predictors of how long a relationship will last (Berscheid and Reis 1998).

So, do "opposites attract"? Not according to the research. We have all heard that people are attracted to their "opposite" in personality, social status, background, and other characteristics. Many of us grow up believing this to be true. However, if the research tells us one thing about interpersonal attraction, it is that with few exceptions we are attracted to people who are similar or even identical to us in socioeconomic status, race, ethnicity, religion, perceived personality traits, and general attitudes and opinions (Taylor et al. 2005; Brehm et al. 2002)."Dominant" people tend to be attracted to other dominant people, not to "submissive" people. Couples tend to have similar opinions about political issues of great importance to them, such as attitudes about abortion, crime, animal rights, and urban violence. Overall, couples tend to exhibit strong cultural or subcultural similarity, not difference.

There are exceptions, of course. We sometimes fall in love with the *exotic*—the culturally or socially different. Novels and movies return endlessly to the story of the young, working class woman who falls in love with a rich, older man, but such a pairing is by far the exception and not the rule. When it comes to long-term relationships, including both friends and lovers (whether heterosexual, gay, or lesbian), humans vastly prefer a great degree of similarity even though, if asked, they might deny it. In fact, the less similar a heterosexual relationship is with respect to race, social class, age, and educational aspirations (how far in school the person wants to go), the quicker the relationship is likely to break up (Silverthorne and Quinsey 2000; Worchel et al. 2000; Berscheid and Reis 1998; Stover and Hope 1993; Hill et al. 1976).

Debunking Society's Myths

Myth: Love is purely an emotional experience that you cannot predict or control.

Sociological perspective: Whom you fall in love with can be predicted beyond chance by such factors as proximity, how often you see the person, how attractive you perceive the person to be, and whether you are similar (not different) to her or him in social class, race–ethnicity, religion, age, educational aspirations, and general attitudes, including political attitudes and beliefs.

Most young romantic relationships, regrettably, come to an end. On campus, relationships tend to break up most often during gaps in the school calendar, such as winter and spring break. Summers are especially brutal on relationships formed during the academic year. Breakups are seldom mutual. Almost always, only one member of the pair wants to break off the relationship, whereas the other wants to keep it going. The sad truth means that the next time someone tells you that their breakup last week was "mutual," you know they are probably lying or deceiving themselves (Taylor et al. 2006).

Theories About Analyzing Social Interaction

Groups, statuses, and roles form a web of social interaction. The interaction people have with one another is a basic element of society. Sociologists have developed different ways of understanding social interaction. Functional theory, discussed in Chapter 1, is one such concept. Here we detail four others: the social construction of reality, ethnomethodology, impression management, and social exchange. The first three theories come directly from the symbolic interaction perspective.

The Social Construction of Reality

What holds society together? This is a basic question for sociologists, one that, as we have seen, has long guided sociological thinking. Sociologists note that society cannot hold together without something that is shared—a shared social reality.

Some sociological theorists have argued convincingly that there is little actual reality beyond that produced by the process of social interaction itself. This is the principle of *the social construction of reality*, the idea that our perception of what is real is determined by the subjective meaning that we attribute to an experience, a principle central to symbolic interaction theory (Berger and Luckmann 1967; Blumer 1969). Hence, there is no objective "reality" in itself. Things do not have their own intrinsic meaning. We subjectively impose meaning on things.

Children do this routinely—impose inherent meaning on things. Upon seeing a marble roll off a table, the child attributes causation (meaning) to the marble: The marble rolled off the table "because it wanted to." Such perceptions carry into adulthood: The man walking down the street who accidentally walks smack into a telephone pole at first thought glares at the pole, as though the pole somehow caused the accident! He inadvertently attributes causation and meaning to an inanimate object—the telephone pole (Taylor et al. 2006).

Considerable evidence exists that people do just that; they force meaning on something when doing so allows them to see or perceive what they want to perceive—even if that perception seems to someone else to be contrary to actual fact. They then come to believe that what they perceived is indeed "fact." A classic and convincing study of this is Hastorf and Cantril's (1954) study of Princeton and Dartmouth students who watched a film of a game of basketball between the two schools. Both sets of students watched the same film. The students were instructed to watch carefully for rule infractions by each team. The results were that the Princeton students reported twice as many rule infractions involving the Dartmouth team than the Dartmouth students saw. The Dartmouth students saw about twice as many rule infractions by Princeton than the Princeton students saw! Remember that they all saw exactly the same game—the same "facts." We see the "facts" we want to see, as a result of the social construction of reality.

As we saw in Chapter 1, our perceptions of reality are determined by what is called the *definition of the situation:* We observe the context in which we find ourselves and then adjust our attitudes and perceptions accordingly. Sociological theorist W.I. Thomas embodies this idea in his well-known dictum that *situations defined as real are real in their consequences* (Thomas 1966/1931). The Princeton and Dartmouth students saw different "realities" depending on what college they were attending, and the consequences (the perceived rule infractions) were very real to them.

The definition of the situation is a principle that can also affect a "factual" event such as whether an emergency room patient is perceived to be dead by the doctors. In his research in the emergency room of a hospital, Sudnow (1967) found that patients who arrived at the emergency room with no discernible heartbeat or breathing were treated differently by the attending physician depending on the patient's age. A person in his or her early twenties or younger was not immediately pronounced "dead on arrival" (DOA). Instead, the physicians spent a lot of time listening for and testing for a heartbeat, stimulating the heart, examining the patient's eyes, giving oxygen, and administering other stimulation to revive the patient. If the doctor obtained no lifelike responses, the patient was pronounced dead. Older patients, however, were on the average less likely to receive such extensive procedures. The older person was examined less thoroughly and often was pronounced dead on the spot with only a stethoscopic examination of the heart. In such instances, how the physicians defined the situation—how they socially constructed reality—was indeed real in its consequence for the patient!

Understanding the social construction of reality helps one see many aspects of society in a new light. Race and gender are significant influences on social experience because people believe them to be so. Indeed, society is constructed based on certain assumptions about the significance of race and gender. These assumptions have guided the formation of social institutions, including what work people do, how families are organized, and how power is exercised.

Ethnomethodology

Our interactions are guided by rules that we follow. These rules are the *norms* of social interaction. Again, what holds society together? Society cannot hold together without norms, but what rules do we follow? How do we know what these rules or norms are? An approach in sociology called *ethnomethodology* is a clever technique for finding out.

Ethnomethodology (Garfinkel 1967), after *ethno* for "people" and *methodology* for "mode of study," is a clever technique for studying human interaction by deliberately disrupting social norms and observing how individuals attempt to restore normalcy. The idea is that to study such norms, one must first break them, because the subsequent behavior of the people involved will reveal just what the norms were in the first place.

Ethnomethodology is based on the premise that human interaction takes place within a consensus and interaction is not possible without this consensus. The consensus is part of what holds society together. According to Garfinkel, this consensus will be revealed by people's *background expectancies,* namely, the norms for behavior that they carry with them into situations of interaction. It is presumed that these expectancies are to a great degree shared, and thus studying norms by deliberately violating them will reveal the norms that most people bring with them into interaction. The ethnomethodologist argues that you cannot simply walk up to someone and ask what norms the person has and uses, because most people will not be able to articulate them. We are not wholly conscious of what norms we use even though they are shared. Ethnomethodology is designed to "uncover" those norms.

Ethnomethodologists often use ingenious procedures for uncovering those norms by thinking up clever ways to interrupt "normal" interaction. William Gamson, a sociology professor, had one of his students go into a grocery store where jelly beans, normally priced at that time at 49 cents per pound, were on sale for 35 cents. The student engaged the saleswoman in conversation about the various candies and then asked for a pound of jelly beans. The saleswoman then wrapped them and asked for 35 cents. The rest of the conversation went like this:

> ***Student:*** Oh, only 35 cents for all those nice jelly beans? There are so many of them. I think I will pay $1 for them.
>
> ***Saleswoman:*** Yes, there are a lot, and today they are on sale for only 35 cents.
>
> ***Student:*** I know they are on sale, but I want to pay $1 for them. I just love jelly beans, and they are worth a lot to me.
>
> ***Saleswoman:*** Well, uh, no, you see, they are selling for 35 cents today, and you wanted a pound, and they are 35 cents a pound.
>
> ***Student:*** (voice rising) I am perfectly capable of seeing that they are on sale at 35 cents a pound. That has nothing to do with it. It is just that I personally feel that they are worth more, and I want to pay more for them.
>
> ***Saleswoman:*** (becoming quite angry) What is the matter with you? Are you crazy or something? Everything in this store is priced more than what it is worth. Those jelly beans probably cost the store only a nickel. Now do you want them or should I put them back?

table 4.2 Theories of Social Interaction

	The Social Construction of Reality	Ethnomethodology	Dramaturgy	Social Exchange Theory	Game Theory
Interprets society as:	organized around the subjective meaning that people give to social behavior	held together through the consensus that people share around social norms; you can discover these norms by violating them	a stage on which actors play their social roles and give impression to those in their "audience"	a series of interactions that are based on estimates of rewards and punishments	a system in which people strategize "winning" and "losing" in their interactions with each other
Analyzes social interaction as:	based on the meaning people give to actions in society	a series of encounters in which people manage their impressions in front of others	enactment of social roles played before a social audience	a rational balancing act involving perceived costs and benefits of a given behavior	calculated risks to balance rewards and punishments

At this point the student became quite embarrassed, paid the 35 cents, and hurriedly left (Gamson and Modigliani 1974).

The point here is that the saleswoman approached the situation with a presumed consensus, a consensus that becomes revealed by its deliberate violation by the student. The puzzled saleswoman took measures to attempt to normalize the interaction, even to *force* it to be normal.

Impression Management and Dramaturgy

Another way of analyzing social interaction is to study impression management, a term coined by symbolic interaction theorist Erving Goffman (1959). **Impression management** is a process by which people control how others perceive them. A student handing in a term paper late may wish to give the instructor the impression that it was not the student's fault but was because of uncontrollable circumstances ("my computer hard drive crashed," "my dog ate the last hard copy," and so on). The impression that one wishes to "give off" (to use Goffman's phrase) is that "I am usually a very diligent person, but today—just today—I have been betrayed by circumstances."

Impression management can be seen as a type of con game. We willfully attempt to manipulate others' impression of us. Goffman regarded everyday interaction as a series of attempts to con the other. In fact, trying in various ways to con the other is, according to Goffman, at the very center of much social interaction and social organization in society: Social interaction is just a big con game.

Perhaps this cynical view is not true of all social interaction, but we do present different "selves" to others in different settings. The settings are, in effect, different stages on which we act as we relate to others. For this reason, Goffman's theory is sometimes called the *dramaturgy model* of social interaction, a way of analyzing interaction that assumes the participants are actors on a stage in the drama of everyday social life. People present different faces (give off different impressions) on different stages (in different situations or different roles) with different others. To your mother, you may present yourself as the dutiful, obedient daughter, which may not be how you present yourself to a friend. Perhaps you think acting like a diligent student makes you seem a jerk, so you hide from your friends that you are really interested in a class or enjoy your homework. Analyzing impression management reveals that we try to con the other into perceiving us as we want to be perceived. The box "Doing Sociological Research: Doing Hair, Doing Class" shows how impression management can be involved in many settings, including the everyday world of the hair salon.

AP Images/The Fort Collins Coloradoan, Rich Abrahamson

Impression management *is the display of how you want others to define you. It can be especially obvious when you feel strong attachments to a particular group or identity.*

doing sociological research

Doing Hair, Doing Class

Research Question: Sociologist Debra Gimlin was curious about a common site for social interaction—hair salons. She noticed that the interaction that occurs in hair salons is often marked by differences in the social class status of clients and stylists. Her research question was, How do women attempt to cultivate the cultural ideals of beauty, and in particular, how is this achieved through the interaction between hair stylists and their clients?

Research Method: She did her research by spending more than 200 hours observing social interaction in a hair salon. She watched the interaction between clients and stylists and conducted interviews with the owner, the staff, and twenty women customers. During the course of her fieldwork, she recorded her observations of the conversations and interaction in the salon, frequently asking questions of patrons and staff. The patrons were mostly middle- and upper-middle class, the stylists, working class; all the stylists were White, as were most of the clients.

Research Results: "Beauty work" as Gimlin calls it, involves the stylist bridging the gap between those who seek beauty and those who define it; her (or his) role is to be the expert in beauty culture, bringing the latest fashion and technique to clients. Beauticians are also expected to engage in some "emotion work"—that is, they are expected to nurture clients and be interested in their lives; often they are put in the position of sacrificing their professional expertise to meet clients' wishes.

According to Gimlin, since stylists typically have lower class status than their clients, this introduces an element into the relationship that stylists negotiate carefully in their routine social interaction. Hairdressers emphasize their special knowledge of beauty and taste as a way of reducing the status differences

A clever study by Albas and Albas (1988) demonstrates just how pervasive impression management is in social interaction. The Albases studied how students interacted with one another when the instructor returned graded papers during class. Some students got good grades ("aces"), others got poor grades ("bombers"), but both employed a variety of devices (cons) to maintain or give off a favorable impression. For example, the aces wanted to show off their grades, but they did not want to appear to be braggarts, so they casually or "accidentally" let others see their papers. In contrast, the bombers hid or covered their papers to hide their poor grades, said they "didn't care" what they got, or simply lied about their grades.

One thing that Goffman's theory makes clear is that social interaction is a very perilous undertaking. Have you ever been embarrassed? Of course you have; we all have. Think of a really big embarrassment that you experienced. Goffman defines embarrassment as a spontaneous reaction to a sudden or transitory challenge to our identity: We attempt to restore a prior perception of our "self" by others. Perhaps you were giving a talk before a class and then suddenly forgot the rest of the talk. Or perhaps you recently bent over and split your pants. Or perhaps you are a man and barged accidentally into a women's bathroom. All these actions will result in embarrassment, causing you to "lose face."

You will then attempt to *restore face* ("save face"), that is, eliminate the conditions causing the embarrassment. You thus will attempt to con others into perceiving you as they might have before the embarrassing incident. One way to do this is to shift blame from the self to some other, for example, claiming in the first example that the teacher did not give you time to adequately memorize the talk; or in the second example, claiming that you will never buy that particular, obviously inferior brand of pants again; or in the third example, claiming that the sign saying "Women's Room" was not clearly visible. All these represent deliberate manipulations (cons) to save face on your part—to restore the other's prior perception of you.

Social Exchange and Game Theory

Another way of analyzing social interaction is through the social exchange model. The *social exchange model* of social interaction holds that our interactions are between themselves and their clients. They also try to nullify the existing class hierarchy by conceiving an alternative hierarchy, not one based on education, income, or occupation but only on the ability to style hair competently. Thus, stylists describe clients as perhaps "having a ton of money," but unable to do their hair or know what looks best on them. Stylists become confidantes with clients, who often tell them highly personal information about their lives. Appearing to create personal relationships with their clients, even though they never see them outside the salon, also reduces status differences.

Conclusions and Implications: Gimlin concludes that beauty ideals are shaped in this society by an awareness of social location and cultural distinctions. As she says, "Beauty is . . . one tool women use as they make claims to particular social statuses" (1996: 525).

Questions to Consider

The next time you get your hair cut, you might observe the social interaction around you and ask how class, gender, and race shape interaction in the salon or barbershop that you use. Try to get someone in class to collaborate with you so that you can compare observations in different salon settings. In doing so, you will be studying how gender, race, and class shape social interaction in everyday life.

1. Would you expect the same dynamic in a salon where men are the stylists?
2. Do Gimlin's findings hold in settings where the customers and stylists are not White or where they are all working class?
3. In your opinion, would Gimlin's findings hold in an African American *men's* barbershop?

Source: Gimlin, Debra. 1996. "Pamela's Place: Power and Negotiation in the Hair Salon." *Gender & Society* 10 (October): 505–526.

determined by the rewards or punishments that we receive from others (Homans 1974; Blau 1986; Taylor et al. 2006; Cook et al. 1998). A fundamental principle of exchange theory is that an interaction that elicits approval from another (a type of reward) is more likely to be repeated than an interaction that incites disapproval (a type of punishment). According to the exchange principle, one can predict whether a given interaction is likely to be repeated or continued by calculating the degree of reward or punishment inspired by the interaction. If the reward for an interaction exceeds the punishment, then a potential for *social profit* exists and the interaction is likely to occur or continue.

Rewards can take many forms. They can include tangible gains such as gifts, recognition, and money, or subtle everyday rewards such as smiles, nods, and pats on the back. Similarly, punishments come in many varieties, from extremes like public humiliation, beating, banishment, or execution, to gestures as subtle as a raised eyebrow or a frown. For example, if you ask someone out for a date and the person says yes, you have gained a reward, and you are likely to repeat the interaction. You are likely to ask the person out again, or to ask someone else out. If you ask someone out, and he or she glares at you and says, "No way!" then you have elicited a punishment that will probably cause you to shy away from repeating this type of interaction with that person.

Social exchange theory has grown partly out of *game theory,* a mathematic and economic theory that predicts that human interaction has the characteristics of a "game," namely, strategies, winners and losers, rewards and punishments, and profits and costs (Nash 1951; Dixit and Sneath 1997; Kuhn and Nasar 2002). Simply asking someone out for a date indeed has a gamelike aspect to it, and you will probably use some kind of strategy to "win" (have the other agree to go out with you) and "get rewarded" (have a pleasant or fun time) at minimal "cost" to you (you don't want to spend a large amount of money on the date or you do not want to get into an unpleasant argument on the date). The interesting thing about game theory is that it sees human interaction as just that: a game. Impression management theory also contains a gamelike element in its hypothesis that human interaction is a big con game. The mathematician John Nash is one of the inventors of game theory and was featured in the movie, *A Beautiful Mind.*

seeing society in everyday life

Social Interaction in an Age of Technology

Just as the invention of new technologies during the Industrial Revolution changed how people worked and lived, so is the Cyberspace Revolution transforming social interaction in society. Only a few years ago, forms of interaction and communication that are now common would have been seen as science fiction. What are some of the changes in social interaction—and, thus, in society—that have come as the result of technological changes?

© Bob Daemmrich/The Image Works

If you look, you will frequently see groups of friends, often young people, gathered together but with everyone talking to someone else on a cell phone. Although new technologies allow people to stay closely in touch, do they also intrude on the intimacy that one might otherwise share in face-to-face interaction?

AP Images/Matt Sayles

Many people have noted that, even with the increasing presence of computer technology, there is still a large racial and class divide in who has access to these various devices. White adults are far more likely, for example, to have Internet access than are Black adults, as are groups in higher income brackets. It remains to be seen whether this will hold up over time, since younger people under age 30 are far more likely to have Internet access than older adults.

AP Images/Joanne Carole

New forms of technology also have introduced new language into society—a form of cultural innovation. If you were asked to "TMB," would you "KWIM"? (If not, see the table in Chapter 2 on page 42).

AP Images/Kevork Djansezian

The new forms of social interaction that computer technologies have created also raise new social issues. For example, what social issues arise from the intrusions on privacy that these new technologies can involve?

Utilizing new technologies is commonplace. The ability to communicate instantly, including around the globe, is a tremendous convenience and can draw distant groups closer together, but do they also produce more alienating social relationships?

© Jeff Greenburg/Photo Edit, Inc.

© Syracuse Newspapers/Michele Gabel/The Image Works

The introduction of new technologies relies on new skills that people learn to be able to work (and play) effectively, but what new forms of inequality might be produced as the result?

Interaction in Cyberspace

When people interact and communicate with one another by means of personal computers—through e-mail, chat rooms, computer bulletin boards, virtual communities, and other computer-to-computer interactions—they are engaging in **cyberspace interaction** (or virtual interaction).

The character of cyberspace interaction is changing as new technologies emerge. Not long ago, nonverbal interaction was absent in cyberspace as people could not "see" what others were like. But with the introduction of video-based cyberspace, such as YouTube and photos on Facebook and MySpace, people can now display still and moving images of themselves. These provide new opportunities for what sociologists would call the presentation of self and impression management in cyberspace. Sometimes this comes with embarrassing consequences. The young college student who displays a semi-nude photo of herself, projecting a sexual presentation of self, may be horrified if one of her parents or a potential employer visits her Facebook site!

Cyberspace interaction is becoming increasingly common among all age, gender, and race groups, although clear patterns are also present in who is engaged in this form of social interaction and how people use it (Table 14.3). Women, for example, still lag behind men in use of the Internet, but young women and Black women use the Internet more than their male peers. Older women lag behind older men in their use. Men are also, in general, more intense users, meaning that they log on more often, spend more time online, and are more likely to use broadband.

Gender differences also prevail in how women and men use the Internet. Women are more likely to use e-mail to write to friends and family, share news, plan events, and forward jokes. And women are more

a sociological eye on the media

Downloading Music—A Community in Cyberspace?

Most of the time, people think of communities as physical places where people live, but sociologists think of communities as social forms where there is sustained and active participation among people with a common identity. In the age of the Internet, the sociological concept of community has now been expanded to include *virtual communities*—those where people may never meet face-to-face, yet still maintain the common sense of identity and social interaction that has traditionally defined communities. Within such virtual communities sociologists find that there are roles, statuses, and *norms*—just as there are in communities more traditionally bound by face-to-face interaction.

Take the example of music collecting and trading. Sociologist Peter Nieckarz has studied the music-trading community through participant observation. He collected data by analyzing information on trading-oriented websites and Internet message boards—some of which were dedicated to specific bands and others, more general trading sites.

He found that there are particular statuses and roles that people play within this virtual community, including *tapers, site operators, good traders, bad traders*, and *newbies*. Each has a unique role: Tapers invest in the equipment and travel to shows; site operators run the servers that allow people to download shows; good and bad traders are defined in terms of whether they uphold the norms of the trading community; and newbies are those just being socialized into the community.

In part, money plays a role in the status system because those with the most resources to invest in equipment or trade heavily have the highest status. But status is also derived from adherence to community norms. Norms of politeness and friendliness govern the transactions that people have when trading. You also become a "bad trader" by violating norms of not "ripping people off," providing timely turnaround on trades, and being honest about the quality of a recoding.

In these ways, music trading and collecting engages a social community—one that can be studied and understood using the same sociological concepts that illuminate other forms of social interaction.

Source: Nieckarz, Peter P., Jr. 2005. "Community in Cyber Space?: The Role of the Internet in Facilitating and Maintaining a Community of Live Music Collecting and Trading." *City & Community* 4 (December): 403–423.

table 4.3 Teens Online: Who Is Most Likely to Create an Online Profile?

	Percentage in Each Group Who Create Online Profiles		Percentage in Each Group Who Create Online Profiles
Sex		**Age by Sex**	
Boys	51%	Boys aged 15–17	57
Girls	58	Girls aged 15–17	70
Age		**Household Income**	
12–14	45	Less than $50,000	55
15–17	64	$50,000 or more	56
Age by Sex		**Race-Ethnicity**	
Boys aged 12–14	46	White, non-Hispanic	53
Girls aged 12–14	44	Non-white	58

Source: Pew Internet and American Life Project: *Teens and Parents Survey.* 2006. www.pewinternet.org

likely to report that e-mail nurtures their relationships. Men, on the other hand, use the Internet more to transact business, and they look for a wider array of information than women do. Men are also more likely to use the Internet for hobbies, including such things as sports fantasy leagues, downloading music, and listening to radio (Pew Internet and American Life Project 2006).

More than half of all young people (those aged 12–17) now use online social networking sites, with older girls among the most frequent users of such sites. The popularity of these sites is also growing rapidly among youth. Again, gender patterns emerge in how young people interact online. Boys are more likely than girls to say they use these sites to make new friends and to flirt; girls use it more to stay in touch with friends they already have (Pew Internet and American Life Project 2006).

It is too early to know the implications of these cyberspace interactions. Some think it will make social life more alienating with people developing weaker social skills and less ability for successful face-to-face interaction. Some studies have noted that people can develop extremely close and in-depth relationships as a result of their interaction in cyberspace (McKenna 2002). The Internet also creates more opportunity for people to misrepresent themselves or even create completely false—or even stolen—identities. But studies find that computer-mediated interactions also follow some of the same patterns that are found in face-to-face interaction. People still "manage" identities in front of a presumed audience; they project images of self to others that are consistent with the identity they have created for themselves, and they form social networks that become the source for evolving identities, just as people do in traditional forms of social interaction (Brignall and Van Valey 2005; Turkle 1995).

In this respect, cyberspace interaction is the application of Goffman's principle of *impression management.* The person can put forward a totally different and wholly created self, or identity. One can give off, in Goffman's terms, any impression one wishes

Internet-based network sites, such as Facebook and MySpace, are transforming how people form and sustain relationships with others.

and, at the same time, know that one's true self is protected by anonymity. This gives the individual quite a large and free range of roles and identities from which to choose. As predicted by symbolic interaction theory, of which Goffman's is one variety, *the reality of the situation grows out of the interaction process itself.* This is a central point of symbolic interaction theory: Interaction creates reality.

Cyberspace interaction has thus resulted in new forms of social interaction in society—in fact, a new social order containing both deviants and conformists. These new forms of social interaction have their own rules and norms, their own language, their own sets of beliefs, and practices or rituals—in short, all the elements of culture, as defined in Chapter 2. For sociologists, cyberspace also provides an intriguing new venue in which to study the connection between society and social interaction.

"On the Internet, nobody knows you're a dog."

4 Chapter Summary

- ***What is society?***

  Society is a system of social interaction that includes both culture and *social organization.* Society includes *social institutions,* or established organized social behavior, and exists for a recognized purpose; *social structure,* the patterned relationships within a society.

- ***What holds society together?***

  According to theorist Emile Durkheim, society with all its complex social organization and culture, is held together, depending on overall type, by *mechanical solidarity* (based on individual similarity) and *organic solidarity* (based on a *division of labor* among dissimilar individuals). Two other forms of social organization also contribute to the cohesion of a society: *gemeinschaft* ("community," characterized by cohesion based on friendships and loyalties) and *gesellschaft* ("society," characterized by cohesion based on complexity and differentiation).

- ***What are the types of societies?***

  Societies across the globe vary in type, as determined mainly by the complexity of their social structures, their division of labor, and their technologies. From least to most complex, they are *foraging, pastoral, horticultural, agricultural* (these four constitute *preindustrial* societies), *industrial,* and *postindustrial* societies.

- ***What are the forms of social interaction in society?***

  All forms of social interaction in society are shaped by the structure of its social institutions. A *group* is a collection of individuals who interact and communicate with each other, share goals and norms, and have a subjective awareness of themselves as a distinct social unit. *Status* is a hierarchical position in a structure; a *role* is the expected behavior associated with a particular status. A *role* is the behavior others expect from a person associated with a particular status. Patterns of social interaction influence nonverbal interaction as well as patterns of attraction and affiliation.

- ***What theories are there about social interaction?***
  Social interaction takes place in society within the context of social structure and social institutions. Social interaction is analyzed in several ways, including the *social construction of reality* (we impose meaning and reality on our interactions with others); *ethnomethodology* (deliberate interruption of interaction to observe how a return to "normal" interaction is accomplished); *impression management* (a person "gives off" a particular impression to "con" the other and achieve certain goals, as in *cyberspace interaction*); and *social exchange* (one engages in gamelike reward and punishment interactions to achieve one's goals).

- ***How is technology changing social interaction?***
  Increasingly people engage with each other through *cyberspace interaction*. Social norms develop in cyberspace as they do in face-to-face interaction, but a person in cyberspace can also manipulate the impression that he or she gives off, thus creating a new "virtual" self.

Key Terms

achieved status 103
ascribed status 103
collective consciousness 96
cyberspace interaction 116
division of labor 97
ethnomethodology 110
gemeinschaft 97
gesellschaft 97
group 101
impression management 111
macroanalysis 94
master status 103
mechanical solidarity 96
microanalysis 94
organic solidarity 97
postindustrial society 100
preindustrial society 98
proxemic communication 106
role 103
role conflict 104
role modeling 104
role set 104
role strain 104
social institution 95
social interaction 94
social organization 95
social structure 96
society 94
status 102
status inconsistency 103
status set 102

Online Resources

Sociology: The Essentials Companion Website

www.thomsonedu.com/sociology/andersen

Visit your book companion website where you will find more resources to help you study and write your research papers. Resources include Suggested Readings, web links, and a MicroCase Online feature that teaches you how to research society. Other resources include Learning Objectives, Internet exercises, quizzing, and flash cards.

is an easy-to-use online resource that helps you study in less time to get the grade you want NOW.

www.thomsonedu.com/login

Need help studying? This site is your one-stop study shop. Take a Pre-Test and Thomson NOW will generate a Personalized Study Plan based on your test results. The Study Plan will identify the topics you need to review and direct you to online resources to help you master those topics. You can then take a Post-Test to determine the concepts you have mastered and what you still need to work on.

5

Chapter five

CHAPTER FIVE

Chapter

Groups and Organizations

Types of Groups

Social Influence in Groups

Formal Organizations and Bureaucracies

Diversity: Race, Gender, and Class in Organizations

Functional, Conflict, and Symbolic Interaction: Theoretical Perspectives

Chapter Summary

Twelve citizens sit together in an elevated enclosure, like a choir loft, and silently watch a drama unfold before them, day after day. They are respectfully addressed by highly paid professionals: lawyers, judges, expert witnesses. Their job, ruling on the innocence or guilt of a defendant or the settlement of a legal claim, was once the prerogative only of kings. Their decision may mean freedom or incarceration, fortune or penury, even life or death.

Juries have been the focus of much research, in part because they fill such a vital role in our society and in part because within this curiously artificial, yet intimate, group of random strangers can be found a wealth of interesting sociological phenomena. Jury verdicts and jury deliberations show the same inescapable influences of status, race, and gender that affect the rest of society. Juries are, in some ways, society in miniature.

During jury selection, in a process called the *voir dire,* lawyers on both sides are entitled to eliminate any potential jurors and no explanation is required. Many lawyers who have great faith in their ability to judge jurors consider jury selection to be the most important part of a trial. By choosing jurors, they are choosing the verdict. Consider some of the folk wisdom clung to by trial lawyers in the past—with varying degrees of accuracy: Farmers believe in strict responsibility, whereas waiters and bartenders are forgiving; avoid the clergy; select married women.

continued

A guideline for Dallas, Texas, prosecutors advised against selecting "Yankees . . . unless they appear to have common sense" (Guinther 1988: 54). High-powered legal teams now make room for a new breed of legal specialist—the *trial consultant,* trained in sociological and psychological techniques, who contributes nothing but juror analysis as part of the jury selection process.

These analyses go beyond simply identifying the bias of a given juror. Juries are *groups,* and groups behave differently from individuals. Understanding group behavior is critical to predicting the performance of a jury. For instance, it is possible to make an educated prediction about who in a jury will become the most influential. Researchers have found that people with high status in society do the most talking in jury deliberations, and other jurors consider them to be the most helpful in reaching a verdict (Hans and Martinez 1994; Berger and Zelditch 1985).

Factions or subgroupings form during jury deliberations, and if jury analysts expect a difficult decision, they can attempt to influence how factionalized juries will resolve their disputes based on sociological and psychological data about small group decision making (Saks and Marti 1997). For instance, jurors are much less likely to defect from large factions than from small ones. The larger the faction, the less willing a juror will be to defy the weight of group opinion.

© John Neubauer/PhotoEdit

This jury of twelve persons is voting on something, perhaps on whether to acquit a person accused of a serious crime. The social pressures in a jury are extremely strong, making the lone "hold-out" person very unlikely.

What does this say about the state of justice in our legal system, when guilt or innocence depends not only on the legal facts but also on sociological aspects of the jury? Like society as a whole, and like organizations and bureaucracies, groups are subject to social influences. Whether a relatively small group, such as a jury, or a large organization, such as a major corporation, people and groups are influenced by sociological forces.

Types of Groups

Each of us is a member of many groups simultaneously. We have relationships in groups with family, friends, team members, and professional colleagues. Within these groups are gradations in relationships: We are generally closer to our siblings (our sisters and brothers) than our cousins; we are intimate with some friends, merely sociable with others. If we count all our group associations, ranging from the powerful associations that define our daily lives to the thinnest connections with little feeling (other pet lovers, other company employees), we will uncover connections to literally hundreds of groups.

What is a group? Recall from Chapter 4, a **group** is two or more individuals who interact, share goals and norms, and have a subjective awareness as "we." To be considered a group, a social unit must have all three characteristics. The hazy boundaries of the definition are necessary. Consider three superficially similar examples: The individuals in a line waiting to board a train are unlikely to have a sense of themselves as one group. A line of prisoners chained together and waiting to board a bus to the penitentiary is more likely to have a stronger sense of common feeling. Finally, and sadly, no doubt the passengers who overpowered the hijacking terrorists on American Airlines Flight 93 on September 11, 2001, subsequently crashing the plane into a Pennsylvania field, became a group for a few moments.

As you remember from the previous chapter, certain gatherings are not groups in the strict sense, but may be *social categories* (for example, teenagers, truck drivers) or *audiences* (everyone watching a movie). The importance of defining a group is not to perfectly decide if a social unit is a group—an unnecessary endeavor—but to help us understand the behavior of people in society. As we inspect groups, we can identify characteristics that reliably predict trends in the behavior of the group and even the behavior of individuals in the group.

The study of groups has application at all levels of society, from the attraction between people who fall in love to the characteristics that make some corporations drastically outperform their competitors. The aggregation of individuals into groups has a transforming power, and sociologists understand the social forces that make these transformations possible.

In this chapter, we move from the *micro* level of analysis (the analysis of groups and face-to-face social influence) to the *macro* level of analysis (the analysis of formal organizations and bureaucracies).

Dyads and Triads: Group Size Effects

Even the smallest groups are of acute sociological interest. A **dyad** is a group consisting of exactly two people. A **triad** consists of three people. This seemingly minor distinction, first scrutinized by the German sociologist **Georg Simmel** (1858–1918), can have critical consequences for group behavior (Simmel 1902). Simmel was interested in discovering the effects of size on groups, and he found that the mere difference between two and three people spawned entirely different group dynamics (the behavior of a group over time).

Imagine two people standing in line for lunch. First one talks, then the other, then the first again. The interaction proceeds in this way for several minutes. Now a third person enters the interaction. The character of the interaction suddenly changes: At any given moment, two people are interacting more with each other than either is with the third. When the third person wins the attention of the other two, a new dyad is formed, supplanting the previous pairing. The group, a triad, then consists of a dyad (the pair that is interacting) plus an *isolate.*

Triadic segregation is what Simmel called the tendency for triads to segregate into a pair and an isolate (a single person). A triad tends to segregate into a *coalition* of the dyad against the isolate. The isolate then has the option of initiating a coalition with either member of the dyad. This choice is a type of social advantage, leading Simmel to coin the principle of *tertius gaudens,* a Latin term meaning "the third one gains." Simmel's reasoning has led to numerous studies of coalition formation in groups.

For example, interactions in a triad often end up as "two against one." You may have noticed this principle of coalition formation in your own conversations. Perhaps two friends want to go to a movie you do not want to see. You appeal to one of them to go instead to a minor league baseball game. She wavers and comes over to your point of view—now you have formed a coalition of two against one. The friend who wants to go to the movies is now the isolate. He may recover lost social ground by trying to form a new coalition by suggesting a new alternative (going bowling or to a different movie). This flipflop interaction may continue for some time, demonstrating another observation by Simmel: A triad is an unstable social grouping, whereas dyads are relatively stable. The minor distinction between dyads and triads is one person but has important consequences because it changes the character of the interaction within the group. Simmel is known as the discoverer of **group size effects**—the effects of group number on group behavior *independently of the personality characteristics of the members themselves.*

Primary and Secondary Groups

Charles Horton Cooley (1864–1929), a famous sociologist of the Chicago School of sociology, introduced the concept of the **primary group,** defined as a group consisting of intimate, face-to-face interaction and relatively long-lasting relationships. Cooley had in mind the family and the early peer group. In his original formulation, *primary* was used in the sense of "first," the intimate group of the formative years (Cooley 1967/1909). The insight that there was an important distinction between intimate groups and other groups proved extremely fruitful. Cooley's somewhat narrow concept of family and childhood peers has been elaborated upon over the years to include a variety of intimate relations as examples of primary groups.

Primary groups have a powerful influence on an individual's personality or self-identity. The effect of family on an individual can hardly be overstated. The weight of peer pressure on school children is particularly notorious. Street gangs are a primary group, and their influence on the individual is significant; in fact, gang members frequently think of themselves as a family. The intense camaraderie formed among Marine Corps units in boot camp and in war, such as the war in Iraq, is another classic example of primary group formation and the resulting intense effect on the individual.

In contrast to primary groups are **secondary groups,** those that are larger in membership, less intimate, and less long-lasting. Secondary groups tend to be less significant in the emotional lives of people. Secondary groups include all the students at a college or university, all the people in your neighborhood, and all the people in a bureaucracy or corporation.

One of the best examples of the primary group is that consisting of parent and child.

doing sociological research

Sharing the Journey

Modern society is often characterized as remote, alienating, and without much feeling of community or belonging to a group. This image of society has been carefully studied by sociologist Robert Wuthnow, who noticed that in the United States people are increasingly looking to small groups as a place where they can find emotional and spiritual support and where they find meaning and commitment, despite the image of society as an increasingly impersonal force.

Research Question: Wuthnow began his research by noting that, even with the individualistic culture of U.S. society, small groups play a major role in this society. He saw the increasing tendency of people to join recovery groups, reading groups, spiritual groups, and a myriad of other support groups. Wuthnow began his research by asking some specific questions, including, "What motivates people to join support groups?" "How do these groups function?" and "What do members like most and least about such groups?" His broadest question, however, was to wonder how the wider society is influenced by the proliferation of small support groups.

Research Methods: To answer these questions, a large research team of fifteen scholars designed a study that included both a quantitative and a qualitative dimension. They distributed a survey to a representative sample of more than 1000 people in the United States. Supplementing the survey were interviews with more than 100 support group members, group leaders, and clergy. The researchers chose twelve groups for extensive study; researchers spent six months to three years tracing the history of these groups, meeting with members and attending group sessions.

Research Results: Based on this research, Wuthnow concludes that the small group movement is funda-

SEE FOR YOURSELF

ANALYZING GROUPS

Identify a *group* of which you are a part. How does one become a member of this group? Who gets included and who gets excluded? Does the group share any unique language or other cultural characteristics (such as dress, jargon, or other group identifiers). Does anyone ever leave the group and, if so, why?

1. Would you describe this group mainly as a *primary* or a *secondary group*? Why?
2. Now think about this group from the perspective of functionalist theory, conflict theory, and symbolic interaction theory. What are the *manifest and latent functions* of the group? Is there a hierarchy within the group? Is there competition between group members? What social meanings do members of the group share?

Secondary groups occasionally take on the characteristics of primary groups. The process can be accelerated in situations of high stress or crisis. For example, when a neighborhood meets with catastrophe, such as a flood, people who may be only acquaintances often come to depend on each other and in the process become more intimate. The secondary group of neighbors becomes, for a time, a primary group. This is precisely what happened in the otherwise highly impersonal neighborhoods in New York City near ground zero of the September 11, 2001, terrorist attack on the World Trade Center: Thousands of people pitched in to help, and as a result, many primary groups formed.

Primary and secondary groups serve different needs. Primary groups give people intimacy, companionship, and emotional support. These human desires are termed **expressive needs** (also called socioemotional needs). Family and friends share and amplify your good fortune, rescue you when you misbehave, and cheer you up when life looks grim. Primary groups are a major influence on social life and an important source of social control. They are also a dominant influence on your likes and dislikes, preferences in clothing, political views, religious attitudes, and other characteristics. Many studies have shown the overwhelming influence of family and friendship groups on religious and political affiliation, as shown in the box "Doing Sociological Research: Sharing the Journey" (Wuthnow 1994).

mentally altering U.S. society. Forty percent of all Americans belong to some kind of small support group. As the result of people's participation in these groups, social values of community and spirituality are undergoing major transformation. People say they are seeking community when they join small groups—whether the group is a recovery group, a religious group, a civic association, or some other small group. People turn to these small groups for emotional support more than physical or monetary support.

Conclusions and Implications: Wuthnow argues that large-scale participation in small groups has arisen in a social context in which the traditional support structures in U.S. society, such as the family, no longer provide the sense of belonging and social integration that they provided in the past. Geographic mobility, mass society, and the erosion of local ties all contribute to this trend. People still seek a sense of community, but they create it in groups that also allow them to maintain their individuality. In voluntary small groups, you are free to leave the group if it no longer meets your needs.

Wuthnow also concludes that these groups represent a quest for spirituality in a society when, for many, traditional religious values have declined. As a consequence, support groups are redefining what is sacred. They also replace explicit religious tenets imposed from the outside with internal norms that are implicit and devised by individual groups. At the same time, these groups reflect the pluralism and diversity that characterize society. In the end, they buffer the trend toward disintegration and isolation that people often feel in mass societies.

Source: Wuthnow, Robert. 1994. *Sharing the Journey: Support Groups and America's New Quest for Community.* New York: Free Press.

Secondary groups serve **instrumental needs** (also called task-oriented needs). Athletic teams form to have fun and win games. Political groups form to raise funds and bend the will of the legislature. Corporations form to make profits, and employees join corporations to earn a living. Needless to say, intimacies can develop in the act of fulfilling instrumental needs, and primary groups may also devote themselves to meeting instrumental needs. The true distinction between primary and secondary groups is in how strongly the participants feel about one another and how dependent they are on the group for sustenance and identity. Both primary and secondary groups are indispensable elements of life in society.

Reference Groups

Primary and secondary groups are groups to which members belong. **Reference groups** are those to which you may or may not belong but use as a standard for evaluating your values, attitudes, and behaviors (Merton and Rossi 1950). Reference groups are generalized versions of role models. They are not "groups" in the sense that the individual interacts within (or in) them. Do you pattern your behavior on that of sports stars, musicians, military officers, or business executives? If so, those models are reference groups for you.

© DoD, Staff Sgt. Dawn Finniss, U.S. Air Force

Identification with a reference group *has a significant influence on one's identity.*

Imitation of reference groups can have both positive and negative effects. Members of a Little League baseball team may revere major league baseball players and attempt to imitate laudable behaviors such as tenacity and sportsmanship. But young baseball fans are also liable to be exposed to tantrums, fights, and tobacco chewing and spitting. This illustrates that the influence of a reference group can be both positive and negative.

Research has shown that identification with reference groups can strongly influence self-evaluation and self-esteem. Before the push for school desegregation began, it was thought that all-Black schools contributed to negative self-evaluation among Black students. Desegregation was expected to raise the self-esteem of Black children (Clark and Clark 1947). The original Clark and Clark study suggested that Black children in all-Black schools preferred playing with White dolls over Black dolls. This behavior was presumed to demonstrate a negative self-evaluation and self-rejection on the part of Black children.

In some cases desegregation did raise the self-esteem of Black children, but later research has also found that identification with a positive reference group was more important than desegregation. When racial or ethnic groups were consistently presented in a positive way, as in later multicultural educational programs designed to increase pride in Black culture, the self-esteem of the children was greater than that of Black children in integrated programs with no multicultural component. The same has been found for Latino children enrolled in Latino cultural awareness programs. Plainly, the representation of racial and ethnic groups in a society can have a striking positive effect on children acquiring their lifetime set of group affiliations (Harris 2006; Zhou and Bankston 2000; Baumeister 1998; Steele 1996, 1992; Steele and Aronson 1995; Banks 1976).

© David R. Frazier/The Image Works

The impersonality of society leads many to join support groups.

In-Groups and Out-Groups

When groups have a sense of themselves as "us," there will be a complementary sense of other groups as "them." The distinction is commonly characterized as *in-groups* versus *out-groups.* The concept was originally elaborated by the early sociological theorist **W.I. Thomas** (Thomas 1931). College fraternities and sororities certainly exemplify "in" versus "out." So do families. The same can be true of the members of your high school class, your sports team, your racial group, your gender, and your social class. Members of the upper classes in the United States occasionally refer to one another as PLUs—"people like us" (Graham 1999; Frazier 1957).

Attribution theory is the principle that we all make inferences about the personalities of others, such as concluding what the other is "really like." These attributions depend on whether you are in the in-group or the out-group. Thomas F. Pettigrew has summarized the research on attribution theory, showing that individuals commonly generate a significantly distorted perception of the motives and capabilities of other people's acts based on whether that person is an in-group or out-group member (Pettigrew 1992; Gilbert and Malone 1995). Pettigrew describes the misperception as **attribution error,** meaning errors made in attributing causes for people's behavior to their membership in a particular group, such as a racial group. Attribution error has several dimensions, all tending to favor the in-group over the out-group. *We tend to perceive people in our in-group positively and those in out-groups negatively regardless of their actual personal characteristics:*

1. When onlookers observe improper behavior by an out-group member, onlookers are likely to attribute the deviance to the disposition of the wrongdoer. Disposition refers to the perceived "true nature" or "inherent nature" of the person, often considered to be genetically determined. *Example:* A White person sees a Hispanic person carrying a knife and, without further information, attributes this behavior to the presumed "inherent tendency" for Hispanics to be violent. The same would be true if a Hispanic person, without additional information, assumed that all Whites have an "inherent tendency" to be racist.
2. When the *same* behavior is exhibited by an in-group member, the perception is commonly held that the act is due to the *situation* of the wrong-

doer, not to the in-group member's inherent disposition. *Example:* A White person sees another White person carrying a knife and concludes, without further information, that the weapon must be carried for protection in a dangerous area.

3. If an out-group member is seen to perform in some laudable way, the behavior is often attributed to a variety of special circumstances, and the out-group member is seen as "the exception."
4. An in-group member who performs in the same laudable way is given credit for a worthy personality disposition.

Typical attribution errors include misperceptions between racial groups and between men and women. If a White police officer shoots a Black or Latino, a White individual, given no additional information, is likely to assume that the victim instigated the shooting and thus "deserved" to be shot. On the other hand, a Black person is more likely to assume that the police officer fired unnecessarily, perhaps because the officer is dispositionally assumed to be a racist (Taylor et al. 2006; Kluegel and Bobo 1993; Bobo and Kluegel 1991).

A related phenomenon has been seen in men's perceptions of women coworkers. Meticulous behavior in a man is perceived positively and is seen by other men as "thorough"; in a woman, the *exact same* behavior is perceived negatively and is considered "picky." Behavior applauded in a man as "aggressive" is condemned in a woman exhibiting the same behavior as "pushy" or "bitchy" (Uleman et al. 1996; Wood 1994).

Social Networks

As already noted, no individual is a member of only one group. Social life is far richer than that. A **social network** is a set of links between individuals, between groups, or between other social units, such as bureaucratic organizations or even entire nations (Aldrich and Ruef 2006; Centeno and Hargittai 2003; Mizruchi 1992). One could say that people belong to several networks. Your group of friends, for instance, or all the people on an electronic mail list to which you subscribe are social networks (Wasserman and Faust 1994).

The network of people you are closest to, rather than those merely linked to you in some impersonal way, is probably most important to you. Numerous research studies indicate that people get jobs via their personal networks more often than through formal job listings, want ads, or placement agencies (Ruef et al. 2003; Petersen et al. 2000; Granovetter 1995, 1974). This is especially true for high-paying, prestigious jobs. Getting a job is more often a matter of who you know than what you know. Who you know, and who they know in turn, is a social network that may have a marked effect on your life and career.

4F Images/Bebeto Matthews

This job candidate does a last-minute check of his resume just before being interviewed by a company representative who contacted the job candidate through a social network.

Networks form with all the spontaneity of other forms of human interaction (Mintz and Schwartz 1985; Wasserman and Faust 1994; Fischer 1981; Knoke 1992). Networks evolve, such as social ties within neighborhoods, professional contacts, and associations formed in fraternal, religious, occupational, and volunteer groups. Networks to which you are only *weakly* tied (you may know only one person in your neighborhood) provide you with access to that network, hence the sociological principle that there is "strength in weak ties" (Petersen et al. 2000; Montgomery 1992; Granovetter 1973).

Networks based on race, class, and gender form with particular readiness. This has been especially true of job networks. The person who leads you to a job is likely to have a similar social background. Recent research indicates that the "old boy network"—any network of White, male corporate executives—is less important than it used to be, although it is certainly not gone. The diminished importance of the old boy network is because of the increasing prominence of women and minorities in business organizations. In fact, among African American and Latino individuals, family can provide network contacts that can lead to jobs and upward mobility (Dominguez and Watkins 2003). Still, as we will see later in this chapter, women and minorities are considerably underrepresented in corporate life, especially in high-status jobs (Padavic and Reskin 2002; Green et al. 1999; Collins-Lowry 1997; Gerson 1993).

Networks can reach around the world, but how big is the world? How many of us, when we discover someone we just met is a friend of a friend, have remarked, "My, it's a small world"? Research into what has come to be known as the *small world problem* has shown that networks make the world a lot smaller than you might think.

Original small-world researchers Travers and Milgram wanted to test whether a document could be routed to a complete stranger more than 1000 miles away using only a chain of acquaintances (Travers and Milgram 1969; Lin 1989; Kochen 1989; Watts and Strogatz 1998; Watts 1999). If so, how many steps would be required? The researchers organized an experiment in which approximately 300 senders were all charged with getting a document to one receiver, a complete stranger. The receiver was a male, Boston stockbroker. The senders were one group of Nebraskans and one group of Bostonians chosen completely at random. Every sender in the study was given the receiver's name, address, occupation, alma mater, year of graduation, wife's maiden name, and hometown. They were asked to send the document directly to the stockbroker if they knew him on a first-name basis. Otherwise, they were asked to send the folder to a friend, relative, or acquaintance known on a first-name basis who might be more likely than the sender to know the stockbroker.

How many intermediaries do you think it took, on average, for the document to get through? (Most people estimate from twenty to hundreds.) The average number of intermediate contacts was only 6.2! However, only about one-third of the documents actually arrived at the target. This was quite impressive, considering that the senders did not know the target person—hence the current expression that any given person in the country is on average only about "six degrees of separation" from any other person. In this sense the world is "small."

This original small-world research has recently been criticized on two grounds: First, only one-third of the documents actually reached the target person. The 6.2 average intermediaries applied only to these completed chains. Thus two-thirds of the initial documents never reached the target person. For these persons, the world was certainly not "small." Second, the sending chains tended to closely follow occupational, social class, and ethnic lines—just as general network theory would predict (Wasserman and Faust 1994). Thus, the world may indeed be "small," but only for people in your social network (Ruef et al. 2003; Kleinfeld 2002; Watts 1999).

A study of Black national leaders by Taylor and associates shows that Black leaders form a very closely knit network, one considerably more closely knit than longer-established White leadership networks (Taylor 1992b; Jackson 2000; Jackson et al. 1995, 1994; Mills 1956; Domhoff 2002; Kadushin 1974; Moore 1979; Alba and Moore 1982). The world is indeed quite "small" for America's Black leadership. Included in the study were Black members of Congress, mayors, business executives, military officers (generals and full colonels), religious leaders, civil rights leaders, media personalities, entertainment and sports figures, and others. The study found that when considering only direct personal acquaintances—not indirect links involving intermediaries—one-fifth of the entire national Black leadership network know each other directly as a friend or close acquaintance. The Black leadership network is considerably more closely connected than White leadership networks. The Black network had greater density. Add only one intermediary, the friend of a friend, and the study estimated that almost three-quarters of the entire Black leadership network are included. Therefore, any given Black leader can generally get in touch with three-quarters of all other Black leaders in the country either by knowing them personally (a "friend") or via only one common acquaintance (a "friend of a friend"). That's pretty amazing when one realizes that the study is considering the population of Black leaders in the entire country.

Social Influence in Groups

The groups in which we participate exert tremendous influence on us. We often fail to appreciate how powerful these influences are. For example, who decides what you should wear? Do you decide for yourself each morning, or is the decision already made for you by fashion designers, role models, and your peers? Consider how closely your hair length, hair styling, and choice of jewelry has been influenced by your peers. Did you invent your baggy pants, your dreadlocks, or your blue blazer? People who label themselves as nonconformists often conform rigidly to the dress code and other norms of their in-group. This was true of the Beatniks in the 1950s, the hippies of the early 1960s and 1970s, the punk rockers of the 1970s and 1980s, and the grunge kids and goths of the 1990s.

After the rebelliousness of youth has faded, the influences of our youth extend to adulthood. The choice of political party among adults (Republican, Democratic, or independent) correlate strongly with the party of one's parents, again demonstrating the power of the primary group. Seven out of ten people vote with the party of their parents, even though these same people insist that they think for themselves when voting (Taylor et al. 2006; Worchel et al. 2000; Jennings and Niemi 1974). Furthermore, most people share the religious affiliation of their parents, although they will insist that they chose their own religion, free of any influence by either parent.

We all like to think we stand on our own two feet, immune to a phenomenon as superficial as group pressure. The conviction that one is impervious to social influence results in what social psychologist Philip Zimbardo calls the *not-me syndrome:* When confronted with a description of group behavior that is disappointingly conforming and not individualistic, most individuals counter that some people may conform, "but not me" (Zimbardo et al. 1977; Taylor et al. 2006). But sociological experiments often reveal a dramatic gulf between what people think they will do and what they actually do. The conformity study by Solomon Asch discussed in the following section is a case in point.

The Asch Conformity Experiment

We learned in the previous section that social influences are evidently quite strong. Are they strong enough to make us disbelieve our own senses? In a classic piece of work known as the Asch conformity experiment, Solomon Asch showed that even simple objective facts cannot withstand the distorting pressure of group influence (Asch 1955, 1951).

Examine the two illustrations in Figure 5.1. Which line on the right is equal in length to the line on the left (line S)? Line B, obviously. Could anyone fail to answer correctly?

In fact, Solomon Asch discovered that social pressure of a rather gentle sort was sufficient to cause an astonishing rise in the number of wrong answers. Asch lined up five students at a table and asked which line in the illustration on the right is the same length as the line on the left. Unknown to the fifth student, the first four were *confederates*—collaborators with the experimenter who only pretended to be participants. For several rounds, the confederates gave correct answers to Asch's tests. The fifth student also answered correctly, suspecting nothing. Then the first student gave a wrong answer. The second student gave the same wrong answer. Third, wrong. Fourth, wrong. Then came the fifth student's turn.

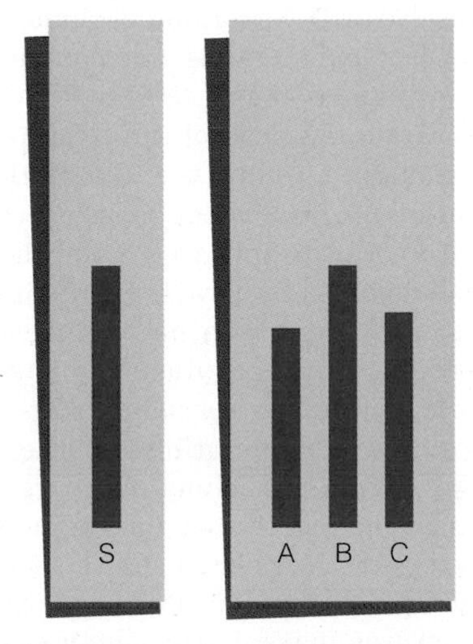

FIGURE 5.1 *Lines from Asch Experiment*

Source: Asch, Solomon, 1955. "Opinion and Social Pressure." *Scientific American 19* (July): 31–35.

In Asch's experiment, fully *one-third* of all students in the fifth position gave the same wrong answer as the confederates at least half the time. Forty percent gave "some" wrong answers. Only one-fourth of the students consistently gave correct answers in defiance of the invisible pressure to conform.

Line length is not a vague or ambiguous stimulus. It is clear and objective. Wrong answers from one-third of all subjects is a very high proportion. The subjects fidgeted and stammered while doing it, but they did it nonetheless. Those who did not yield to group pressure showed even more stress and discomfort than those who yielded to the (apparent) opinion of the group.

Would you have gone along with the group? Perhaps, perhaps not. Sociological insight grows when we acknowledge the fact that fully one-third of all participants will yield to the group. The Asch experiment has been repeated many times over the years, with students and nonstudents, old and young, in groups of different sizes, and in different settings (Worchel et al. 2000; Cialdini 1993). The results remain essentially the same. One-third to one-half of the participants make a judgment contrary to fact, yet in conformity with the group.

The Milgram Obedience Studies

What are the limits of social pressure? In terms of moral and psychological issues, judging the length of a line is a small matter. What happens if an authority figure demands obedience—a type of conformity—even if the task is something the test subject (the person) finds morally wrong and reprehensible? A chilling answer emerged from the now famous Milgram Obedience Studies, done from 1960 through 1974 by Stanley Milgram (Milgram 1974).

In this study, a naive research subject entered a laboratory-like room and was told that an experiment on learning was to be conducted. The subject was to act as a teacher, presenting a series of test questions to another person, the learner. Whenever the learner gave a wrong answer, the teacher would administer an electric shock.

The test was relatively easy. The teacher read pairs of words to the learner, such as

blue box
nice house
wild duck

The teacher then tested the learner by reading a multiple-choice answer, such as

blue sky ink box lamp

(a)

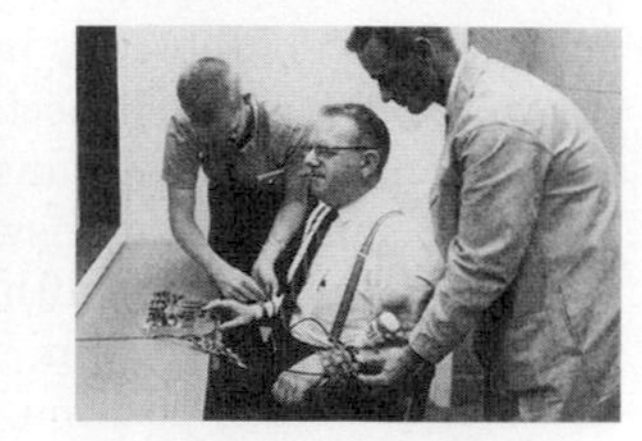

(b)

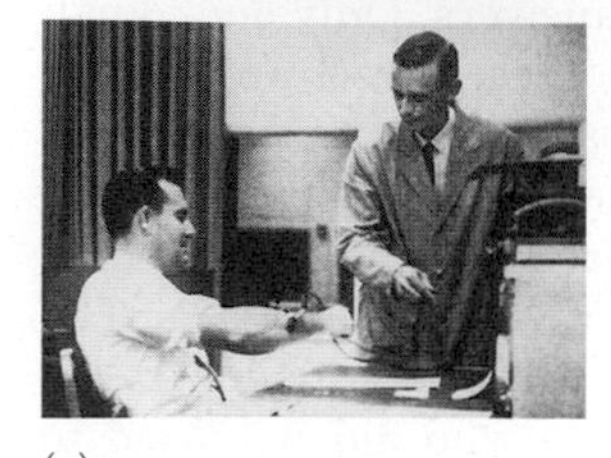

(c)

(d)

FIGURE 5.2 *Milgram's Setup*

These photographs show how intimidating—and authoritative—the Milgram experiment must have been. The first picture (a) shows the formidable-looking shock generator. The second (b) shows the role player, who pretends to be getting the electric shock, being hooked up. The third (c) shows an experimental subject (seated) and the experimenter (in lab coat, standing). The fourth picture (d) shows a subject terminating the experiment prematurely, that is, before giving the highest shock level (voltage). A large majority (65 percent) of subjects did not do this and actually went all the way to the maximum shock level.

Source: Milgram, Stanley. 1974. *Obedience to Authority: An Experimental View.* New York: Harper & Row, p. 25.

The learner had to recall which term completed the pair of terms given originally, in this case, blue box.

If the learner answered incorrectly, the teacher was to press a switch on the shock machine, a formidable-looking device that emitted an ominous hum when activated (see Figure 5.2). For each successive wrong answer, the teacher was to increase the intensity of the shock by 15 volts.

The machine bore labels clearly visible to the teacher: Slight Shock, Moderate Shock, Strong Shock, Very Strong Shock, Intense Shock, Extreme Intensity Shock, Danger: Severe Shock, and lastly, XXX at 450 volts. As the voltage rose, the learner responded with squirming, groans, then screams.

The experiment was rigged. The learner was a confederate. No shocks were actually delivered. The true purpose of the experiment was to see if any "teacher" would go all the way to 450 volts. If the subject tried to quit, the experimenter responded with a sequence of prods:

"Please continue."

"The experiment requires that you continue."

"It is absolutely essential that you continue."

"You have no other choice, you *must* go on."

In the first experiment, fully 65 percent of the volunteer subjects ("teachers") went *all the way* to 450 volts on the shock machine.

Milgram himself was astonished. Before carrying out the experiment, he had asked a variety of psychologists, sociologists, psychiatrists, and philosophers to guess how many subjects would actually go all the way to 450 volts. The opinion of these consultants was that only one-tenth of 1 percent (one in one thousand) would actually do it.

What would you have done? Remember the "not-me" syndrome. Think about the experimenter saying, "You have no other choice, you must go on." Most people claim they would refuse to continue as the voltage escalated. The importance of this experiment derives in part from how starkly it highlights the difference between what people *think* they will do and what they *actually* do.

Milgram devised a series of additional experiments in which he varied the conditions to find out what would cause subjects not to go all the way to 450 volts. He moved the experiment from an impressive laboratory to a dingy basement to counteract some of the tendency for people to defer to a scientist conducting a scientific study. One learner was instructed to complain of a heart condition. Still, well over half of the subjects delivered the maximum shock level. Speculating that women might be more humane than men (all prior experiments used only male subjects), Milgram did the experiment again using only women subjects. The results? Exactly the same. Class background made no difference. Racial and ethnic differences had no detectable effect on compliance rate.

At the time that the Milgram experiments were conceived, the world was watching the trial in Jerusalem of World War II Nazi Adolf Eichmann. Millions of Jews, Gypsies, homosexuals, and communists were murdered between 1939 and 1945 by the Nazi party, led by Adolf Hitler. As head of the Gestapo's "Jewish section," Eichmann oversaw the deportation of Jews to concentration camps and the mass executions that followed. Eichmann disappeared after the war, was abducted in Argentina by Israeli agents in 1961, and transported to Israel, where he was tried and ultimately hanged for crimes against humanity.

The world wanted to see what sort of monster could have committed the crimes of the Holocaust, but a jarring picture of Eichmann emerged. He was slight and mild mannered, not the raging ghoul that everyone expected. He insisted that although he had indeed been a chief administrator in an organization whose product was mass murder, he was guilty only of doing what he was told to do by his superiors. He did not hate Jews, he said. In fact, he had a Jewish half-cousin whom he hid and protected. He claimed, "I was just following orders."

How different was Adolph Eichmann from the rest of us? The political theorist Hannah Arendt dared to suggest in her book *Eichman in Jerusalem* (1963) that evil on a giant scale is banal. It is not the work of monsters, but an accident of civilization. Arendt argued that to find the villain, we need to look into ourselves.

The Iraqi Prisoners at Abu Ghraib: Research Predicts Reality?

We have just learned that ordinary people will do horrible things to other humans simply because of the influence of the group, because of an authority figure, or because of a combination of both. This has been the lesson of the Asch studies and the Milgram studies. Recent events in the world have once again shown vividly and clearly how accurate such sociological and psychological experiments are in the prediction of actual human behavior.

In the spring of 2004, it was revealed that American soldiers who were military police guards at a prison in Iraq (the prison was named Abu Ghraib) had engaged in severe torture of Iraqi prisoners of war. The torture included sexual abuse of the prisoners—having male prisoners simulate sex with other male prisoners, positioning their mouths next to the genitals of another male prisoner, being forced to masturbate in view of others, and other such acts. Still other acts of torture involved physical abuse such as beatings, stomping on the fingers of prisoners (thus fracturing them), and a large number of other physical acts of torture, including bludgeoning, some allegedly resulting in deaths of prisoners. Such tortures are clearly outlawed by the Geneva Conventions and by clearly stated U.S. principles of war. Both male and female guards participated in these acts of torture, and although most of the Iraqi victims were male, some were female.

The guards later claimed that they were simply following orders, either orders directly given or indirectly assumed. President George W. Bush and then Secretary of Defense Donald H. Rumsfeld both claimed that the acts of torture were merely the acts of a "corrupt few" and that the vast majority of American soldiers would never engage in such horrible acts.

Now consider what we know from research. The Milgram studies strongly suggest that many ordinary soldiers—who were not at all "corrupt," at least not more than average—would indeed engage in these acts of torture, particularly if they believed that they were under orders to do so, *or* if they believed that they would not be punished in any way if they did. The American soldiers must bear a significant portion of the responsibility for their own behavior. *Nonetheless, the causes of the soldiers' behaviors lie not in their personalities (their "natures") but in the social structure and group pressures of the situation.*

Evidently, the soldiers (guards) in the Abu Ghraib prison may not have received orders to torture prisoners. A now classic study of a simulated prison by Haney, Banks, and Zimbardo (1973) shows this quite clearly. In this study, Stanford University students were told by an experimenter to enter a dungeon-like basement. Half were told to pretend to be guards (to role-play being a guard) and half were told that they were prisoners (to role-play being a prisoner). Which students were told what was *randomly determined.*

After two or three days, the guards, completely on their own, began to act very sadistically and brutally toward the prisoners—having them strip naked, simulate sex, act subservient, and so on. Interestingly, the prisoners for the most part did just what the guards wanted them to do, no matter how unpleasant the requested act! The experiment was so scary that the researchers terminated the experiment after six days—more than one week early.

Remember that this study was conducted in 1973—thirty-one years *before* Abu Ghraib. Yet this simulated prison study predicted quite precisely how both "guards" and "prisoners" would act in a prison situation. Group influence effects uncovered by the Asch as well as the Milgram studies ruled in both the simulated prison of 1973 as well as the only too real Iraq prison of 2004.

Debunking Society's Myths

Myth: People are just individuals who make up their own minds about how to behave.

Sociological perspective: The Asch, Milgram, and simulated prison experiments conclusively show that people are profoundly influenced by group pressure, often causing them to make up their minds contrary to objective fact and even to deliberately cause harm to another person.

Groupthink

Wealth, power, and experience are apparently not enough to save us from social influences. **Groupthink,** as described by I.L. Janis, is the tendency for group members to reach a consensus opinion, even if that decision is downright stupid (Janis 1982).

Janis reasoned that because major government policies are often the result of group decisions, it would be fruitful to analyze group dynamics that operate at the highest level of government—for instance, in the Office of the President of the United States. The president makes decisions based upon group discussions with his advisers. The president is human and thus susceptible to group influence. To what extent have past presidents and their advisers been influenced by group decision making instead of just the facts?

Janis investigated five ill-fated decisions, all the products of group deliberation:

- The decision of the Naval High Command in 1941 *not* to prepare for the attack on Pearl Harbor by Japan, which occurred anyway.
- President Harry Truman's decision to send troops to North Korea in 1951.
- President John F. Kennedy's attempt to overthrow Cuba by launching the ill-fated invasion at the Bay of Pigs in 1962.
- President Lyndon B. Johnson's decision in 1967 to increase the number of U.S. troops in Vietnam.
- The fateful decision by President Richard M. Nixon's advisers in 1972 to break into the Democratic Party headquarters at the Watergate apartment complex, launching the famed Watergate affair.

All the preceding were group decisions, and all were absolute fiascoes. For example, the Bay of Pigs invasion was a major humiliation for the United States—a covert outing so ill conceived it is hard to imagine how it survived discussion by a group of foreign policy experts. Fifteen hundred Cuban exiles trained by the CIA to parachute into heavily armed Cuba landed in an impassibly dense 80-mile swamp far from their planned drop zone with inadequate weapons and incorrect maps. A sea landing was demolished by well-prepared, warned defenders.

Janis discovered a common pattern of misguided thinking in his investigations of presidential decisions. He surmised that outbreaks of groupthink had several things in common:

1. *An illusion of invulnerability.* "With such a brilliant team, and such a nation, how could any plan fail?" thought those in the group.
2. *A falsely negative impression of those who are antagonists to the group's plans.* Fidel Castro was perceived to be clownish, and Cuban troops were supposed to be patsies. In truth, the defenders at the Bay of Pigs were actually highly trained commandoes. Castro has remained in power to this day, several decades after the failed invasion.
3. *Discouragement of dissenting opinion.* As groupthink takes hold, dissent is equated with disloyalty. This can discourage dissenters from voicing their objections.
4. *An illusion of unanimity.* In the aftermath, many victims of groupthink recall their reservations, but at the moment of decision there is a prevailing sense that the entire group is in complete agreement.

We now might ask if groupthink influenced the torture of Iraq prisoners at the Abu Ghraib prison. The actions of the military guards there were, it seems at least in part, directly or indirectly the result of high-level group decisions among presidential advisors. Groupthink is not inevitable when a team gathers to make a decision, but it is common and appears in all sorts of groups, from student discussion groups to the highest councils of power (Flowers 1977; McCauley 1989; Aldag and Fuller 1993; Paulus et al. 2001; Kelley et al. 1999).

Debunking Society's Myths

Myth: A group of experts brought together in a small group will solve a problem according to their collective expertise.

Sociological perspective: Groupthink can lead even the most qualified people to make disastrous decisions because people in groups in the United States tend to seek to consensus at all costs.

Risky Shift

The term *groupthink* is commonly associated with group decision making with consequences that are not merely unexpected but disastrous. Another group phenomenon, **risky shift** (also called *polarization shift*), may help explain why the products of groupthink are frequently calamities. Have you ever found yourself in a group engaged in a high-risk activity that you would not do alone? When you created mischief as a child, were you not usually part of a group? If so, you were probably in the thrall of risky shift—the general tendency for groups to be more risky than individuals taken singly.

Risky shift was first observed by James Stoner (1961). Stoner gave study participants descriptions of a situation involving risk, such as one in which persons seeking a job must choose between job security versus a potentially lucrative but risky advancement. The participants were then asked to decide how much risk the person should take. Before performing his study, Stoner believed that individuals in a group would take less risk in a group than individuals alone, but he found the opposite: After his groups had

thinking sociologically

Think of a time when you engaged in some risky behavior. What group were you part of, and how did the group influence your behavior? How does this illustrate the concept of *risky shift*? Is there more risky shift with more people in the group? If so, this would illustrate a *group size effect*.

Streaking, or running nude in a public place, is more common as a group activity than as a strictly individual one. This illustrates how the group can provide the persons in it with deindividuation, *or merging of self with group. This allows the individual to feel less responsibility or blame for his or her actions, thus convincing herself or himself that the group must share the blame.*

engaged in open discussions, they favored greater risk than they would before discussion.

His research stimulated hundreds of studies using males and females, different nationalities, different tasks, and other variables (Pruitt 1971; Blaskovitch 1973; Johnson et al. 1977; Hong 1978; Worchel et al. 2000; Taylor et al. 2006). The results are complex. Most but not all group discussion leads to greater risk-taking. In subcultures that value caution above daring, as in some work groups of Japanese and Chinese firms, group decisions are less risky after discussion than before. The shift can occur in either direction, driven by the influence of group discussion, but there is generally some kind of shift, in one direction or the other, rather than no shift at all (Kerr 1992).

What causes risky shifts? The most convincing explanation is that deindividuation occurs. **Deindividuation** is the sense that one's self has merged with a group. In terms of risk-taking, one feels that responsibility (and possibly blame) is borne not only by oneself but also by the group. This seems to have happened among the American prison guards who tortured prisoners at Abu Ghraib prison: Each guard could convince himself or herself that responsibility, hence blame, was to be borne by the group as a whole.

The greater the number of people in a group, the greater the tendency toward deindividuation. In other words, deindividuation is a *group size effect.* As groups get larger, trends in risk-taking are amplified.

Formal Organizations and Bureaucracies

Groups, as we have seen, are capable of greatly influencing individuals. The study of groups and their effects on the individual represent an example of *microanalysis,* to use a concept introduced in Chapter 4. In contrast, the study of formal organizations and bureaucracies, a subject to which we now turn, represents an example of *macroanalysis.* The focus on groups drew our attention to the relatively small and less complex, whereas the focus on organizations draws our attention to the relatively large and structurally more complex.

A **formal organization** is a large secondary group, highly organized to accomplish a complex task or tasks and to achieve goals efficiently. Many of us belong to various formal organizations: work organizations, schools, and political parties, to name a few. Organizations are formed to accomplish particular tasks and are characterized by their relatively large size, compared with a small group such as a family or a friendship circle. Often organizations consist of an array of other organizations. The federal government is a huge organization comprising numerous other organizations, most of which are also vast. Each organization within the federal government is also designed to accomplish specific tasks, be it collecting your taxes, educating the nation's children, or regulating the nation's transportation system and national parks.

Organizations develop cultures and routine practices. The culture of an organization may be reflected in certain symbols, values, and rituals. Some organizations develop their own language and styles of dress. The norms can be subtle, such as men being expected to wear long-sleeve shirts and ties or women being expected to wear stockings, even on hot summer days. It does not take explicit rules to regulate this behavior; comments from coworkers or bosses may be enough to enforce such organizational norms. Some work organizations have instituted a practice called "casual day" or "dress down day"—one day per week, usually Friday, when workers can dress less formally.

Organizations tend to be persistent, although they are also responsive to the broader social environment where they are located (DiMaggio and Powell 1991). Organizations are frequently under pressure to respond to changes in the society by incorporating new practices and beliefs into their structure. Business corporations, as an example, have had to respond to increasing global competition; they do so by expanding into new international markets, developing a globally focused workforce, and trimming costs by *downsizing,* that is, by eliminating workers and various layers of management.

Another recent response to increased global competition is *outsourcing*—having manufacturing tasks ordinarily performed by the home company (such as the manufacture of athletic shoes or soccer balls) performed instead by foreign workers.

Organizations can be tools for innovation, depending on the organization's values and purpose. Rape crisis centers are examples of organizations that originally emerged from the women's movement because of the perceived need for services for rape victims. Rape crisis centers have, in many cases, changed how police departments and hospital emergency personnel respond to rape victims. By advocating changes in rape law and services for rape victims, rape crisis centers have generated change in other organizations as well (Schmitt and Martin 1999; Fried 1994).

Types of Organizations

Sociologists Blau and Scott (1974) and Etzioni (1975) classify formal organizations into three categories distinguished by their types of membership affiliation: normative, coercive, and utilitarian.

Normative Organizations. People join *normative organizations* to pursue goals that they consider worthwhile. They obtain personal satisfaction but no monetary reward for membership in such an organization. In many instances, people join the normative organization for the social prestige that it offers. Many are service and charitable organizations and are often called *voluntary organizations*. They include organizations such as the PTA, Kiwanis clubs, political parties, religious organizations, the National Association for the Advancement of Colored People (NAACP), B'nai B'rith, La Raza, and other similar voluntary organizations that are concerned with specific issues. Civic and charitable organizations, such as the League of Women Voters, and political organizations, such as the National Women's Political Caucus, reflect the fact that for decades women have been excluded from traditionally all-male voluntary organizations and political networks, such as the Kiwanis and Lions clubs. Like other service and charitable organizations, these groups have been created to meet particular needs, ones that members see as not being served by other organizations.

Gender, class, race, and ethnicity all play a role in who joins what voluntary organization. Social class is reflected in the fact that many people do not join certain organizations simply because they cannot afford to join. Membership in a professional organization, as one example, can cost hundreds of dollars each year. Those who feel disenfranchised, however, may join grassroots organizations—voluntary organizations that spring from specific local needs that people think are unmet. Tenants may form an organization to protest rent increases or lack of services, or a new political party may emerge from people's sense of alienation from existing party organizations. African Americans, Latinos, and Native Americans have formed many of their own voluntary organizations—in part because of their historical exclusion from traditional White voluntary organizations—which are now vibrant ongoing organizations in their own right (such as the African American organizations Delta Sigma Theta and Alpha Kappa Alpha sororities and the fraternities Alpha Phi Alpha, Kappa Alpha Psi, and Omega Psi Phi; see Giddings 1994).

The NAACP, founded in 1909 by W.E.B. Du Bois (recall Chapter 1), and the National Urban League are two other large national organizations that have historically fought racial oppression on the legal and urban fronts, respectively. La Raza Unida, a Latino organization devoted to civic activities as well as combating racial–ethnic oppression, has a large membership, with Latinas holding major offices. In fact, such voluntary organizations dedicated to the causes of people of color have in recent years had more women in leadership positions than have many standard, White organizations. Similarly, Native American voluntary organizations have boasted increasing numbers of women in leadership positions (Feagin and Feagin 1993; Snipp 1996, 1989).

Coercive Organizations. *Coercive organizations* are characterized by membership that is largely involuntary. Prisons are an example of organizations that people are coerced to "join" by virtue of punishment for their crime. Similarly, mental hospitals are coercive organizations: People are placed in them, often involuntarily, for some form of psychiatric treatment. In many respects, prisons and mental hospitals are similar in their treatment of inmates or patients. They both have strong security measures such as guards, locked and barred windows, and high walls (Goffman 1961; Rosenhan 1973). Sexual harassment and sexual victimization are common in both prisons and mental hospitals (Andersen 2003; Chesney-Lind 1992).

The sociologist Erving Goffman has described coercive organizations as total institutions. A **total institution** is an organization that is cut off from the rest of society and one in which resident individuals are subject to strict social control (Goffman 1961). Total institutions include two populations: the "inmates" and the staff. Within total institutions, the staff exercises complete power over inmates, for example, nurses over mental patients and guards over prisoners. The staff administers all the affairs of everyday life, including basic human functions such as eating and sleeping. Rigid routines are characteristic of total institutions, thus explaining the common complaint by those in hospitals that they cannot sleep because nurses repeatedly enter their rooms at night, regardless of whether the patient needs treatment.

Utilitarian Organizations. The third type of organization named is *utilitarian.* These are large organizations, either for-profit or nonprofit, that individuals join for specific purposes, such as monetary reward. Large business organizations that generate profits (in the case of for-profit organizations) and salaries and wages for the organization's employees (as with either for-profit or nonprofit organizations) are utilitarian organizations. Examples of large, for-profit organizations include General Motors, Microsoft, Amazon.com, and Procter & Gamble. Examples of large nonprofit organizations that pay salaries to employees are colleges, universities, and the Educational Testing Service (ETS).

Bureaucracy

As formal organizations develop, many become a **bureaucracy,** a type of formal organization characterized by an authority hierarchy, a clear division of labor, explicit rules, and impersonality. Bureaucracies are notorious for their unwieldy size and complexity as well as their reputation for being remote and cumbersome organizations that are highly impersonal and machinelike in their operation. The federal government is a good example of a cumbersome bureaucracy that many believe is ineffective because of its sheer size. Numerous other formal organizations have developed into huge bureaucracies: IBM, Disney, many universities, hospitals, state motor vehicle registration systems, and some law firms. Other formal organizations, such as Enron and WorldCom, quickly developed into large bureaucracies, then subsequently collapsed under fraudulent accounting procedures.

The early sociological theorist **Max Weber** (1947/1925) analyzed the classic characteristics of the bureaucracy. These characteristics represent what he called the *ideal type bureaucracy*—a model rarely seen in reality but which defines the typical characteristics of a social form. The characteristics of bureaucracies are as follows.

1. **High degree of division of labor and specialization.** The notion of the specialist embodies this criterion. Bureaucracies ideally employ specialists in the various positions and occupations, and these specialists are responsible for a specific set of duties. Sociologist Charles Perrow (1994, 1986) notes that many modern bureaucracies have hierarchical authority structures and an elaborate division of labor.
2. **Hierarchy of authority.** In bureaucracies, positions are arranged in a hierarchy so that each is under the supervision of a higher position. Such hierarchies are often represented in an *organizational chart,* a diagram in the shape of a pyramid that shows the relative rank of each position plus the lines of authority between each. These lines of authority are often called the "chain of command," and they show not only who has authority, but also who is responsible to whom and how many positions are responsible to a given position.
3. **Rules and regulations.** All the activities in a bureaucracy are governed by a set of detailed rules and procedures. These rules are designed, ideally, to cover almost every possible situation and problem that might arise, including hiring, firing, salary scales, and rules for sick pay and absences.
4. **Impersonal relationships.** Social interaction in the bureaucracy is supposed to be guided by *instrumental* criteria, such as the organization's rules, rather than by *expressive needs,* such as personal attractions or likes and dislikes. The ideal is that the objective application of rules will minimize matters such as personal favoritism—giving someone a promotion simply because you like him or her or firing someone because you do not like him or her. Of course, as we will see, sociologists have pointed out that bureaucracy has another face—the social interaction that keeps the bureaucracy working and often involves interpersonal friendships and social ties, typically among people taken for granted in these organizations, such as the support staff.
5. **Career ladders.** Candidates for the various positions in the bureaucracy are supposed to be selected on the basis of specific criteria, such as education, experience, and standardized examinations. The idea is that advancement through the organization becomes a career for the individual.
6. **Efficiency.** Bureaucracies are designed to coordinate the activities of many people in pursuit of organizational goals. Ideally, all activities have been designed to maximize this efficiency. The whole system is intended to keep social–emotional relations and interactions at a minimum and instrumental interaction at a maximum.

Bureaucracy's Other Face

All the characteristics of Weber's "ideal type" are general defining characteristics. Rarely do actual bureaucracies meet this exact description. A bureaucracy has, in addition to the ideal characteristics of structure, an *informal structure.* This includes social interactions, even network connections, in bureaucratic settings that ignore, change, or otherwise bypass the formal structure and rules of the organization. This informal structure often develops among those who are taken for granted in organizations, such as secretaries. Sociologist Charles Page (1946) coined the phrase *bureaucracy's other face* to describe this condition.

This other face is informal culture. It has evolved over time as a reaction to the formality and impersonality of the bureaucracy. Thus, secretaries will sometimes "bend the rules a bit" when asked to do something more quickly than usual for a boss they like and bend the rules in another direction for a boss they do not like by slowing down or otherwise sabotaging the boss's work. Researchers have noted, for example, that secretaries and assistants have more authority than their job titles—and salaries—suggest. As a way around the cumbersome formal communication channels within the organization, the informal network, or "grapevine," often works better, faster, and sometimes even more accurately than the formal channels. As with any culture, the informal culture in the bureaucracy has its own norms or rules. One is not supposed to "stab friends in the back," such as by "ratting on" them to a boss or spreading a rumor about them that is intended to get them fired. Yet just as with any norms, there is deviation from the norms, and "backstabbing" and "ratting" does happen.

Bureaucracy's other face can also be seen in the workplace subcultures that develop, even in the largest bureaucracies. Some sociologists interpret the subcultures that develop within bureaucracies as people's attempts to humanize an otherwise impersonal organization. Keeping photographs of family and loved ones in the office, placing personal decorations on one's desk (if you are allowed), and organizing office parties are some ways people resist the impersonal culture of bureaucracies. Of course, this informal culture can also become exclusionary, increasing the isolation that some workers feel at work. Gay and lesbian workers may feel left out when other workers gossip about people's heterosexual dates; minority workers may be excluded from the casual conversations in the workplace that connect nonminority people to one another.

The informal norms that develop within the modern day bureaucracy often cause worker productivity to go up or down, depending on the norms and how they are informally enforced. The classic 1930s' Hawthorne Studies, so named because they were carried out at the Western Electric telephone plant in Hawthorne, Illinois (Roethlisberger and Dickson 1939), discovered that small groups of workers developed their own ideas—their own norms—about how much work they should produce each day. If someone produced too many completed tasks in a day, he would make the rest of the workers "look bad" and run the risk of having the organization raise its expectations of how much work the group might be expected to produce. Because of this, anyone producing too much was informally labeled a "rate buster," and that person was punished by some act, such as punches on the shoulder (called "binging"), or by group ridicule (called "razzing"). By the same token, one could be accused of producing too little, in which case he was labeled a "chiseler" and punished in the same way by either binging or razzing. This informal culture of bureaucracy's other face continues today in a manner similar to the culture initially discovered in the early Hawthorne studies (Perrow 1994, 1986).

Problems of Bureaucracies

In contemporary times, problems have developed that grow out of the nature of the complex bureaucracy. Two problem areas already discussed are the occurrence of risky shift in work groups and the development of groupthink. Additional problems include a tendency to *ritualism* and the potential for *alienation* on the part of those within the organization.

Ritualism. Rigid adherence to rules can produce a slavish following of them, regardless of whether it accomplishes the purpose for which the rule was originally designed. The rules become ends in themselves rather than means to an end: This is ritualism.

Two now classic examples of the consequences of *organizational ritualism* have come to haunt us: the explosion of the space shuttle *Challenger* on January 28, 1986, and, to our horror, the breakup of yet another space shuttle, the *Columbia,* on February 1, 2003. People in the United States became bound together at the moment of the *Challenger* accident. Many remember where they were and exactly what they were doing when they heard about the tragedy. The failure of the essential O-ring gaskets on the solid fuel booster rockets of the *Challenger* shuttle caused the catastrophic explosion. It was revealed later that the O-rings became brittle at below-freezing temperatures, as was the temperature at the launch pad the evening before the *Challenger* lifted off.

Why did the managers and engineers at NASA (National Aeronautics and Space Administration) allow the shuttle to lift off given these prior conditions? The managers had all the information about the O-rings before the launch. Furthermore, engineers had warned them against the danger. In a detailed analysis of the decision to launch, sociologist Diane Vaughan (1996) uncovered both risky shift and organizational ritualism within the organization. The NASA insiders, confronted with signals of danger, proceeded as if nothing was wrong when they were repeatedly faced with the evidence that something was indeed *very* wrong. They in effect *normalized* their own behavior so that their actions became acceptable to them, representing nothing out of the ordinary. This is an example of organizational ritualism, as well as what Vaughan calls the "normalization of deviance."

Unfortunately, history repeated itself on February 1, 2003, when the space shuttle *Columbia,* upon its return from space, broke up in a fiery descent into the atmosphere above Texas, killing all who were

The horror of the explosion of the space shuttle Challenger in 1986 is seen in the faces of the observers here. All seven astronauts died in the explosion (top right photo). Sociologist Diane Vaughan (1996) attributes the disaster to an ill-formed launch decision in the bureaucracy of NASA based on group interaction phenomena such as risky shift, ritualism, groupthink, and the normalization of deviance. Tragedy struck again in February 2003, when the space shuttle Columbia broke up upon reentry into the atmosphere, killing all seven of the astronauts on board (bottom right photo).

aboard. The evidence shows that a piece of hard insulating foam separated from an external fuel tank during launch and struck the shuttle's left wing, damaging it and dislodging its heat-resistant tiles that are necessary for reentry. The absence of these tiles caused a burn-up upon reentry into the atmosphere. With eerie similarity to the earlier 1986 *Challenger* accident, a recent research report concludes that a "flawed institutional culture" and—citing sociologist Diane Vaughan—a normalization of deviance accompanying a gradual erosion of safety margins were among the causes of the *Columbia* accident (Schwartz and Wald 2003, and a Spring 2007 report on CNN).

No single individual was at fault in either accident. The story is not one of evil but rather of the ritualism of organizational life in one of the most powerful bureaucracies in the United States. It is a story of rigid group conformity within an organizational setting and of how deviant behavior is redefined, that is, socially constructed, to be perceived as normal. Organizational procedures in this case, or rituals, so dominate, the means toward goals become the goals themselves. A critical decrease in overall safety and dramatic increase in risk are the results. Vaughan's analysis is a powerful warning about the hidden hazards of bureaucracy in a technological age.

Alienation. The stresses on rules and procedures within bureaucracies can result in a decrease in the overall cohesion of the organization. This often psychologically separates a person from the organization and its goals. This state of *alienation* results in increased turnover, tardiness, absenteeism, and overall dissatisfaction with the organization.

Alienation can be widespread in organizations where workers have little control over what they do or where workers themselves are treated like machines—employed on an assembly line, doing the same repetitive action for an entire work shift. Alienation is not restricted to manual labor, however. In organizations where workers are isolated from others, where they are expected only to implement rules, or where they think they have little chance of advancement, alienation can be common. As we will see, some organizations have developed new patterns of work to try to minimize worker alienation and thus enhance their productivity.

The McDonaldization of Society

Sometimes the problems and peculiarities of bureaucracy can have effects on the total society. This has been the case with what George Ritzer (2002) has called the *McDonaldization of society,* a term coined from the well-known fast-food chain. In fact, one study (Schlosser 2001) concluded that each month, 90 percent of the children in the United States between ages 3 and 9 visit McDonald's! Ritzer noticed that the principles that characterize fast-food organizations are increasingly dominating more aspects of U.S. society, indeed, of societies around the world. McDonaldization refers to the increasing and ubiquitous presence of the fast-food model in most organizations that shape daily life. Work, travel, leisure, shopping, health care, education, and politics have all become subject to McDonaldization. Each industry is based on a principle of high and efficient productivity, which translates into a highly rational social organization, with workers employed at low pay but with customers experiencing ease, convenience, and familiarity.

Ritzer argues that McDonald's has been such a successful model of business organization that other industries have adopted the same organizational characteristics, so much so that their nicknames associate them with the McDonald's chain: McPaper for *USA Today,* McChild for child-care chains like KinderCare, and McDoctor for the drive-in clinics that deal quickly and efficiently with minor health and dental problems.

Based in part upon Max Weber's concept of the ideal bureaucracy mentioned earlier, Ritzer identifies four dimensions of the McDonaldization process: efficiency, calculability, predictability, and control:

1. **Efficiency** means that things move from start to finish in a streamlined path. Steps in the production of a hamburger are regulated so that each hamburger is made exactly the same way—hardly characteristic of a home-cooked meal. Business can be even more efficient if the customer does the work once done by an employee. In fast-food restaurants, the claim that you can "have it your way" really means that you assemble your own sandwich or salad.
2. **Calculability** means there is an emphasis on the quantitative aspects of products sold—size, cost, and the time it takes to get the product. At McDonald's, branch managers must account for the number of cubic inches of ketchup used per day; likewise, ice cream scoopers in chain stores measure out predetermined and exact amounts of ice cream. Workers are monitored for how long it takes them to complete a transaction; every bit of food and drink is closely monitored by computer, and everything has to be accounted for.
3. **Predictability** is the assurance that products will be exactly the same, no matter when or where they are purchased. Eat an Egg McMuffin in New York, and it will likely taste just the same as an Egg McMuffin in Los Angeles or Paris!
4. **Control** is the primary organizational principle that lies behind McDonaldization. Behavior of the customers and workers is reduced to a series of machinelike actions. Ultimately, efficient technologies replace much of the work that humans once performed.

AP Images/Tony Dejak

Evidence of the "McDonaldization of society" can be seen everywhere, perhaps including on your own campus. Shopping malls, food courts, sports stadiums, even cruise ships reflect this trend toward standardization.

McDonaldization clearly brings many benefits. There is a greater availability of goods and services to a wide proportion of the population; instantaneous service and convenience to a public with less free time; predictability and familiarity in the goods bought and sold; and standardization of pricing and uniform quality of goods sold, to name a few benefits. However, this increasingly rational system of goods and services also spawns irrationalities. For example, the majority of workers at McDonald's lack full-time employment, have no worker benefits, have no control over their workplace, and quit on average after only four or five months (Schlosser 2001). Ritzer argues that, as we become more dependent on the familiar and that which is taken for granted, there is the danger of dehumanization. People lose their creativity, and there is little concern with the quality of goods and services, thereby disrupting something fundamentally human—the capacity for error, surprise, and imagination.

Notice, as you go through your daily life the extraordinary presence of the process Ritzer has observed.

In what areas of life do you see the process of McDonaldization? How has it influenced the national culture? In a society with as much cultural diversity as the United States, what does it mean that we have developed a system for the production and distribution of goods that relies on such uniformity? These questions will help you see how McDonaldization has permeated U.S. society and will help you think about formal organizations as a sociologist would. You might ask where you see evidence of McDonaldization on your campus, your job and in your community.

Diversity: Race, Gender, and Class in Organizations

The hierarchical structuring of positions within organizations results in the concentration of power and influence with a few individuals at the top. Since organizations tend to reflect patterns within the broader society, this hierarchy, like that of society, is marked by inequality in race, gender, and class relations. Although the concentration of power in organizations is incompatible with the principles of a democratic society (Perrow 1994, 1986), discrimination against women and minorities still occurs. There have been widespread disparities in the promotion rates for White and Black workers, with Whites more likely to be promoted and promoted more quickly—a pattern repeated in many work organizations (Eichenwald 1996; McGuire and Reskin 1993; Collins 1989).

Traditionally, within organizations the most powerful positions are held by White men of upper social class status. Women and minorities, on average, occupy lower positions in the organization. Although very small numbers of minorities and women do get promoted, there is typically a "glass ceiling" effect, meaning that women and minorities may be promoted but only up to a point. The glass ceiling acts as a barrier to the promotion of women and minorities into higher ranks of management, as discussed further in Chapter 10. What then are the barriers that prevent more inclusiveness in the higher ranks of organizations?

Sociological research finds that organizations are sensitive to the climate in which they operate. The more egalitarian the environment in which a firm operates, the more equitable is its treatment of women and minorities.

Yet studies find that patterns of race and gender discrimination persist throughout organizations, even when formal barriers to advancement have been removed. Many minorities are now equal to Whites in education, particularly in organizational jobs that require advanced graduate degrees such as the master of business administration (MBA). Still, White men in organizations are more likely to receive promotions than African American, Hispanic, and Native American workers *with the same education* (DeWitt 1995; Zwerling and Silver 1992). The same thing often happens to both White and minority women in organizations: Women are less likely to receive promotions than a White male with the same education, and sometimes even less education. Studies consistently find that women are held to higher promotion standards than men; for men, the longer they are in a position in an organization, the more likely they will be promoted, but the same is not true for women

Few organizational boards and executive committees contain minorities and women, as does this one.

(Smith 1999). Studies find that women change jobs more frequently within organizations than do men, but these tend to be lateral moves. For men, job changes are more likely to mark a jump from a lower level to a higher level in the organization, thus constituting a promotion.

Things work the same way with respect to people being discharged or fired. Studies show quite clearly that Black federal employees (men and women) were more than *twice as likely* to be dismissed than their White counterparts (DeWitt 1995; Zwerling and Silver 1992). This disparity occurred regardless of education and regardless of occupational category, pay level, type of federal agency, age, performance rating, seniority, or attendance record. The main reasons cited in the studies for such disparities were lack of network contacts with the old boy network and racial bias within the organization. These studies were particularly well designed because they took into account quite a few factors besides race–ethnicity that are often given as reasons for not promoting minorities. The studies showed quite conclusively that with all factors taken into account, it is still race that is an important reason for the way people are treated in organizations today. The studies strongly suggest that racism still thrives in the bureaucracy.

Debunking Society's Myths

Myth: Programs designed to enhance the number of women and minorities in organizational leadership are no longer needed because discriminatory barriers have been removed.

Sociological perspective: Research continues to find significant differences in the promotion rates for women and minorities in most organizational settings. Even with the removal of formal discriminatory barriers, organizational practices persist that block the mobility of these workers.

A classic study by Rosabeth Moss Kanter (1977) shows how the structure of organizations leads to obstacles in the advancement of groups that are tokens—rarely represented—in the organizational environment. Kanter demonstrated how the hierarchical structure of the bureaucracy negatively affects both minorities and women who are underrepresented in the organization. In such cases, they represent *token* minorities or women; they feel put "out front" and under the all-too-watchful eyes of their superiors. As a result, they often suffer severe stress (Jackson 2000; Jackson et al. 1995, 1994; Spangler et al. 1978; Kanter 1977; Yoder 1991). Their coworkers often accuse them of getting the position simply because they are women, minorities, or both. It is a widespread phenomenon in universities and colleges that not only minorities, but women as well, are often accused of being admitted simply because of their gender and/or their race, even in instances when this has clearly not been the case. This is stressful for the person and shows that tokenism can have very negative consequences.

Social class, in addition to race and gender, plays a part in determining people's place within formal organizations. Middle- and upper-class employees in organizations make higher salaries and wages and are more likely to get promoted than are people of lower social class status, even for individuals who are of the same race or ethnicity. This even holds for people coming from families of lower social class status who are as well educated as their middle- and upper-class coworkers. Thus, their lower salaries and lack of promotion cannot necessarily be attributed to a lack of education. In this respect, their treatment in the bureaucracy only perpetuates rather than lessens the negative effects of the social class system in the United States.

The social class stratification system in the United States produces major differences in the opportunities and life chances of individuals, and the bureaucracy simply carries these differences forward. Class stereotypes also influence hiring practices in organizations. Personnel officers look for people with "certain demeanors," a code for those who convey middle-class or upper middle-class standards of dress, language, manners, and so on, which some people may be unable to afford or may not possess.

Patterns of race, class, and gender inequality in organizations persist at the same time that many organizations have become more aware of the need for a more diverse workforce. Responding to the simple fact of more diversity within the working-age population, organizations have developed human relations experts who work within organizations to enhance sensitivity to diversity. Such diversity training has become commonplace in many large organizations.

Functional, Conflict, and Symbolic Interaction: Theoretical Perspectives

All three major sociological perspectives—functional, conflict, and symbolic interaction—are exhibited in the analysis of formal organizations and bureaucracies (see Table 5.1). The functional perspective, based in this case on the early writing of Max Weber, argues that certain functions, called *eufunctions* (that is, positive functions) characterize bureaucracies and

table 5.1 Theoretical Perspectives on Organizations

	Functionalist Theory	Conflict Theory	Symbolic Interaction Theory
Central Focus	Positive functions (such as efficiency) contribute to unity and stability of the organization	Hierarchical nature of bureaucracy encourages conflict between superior and subordinate, men and women, and people of different racial or class backgrounds	Stresses the role of self in the bureaucracy and how the self develops and changes
Relationship of Individual to the Organization	Individuals, like parts of a machine, are only partly relevant to the operation of the organization	Individuals subordinated to systems of power and experience stress and alienation as a result	Interaction between superiors and subordinates forms the structure of the organization
Criticism	Hierarchy can result in dysfunctions such as ritualism and alienation	De-emphasizes the positive ways that organizations work	Tends to downplay overall social organization

contribute to their overall unity. The bureaucracy exists to accomplish these eufunctions, such as efficiency, control, impersonal relations, and a chance for the individual to develop a career within the organization. As we have seen, however, bureaucracies develop the "other face" (informal interaction and culture, as opposed to formal or bureaucratic interaction and culture) as well as the problems of ritualism and alienation of the person from the organization. These latter problems are called *dysfunctions* (negative functions), which have the consequence of contributing to the disunity and lack of harmony in the bureaucracy.

The conflict perspective argues that the hierarchical or stratified nature of the bureaucracy in effect encourages rather than inhibits conflict among the individuals within it. These conflicts are between superior and subordinate, as well as between racial and ethnic groups, men and women, and people of different social class backgrounds, hampering the smooth and efficient running of the bureaucracy.

Consider the symbolic interaction perspective as underlying two management theories, those of Argyris (1990) and of Ouchi (1981). Symbolic interaction stresses the role of the self in any group and especially how the self develops as a product of social interaction. Argyris's theory advocates increased involvement of the self within the organization as a way of "actualizing" the self. This helps reduce the disconnection between individual and organization as well as other organizational problems and dysfunctions. Ouchi's theory argues that increased interaction between superior and subordinate, based on the Japanese organization model of executives "walking around" and interacting more on a primary group basis, will reduce organizational dysfunction.

5 Chapter Summary

- ***What are the types of groups?***

Groups are a fact of human existence and permeate virtually every facet of our lives. Group size is important, as is the otherwise simple distinction between dyads and triads. *Primary groups* form the basic building blocks of social interaction in society. *Reference groups* play a major role in forming our attitudes and life goals, as do our relationships with in-groups and out-groups. *Social networks* partly determine things such as who we know and the kinds of jobs we get. Networks based on race–ethnicity, social class, and other social factors are extremely closely connected or very dense.

- ***How strong is social influence?***

The social influence groups exert on us is tremendous, as seen by the Asch conformity experiments. The Milgram experiments demonstrated that the interpersonal influence of an authority figure can cause an individual to act against his or her deep convictions. The recent torture and abuse of Iraqi prisoners of war by American soldiers as prison guards serves as testimony to the powerful effects of both social influence and authority structures.

- ***What is the importance of groupthink and risky shift?***
  Groupthink can be so pervasive that it adversely affects group decision making. *Risky shift* similarly often compels individuals to reach decisions that are at odds with their better judgment.

- ***What are the types of formal organizations and bureaucracies, and what are some of their problems?***
  There are several types of *formal organizations,* such as normative, coercive, or utilitarian. Weber typified *bureaucracies* as organizations with an efficient division of labor, an authority hierarchy, rules, impersonal relationships, and career ladders. Bureaucratic rigidities often result in organizational problems such as ritualism and resulting "normalization of deviance," which may have been significantly responsible for the space shuttle *Challenger* explosion in 1986, and the space shuttle *Columbia* breakup in 2003. The "McDonaldization of society" has resulted in greater efficiency, calculability, and control in many industries, probably at the expense of some individual creativity.

- ***What are the problems of diversity in organizations?***
  Formal organizations perpetuate inequality on the basis of race–ethnicity, gender, and social class. Even today Blacks, Hispanics, and Native Americans are less likely to get promoted and more likely to get fired than Whites of comparable education and other qualifications. Women experience similar effects of inequality, especially negative effects of tokenism, such as stress and lowered self-esteem. Finally, persons of less than middle-class origins make less money and are less likely to get promoted than a middle-class person of comparable education. Diversity training has been introduced into some organizations as an attempt to combat such problems.

- ***What do the functional, conflict, and symbolic interaction theories say about organizations?***
  Functional, conflict, and symbolic interaction theories highlight and clarify the analysis of organizations by specifying both organizational functions and dysfunctions (*functional theory*); by analyzing the consequences of hierarchical, gender, race, and social class conflict in organizations (*conflict theory*); and, finally, by studying the importance of social interaction and integration of the self into the organization (*symbolic interaction theory*).

Key Terms

attribution error 126
attribution theory 126
bureaucracy 135
deindividuation 133
dyad 123
expressive needs 124
formal organization 133
group 122
group size effect 123
groupthink 131
instrumental needs 125
primary group 123
reference group 125
risky shift 132
secondary group 123
social network 127
total institution 134
triad 123
triadic segregation 123

Online Resources

Sociology: The Essentials Companion Website

www.thomsonedu.com/sociology/andersen

Visit your book companion website where you will find more resources to help you study and write your research papers. Resources include Suggested Readings, web links, and a MicroCase Online feature that teaches you how to research society. Other resources include Learning Objectives, Internet exercises, quizzing, and flash cards.

is an easy-to-use online resource that helps you study in less time to get the grade you want NOW.

www.thomsonedu.com/login

Need help studying? This site is your one-stop study shop. Take a Pre-Test and Thomson NOW will generate a Personalized Study Plan based on your test results. The Study Plan will identify the topics you need to review and direct you to online resources to help you master those topics. You can then take a Post-Test to determine the concepts you have mastered and what you still need to work on.

6
Chapter six
CHAPTER SIX

Deviance and Crime

In the early 1970s, an airplane carrying forty members of an amateur rugby team crashed in the Andes Mountains in South America. The twenty-seven survivors were marooned at 12,000 feet in freezing weather and deep snow. There was no food except for a small amount of chocolate and some wine. A few days after the crash, the group heard on a small transistor radio that the search for them had been called off.

Scattered in the snow were the frozen bodies of dead passengers. Preserved by the freezing weather, these bodies became, after a time, sources of food. At first, the survivors were repulsed by the idea of eating human flesh, but as the days wore on, they agonized over the decision about whether to eat the dead crash victims, eventually concluding that they had to eat if they were to live.

In the beginning, only a few ate the human meat, but soon the others began to eat, too. The group experimented with preparations as they tried different parts of the body. They developed elaborate rules about how, what, and whom they would eat. Some could not bring themselves to cut the meat from the human body, but would slice it once someone else had cut off large chunks. They all refused to eat certain parts—the lungs, skin, head, and genitals.

continued

six chapter six

CHAPTER SI

After two months, the group sent out an expedition of three survivors to find help. The group was rescued, and the world learned of their ordeal. Their cannibalism (the eating of other human beings) was generally accepted as something they had to do to survive. Although people might have been repulsed by the story, the survivors' behavior was understood as a necessary adaptation to their life-threatening circumstances. The survivors also maintained a sense of themselves as good people even though what they did profoundly violated ordinary standards of socially acceptable behavior in most cultures in the world (Read 1974; Miller 1991; Henslin 1993).

Was the behavior of the Andes crash survivors socially deviant? Were the people made crazy by their experience, or was this a normal response to extreme circumstances?

Compare the Andes crash to another case of human cannibalism. In 1991, in Milwaukee, Wisconsin, Jeffrey Dahmer pled guilty to charges of murdering at least fifteen men in his home. Dahmer lured the men—eight of them African American, two White, and one a fourteen-year-old Laotian (Asian) boy—to his apartment, where he murdered and dismembered them, then cooked and ate some of their body parts. For those he considered most handsome, he boiled the flesh from their heads so that he could save and admire their skulls. Dahmer was seen as a total social deviant, someone who violated every principle of human decency. Even hardened criminals were disgusted by Dahmer. In fact, he was killed in prison by another inmate in 1994.

Why was Dahmer's behavior considered so deviant when that of the Andes survivors was not? The answer can be found by looking at the situation in which these behaviors occurred. For the Andes survivors, eating human flesh was essential for survival. For Dahmer, however, it was murder. From a sociological perspective, the deviance of cannibalism resides *not just in the act itself* but also the social context in which it occurs. The exact same behavior—eating other human beings—is considered reprehensible in one context and acceptable in another. That is the essence of the sociological explanation: The nature of deviance is not only in the personality of the deviant person, nor is it inherently in the deviant act itself, but instead it is a significant part and product of the social structure.

Defining Deviance

Sociologists define **deviance** as *behavior that is recognized as violating expected rules and norms.* Deviance is more than simple nonconformity; it is behavior that departs significantly from social expectations. In the sociological perspective on deviance, there is subtlety that distinguishes it from our commonsense understanding of the same behavior:

- *The sociological definition of deviance stresses social context, not just individual behavior.* Sociologists see deviance in terms of group processes, definitions, and judgments, not just as unusual individual acts.
- *The sociological definition of deviance recognizes that not all behaviors are judged similarly by all groups.* What is deviant to one group may be normative (not deviant) to another.
- *The sociological definition of deviance recognizes that established rules and norms are* socially created, *not just morally decided or individually imposed.* Sociologists emphasize that deviance lies not just in behavior itself, but in the social responses of groups to behavior by others.

Sociological Perspectives on Deviance

Strange, unconventional, or nonconformist behavior is often understandable in its sociological context. Consider suicide. Are people who commit suicide mentally disturbed, or might their behavior be explained? Are there conditions under which suicide is acceptable behavior? Would one who commits suicide in the face of a terminal illness be judged differently from a despondent person who jumps from a window? These are the kinds of questions asked by sociologists who study deviance.

Once you have a sociological perspective on deviance, you are likely to see things a little differently when you observe someone behaving in an unusual way. Sociologists sometimes use their understanding of deviance to explain otherwise ordinary events—such as tattooing and body piercing (Irwin 2001; Vail 1999), eating disorders (Sharp et al. 2000; Lovejoy 2001; Thompson 1994), or drug and alcohol use (Inciardi 2001; Humphries 1999; Logio-Rau 1998).

Sociologists distinguish two types of deviance: formal and informal. *Formal deviance* is behavior that breaks laws or official rules. Crime is an example. There are formal sanctions against formal deviance, such as imprisonment and fines. *Informal deviance* is behavior that violates customary norms (Schur 1984). Although such deviance may not be specified in law, it is judged to be deviant by those who uphold the society's norms. A good example is body piercing.

Although there are no laws against such a practice, it violates older norms about dress and appearance and is judged by many outside the youth culture to be socially deviant, even though it is quite fashionable for others.

The study of deviance can be divided into the study of why people violate laws or norms and the study of how society reacts. This reaction includes the labeling process by which deviance comes to be recognized as such. *Labeling theory* is discussed in detail later, but it is important to point out here that the meaning of deviance is not just in the breaking of norms or rules; it also includes how people react to those behaviors. This societal reaction to deviant behavior suggests that social groups actually *create* deviance "by making the rules whose infraction constitutes deviance, and by applying those rules to particular people and labeling them as outsiders" (Becker 1963: 9).

The Context of Deviance. Some situations are more conducive to the formation of deviant behavior. Even the most unconventional behavior can be understood if we know the context in which it occurs. Behavior that is deviant in one circumstance may be normal in another, or behavior may be ruled deviant only when performed by certain people. For example, people who break gender stereotypes may be judged as deviant even though their behavior is considered normal for the other sex. Heterosexual men and women who kiss in public are the image of romance; lesbians and gay men who even dare to hold hands in public are often seen as flaunting their sexual orientation.

The definition of deviance can also vary over time. Acquaintance rape (formerly called "date rape"), for example, was not considered social deviance until recently. Women have been presumed to mean yes when they said no, and men were expected to "seduce" women through aggressive sexual behavior. Even now, women who are raped by someone they know may not think of it as rape. If they do, they may find that prosecuting the offender is difficult because others do not think of it as rape. Studies have found that students think that acquaintance rape is justified if the victim is wearing provocative clothing (Workman and Freeburg 1999; Johnson 1995; Cassidy and Hurrell 1995). People with more traditional attitudes about gender roles are more likely to excuse men's aggression in acquaintance rape and to define the situation as something other than rape. These examples show that the definition of deviance derives not only from what one does, but also from who does it, when, and where.

The sociologist Emile Durkheim argued that one reason acts of deviance are publicly punished is that the social order is threatened by deviance. Judging those behaviors as deviant and punishing them confirms general social standards. Therein lies the value of widely publicized trials, public executions, or the historical practice of displaying a wrongdoer in the stocks, which held one's feet fast, or the pillory, which held the hands and head. Passersby were permitted to hurl both stones and large rocks at those persons so immobilized. The punishment affirms the collective beliefs of the society, reinforces social order, and inhibits future deviant behavior, especially as defined by those with the power to judge others.

Durkheim argued that societies actually *need* deviance to know what presumably normal behavior is. In this sense, Durkheim considered deviance "functional" for society (Durkheim 1951/1897; Erikson 1966). You could observe Durkheim's point in the

Debunking Society's Myths

Myth: Deviance is bad for society because it disrupts normal life.

Sociological perspective: Deviance tends to stabilize society. By defining some forms of behavior as deviant, people are affirming the social norms of groups. In this sense, society actually to some extent *creates* deviance.

RAWA/WorldPicture News

This widely distributed photo of a woman being executed by the Taliban in Afghanistan illustrates the extreme sanctions that can be brought against those defined as deviant by a powerful group. In this case, the photo also mobilized world condemnation of the Taliban regime for its treatment of women.

aftermath of the terrorist attacks of September 11, 2001. Horrified by the sight of hijacked planes flying into the World Trade Center and the Pentagon and crashing in a Pennsylvania field, U.S. citizens responded through publicly demonstrating strong patriotism. *Durkheim would interpret these terrorist acts as deviance producing strong social solidarity.* This was one of Durkheim's most important insights: Deviance produces social solidarity. Instead of breaking society up, deviance produces a pulling together, or social solidarity.

The Influence of Social Movements. The perception of deviance may also be influenced by social movements, which are networks of groups that organize to support or resist changes in society (see Chapter 16). With a change in the social climate, formerly acceptable behaviors may be newly defined as deviant. Smoking, for instance, was once considered glamorous, sexy, and "cool." Now, smokers are widely scorned as polluters and, despite strong lobbying by the tobacco industry, regulations against smoking have proliferated.

Whereas in 1987 only 17 percent of the public thought that smoking should be banned in restaurants, by 2006 over half (60 percent) thought so (Gallup Organization 2006). The increase in public disapproval of smoking results as much from social and political movements as it does from the known health risks. The success of the antismoking movement has come from the mobilization of constituencies able to articulate to the public that smoking is dangerous. Note that the key element here is the ability of people to mobilize—not just the evidence of risk. In other words, there has to be a social response for deviance to be defined as such; scientific evidence of harm in and of itself is not enough.

© Jeff Greenberg/PhotoEdit

Once considered "cool", smokers are now considered to be deviants, scorned as polluters, and often banished to outside office buildings, as here.

Social movements like the gay and lesbian movement can also organize to remove the deviant label from certain behaviors. Gay and lesbian behavior traditionally has been defined as deviant, but this movement has encouraged people to see gay and lesbian relationships as legitimate. By organizing to legalize gay marriage or to extend employment benefits to domestic partners, people in this movement have not only advocated new rights for lesbian women and gay men but have challenged the public label of gays and lesbians as deviant.

The Social Construction of Deviance. Perhaps because it violates social conventions or because it sometimes involves unusual behavior, deviance captures the public imagination. Commonly, however, the public understands deviance as the result of individualistic or personality factors. Many people see deviants as crazy, threatening, "sick," or in some other ways inferior, but sociologists see deviance as influenced by society—by the same social processes and institutions that shape all social behavior.

Deviance, for example, is not necessarily irrational or "sick" and may be a positive and rational adaptation to a situation. Think of the Andes survivors discussed in this chapter's opener. Was their action (eating human flesh) irrational, or was it an inventive and rational response to a dreadful situation? To use another example, are gangs the result of the irrational behavior of maladjusted youth, or are they predictable responses to a social situation?

Sociological studies of gangs in the United States shed light on this question. The family situations of gang members are often problematic, although girls in gangs tend to be more isolated from their families than are boys in gangs (Fleisher 2000; Esbensen-Finn et al. 1999). Given the class, race, and gender inequality faced by minority youth, many turn to gangs for the social support they lack elsewhere (Walker-Barnes and Mason 2001; Moore and Hagedorn 1996). For example, some young, poor Puerto Rican girls live in relatively confined social environments with little opportunity for educational or occupational advancement. Their community expects them to be "good girls" and to remain close to their families. Joining a gang is one way to reject these restrictive roles (Messerschmidt 1997; Campbell 1987). Are these young women irrational or just doing the best they can to adapt to their situation? Sociologists interpret their behavior as an understandable adaptation to conditions of poverty, racism, and sexism.

In some subcultures or situations, deviant behavior is encouraged and praised. Have you ever been egged on by friends to do something that you thought was deviant or have you done something you knew was wrong? Many argue that the reason so many college students drink excessively is that the student subculture encourages them—even though students know it is harmful. Similarly, the juvenile delinquent regarded by school authorities as defiant and obnoxious is rewarded and praised by peers for the very behaviors that school authorities loathe. Much deviant behavior occurs, or escalates, because of the social support from the group in which it develops.

Some behavior patterns defined as deviant are also surprisingly similar to so-called normal behavior. Is a heroin addict who buys drugs with whatever money he can find so different from a business executive who spends a large proportion of his discretionary income on alcohol? Each may establish a daily pattern that facilitates drug use; each may select friends based on shared interests in drinking or taking drugs; and each may become so physically, emotionally, and socially dependent on their "fix" that life seems unimaginable without it. Which of the two is more likely to be considered deviant?

The point is that deviance is both created and defined within the social context. It is not just weird, pathological, irrational, off-putting, or unconventional behavior. Sociologists who study deviance understand it in the context of social relationships and society. They define deviance in terms of existing social norms and the social judgments people make about one another. Sometimes behavior that is deviant in one context may be perfectly normal in another (for example, men wearing earrings or women wearing boxer shorts). Indeed, deviant behavior can sometimes be indicative of changes that are taking place in the cultural folkways. Thus, whereas only a few years ago body piercing and tattooing were associated with gangs and disrespectable people, now it is considered fashionable among young, middle-class people—even though to some it is still a mark of deviance.

Psychological Explanations of Deviance

Sociologists explain deviance as more than a person's individual personality. Psychologists, on the other hand, explain deviance by emphasizing individual personality factors as the underlying cause of deviant behavior. Sociologists have criticized this perspective, not because it is wrong, but because it is incomplete. By locating causes of deviance within individuals, many psychological explanations tend to overlook the context in which deviance is produced. Individual motivation simply does not explain the social patterns that sociologists observe in studying deviance. Why is deviance more common in some groups than others? Why are some more likely to be labeled deviant than others, even if they engage in the exact same behavior? How is deviance related to patterns of inequality in society? To answer these questions requires a sociological explanation. Sociologists do not ignore individual psychology but integrate it into an explanation of deviance that focuses on the social conditions surrounding the behavior, going beyond explanations of deviance that root it in the individual personality.

Commonly, people will say that someone who commits a very deviant act is "sick." This common explanation is what sociologists call the **medicalization of deviance** (Conrad and Schneider 1992). Medicalizing deviance attributes deviant behavior to a "sick" state of mind, where the solution is to "cure" the deviant through therapy or other psychological treatment.

An example is found in alcoholism. There is some evidence that there may be a genetic basis to alcoholism, and certainly alcoholism must be understood at least in part in medical terms, but viewing alcoholism *solely* from a medical perspective ignores the social causes that influence the development and persistence of this behavior. Practitioners know that medical treatment alone does not solve the problem. The social relationships, social conditions, and social habits of alcoholics must be altered, or the behavior is likely to recur.

> **thinking sociologically**
>
> **Ask some of your friends to explain why rape occurs. What evidence of the *medicalization of deviance* exists in your friends' answers?**

Sociologists criticize the medicalization of deviance for ignoring the effects of social structures on the development of deviant behavior. From a sociological perspective, deviance originates in society, not just in individuals. Changing the incidence of deviant behavior requires changes in society in addition to changes in individuals. Deviance, to most sociologists, is not a pathological state but an *adaptation to the social structures* in which people live. Factors such as family background, social class, racial inequality, and the social structure of gender relations in society produce deviance, and these factors must be considered in order to explain deviance.

Sociological Theories of Deviance

Sociologists have drawn on several major theoretical traditions to explain deviant behavior, including functionalism, conflict theory, and symbolic interaction theory.

Functionalist Theories of Deviance

Recall that functionalism is a theoretical perspective that interprets all parts of society, even those that may seem dysfunctional, as instead contributing to the stability of the whole. At first glance, deviance seems to be dysfunctional for society. Functionalist theorists argue otherwise (see Table 6.1). They contend that deviance is functional *because it creates*

table 6.1 Sociological Theories of Deviance

Functionalist Theory	Symbolic Interaction Theory	Conflict Theory
Deviance creates social cohesion.	Deviance is a learned behavior, reinforced through group membership.	Dominant classes control the definition of and sanctions attached to deviance.
Deviance results from structural strains in society.	Deviance results from the process of social labeling, regardless of the actual commission of deviance.	Deviance results from social inequity in society.
Deviance occurs when people's attachment to social bonds is diminished.	Those with the power to assign deviant labels themselves produce deviance.	Elite deviance and corporate deviance goes largely unrecognized and unpunished.

social cohesion. Branding certain behaviors as deviant provides contrast with behaviors that are considered normal, giving people a heightened sense of social order. Norms are meaningless unless there is deviance from them; thus, deviance is necessary to clarify what society's norms are. Group coherence then comes from sharing a common definition of legitimate, as well as deviant, behavior. The collective identity of the group is affirmed when those who are defined as deviant are ridiculed or condemned by group members (Erikson 1966).

To give an example, think about how many people define gay men as deviant. Although lesbians and gay men have rejected this label, labeling homosexuality as deviant is one way of affirming the presumed normality of heterosexual behavior. Labeling someone else an outsider is, in other words, a way of affirming one's "insider" identity.

Durkheim: The Study of Suicide. The functionalist perspective on deviance stems originally from the work of **Emile Durkheim.** Recall that one of Durkheim's central concerns was how society maintains its coherence (or social order). Durkheim saw deviance as functional for society because it produces solidarity among society's members. He developed his analysis of deviance in large part through his analysis of suicide. Through this work, he discovered a number of important sociological points. First, he criticized the usual psychological interpretations of why people commit suicide, turning instead to sociological explanations with data to back them up. Second, he emphasized the role of social structure in producing deviance. Third, he pointed to the importance of people's social attachments to society in understanding deviance. Finally, he elaborated the functionalist view that deviance provides the basis for social cohesion. His studies of suicide illustrate these points.

Durkheim was the first to argue that the causes of suicide were to be found in social factors, not individual personalities. Observing that the rate of suicide in a society varied with time and place, Durkheim looked for causes linked to these factors other than emotional stress. Durkheim argued that suicide rates are affected by the different social contexts in which they emerge. He looked at the degree to which people feel integrated into the structure of society and their social surroundings as social factors producing suicide.

Durkheim analyzed three types of suicide: anomic suicide, altruistic suicide, and egoistic suicide. **Anomie,** as defined by Durkheim, is the condition that exists when social regulations in a society break down; the controlling influences of society are no longer effective, and people exist in a state of relative normlessness. The term *anomie* refers not to an individual's state of mind, but instead to social conditions.

Anomic suicide occurs when the disintegrating forces in the society make individuals feel lost or alone. Teenage suicide is often cited as an example of anomic suicide. Studies of college campuses, for example, trace the cause of campus suicides to feelings of depression and hopelessness (Langhinrichsen-Rohling et al. 1998). These feelings are, however, more likely to arise in certain sociological contexts. Thus, suicide is more likely committed by those who have been sexually abused as children or by those whose parents are alcoholics (Thakkar et al. 2000; Bryant and Range 1997).

Altruistic suicide occurs when there is excessive regulation of individuals by social forces. An example is someone who commits suicide for the sake of a religious or political cause. For example, after hijackers took control of four airplanes—crashing two into the World Trade Center in New York, one into the Pentagon, and, through the intervention of passengers, one in a Pennsylvania field—many wondered how anyone could do such a thing, killing themselves in the process. Although sociology certainly does not excuse such behavior, it can help explain it. Terrorists and *suicide bombers,* such as those now familiar to us from the U.S.–Iraq war, are so regulated by their extreme beliefs that they are willing to die to kill as many people as possible to achieve their goals. As Durkheim argued, altruistic suicide results when individuals are excessively dominated by the expectations of their social group. People who commit altruistic suicide subordinate themselves to collective expectations, even when death is the result.

Egoistic suicide occurs when people feel totally detached from society. This helps explain the high rate of suicide among the elderly in the United States. People over seventy-five years of age have one of the highest rates of suicide, presumably because the elderly lose many of their functional ties to society (National Center for Health Statistics 2004). Ordinarily, people are integrated into society by work roles, ties to family and community, and other social bonds. When these bonds are weakened through retirement or loss of family and friends, the likelihood of egoistic suicide increases. Elderly people who lose these ties are the most susceptible to egoistic suicide. Suicide is also more likely to occur among people who are not well integrated into social networks (Berkman et al. 2000). Thus, it should not be surprising that women have lower suicide rates than men (National Center for Health Statistics 2006). Sociologists explain this fact as a result of men being less embedded in social relationships of care and responsibility than women (Watt and Sharp 2001).

Durkheim's major point is that suicide is a social, not just an individual, phenomenon. Recall from Chapter 1 that Durkheim sees sociology as the discovery of the social forces that influence human behavior. As individualistic as suicide might seem, Durkheim discovered the influence of social structure even here (see Map 6.1).

In the spring of 2007, Seung-Hui Cho, a college student, armed with two large-capacity semiautomatic pistols, shot and killed thirty-one people at Virginia Tech University, wounded fourteen others, and then killed himself, bringing the total killed to thirty-two dead (see the box, "The Virginia Tech Student Massacre: Carnage with Egoistic Suicide?"). This act shocked the nation, and the media instantly dubbed it the largest mass rampage killing in U.S. history. No doubt sociologists (such as Katherine Newman 2006) will see a list of social–structural elements that are common to both the Virginia Tech rampage and similar "school shootings" in the past, such as the rampage killings at Columbine High School in Littleton, Colorado, in 1999. Both acts were committed by individuals who could be characterized as extremely socially isolated and utterly outside a network of peers. All three perpetrators (Cho in the case of Virginia Tech and Dylan Klebold and Eric Harris in the case of Columbine High School) were extreme social isolates, and all three committed suicide immediately after their carnage. In Durkheim's sense, all three instances represented examples of egoistic suicide given the attributes of social isolation, lack of integration into society, troubled individual histories, including severe psychological disturbance, and a known narcissistic desire to "make their mark" in history, such by killing the largest number of individuals possible in a single attack. Diaries left behind by Mr. Cho after his suicide showed without question that he wanted precisely this kind notoriety.

Strong ties among the Navajo produce social integration, resulting in the fact that the Navajo have one of the lowest suicide rates of any group in the United States, and also lowest among other Native American tribal groups.

Merton: Structural Strain Theory. The functionalist perspective on deviance has been further elaborated by the sociologist **Robert Merton** (1910–2003). Merton's **structural strain theory** traces the origins of deviance to the tensions caused by the gap between cultural goals and the means people have available to achieve those goals. Merton noted that societies are characterized by both culture and social structure. Culture establishes goals for people in society; social structure provides, or fails to provide, the means for people to achieve those goals. In a well-integrated society, according to Merton, people use accepted means to achieve the goals society establishes. In other words, the goals and means of the society are in balance. When the means are out of balance with the goals, deviance is likely to occur. According to Merton, this imbalance, or disjunction, between cultural goals and structurally available means can actually compel the individual into deviant behavior (Merton 1968).

To explain further, a collective goal in U.S. society is the achievement of economic success. The legitimate means to achieve such success are education and jobs, but not all groups have equal access to those

map 6.1 MAPPING AMERICA'S DIVERSITY

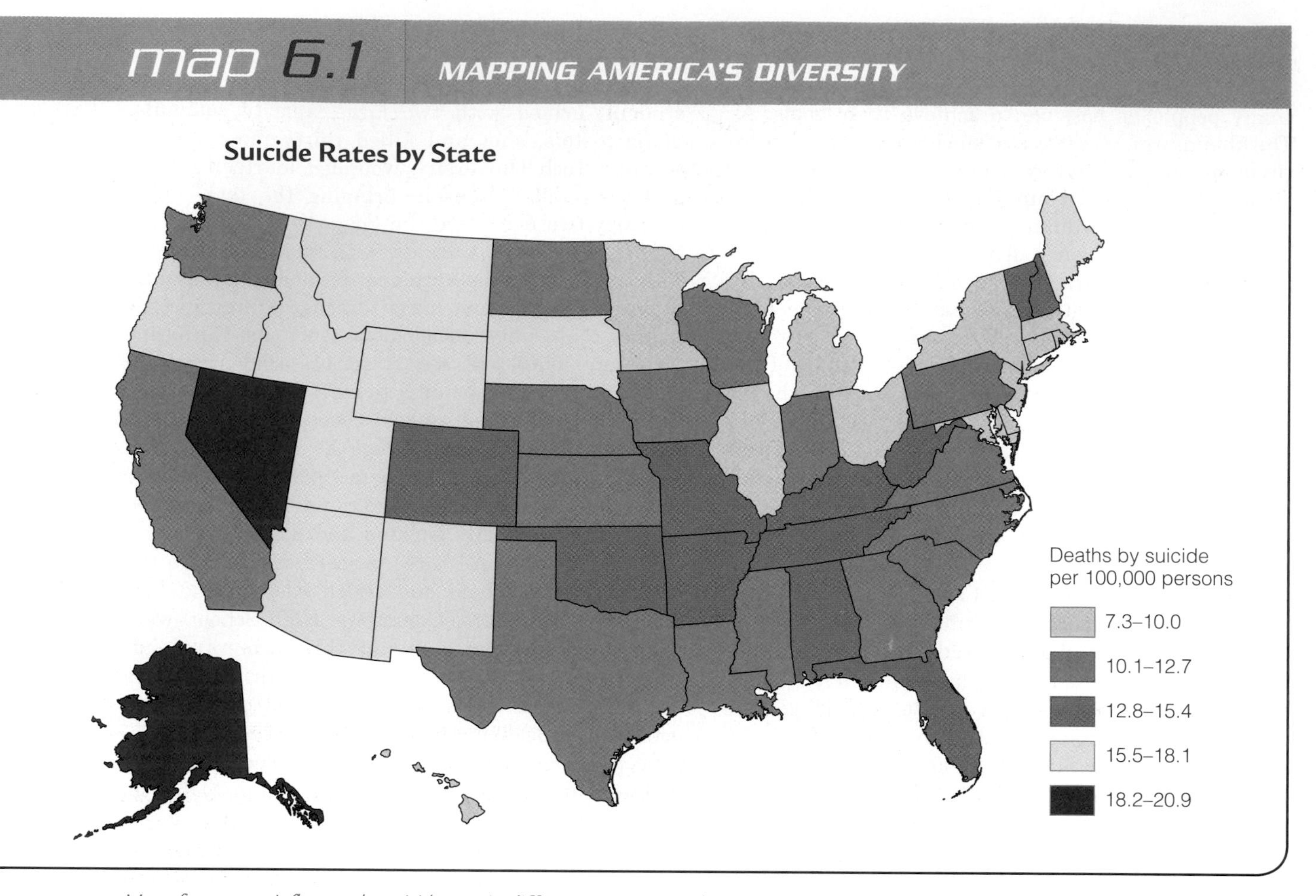

Many factors can influence the suicide rate in different contexts. As discussed in the text, suicides can be caused by a multiple of structural and cultural factors, and sometimes these factors may be differently distributed by state or region. What are some of the social facts about the different states and regions that might affect the different rates of suicide you see in this map? What, in particular, might you guess about such social facts characterizing the states with the highest suicide rates?

Data: U.S. Census Bureau, 2007. *The 2007 Statistical Abstract: National Data Book.* Washington, DC: U.S. Government Printing Office.

means. The result is structural strain that produces deviance. According to Merton, lower-class individuals are most likely to experience these strains because they internalize the same goals and values of the rest of society but have blocked opportunities for success. Structural strain theory therefore helps explain the high correlation that exists between unemployment and crime.

Figure 6.1 illustrates how strain between cultural goals and structurally available means can produce deviance. *Conformity* is likely to occur when the goals are accepted and the means for attaining the goals are made available to the individual by the social structure. If this does not occur, then cultural–structural stain exists, and at least one of four possible forms of deviance is likely to result: innovative deviance, ritualistic deviance, retreatism deviance, or rebellion.

Consider the case of female prostitution: The prostitute has accepted the cultural values of dominant society—obtaining economic success and material wealth. Yet if she is poor, then the structural means to attain these goals are less available to her and turning to prostitution—a type of *innovative deviance*—is likely to result. The stockbroker who engages in illegal insider trading constitutes another example of innovative deviance: The cultural goal

	Cultural goals accepted?	Institutionalized means toward goal available?
Conformity	Yes	Yes
Innovative deviance	Yes	No
Ritualistic deviance	No	Yes
Retreatism deviance	No	No
Rebellion	No (old goals) Yes (new goals)	No (old means) Yes (new means)

FIGURE 6.1 *Merton's Structural Strain Theory*

The Virginia Tech Student Massacre: Carnage with Egoistic Suicide?

The relative calm of the campus at Virginia Tech University in Blacksburg, Virginia, was shattered in the spring of 2007 by Seung-Hui Cho (given as Cho Seung-Hui in some sources). Carrying two semiautomatic handguns, Cho systematically shot and killed thirty-one people and wounded fourteen others, carefully pausing several times to reload his high-capacity Glock 19 pistols. He then killed himself, bringing the total dead to thirty-two. Police were later struck with the high degree of calculation that went into Cho's mayhem.

This school shooting unfortunately is characterized by three important ingredients that have characterized school shootings in the past and that also contain elements of a "classic" *egoistic suicide,* as originally analyzed by Durkheim:

1. *Extreme social isolation of the perpetrator.* Cho socially interacted with virtually none of his classmates. He was so isolated that students living on his floor right next to him knew absolutely nothing about him and had no contact with him either inside or outside the classroom. In the elaborate diaries he left behind, he called students around him "rich brats" who "never felt a single ounce of pain." His diaries provided considerable evidence that he was also extremely isolated from his parents and his siblings.
2. *A sense of grandiosity on the part of the perpetrator.* This is the "egoistic" element of the perpetrator's behavior. Past studies have shown that those who have engaged in rampage killings, particularly school shootings, have had a greatly exaggerated sense of their role in society in that they see themselves as having the ability to exert strong control over all others. This may explain, at least in part, why perpetrators such as Cho set out to purposely kill as many human beings as possible in the shortest time available, and then eliminate themselves before "getting caught." (Columbine High School shooters Dylan Klebold and Eric Harris also revealed in their own diaries a sense of self-grandiosity as well as a desire to kill as many students as possible.)
3. *Severe mental disorder.* The perpetrator is clearly mentally disturbed. In Cho's case, before his rampage he had been diagnosed by a psychiatrist as "mentally ill and in need of hospitalization," and he was also seen as "incapable of volunteering or unwilling to volunteer for treatment." Despite this diagnosis, Cho was not forced into treatment, nor was he officially targeted to receive mental counseling. Psychiatrists often do not hospitalize suspected mental patients, even those identified (labeled?) as dangerous, on the grounds that not everyone who is so diagnosed will actually go out and kill others.

The presence of each of these three elements in Cho's case sheds some light on the nature of egoistic suicide: It reflects both individual (personality) causes as well as social circumstances (such as extreme isolation from a social structure and social networks) that *combine* to compel the individual into severely deviant behavior.

A final observation needs to be added. The first person killed by Cho was a woman. The woman had had very little, if any, prior contact with Cho. This nonetheless erroneously led the media and the campus police to suspect that the killing was merely a "lovers' quarrel" of some sort—a "domestic dispute" according to University President Steger. It was thus believed that the gunman had fled the campus and would not return—a wholly erroneous conclusion. The assumption unfortunately appeared to be that a man killing his girlfriend is no big deal: There is supposedly no real danger to others in such an act. This thus demonstrates how *a priori* institutionalized sexism can result in a highly unfortunate and horribly incorrect conclusion.

AP Images/Carolyn Kaster

Graduation took place on a somber note at Virginia Tech University despite the rampage killings that had taken place on the campus.

Source: Time Magazine, April 30, 2007; *Mother Jones,* April 2007, http://www.motherjones.com/news/update/2007/04virginiatech women.html; and Newman, Katherine S. 2004. *Rampage: The Social Roots of School Shootings* (with Cybelle Fox, David Harding, Jal Mehta, and Wendy Roth). New York: Basic Books.

(wealth) is accepted, but nontraditional means (insider trading) are available and used.

Other forms of deviance also represent this disjunction, or strain, between goals and means. *Retreatism deviance* becomes likely when neither the goals nor the means are available. Examples of retreatism are the severe alcoholic or the homeless person or the hermit. *Ritualistic deviance* is illustrated in the case of some eating disorders among college women, such as *bulimia* (purging one's self after eating). The cultural goal of extreme thinness is perceived as unattainable even though the means for trying to attain it are plentiful, for example, good eating habits and proper diet methods (Sharp 2000). Finally, *rebellion* as a form of deviance is likely to occur when new goals are substituted for more traditional ones and also new means are undertaken to replace older ones, as by force or armed combat. Many right-wing extremist groups, such as the American Nazi Party, "skinheads," and the Ku Klux Klan (KKK), are examples of this type of deviance.

Social Control Theory. Taking functionalist theory in another direction, Travis Hirschi has developed social control theory to explain deviance. **Social control theory,** a type of functionalist theory, suggests that deviance occurs when a person's (or group's) attachment to social bonds is weakened (Hirschi 1969; Gottfredson and Hirschi 1995, 1990). According to this view, people internalize social norms because of their attachments to others. People care what others think of them and therefore conform to social expectations because they accept what people expect. You can see here that social control theory, like the functionalist framework from which it stems, assumes the importance of the socialization process in producing conformity to social rules. When that conformity is broken, deviance occurs.

Social control theory assumes there is a common value system within society, and breaking allegiance to that value system is the source of social deviance. This theory focuses on how deviants are attached (or not) to common value systems and what situations break people's commitment to these values. Social control theory suggests that most people probably feel some impulse toward deviance at times but that the attachment to social norms prevents them from actually participating in deviant behavior. Sociologists find that juveniles whose parents exercise little control over violent behavior and who learn violence from aggressive peers are most likely to engage in violent crimes (Heimer 1997), as was the case with the two teenagers who killed twelve students and a teacher at Columbine High School.

Functionalism: Strengths and Weaknesses. Functionalism emphasizes that social structure, not just individual motivation, produces deviance. Functionalists argue that social conditions exert pressure on individuals to behave in conforming or nonconforming ways. Types of deviance are linked to one's place in the social structure; thus, a poor person blocked from economic opportunities may use armed robbery to achieve economic goals, whereas a stockbroker may use insider trading to achieve the same. Functionalists acknowledge that people choose whether to behave in a deviant manner but believe that they make their choice from among socially prestructured options. The emphasis in functionalist theory is on social structure, not individual action. In this sense, functionalist theory is highly sociological.

Functionalists also point out that what appears to be dysfunctional behavior may actually be functional for the society. An example is the fact that most people consider prostitution to be dysfunctional behavior; from the point of view of an individual, that is true: It demeans the women who engage in it, puts them at physical risk, and subjects them to sexual exploitation. From the view of functionalist theory, however, prostitution supports and maintains a social system that links women's gender roles with sexuality, associates sex with commercial activity, and defines women as passive sexual objects and men as sexual aggressors. In other words, what appears to be deviant may actually serve various purposes for society.

Critics of the functionalist perspective argue that it does not explain how norms of deviance are first established. Despite its analysis of the ramifications of deviant behavior for society as a whole, functionalism does little to explain why some behaviors are defined as normative and others as illegitimate. Questions such as who determines social norms and on whom such judgments are most likely to be imposed are seldom asked by anyone using a functionalist perspective. Functionalists see deviance as having stabilizing consequences in society, but they tend to overlook the injustices that labeling someone deviant can produce. Others would say that the functionalist perspective too easily assumes that deviance has a positive role in society; thus, functionalists rarely consider the differential effects that the administration of justice has on various social groups. The tendency in functionalist theory to assume that the system works for the good of the whole too easily ignores the inequities in society and how these inequities are reflected in patterns of deviance. These issues are left for sociologists who work from the perspectives of conflict theory and symbolic interaction.

Conflict Theories of Deviance

Recall that conflict theory emphasizes the unequal distribution of power and resources in society. It links the study of deviance to social inequality. Based on the work of Karl Marx (1818–1883; see Chapter 1), conflict theory sees a dominant class as controlling the resources of society and using its power to create the institutional rules and belief systems that support its power. Like functionalist theory, conflict theory is a *macrostructural* approach; that is, both theories look at the structure of society as a whole in developing explanations of deviant behavior.

Because some groups of people have access to fewer resources in capitalist society, they are forced into crime to sustain themselves. Conflict theory posits that the economic organization of capitalist societies produces deviance and crime. The high rate of crime among the poorest groups, especially economic crimes such as theft, robbery, prostitution, and drug selling, are a result of the economic status of these groups. Rather than emphasizing values and conformity as a source of deviance as do functional analyses, conflict theorists see crime in terms of power relationships and economic inequality (Grant and Martínez 1997).

The upper classes, conflict theorists point out, can also better hide crimes they commit because affluent groups have the resources to mask their deviance and crime. As a result, a working-class man who beats his wife is more likely to be arrested and prosecuted than an upper-class man who engages in the same behavior. In addition, those with greater resources can afford to buy their way out of trouble by paying bail, hiring expensive attorneys, or even resorting to bribes.

Corporate crime is crime committed within the legitimate context of doing business. Conflict theorists expand our view of crime and deviance by revealing the significance of such crimes. They argue that appropriating profit based on exploitation of the poor and working class is inherent in the structure of capitalist society. **Elite deviance** refers to the wrongdoing of wealthy and powerful individuals and organizations (Simon 2007). Elite deviance includes what early conflict theorists called *white-collar crime* (Sutherland 1940; Sutherland and Cressey 1978). Elite deviance includes tax evasion, illegal campaign contributions, corporate scandals such as fraudulent accounting practices that endanger or deceive the public but profit the corporation or individuals within it, and even government actions that abuse the public trust.

The ruling groups in society develop numerous mechanisms to protect their interests according to conflict theorists who argue that law, for example, is created by elites to protect the interests of the dominant class. Thus, law, supposedly neutral and fair in its form and implementation, works in the interest of the most well to do (Weisburd et al. 2001, 1991; Spitzer 1975). Another way that conflict theorists see dominant groups as using their power is through the excessive regulation of populations that are a potential threat to affluent interests. Periodically sweeping the homeless off city streets, especially when there is a major political event or other elite event occurring, is a good example.

Conflict theory emphasizes the significance of social control in managing deviance and crime. **Social control** is the process by which groups and individuals within those groups are brought into conformity with dominant social expectations. Social control, as we saw in Chapter 3, can take place simply through socialization, but dominant groups can also control the behavior of others through marking them as deviant. An example is the historic persecution of witches during the Middle Ages in Europe and during the early Colonial period in America (Ben-Yehuda 1986; Erikson 1966).

From the point of view of conflict theory, social control agents play a significant role in defining deviant behavior.

Witches often were women who were healers and midwives—those whose views were at odds with the authority of the exclusively patriarchal hierarchy of the church, then the ruling institution. "Witch hunt" is a term still used today to refer to the aggressive pursuit of those who dissent from prevailing political and social norms.

One implication of conflict theory, especially when linked with labeling theory, is that the power to define deviance confers an important degree of social control. **Social control agents** are those who regulate and administer the response to deviance, such as the police and mental health workers. Members of powerless groups may be defined as deviant for even the slightest infraction against social norms, whereas others may be free to behave in deviant ways without consequence. Oppressed groups may actually engage in more deviant behavior, but it is also true that they have a greater likelihood of being labeled deviant and incarcerated or institutionalized, whether or not they have actually committed an offense. This is evidence of the power wielded by social control agents.

When powerful groups hold stereotypes about other groups, the less powerful people are frequently assigned deviant labels. As a consequence, the least powerful groups in society are subject most often to social control. You can see this in the patterns of arrest data. Poor people are more likely to be considered criminals and therefore more likely to be arrested, convicted, and imprisoned than middle- and upper-class people. The same is true of Latinos, Native Americans, and African Americans. Sociologists point out that this does not necessarily mean that these groups are somehow more criminally prone; rather, they take it as evidence of the differential treatment of these groups by the criminal justice system.

Conflict Theory: Strengths and Weaknesses. The strength of conflict theory is its insight into the significance of power relationships in the definition, identification, and handling of deviance. It links the commission, perception, and treatment of crime to inequality in society and offers a powerful analysis of how the injustices of society produce crime and result in different systems of justice for disadvantaged and privileged groups. Not without its weaknesses, however, critics point out that laws protect most people, not just the affluent, as conflict theorists argue.

In addition, although conflict theory offers a powerful analysis of the origins of crime, it is less effective in explaining other forms of deviance. For example, how would conflict theorists explain the routine deviance of middle-class adolescents? They might point out that much of middle-class deviance is driven by consumer marketing. Profits are made from the accoutrements of deviance—rings in pierced eyebrows, "gangsta" rap music, and so on—but economic interests alone cannot explain all the deviance observed in society. As Durkheim argued, deviance is functional for the whole of society, not just those with a major stake in the economic system.

Symbolic Interaction Theories of Deviance

Whereas functionalist and conflict theories are *macrosociological* theories, certain *microsociological* theories of deviance look directly at the interactions people have with one another as the origin of social deviance. *Symbolic interaction theory* holds that people behave as they do because of the meanings people attribute to situations (see Chapter 1). This perspective emphasizes the meanings surrounding deviance, as well as how people respond to those meanings. Symbolic interaction emphasizes that deviance originates in the interaction between different groups and is defined by society's reaction to certain behaviors. Symbolic interactionist theories of deviance originate in the perspective of the Chicago School of sociology.

W.I. Thomas and the Chicago School. **W.I. Thomas** (1863–1947), one of the early sociologists from the University of Chicago, was among the first to develop a sociological perspective on social deviance. Thomas explained deviance as *a normal response to the social conditions in which people find themselves.* He called this perspective *situational analysis,* meaning that people's actions and the subjective meanings attributed to these actions, including deviant behavior, must be understood in social, not individualized, frameworks. Much influenced by his women students in the Chicago School (Deegan 1990), Thomas was one of the first to argue that delinquency was caused by the social disorganization brought on by slum life and urban industrialism; he saw it as a problem of social conditions, not individual character.

Differential Association Theory. Thomas's work laid the foundation for a classic theory of deviance: differential association theory. **Differential association theory** interprets deviance, including criminal behavior, as behavior one learns through interaction with others (Sutherland 1940; Sutherland and Cressey 1978). Edwin Sutherland argued that becoming a criminal or a juvenile delinquent is a matter of learning criminal ways within the primary groups to which one belongs. To Sutherland, people become criminals when they are more strongly socialized to break the law than to obey it. Differential association theory emphasizes the interaction people have with their peers and others in their environment. Those who "differentially associate" with delinquents, deviants, or criminals learn to value deviance. The greater the frequency, duration, and intensity of their immersion in deviant environments, the more likely it is that they will become deviant.

Consider the career path of con artists and hustlers. Hustlers seldom work alone. Like any skilled worker, they have to learn the "tricks of the trade." A new recruit becomes part of a network of other hustlers who teach the recruit the norms of the deviant culture (Prus and Sharper 1991). Crime also tends to run in families. This does not necessarily mean that crime is passed on in genes from parent to child. It means that youths raised in deviant families are more likely socialized to become deviant themselves (Miller 1986). Differential association theory offers a compelling explanation for how deviance is culturally transmitted—that is, people pass on deviant expectations through the social groups in which they interact, of which the family is but one.

Critics of differential association theory have argued that this perspective tends to blame deviance on the values of particular groups. Differential association has been used, for instance, to explain the higher rate of crime among the poor and working class, arguing that this higher rate of crime occurs because they do not share the values of the middle class. Such an explanation, critics say, is class-biased, because it overlooks the deviance that occurs in the middle-class culture and among elites. Disadvantaged groups may share the values of the middle class but cannot necessarily achieve them through legitimate means (a point, you will remember, made by Merton's structural strain theory). Still, differential association theory offers a good explanation of why deviant activity may be more common in some groups than others, and it emphasizes the significant role that peers play in encouraging deviant behavior.

Labeling Theory. **Labeling theory** interprets the responses of others as the most significant factor in understanding how deviant behavior is both created and sustained (Becker 1963). Labeling theory stems from the work of W.I. Thomas, who it will be recalled wrote, "If men define situations as real, they are real in their consequences" (Thomas and Thomas 1928: 572). A *label* is the assignment or attachment of a deviant identity to a person by others, including by agents of social institutions. Therefore, peoples' reactions, not the action itself, produce deviance as a result of the labeling process. Once applied, the deviant label is difficult to shed.

Linked with conflict theory, labeling theory shows how those with the power to label an act or a person deviant and to impose sanctions—such as police, court officials, school authorities, experts, teachers, and official agents of social institutions—wield great power in determining societal understandings of deviance. When they apply the "deviant" label, it sticks. Furthermore, because deviants are handled through bureaucratic organizations, bureaucratic workers "process" people according to rules and procedures, seldom questioning the basis for those rules or willing or able to challenge them (Cicourel 1968; Kitsuse and Cicourel 1963; Margolin 1992; Montada and Lerner 1998).

Once the label is applied, it is difficult for the deviant to recover a nondeviant identity. To give an example, once a social worker or psychiatrist labels a client mentally ill, that person will be treated as mentally ill, regardless of his or her actual mental state. Pleas by the accused that he or she is mentally sound are typically taken as further evidence of mental illness! A person's anger and frustration about the label are taken as further support for the diagnosis. Once labeled, a person may have great difficulty changing his or her classification; the label itself has consequences.

A person need not have actually engaged in deviant behavior to be labeled deviant; yet, once applied, the label has social consequences. Labeling theory helps explain why convicts released from prison have such high rates of *recidivism* (return to criminal activities). Convicted criminals are formally and publicly labeled wrongdoers. They are treated with suspicion ever afterward and have great difficulty finding legitimate employment: The label "ex-con" defines their future options.

It is in fact exceedingly difficult for an ex-con to find employment after release from prison, and it is especially difficult if the person is male and Black or Hispanic. In a clever study, Pager (2005, 2003) had pretrained role-players pose as ex-cons looking for a job. These role-players then went on to the job market and were interviewed for various jobs; all of them used the same preset script during the interview. The idea of the study was to see how many would be invited back for another interview. The results were staggering: A very small percentage of the male Black and Hispanic role-players were invited back for interviews yet a large percentage of the male White role-players were. These differences could not be attributed to differences in interaction displayed during the interview, because everyone used the exact same pre-prepared script.

Labeling theory points to a distinction often made by sociologists between primary, secondary, and tertiary deviance. *Primary deviance* is the actual violation of a norm or law. *Secondary deviance* is the behavior that results from being labeled deviant, regardless of whether the person has engaged in deviance. A student labeled a "troublemaker," for example, might accept this identity and move from being merely mischievous to engaging in escalating delinquent acts. In that case, the person at least partly accepts the deviant label and acts in accordance with that role. *Tertiary deviance* occurs when the deviant fully accepts the deviant role but rejects the stigma associated with it, as when lesbians and gays proudly display their identity (Lemert 1972; Kitsuse 1980).

Both social class and the role of prisons in society play an important role in the creation of secondary deviance. Researchers Bruce Western (2006) and Jeffrey Reiman (2004) note that the prison system in the United States is in effect designed to *train and socialize* prisoners into a career of secondary deviance and to tell the public that crime is a threat primarily from the poor (see the box, "Understanding Diversity: The Rich Get Richer and the Poor Get Prison"). Reiman sees that the goal of the prison system is not to reduce crime, but to impress upon the public that crime is inevitable and that it originates only from the lower classes. Prisons accomplish this, even if unintentionally, by demeaning prisoners, not training them in marketable skills, and stigmatizing them as different from "decent citizens." As a consequence, the person will never be able to pay his or her debt to society and the prison system has created the very behavior it intended to eliminate.

Labeling theory suggests that deviance refers not just to something one does, but to something one becomes. **Deviant identity** is the definition a person has of himself or herself as a deviant. Most often, deviant identities emerge over time (Lemert 1972;

Simon 2007). A drug addict, for example, may not think of herself as a junkie until she realizes she no longer has nonusing friends. The formation of a deviant identity, like other identities, involves a process of social transformation in which a new self-image and new public definition of a person emerges. This is a process that involves how people view the deviant and how the deviant views himself or herself. Studies of tattoo "collectors" (that is, those who are very heavily tattooed) find, for example, that if collectors first learn to interpret tattooing as a desirable thing, they then begin to feel connected to a subculture of other collectors and eventually come to see their tattoos as part of themselves (Irwin 2001; Vail 1999; Montada and Lerner 1998; Scheff 1984).

thinking sociologically

Perform an experiment by doing something deviant for a period, such as carry around a teddy bear doll and treat it as a live baby; or, stand in the street and look into the air, as though you are looking at something up there. Make a record of how others respond to you, and then ask yourself how *labeling theory* is important to the study of *deviance*. Then take your experiment a step further and ask yourself how people's reactions to you might have differed had you been of another race or gender. You might want to structure this question into your experiment by teaming up with a classmate of another race or gender. You could then compare responses to the same behavior by both of you. *A note of caution:* Do not do anything illegal or dangerous; even the most seemingly harmless acts of deviance can generate strong (and sometimes hostile) reactions, so be careful in planning your experiment!

Deviant Careers. In the ordinary context of work, a career is the sequence of movements a person makes through different positions in an occupational system (Becker 1963). A **deviant career** is the sequence of movements people make through a particular subculture of deviance. Deviant careers can be studied sociologically, like any other career. Within deviant careers, people are socialized into new "occupational" roles and encouraged, both materially and psychologically, to engage in deviant behavior. The concept of a deviant career emphasizes that there is a progression through deviance: Deviants are recruited, given or denied rewards, and promoted or demoted. As with legitimate careers, deviant careers involve an evolution in the person's identity, values, and commitment over time. Deviants, like other careerists, may have to demonstrate their commitment to the career to their superiors, perhaps by passing certain tests of their mettle, such as when a gang expects new members to commit a crime, perhaps even shoot someone.

The concept of a deviant career helps explain why being caught and labeled deviant may actually reinforce, rather than deter, one's commitment to a deviant career. For example, hospitalized mental patients are often rewarded with comfort and attention for "acting sick" but are punished when they act normally—for instance, if they rebel against the boredom and constraints of institutionalization. Acting the sick role will foster their career as "a mentally ill person" (Scheff 1984, 1966).

A first arrest on weapons charges may be seen as a rite of passage that brings increased social status among peers. Whereas those outside the deviant group may think that arrest is a deterrent to crime, it may actually encourage a person to continue along a deviant path. Punishments administered by the authorities may even become badges of honor within a deviant community. Similarly, labeling a teenager "bad" for behavior that others think is immoral may actually encourage the behavior to continue because the juvenile may take this as a sign of success as a deviant.

Like anyone else, deviants may experience career mobility—that is, they may move up or down in rank within the deviant community. Male prostitution, for example, is a career organized around a hierarchy of illicit sexual services. Men or boys are recruited into prostitution at different ranks: as street hustlers, bar hustlers, or escorts. Some may become specialists in sadomasochism or in cross-dressing. Newcomers often acquire mentors who educate them in the deviant lifestyle. Someone who "learns the ropes" is more likely to continue in a deviant career (Luckenbill 1986).

Deviant Communities. The preceding discussion continues to indicate an important sociological point: Deviant behavior is not just the behavior of maladjusted individuals; it often takes place within a group context and involves group response. Some groups are actually organized around particular forms of social deviance; these are called **deviant communities.**

Like subcultures and countercultures, deviant communities maintain their own values, norms, and rewards for deviant behavior. Joining a deviant community closes one off from conventional society and tends to solidify deviant careers because the deviant individual receives rewards and status from the in-group. Disapproval from the out-group may only enhance one's status within. Deviant communities also create a worldview that solidifies the deviant identity of their members. They may develop symbolic systems such as emblems, forms of dress, publications, and other symbols that promote their identity as a deviant group. Gangs wear their "colors"; prostitutes have their own vocabulary of *tricks* and *johns;* skinheads have their insignia and music. All are examples of deviant communities. Ironically, subcultural norms

© Mark Peterson/CORBIS Saba

Some deviance develops in deviant communities, such as the "skinheads" shown here marching in a Ku Klux Klan rally protesting the Martin Luther King, Jr. holiday. Such right-wing extremist groups have become more common in recent years.

and values reinforce the deviant label both inside and outside the deviant group, thereby reinforcing the deviant behavior.

Some deviant communities are organized specifically to provide support to those in presumed deviant categories. Groups such as Alcoholics Anonymous, Weight Watchers, and various twelve-step programs help those identified as deviant overcome their deviant behavior. These groups, which can be quite effective, accomplish their mission by encouraging members to accept their deviant identity as the first step to recovery.

A Problem with Official Statistics. Because labeling theorists see deviance as produced by those with the power to assign labels, they question the value of official statistics as indicators of the true extent of deviance. *Reported rates of deviant behavior are themselves the product of socially determined behavior, specifically the behavior of identifying what is deviant.* Official rates of deviance are produced by people in the social system who define, classify, and record certain behaviors as deviant and others as legitimate. Labeling theorists are more likely to ask how behavior becomes labeled deviant than they are to ask what motivates people to become deviant (Best 2001; Kitsuse and Cicourel 1963).

In the aftermath of terrorist attacks on the World Trade Center, officials debated whether to count the deaths of thousands as murder or as a separate category of terrorism. The decision would change the official rate of deviance by inflating or deflating the reported crime rate of murder in New York City in that year. In the end, these deaths were not counted in the murder rate. Labeling theorists think that official rates of deviance do not necessarily reflect the actual commission of crimes or deviant acts; instead, the official rates reflect social judgments.

Likewise, when AIDS (acquired immune deficiency syndrome) first emerged, it was highly stigmatized because of its perceived association with gay men. Obituaries of AIDS victims seldom noted that the death was a result of AIDS. More typically, obituaries reported only that the person had died following a "long illness."

In another example, official rape rates are underestimates of the actual extent of rape, largely due to victims' reluctance to report. Also, some rapes are less likely to be "counted" as rape by police, such as if the victim is a prostitute, was drunk at the time of the assault, or had a previous relationship with the assailant. Moreover, rapes resulting in death are classified as homicides and therefore do not appear in the official statistics on rape. There is also enough evidence of discriminatory treatment by the police to know that official statistics showing higher arrest rates for African American men tell us as much about police behavior as about the behavior of African American men (Babbie 2003; DeFleur 1975). Given such problems, any official statistics must be interpreted with caution.

Labeling Theory: Strengths and Weaknesses. The strength of labeling theory is its recognition that the judgments people make about presumably deviant behavior have powerful social effects. Labeling theory does not, however, explain why deviance occurs in the first place. It may illuminate the consequences of a young man's violent behavior, but it does not explain the actual origins of the behavior. Put bluntly, it does not explain why some people become deviant and others do not. This shortcoming in the analysis of deviance has been carefully scrutinized by conflict theorists who place their analysis of deviance within the power relationships of race, class, and gender.

Forms of Deviance

Although there are many forms of deviance, the sociology of deviant behavior has focused heavily on subjects such as mental illness, social stigmas, and crime. As we review each, you will also see how the different sociological theories about deviance contribute to understanding each subject. In addition, you will see how the social context of race, class, and gender relationships shape these different forms of deviance. Race, class, and gender are not just individual attributes; they are patterns of relationships supported by social institutions and social ideologies. Consequently, they are an important part of the social context in which different forms of deviance emerge and from which people make judgments about who is deviant and who is not.

Mental Illness

Sociological explanations of mental illness look to the social systems in which mental illness is defined, identified, and treated, even though it is typical for many to think of mental illness only in psychological terms. This has several implications for understanding mental illness. Functionalist theory suggests, for example, that by recognizing mental illness, society also upholds normative values about more conforming behavior. Symbolic interaction theory tells us that mentally ill people are not necessarily "sick" but are the victims of societal reactions to their behavior. Some, such as psychiatrist Thomas Szasz, go so far as to say there is no such thing as mental illness, only people's reactions to unusual behavior. From this point of view, people learn faulty self-images and then are cast into the role of patient when they are treated by therapists. Once someone is labeled as a "patient," he or she is forced into the "sick" role, as expected by those who reinforce it, and it becomes difficult to get out of the role (Szasz 1974).

Labeling theory, combined with conflict theory, suggests that those people with the fewest resources are most likely to be labeled mentally ill. Women, racial minorities, and the poor all suffer higher rates of reported mental illness and more serious disorders than are groups of higher social and economic status. Furthermore, research over the years has consistently shown that middle- and upper-class persons are more likely to receive some type of psychotherapy for their illness. Poorer individuals and minorities are more likely to receive only physical rehabilitation and medication, with no accompanying psychotherapy (originally noted by Hollingshead and Redlich 1958, but an observation that pertains just as well to the present; Simon 2007).

Sociologists give two explanations for the correlation between social status and mental illness. On the one hand, the stresses of being in a low-income group, being a racial minority, or being a woman in a sexist society contribute to higher rates of mental illness; the harsher social environment is a threat to mental health. On the other hand, the same behavior that is labeled mentally ill for some groups may be tolerated and not so labeled in others. For example, behavior considered crazy in a homeless woman (who is likely to be seen as "deranged") may be seen as merely eccentric or charming when exhibited by a rich person.

Patterns of mental illness also reflect gender relations in society. Women have higher rates of mental illness than men, although men and women differ in the kinds of mental illnesses they experience (Horton 1995). Most sociologists conclude that the higher rates of mental illness for women stem from the roles women are forced to play in society. Learned personality traits developed by women are joined by other sources of stress that can lead to mental illness (Walters 1993). Poverty, unhappy marriages, physical and sexual abuse, the stress of rearing children, and even a high percentage of time doing household work all contribute to higher rates of mental illness for women (Elliott 2001). As labeling theory would predict, gender stereotypes mean that women are more likely than men to be labeled mentally ill. The frequency with which physicians label women's complaints to be "psychologically grounded" is evidence of this tendency.

© Keith Erskine/Alamy

This man's extreme spiked hairdo shows deviant behavior in society but rigid conformity to the norms of the peer group or deviance community.

Debunking Society's Myths

Myth: Mental illness is an abnormality best studied by psychologists and physicians.

Sociological perspective: Mental illness follows patterns associated with race, class, and gender relations in society and is subject to a significant labeling effect. Those who study and treat mental illness benefit from combining a sociological perspective with both medical and psychological knowledge.

Social Stigmas

A **stigma** is an attribute that is socially devalued and discredited. Some stigmas result in people being labeled deviant. The experiences of people who are disabled, disfigured, or in some other way stigmatized are studied in much the same way as other forms of social deviance. Like other deviants, people with stigmas are stereotyped and defined only in terms of their presumed deviance.

Think, for example, of how disabled people are treated in society. Their disability can become a **master status** (Chapter 4), a characteristic of a person that overrides all other features of the person's identity (Goffman 1963). Physical disability can become a master status when other people see the disability as the defining feature of the person; a person with a disability becomes "that blind woman" or "that paralyzed guy." People with a particular stigma are often seen to be all alike. This may explain why stigmatized individuals of high visibility are often expected to represent the whole group.

People who suddenly become disabled often have the alarming experience of their new master status rapidly erasing their former identity. They may be treated and seen differently by people they know. A master status may also prevent people from seeing other parts of a person. A person with a disability may be assumed to have no meaningful sex life, even if the disability is unrelated to sexual ability or desire. Sociologists have argued that the negative judgments made about people with stigmas tend to confirm the "usualness" of others (Goffman 1963b: 3). For example, when welfare recipients are stigmatized as lazy and undeserving of social support, others are indirectly promoted as industrious and deserving.

Stigmatized individuals are measured against a presumed norm and may be labeled, stereotyped, and discriminated against. In Goffman's words, people with stigmas are perceived to have a "spoiled identity." Seen by others as deficient or inferior, they are caught in a role imposed by the stigma. They may respond by trying to hide their stigma or by blaming others. What happens, for example, to those who have a sexually transmitted disease? Particularly because this is associated with sexual immorality, there is typically shame and embarrassment; those with a sexually transmitted disease may try to conceal that they have it (Nack 2000).

Sometimes, people with stigmas bond with others, perhaps even strangers. This can involve an acknowledgment of "kinship" or affiliation that can be as subtle as an understanding look, a greeting that makes a connection between two people, or a favor extended to a stranger who the person sees as sharing the presumed stigma. Public exchanges are common between various groups that share certain forms of disadvantage, such as people with disabilities, lesbians and gays, or members of other minority groups.

Substance Abuse: Drugs and Alcohol

As with mental illness and stigmas, sociologists study the social factors that influence drug and alcohol use. Who uses what and why? How are users defined by others? These questions guide sociological research on substance abuse.

One of the first things to ask when thinking about drugs and alcohol is why using one substance is considered deviant and stigmatizing and using another is not. How do such definitions of deviance change over time? Until recently, cigarette smoking was considered normative—indeed, glamorous and sexy. Now, although smoking is still common, it has become more of a negative stigma. Some might say that this change resulted from the known risks of nicotine addiction. But just knowing the risks of smoking is not enough to define it as deviant behavior. Sociologists study how social groups have mobilized to define smoking as deviant, and they analyze how the tobacco industry has navigated through the climate of public opinion to maintain the industry's profits (Kall 2002; Brown 2000).

Like cigarettes, alcohol is also a legal drug. Whether one is labeled an alcoholic depends in large part on the social context in which one drinks, not solely on the amount of alcohol consumed. For years, the businessman's lunch where executives drank two or three martinis was viewed as normative. Drinking from a bottle in a brown bag on the street corner is considered highly deviant; having martinis in a posh bar is seen as cool.

Sociological understandings challenge views of drug and alcohol use as stemming from individual behaviors that lead to substance abuse. Patterns of use vary by factors such as age, gender, and race (see Figure 6.2). Age is one significant predictor of illegal drug use. Young people are more likely to use marijuana and cocaine and binge drink than are people over age twenty-five. Alcohol is more likely to be used by those aged eighteen to thirty-four, although the difference here is less than for other drugs.

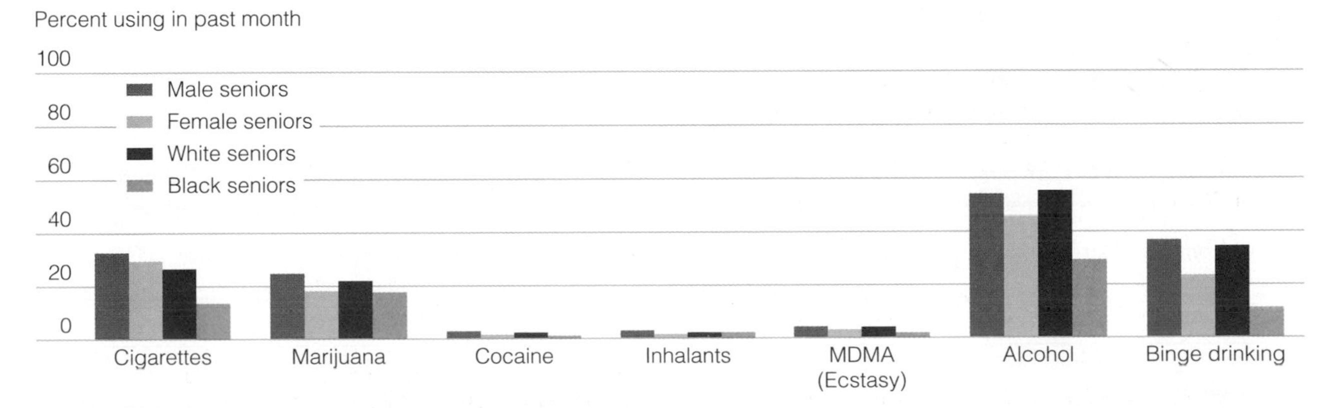

FIGURE 6.2 *Use of Selected Substances by High School Seniors*

National Center for Health Statistics. 2006. *Health United States 2006*. Hyattsville, MD: U.S. Department of Health and Human Services.

Crime and Criminal Justice

The concept of deviance in sociology is a broad one, encompassing many forms of behavior—both legal and illegal, ordinary and unusual. **Crime** is one form of deviance, specifically behavior that violates particular criminal laws. Not all deviance is crime. Deviance becomes crime when it is designated by the institutions of society as violating a law or laws. *Deviance* is behavior that is recognized as violating rules and norms of society. Those rules may be formal laws, in which case the deviant behavior would be called *crime,* or informal customs or habits, in which case the deviant behavior would not be called *crime.*

Criminology is the study of crime from a scientific perspective. Criminologists include social scientists such as sociologists who stress the societal causes of crime, psychologists who stress the personality causes of crime, and political scientists who view crime as being both caused and regulated by the powerful institutions in society. All the theoretical perspectives on deviance that we examined earlier contribute to our understanding of crime (see Table 6.2). According to the functionalist perspective, crime may be necessary to hold society together—a profound hypothesis. By singling out criminals as socially deviant, others are defined as good. The nightly reporting of crime on television is a demonstration of this sociological function of crime. Conflict theory suggests that disadvantaged groups are more likely to become criminal; it also sees the well-to-do as better able to hide their crimes and less likely to be punished. Symbolic interaction helps us understand how people learn to become criminals or come to be accused of criminality even when they may be innocent. Each perspective traces criminal behavior to social conditions rather than only to the intrinsic tendencies or personalities of individuals.

Measuring Crime: How Much Is There?

Is crime increasing in America? One would certainly think so from watching the media. Images of violent crime abound and give the impression that crime is a constant threat and is on the rise. Data on crime actually show that violent crime peaked in 1990, but *decreased* through the 1990s and leveled off a bit for until 2005 (see Figure 6.3). Assault and robbery, in particular, decreased quite significantly through the 1990s. Murder and rape remained more constant, though they too have shown some decline since the 1990s.

Data about crime come from the Federal Bureau of Investigation (FBI) based on reports from police departments across the nation. The data are distributed annually in the *Uniform Crime Reports* and are the basis for official reports about the extent of crime and its rise and fall over time. These data show that, although media reporting of crime has remained high and about the same, the officially reported rate of crime has decreased.

A second major source of crime data is the *National Crime Victimization Surveys* published by the Bureau of Justice Statistics in the U.S. Department of Justice. These data are based on surveys in which national samples of people are periodically asked if they have been the victims of one or more criminal acts. These surveys also show that violent crime, including rape, assault, robbery, and murder, declined by 15 percent over the 1990s.

Both of these sources of data—the *Uniform Crime Reports* and the *National Crime Victimization Surveys*—are subject to the problem of underreporting. About half to two-thirds of all crimes may not be reported to police, meaning that much crime never shows up in the official statistics. Certain serious crimes, such as rape, are significantly underreported, as we have already noted. Victims may be too upset to report a rape to the police or they may believe that

table 6.2 Sociological Theories of Crime

Functionalist Theory	Symbolic Interaction Theory	Conflict Theory
Societies require a certain level of crime in order to clarify norms.	Crime is behavior that is learned through social interaction.	The lower the social class, the more the individual is *forced* into criminality.
Crime results from social structural strains within society.	Labeling criminals tends to reinforce rather than deter crime.	Inequalities in society by race, class, gender, and other forces tend to produce criminal activity.
Crime may be functional to society, thus difficult to eradicate.	Institutions with the power to label, such as prisons, actually produce rather than lessen crime.	Reducing social inequalities in society with reduce crime.

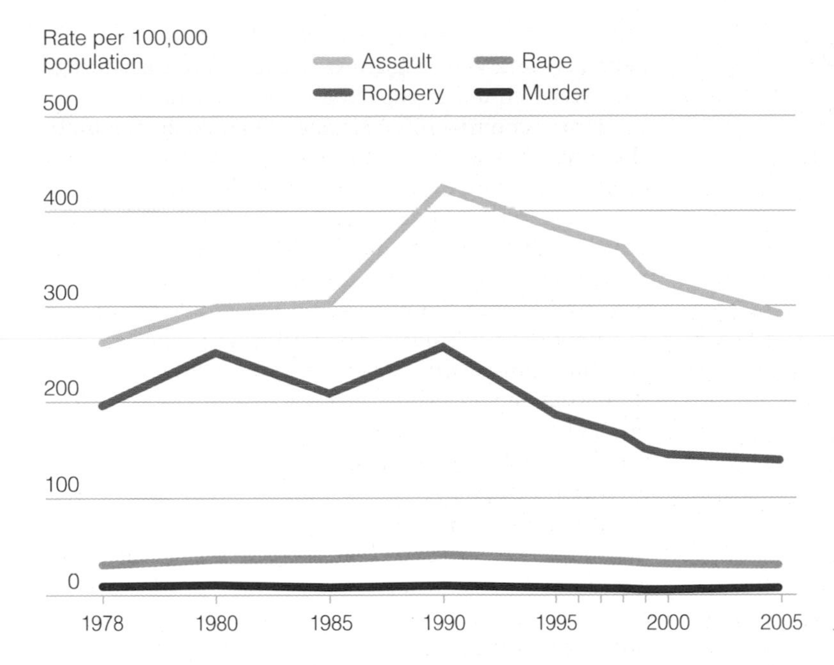

FIGURE 6.3 *Violent Crime in the United States*

Source: Federal Bureau of Investigation, 2007. *Uniform Crime Reports.* Washington, DC: U.S. Department of Justice. **http://www.fbi.gov**

[a sociological eye on the media

Images of Violent Crime

The media routinely drive home two points to the consumer: first, that violent crime is always high and may be increasing over time; and second, that there is much *random* violence constantly around us. The media bombard us with stories of "wilding," in which bands of youths kill random victims. Many of us think road rage is extensive (which it is not) and completely random. Most of us are now aware of violence in some high schools where students armed with automatic weapons kill their fellow students and indeed in some colleges as well, such as Virginia Tech (see the box "The Virginia Tech Student Massacre: Carnage with Egoistic Suicide"). The media vividly and routinely report such occurrences as pointless, random, and probably increasing.

The evidence shows that violent crime in the United States, although it increased during the 1970s and 1980s, nonetheless began to decrease in 1990 and continues to decrease nationally through the present. For example, both robbery and physical assaults have declined dramatically since 1990 (see Figure 6.3). Yet according to research (Best 1999; Glassner 1999), the media have consistently given a picture that violent crime has increased during this same period and, furthermore, that the violence is completely unpatterned and random.

No doubt there are occasions when victims are indeed picked at random. But the statistical rule of randomness could not possibly explain what has come to be called *random violence,* a vision of patternless chaos that is advanced by the media. If randomness truly ruled, *then each of us would have an equal chance of being a victim*—and of being a criminal. This is assuredly not the case. As Best notes, the notion of random violence, and the notion that it is increasing, ignores virtually everything that criminologists, psychologists, sociologists, and extensive research studies know about crime: It is highly patterned and significantly predictable, beyond sheer chance, by taking into account the social structure, social class, location, race–ethnicity, gender, labeling, age, whom one's family members are, and other such variables and forces in society that affect both criminal and victim. The broad picture, then, is clearly not conveyed in the media: Criminal violence is not increasing, but decreasing, and it is not random, but highly patterned and even predictable.

SEE FOR YOURSELF

CRIME IN THE MEDIA

Over the next three days, watch your local television news, keeping a written record of the racial identity of every criminal noted in the news. Also keep a record of what kind of crime is being reported. At the end of your observation period, count how many images of each radical group you have observed and the type of crime reported. Explain what your results suggest about racial stereotyping and crime reporting in the media.

the police will not believe a rape has occurred. Equally significant, the victim may not want to undergo the continued emotional stress of an investigation and trial. Recall from earlier in this chapter that certain kinds of noncriminal deviance, such as suicide, are also underreported, particularly by upper-income families, because of embarrassment to the deceased person's family.

Another problem arises in the attempt to measure crime by means of official statistics. The FBI's *Uniform Crime Reports* stress what are called **index crimes,** which include the violent crimes of murder, manslaughter, rape, robbery, and aggravated assault, plus property crimes of burglary, larceny-theft, and motor vehicle theft. These crimes are committed mostly by individuals who are disproportionately minority and poor. Statistics based on these offenses do not reflect the crimes that tend to be committed by middle-class and upper-class persons, such as tax violations, insider trading, embezzlement, fraudulent accounting practices, and other so-called elite crimes. The official statistics provide a relatively inflated picture for index crimes but an underreported picture of elite crimes, giving a biased picture of crime. A final result is, unfortunately, that the public sees the stereotyped "criminal" as a lower-class person, most likely an African American or Latino male, not as a middle- or upper-class White person who has committed tax fraud. The official statistics give biased support to the stereotype. This in turn perpetuates the public belief that the "typical" criminal is lower class and minority instead of upper class and nonminority. Criminals, however, can be either.

Personal and Property Crimes. The *Uniform Crime Reports* are subject to the same biases in official statistics mentioned earlier, but they are the major source of information on patterns of crime and arrest, with crimes classified into four categories. **Personal crimes** are violent or nonviolent crimes directed against people. Included in this category are murder, aggravated assault, forcible rape, and robbery. As we see in Figure 6.3, aggravated assault is the most frequently reported personal crime.

Hate crimes refer to assaults and other malicious acts (including crimes against property) motivated by various forms of social bias, including that based on race, religion, sexual orientation, ethnic/national origin, or disability. This form of crime has been increasing in recent years, especially against gays and lesbians (Jenness and Broad 2002).

Property crimes involve theft of property without threat of bodily harm. These include burglary (breaking and entering), larceny (the unlawful taking of property, but without unlawful entry), auto theft, and arson. Property crimes are the most frequent criminal infractions.

Finally, **victimless crimes** violate laws but are not listed in the FBI's serious crime index. These include illicit activities, such as gambling, illegal drug use, and prostitution, in which there is no complainant. Enforcement of these crimes is typically not as rigorous as the enforcement of crimes against persons or property, although periodic crackdowns occur, such as the current policy of mandatory sentencing for drug violations.

Elite and White-Collar Crime. Sociologists use the term *white-collar crime* to refer to criminal activities by people of high social status who commit their crimes in the context of their occupation (Sutherland and Cressey 1978). White-collar crime includes activities such as embezzlement (stealing funds from one's employer), involvement in illegal stock manipulations (insider trading), and a variety of violations of income tax law, including tax evasion. Also included are manipulations of accounting practices to make one's company appear profitable, thus artificially increasing the value of the company's stock.

White-collar crime seldom generates great concern in the public mind—far less than the concern directed at street crime. In terms of total dollars, however, white-collar crime is even more consequential for society. Scandals involving prominent white-collar criminals come to the public eye occasionally, such as the insider stock sale and subsequent cover-up by media personality Martha Stewart in 2002, for which she was sentenced to five months in prison in 2004. Nonetheless, white-collar crime is generally the least investigated and least prosecuted.

Some would argue, for example, that tobacco executives are guilty of crimes, given the known causal relationship between smoking and lung disease. From a sociological point of view, one interesting question that stems from studies of deviant behavior is how those who engage in deviance "normalize" their behavior. Most tobacco executives, for example, likely believe they are doing nothing wrong; they are likely to believe they are pursuing good business practices even though that business may have serious consequences for public health and safety (Rosenblatt 1994). From the perspective of conflict theory, sociologists argue that class bias lies at the heart of what is perceived as criminal behavior.

Debunking Society's Myths

Myth: The only *real* crimes are offenses such as robbery, assault, murder, and the like.

Sociological perspective: Many other types of activities are true crimes; they are in violation of the law. Examples of what people do not usually consider to be crimes are "elite" crimes such as income tax evasion, insider trading, embezzlement, and faulty business accounting practices. Although such crimes have been in the headlines more often in the past few years, such crimes are still prosecuted less frequently and less aggressively than crimes of the "street" among those of relatively lower social class position.

Organized Crime

The structure of crime and criminal activity in the United States often takes on an organized, almost institutional character. This is crime in the form of mob activity and racketeering and is called organized crime. Also, there are crimes committed by bureaucracies, known as corporate crime. Both types of crime are so highly organized, complex, and sophisticated that they take on the nature of social institutions.

Organized crime is crime committed by structured groups typically involving the provision of illegal goods and services to others. Organized crime syndicates are typically stereotyped as the Mafia, but the term can refer to any group that exercises control over large illegal enterprises, such as the drug trade, illegal gambling, prostitution, weapons smuggling, or money laundering. These organized crime syndicates are often based on racial or ethnic ties as well as family and kinship ties, with different groups dominating and replacing each other in different criminal "industries."

A key concept in sociological studies of organized crime is that these industries are organized along the same lines as legitimate businesses; indeed, organized crime has taken on a corporate form (Carter 1999). There are likely senior partners who control the profits of the business, workers who manage and provide the labor for the business, and clients who buy the services that organized crime provides. In-depth studies of the organized-crime underworld are difficult, owing to its secretive nature and dangers. As organized crime has moved into seemingly legitimate corporate organizations, it is even more difficult to trace, although some sociologists have penetrated

Abbot Genser/© HBO/courtesy Everett Collection

This dinner gathering of HBO's "Sopranos" gang illustrates both the hierarchy and group cohesion that characterizes organized crime.

underworld networks and provided fascinating accounts of how these crime networks are organized (Block and Scarpitti 1993; Carter 1999). Movies such as *The Godfather* and *Goodfellas,* as well as the popular TV series *The Sopranos,* have tended to glamorize organized crime.

The dons, or godfathers, that head crime organizations may lead relatively quiet lives in the suburbs and are good family men who attend religious services and spend time with their children. Traditionally, women have been excluded from meaningful leadership roles in organized crime, which is distinguished from other kinds of crime by its rigid hierarchy of godfathers, bosses, captains, underlings, hit men, and the like. The financial success of organized crime depends upon monopolistic control of prostitution and drug dealing; infiltration of legitimate business monopolies, such as waste and garbage removal; and the use of torture and murder for enforcement. *Racketeering,* the extortion of money from legitimate small and large businesses on a regular basis, is another widespread and well-organized undertaking. *Extortion* is accomplished by forcing businesspeople to buy "protection" for their businesses or insisting that they purchase products that they do not want or need (Scarpitti et al. 1997; Carter 1999).

Corporate Crime and Deviance: Doing Well, Doing Time

Corporations and even entire governments may engage in deviance—behavior that can be very costly to society. Sociologists estimate that the costs of corporate crime may be as high as $200 billion every year, dwarfing the take from street crime (roughly $15 billion), which most people imagine is the bulk of criminal activity. Tax cheaters in business alone probably skim $50 billion a year from the IRS, three times the value of street crime. Taken as a whole, the cost of corporate crime is almost 6000 times the amount taken in bank robberies in a given year and 11 times the total amount for all theft in a year (Reiman 2004).

Corporate crime and deviance is wrongdoing that occurs within the context of a formal organization or bureaucracy and is actually sanctioned by the norms and operating principles of the bureaucracy (Simon 2007). This can occur within any kind of organization—corporate, educational, governmental, or religious. It exists once deviant behavior becomes institutionalized in the routine procedures of an organization. The recent scandals involving sexual assaults of youths by Catholic priests and the attempted coverups by assigning offending priests to different parishes in different towns or states, constitute examples of corporate, or organizational, crimes. Individuals within the organization may participate in the deviant behavior with little awareness that their behavior is illegitimate. In fact, their actions are likely to be defined as in the best interests of the organization—business as usual. New members who enter the organization learn to comply with the organizational expectations or leave.

In the 1980s, Beech Nut baby foods proudly claimed that their "nutritionists prepare fresh-tasting vegetables, meats, fruits, cereals, and juices without artificial flavoring." Not only was there no artificial flavoring in the company's apple juice, there were no apples. Beech Nut was selling sugar water colored brown to resemble apple juice (Ermann and Lundman 1992). No one ordered plant operations to make fake juice. Instead, the owners insisted that the plant make a stronger return on its investment by cutting corners, and the pursuit of higher profits resulted in corporate crime. Most of the people in the production line probably never knew what was happening.

Sociological studies of corporate deviance show that this form of deviance is embedded in the ongoing and routine activities of organizations (Punch 1996; Lee and Ermann 1999). They represent cases of what in Chapter 5 was called the *normalization of deviance.* Instead of conceptualizing organizational deviance as merely the behavior of bad individuals, sociologists see it as the result of employees following rules and making decisions in more ordinary ways.

A case in point was the stock trading and accounting practices of the Enron Corporation of Houston, Texas. These deviant practices led to the downfall of that organization early in 2002 and the arrest and conviction of its president and CEO, Jeffrey Skilling and Kenneth Lay (since deceased) in 2006. In the summer of 2001, company executives who found company profits and thus their own personal company stock holdings declining in value quickly sold off their own stock and pocketed the resulting money before the stocks declined further in value. At the same time, they forbade their own rank-and-file employees from selling their own company stock. The stock held by these unfortunate employees declined to almost nothing over the next several months, thus wiping out the retirement accounts, or "nest eggs," of hundreds of Enron employees. Furthermore, the Enron executives enlisted their own accounting firm, the nationally known firm of Arthur Andersen, to cover up these practices and "cook the books" to conceal illegal stock transactions. Since then the Andersen firm has been hauled into court.

Another recent example of corporate and accounting malfeasance involved the WorldCom Corporation, a telecommunications company of worldwide repute. The company engaged in a multimillion dollar accounting fraud that disguised mounting losses from early in the year 2000 through the summer of 2002. By incorrectly reporting its operating expenses as though they were capital gains—again, employing illegal accounting—the company was able to inflate the value of its own stock even though its own finances were rapidly deteriorating (Eichenwald 2002; Costello 2004).

Race, Class, Gender, and Crime

Arrest data show a clear pattern of differential arrests along lines of race, gender, and class. To sociologists the central question posed by such data is whether this reflects actual differences in the extent of crime among different groups or whether this reflects differential treatment by the criminal justice system. The answer is "both" (D'Alissio and Stolzenberg 2003).

Certain groups are more likely to commit crime than others, because crime is distinctively linked to patterns of inequality in society. Unemployment, for example, is one correlate of crime, as is poverty. In a direct test of the association between inequality and crime, sociologist Ramiro Martinez, Jr. (2002, 1996) has explored the connection between rates of violence in Latino communities and the degree of inequality in 111 U.S. cities. His research shows a clear link between the likelihood of lethal violence and the socioeconomic conditions for Latinos in these different cities.

Sociologists use the analysis of socioeconomic conditions to explain the commission of crime, but they also make the important point that the process of prosecution by the criminal justice system is significantly related to patterns of race, gender, and class inequality. We see this in the bias of official arrest statistics, in treatment by the police, in patterns of sentencing, and in studies of imprisonment.

Race, Class, and Crime. One of the most important areas of sociological research on crime is the relationship between social class, crime, and race. Arrest statistics show a strong correlation between social class and crime, the poor being more likely than others to be arrested for crimes. Does this mean that the poor commit more crimes? To some extent, yes. Sociologists have demonstrated a strong relationship between unemployment, poverty, and crime (Scarpitti et al. 1997; Hagan 1993; Britt 1994; M. Smith et al. 1992). The reason is simple: Those who are economically deprived often see no alternative to crime—as Merton's structural strain theory would predict.

Moreover, law enforcement is concentrated in lower income and minority areas. People who are better off are further removed from police scrutiny and better able to hide their crimes. Although white-collar crime costs society more than street crime, it often goes undetected. When and if white-collar criminals are prosecuted and convicted, they typically receive light sentences—often a fine or community service instead of imprisonment. Middle- and upper-income people may be perceived as being less in need of imprisonment because they likely have a job and high-status people to testify for their good character. White-collar crime is simply perceived as less threatening than crimes by the poor. Class also predicts who most likely will be victimized by crime, with those at the highest ends of the socioeconomic scale least likely to be victims of violent crime (see Figure 6.4).

Paralleling the correlation between class and crime shown by arrest data is a strong relationship between race and crime. Minorities constitute 25 percent of the population of the United States but are more than 33 percent of the people arrested for property crimes and almost 50 percent of those arrested

AP Images/David J. Phillip

Daniel Acker/Bloomberg News/Landov

On the left, Andrew S. "Fast Andy" Fastow, former chief financial officer of the Enron corporation, is being taken to court by FBI agents. He was given a six-year jail sentence in September 2006. Martha Stewart, above right, the media/household planning tycoon, has spent five months in prison for stock fraud.

The Rich Get Richer and the Poor Get Prison

Jeffrey H. Reiman (2007) notes that the prison system in the United States, instead of serving as a way to rehabilitate criminals, is in effect designed to train and socialize inmates into a career of crime. It is also designed in such a way as to assure the public that crime is a threat primarily from the poor and that it originates at the lower rungs of society. This observation has also been made by Bruce Western (2006). Reiman and Western note that prisons contain elements that seem designed to accomplish this view. One can "construct" a prison that ends up looking like a U.S. prison. First, continue to label as criminals those who engage in crimes that have no unwilling victim, such as prostitution or gambling. Second, give prosecutors and judges broad discretion to arrest, convict, and sentence based on appearance, dress, race, and apparent social class. Third, treat prisoners in a painful and demeaning manner, as one might treat children. Fourth, make certain that prisoners are not trained in a marketable skill that would be useful upon their release. And, finally, assure that prisoners will forever be labeled and stigmatized as different from "decent citizens," even after they have paid their debt to society. Once an ex-con, always an ex-con. One has thus socially constructed a U.S. prison, an institution that will continue to generate the very thing that it claims to eliminate.

Sources: Reiman, Jeffrey H. 2007. *The Rich Get Richer and the Poor Get Prison.* 8th edition. Boston, MA: Allyn and Bacon; Western, Bruce. 2006. *Punishment and Inequality in America.* New York: Russell Sage Foundation.

for violent crimes. African Americans and Hispanics are more than twice as likely to be arrested for a crime than are Whites. Native Americans and Asians are exceptions, with both groups having relatively low rates of arrest for crime (Federal Bureau of Investigation 2006).

These data may seem to reinforce racial stereotypes, but sociologists have learned not to take these statistics at face value. Instead, they consider how poor and rich communities are policed as well as the social origins of crime to explain the differences in criminal behavior among groups. What do they find? Police have wide latitude in deciding when to enforce laws and make arrests. Their discretion is greatest when dealing with minor offenses, such as disorderly conduct. Sociological research has shown that police discretion is strongly influenced by class and race judgments, just as labeling theory would predict (Avakame et al. 1999). The police are more likely to arrest people who they perceive as troublemakers. They also are more likely to make arrests when the person complaining is White. Finally, minority communities are policed much more intensively, which leads to more frequent arrests of the residents.

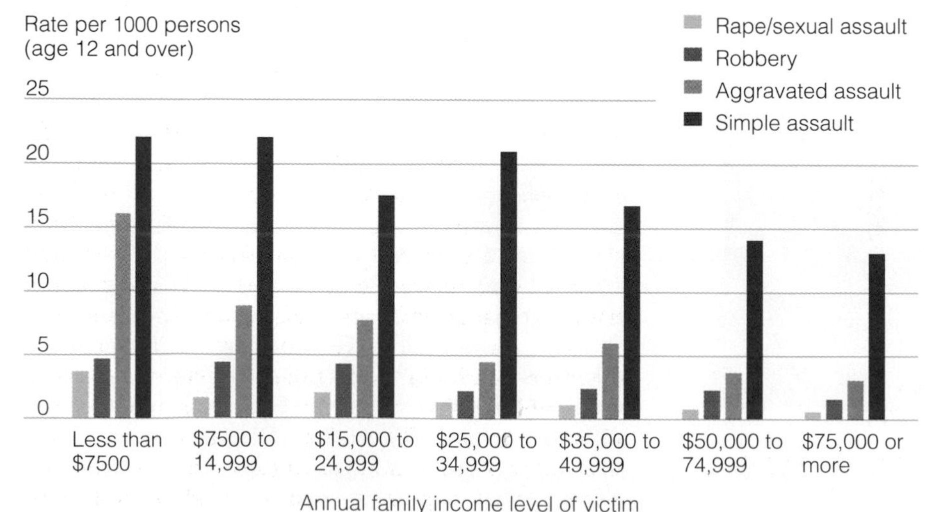

FIGURE 6.4 *Victimization by Crime: A Class Phenomenon*

U.S. Bureau of Statistics. 2006. Washington, DC: U.S. Department of Justice.

A research report on juvenile offenders demonstrates that Black and Latino youths with no prior criminal record are treated far more severely in the juvenile justice system than Whites of comparable social class who also have no prior criminal record. Minority youths are more likely to be arrested, held in jail, sent to juvenile or adult court for trial, convicted, and given longer prison terms. The racial disparities in the juvenile court system are magnified with each additional step into the justice system. In some cases, the racial disparities are stunning. For example, the report notes that 25 percent of arrested White youths are sent to prison, but *nearly 60 percent* of arrested Black youths are imprisoned. That is truly a wide racial disparity. The report concludes that these racial disparities lie not in overt discrimination on the part of prosecutors, judges, and other court personnel, but instead in the stereotypes that these decision makers rely on at each point in the juvenile justice system. Being Black, wearing low-hung baggy pants, and sporting dreadlocks is likely to get a person quickly through the various stages of the juvenile justice system and then quickly into prison (Butterfield 2000; Western 2006).

Bearing in mind the factors that affect the official rates of arrest and conviction—bias of official statistics, influence of powerful individuals, discrimination in patterns of arrest, differential policing—there remains evidence that the actual commission of crime varies by race. Why? Again, sociologists find a compelling explanation in social structural conditions. Racial minority groups are far more likely than Whites to be poor, unemployed, and living in single-parent families. These social facts are all predictors of a higher rate of crime. Note, too, as Figure 6.5 shows, that African Americans are generally more likely to be victimized by crime.

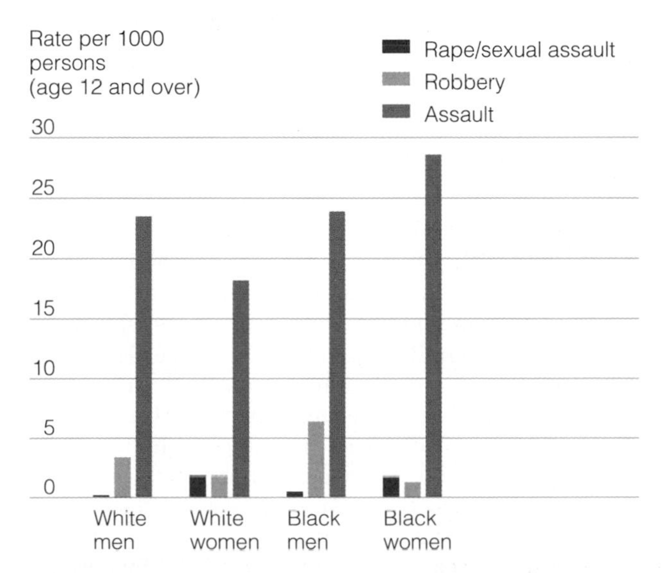

FIGURE 6.5 *Crime Victimization (by race and gender)*

Gender and Crime. Until recently, most sociological research on crime and deviance focused on men. Women's crime was seen as uninteresting or unimportant and studied only with the stereotyped vision of women as accomplices to men or as prostitutes. Newer research focuses on the analysis of crime and deviance among women (Belknap 2001).

Generally, women commit fewer crimes than men. Why? Although the number of women arrested for crime has increased slightly in recent years, the numbers are still small relative to men, except for a few crimes, such as fraud, embezzlement, and prostitution. Some argue that women's lower crime participation reflects their socialization into less risk-taking roles; others say that women commit crimes that are extensions of their gender roles—this would explain why the largest number of arrests of women are for crimes such as shoplifting, credit card fraud, and passing bad checks.

Nonetheless, women's participation in crime has been increasing in recent years. Sociologists relate this to several factors. Women are now more likely to be employed in jobs that present opportunities for crimes, such as property theft, embezzlement, and fraud. Violent crime by women has also increased notably since the early 1980s, possibly because the images that women have of themselves are changing, making new behaviors possible. Most significant, crime by women is related to their continuing disadvantaged status in society. Just as crime is linked to socioeconomic status for men, so it is for women (Belknap 2001; Miller 1986).

Despite recent achievements, many women remain in disadvantaged, low-wage positions in the labor market. At the same time, changes in the social structure of families mean that more women are economically responsible for their children without the economic support of men. Disadvantaged women may turn to illegitimate means of support, a trend that may be exacerbated by reductions in welfare support.

Women are also less likely than men to be victimized by crime, although victimization by crime among women varies significantly by race and age. Black women are much more likely than White women to be victims of violent crime; young Black women are especially vulnerable. Divorced, separated, and single women are more likely than married women to be crime victims. Regardless of their actual rates of victimization, women are more fearful of crime than men. Minority women and widowed, separated, and divorced women are the most fearful. Women's fear of crime increases with age even though the likelihood of victimization decreases with age, a fact that researchers attribute to the elderly's increased sense of vulnerability (Joseph 1997; Weinrath and Gartrell 1996; Gordon and Riger 1989). Women's fear of crime, sociologist Esther Madriz argues, results from an ideology that depicts women as needing protection and identifies public spaces as reserved for men. Women's fear of crime can be seen as a system of social control that keeps women from enjoying the rights of full citizenship (Madriz 1997).

For all women, victimization by rape is probably the greatest fear. Although rape is the most underreported crime, until recently, it has been one of the fastest growing—something criminologists explain as the result of a greater willingness to report *and* an actual increase in the extent of rape (Federal Bureau of Investigation 2006). More than 200,000 rapes (including attempted rapes) are reported to the police annually. Officials estimate that this is probably only about one in four of all rapes committed. Many women are reluctant to report rape because they fear the consequences of having the criminal justice system question them. Rape victims are least likely to report the assault when the assailant is someone known to them, even though a large number of rapes are committed by someone the victim knows.

A disturbingly frequent form of rape is *acquaintance rape*—rape committed by an acquaintance or someone the victim has just met. The extent of acquaintance rape is difficult to measure. The Bureau of Justice Statistics finds that 3 percent of college women experience rape or attempted rape in a given college year; 13 percent report being stalked (Fisher et al. 2001). Research finds that acquaintance rape is linked to men's acceptance of various rape myths (such as believing that a woman's "no" means "yes"), the use of alcohol, and the peer support that men receive in some all-male groups and organizations, such as fraternities (Taylor et al. 2006; Ullman et al. 1999; Boeringer 1999; Belknap et al. 1999).

Sociological research on rape shows that it is clearly linked to gender relations in society. Rape is an act of aggression against women, and as many sociologists have argued, it stems from learned gender roles that teach men to be sexually aggressive. This is reflected in sociological research on convicted rapists who think they have done nothing wrong and who believe, despite having overpowered their victims, that the women asked for it (Taylor et al. 2006; Scully 1990). Sociologists have argued that the causes of rape lie in women's status in society—that women are treated as sexual objects for men's pleasure. The relationship between women's status and rape is also reflected in data revealing who is most likely to become a rape victim.

African American women, Latinas, and poor women have the highest likelihood of being raped, as do women who are single, divorced, or separated. Young women are also more likely to be rape victims than older women (U.S. Bureau of Justice Statistics 2006). Sociologists interpret these patterns to mean that the most powerless women are also most subject to this form of violence.

The Criminal Justice System: Police, Courts, and the Law

Sociological studies consistently find patterns of differential treatment by the institutions that respond to deviance and crime in society. Whether it is in the police station, the courts, or prison, the factors of race, class, and gender are highly influential in the administration of justice in this society. Those in the most disadvantaged groups are more likely to be defined and identified as deviant independently of their behavior and, having encountered these systems of authority, are more likely to be detained and arrested, found guilty, and punished.

Debunking Society's Myths

Myth: The criminal justice system treats all people according to the neutral principles of law.

Sociological perspective: Race, class, and gender continue to have an influential role in the administration of justice. For example, even when convicted of the same crime as Whites, African American and Latino male defendants with the *same* prior arrest record as Whites are more likely to be arrested, sentenced, and to be sentenced for longer terms than White defendants.

The Policing of Minorities. There is little question that minority communities are policed more heavily than White neighborhoods; moreover, policing in minority communities has a different effect from that in White, middle-class communities. To middle-class Whites, the presence of the police is generally reassuring, but for African Americans and Latinos an encounter with a police officer can be terrifying. Regardless of what they are doing at the time, minority people, men in particular, are perceived as a threat, especially if they are observed in communities where they "don't belong."

Racial profiling has recently come to the public's attention, although it is a practice that has a long history. Often referred to half in jest by African Americans as the offense of "DWB," or "driving while Black," **racial profiling** on the part of a police officer is the use of race alone as the criterion for deciding whether to stop and detain someone on suspicion of their having committed a crime. The police argue that racial profiling is justified because a high proportion of Blacks and Hispanics commit crimes. Although the crime rate for Blacks and Hispanics is higher than that of Whites, race is a particularly bad basis for suspicion because the vast majority of Blacks and Hispanics, like the vast majority of Whites, do not commit any crime at all. Annually at least 90 percent of all African Americans are *not* arrested. That means on any given day, there is roughly a 90 percent probability that an African American in a car has not committed a crime, which leaves a 10 percent chance that the African American in the car has actually committed a crime. Nonetheless, *eight out of every ten* automobile searches carried out by state troopers on the New Jersey Turnpike over ten years were conducted on vehicles driven by Blacks and Hispanics; the vast majority of these searches turned up no evidence of contraband or crimes of any sort (Kocieniewski and Hanley 2000; Cole 1999).

Racial minorities are more likely than the rest of the population to be victims of excessive use of force by the police, also called police brutality. Studies show that most cases of police brutality involve minority citizens and that there is usually no penalty for the officers involved. Moreover, showing a disrespectful attitude is just as likely to generate police brutality as posing a serious bodily threat to the police (Lersch and Feagin 1996). Increasing the number of minority police officers has some effect on how the police treat minorities. Cities where African Americans head the police department show a concurrent decline in police brutality complaints, an increase in minority police and minority recruitment, and, in some cases, a decrease in crime. Simply increasing the number of African Americans in police departments does not, however, reduce crime dramatically, because it does not change the material conditions that create crime to begin with (Cashmore 1991).

Race and Sentencing. What happens once minority citizens are arrested for a crime? On arraignment, bail is set higher for African Americans and Latinos than for Whites, and minorities have less success with plea bargains. Extensive research finds that, once on trial, minority defendants are found guilty more often than White defendants. At sentencing, African Americans and Hispanics are likely to get longer sentences than Whites, even when they have the same number of prior arrests and socioeconomic background as Whites. Young African American men, as well as Latino men, are sentenced more harshly than any other group, and once sentenced they are less likely to be released on probation (Western 2006; Steffensmeier and Demuth 2000; Mauer 1999; Steffensmeier et al. 1998; Chambliss and Taylor 1989; Bridges and Crutchfield 1988). Any number of factors influence judgments about sentencing, including race of the judge, severity of the crime, race of the victim, and the gender of the defendant, but throughout these studies, race is shown to consistently matter—and matter a lot.

Racial discrimination is particularly evident with regard to the death penalty. Currently, of the approximately 3500 prisoners on death row, 44 percent are Black (U.S. Bureau of Justice Statistics 2006). Research shows that when Whites and minorities commit the same crime against a White victim, minorities are more likely to receive the death penalty. African American men convicted of raping White women are the group most likely to receive the death penalty. African Americans who kill Whites are also three times more likely to get the death penalty than those who kill African Americans, regardless of the race of the perpetrator (Paternoster and Brame 2003; Keil and Vito 1995).

Prisons: Rehabilitation or Racialized Mass Incarceration? Racial minorities account for *more than half* of the federal and state male prisoners in the United States. Blacks have the highest rates of imprisonment, followed by Hispanics, then Native Americans and Asians. (Native Americans and Asian Americans together are less than 1 percent of the total prison population.) Hispanics are the fastest growing minority group in prison (U.S. Bureau of Justice Statistics 2006). Native Americans, though a small proportion of the prison population, are still overrepresented in prisons. In theory, the criminal justice system is supposed to be unbiased, able to objectively weigh guilt and innocence. The reality is that the criminal justice system reflects the racial and class stratification and biases in society.

The United States and Russia have the highest rate of incarceration in the world (see Figure 6.6). In the United States, the rate of imprisonment has been rapidly growing (see Figure 6.7). By all signs, the population of state and federal prisons continues to grow, with the population in prisons exceeding the capacity of the facilities. The cost to the nation of keeping people behind bars is at least $160 billion (U.S. Bureau of Justice Statistics 2006).

Given the preponderance of Blacks and Hispanics in U.S. prisons, sociologist Bruce Western, in an extensive study, has coined the term *racialized mass incarceration* to describe this condition. Although it is certainly true that, as we have already noted, Blacks and Hispanics themselves commit proportionately more crime than Asians and Native American Indians, it is nonetheless *also* demonstrably true that the structure of the U.S. criminal justice system disproportionately *propels* Blacks and Hispanics into prison at a greater rate than same-aged Whites who have the same prior criminal record. Western gives at least three reasons for this: First, jobless (unemployment) rates have been much higher for Blacks and Hispanics; second, both federal and state Republican administrations have traditionally been more hostile to persons of color who run afoul of the criminal justice system than to Whites who do so; and, third, the "three strikes" law (giving mandatory long sentences if caught three times for the same drug offense)

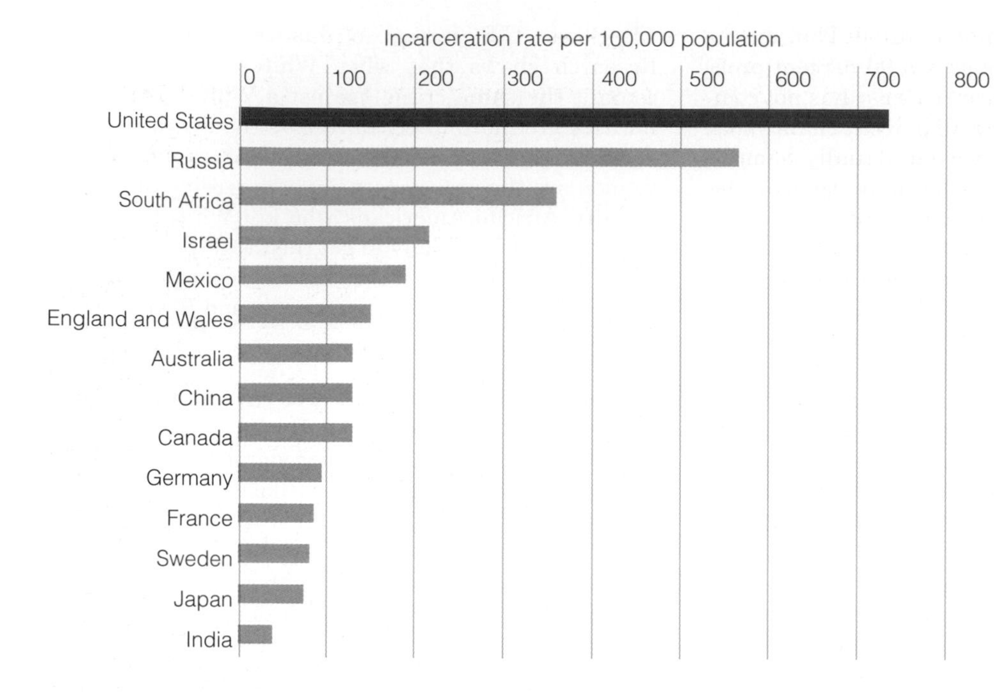

FIGURE 6.6 *Incarceration Rates for Selected Nations*

The Sentencing Project. 2006. Washington, DC. www.sentencingproject.com

disproportionately affects minorities of color. The situation is so severe and there are so many minority persons now in prison that sociologist Bruce Western calls them a *new color-caste* in U.S. society—in other words, a society unto itself (Western 2006).

The picture of incarceration in the United States seems contradictory. The overall violent crime rate has declined (see Figure 6.3); this would cause us to expect that the rate of admissions to prison would also decline. Yet at the same time the numbers of individuals in state and federal prisons have been increasing (Figure 6.7). Why is there such growth in the prison population when the crime rate has been declining? As already noted, a major reason for the increasing number of individuals behind bars is the increased enforcement of drug offenses and the mandatory sentencing that has been introduced. Nearly one-quarter of those in state prisons are serving a drug sentence. Sixty percent of federal prisoners are serving drug sentences, more than double that of 1980.

The number of women behind bars has also increased at a faster rate than for men, although the number of women in prison is small by comparison. Women are only 8 percent of all state and federal prisoners (U.S. Bureau of Justice Statistics 2005).

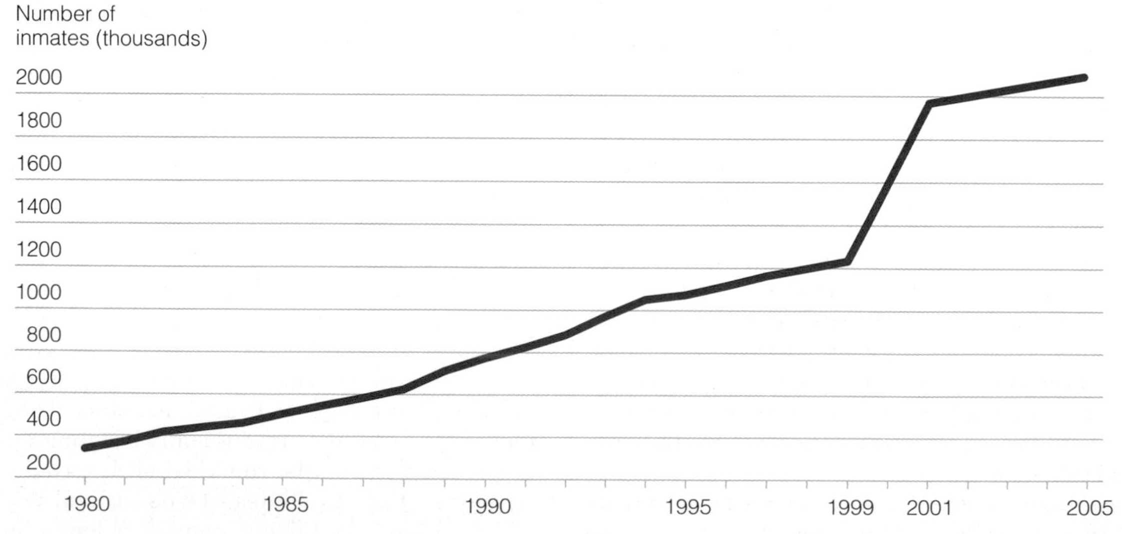

FIGURE 6.7 *State and Federal Prison Population, 1980–2004*

Source: U.S. Bureau of Justice Statistics. 2007. "Prison and Jail Inmates." **www.ojp.usdoj.gov/bjs/prisons.html**

Like men, three-fourths of women in federal prisons are there because of drug offenses; often they have participated in these crimes by going along with the behavior of their boyfriends (Miller 1986). The typical woman in prison is a poor, young minority who dropped out of high school, is not married, and has two or more children. Of all women prisoners, about *two-thirds* have been victims of sexual abuse. Women prisoners are also more likely than men to be positive for HIV-infection (Greenfeld and Snell 1999).

Women in prison face unique problems, in part because they are in a system designed for men and run mostly by men, which tends to ignore the particular needs of women. For example, 25 percent of the women entering prison are pregnant or have just given birth, but they often get no prenatal or obstetric care. Male prisoners are trained for such jobs as auto mechanics, whereas women are more likely to be trained in relatively lower-status jobs such as beauticians and launderers. The result is that few women offenders are rehabilitated by their experience in prison.

The United States, then, is putting offenders in prison at a record pace. Is crime being deterred? Are prisoners being rehabilitated? Or are they simply being *warehoused*—put on a shelf? If the deterrence argument were correct, we would expect that increasing the risk of imprisonment would lower the rate of crime. For example, we would expect drug use to decline as enforcement of drug laws increases. In the past few years, there has been a marked increase in drug law enforcement but not the expected decrease in drug use. Using drugs as an example, then, it appears that the threat of imprisonment does not deter crime (Mauer 1999).

There is also little evidence that the criminal justice system rehabilitates offenders. Using drugs as the example again, only about 20 percent of those imprisoned for drug offenses ever receive drug treatment. Although the nation is getting "tough on crime," it is doing little to see that offenders do not continue to commit drug crimes once they are released from prison.

If the criminal justice system fails to reduce crime, what does it do? Some sociologists contend that the criminal justice system is not meant to reduce crime but has other functions, namely to reinforce an image of crime as a threat from the poor and from racial groups. The prison experience is a demeaning one, poorly suited to training prisoners in marketable skills or to letting them repay their debt to society. Rather than teaching prisoners self-control and self-direction, prisons deny inmates the least control over their everyday life (Reiman 2004). In the end, prisons seem, at least in some cases, to refine criminals rather than rehabilitate them (see the box, "Understanding Diversity: The Rich Get Richer and the Poor Get Prison").

Terrorism as International Crime: A Global Perspective

Crime now crosses international borders and has become global. Terrorism is an example of the globalization of crime. The morning of Tuesday, September 11, 2001, was not expected to be any different from other early fall mornings. However, at approximately 8:45 A.M., the world stood still with images of a civilian airplane crashing into the north tower of the World Trade Center in New York City. Many at that moment assumed that the incident was nothing more than an accident by a pilot who might have lost consciousness or a plane that had a mechanical failure. This was a short-lived assumption. At 9:03 A.M., a second plane crashed into the south tower of the World Trade Center, leaving no doubt that this constituted a terrorist attack on the United States. This was confirmed a short time later by reports that another airplane had crashed into the Pentagon, and still another hijacked plane had crashed in Pennsylvania. Virtually everyone in the United States now remembers where they were and what they were doing during those horrible moments.

The FBI includes *terrorism* in its definition of crime, seeing it as violent action to achieve political ends (White 2002). Thus, terrorism is a crime that violates both international and domestic laws. It is a crime that crosses national borders, and its understanding requires a global perspective. The origins of modern terrorism are historically traceable from western European ideology to the Russian Revolution, and then to the incorporation of Russian Revolutionary

AP Images

Threats of terrorism, such as bioterrorism, have resulted in increased security and countermeasures, particularly in urban areas.

thought in the nationalist struggle in Ireland. Contrary to what many think, the origins of international terrorism go back farther in history than the Israel–Palestine conflict and farther back than Osama bin Laden's terrorist organization, al Qaeda.

Terrorism is globally linked to other forms of international crime. A case in point was Osama bin Laden's role in the international opium trade. Afghanistan, where bin Laden's terrorist al Qaeda organization was headquartered, was the world's largest grower of opium-producing poppies. It is suspected that profits from the international drug trade helped finance the September 11 terrorist attacks. Therefore, a global perspective on crime involves recognizing the global basis of some international crime networks that cross national borders (Binns 2003).

Many nations have long experienced terrorism in the form of bombings, hijackings, suicide attacks, and other terrorist crimes. But the attacks of September 11 focused the world's attention on the problem of terrorism in new ways, including increased fears of **bioterrorism**—the form of terrorism involving the dispersion of chemical or biological substances intended to cause widespread disease and death. Fears of bioterrorism were exacerbated in the United States in 2002 with a threat of the spread of anthrax. This exceptionally deadly bacteria capable of causing virtually instant death upon inhaling the spores was found in the mail of offices of Congress, the Postal Service, the Supreme Court, and other locations, resulting in the deaths of several persons.

Another form of terrorism, and thus cause for international concern, is **cyberterrorism,** the use of the computer to commit one or more terrorist acts. Terrorists may use computers in a number of ways. Data-destroying computer viruses may be implanted electronically in an enemy's computer. Another use would be to employ "logic bombs" that lie dormant for years until they are electronically instructed to overwhelm a computer system. The use of the Internet to serve the needs of international terrorists has already become a reality (Jucha 2002, 2007).

Without understanding the political, economic, and social relations from which terrorist groups originate, terrorist acts seem like the crazed behavior of violent single individuals. Although sociologists in no way excuse such acts, they look to the social structure of conflicts from which terrorism emerges as the cause of such criminal and deviant behavior. Even then, terrorism is not only the work of international extremist groups, as witnessed by the bombing of the Oklahoma City federal buildings in 1995. For this terrorist act, two White, male U.S. citizens were tried and convicted, and one of them was executed. Terrorism, whether domestic or international, is best understood not only as individual insanity, but also as a politically, economically, and socially oriented form of violence.

Chapter Summary

- ***What is the difference between deviance and crime?***

  Deviance is behavior that violates norms and rules of society, and *crime* is a type of deviant behavior that violates criminal law. *Criminology* is the study of crime from a scientific perspective.

- ***How do sociologists conceptualize and explain deviance and crime?***

  Deviance is behavior that is recognized as violating expected rules and norms and that should be understood in the social context in which it occurs. Psychological explanations of deviance place the cause of deviance primarily within the individual. Sociologists emphasize the total social context in which deviance occurs. Sociologists see deviance more as the result of group, not individual, behavior.

- ***What does sociological theory contribute to the study of deviance and crime?***

  Functionalist theory sees both deviance and crime as functional for the society because it affirms what is acceptable by defining what is not. *Structural strain theory,* a type of functionalist theory, predicts that societal inequalities actually force and compel the individual into deviant and criminal behavior. *Conflict theory* explains deviance and crime as a consequence of unequal power relationships and inequality in society. *Symbolic interaction theory* explains deviance and crime as the result of meanings people give to various behaviors. *Differential association theory,* a type of symbolic interaction theory, interprets deviance as behavior learned through social interaction with other deviants. *Labeling theory,* also a type of symbolic interaction theory, argues that societal reactions to behavior produce deviance, with some groups having more power than others to assign deviant labels to people.

- ***What are the major forms of deviance?***
  Mental illness, *stigma,* and substance abuse are major forms of deviance studied by sociologists, although deviance comprises many different forms of behavior. Sociological explanations of mental illness focus on the social context in which mental illness develops and is treated. Social *stigmas* are attributes that are socially devalued. Substance abuse includes alcohol and drug abuse but is not limited to these two forms.

- ***What are the connections between inequality, deviance, and crime?***
  Sociological studies of crime analyze the various types of crimes, such as *elite crime, organized crime, corporate crime,* and *personal* and *property crimes.* Many types of crimes are underreported, such as rape and certain elite and corporate crimes. Sociologists study the conditions, including race, class, and gender inequality, that produce crime and shape how different groups are treated by the criminal justice system, such as showing group differences in sentencing.

- ***How is crime related to race, class, and gender?***
  In general, crime rates for a variety of crimes are higher among minorities than among Whites, among poorer persons than among middle- or upper-class persons, and among men than among women. Women, especially minority women, are more likely to be victimized by serious crimes such as rape or violence from a spouse or boyfriend.

- ***How is globalization affecting the development of deviance and crime?***
  International *terrorism* is a crime, and crime is thus global. Other global crimes of significance are *bioterrorism* and *cyberterrorism.* Osama bin Laden's al Qaeda organization, centered in Afghanistan, was central to the international drug trade. Thus, crimes are clearly not just the acts of a crazed small group of individuals but the result of structural and cultural conditions.

Key Terms

altruistic suicide 150
anomic suicide 150
anomie 150
bioterrorism 174
crime 162
criminology 162
cyberterrorism 174
deviance 146
deviant career 158
deviant communities 158
deviant identity 157
differential association theory 156
egoistic suicide 151
elite deviance 155
hate crime 164
index crimes 164
labeling theory 157
master status 160
medicalization of deviance 149
organized crime 165
personal crimes 164
property crimes 164
racial profiling 170
social control 155
social control agents 155
social control theory 154
stigma 160
structural strain theory 151
victimless crimes 164

Online Resources

Sociology: The Essentials Companion Website

www.thomsonedu.com/sociology/andersen
Visit your book companion website where you will find more resources to help you study and write your research papers. Resources include Suggested Readings, web links, and a MicroCase Online feature that teaches you how to research society. Other resources include Learning Objectives, Internet exercises, quizzing, and flash cards.

is an easy-to-use online resource that helps you study in less time to get the grade you want NOW.

www.thomsonedu.com/login
Need help studying? This site is your one-stop study shop. Take a Pre-Test and Thomson NOW will generate a Personalized Study Plan based on your test results. The Study Plan will identify the topics you need to review and direct you to online resources to help you master those topics. You can then take a Post-Test to determine the concepts you have mastered and what you still need to work on.

7
Chapter seven
CHAPTER SEVEN
Chapter

Social Class and Social Stratification

[**One afternoon in** a major U.S. city, two women go shopping. They are friends—wealthy, suburban women who shop for leisure. They meet in a gourmet restaurant and eat imported foods while discussing their children's private schools. After lunch, they spend the afternoon in exquisite stores—some of them large, elegant department stores; others, intimate boutiques where the staff know them by name. When one of the women stops to use the bathroom in one store, she enters a beautifully furnished room with an upholstered chair, a marble sink with brass faucets, fresh flowers on a wooden pedestal, shining mirrors, an ample supply of hand towels, and jars of lotion and soaps. The toilet is in a private stall with solid doors. In the stall there is soft toilet paper and another small vase of flowers.

The same day, in a different part of town, another woman goes shopping. She lives on a marginal income earned as a stitcher in a textiles factory. Her daughter badly needs a new pair of shoes because she has outgrown last year's pair. The woman goes to a nearby discount store where she hopes to find a pair of shoes for under $15, but she dreads the experience. She knows her daughter would like other new things—a bathing suit for the summer, a pair of jeans, and a blouse. But this summer the daughter will have to wear hand-me-downs because medical bills over the winter have depleted the little money left after food and rent. For the mother, shopping is not recreation but a bitter chore reminding her of the things she is unable to get for her daughter.

continued

Mitchell Funk/Getty Images/Photographer's Choice

chapter seven

CHAPTER SE

While this woman is shopping, she, too, stops to use the bathroom. She enters a vast space with sinks and mirrors lined up on one side of the room and several stalls on the other. The tile floor is gritty and gray. The locks on the stall doors are missing or broken. Some of the overhead lights are burned out, so the room has dark shadows. In the stall, the toilet paper is coarse. When the woman washes her hands, she discovers there is no soap in the metal dispensers. The mirror before her is cracked. She exits quickly, feeling as though she is being watched.

Two scenarios, one society. The difference is the mark of a society built upon class inequality. The signs are all around you. Think about the clothing you wear. Are some labels worth more than others? Do others in your group see the same marks of distinction and status in clothing labels? Do some people you know never seem to wear the "right" labels? Whether it is clothing, bathrooms, schools, homes, or access to health care, the effect of class inequality is enormous, giving privileges and resources to some and leaving others struggling to get by.

Great inequality divides society. Nevertheless, most people think that in the United States equal opportunity exists for all. The tendency is to blame individuals for their own failure or attribute success to individual achievement. Many people think the poor are lazy and do not value work. At the same time, the rich are often admired for their supposed initiative, drive, and motivation. Neither is an accurate portrayal. There are many hard-working individuals who are poor, and most rich people have inherited their wealth rather than earned it themselves.

Observing and analyzing class inequality is fundamental to sociological study. What features of society cause different groups to have different opportunities? Why is there such an unequal allocation of society's resources? Sociologists respect individual achievements but have found the greatest cause for the disparities in material success is the organization of society. Instead of understanding inequality as the result of individual effort, sociologists thus study the social structural origins of inequality.

Social Differentiation and Social Stratification

All social groups and societies exhibit social differentiation. **Status,** as we have seen earlier, is a socially defined position in a group or society. **Social differentiation** is the process by which different statuses develop in any group, organization, or society. Think of a sports organization. The players, the owners, the managers, the fans, the cheerleaders, and the sponsors all have a different status within the organization. Together they constitute a whole social system, one that is marked by social differentiation.

Status differences can become organized into a hierarchical social system. Social stratification is a relatively fixed, hierarchical arrangement in society by which groups have different access to resources, power, and perceived social worth. **Social stratification** is a system of structured social inequality. To use a sports example, owners control the resources of the teams; players, although they may earn high salaries, do not control the team resources and have less power than the owners and managers. Sponsors (including individuals and corporations) provide the

table 7.1 Inequality in the United States

- One in six children in the United States live in poverty, including 34 percent of African American children, 28 percent of Hispanic children, 10 percent of White children, and 10 percent of Asian American children (DeNavas-Walt et al. 2006).
- The rate of poverty among people in the United States has been steadily increasing since 2000 (DeNavas-Walt et al. 2006).
- Among women heading their own households, 31 percent live below the poverty line (DeNavas-Walt et al. 2006).
- One percent of the U.S. population controls 33 percent of the total wealth in the nation; the bottom 20 percent owe more than they own (Mishel et al. 2005).
- When Leona Helmsley (the hotel financier) died, she left a $12 million inheritance to her dog, with the provision that the dog should eventually be buried in a mausoleum with Helmsley—a mausoleum that would be steam-cleaned once per week in perpetuity.
- The average CEO of a major company has a salary of $13.1 million dollars per year; workers earning the minimum wage make $10,712 per year if they work 40 hours a week for 52 weeks and hold only one job (Lavelle 2001).

resources on which this system of stratification rests; fans are merely observers who pay to watch the teams play. Sports organizations include a system of stratification because the groups that constitute the organization are arranged in a hierarchy where some have more resources and power than others.

All societies seem to have a system of social stratification, although they vary in the degree and complexity of stratification. Some societies stratify only along a single dimension, such as age, keeping the stratification system relatively simple. Most contemporary societies are more complex, with many factors interacting to create social strata. In the United States, social stratification is strongly influenced by class, which is in turn influenced by matters such as one's occupation, income, and education, along with race, gender, and other influences such as age, region of residence, ethnicity, and national origin.

Estate, Caste, and Class

Stratification systems can be broadly categorized into three types: estate systems, caste systems, and class systems. In an **estate system** of stratification, the ownership of property and the exercise of power is monopolized by an elite who have total control over societal resources. Historically, such societies were feudal systems where classes were differentiated into three basic groups—the nobles, the priesthood, and the commoners. Commoners included peasants (usually the largest class group), small merchants, artisans, domestic workers, and traders. The nobles controlled the land and the resources used to cultivate the land, as well as all the resources resulting from peasant labor.

Estate systems of stratification are most common in agricultural societies. Although such societies have been largely supplanted by industrialization, there are still examples of societies that have a small but powerful landholding class ruling over a population that works mainly in agricultural production. Unlike the feudal societies of the European Middle Ages, however, contemporary estate systems of stratification display the influence of international capitalism. The "noble class" comprises not knights who conquered lands in war, but international capitalists or local elites who control the labor of a vast and impoverished group of people, such as in some South American societies where landholding elites maintain a dictatorship over peasants who labor in agricultural fields.

In a **caste system,** one's place in the stratification system is an *ascribed status* (see Chapter 4), meaning it is a quality given to an individual by circumstances of birth. The hierarchy of classes is rigid in caste systems and is often preserved through formal law and cultural practices that prevent free association and movement between classes.

The system of *apartheid* in South Africa was a stark example of a caste system. Under apartheid, the travel, employment, associations, and place of residence of Black Africans were severely restricted. Segregation was enforced using a pass system in which Blacks in White areas were obliged to account for themselves to White authorities. Interracial marriage was illegal, and Black Africans were prohibited from voting. The apartheid system was overthrown by election in 1992.

In *class systems,* stratification exists, but a person's placement in the class system can change according to personal achievements; that is, class depends to some degree on *achieved status,* defined as status that is earned by the acquisition of resources and power, regardless of one's origins. Class systems are more open than caste systems because position does not depend strictly on birth, and classes are less rigidly defined than castes because the divisions are blurred by those who move between one class and the next.

Social class differences make it seem as if some people are practically living in two different societies.

Social class influences many opportunities and practices in society, including the leisure that different groups experience.

© David R. Frazier/PhotoEdit
© Stock Image/Alamy
© Jeff Greenberg/Peter Arnold, Inc.

Despite the potential for movement from one class to another, in the class system found in the United States, class placement still depends heavily on one's social background. Although ascription (the designation of ascribed status according to birth) is not the basis for social stratification in the United States, the class a person is born into has major consequences for that person's life. Patterns of inheritance; access to exclusive educational resources; the financial, political, and social influence of one's family; and similar factors all shape one's likelihood of achievement. Although there is no formal obstacle to movement through the class system, individual achievement is very much shaped by an individual's class of origin.

thinking sociologically

Take a shopping trip to different stores and observe the appearance of stores serving different economic groups. What kinds of bathrooms are there in stores catering to middle-class clients? The rich? The working class? The poor? Which ones allow the most privacy or provide the nicest amenities? What fixtures are in the display areas? Are they simply utilitarian with minimal ornamentation, or are they opulent displays of consumption? Take detailed notes of your observations, and write an analysis of what this tells you about *social class* in the United States.

Defining Class

In common terms, *class* refers to style or sophistication. In sociological use, **social class** (or **class**) is the social structural position groups hold relative to the economic, social, political, and cultural resources of society. Class determines the access different people have to these resources and puts groups in different positions of privilege and disadvantage. Each class has members with similar opportunities who tend to share a common way of life. Class also includes a cultural component in that class shapes language, dress, mannerisms, taste, and other preferences. *Class is not just an attribute of individuals; it is a feature of society.*

The social theorist Max Weber described the consequences of stratification in terms of **life chances,** meaning the opportunities that people have in common by virtue of belonging to a particular class. Life chances include the opportunity for possessing goods, having an income, and having access to particular jobs. Life chances are also reflected in the quality of everyday life. Whether you dress in the latest style or wear another person's discarded clothes, have a vacation in an exclusive resort, take your family to the beach for a week, or have no vacation at all—these life chances are the result of being in a particular class.

Class is a structural phenomenon; it cannot be directly observed. Nonetheless, you can "see" class through various displays that people project—often unintentionally—about their class status. What clothing do you wear? Do some worn objects project higher class status than others? How about cars? What class status is displayed through the car you drive—or, for that matter, whether you even have a car or use a bus to get to work. In these and myriad other ways, class is projected to others as a symbol of our presumed worth in society.

Social class can be observed in the everyday habits and presentations of self that people project. Common objects, such as clothing and cars, become symbols of one's class status. As such, they can be ranked not only in terms of their economic value, but also in terms of the status that various brands and labels

SEE FOR YOURSELF

STATUS SYMBOLS IN EVERYDAY LIFE

You can observe the everyday reality of social class by noting the status that different ordinary objects have within the context of a class system. Make a list of every car brand you can think of—or, if you prefer, every clothing label. Then rank your list with the highest status brand (or label) at the top of the list, going down to the lowest status. Then answer the following questions:

1. Where does the presumed value of this object come from? Does the value come from the actual cost of producing the object or something more subjective?
2. Do people make judgments about people wearing or driving the different brands you have noted? What judgments do they make? Why?
3. What consequences do you see (positive and negative) of the ranking you have observed? Who benefits from the ranking and who does not?

What does this exercise reveal about the influence of status symbols in society?

carry. The interesting thing about social class is that a particular object may be quite ordinary, but with the right "label" it becomes a *status symbol* and thus becomes valuable. Take the example of Vera Bradley bags. These paisley bags are made of ordinary cotton with batting. Not long ago, such cloth was cheap and commonplace, associated with rural, working class women. If such a bag were sewn and carried by a poor person living on a farm, the bag (and perhaps the person!) would be seen as ordinary, almost worthless. But, transformed by the right label (and some good marketing), Vera Bradley bags have become status symbols, selling for a high price (often a few hundred dollars—a price one would never pay for a simple cotton purse). Presumably having such a bag denotes the status of the person carrying it. (See also the box "See for Yourself: Class Symbols in Everyday Life.")

Because sociologists cannot isolate and measure social class directly, they use other indicators to serve as measures of class. A prominent indicator of class is income; other common indicators are education, occupation, and place of residence. These indicators alone do not define class, but they are often accurate measures of the class standing of a person or group. We will see that these indicators tend to be linked. A good income, for example, makes it possible to afford a house in a prestigious neighborhood and an exclusive education for one's children. In the sociological study of class, indicators such as income and education have had enormous value in revealing the outlines and influences of the class system.

The Class Structure of the United States

The class structure of the United States is elaborate, arising from the interactions of old wealth, new wealth, intensive immigration, a culture of entrepreneurship and individualism, and in recent times, vigorous globalization. One can conceptualize the class system as a ladder, with different class groups arrayed up and down the rungs, each rung corresponding to a different level in the class system. Conceptualized this way, social class is the common position groups hold in a status hierarchy (Wright 1979; Lucal 1994); class is indicated by factors such as levels of income, occupational standing, and educational attainment. People are relatively high or low on the ladder depending on the resources they have, whether those resources are education, income, occupation, or any of the other factors known to influence people's placement (or ranking) in the stratification system. Indeed, an abundance of sociological research has

Although "rags to riches" stories are common and there are examples of ordinary people who have become fabulously wealthy, such social mobility is more the exception than the rule.

seeing society in everyday life

Social Inequality

If you use your sociological imagination, you will begin to see the signs of social inequality in many places, including common place locations, such as shopping, but also in events that are out of the ordinary, such as natural disasters. Sociologists Christine Williams noted this in her research for the book, *Inside Toyland,* where she studied the influence of class, race, and gender in the different experiences of shopping in a "big box" toy store compared with shopping in a boutique toy store. Where do you see class inequality in your everyday life?

© Jeff Greenburg/PhotoEdit

AP Images/Don Ryan

How does the experience of shopping vary by social class?

Of course, everybody loves a bargain, but social class inequality means that people will have very different experiences. For some, shopping is a cruel reminder of their class status; for others, it may be a form of leisure and a mark of their higher status. For the upper class shopper, shopping is likely to be more personal and accommodating.

How are status symbols related to the process of globalization?

Although globalization may not be immediately visible, many of the products by which people in one place accrue status are linked to the low status of workers in other nations. This is known as the global assembly line.

AP Images/Richard Vogel

© Mark Wilson/Getty Images

AP Images/Dave Martin

Don't natural disasters affect all people?

Hurricane Katrina showed the public that, although a natural disaster can affect anyone, the impact of such disasters falls especially hard on people with the fewest resources. Katrina unmasked the deadly reality of race and poverty in the United States.

AP Images/Alex Brandon

How have race and class affected the rebuilding of New Orleans?

Although everyone affected by the devastation of Hurricane Katrina has struggled with the enormity of rebuilding, rebuilding has been slow to even begin in the city's poorest section, the Ninth Ward.

AP Images/Cheryl Gerber

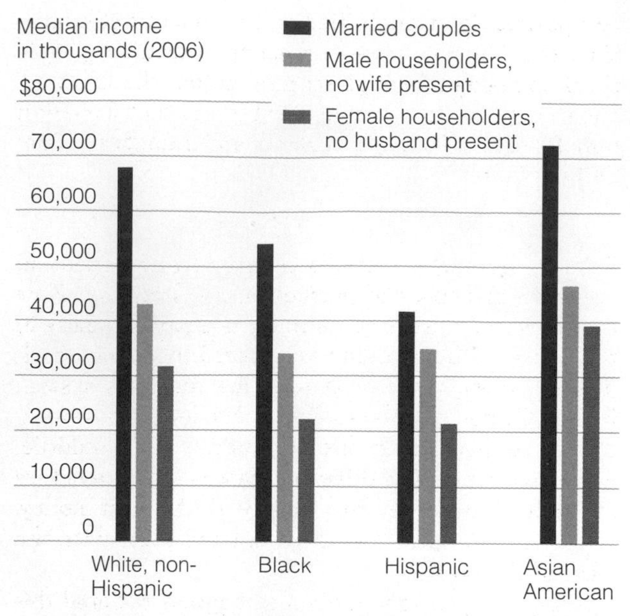

FIGURE 7.1 *Median Annual Income by Race and Household Status*

As illustrated in this graph, married couple households have the highest median income in all racial-ethnic groups; female-headed households, the least. The only households that reach median income status are White (non-Hispanic), African American, and Asian married households.

Source: U.S. Census Bureau. 2006. *Historical Income Tables,* Table F-8. www.census.gov

stemmed from the concept of **status attainment,** the process by which people end up in a given position in the stratification system. Status attainment research describes how factors such as class origins, educational level, and occupation produce class location.

This laddered model of class suggests that in the United States stratification is hierarchical but somewhat fluid. Different gradients in the stratification system are not fixed as they might be in a society where one's placement is completely a matter of the class into which one is born. In a relatively open class system such as the United States, people's achievements do matter, although the extent to which people rise rapidly and dramatically through the stratification system is less than the popular imagination envisions. Some people do begin from modest origins and amass great wealth and influence (celebrities such as Bill Gates, Oprah Winfrey, and millionaire athletes), but these are the exceptions, not the rule. Some people move down in the class system, but as we will see, most people remain relatively close to their class of origin. When people rise or fall in the class system, the distance they travel is usually relatively short, as we will see further in the section on social mobility.

The image of stratification as a laddered system, with different gradients of social standing, emphasizes that one's **socioeconomic status (SES)** is derived from certain factors. Income, occupational prestige, and education are the three measures of socioeconomic status that have been found to be most significant in determining people's placement in the stratification system. **Income** is the amount of money a person receives in a given period. As we will see, income is distinct from **wealth** (or net worth), which is the total value of what one owns, minus one's debts. The **median income** for a society is the midpoint of all household incomes. In other words, half of all households earn more than the median income; half earn less. In the laddered model of class, those bunched around the median income level are considered middle class. In 2005, median household income in the United States was $46,236 (see Figure 7.1; DeNavas-Walt et al. 2006).

SEE FOR YOURSELF

INCOME DISTRIBUTION—SHOULD GRADES BE THE SAME?

Figure 7.2 shows you the income distribution with the United States. Imagine that grades in your class were distributed based on the same curve. Let's suppose that after students arrived in class and sat down, different groups received their grades based on where they were sitting in the room. Only students in the front receive As; the back, Ds and Fs. The middle of the room gets the Bs and Cs. Write a short essay answering the following questions based on this hypothetical scenario.

1. How many students would receive As, Bs, Cs, Ds, and Fs?
2. Would it be fair to distribute grades this way? Why or why not?
3. Which groups in the class might be more likely to support such a distribution? Who would think the system of grade distribution should be changed?
4. What might different groups do to preserve or change the system of grade distribution? What if you really needed an A, but got one of the Fs? What might you do?
5. Are there circumstances in actual life that are beyond the control of people and that shape the distribution of income?
6. How is social stratification maintained by the beliefs that people have about merit and fairness?

Adapted from: Brislen, William, and Clayton D. Peoples. 2005. "Using a Hypothetical Distribution of Grades to Introduce Social Stratification." *Teaching Sociology* 33 (January): 74–80.

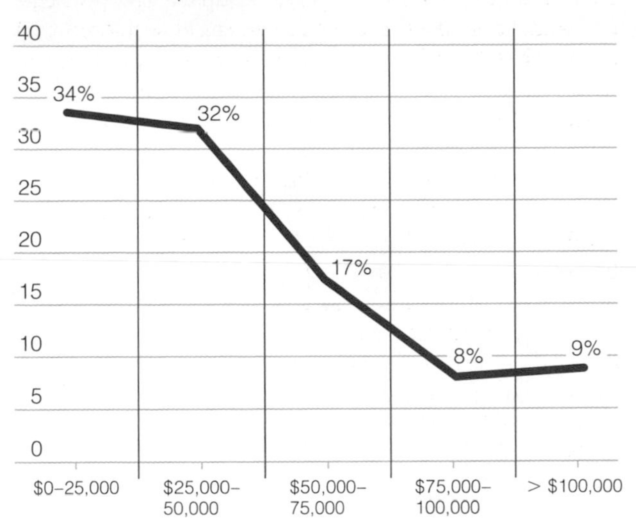

FIGURE 7.2 *Income Distribution in the United States*

This graph shows the percentage of the population that falls into each of the five income brackets.

Source: DeNavas-Walt, Carmen, Bernadette Proctor, and Cheryl Lee. 2006. *Income, Poverty, and Health Insurance Coverage in the United States: 2005.* Washington, DC: U.S. Census Bureau, #P60-231.

In the laddered model of class, those bunched around the median income level are considered middle class. Sociologists may argue about which income brackets constitute middle-class standing, but they agree that income is a significant indicator of class, although not the only one.

Occupational prestige is a second important indicator of socioeconomic status. **Prestige** is the value others assign to people and groups. **Occupational prestige** is the subjective evaluation people give to jobs. To determine occupational prestige, sociological researchers typically ask nationwide samples of adults to rank the general standing of a series of jobs. These subjective ratings provide information about how people perceive the worth of different occupations. People tend to rank professionals, such as physicians, professors, judges, and lawyers highly, with occupations such as electrician, insurance agent, and police officer falling in the middle. Occupations with low occupational prestige are maids, garbage collectors, and shoe-shiners (Nakao and Treas 2000, Davis and Smith 1984). These rankings do not reflect the worth of people within these positions but are indicative of the judgments people make about the worth of these jobs.

The final major indicator of socioeconomic status is **educational attainment,** typically measured as the total years of formal education. The more years of education attained, the more likely a person will have higher class status. The prestige attached to occupations is strongly tied to the amount of education the job requires—the more education people think is needed for a given occupation, the more occupational prestige people attribute to that job (Blau and Duncan 1967; MacKinnon and Langford 1994; Ollivier 2000).

Layers of Social Class

Taken together, income, occupation, and education are good measures of people's social standing. How then do sociologists understand the array of classes in the United States? Using a laddered model of stratification, most sociologists describe the class system in the United States as being divided into several classes: upper, upper middle, middle, lower middle, and lower class. The different classes are arrayed up and down the ladder, with those with the most money, education, and prestige at the top and those with the least at the bottom.

Map 7.1 provides a visual image of regional differences in the distribution of income by showing median income in different parts of the country, organized by county.

In the United States, the *upper class* owns the major share of corporate and personal wealth; it includes those who have held wealth for generations as well as those who have recently become rich. Only a very small proportion of people actually constitute the upper class, but they control vast amounts of wealth and power in the United States. Those in this class are elites who exercise enormous control throughout society. Some wealthy individuals can wield as much power as entire nations (Friedman 1999).

Despite social myths to the contrary, the best predictor of future wealth is the family into which you are born. Each year, the business magazine *Forbes* publishes a list of the 400 wealthiest families and individuals in the country. By 2005, for the first time in history, you had to have at least $1 billion to be on the list! Bill Gates, the richest person on this list, has an estimated worth of $53 billion. Of all the wealth represented on the *Forbes 400* list, most is inherited, although since the 1990s there has been some increase in the number of people on the list with self-created wealth (Miller and Seraphin 2006).

Those in the upper class with newly acquired wealth are known as the *nouveau riche.* Luxury vehicles, high-priced real estate, and exclusive vacations may mark the lifestyle of the newly rich. However, although they may have vast amounts of money, they are often not accepted into "old rich" circles.

The *upper middle class* includes those with high incomes and high social prestige. They tend to be well-educated professionals or business executives. Their earnings can be quite high indeed, even millions of dollars a year. It is difficult to estimate exactly how many people fall into this group because of the difficulty of drawing lines between the upper, upper

map 7.1 MAPPING AMERICA'S DIVERSITY

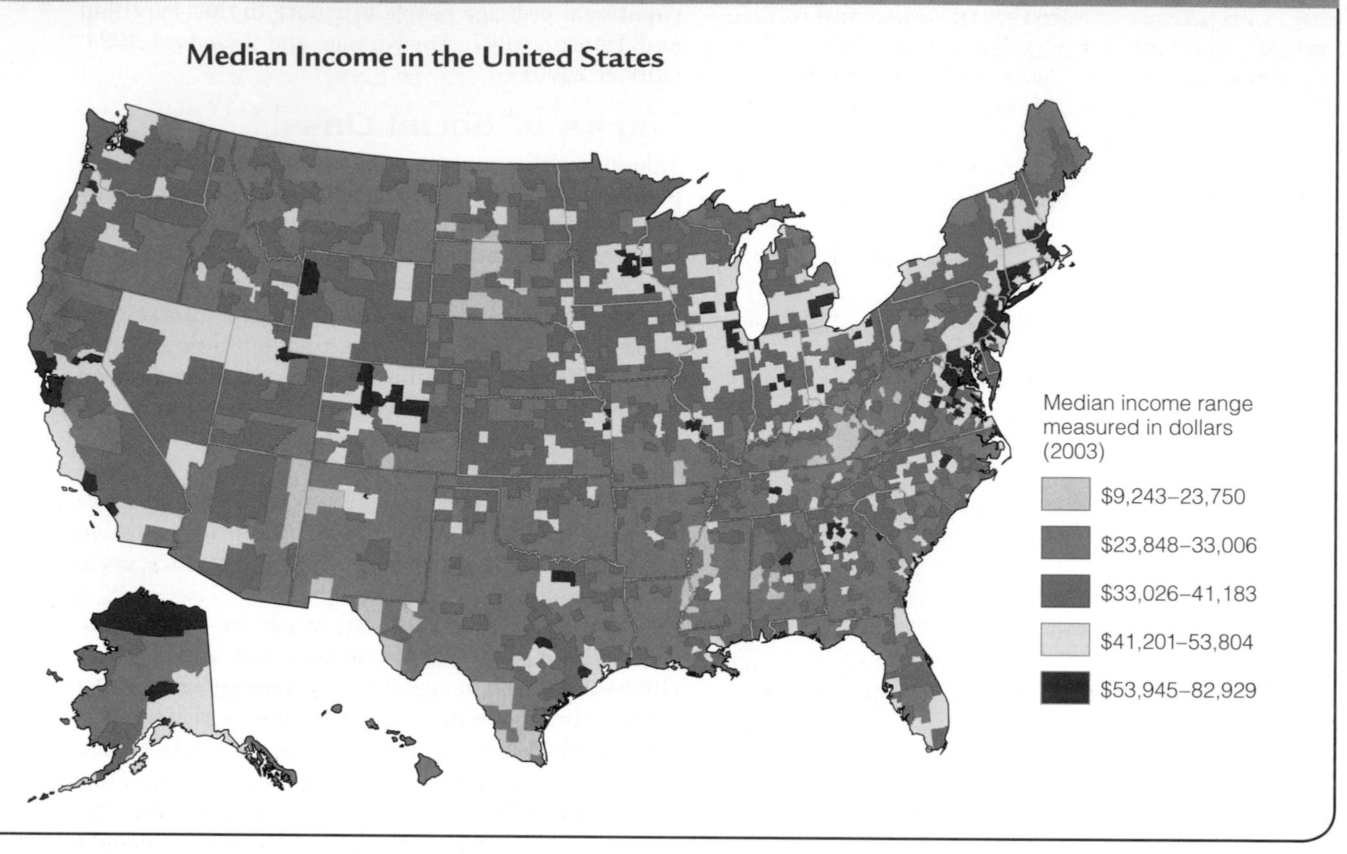

If you look closely at this map, you will see that median income tends to be higher in more urban areas. Thus, in 2003, median income inside metropolitan areas was $46,060 and outside such areas, $35,112. What the map does not show, however, are differences within cities. Median income inside central cities is substantially lower ($37,174) than the median income within metropolitan areas, but out of the center city—that is, in suburban areas—it is $51,737. Given this, what do you conclude about the significance of residence in the structure of the class system?

Data: U.S. Census Bureau. 2004. *American FactFinder.* **www.census.gov**

middle, and middle classes. Indeed, the upper middle class is often thought of as "middle class" because their lifestyle sets the standard to which many aspire, but this lifestyle is actually unattainable by most. A large home full of top-quality furniture and modern appliances, two or three luxury cars, vacations every year (perhaps a vacation home), high-quality college education for one's children, and a fashionable wardrobe are simply beyond the means of a majority of people in the United States.

The *middle class* is hard to define in part because being "middle class" is more than just economic position. A very large portion of Americans identify themselves as middle class even though they vary widely in lifestyle and in resources at their disposal. But the idea that the United States is an open class system leads many to think that the majority have a middle-class lifestyle; thus, the middle class becomes the ubiquitous norm even though many who call themselves middle class have a tenuous hold on this class position.

Debunking Society's Myths

Myth: Unlike other societies where there are distinct social classes, most Americans are middle class and share similar values and life chances.

Sociological perspective: Although middle-class values form the dominant culture in the United States, this assertion overlooks the extent to which class differentiates life chances. Elites control most of the wealth in the United States, and many in the middle class find their class position to be tenuous. Though often ignored in the popular mind, the working class forms a substantial portion of the class structure.

The *lower middle class* includes workers in the skilled trades and low-income bureaucratic workers, many of whom may actually think of themselves as middle class. Also known as the *working class,* this class includes blue-collar workers (those in skilled trades who do manual labor) and many service workers, such as secretaries, hair stylists, food servers, police, and firefighters. A medium to low income, education, and occupational prestige define the lower middle class relative to the class groups above it. The term *lower* in this class designation refers to the relative position of the group in the stratification system, but it has a pejorative sound to many people, especially to people who are members of this class, many of whom think of themselves as middle class.

The *lower class* is composed primarily of the displaced and poor. People in this class have little formal education and are often unemployed or working in minimum-wage jobs. People of color and women make up a disproportionate part of this class. The poor include the *working poor*—those who work at least 27 hours a week but whose wages fall below the federal poverty level. Six percent of all working people now live below the poverty line—a proportion that has generally increased over time. Among white workers, men and women are nearly equivalent in the likelihood of being among the working poor, but among African American workers, Black women are twice as likely as Black men to be poor (Bureau of Labor Statistics 2005).

The concept of the **urban underclass** has been added to the lower class (Wilson 1987). The underclass includes those who are likely to be permanently unemployed and without much means of economic support. The underclass has little or no opportunity for movement out of the worst poverty. Rejected from the economic system, those in the underclass may become dependent on public assistance or illegal activities. Structural transformations in the economy have left large groups of people, especially urban minorities, in these highly vulnerable positions. The growth of the urban underclass has exacerbated the problems of urban poverty and related social problems (Wilson 1996, 1987).

Class Conflict

Sociologists have also analyzed class according to the perspective of conflict theory, derived from the early work of Karl Marx. This perspective defines classes in terms of their structural relationship to other classes and their relationship to the economic system. The analysis of class from this sociological perspective interprets inequality as resulting from the unequal distribution of power and resources in society (see Chapter 1). Instead of seeing class simply as a ladder, sociologists who work from a conflict perspective see the classes as facing off against each other, with elites exploiting and dominating others. The key idea in this model is that class is not simply a matter of what individuals possess in terms of income and prestige; instead, class is defined by the relationship of the classes to the larger system of economic production (Vanneman and Cannon 1987; Wright 1985).

From a conflict perspective, the position of the middle class in society is unique. The middle class, or the *professional–managerial class,* includes managers, supervisors, and professionals. Members of this group have substantial control over other people, primarily through their authority to direct the work of others, impose and enforce regulations in the workplace, and determine dominant social values. Although, as Marx argued, the middle class is controlled by the ruling class, members of this class tend to identify with the interests of the elite. The professional–managerial class, however, is caught in a contradictory position between elites and the working class. Like elites, those in this class have some control over others, but like the working class, they have minimal control over the economic system (Wright 1979). As capitalism progresses, Karl Marx argued that more and more of those in the middle class drop into the working class as they are pushed out of managerial jobs into working-class jobs or as

© Tony Freeman/PhotoEdit

© Annie Dovie

The class status of different groups can readily be seen by looking at the context of people's daily lives, as shown in these two shopping areas.

professional jobs become organized more along the lines of traditional working-class employment.

Has this happened? Not to the extent Marx predicted. He thought that ultimately there would be only two classes—the capitalist and the proletariat. To some extent, however, this is occurring. Classes have become more polarized, with the well-off accumulating even more resources and the middle class seeing their income as either flat or falling, measured in constant dollars (Mishel et al. 2005). Rising levels of debt among the middle class have contributed to this growing inequality. Many now have a fragile hold on being middle class: The loss of a job, a family emergency (such as the death of a working parent), divorce, disability, or a prolonged illness can quickly leave middle- and working-class families in a precipitous financial state. At the same time, corporate mergers, tax policies that favor the rich, a decline in corporate taxes, and sheer greed are concentrating more wealth in the hands of a few.

Members of the working class have little control over their own work lives; instead, they generally have to take orders from others. This concept of the working class departs from traditional blue-collar definitions of working-class jobs because it includes many so-called "white collar workers" (secretaries, salespeople, and nurses), any group working under the rules imposed by managers. The middle class may exercise some autonomy at work, but the working class has little power to challenge decisions of those who supervise them, except insofar as they can organize collectively, as in unions, strikes, or other collective work actions.

© Jim West/The Image Works

Labor unions, traditionally dominated by White men in the skilled trades, are not only more diverse, but also represent workers in occupations typically thought of as "white collar" work.

Whether you use a laddered model of class or a class conflict perspective, you can see that the class structure in the United States is hierarchical. Class position gives different people access to jobs, income, education, power, and social status—all of which bestow further opportunities on some and deprive others of success. People sometimes move from one class to another (although this is not the norm), but the class structure is a system with boundaries built into it, generating class conflict. The middle and working classes shoulder much of the tax burden for social programs, producing resentment by these groups toward the poor. At the same time, corporate taxes have declined and tax loopholes for the rich have increased—an indication of the privilege that is perpetuated by the class system. Whatever features of the class system different sociologists study, they see class stratification as a dynamic process—one involving the interplay of access to resources, judgments about different groups, and the exercise of power by a few.

The Distribution of Wealth and Income

One thing that is clear about the U.S. class structure is that there is enormous class inequality in this society and it is growing. Elites control an enormous share of the wealth and exercise tremendous control over others. The gap between the rich and the poor is also increasing, and many in the middle and working class find their class standing slipping. Figure 7.3 shows the increasing gap that has developed between the upper classes and everyone else in recent years. Income growth has been greatest for those at the top end of the population—the top 20 percent and the top 5 percent of all income groups. For everyone else, income growth has remained relatively flat.

Contributing to the income gap is very high compensation for CEOs of major companies. In the top 500 companies, CEO compensation averages $13.5 million per year—with unprecedented growth in the last twenty years even when workers' income has remained flat. Weighed against average income for all full-time workers, CEOs are earning 350 times what workers earn (AFL-CIO 2002; DeCarlo 2006). This growth has been augmented by government policies that have provided tax breaks for the highest earners. When you add in wealth, as we will see below, you start to see the full picture of inequality in the United States, verifying the popular adage that the rich are getting richer and the poor, poorer. As the classes become more polarized, the idea that the United States is primarily a middle-class society could evaporate.

When discussing patterns of stratification, it is important to distinguish wealth and income. *Wealth* is

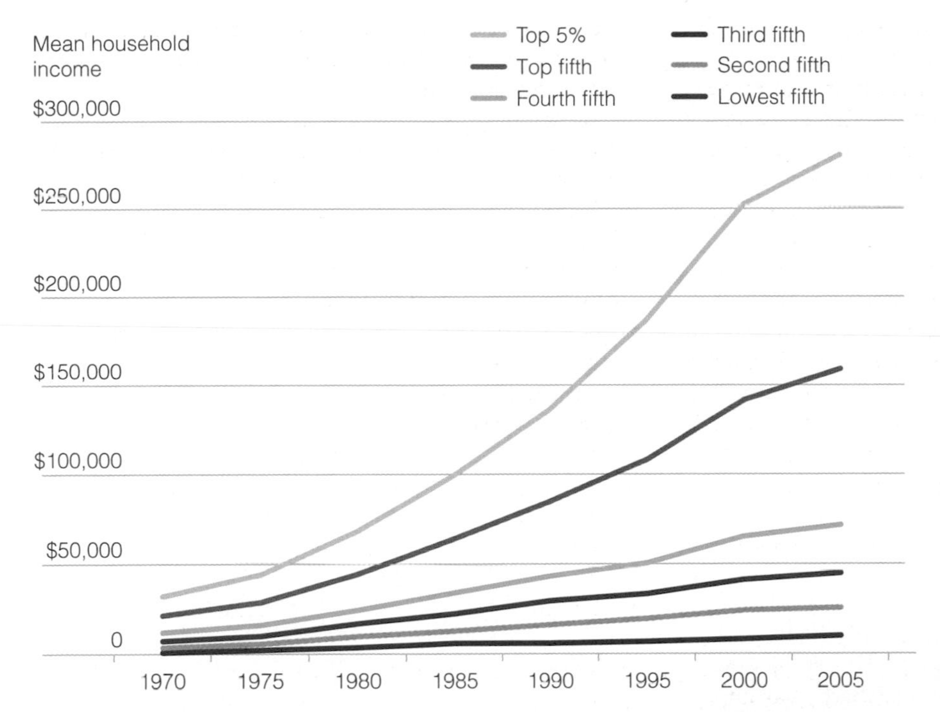

FIGURE 7.3 *Mean Household Income Received by Different Income Groups, 1970–2005*

As you can see in this graph, for most income groups in the United States, income (adjusted for rate of inflation) has remained relatively flat in recent years. The top groups, however, have seen huge increases in income—a trend resulting in growing inequality in the United States.

Source: U.S. Census Bureau. 2006. *Historical Income Tables—Households*, Table H-3.

the monetary value of everything one actually owns. It is calculated by adding all financial assets (stocks, bonds, property, insurance, savings, value of investments, and so on) and subtracting debts, which gives a dollar amount that is one's *net worth*. Wealth allows you to accumulate assets over generations, giving advantages to subsequent generations that they might not have had on their own. *Income* is the amount of money brought into a household from various sources (wages, investment income, dividends, and so on) during a given period. Unlike income, wealth is cumulative—that is, its value tends to increase through investment; it can be passed on to the next generation, giving those who inherit wealth a considerable advantage in accumulating more resources.

To understand the significance of wealth compared to income in determining class location, imagine two college graduates graduating in the same year, from the same college, with the same major and same grade point average. Imagine further that upon graduation, both get jobs with the same salary in the same organization. Yet, in one case parents paid all the student's college expenses and gave her a car upon graduation. The other student worked while in school and graduated with substantial debt from student loans. This student's family has no money with which to help support the new worker. Who is better off? Same salary, same credentials, but wealth (even if modest) matters. It gives one person an advantage—one that will be played out many times over as the young worker buys a home, finances her own children's education, and possibly inherits additional assets.

Where is all the wealth? The wealthiest 1 percent own 33 percent of all net worth; the bottom 80 percent

AP Images/Steve Helber/POOL

Working-class heroes: The miners who were dramatically rescued from the Quecreek mine in the summer of 2001 signed a deal to tell their story in a made-for-TV movie. This will give them the equivalent of three years' pay each. (They earn about $40,000 per year, counting overtime). That is a pittance compared to the head of the group that owns the mine: He earns about $18 million a year in salary and bonuses—making in five minutes what it would take the miners a year to earn!

doing sociological research

The Fragile Middle Class

Research Question: The hallmark of the middle class in the United States is its presumed stability. Home ownership, a college education for children, and other accoutrements of middle-class status (nice cars, annual vacations, an array of consumer goods)—these are the symbols of middle-class prosperity. But the rising rate of bankruptcy among the middle class shows that the middle class is not as secure as it is presumed to be (see Figure 7.4). Personal bankruptcy has risen dramatically in recent years—more than one million filings for bankruptcy per year. How can this be happening in such a prosperous society? This is the question examined by Teresa Sullivan, Elizabeth Warren, and Jay Lawrence Westbrook in their study of bankruptcy and debt among the middle class.

Research Method: They based their study on an analysis of official records of bankruptcy in five states, as well as on detailed questionnaires given to individuals who filed for bankruptcy.

Research Results: Their findings debunk the idea that bankruptcy is most common among poor people. Instead, they found bankruptcy is mostly a middle-class phenomenon representing a cross-section of those in this class (meaning that those who are bankrupt are matched on the demographic characteristics of race, age, and gender with others in the middle class). They also debunk the notion that bankruptcy is rising because it is so easy to file. Rather, they found many people in the middle class so overwhelmed with debt that they cannot possibly pay it off. Most often people file for bankruptcy as a result of job loss and lost wages. But divorce, medical problems, housing expenses, and credit card debt also drive many to bankruptcy court.

Conclusions and Implications: Sullivan and her colleagues explain the rise of bankruptcy as stemming from structural factors in society that fracture the stability of the middle class. The volatility of jobs under modern capitalism is one of the biggest factors, but add to this the "thin safety net"—no health insurance for many, but rising medical costs. Also, the American dream of owning one's own home means many are "mortgage poor"—extended beyond their ability to keep up.

In addition, the United States is a credit-driven society. Credit cards are routinely mailed to people in the middle class, encouraging them to buy beyond their means. You can now buy virtually anything on credit: cars, clothes, doctor's bills, entertainment, groceries. You can even use one credit card to pay off other credit cards. Indeed, it is difficult to live in this society without credit cards. Increased debt is the result. Many are simply unable to keep up with compounding interest and penalty payments, and debt takes on a life of its own as consumers cannot keep up with even the interest payments on debt.

Sullivan, Warren, and Westbrook conclude that increases in debt and uncertainty of income combine to produce the fragility of the middle class. Their research shows that "even the most secure family may be only a job loss, a medical problem, or an out-of-control credit card away from financial catastrophe" (2000: 6).

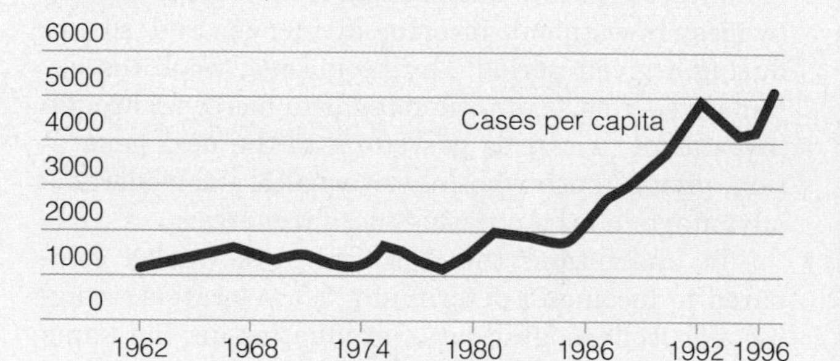

FIGURE 7.4 *Bankruptcy Cases per Million Adults*

Questions to Consider

1. Have you ever had a credit card? If so, how easy was it to get? Is it possible to get by without a credit card?
2. What evidence do you see in your community of the fragility or stability of different social class groups?

Source: Reprinted with permission from T.A. Sullivan, E. Warren, and J.L. Westbrook, *The Fragile Middle Class: Americans in Debt.*

thinking sociologically

Suppose that you wanted to reduce inequality in the United States. Since you know that the transmission of *wealth* is one of the bases of *stratification,* would you be willing to eliminate the right to inherit property to achieve greater equality?

control only 16 percent. The top 1 percent also own almost half of all stock; the bottom 80 percent own only 6 percent of total stock holdings. Moreover, there has been an increase in the concentration of wealth since the 1980s (Mishel et al. 2005). As just one example, John D. Rockefeller is typically heralded as one of the wealthiest men in U.S. history. But comparing Rockefeller with Bill Gates, in the value of today's dollars, Gates has already surpassed Rockefeller's riches (Myerson 1998).

In contrast to the vast amount of wealth and income controlled by elites, a very large proportion of Americans have hardly any financial assets once debt is subtracted. Figure 7.5 shows the net worth of different income brackets in the United States, and you can see that most of the population have very low net worth. Although you cannot tell by looking at Figure 7.5, another 18 percent have zero or negative net worth—usually because their debt exceeds their assets, and levels of debt and bankruptcy for U.S. households are now at an all-time high (Sullivan et al. 2000; Mishel et al. 2005). The American dream of owning a home, a new car, taking annual vacations, and sending one's children to good schools—not to mention saving for a comfortable retirement—is increasingly unattainable by many. When you see the amount of income and wealth controlled by a small segment of the population, a sobering picture of class inequality emerges.

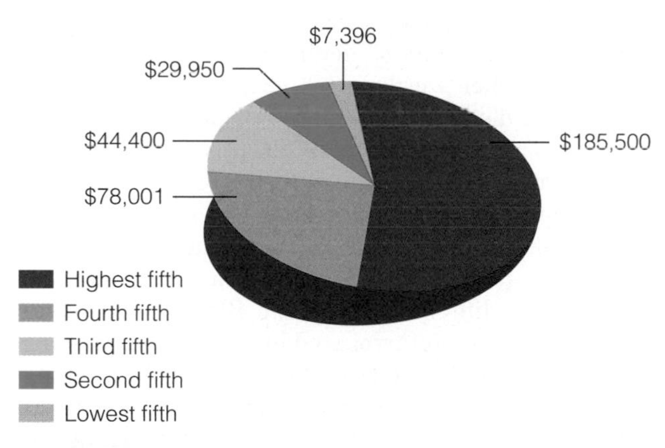

FIGURE 7.5 *Median Net Worth by Household Income*

Recall that one's net worth is the value of everything owned minus one's debt. You can see here the vast differences in wealth across different income groups. Remember that median is the midpoint; thus, in each of these groups, half of those in the group have more wealth; half, less.

Source: Orzechowski, Shawna, and Peter Sepielli.2003. *Net Worth and Asset Ownership of Households: 1998 and 2000.* Washington, DC: U.S. Census Bureau, P70-88.

Popular legends, however, extol the possibility of anyone becoming rich in the United States. The well-to-do are admired not just for their style of life, but also for their supposed drive and diligence. The admiration for those who rise to the top makes it seem like anyone who is clever enough and works hard can become fabulously rich. Despite the prominence of rags-to-riches stories in American legend, most wealth in this society is inherited. A few individuals make their way into the elite class by virtue of their own success, but this is rare. The upper class is also overwhelmingly White and Protestant. The wealthy also exercise tremendous political power by funding lobbyists, exerting their social and personal influence on other elites, and contributing heavily to political campaigns. Studies of elites also find that they tend to be politically quite conservative (Burris 2000; Zweigenhaft and Domhoff 2006). They travel in exclusive social networks that tend to be open only to those in the upper class. They tend to intermarry, their children are likely to go to expensive schools, and they spend their leisure time in exclusive resorts.

Race also influences the pattern of wealth distribution in the United States; for every dollar of wealth held by White Americans, Black Americans have only 26 cents (Oliver and Shapiro 1995; Conley 1999). At all levels of income, occupation, and education, Black families have lower levels of wealth than similarly situated White families. Being able to draw on assets during times of economic stress means that families with some resources can better withstand difficult times than those without assets. Even small assets, such as home ownership or a savings account, provide protection from crises such as increased rent, a health emergency, or unemployment. Because the effects of wealth are intergenerational—that is, they accumulate over time—just providing equality of opportunity in the present does not address the differences in class status that Black and White Americans experience (Oliver and Shapiro 1995; Shapiro 2005).

Debunking Society's Myths

Myth: Mothers on welfare have more children to increase the size of their welfare checks.

Sociological perspective: No causal relationship exists between the size of welfare benefits and the number of births by welfare recipients. "Family cap" policies now in place prohibit increasing welfare benefits with the birth of an additional child.

What explains the disparities in wealth by race? Wealth accumulates over time. Thus, government policies in the past have prevented Black Americans from being able to accumulate wealth. Discriminatory housing policies, bank lending policies, tax codes, and so forth have disadvantaged Black Americans, resulting in the differing assets Whites and Blacks in general hold now. Even though some of these discriminatory policies have ended, many continue. Either way, their effects persist, resulting in what sociologists Melvin Oliver and Thomas Shapiro call the *sedimentation of racial inequality.*

Understanding the significance of wealth in shaping life chances for different groups also challenges the view that all Hispanics have similar experiences and wealth. Cuban Americans and Spaniards are similar to Whites in their wealth holdings, whereas Mexicans, Puerto Ricans, Dominicans, and other Hispanic groups more closely resemble African Americans on the various indicators of wealth and social class. Likewise, one can better understand differences in class status among Asian American groups by carefully considering the importance not just of income, education, and occupation, but also patterns in the net assets of different groups (Oliver and Shapiro 2001).Without significant wealth holdings, families of any race are less able to transmit assets from previous generations to the next generation, one main support of social mobility.

Despite beliefs to the contrary, class divisions in the United States are real and becoming more pronounced. The elite are becoming better off, whereas the positions of the middle class, working class, and poor seem to be worsening. Indeed, many middle-class people feel that their way of life is slipping away. Many factors have contributed to the declining fortunes of the lower and middle classes in the United States, including the profound effects of national and global economic change. Reductions in state and federal spending have eliminated many government jobs—jobs that have traditionally been the route to middle-class standing for many workers. Under economic restructuring manufacturing jobs are being eliminated in the United States, with those formerly employed in this sector having to take lower-wage jobs to survive. At the same time, recent changes in tax policy heavily favor the already wealthy. Coupled with reductions in various federal support programs, the degree of class polarization is amplified.

The new economy has had mixed results for different groups (Andersen 2000). Income levels for women have increased, but they have decreased for men, except for the top 30 percent of earners. However, since 1980, women in the bottom 10 percent of wage earners have also seen their wages decline. These are aggregate numbers; they reflect trends in the overall population and show the varying impact of economic restructuring on different groups of people (Kilborn and Clemetson 2002; Mishel et al. 2005). Whether these trends reverse or continue remains to be seen.

The tax structure has also distributed benefits unevenly, leading to some of the resistance to supporting social programs that would be subsidized through federal taxes. Corporations are benefiting the most from the tax structure, as corporate taxes have fallen in recent years. While most Americans are paying more in federal tax than ever before (an increase from 13 to 15 cents per dollar of income earned since 1990), corporate taxes since 1990 have fallen from 26 cents on the dollar to 20 cents, even though corporate profits were up 252 percent in that period (Johnston 2000). Individuals at the upper ends of the class system have also been able to take advantage of numerous tax benefits and loopholes, reducing their tax burden, whereas the burden on the middle classes has increased. Understanding the differential impact of changes in the economy is an important part of analyzing the dynamics of social stratification, as well as contemporary politics.

Diverse Sources of Stratification

Class is only one basis for stratification in the United States. Factors such as age, ethnicity, and national origin have a tremendous influence on stratification. Race and gender are two primary influences in the stratification system in the United States. In fact, analyzing class without also analyzing race and gender can be misleading. Race, class, and gender, as we are seeing throughout this book, are overlapping systems of stratification that people experience simultaneously. A working-class Latina, for example, does not experience herself as working class at one moment, Hispanic at another moment, and a woman the next. At any given point in time, her position in society is the result of her race, class, *and* gender status. In other words, class position is manifested differently, depending on one's race and gender, just as gender is experienced differently depending on one's race and class and race is experienced differently depending on one's gender and class. Depending on one's circumstances, race, class, or gender may seem particularly salient at a given moment in a person's life. For example, a Black middle-class man stopped and interrogated by police when driving through a predominantly White middle-class neighborhood may at that moment feel his racial status as his single most outstanding characteristic, but at all times his race, class, and gender influence his life chances. As social categories, race, class, and gender shape all people's experience in this society—not just those who are disadvantaged (Andersen and Collins 2007).

Class also significantly differentiates group experience within given racial and gender groups. Latinos, for example, are broadly defined as those who trace their origins to regions originally colonized by Spain. The ancestors of this group include both White Spanish colonists and the natives who were enslaved on Spanish plantations. Today, some Latinos identify as

White, others as Black, and others by their specific national and cultural origins. The very different histories of those categorized as Latino are matched by significant differences in class. Some may have been schooled in the most affluent settings; others may be virtually unschooled. Those of upper-class standing may have had little experience with prejudice or discrimination; others may have been highly segregated into barrios and treated with extraordinary prejudice. Latinos who live near each other geographically in the United States and who are the same age and share similar ancestry may have substantially different experiences based on their class standing (Massey 1993). Neither class, race, nor gender, taken alone, can be considered an adequate indicator of different group experiences, as shown in the box "Understanding Diversity: Latino Class Experience."

The Race–Class Debate. The relationship between race and class is much debated among sociologists. The Black middle class goes all the way back to the small numbers of free Blacks in the eighteenth and nineteenth centuries (Frazier 1957), expanding in the twentieth century to include those who were able to obtain an education and become established in industry, business, or a profession. Although wages for Black middle-class and professional workers never matched those of Whites in the same jobs, within the Black community the Black middle class has had relatively high prestige. Many sociologists conclude that the class structure among African Americans has existed alongside the White class structure, separate and different.

In recent years, both the African American and Latino middle classes have expanded, primarily as the result of increased access to education and middle-class occupations for people of color (Higginbotham 2001; Pattillo-McCoy 1999). Although middle-class Blacks and Latinos may have economic privileges that others in these groups do not have, their class standing does not make them immune to the negative effects of race. Asian Americans also have a

Latino Class Experience

Latinos in the United States are a diverse population of many different groups, each with many different national origins, histories, and cultural backgrounds. This diversity is represented by the fact that Latinos do not agree among themselves on what they should be called; some prefer "Latino," others "Hispanic," and some prefer to be identified by their cultural origins, as in Chicanos, Cuban Americans, or Puerto Riqueños. Some think of themselves as White. Generational differences further add to the diversity among Latinos.

Sociologist Douglas Massey reminds us of this diversity in thinking about the diverse class experiences of Latinos. Massey writes that as a result of different histories, Latinos live in different socioeconomic circumstances:

> *They may be fifth-generation Americans or new immigrants just stepping off the jetway. Depending on when and how they got to the United States, they may also know a long history of discrimination and repression or they may see the United States as a land of opportunity where origins do not matter. They may be affluent and well educated or poor and unschooled; they may have no personal experience of prejudice or discrimination, or they may harbor stinging resentment at being called a "spic" or being passed over for promotion because of their accent.* (1993: 7–8)

Massey also notes that levels of residential segregation and poverty vary across different Latino groups. He notes that as socioeconomic levels rise and as immigrants reach second, third, and later generations within the United States, the degree of segregation for Latinos progressively falls. Puerto Ricans have higher levels of segregation than other Latino groups, as well as some of the highest rates of poverty. The route of Latinos into this country also contributes to their class position. Some Latinos are indigenous to the United States—that is, their families owned land in the American Southwest that was colonized by White settlers; others are recent immigrants who enter the economy as low-wage workers with little opportunity for upward mobility.

Diversity among Latinos has many implications for understanding the experience of this population. Some are caught in the economic underclass, others are middle class; a few are among the nation's elites. Massey reminds us that factors such as class, historical origins, residential segregation, race, and migration patterns must be carefully analyzed to understand Latino experiences.

Source: Massey, Douglas S. 1993. "Latino Poverty Research: An Agenda for the 1990s." *Social Science Research Council Newsletter* 47 (March): 7–11.

significant middle class, but they have also been stereotyped as the most successful minority group because of their presumed educational achievement, hard work, and thrift. This stereotype is referred to as the *myth of the model minority* and includes the idea that a minority group must adopt alleged dominant group values to succeed. This myth about Asian Americans obscures the significant obstacles to success that Asian Americans encounter, and it ignores the hard work and educational achievements of other racial and ethnic groups. The idea that Asian Americans are the "model minority" also obscures the high rates of poverty among many Asian American groups (Lee 1996).

Despite recent successes, many in the Black middle class have a tenuous hold on this class status. The Black middle class remains as segregated from Whites as the Black poor, and continuing racial segregation in neighborhoods means that Black middle-class neighborhoods are typically closer to Black poor neighborhoods than the White middle-class neighborhoods are to White poor ones. This exposes many in the Black middle class to some of the same risks as those in poverty. This is not to say that the Black middle class has the same experience as the poor, but it challenges the view that the Black middle class "has it all" (Pattillo-McCoy 1999). Furthermore, Black Americans are still much more likely to be working class than middle class; they are also more likely to be working class than are Whites (Horton et al. 2000).

The Influence of Gender and Age. The effects of gender further complicate the analysis of class. In the past, women were thought to derive their class position from their husbands or fathers, but sociologists now challenge this assumption. Measured by their own income and occupation, the vast majority of women would likely be considered working class. The median income for women, even among those employed full time, is far below the national median income level. In 2005, when median income for men working year-round and full time was $41,386, the median income of women working year-round, full time was $31,858 (DeNavas-Walt et al. 2006). The vast majority of women work in low-prestige and low-wage occupations, even though women and men have comparable levels of educational attainment. Women's class status is abundantly clear in the aftermath of a divorce, when women's income drops significantly and men's increases (Peterson 1996a, 1996b; Weitzman 1996; Weitzman 1985).

Age, too, is a significant source of stratification, as we saw in Chapter 3. Just being born in a particular generation can have a significant influence on one's life chances. The fears of today's young, middle-class people that they will be unable to achieve the lifestyles of their parents show the effect that being in a particular generation can have on one's life chances.

The age group most likely to be poor are children, 17.6 percent of whom live in poverty in the United States. This represents a change from the recent past when the aged were the most likely to be poor (see Figure 7.6). Although many elderly people are now poor (10 percent of those 65 and over), far fewer in this age category are poor than was the case not many years ago (DeNavas-Walt et al. 2006). This shift reflects the greater affluence of the older segments of the population—a trend that is likely to continue as the current large cohort of middle-aged, middle-class Baby Boomers grow older.

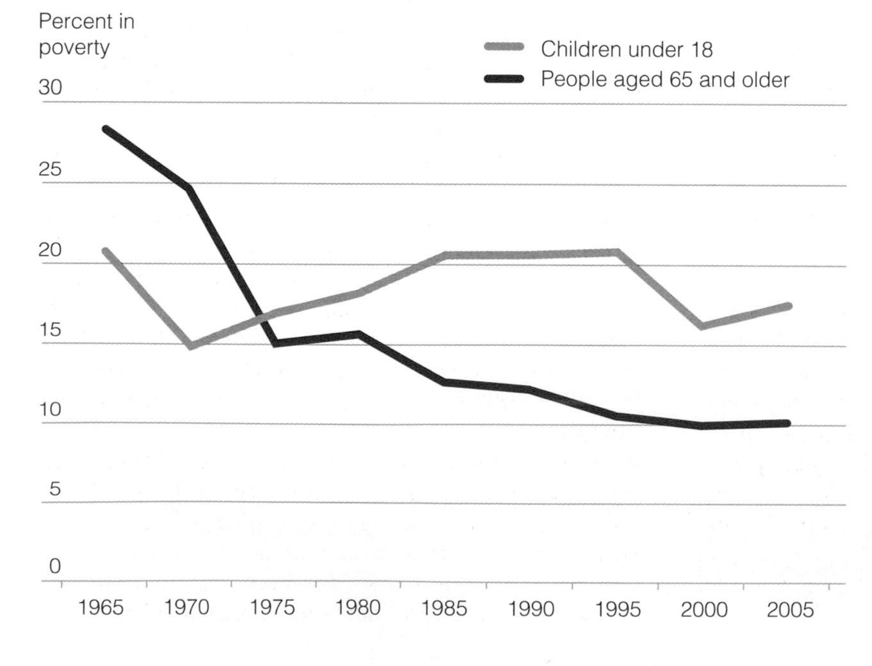

FIGURE 7.6 *Poverty among the Old and the Young, 1965–2005*

Source: DeNavas-Walt, Carmen, Bernadette Proctor, and Cheryl Lee. 2006. *Income, Poverty, and Health Insurance Coverage in the United States: 2005.* Washington, DC: U.S. Census Bureau, P60-231.

Social Mobility: Myths and Realities

There is a general belief in the United States that anyone can, by his or her own labor, move relatively freely through the class system. The assumption is that the United States class system is a **meritocracy**—that is, a system in which one's status is based on merit or accomplishments, not other social characteristics. As the word suggests, in a meritocracy, people move up and down through the class system based on merit not based on other characteristics. Is this the case in the United States?

Defining Social Mobility

Social mobility is a person's movement over time from one class to another. Social mobility can be up or down, although the American dream emphasizes upward movement. Mobility can be either *inter*generational, occurring between generations, as when a daughter rises above the class of her mother or father; or *intra*generational, occurring within a generation, as when a person's class status changes as the result of business success (or disaster).

Societies differ in the extent to which social mobility is permitted. Some societies are based on *closed class systems,* in which movement from one class to another is virtually impossible. In a caste system, for example, mobility is strictly limited by the circumstances of one's birth. At the other extreme are *open class systems,* in which placement in the class system is based on individual achievement, not ascription. In open class systems, there are relatively loose class boundaries, high rates of class mobility, and weak perceptions of class difference.

The Extent of Social Mobility

Does social mobility occur in the United States? Social mobility is much more limited than people believe. Success stories of social mobility do occur, but research finds that experiences of mobility over great distances are rare, certainly far less than believed. Most people remain in the same class as their parents, and many drop to a lower class. The social mobility that does exist is greatly influenced by education. African Americans, as well as immigrant groups, are often strongly committed to social mobility through education; increases in educational attainment for African Americans account for a considerable portion of the gains they have made (Smith 1989). But most of the time, among all groups, people remain in the class where they started. What mobility exists is typically short in distance. Evidence now suggests that mobility between generations may be becoming even more rigid than in the past (Sawhill and McClanahan 2006).

Social mobility is influenced most by factors that affect the whole society, not by individual characteristics. In other words, when mobility occurs, it is usually because of changes in the occupational system, economic cycles, and demographic factors, such as the number of college graduates in the labor force (Beller and Hout 2006). Social mobility in the United States is not impossible. Indeed, many have immigrated to this nation with the knowledge that their life chances are better here than in their countries of origin. But, in sum, social mobility is much more limited than the American dream of mobility suggests.

Class Consciousness

Because of the widespread belief that mobility is possible, people in the United States, compared to many other societies, tend not to be very conscious of class. **Class consciousness** is the perception that a class structure exists along with a feeling of shared identification with others in one's class—that is, those with whom you share life chances (Centers 1949). Notice that there are two dimensions to the definition of class consciousness: the idea that a class structure exists and one's class identification.

There has been a long-standing argument that Americans are not very class conscious because of the belief that upward mobility is possible. Images of opulence also saturate popular culture, making it seem that such material comforts are available to anyone. The faith that upward mobility is possible ironically perpetuates inequality since, if people believe that everyone has the same chances of success, they are likely to think that whatever inequality exists must be fair—or the result of individual success and failure.

Class inequality in any society is usually buttressed by ideas that support (or actively promote) inequality. Beliefs that people are biologically, culturally, or socially different can be used to justify the higher position of some groups. If people believe these ideas, the ideas provide legitimacy for the system. Karl Marx used the term **false consciousness** to describe the class consciousness of subordinate classes who had internalized the view of the dominant class. Marx argued that the ruling class controls subordinate classes by infiltrating their consciousness with belief systems that are consistent with the interests of the ruling class. If people accept these ideas, which justify inequality, they need not be overtly coerced into accepting the roles designated for them by the ruling class.

Class consciousness in the United States has been higher at some times than others. Now, 46 percent of the public identifies as working class; 46 percent as middle class; 5 percent as lower class; and 3 percent as upper class (National Opinion Research Center 2004). There have been times when class

a sociological eye on the media

Reproducing Class Stereotypes

The media have a substantial impact on how people view the social class system and different groups within it. Especially because people tend to live and associate with people in their own class, how they see others can be largely framed by the portrayal of different class groups in the media. Research has found this to be true and, in addition, has found that mass media have the power to shape public support for policies for public assistance.

To begin with, the media overrepresent the lifestyle of the most comfortable classes. It is the rare family that can afford the home decor and fashion depicted even in soap operas, ironically most likely watched by those in the working class. Media portrayals, such as those found on television talk shows—as well as sports—tend to emphasize stories of upward mobility. When the working class is depicted, it tends to be shown as deviant—reinforcing class antagonism and giving viewers a sense of moral and "class superiority" (Gersch 1999).

© John Burke/Index Stock

Content analyses of the media also find that the poor are largely invisible in the media (Mantsios 2001). Those poor people who are depicted in television and magazines are more often portrayed as Black than is actually the case, leading people to overestimate the actual number of the Black poor. The elderly and working poor are rarely seen (Clawson and Trice 2000; Gilens 1996). Representations of welfare overemphasize themes of dependency, especially when the portrayal is of African Americans. Women are also more likely than men to be represented as dependent (Misra et al. 2003). And rarely are welfare activists shown as experts; rather, public officials are typically given the voice of authority (Ryan 1996). One result is that the media end up framing the "field of thinkable solutions to public problems" (Sotirovic 2000, 2001), but do so within a context that ignores the social structural context of social issues.

Further Resources: See the film, *Class Dismissed.: How TV Frames the Working Class,* Media Education Foundation. www.mediaed.org

Sources: Gersch, Beate. 1999. "Class in Daytime Talk Television." *Peace Review* 11 (June): 275–281; Sotirovic, Mira. 2001. "Media Use and Perceptions of Welfare." *Journal of Communication* 51 (December): 750–774; Sotirovic, Mira. 2000. "Effects of Media Use on Audience Framing and Support for Welfare." *Mass Communication & Society* 2–3 (Spring–Summer): 269–296; Bullock, Heather E., Karen Fraser, and Wendy R. Williams. 2001. "Media Images of the Poor." *The Journal of Social Issues* 57 (Summer): 229–246; Clawson, Rosalee A., and Rakuya Trice. 2000. "Poverty as We Know It: Media Portrayals of the Poor." *The Public Opinion Quarterly* 64 (Spring): 53–64; Gilens, Martin. 1996. "Race and Poverty in America: Public Misperceptions and the American News Media." *Public Opinion Quarterly* 60 (Winter): 515–541; Misra, Joy, Stephanie Moller, and Marina Karides. 2003. "Envisioning Dependency: Changing Media Depictions of Welfare in the 20th Century." *Social Problems* 50 (November): 482–504; Ryan, Charlotte. 1996. "Battered in the Media: Mainstream News Coverage of Welfare Reform." *Radical America* 26 (August): 29–41.

consciousness was higher, such as during the labor movement of the 1920s and 1930s. Then, working-class people had a very high degree of class consciousness and mobilized on behalf of workers' rights. But now the formation of a relatively large middle class and a relatively high standard of living militates against class discontent. Racial and ethnic divisions also make strong alliances within various classes less stable. The growing inequality of today could result in a higher degree of class consciousness, but this has not yet developed into a significant class-based movement for change.

Why Is There Inequality?

Stratification occurs in all societies. Why? This question originates in classical sociology in the works of Karl Marx and Max Weber, theorists whose work continues to inform the analysis of class inequality today.

Karl Marx: Class and Capitalism

Karl Marx (1818–1883) provided a complex and profound analysis of the class system under capitalism—an analysis that, although more than 100 years old, continues to inform sociological analyses and has been the basis for major world change. Marx defined classes in relationship to *the means of production,* defined as the system by which goods are produced and distributed. In Marx's analysis, two primary classes exist under capitalism: the *capitalist class,* those who own the means of production, and the *working class* (or proletariat), those who sell their labor for wages. There are further divisions within these two classes: the *petty bourgeoisie,* small business owners and managers (those whom you might think of as middle class) who identify with the interests of the capitalist class but do not own the means of production, and the *lumpenproletariat,* those who have become unnecessary as workers and are then discarded. (Today, these would be the underclass, the homeless, and the permanently poor.)

Marx thought that with the development of capitalism, the capitalist and working class would become increasingly antagonistic (something he referred to as class struggle). As class conflicts became more intense, the two classes would become more polarized, with the petty bourgeoisie becoming deprived of their property and dropping into the working class. This analysis is still reflected in contemporary questions about whether the classes are becoming more polarized, with the rich getting richer and everyone else worse off, as we have seen.

In addition to the class struggle that Marx thought would characterize the advancement of capitalism, he also thought that capitalism was the basis for other social institutions. Capitalism is the *infrastructure* of society, with other institutions (such as law, education, the family, and so forth) reflecting capitalist interests. Thus, according to Marx, the law supports the interests of capitalists; the family promotes values that socialize people into appropriate work roles; and education reflects the interests of the capitalist class. Over time, capitalism increasingly penetrates society, as we can clearly see with the corporate mergers that characterize modern life and the predominance of capitalist values in society's institutions.

Why do people support such a system? Here is where ideology plays a role. **Ideology** refers to belief systems that support the status quo. According to Marx, the dominant ideas of a society are promoted by the ruling class, as we will explore further in the section on class consciousness. Through their control of the communications industries in modern society, the ruling class is able to produce ideas that buttress their interests.

Much of Marx's analysis boils down to the consequences of a system based on the pursuit of profit. If goods were exchanged at the cost of producing them, no profit would be produced. Capitalist owners want to sell commodities for more than their actual value—more than the cost of producing them, including materials and labor. Because workers contribute value to the system and capitalists extract value, Marx saw capitalist profit as the exploitation of labor. Marx believed that as profits became increasingly concentrated in the hands of a few capitalists, the working class would become increasingly dissatisfied. The basically exploitative character of capitalism, according to Marx, would ultimately lead to its destruction as workers organized to overthrow the rule of the capitalist class. *Class conflict* between workers and capitalists, he argued, was inescapable, with revolution being the inevitable result. Perhaps the class revolution that Marx predicted has not occurred, but the dynamics of capitalism that he analyzed are unfolding before us.

At the time Marx was writing, the middle class was small and consisted mostly of small business owners and managers. Marx saw the middle class as dependent on the capitalist class, but exploited by it, because the middle class did not own the means of production. He saw middle-class people as identifying with the interests of the capitalist class because of the similarity in their economic interests and their dependence on the capitalist system. Marx believed that the middle class failed to work in its own best interests because it falsely believed that it benefited from capitalist arrangements. Marx thought that in the long run the middle class would pay for their misplaced faith when profits became increasingly concentrated in the hands of a few and more and more of the middle class dropped into the working class. Because he did not foresee the emergence of the large and highly differentiated middle class we have today, not every part of Marx's theory has proved true. Still, his analysis provides a powerful portrayal of the forces of capitalism and the tendency for wealth to belong to a few, whereas the majority work only to make ends meet. He has also influenced the lives of billions of people under self-proclaimed Marxist systems that were created in an attempt, however unrealized, to overcome the pitfalls of capitalist society.

Max Weber: Class, Status, and Party

Max Weber (1864–1920) agreed with Marx that classes were formed around economic interests, and he agreed that material forces (that is, economic forces) have a powerful effect on people's lives. However, he disagreed with Marx that economic forces are the primary dimension of stratification. Weber saw three dimensions to stratification:

- *class* (the economic dimension);
- *status* (or prestige, the cultural and social dimension); and
- *party* (or power, the political dimension).

Weber is thus responsible for a *multidimensional view* of social stratification because he analyzed the connections between economic, cultural, and political systems. Weber pointed out that, although the economic, social, and political dimensions of stratification are usually related, they are not always consistent. A person could be high on one or two dimensions but low on another. A major drug dealer is an example: high wealth (economic dimension) and power (political dimension) but low prestige (social dimension), at least in the eyes of the mainstream society, even if not in other circles.

Weber defined *class* as the economic dimension of stratification—how much access to the material goods of society a group or individual has, as measured by income, property, and other financial assets. A family with an income of $100,000 per year clearly has more access to the resources of a society than a family living on an income of $40,000 per year. Weber understood that a class has common economic interests and that economic well-being was the basis for one's life chances. But, in addition, he thought that people were also stratified based on their status and power differences.

Status, to Weber, is the prestige dimension of stratification—the social judgment or recognition given to a person or group. Weber understood that class distinctions are linked to status distinctions—that is, those with the most economic resources tend to have the highest status in society, but not always. In a local community, for example, those with the most status may be those who have lived there the longest, even if newcomers arrive with more money. Although having power is typically related to also having high economic standing and high social status, this is not always the case, as you saw with the example of the drug dealer.

Finally, *party* (or what we would now call power) is the political dimension of stratification. It is the capacity to influence groups and individuals even in the face of opposition. Power is also reflected in the ability of a person or group to negotiate their way through social institutions. An unemployed Latino man wrongly accused of a crime, for instance, does not have much power to negotiate his way through the criminal justice system. By comparison, business executives accused of corporate crime can afford expensive lawyers and, thus, frequently go unpunished or, if they are found guilty, serve relatively light sentences in comparatively pleasant facilities. Again, Weber saw power as linked to economic standing, but he did not think that economic standing was always the determining cause of people's power.

Marx and Weber explain different features of stratification. Both understood the importance of the economic basis of stratification, and they knew the significance of class for determining the course of one's life. Marx saw people as acting primarily out of economic interests. Weber refined the sociological analyses of stratification to account for the subtleties that can be observed when you look beyond the sheer economic dimension to stratification, stratification being the result of economic, social, and political forces. Together, Marx and Weber provide compelling theoretical grounds for understanding the contemporary class structure.

Functionalism and Conflict Theory: The Continuing Debate

Marx and Weber were trying to understand why differences existed in the resources and power that different groups in society hold. The question persists of why there is inequality. Two major frameworks in sociological theory—functionalist and conflict theory—take quite different approaches to understanding inequality (see Table 7.2).

The Functionalist Perspective on Inequality

Functionalist theory views society as a system of institutions organized to meet society's needs (see Chapter 1). The functionalist perspective emphasizes that the parts of society are in basic harmony with each other; society is characterized by cohesion, consensus, cooperation, stability, and persistence (Parsons 1951a; Merton 1957; Eitzen and Baca Zinn 2007). Different parts of the social system complement one another and are held together through social consensus and cooperation. To explain stratification, functionalists propose that the roles filled by the upper classes—such as governance, economic innovation, investment, and management—are essential for a cohesive and smoothly running society and hence are rewarded in proportion to their contribution to the social order (Davis and Moore 1945).

According to the functionalist perspective, social inequality serves an important purpose in society: It motivates people to fill the different positions in society that are needed for the survival of the whole. Functionalists think that some positions in society are more important than others and require the most talent and training. Rewards attached to those positions (such as higher income and prestige) ensure that people will make the sacrifices needed to acquire the training for functionally important positions (Davis and Moore 1945). Higher class status thus comes to those who acquire what is needed for success (such as education and job training). In other words, functionalist theorists see inequality as based on a reward system that motivates people to succeed.

table 7.2 *Functional and Conflict Theories of Stratification*

Interprets	Functionalism	Conflict Theory
Inequality	Inequality serves an important purpose in society by motivating people to fill the different positions in society that are needed for the survival of the whole.	Inequality results from a system of domination and subordination where those with the most resources exploit and control others.
Class structure	Differentiation is essential for a cohesive society.	Different groups struggle over societal resources and compete for social advantage.
Reward system	Rewards are attached to certain positions (such as higher income and prestige) as a way to ensure that people will make the sacrifices needed to acquire the training for functionally important positions in society.	The more stratified a society, the less likely that society will benefit from the talents of all its citizens, since inequality prevents the talents of those at the bottom from being discovered and used.
Classes	Some positions in society are more functionally important than others and are rewarded because they require the greatest degree of talent and training.	Classes exist in conflict with each other as they vie for power and economic, social, and political resources.
Life chances	Those who work hardest and succeed have greater life chances.	The most vital jobs in society—those that sustain life and the quality of life—are usually the least rewarded.
Elites	The most talented are rewarded in proportion to their contribution to the social order.	The most powerful reproduce their advantage by distributing resources and controlling the dominant value system.
Class consciousness/ideology	Beliefs about success and failure confirm status of those who succeed.	Elites shape societal beliefs to make their unequal privilege appear to be legitimate and fair.
Social mobility	Upward mobility is possible for those who acquire the necessary talents and tools for success (such as education and job training).	There is blocked mobility in the system because the working class and poor are denied the same opportunities as others.
Poverty	Poverty serves economic and social functions in society.	Poverty is inevitable because of the exploitation built into the system.
Social policy	Because the system is basically fair, social policies should only reward merit.	Because the system is basically unfair, social policies should support disadvantaged groups by redirecting society's resources for a more equitable distribution of income and wealth.

The Conflict Perspective on Inequality. Conflict theory also sees society as a social system, but unlike functionalism, conflict theory interprets society as being held together through conflict and coercion. From a conflict-based perspective, society comprises competing interest groups, some with more power than others. Different groups struggle over societal resources and compete for social advantage. Conflict theorists argue that those who control society's resources also hold power over others. The powerful are also likely to act to reproduce their advantage and try to shape societal beliefs to make their privileges appear to be legitimate and fair. In sum, conflict theory emphasizes the friction in society rather than the coherence and sees society as dominated by elites.

From the perspective of conflict theory, derived largely from the work of Karl Marx, social stratification is based on class conflict and blocked opportunity. Conflict theorists see stratification as a system of domination and subordination in which those with the most resources exploit and control others. They also see the different classes as in conflict with each other, with the unequal distribution of rewards reflecting the class interests of the powerful, not the survival needs of the whole society (Eitzen and Baca Zinn 2007). According to the conflict perspective, inequality provides elites with the power to distribute resources, make and enforce laws, and control value systems; elites use these powers in ways that reproduce inequality. Others in the class structure, especially the working class and the poor, experience blocked mobility.

Conflict theorists argue that the consequences of inequality are negative. From a conflict point of view, the more stratified a society, the less likely that society will benefit from the talents of its citizens; inequality limits the life chances of those at the bottom, preventing their talents from being discovered and used. To the waste of talent is added the restriction of human creativity and productivity.

The Debate between Functionalist and Conflict Theory. Implicit in the argument of each perspective is criticism of the other perspective. Functionalism assumes that the most highly rewarded jobs are the most important for society, whereas conflict theorists argue that some of the most vital jobs in society—those that sustain life and the quality of life, such as farmers, mothers, trash collectors, and a wide range of other laborers—are usually the least rewarded. Conflict theorists also criticize functionalist theory for assuming that the most talented get the greatest rewards. They point out that systems of stratification tend to devalue the contributions of those left at the bottom and to underutilize the diverse talents of all people (Tumin 1953). In contrast, functionalist theorists contend that the conflict view of how economic interests shape social organization is too simplistic. Conflict theorists respond by arguing that functionalists hold too conservative a view of society and overstate the degree of consensus and stability that exists.

The debate between functionalist and conflict theorists raises fundamental questions about how people view inequality. Is it inevitable? How is inequality maintained? Do people basically accept it? This debate is not just academic. The assumptions made from each perspective frame public policy debates. Whether the topic is taxation, poverty, or homelessness, if people believe that anyone can get ahead by ability alone, they will tend to see the system of inequality as fair and accept the idea that there should be a differential reward system. Those who tend toward the conflict view of the stratification system are more likely to advocate programs that emphasize public responsibility for the well-being of all groups and to support programs and policies that result in more of the income and wealth of society going toward the needy.

Following Hurricane Katrina, many residents, particularly the poor, were forced to live on a raised section of Interstate 10 in New Orleans as the waters continued to rise.

Poverty

Many people in the United States were shocked when, following the devastation of Hurricane Katrina in New Orleans and the Gulf Coast, thousands of poor, mostly African American, people were seen on national TV in horrific circumstances, struggling to stay alive and visibly poor. Katrina uncovered one of the faces of poverty in the United States. For many people it was surprising to see conditions that are normally associated with poor, underdeveloped nations right here in the United States—in one of our major and beloved cities.

The truth is that, despite the relatively high average standard of living in the United States, poverty afflicts millions of people. And the particulars of poverty are deeply related to the social structures of class, race, and gender. In New Orleans, for example, when Katrina hit, 28 percent of the population was poor—twice the national rate. Two-thirds of families headed by women in New Orleans were poor—also twice the national rate (U.S. Census Bureau 2004). Although a disaster like Katrina can hurt anyone—and did harm people of differing social and economic statuses—these disasters are not just natural; they also have sociological dimensions (Bobo and Dawson 2006; Hartman and Squires 2006).

thinking sociologically

Using the current federal *poverty line* ($19,874) as a guide, develop a monthly budget that does not exceed this income level and that accounts for all of your family's needs. For purposes of this exercise, assume that you head a family of four, the figure on which this poverty threshold is based. Base your budget on the actual costs of such things in your locale (rent, food, transportation, utilities, clothing, and so forth). Don't forget to account for taxes (state, federal, and local), health care expenses, your children's education, car repairs, and so on. What does this exercise teach you about those who live below the poverty line?

The poverty that Katrina unmasked was not, of course, news to social scientists who have long analyzed the extent of poverty in America and how it affects society's problems. Poor health care, failures in the education system, and crime are all related to poverty. Who is poor, and why is there so much poverty in an otherwise relatively affluent society?

The federal government has established an official definition of poverty used to determine eligibility for government assistance and to measure the extent of poverty in the United States. The **poverty line** is the amount of money needed to support the basic needs of a household, as determined by government; below this line, one is considered officially poor. To determine the poverty line, the Social Security Administration takes a low-cost food budget (based on dietary information provided by the U.S. Department of Agriculture) and multiplies by a factor of three, assuming that a family spends approximately one-third of its budget on food. The resulting figure is the official poverty line, adjusted slightly each year for increases in the cost of living. In 2005, the official poverty line for a family of four was $19,874. Although a cutoff point is necessary to administer antipoverty programs, this definition of poverty can be misleading. A person or family earning $1 above the cutoff point would not be officially categorized as poor.

Who Are the Poor?

There are now more than 37 million poor people in the United States, representing 12.6 percent of the population. Since the 1950s, poverty has declined in the United States; it increased sharply from about 1978 until the mid-1990s, declined, and has been steadily on the rise again since 2000. Although the majority of the poor are White, disproportionately high rates of poverty are also found among Asian Americans, Native Americans, Black Americans, and Hispanics. One-third of Native Americans, 24 percent of African Americans, 22 percent of Hispanics, 10 percent of Asians and Pacific Islanders, and 11 percent of non-Hispanic Whites are poor (DeNavas-Walt et al. 2006). Among Hispanics, there are further differences among groups. Puerto Ricans—the Hispanic group with the lowest median income—have been most likely to suffer increased poverty, probably because of their concentration in the poorest segments of the labor market and their high unemployment rates (Tienda and Stier 1996; Hauan et al. 2000). Asian American poverty has also increased substantially in recent years, particularly among the most recent immigrant groups, including Laotians, Cambodians, Vietnamese, Chinese, and Korean immigrants; Filipino and Asian Indian families have lower rates of poverty (Lee 1994).

The vast majority of the poor have always been women and children, but their proportion of the poor has been increasing in recent years. The term **feminization of poverty** refers to the increasing proportion of the poor who are women and children. This trend results from several factors, including the dramatic growth of female-headed households, a decline in the proportion of the poor who are elderly (not matched by a decline in the poverty of women and children), and continuing wage inequality between women and men. The large number of poor women is associated with a commensurate large number of poor children. By 2005, 18 percent of all children (those under age 18) in the United States were poor, including 9.5 percent of non-Hispanic White children, 34.5 percent of Black children, 28 percent of Hispanic children, and 11.5 percent of Asian American children (DeNavas-Walt 2006).

One-third of all families headed by women are poor (see Figure 7.7). In recent years, wages for young workers have declined; since most unmarried mothers are quite young, there is a strong likelihood that their children will be poor. Because of the divorce rate and generally little child support provided by men, women are also increasingly likely to be without the contributing income of a spouse and for longer periods of their lives. Women are more likely than men to live with children and to be financially responsible for them. However, women without children also suffer a high poverty rate, compounded in recent times by the fact that women now live longer than before and are less likely to be married than in previous periods.

Budget cutbacks in federal support programs and in federal employment have also contributed to the feminization of poverty. Almost two-thirds of the poor receive no food stamps; 80 percent get no housing assistance; less than half receive Medicaid, which is the federal health care system for the poor (U.S. Census

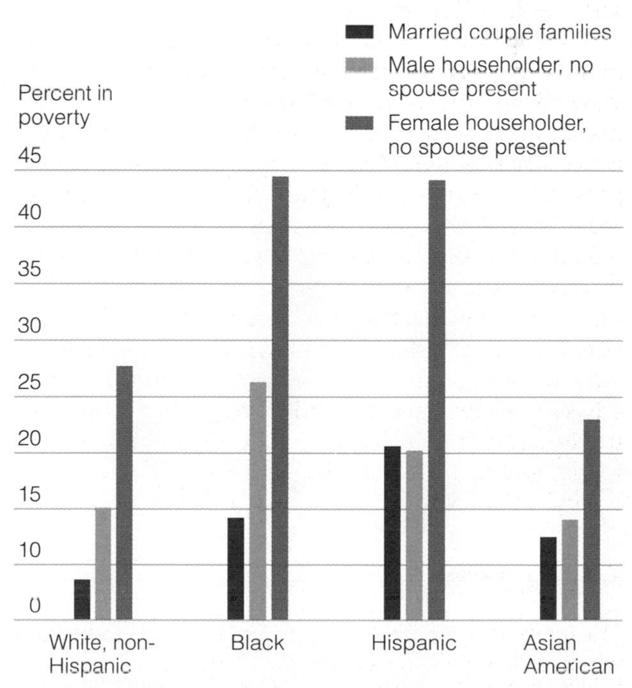

FIGURE 7.7 *Poverty Status by Family Type and Race*

Source: U.S. Census Bureau. 2006. *Historical Income Tables—Households,* Table POV-02.

Bureau 2007). Government budget reductions also disproportionately affect women because women are more reliant than men on public sector jobs. All told, government support for poor women and their children is negligible, despite the fact that the system is reviled as overly generous and producing sloth and dependence.

The poor are not a one-dimensional group. They are racially diverse, including Whites, Blacks, Hispanics, Asian Americans, and Native Americans. They are diverse in age, including not just children and young mothers, but men and women of all ages, and especially a substantial number of the elderly, many of whom live alone. They are also geographically diverse, to be found in areas east and west, south and north, urban and rural, although, as Map 7.2 shows, poverty rates are generally higher in the South and Southwest. One marked change in poverty is the growth of poverty in suburban areas to over 9 percent of all poverty. At the same time, half of the poor live inside central cities. But the focus on urban poverty should not cause you to lose sight of the extent of rural poverty as well.

Among the poor are the thousands of homeless. It is difficult to estimate the number of homeless people. Depending on how one defines and measures homelessness, the estimates vary widely. If you count the number of homeless on any given night, there may be about 444,000 to 842,000 homeless people (depending on the month measured), but measuring those experiencing homelessness over a period of one year, the estimates jump to 2.3 to 3.5 million people (Urban Institute 2000; National Coalition for the Homeless 2006).

Whatever the actual numbers of homeless people, there has been an increase in homelessness over the past two decades, not even counting those displaced by Hurricane Katrina in 2005. Families are the fastest growing segment of the homeless—40 percent. Families with children are the fastest growing segment of the homeless population. Moreover, half of the women with children who are

map 7.2 MAPPING AMERICA'S DIVERSITY

Poverty in the United States

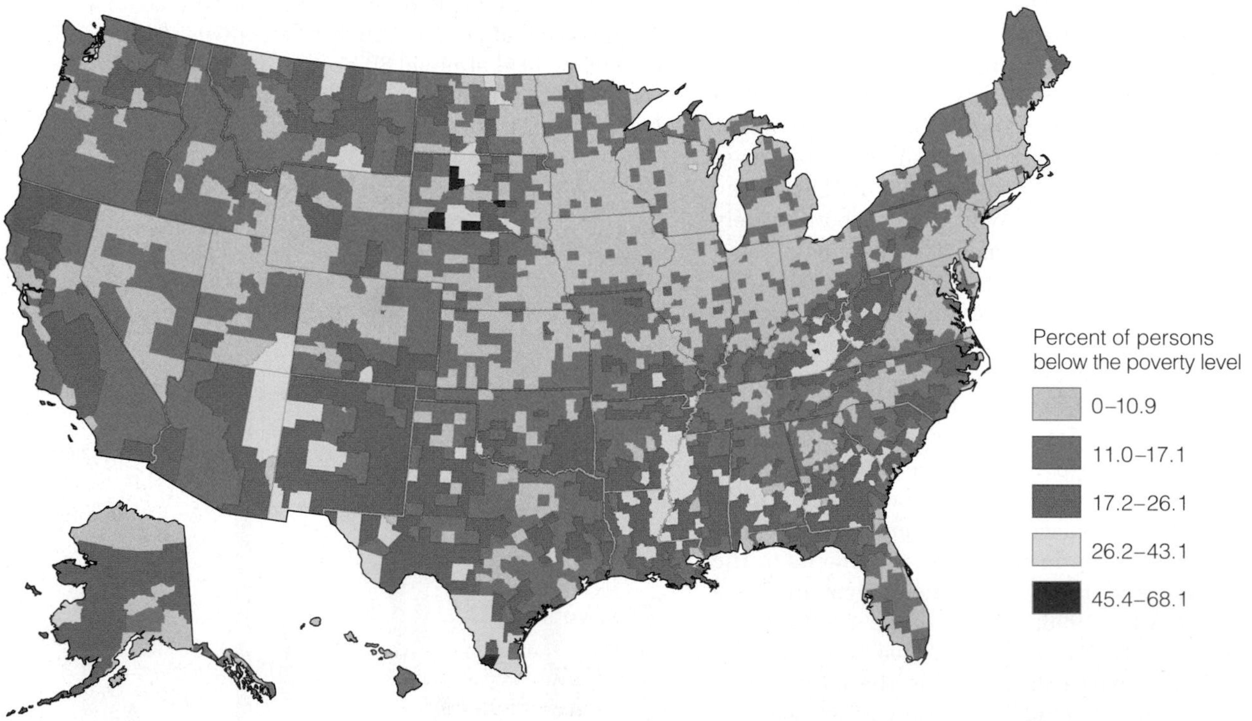

Percent of persons below the poverty level
0–10.9
11.0–17.1
17.2–26.1
26.2–43.1
45.4–68.1

This map shows regional differences in poverty rates (that is, the percentage of poor in different counties). As you can see, poverty is much higher in the South (14.1 percent) than in the West (12.6 percent), Northeast (11.3 percent), and Midwest (10.7 percent). Various social factors explain different rates of poverty, including regional labor markets, the degree of urbanization, immigration patterns, and population composition, among other factors. What do you think the major causes of poverty are in your region?

Data: U.S. Census Bureau. 2004. *American FactFinder.* **www.census.gov**

homeless have fled from domestic violence (National Coalition against Domestic Violence 2001; Zorza 1991). Among homeless people, close to half are African American; another 35 percent are White; 13 percent, Hispanic; 2 percent, Native American; and 1 percent, Asian (National Law Center on Homelessness and Poverty 2004; National Coalition for the Homeless 2006).

There are many reasons for homelessness. The great majority of the homeless are on the streets because of unemployment and/or eviction. Reductions in federal support for affordable housing have left many with no place to live. Combined with eroding work opportunities (particularly in jobs with decent benefits), inadequate housing for low-income people, and reductions in public assistance, many people have no choice but to live on the street. Add to that problems of inadequate health care, domestic violence, and addiction, and you begin to understand the factors that create homelessness. Some of the homeless (about 20 to 25 percent) are mentally ill; the movement to get mental patients out of institutional settings has left many without mental health resources that might help them (National Coalition for the Homeless 2006).

Causes of Poverty

Most agree that poverty is a serious social problem. There is far less agreement on what to do about it. Public debate about poverty hinges on disagreements about its underlying causes. Two points of view prevail: Some blame the poor for their own condition, and some look to social structural causes to explain poverty. The first view, popular with the public and many policy makers, is that poverty is caused by the cultural habits of the poor. According to this point of view, behaviors such as crime, family breakdown, lack of ambition, and educational failure generate and sustain poverty, a syndrome to be treated by forcing the poor to fend for themselves. The second view is a more structural view, with sociologists seeing poverty as rooted in the structure of society, not in the morals and behaviors of individuals.

Blaming the Victim: The Culture of Poverty. Blaming the poor for being poor stems from the myth that success requires only individual motivation and ability. Many in the United States adhere to this view and hence have a harsh opinion of the poor. This attitude is also reflected in U.S. public policy concerning poverty, which is rather ungenerous compared with other industrialized nations. Those who blame the poor for their own plight typically argue that poverty is the result of early childbearing, drug and alcohol abuse, refusal to enter the labor market, and crime. Such thinking puts the blame for poverty on individual choices, not on societal problems. In other words, it blames the victim, not the society, for social problems (Ryan 1971).

The **culture of poverty** argument attributes the major causes of poverty to the absence of work values and the irresponsibility of the poor. In this light, poverty is seen as a dependent way of life that is transferred, like other cultures, from generation to generation. The culture of poverty argument has now been adapted by many policy makers to argue that the actual causes of poverty are found in the breakdown of major institutions, including the family, schools, and churches.

Is the culture of poverty argument true? To answer this question, we might ask: Is poverty transmitted across generations? Researchers have found only mixed support for this assumption. Many of those who are poor remain poor for only one or two years; only a small percentage of the poor are chronically poor. More often, poverty results from a household crisis, such as divorce, illness, unemployment, or parental death. People tend to cycle in and out of poverty. The public stereotype that poverty is passed through generations is thus not well supported by the facts.

A second question is, Do the poor want to work? The persistent public stereotype that they do not is central to the culture of poverty thesis. This attitude presumes that poverty is the fault of the poor, that if they would only change their values and adopt the American work ethic, then poverty would go away. What is the evidence for these claims?

Most of the able-bodied poor *do* work, even if only part-time. Moreover, as we saw above, the number of workers who constitute the *working poor* has increased. You can see why this is true when you calculate the income of someone working full-time for minimum wage. Someone working forty hours per week, fifty-two weeks per year, at minimum wage will have an income far below the poverty line. This is the major reason that many have organized a *living wage campaign,* intended to raise the federal minimum wage to provide workers with a decent standard of living.

Current policies that force those on welfare to work also tend to overlook how difficult it is for poor people to retain the jobs they get. Prior to welfare reform in the mid-1990s, poor women who went off welfare to take jobs often found they soon had to return to welfare because the wages they earned were not enough to support their families. Leaving welfare often means losing health benefits, yet incurring increased living expenses. The jobs that poor people find often do not lift them out of poverty. In sum, attributing poverty to the values of the poor is both unproven and a poor basis for public policy (Albelda and Tilly 1996; Catanzarite and Ortiz 1996).

Structural Causes of Poverty. From a sociological point of view, the underlying causes of poverty lie in the economic and social transformations taking place in the United States. Careful scholars do not attribute poverty to a single cause. There are many causes. Two of the most important are the *restructuring of the economy,* which has resulted in diminished earning power and increased unemployment, and *the status of women in the family and the labor market,* which has contributed to women being overrepresented among the poor. Add to these underlying conditions the federal policies in recent years that have *diminished social support for the poor* in the form of welfare, public housing, and job training. Given these reductions in federal support, it is little wonder that poverty is so widespread.

The restructuring of the economy has caused the disappearance of manufacturing jobs, traditionally an avenue of job security and social mobility for many workers, especially African American and Latino workers (Baca Zinn and Eitzen 2007; Wilson 1996). The working class has been especially vulnerable to these changes. Economic decline in those sectors of the economy where men have historically received good pay and good benefits has meant that fewer men are the sole support for their families. Most families now need two incomes to achieve a middle-class way of life. The new jobs that are being created fall primarily in occupations that offer low wages and few benefits; they also tend to be filled by women, especially women of color, leaving women poor and men out of work (McCall 2001; Browne 1999; Andreasse 1997). Such jobs offer little chance to get out of poverty. New jobs are also typically located in neighborhoods far away from the poor, creating a mismatch between the employment opportunities and the residential base of the poor.

Declining wage rates caused by transformations taking place within the economy fall particularly hard on young people, women, and African Americans and Latinos, who are the groups most likely to be among the working poor (Fine and Weis 1998). The high rate of poverty among women is also strongly related to women's status in the family and the labor market. Divorce is one cause of poverty, although minority women are more likely than White women to be poor even within marriage (Catanzarite and Ortiz 1996). Women's child-care responsibilities make working outside the home on marginal incomes difficult. Many women with children cannot manage to work outside the home, because it leaves them with no one to watch their children. More women now depend on their own earnings to support themselves, their children, and other dependents. Whereas unemployment has always been considered a major cause of poverty among men, for women, wage discrimination has a major role. Notice that the median income for all women ($23,074 in 2005) is not that far above the poverty line.

The persistence of poverty also increases tensions between different classes and racial groups. William Julius Wilson, one of the most noted analysts of poverty and racial inequality, has written, "The ultimate basis for current racial tension is the deleterious effect of basic structural changes in the modern American economy on Black and White lower-income groups, changes that include uneven economic growth, increasing technology and automation, industry relocation, and labor market segmentation" (1978: 154). This demonstrates the power of sociological thinking by convincingly placing the causes of both poverty and racism in their societal context, instead of the individualistic thinking that tends to blame the poor for their plight.

Welfare and Social Policy

Current welfare policy is covered by the 1996 Personal Responsibility and Work Reconciliation Act (PRWRA). This federal policy eliminated the longstanding welfare program titled Aid to Families with Dependent Children (AFDC), which was created in 1935 as part of the Social Security Act. Implemented during the Great Depression, AFDC was meant to assist poor mothers and their children. It acknowledged that some people are victimized by economic circumstances beyond their control and deserve assistance. For much of its lifetime, this law supported mostly White mothers and their children; not until the 1960s did welfare come to be identified with Black families.

The new welfare policy gives block grants to states to administer their own welfare programs through the program called **Temporary Assistance for Needy Families (TANF).** TANF stipulates a lifetime limit of five years for people to receive aid and requires all welfare recipients to find work within two years—a policy known as *workfare.* Those who have not found work within two months of receiving welfare can be required to perform community service jobs for free.

In addition, welfare policy denies payments to unmarried teen parents under eighteen years of age unless they stay in school and live with an adult. It also requires unmarried mothers to identify the fathers of their children or risk losing their benefits (Hays 2003; Edin and Kefalas 2005). These broad guidelines are established at the federal level, but individual states can be more restrictive, as many have been. At the heart of welfare reform is the idea that public assistance creates dependence, discouraging people from seeking jobs. The very title of the new law, emphasizing personal responsibility and work, suggests that poverty is the fault of the poor. Low-income women, for example, are stereotyped as just wanting to have babies to increase the size of their welfare checks—even though research finds no support for this idea (Edin and Kefalas 2005)!

Is welfare reform working? Many claim that welfare reform is working because, since passage of the new law, the welfare rolls have shrunk. But having fewer people on welfare does not mean poverty is reduced; in fact, as we have seen, poverty has actually increased since passage of welfare reform. Having fewer people on the rolls can simply mean that people are without a safety net.

Many studies also find that low-wage work does not lift former welfare recipients out of poverty (Hays 2003). Critics of the current policy also argue that forcing welfare recipients to work provides a cheap labor force for employers and potentially takes jobs from those already employed. In the first few years of welfare reform, the nation was also in the midst of an economic boom; jobs were thus more plentiful. But in an economic downturn, those who are on aid or in marginal jobs are vulnerable to economic distress, particularly given the time limits now placed on receiving public assistance (Albelda and Withorn 2002).

Debunking Society's Myths

Myth: Marriage is a good way to reduce women's dependence on welfare.

Reality: Forcing women to marry encourages women's dependence on men and punishes women for being independent. Research indicates that poor women place a high value on marriage and want to be married, but also understand that men's unemployment and instability makes their ideal of marriage unattainable (Edin and Kefalas 2005). In addition, large numbers of women receiving welfare have been victims of domestic violence.

Research done to assess the impact of a changed welfare policy is relatively recent. Politicians brag that welfare rolls have shrunk, but reduction in the welfare rolls is poor measure of the true impact of welfare reform because this would be true simply because people are denied benefits. And because welfare has been decentralized to the state level, studies of the impact of current law must be done on a state-by-state basis. Such studies are showing that those who have gone into workfare programs most often earn wages that keep them below the poverty line. Although some states report that family income has increased following welfare reform, the increases are slight. More people have been evicted because of falling behind on rent. Families also report an increase in other material hardships, such as phones and utilities being cut off. Marriage rates among former recipients have not changed, although more now live with nonmarital partners, most likely as a way of sharing expenses. The number of children living in families without either parent has also increased, probably because parents had to relocate to find work. In some states, the numbers of people neither working nor receiving aid also increased (Lewis et al. 2002; Acker et al. 2002; Bernstein 2002).

The public debate about welfare rages on, often in the absence of informed knowledge from sociological research and almost always without input from the subjects of the debate, the welfare recipients themselves. Although stigmatized as lazy and not wanting to work, those who have received welfare actually believe that it has negative consequences for them, but they say they have no other viable means of support. They typically have needed welfare when they could not find work or had small children and were without child care. Most were forced to leave their last job because of layoffs or firings or because the work was only temporary. Few left their jobs voluntarily.

Welfare recipients also say that the welfare system makes it hard to become self-supporting, because the wages one earns while on welfare are deducted from an already minimal subsistence. Furthermore, there is not enough affordable day care for mothers to leave home and get jobs. The biggest problem they face in their minds is lack of money. Contrary to the popular image of the conniving "welfare queen," welfare recipients want to be self-sufficient and provide for

*Is It True?**

	True	False
1. Income growth has been greatest for those in the middle class in recent years.		
2. The average American household has most of its wealth in the stock market. Poverty in U.S. suburbs is increasing.		
3. Social mobility is greater in the United States than in any other Western nation.		
4. The majority of welfare recipients are African American.		
5. Poor, teen mothers do not have the same values about marriage as middle-class people.		
6. Welfare reform has reduced the welfare rolls.		
7. Old people are the most likely to be poor.		
8. Poverty in U.S. suburbs is increasing.		

*** The answers can be found on page 206.**

Is It True? [Answers]

1. FALSE. Income growth has been highest in the top 20 percent of income groups; most others have seen their incomes remain relatively flat (DeNavas-Walt et al. 2006).
2. FALSE. Eighty percent of all stock is owned by a small percentage of people. For most people, homeownership is the most common financial asset (minus mortgage debt; Orzechowski and Sepielli 2003).
3. FALSE. The U.S. has lower rates of social mobility than Canada, Sweden, and Norway and ranks near the middle in comparison with other Western nations (Beller and Hout 2006).
4. FALSE. The majority of those receiving welfare (TANF) are white (DeNavas-Walt et al. 2006).
5. FALSE. Research finds that poor, teen mothers value marriage and want to be married, but they associate it with economic security, which they do not think they can achieve (Edin and Kefalas 2005).
6. TRUE. Welfare reform has shrunk the number of people receiving welfare, but at the same time, poverty has actually increased (Hays 2003; DeNavas-Walt 2006).
7. FALSE. Although those over 65 used to be the most likely to be poor, poverty among the elderly has declined; the most likely to be poor are children (under 18 years of age).
8. TRUE. Although most of the poor live inside metropolitan areas, poverty in suburban areas has been increasing.

their families, but they face circumstances that make this very difficult to do. Indeed, studies of young, poor mothers find that they place a high value on marriage, but they do not think they—or their boyfriends—have the means to achieve the marriage ideals they cherish (Hays 2003; Edin and Kefalas 2005).

Another popular myth about welfare is that people use their welfare checks to buy things they do not need. But research finds that when former welfare recipients find work, their expenses actually go up. Although they may have increased income, their expenses (in the form of child care, clothing, transportation, lunch money, and so forth) increase, leaving them even less disposable income. Moreover, studies find that low-income mothers who buy "treats" for their children (name-brand shoes, a movie, candy, and so forth) do so because they want to be good mothers (Edin and Lein 1997).

Other beneficiaries of government subsidies have not experienced the same kind of stigma. Social Security supports virtually all retired people, yet they are not stereotyped as dependent on federal aid, unable to maintain stable family relationships, or insufficiently self-motivated. Spending on welfare programs is also a pittance compared with the spending on other federal programs. Sociologists conclude that the so-called welfare trap is not a matter of learned dependency, but a pattern of behavior forced on the poor by the requirements of sheer economic survival (Hays 2003; Edin and Kefalas 2005).

7 Chapter Summary

- ***What different kinds of stratification systems exist?***

  Social stratification is a relatively fixed hierarchical arrangement in society by which groups have different access to resources, power, and perceived social worth. All societies have systems of stratification, although they vary in composition and complexity. *Estate systems* are those where power and property are held by a single elite class; in *caste systems,* placement in the stratification is by birth, and in *class systems,* placement is determined by achievement.

- ***How do sociologists define class?***

  Class is the social structural position groups hold relative to the economic, social, political, and cultural resources of society. It is highly significant in determining one's *life chances.*

- ***How is the class system structured in the United States?***

  Social class can be seen as a hierarchy, like a ladder, where income, occupation, and education are indicators of class. *Status attainment* is the process by which people end up in a given position in this hierarchy. *Prestige* is the value assigned to people and groups by others within this hierarchy. Classes are also organized around common interests and exist in conflict with one another.

- ***Is there social mobility in the United States?***
  Social mobility is the movement between class positions. Education gives some boost to social mobility, but social mobility is more limited than people believe; most people end up in a class position very close to their class of origin. *Class consciousness* is both the perception that a class structure exists and the feeling of shared identification with others in one's class. The United States has not been a particularly class-conscious society because of the belief in upward mobility.

- ***What analyses of social stratification do sociological theorists provide?***
  Karl Marx saw class as primarily stemming from economic forces; Max Weber had a multidimensional view of stratification, involving economic, social, and political dimensions. Functionalists argue that social inequality motivates people to fill the different positions in society that are needed for the survival of the whole, claiming that the positions most important for society require the greatest degree of talent or training and are, thus, most rewarded. Conflict theorists see social stratification as based on class conflict and blocked opportunity, pointing out that those at the bottom of the stratification system are least rewarded because they are subordinated by dominant groups.

- ***How do sociologists explain why there is poverty in the United States?***
  The *culture of poverty thesis* is the idea that poverty is the result of the cultural habits of the poor that are transmitted from generation to generation, but sociologists see poverty as caused by social structural conditions, including unemployment, gender inequality in the workplace, and the absence of support for child care for working parents.

- ***What current policies address the problem of poverty?***
  Current welfare policy, adopted in 1996, provides support through individual states, but recipients are required to work after two years of support and have a lifetime limit of five years' support.

Key Terms

caste system 179
class 180
class consciousness 195
culture of poverty 203
educational attainment 185
estate system 179
false consciousness 195
feminization of poverty 201
ideology 197
income 184
life chances 180
median income 184
meritocracy 195
occupational prestige 185
poverty line 201
prestige 185
social class 180
social differentiation 178
social mobility 195
social stratification 178
socioeconomic status 184
status 178
status attainment 184
Temporary Assistance for Needy Families (TANF) 204
urban underclass 187
wealth 184

Online Resources

Sociology: The Essentials Companion Website

www.thomsonedu.com/sociology/andersen

Visit your book companion website where you will find more resources to help you study and write your research papers. Resources include Suggested Readings, web links, and a MicroCase Online feature that teaches you how to research society. Other resources include Learning Objectives, Internet exercises, quizzing, and flash cards.

is an easy-to-use online resource that helps you study in less time to get the grade you want NOW.

www.thomsonedu.com/login

Need help studying? This site is your one-stop study shop. Take a Pre-Test and Thomson NOW will generate a Personalized Study Plan based on your test results. The Study Plan will identify the topics you need to review and direct you to online resources to help you master those topics. You can then take a Post-Test to determine the concepts you have mastered and what you still need to work on.

8
Chapter eight
CHAPTER EIGHT
Chapter

Global Stratification

"It takes a village to raise a child," the saying goes. But it also seems to take a world to make a shirt—or so it seems from looking at the global dimensions of the production and distribution of goods. Try this simple experiment: Look at the labels on your clothing. (If you do this in class, try to do so without embarrassing yourself and others!) What do you see? "Made in Indonesia," "Made in Vietnam," "Made in Malawi," all indicating the linkage of the United States to systems of production around the world. The popular brand Nike, as just one example, has not a single factory in the United States, although its founder and chief executive officer is one of the wealthiest people in America. Nike products are manufactured mostly in Southeast Asia.

Taking your experiment further, ask yourself: Who made your clothing? A young person trying to lift his or her family out of poverty? Might it have been a child? In many areas of the world, one in five children under age 15 work (International Labour Organization 2002). What countries benefit most from this system of production? Answering these questions reveals much about the interconnection among countries in the *global stratification* system, a system in which the status of the people in one country is intricately linked to the status of the people in others.

continued

chapter eight

CHAPTER EI

eight

Recall from Chapter 1 that C. Wright Mills identified the task of sociology as seeing the social forces that exist beyond the individual. This is particularly important when studying global stratification. The person in the United States (or western Europe or Japan) who thinks he or she is expressing individualism by wearing the latest style is actually part of a global system of inequality. The adornments available to that person result from a whole network of forces that produce affluence in some nations and poverty in others.

Dominant in the system of global stratification are the United States and other wealthy nations. Those at the top of the global stratification system have enormous power over the fate of other nations. Although world conflict stems from many sources, including religious differences, cultural conflicts, and struggles over political philosophy, the inequality between rich and poor nations causes much hatred and resentment. One cannot help but wonder what would happen if the differences between the wealth of some nations and the poverty of others were smaller. In this chapter we examine the dynamics and effects of global stratification.

Global Stratification

In the world today, there are not only rich and poor people, but also rich and poor countries. Some countries are well-off, some countries are doing so-so, and a growing number of countries are poor and getting poorer. There is, in other words, a system of **global stratification** in which the units we are considering are countries, much like a system of stratification within countries in which the units are individuals or families. Just as we can talk about the upper-class or lower-class individuals within a country, we can also talk of the equivalent upper-class or lower-class countries in this world system. One manifestation of global stratification is the great inequality in life chances that differentiates nations around the world. The United Nations' **human development index,** a compilation of data indicating various levels of national well-being, shows the great inequities that global stratification brings (see Map 8.1). Simple measures of well-being, including life expectancy, infant mortality, and access to health services, reveal the consequences of a global system of inequality. And, the gap between the rich and poor is sometimes greater in nations where the average person is least well off. No longer can these nations be understood without considering the global system of stratification of which they are a part.

The effects of the global economy on inequality have become increasingly evident, as witnessed by public concerns about jobs being sent overseas. A coalition of unions, environmentalists, and other groups has also emerged to protest global trade policies that they believe threaten jobs and workers' rights in the United States, as well as contributing to environmental degradation. Such policies also encourage further McDonaldization (see Chapter 5). Thus, popular stores such as The Gap and Niketown have often been targets of political protests because they symbolize the expansion of global capitalism. Protestors see the growth of such stores as eroding local cultural values and spreading the values of unfettered consumerism around the globe. Protests over world trade policies have also emerged in a student-based movement against companies that manufacture apparel with college logos.

The relative affluence of the United States means that U.S. consumers have access to goods produced around the world. A simple thing, such as a child's toy, can represent this global system. For many young girls in the United States, Barbie™ is the ideal of fashion and romance. Young girls may have not just one Barbie, but several, each with a specific role and costume. Cheaply bought in the United States, but produced overseas, Barbie is manufactured by those probably not much older than the young girls who play with her and who would need all of their monthly pay to buy just one of the dolls that many U.S. girls collect by the dozens (Press 1996: 12).

The manufacturing of toys and clothing are examples of the global stratification that links the United States and other parts of the world. When companies export jobs, workers in the United States lose them. For example, note that in 1973, more than 56,000 U.S. workers were employed in toy factories. Now, even though the market has become glutted with the latest popular items, only 27,000 workers work in toy factories. The companies that make the toys amass profits, but U.S. workers lose jobs and then blame foreign workers for taking them. Major toy manufacturers have adopted policies that prohibit the exploitation of child labor, but clearly it is more profitable for companies to take their labor overseas where workers are paid far less and get fewer benefits than many workers in the United States.

Rich and Poor

One dimension of stratification between countries is wealth. Enormous differences exist between the wealth of the countries at the top of the global stratification system and the wealth of the countries at the bottom. Although there are different ways to measure the wealth of nations, one of the most common is to use the per capita **gross national income** (per capita GNI). The GNI measures the total output of goods and services produced by residents of a country each year plus the income from nonresident sources, divided by the size of the population. This does not truly reflect what individuals or families receive in wages or pay; it is simply each person's annual share of their country's income, should, in theory, the proceeds be shared equally.

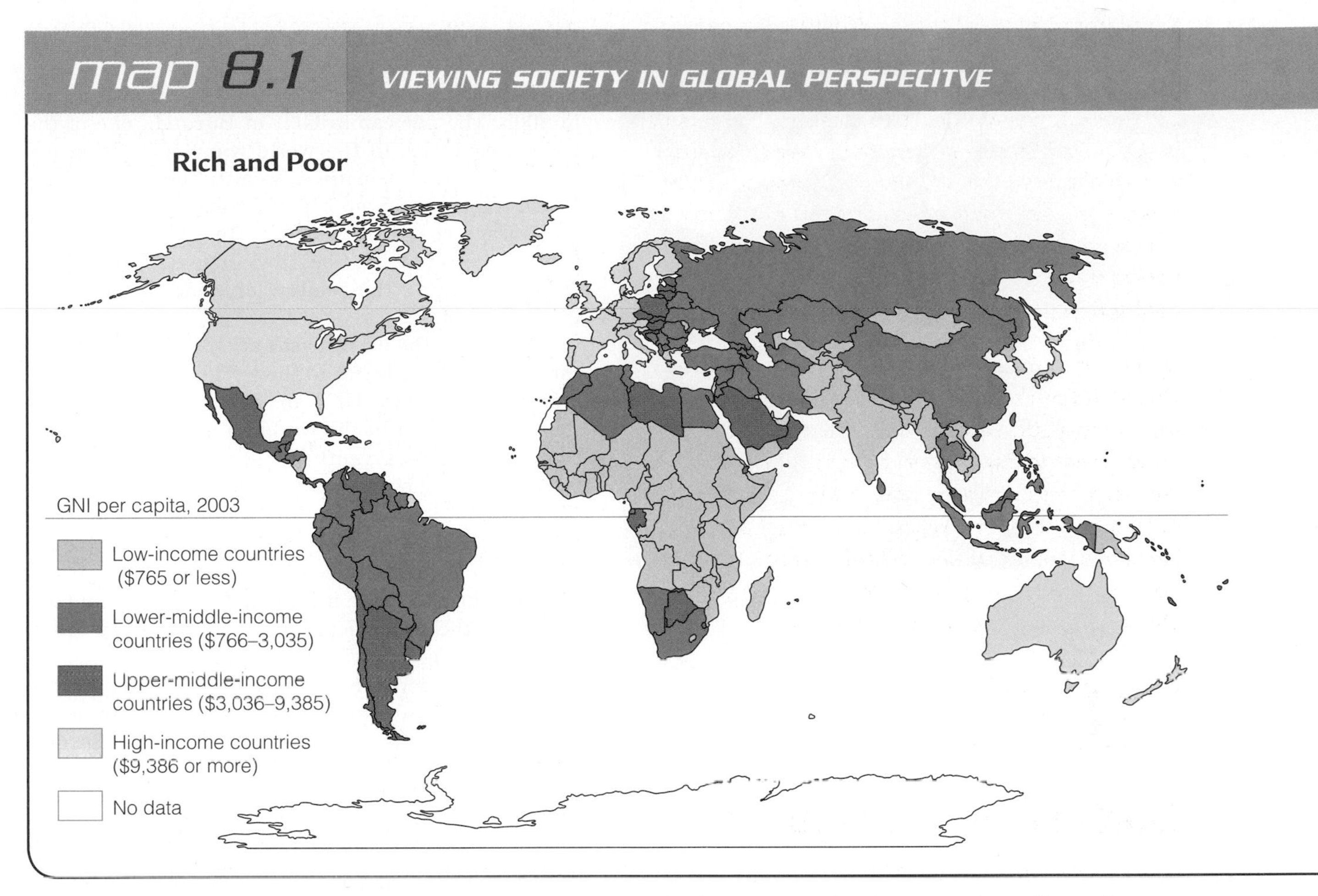

Most nations are linked in a world system that produces wealth for some and poverty for others. The GNI (gross national income), depicted here on a per capita basis for most nations in the world, is an indicator of the wealth and poverty of nations. Identify one of the nations represented here. Would you say this nation is a core, semiperipheral, or peripheral country in the global economy? Why?

Source: World Bank. 2007. *World Bank Atlas.* Washington, DC: World Bank. **www.worldbank.org.** Used by permission.

Global stratification often means that consumption in the more affluent nations is dependent on cheap labor in other less affluent nations.

© DPA/DJD/The Image Works

Many common products marketed in the United States are produced in a global economy.

Per capita GNI is reliable only in countries that are based on a cash economy. It does not measure informal exchanges or bartering in which resources are exchanged without money changing hands. These noncash transactions are not included in the GNP calculation, but they are more common in developing countries. As a result, measures of wealth based on the GNI, or other statistics that count cash transactions, are less reliable among the poorer countries and may underestimate the wealth of the countries at the lower end of the economic scale.

The per capita GNI of the United States, which is one of the wealthier nations in the world (though not the wealthiest on a per capita basis) was $43,740 in 2005. The per capita GNI in Burundi, one of the poorest countries in the world was $100. Using per capita GNI as a measure of wealth, the average person in the United States is 437 times wealthier than an average citizen of Burundi. In Afghanistan, the GNI was estimated to be $875 or less (World Bank 2007), indicating the relative affluence of people in the United States.

Which are the wealthiest nations? Figure 8.1 lists the ten richest countries in the world measured by the annual per capita GNI in 2005. Luxembourg is the richest nation in the world on a per capita basis; the United States, seventh. Of course, Luxembourg has a tiny population compared with the United States. Note that most of the wealthy countries are in western Europe. They are mostly industrialized countries (or support such countries through such businesses as banking), and they are mostly urban. These countries represent the equivalent of the upper class—even though many people within them are poor.

Now consider the ten poorest countries in the world, also shown in Figure 8.1 and again using per capita GNI as the measure of wealth. Several countries are even poorer than these countries, but they are not often listed in official data because they are too poor to report reliable statistics. Most of the world's poorest countries are in eastern or central Africa. These countries have not become industrialized, are largely rural, have high fertility rates, and still depend heavily on subsistence agriculture. Based on the dimension of wealth, these countries rank at the bottom of the global stratification system.

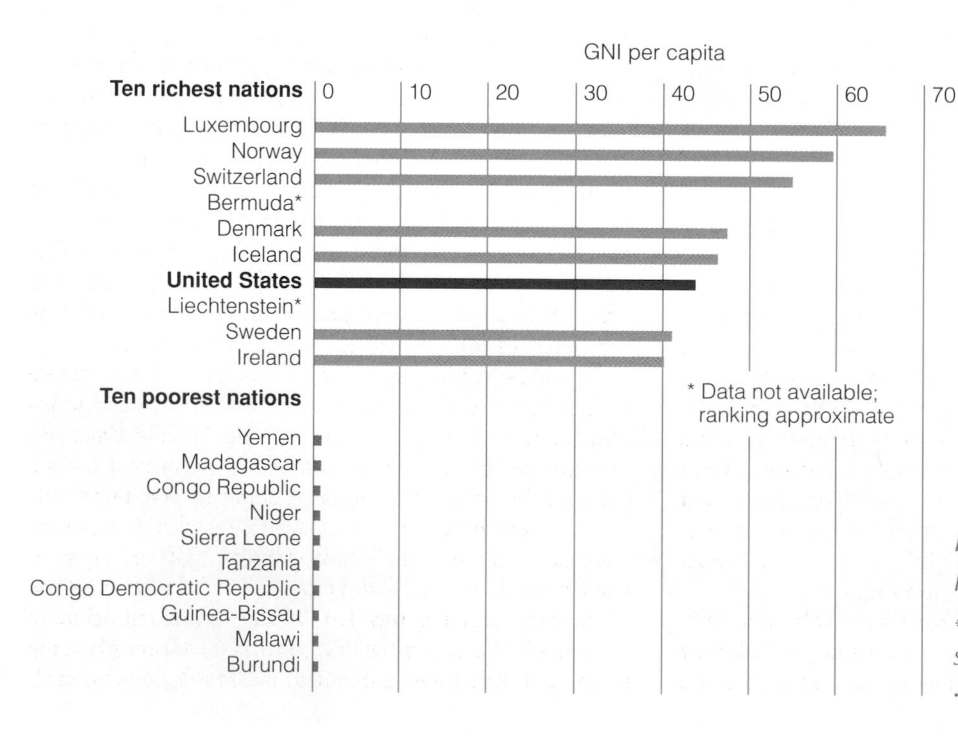

FIGURE 8.1 *The Rich and the Poor: A World View**

*GNI per capita, measured in U.S. dollars. *Source:* World Bank. 2007. **www.worldbank.org**. Used by permission.

© Reuters/CORBIS

Global stratification means not only that enormous differences exist in the relative well-being of different countries, but also within nations. Large numbers of people live in poverty such as in this refugee camp at Mazar-i-Sharif in Afghanistan—one of the poorest nations in the world.

Clearly, many countries in the world are very poor, whereas other countries are rich. This does not mean that all people in rich countries are rich or that all people in poor countries are poor. But on the average, people in poor countries are much worse off than people in rich countries. In many poor countries, the life of an average citizen is desperate; these countries also have the largest populations. In a world with a population of nearly six billion, more than three billion—more than half the world's population—live in the poorest forty-five countries. Often poor nations are rich with natural resources but are exploited for such resources by more powerful nations. We will look more closely at the nature and causes of world poverty later in this chapter.

The Core and Periphery

Global stratification involves nations in a large and integrated network of both economic and political relationships. **Power**—meaning the ability of a country to exercise control over other countries or groups of countries—is a significant dimension of global stratification. Countries can exercise several kinds of power over other countries, including military, economic, and political power. The **core countries** have the most power in the world economic system. These countries control and profit the most from the world system, and thus they are the "core" of the world system. These include the powerful nations of Europe, the United States, Australia, and Japan.

Surrounding the core countries, both structurally and geographically, are the **semiperipheral countries** that are semi-industrialized and, to some degree, represent a kind of middle class (such as Spain, Turkey, and Mexico). They play a middleman role, extracting profits from the poor countries and passing those profits on to the core countries. At the bottom of the world stratification system, in this model, are the **peripheral countries.** These are the poor, largely agricultural countries of the world. Even though they are poor, they often have important natural resources that are exploited by the core countries. This exploitation, in turn, keeps them from developing and perpetuates their poverty. Often these nations are politically unstable, and, though they exercise little world power, political instability can cause a crisis for core nations that depend on their resources. Military intervention by the United States or European nations is often the result.

This categorizing system emphasizes the power of each country in the world economic system. Another way that these countries are sometimes labeled is as first-, second-, and third-world nations. This language grows out of the politics of the Cold War and reflects the political and economic dimensions of global stratification. **First-world countries** consist of the industrialized capitalist countries of the world, including the United States, New Zealand, Australia, Japan, and the countries of western Europe. They are industrialized and have a market-based economy and a democratically elected government. The **second-world countries** are socialist countries, which included the former Soviet Union, China, Cuba, North Korea, and, prior to the fall of the Berlin Wall, the eastern European nations. During the Cold War, these countries had a communist-based government and a state-managed economy, as some still do. Although less developed than the first-world countries, the second-world countries tried to provide citizens with services such as free education, health care, and low-cost housing, consistent with the principles of socialism, but poverty often prevented them from doing so. Moreover, their governments were not democratically elected. **Third-world countries** in this scheme are the countries that are poor, underdeveloped, largely rural, and with high levels of poverty. Many of the governments of the third-world countries are autocratic dictatorships (ruled by one person with absolute authority), though not all. Though these countries are generally poor, wealth is concentrated in the hands of only a few elite.

Because this system of categorization was based on the logic of the Cold War, it has changed. For instance, the oil-rich countries of the Middle East are not part of the first or second world, according to this scheme, but they also do not belong in the same category as the poor countries of Africa and Asia because they have considerably more wealth. The collapse of the Soviet Union and the change in the governments of eastern Europe has led to the transformation of almost all the second-world countries. Although some countries still have a communist-based government,

doing sociological research

Servants of Globalization: Who Does the Domestic Work?

Research Question: International migration is becoming an increasingly common phenomenon. Women are one of the largest groups to experience migration, often leaving poor nations to become domestic workers in wealthier nations. What are these women's experiences in the context of global stratification? This is what Rhacel Salazar Parreñas wanted to know.

Research Method: Parreñas studied two communities of Filipina women, one in Los Angeles and one in Rome, Italy, conducting her research through extensive interviewing with Filipina domestic workers in these two locations. She supplemented the interviews with participant observation in church settings, after-work social gatherings, and in employers' homes. The interviews were conducted in English and Tagalog—sometimes a mixture of both.

Research Results: Parreñas found that Filipina domestics experienced many status inconsistencies. They were upwardly mobile in terms of their home country but were excluded from the middle-class Filipino communities in the communities where they lived. Thus, they experienced feelings of social exclusion in addition to being separated from their own families.

Conclusion and Implications: The women Parreñas studied are part of a new social form: *transnational families*—that is, families whose members live across the borders of nations. These Filipinas provide the labor for more affluent households while their own lives are disrupted by these new global forces. As global economic restructuring evolves, it may be that more and more families will take this form as they adapt to changing economic and social conditions.

Questions to Consider

1. Are there domestic workers in your community who provide child care and other household work for middle- and upper-class households? What is the race, ethnicity, nationality, and gender of these workers? What does this tell you about the division of labor in domestic work and its relationship to global stratification?
2. Why do you think domestic labor is so underpaid and undervalued? Are there social changes that might result in a reevaluation of the value of this work?

Source: Parreñas, Rhacel Salazar. 2001. *Servants of Globalization: Women. Migration and Domestic Work.* Stanford, CA: Stanford University Press.

SEE FOR YOURSELF

THE GLOBAL ECONOMY OF CLOTHING

Look at the labels in your clothes and note where your clothing was made. Where are the products bearing your college logos manufactured and sold? Who profits from the distribution of these goods? What does this tell you about the relationship of *core, semiperipheral,* and *peripheral* countries within world systems theory? What further information would reveal the connections between the country where you live and the countries where your clothing is made and distributed?

many of them, including China, are moving toward a market economic system. Still, like the terms *core* and *periphery,* the terms *first-, second-,* and *third-world* are useful in denoting the relationship of different countries to the world economy and global stratification.

Race and Global Inequality

Along with class inequality, there is a racial component to world inequality, which can be seen in several ways. The rich core countries, those that dominate the world system, are largely European, with the exception of the United States and Japan. In Europe and the United States, the population is mostly White; in the poor countries of the world, mostly in Africa,

Asia, or South America, the populations are largely people of color. On average, there are vast differences in life chances and lifestyle between the countries of the world with White populations and the countries of the world with Black populations.

Exploitation of the human and natural resources of regions populated by people of color has characterized the history of Western capitalism, with people of color being dominated by Western imperialism and colonialism. The inequities that have resulted are enormous. Patterns of malnutrition and hunger show these inequities. More than one billion people in the world suffer from malnutrition and hunger. The vast majority of these people are people of color—that is, those not of European descent (Uvin 1998).

How did this racial inequality come about? On the surface, global capitalism is not explicitly racist, as were earlier forms of industrial capitalism. Yet, in fact, it is the rapid expansion of the global capital system that has led to the increase in racial inequality between nations. In the new capitalist system, a new **international division of labor** has emerged that is not tied to place, but can employ cheap labor anywhere. Cheap labor is usually found in non-Western countries. The exploitation of cheap labor has created a poor and dependent workforce that is mostly people of color. The profits accrue to the wealthy owners, who are mostly White, resulting in a racially divided world. Some have argued further that multinational corporations' exploitation of the poor peripheral nations has forced an exodus of unskilled workers from the impoverished nations to the rich nations. The flood of third-world refugees into the industrialized nations is thereby increasing racial tensions, fostering violence, and destroying worker solidarity (Sirvananadan 1995).

South Africa, the United States, and Brazil each developed different sets of racial categories. Although all three countries have many people of mixed descent, race is defined differently in each place. In South Africa, the particular history of Dutch and English colonialism led to strongly drawn racial categories that defined people as "White," "coloured" (those of mixed descent), or "native/Bantu" (all black South Africans). However, in the United States, given its history of slavery, the "one drop" rule was used, which defined anyone with any African heritage as Black, thus ruling out any category of mixed race.

Brazil is yet a different case. The Brazilian elite declared Brazil a racial democracy at the early stages of national development. Racial differences were thought not to matter. Yet, instead of creating an egalitarian society free of racism, this resulted in Afro-Brazilians being of lower social status while Euro-Brazilians remain at the highest social status. Thus, color continues to matter because it stratifies people (Marx 1997; Frederickson 2003; Telles 2004).

Theories of Global Stratification

How did world inequality occur? Sociological explanations of world stratification generally fall into three camps: modernization theory, dependency theory, and world systems theory, each explained here (see Table 8.1).

table 8.1 Theories of Global Stratification

	Modernization Theory	Dependency Theory	World Systems Theory
Economic Development	Arises from relinquishing traditional cultural values and embracing new technologies and market-driven attitudes and values	Exploits the least powerful nations to the benefit of wealthier nations that then control the political and economic systems of the exploited countries	Has resulted in a single economic system stemming from the development of a world market that links core, semiperipheral, and peripheral nations
Poverty	Results from adherence to traditional values and customs that prevent societies from competing in a modern global economy	Results from the dependence of low-income countries on wealthy nations	Is the result of core nations extracting labor and natural resources from peripheral nations
Social Change	Involves increasing complexity, differentiation, and efficiency	Is the result of neocolonialism and the expansion of international capitalism	Leads to an international division of labor that increasingly puts profit in the hands of a few while exploiting those in the poorest and least powerful nations

Modernization Theory

Modernization theory views the economic development of countries as stemming from technological change. According to this theory, a country becomes more "modernized" by increased technological development, and this technological development is also dependent on other countries. Modernization theory was initially developed in the 1960s to explain why some countries had achieved economic development and why some had not (Rostow 1978).

Modernization theory sees economic development as a process by which traditional societies become more complex and differentiated. For economic development to occur, modernization theory predicts, countries must change their traditional attitudes, values, and institutions. Economic achievement is thought to derive from attitudes and values that emphasize hard work, saving, efficiency, and enterprise. These values are said by the theory to be found in modern (developed) countries, but lacking in traditional societies. Modernization theory suggests that nations remain underdeveloped when traditional customs and culture discourage individual achievement and kin relations dominate.

As an outgrowth of functionalist theory, modernization theory derives some of its thinking from the work of Max Weber. In *The Protestant Ethic and the Spirit of Capitalism* (1958 [1904]), Weber saw the economic development that occurred in Europe during the Industrial Revolution as a result of the values and attitudes of Protestantism. The Industrial Revolution took place in England and northern Europe, Weber argued, because the people of this area were hardworking Protestants who valued achievement and believed that God helped those who helped themselves.

Modernization theory is similar to the argument of the culture of poverty, which sees people as poor because they have poor work habits, engage in poor time management, are not willing to defer gratification, and do not save or take advantage of educational opportunities (see Chapter 7). Countries are poor, in other words, because they have poor attitudes and poor institutions.

Modernization theory can partially explain why some countries have become successful. Japan is an example of a country that has made huge strides in economic development in part because of a national work ethic (McCord and McCord 1986). But the work ethic alone does not explain Japan's success. In sum, modernization theory may partially explain the value context in which some countries become successful and others do not, but it is not a substitute for explanations that also look at the economic and political context of national development. It also rests on an arrogant perspective that the United States and other more economically developed nations have superior values compared to other nations. Critics point out that this perspective blames countries for being poor when other causes of their status in the world may be outside their control. Whether a country develops or remains poor may be the result of other countries exploiting the less powerful. Modernization theory does not sufficiently take into account the interplay and relationships between countries that can affect a country's economic or social condition.

Developing countries, modernization theory says, are better off if they let the natural forces of competition guide world development. Free markets, according to this perspective, will result in the best economic order. But, as critics argue, markets do not develop independently of government's influence. Governments can spur or hinder economic development, especially as they work with private companies, to enact export strategies, restrict imports, or place embargoes on the products of nonfavored nations.

Dependency Theory

Although market-oriented theories may explain why some countries are successful, they do not explain why some countries remain in poverty or why some countries have not developed. It is necessary to look at issues outside the individual countries and to examine the connections between them. Drawing on the fact that many of the poorest nations are former colonies of European powers, another theory of world stratification focuses on the processes and results of European colonization and imperialism. This theory, called **dependency theory,** focuses on explaining the persistence of poverty in the world. It holds that the poverty of the low-income countries is a direct result of their political and economic dependence on the wealthy countries. Specifically, dependency theory argues that the poverty of many countries is a result of exploitation by powerful countries. This theory is derived from the work of Karl Marx, who foresaw that a capitalist world economy would create an exploited class of dependent countries, just as capitalism within countries had created an exploited class of workers.

Dependency theory begins by examining the historical development of this system of inequality. As the European countries began to industrialize in the 1600s, they needed raw materials for their factories and they needed places to sell their products. To accomplish this, the European nations colonized much of the world, including most of Africa, Asia, and the Americas. **Colonialism** is a system by which Western nations became wealthy by taking raw materials from colonized societies and reaping profits from products finished in the homeland. Colonialism worked best for the industrial countries when the colonies were kept undeveloped to avoid competition with the home country. For example, India was a British colony from 1757 to 1947. During that time, Britain bought cheap cotton from India, made it into cloth in British mills, and then sold the cloth back to India, making large profits. Although India was able to make cotton into cloth at a much cheaper cost than the British, and very fine cloth at that, the British nonetheless did not allow India to develop its cotton industry. As long as India was dependent on Britain, Britain became wealthy and India remained poor.

Under colonialism, dependency was created by the direct political and military control of the poor countries by powerful developed countries. Most colonial powers were European countries, but other countries, particularly Japan and China, had colonies as well. Colonization came to an end soon after the Second World War, largely because it became too expensive to maintain large armies and administrative staffs in distant countries. As a result, according to dependency theory, the powerful countries turned to other ways to control the poor countries and keep them dependent. The powerful countries still intervene directly in the affairs of the dependent nations by sending troops or, more often, by imposing economic or political restrictions and sanctions. But other methods, largely economic, have been developed to control the dependent poor countries, such as price controls, tariffs, and, especially, the control of credit.

The rich industrialized nations, according to dependency theory, are able to set prices for raw material produced by the poor countries at very low levels so that the poor countries are unable to accumulate enough profit to industrialize. As a result, the poor, dependent countries must borrow from the rich countries. However, debt creates only more dependence. Many poor countries are so deeply indebted to the major industrial countries that they must follow the economic edicts of the rich countries that loaned them the money, thus increasing their dependency. This form of international control has sometimes been called **neocolonialism,** a form of control of the poor countries by the rich countries but without direct political or military involvement.

Multinational corporations are companies that draw a large share of their profits from overseas investments and that conduct business across national borders. They play a role in keeping the dependent nations poor, dependency theory suggests. Although their executives and stockholders are from the industrialized countries, multinational corporations recognize no national boundaries and pursue business where they can best make a profit. Multinationals buy resources where they can get them cheapest, manufacture their products where production and labor costs are lowest, and sell their products where they can make the largest profits.

Many critics fault companies for perpetuating global inequality by taking advantage of cheap overseas labor to make large profits for U.S. stockholders. Companies are, in fact, doing what they should be doing in a market system: trying to make a profit. Nonetheless, dependency theory views the practices of multinationals as responsible for maintaining poverty in the poor parts of the world.

One criticism of dependency theory is that many poor countries were never colonies, for example, Ethiopia. Some former colonies have also done well. Two of the greatest postwar success stories of economic development are Singapore and Hong Kong. Both of these countries were British colonies—Hong Kong until 1997—and clearly dependent on Britain, yet they have had successful economic development precisely because of their dependence on Britain. Other former colonies are also improving economically, including India.

The gap between the rich and poor worldwide can be staggering. At the same time that many struggle for mere survival, others enjoy the pleasantries of a gentrified lifestyle.

World Systems Theory

Modernization theory examines the factors internal to an individual country, and dependency theory looks to the relationship between countries or groups of countries. Another approach to global stratification is called **world systems theory.** Like the dependency theory, this theory begins with the premise that no nation in the world can be considered in isolation. Each country, no matter how remote, is tied in many ways to the other countries in the world. However, unlike dependency theory, world systems theory argues that there is a world economic system that must be understood as a single unit, not in terms of individual countries or groups of countries. This theoretical approach derives to some degree from the work of the dependency theorists and is most closely associated with the work of Immanuel Wallerstein in *The Modern World System* (1974) and *The Modern World System II* (1980). According to this theory, the level of economic development is explained by understanding each country's place and role in the world economic system.

This world system has been developing since the sixteenth century. The countries of the world are tied together in many ways, but of primary importance are the economic connections in the world markets of goods, capital, and labor. All countries sell their products and services on the world market and buy products and services from other countries. However, this is not a market of equal partners. Because of historical and strategic imbalances in this economic system, some countries are able to use their advantage to create and maintain wealth, whereas other countries that are at a disadvantage remain poor. This process has led to a global system of stratification in which the units are not people, but countries.

World systems theory sees the world divided into three groups of interrelated nations: core or first-world countries, semiperiperial or second-world countries, and peripheral or third-world countries. This world economic system has resulted in a modern world in which some countries have obtained great wealth and other countries have remained poor. The core countries control and limit the economic development in the peripheral countries so as to keep the peripheral countries from developing and competing with them on the world market; thus the core countries can continue to purchase raw materials at a low price.

Although world systems theory was originally developed to explain the historical evolution of the world system, modern scholars now focus on the international division of labor and its consequences. This approach is an attempt to overcome some of the shortcomings in world systems theory by focusing on the specific mechanism by which differential profits are attached to the production of goods and services in the world market. A tennis shoe made by Nike is designed in the United States; uses synthetic rubber made from petroleum from Saudi Arabia; is sewn in Indonesia; is transported on a ship registered in Singapore, which is run by a Korean management firm using Filipino sailors; and is finally marketed in Japan and the United States. At each of these stages, profits are taken, but at very different rates.

thinking sociologically

What are the major industries in your community? Find out in what parts of the world they do business, including where their product is produced. How does the *international division* of labor affect jobs in your region?

World systems theorists call this global production process a **commodity chain,** the network of production and labor processes by which a product becomes a finished commodity. By following a commodity through its production cycle and seeing where the profits go at each link of the chain, one can identify which country is getting rich and which country is being exploited.

The development of a global economy results in much international migration as diverse groups seek opportunities in new parts of the world. One result is the presence of immigrant enclaves in world cities, such as Pakistani immigrants in London.

© David Bacon/The Image Works

Public debates over immigration policy have mobilized many who point out that immigration has long been a part of our national heritage.

World systems theory helps explain the growing phenomenon of international migration. An international division of labor means that the need for cheap labor in some of the industrial and developing nations draws workers from poorer parts of the globe. International migration is also the result of refugees seeking asylum from war-torn parts of the world or from countries where political oppression, often against particular ethnic groups, forces some to leave. The development of a world economy, however, is resulting in large changes in the composition of populations around the globe. **World cities,** that is, those that are closely linked through the system of international commerce, have emerged. Within these cities, families and their surrounding communities often form *transnational communities,* that is, communities that may be geographically distant but socially and politically close. Linked through various communication and transportation networks, transnational communities share information, resources, and strategies for coping with the problems of international migration.

International migration, sometimes legal, sometimes not, has radically changed the racial and ethnic composition of populations not only in the United States, but also in many European and Asian nations (Rodriquez 1999; Light et al. 1998). Some of those who migrate internationally are professional workers, but many others remain in the lowest segments of the labor force where, although their work is critical to the world economy, they are treated with hostility and suspicion, discriminated against, and stereotyped as undeserving and threatening. In many nations, including the United States, this has led to numerous political tensions over immigration, even while the emergence of migrant groups in world cities is now a major feature of the urban landscape (White 1998).

There are a number of criticisms of world systems theory. Certainly, it is useful to see the world as an interconnected set of economic ties between countries and to understand that these ties often result in the exploitation of poor countries. However, it is not at all clear that the system always works to the advantage of the core countries and to the detriment of the peripheral countries. For one, countries that were once at the center of this world system no longer occupy such a lofty position—England, for example. In addition, the world economic system does not always work to the detriment of the peripheral countries and the benefit of core countries. Peripheral countries often benefit by housing low-wage factories, and the core countries are sometimes hurt when jobs move overseas. Also, low-wage sweatshops are found in all nations, not just the peripheral countries. Nonetheless, world systems theory has provided a powerful tool for understanding global inequality.

Consequences of Global Stratification

It is clear that some nations are wealthy and powerful and some are poor and powerless. What are the consequences of this world stratification system? Table 8.2 shows some of the basic indicators of national well-being for selected nations. You can see that there are considerable differences in the quality of life in these different places in the world.

Population

One of the biggest differences in rich and poor nations is population. The poorest countries comprise three billion people—over half the world's population (World Bank 2007). The poorest countries also have the highest birthrates and the highest death rates. The total *fertility rate,* how many live births a woman will have over her lifetime at current fertility rates, shows that in the poorest countries women on the average have almost five children. Because of this high fertility rate, the populations of poor countries are growing faster than the populations of wealthy countries; these countries therefore also have a high proportion of young children.

In contrast, the richest countries have a total population of approximately one billion people—only 15 percent of the world's population. The populations of the richest countries are not growing nearly as fast as the populations of the poorest countries. In the richest countries, women have about two children over their lifetime, and the populations of these countries are growing by only 1.2 percent. Many of the richest countries, including most of the countries of Europe, table 8.2 *Quality of Life: A Comparative Perspective*

	Life Expectancy (years)	Infant Mortality (per 1,000 births)	Adult Literacy (% of population over 15)	Child Malnutrition (% under weight)	Access to Safe Water (percent of population)
Afghanistan	45 yrs	165	29%	n/a	40%
Iran	71	32	77	n/a	94
Iraq	61	115	60	12%	81
Mexico	75.1	22.6	91	n/a	97
United States	77.9	6.9	95	1	100%

n/a = not available.
Source: World Bank. 2007. **www.worldbank.org**; U.S. Census Bureau. 2006. *Statistical Abstract of the United States.* Washington, DC: U.S. Department of Commerce.

are actually experiencing population declines. With a low fertility rate, the rich countries have proportionately fewer children, but they also have proportionately more elderly, which can also be a burden on societal resources. Different from the poorest nations, the richest ones are largely urban.

Rapid population growth as a result of high fertility rates can make a large difference in the quality of life of the country. Countries with high birthrates are faced with the challenge of having too many children and not enough adults to provide for the younger generation. Public services, such as schools and hospitals, are strained in high-birthrate countries, especially because these countries are poor to begin with. However, very low birthrates, as many rich countries are now experiencing, can also lead to problems. In countries with low birthrates, there are often not enough young people to meet labor force needs and workers must be imported from other countries.

Although the data clearly show that poor countries have large populations and high birthrates and rich countries have smaller populations and low birthrates, does this mean that the large population results in the low level of wealth of the country or that high fertility rates keep countries poor?

Scholars are divided on the relationship between the rate of population growth and economic development (Cassen 1994; Demeny 1991). Some theorize that rapid population growth and high birthrates lead to economic stagnation and that too many people keep a country from developing, thus miring the country in poverty (Ehrlich 1990). However, other researchers point out that some countries with very large populations have become developed (Coale 1986). After all, the United States has the third largest population in the world at 281 million people, yet it is one of the richest and most developed nations in the world. China and India, the two nations in the world with the largest populations, are also showing significant economic development. Scholars now believe that even though in some situations large population and high birthrates can impede economic development, in general, fertility levels are affected by levels of industrialization, not the other way around. That is, as countries develop, their fertility levels decrease and their population growth levels off (Hirschman 1994; Watkins 1987).

Health and Environment

Significant differences are also evident in the basic health standards of countries, depending on where they are in the global stratification system. The high-income countries have lower childhood death rates, higher life expectancies, and fewer children born underweight. In addition, most of the people in the high-income countries, but not all, have clean water and access to adequate sanitation. People born today in wealthy countries can expect to live about seventy-seven years, and women outlive men by several years. Except for some isolated or poor areas of the rich countries, almost all people have access to clean water and acceptable sewer systems.

In the poorest countries the situation is completely different. Many children die within the first five years of life, people live considerably shorter lives, and fewer people have access to clean water and adequate sanitation. In the low-income countries, the problems of sanitation, clean water, childhood death rates, and life expectancies are all closely related. In many of the poor countries, drinking water is contaminated from poor or nonexistent sewage treatment. This contaminated water is then used to drink, to clean eating utensils, and to make baby formula. For adults, the waterborne illnesses such as cholera and dysentery sometimes cause severe sickness, but

Increased awareness of the impact of globalization has generated a protest movement with an unusual alliance between those concerned about the loss of jobs and those concerned with the impact of globalization on the environment.

seldom result in death. However, children under age 5, and especially those under the age of 1, are highly susceptible to the illnesses carried in contaminated water. A common cause of childhood death in countries with low incomes is dehydration brought on by the diarrhea contracted by drinking contaminated water.

Degradation of the environment is a problem that affects all nations, which are linked in one vast environmental system. But global stratification also means that some nations suffer at the hands of others. Overdevelopment is resulting in deforestation. The depletion of this natural resource is most severe in South America, Africa, Mexico, and Southeast Asia (World Bank 2007). On the other hand, the overproduction of "greenhouse gas," emission of carbon dioxide from the burning of fossil fuels, is most severe in the United States, Canada, Australia, parts of western Europe and Russia, and, increasingly, China—places that use the most energy.

Although high-income countries have only 15 percent of the world population, together they use more than half of the world's energy. The United States alone uses one-quarter of the world's energy, though it holds only 4 percent of the world's population (see Figure 8.2). Safe water is also crucial; more than a billion people do not have access to safe water. Moreover, water supplies are declining, a problem that will only be exacerbated by population growth and economic development. The World Bank has, in fact, warned that one-half of the world's population will face severe water shortage by the year 2025 (World Bank 2004). Clearly, global stratification has some irreversible environmental effects that are felt around the globe.

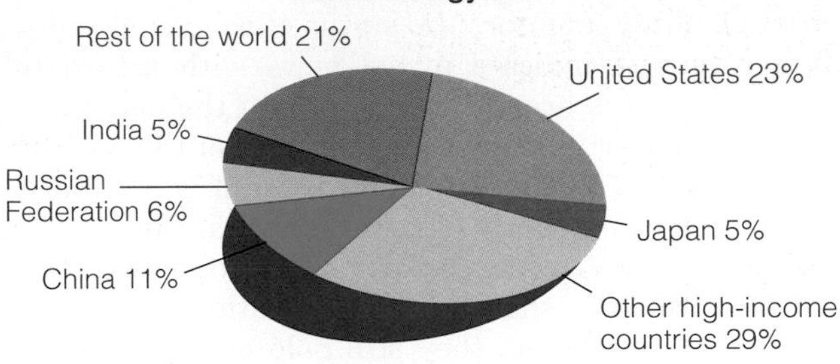

FIGURE 8.2 *Who Uses the World's Energy?*

Source: World Bank. 2004. *World Bank Atlas.* New York: The World Bank. **www.worldbank.org.** Used by permission.

Education and Illiteracy

In the high-income nations of the world, education is almost universal, and the vast majority of people have attended school at least at some level. Literacy and school enrollment are now taken for granted in the high-income nations, although people in these wealthy nations who do not have a good education stand little chance of success. In the middle- and lower-income nations, the picture is quite different. Elementary school enrollment, virtually universal in wealthy nations, is less common in the middle-income nations and even less common in the poorest nations.

How do people survive who are not literate or educated? In much of the world, education takes place outside formal schooling. Just because many people in the poorer countries never go to school, this does not mean that they are ignorant or that they are uneducated. Most of the education in the world takes place in family settings, in religious congregations, or in other settings where elders teach the next generation the skills and knowledge they need to survive. This type of informal education often includes basic literacy and math skills that people in these poorer countries need for their daily lives.

The disadvantage of this informal and traditional education is that, although it prepares people for their traditional lives, it often does not give them the skills and knowledge needed to operate in the modern world. In an increasingly technological world, this can perpetuate the underdeveloped status of some nations.

Gender Inequality

The position of a country in the world stratification system also affects gender relations within different countries. Poverty is usually felt more by women than by men. Although gender inequality has not been achieved in the industrialized countries, compared with women in other parts of the world, women in the wealthier countries are much better off.

The United Nations is one of the organizations that carefully monitors the status of women globally. Their reports indicate mixed news with regard to women's status around the world. On the one hand, women's poverty has declined in some of the nations where it has been extreme, particularly in India, China, and some parts of Latin America. But in sub-Saharan Africa, women's poverty has increased. And though women's share of representation in governments has increased, they still hold only 16 percent of parliamentary seats worldwide. Also, although women have achieved near equity in levels of primary education, there are large gaps in the status of women and men in secondary and higher education—a fact that has huge implications for the work women do in a global economy that increasingly demands educational attainment (United Nations 2005a).

Perhaps most distressing is the global extent of violence against women. Violence takes many forms, including violence within the family, rape, sexual harassment, sex trafficking and prostitution, and state-based violence, among other things. The United Nations has concluded that, "Violence against women persists in every country in the world as a pervasive violation of human rights and a major impediment to achieving gender inequality" (United Nations 2006a: 9). Several factors put women at risk of violence—factors that range from individual level risk factors (such as a history of abuse as a child and substance abuse) to societal level factors, such as gender roles that entrench male dominance and societal norms that tolerate violence as a means of conflict resolution (see Table 8.3). Clearly, the inequalities that mark global stratification have particularly deleterious effects for the world's women.

War and Terrorism

The consequences of global stratification are also found in the international conflicts that bring war and an increased risk of terrorism. Although global inequality is certainly not the only cause of such problems, it contributes to the instability of world peace and the threat of terrorism. Global stratification generates inequities in the distribution of power between nations. Moreover, globalization has created a world-based capitalist class with unprecedented wealth and power. This is a class that now crosses national borders, thus some have defined it as a "transnational capitalist class" (Langman and Morris 2002). Coupled with the enormous poverty that exists, the visibility of this class and its association with Western values leads to resentment and conflict. Furthermore, attempts by wealthier nations to control access to the world's natural resources, such as oil, generate much political conflict. Thus, the same power and affluence that makes the United States a leader throughout the world makes it a target by those who resent its dominance.

In the Middle East, for example, oil production has created prosperity for some and exposed people in these nations to the values of Western culture. When people from different nations, such as those in the Middle East, study at U.S. universities and travel on business or vacations, they are exposed to Western values and patterns of consumption. As one commentator has noted, "Even those who have remained at home have not escaped exposure to Western culture. In most of the countries of the modern Middle-East western cultural influences are pervasive. They see western television programs, they watch western movies, they listen to western music, frequently wear

table 8.3 Risk Factors for Violence against Women: A Global Analysis

The United Nations has studied the frequent use of violence against women in the world and identified the factors that put women at risk. These factors are found at various levels.

Individual Level:	*Community Level:*
• Frequent use of alcohol and drugs • Membership in marginalized communities • Low educational or economic status • History of abuse as a child • Witnessing marital violence in the home	• Women's isolation and lack of social support • Community attitudes that tolerate and legitimate male violence • High levels of social and economic inequality, including poverty
Family/Relationships Level:	***Societal Level:***
• Male control of wealth • Male control of decision making • History of marital violence • Significant disparities in economic, educational, or employment status	• Gender roles that entrench male dominance and women's subordination • Tolerance of violence as a means of conflict resolution • Inadequate laws and policies to prevent and punish violence • Limited awareness and sensitivity on the part of officials and social service providers

Source: United Nations. 2006a. *In-Depth Study on All Forms of Violence against Women.* New York: United Nations.

Global conflicts, including war, can result from the inequality that global stratification produces.

western clothes and visit western web sites. Even western foods are locally available. McDonald's are now found in many of the major cities" (Bailey 2003: 341). Moreover, the sexual liberalism of Western nations and the relative equality of women also add to the volatile mix of nations clashing (Norris and Inglehart 2002).

As a result, some traditional leaders, including religious clerics, define Western culture as a source of degeneracy. Countries such as the United States, where consumerism is rampant, then become the target of those who see this as a threat to their traditional way of life (Ehrlich and Liu 2002). In this sense, global stratification and the dominance of Western culture are inseparable (Bailey 2003). Understood in this way, terrorism is not just a question of clashing religious values (although that is a contributing factor), but it also stems from the global dominance of some nations over others. This is why those who commit atrocious acts, like flying jets into the World Trade Center towers, can define themselves as fighting for a righteous cause.

Terrorism can be defined as premeditated, politically motivated violence perpetrated against noncombatant targets by persons or groups who use their action to try to achieve their political ends (White 2002). Terrorism can be executed through violence or threats of violence and can be executed through various means—suicide bombs, biochemical terror, cyberterror, or other means. Because terrorists operate outside the bounds of normative behavior, it is very difficult to prevent. Although rigid safeguards can be put in place, such safeguards also threaten the freedoms that are characteristics of open, democratic societies. The fact that terrorism is so difficult to stop contributes to the fear that it is intended to generate.

Inequality is also connected to the context in which terrorism emerges. A study of Al Qaeda terrorists finds that the leaders tend to come from middle-class backgrounds, though they often use those who are young, poorly educated, and economically disadvantaged to carry out suicide missions. Families of suicide bombers often receive large cash payments; at the same time they can feel they have served a sacred cause (Stern 2003). The fact that one-third of Iraqis now live in poverty—a change from having a thriving largely middle-class economy in the 1970s and 1980s—helps explain the high rates of violence within Iraq now (United Nations News Center 2007). This also suggests that improving the lives of those who now feel collectively humiliated could provide some protection against terrorism.

World Poverty

One fact of global inequality is the growing presence and persistence of poverty in many parts of the world. There is poverty in the United States, but very few people in the United States live in the extreme levels of deprivation found in some of the poor countries of the world, as seen in Map 8.2.

In the United States, the poverty level is determined by the yearly income for a family of four that is considered necessary to maintain a suitable standard of living. Twelve and a half percent of Americans live in poverty (DeNavas-Walt et al. 2006). This definition of poverty in the United States identifies **relative poverty:** The households in poverty in the United States are poor compared with other Americans, but when one looks at other parts of the world, an income at the U.S. poverty line of $19,874 would make a family very well off.

The United Nations measures world poverty in two ways. **Absolute poverty** is the situation in which people live on less than $1 per day. **Extreme poverty** is defined as the situation in which people live on less than $275 a year, that is, on less than 75 cents a day. There are six hundred million people who live at or below this extreme poverty level.

However, money does not tell the whole story because many people in the poor countries do not always deal in cash. In many countries, people survive by raising crops for personal consumption and by bartering or trading services for food or shelter. These activities do not show up in calculations of poverty levels that use amounts of money as the measure. As a result, the United Nations Development Program also defines what it calls the human poverty index.

The **human poverty index** is a multidimensional measure of poverty, meant to indicate the map 8.2 VIEWING SOCIETY IN GLOBAL PERSPECTIVE

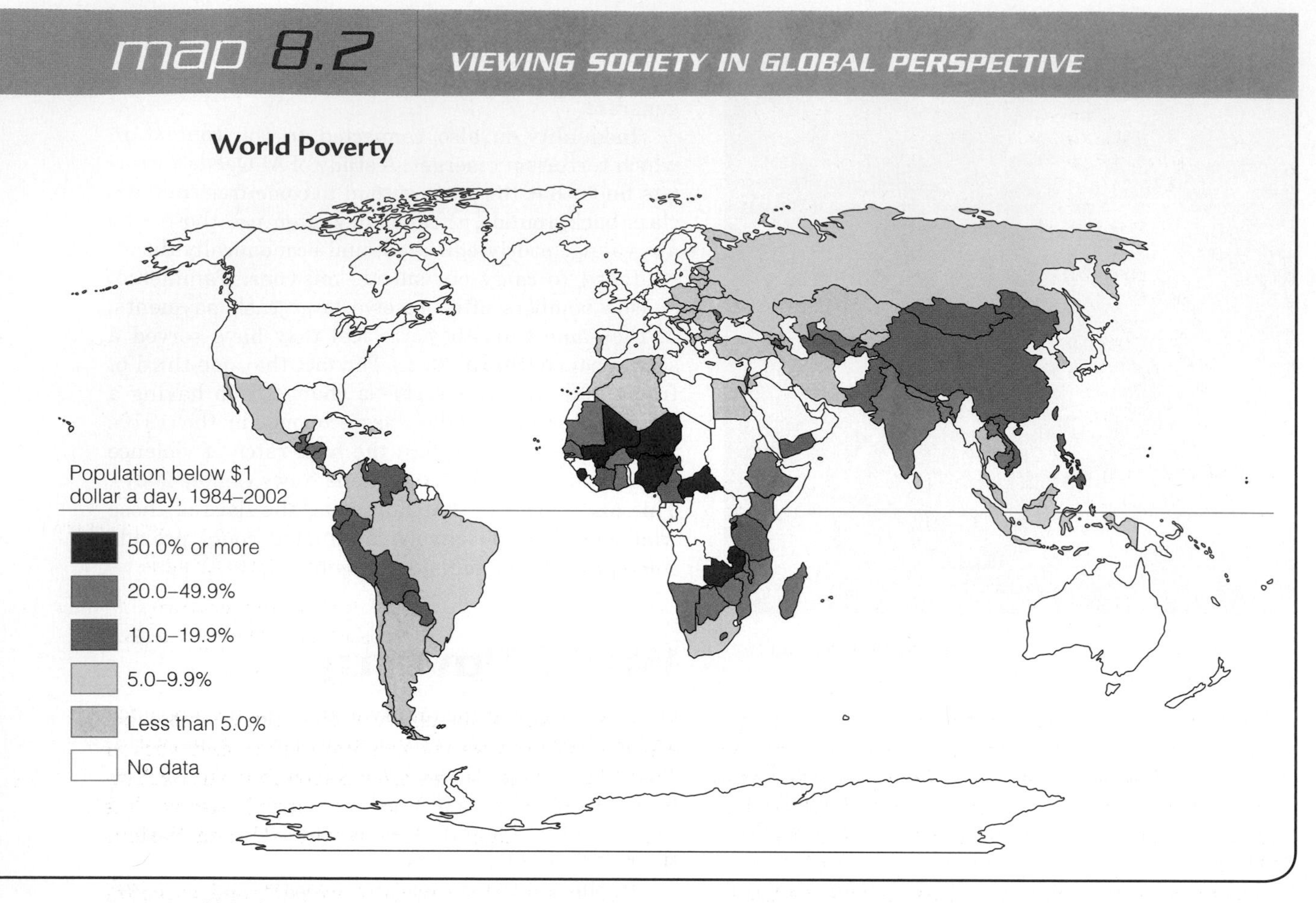

Source: World Bank. 2007. *World Bank Atlas.* Washington, DC: World Bank. **www.worldbank.org.** Used by permission.

degree of deprivation in four basic dimensions of human life: a long and healthy life, knowledge, economic well-being, and social inclusion. Different specific indicators of these different dimensions are used to measure poverty in the industrialized and developing countries because what constitutes these different dimensions of life can vary substantially in such different environments. In developing countries, the indicators are

- the percentage of people born not expected to live to age 40;
- the adult illiteracy rate;
- the proportion of people lacking access to health services and safe water; and
- the percentage of children under age 5 who are moderately or severely underweight.

In industrialized countries, the human poverty index is measured by

- the proportion of people not expected to live to age 60;
- the adult functional illiteracy rate;
- the incidence of income poverty (because income is the largest source of economic provisioning in industrialized countries); and
- long-term unemployment rates.

Figure 8.3 compares the human poverty index in select developing and industrialized nations (United Nations 2000a).

Who Are the World's Poor?

Using the United Nations' definition of absolute poverty (those whose level of consumption falls below $1 per day), one billion people, about one-fifth of the world's population, live in poverty. Another 1.5 billion live on $1–$2 per day, resulting in more than 40 percent of the world's population forming what the United Nations calls a *global underclass.* The good news is that the number of people living in poverty is declining, with most of the reduction attributable to progress in East Asia, particularly in the People's Republic of China. At the same time, poverty in other areas is increasing, including in eastern Europe and central Asia. Sub-Saharan Africa has the highest incidence of poverty of anywhere in the world, despite

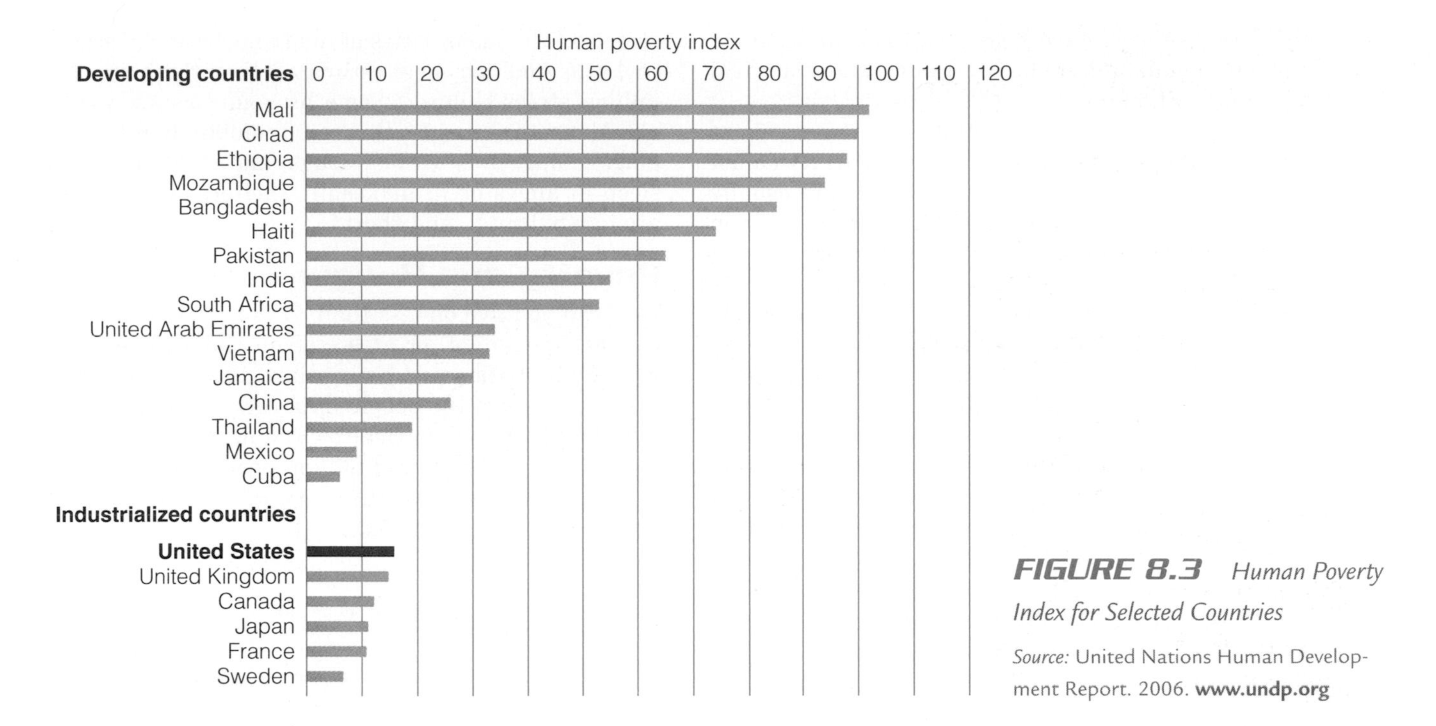

FIGURE 8.3 *Human Poverty Index for Selected Countries*

Source: United Nations Human Development Report. 2006. **www.undp.org**

the rich natural resources of this region. Almost half of the population in this region live in poverty. As a result, infant mortality here is high, life expectancy is low, school enrollment is low, and there are very high death rates due to AIDS (United Nations 2005b).

The character of poverty differs around the globe. In Asia, the pressures of large population growth leave many without sustainable employment. And, as manufacturing has become less labor intensive with more mechanized production, the need for labor in certain industries has declined. Even though new technologies provide new job opportunities, they also create new forms of illiteracy as many people have neither the access nor the skills to use information technology. In sub-Saharan Africa, the poor live in marginal areas where poor soil, erosion, and continuous warfare have created extremely harsh conditions. Political instability and low levels of economic productivity contribute to the high rates of poverty. Solutions to world poverty in these different regions require sustainable economic development, as well as an understanding of the diverse regional factors that contribute to high levels of poverty.

Women and Children in Poverty

There is no country in the world in which women are treated as well as men. As with poverty in the United States, women bear a larger share of the burden of world poverty. Some have called this *double deprivation*—in many of the poor countries women suffer because of their gender and because they disproportionately carry the burden of poverty. For instance, in situations of extreme poverty, women have the burden of taking on much of the manual labor because in many cases the men have left to find work or food. The United Nations concludes that strengthening women's economic security through better work is essential for reducing world poverty (United Nations 2005a).

Because of their poverty, women tend to suffer greater health risks than men. Although women outlive men in most countries, the difference in life expectancy is *less* in the countries in poverty. This is explained by several factors. For one, fertility rates are higher in poor countries. Giving birth is a time of high risk to women, and women in poor countries with poor nutrition, poor maternal care, and the lack of trained birth attendants are at higher risk of dying during and after the birth.

High fertility rates are also related to the degree of women's empowerment in society—an often neglected aspect of the discussion between fertility and poverty. Societies where women's voices do not count for much tend to have high fertility rates as well as other social and economic hardships for women, including lack of education, job opportunities, and information about birth control. Empowering women through providing them employment, education, property, and voting rights can have a strong impact on reducing the fertility rate (Sen 2000).

Women also suffer in some poor countries because of traditions and cultural norms. Most (though not all) of the poor countries are patriarchal, meaning that men control the household. As a result, in some situations of poverty, the women eat after the men, and boys are fed before girls. In conditions of extreme poverty, baby boys may also be fed before baby girls because boys have higher status than girls. As a result, female infants have a lower rate of survival than male infants.

A distressing number of children in the world are also poor (see Figure 8.4). Children in poverty do not have the luxury of an education. Schools are usually few or nonexistent in poor areas of the world, and families are so poor that they cannot afford to send their children to school. Children from a very early age are required to help the family survive by working or performing domestic tasks such as fetching water. In extreme situations, children at a young age work as beggars, young boys and girls are sold to work in sweatshops, and young girls are sold into prostitution by their families. This may seem unusually cruel and harsh by Western standards, but it is difficult to imagine the horror of starvation and the desperation that many families in the world must feel that would force them to take such measures to survive. In poor countries, families feel they must have more children for their survival, yet having more children perpetuates the poverty. The United Nations estimates that there are 211 million children between age 5 and 14 in the paid labor force throughout the world. Most of the children, 127 million, are in Asia, and 48 million are in sub-Saharan Africa (International Labour Organization 2002). Many of these children work long hours in difficult conditions and enjoy few freedoms, making products (soccer balls, clothing, and toys, for example) for those who are much better off.

Another problem in the very poor areas of the world is homeless children (Mickelson 2000). In many situations, families are so poor that they can no longer care for their children, and the children must go without education out on their own, even at young ages. Many of these homeless children end up in the streets of the major cities of Asia and Latin America. In Latin America, it is estimated that there are 13 million street children, some as young as six years old. Alone, they survive through a combination of begging, selling, prostitution, drugs, and stealing. They sleep in alleys or in makeshift shelters. Their lives are harsh, brutal, and short.

Poverty and Hunger

How can you live on less than $1 a day? The answer is that you cannot, or at least you cannot live very well. Malnutrition and hunger are growing problems because many of the people in poverty cannot find or afford food. The World Health Organization estimates that about eight hundred million people in the world are malnourished, which leads to disease (World Health Organization 2000).

Debunking Society's Myths

Myth: There are too many people in the world, and there is simply not enough food to go around.

Sociological perspective: Growing more food will not end hunger. If systems of distributing the world's food were more just, hunger could be reduced.

Hunger results when there is not enough to eat to feed a designated area (such as a region or country). It may be that there is an inadequate supply of food or that households simply cannot afford to purchase enough food to feed themselves. Hunger stifles the mental and physical development of children and leads to disease and death. The trend in the world

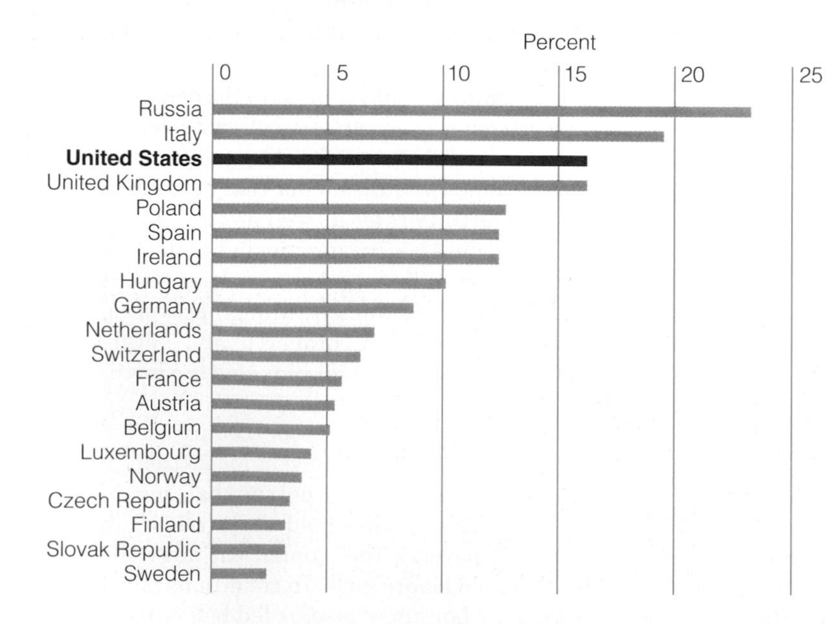

FIGURE 8.4 *Child Poverty in the Wealthier Nations*

Source: Velminckx, Koen, and Timothy Smeeding, eds. 2001. *Child Well-Being, Child Poverty, and Child Policy in Modern Nations.* Tonawanda, NY: University of Toronto Press.

[understanding diversity

War, Childhood, and Poverty

Surgeons were forced to amputate both of Ali Ismaeel Abbas's arms after an errant U.S. bomb slammed into his Baghdad home during the opening phase of the Iraq war. Pictures of the 12-year-old, who lost his parents in the attack, soon appeared on TV screens and in newspapers around the world. Since then, Abbas, who was treated in Kuwait, has come to represent a grim reality: all too often the victims of war are innocent children. (McClelland 2003: 20)

In the past ten years alone, UNICEF estimates that over two million children have died in war, with even more injured, disabled, orphaned, or forced into refugee camps (Machel 1996). One estimate is that, of all the victims of war, 90 percent are civilian—half of those, children (McClelland 2003). Young children are often exploited as combatants or may be used as human shields.

In the aftermath of war children are also highly vulnerable to outbreaks of disease. In Iraq, following the war in 2003, many children died of diseases such as anemia and diarrhea—diseases that can be prevented. Children in Iraq were already living under extreme hardship under the regime of Saddam Hussein. Economic sanctions against Iraq during his regime also produced high infant mortality because of food shortages.

The United Nations has passed resolutions prohibiting the use of children under age 18 in combat. It has linked the threats to children from violence with high rates of poverty around the world. Although reducing poverty would not eliminate the threat of war, it would go a long way toward improving children's lives in war-torn regions.

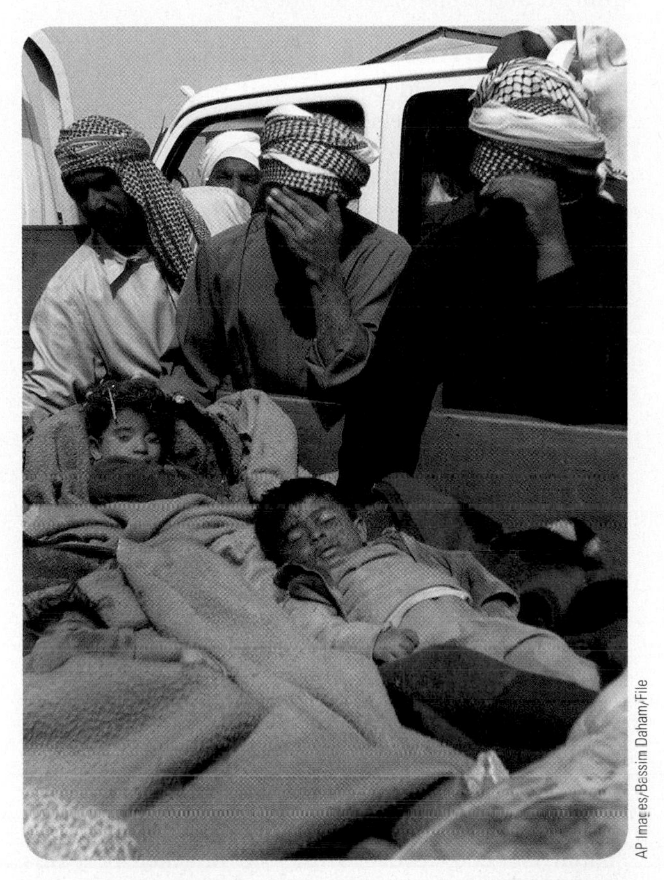

AP Images/Bessim Daham/File

War, though it may seem remote to some, affects millions both in the United States and in war-torn countries. Many of those most affected are children. Here relatives mourn the death of children killed during a U.S. raid in Tikrit, Iraq in 2006.

has been a reduction in the number of malnourished people, with the number of malnourished people expected to drop to under six hundred million by the year 2015. This is encouraging, but it is also short of the goal for reducing hunger that the World Health Organization has hoped for (World Health Organization 2000). Although the food supply is plentiful in the world, and is actually increasing faster than the population, the rate of malnutrition is still dangerously high.

Why are people hungry? Is there not enough food to feed all the people in the world? In fact, plenty of food is grown in the world. The world's production of wheat, rice, corn, and other grains is sufficient to adequately feed all the people in the world. Much grain grown in the United States is stored and not used. The problem is that the surplus food does not get to the truly needy. The people who are starving lack what they need for obtaining adequate food, such as arable land or a job that would pay a living wage.

In many cases, in the past people grew food crops and were able to feed themselves, but today much of the best land has been taken over by agribusinesses that grow cash crops, such as tobacco or cotton, and subsistence farmers have been forced onto marginal lands on the flanks of the desert where conditions are difficult and crops often do not grow.

Causes of World Poverty

What causes world poverty, and why are so many people so desperately poor and starving? More to the point, why is poverty decreasing in some areas but increasing in others? We do know what does *not* cause poverty. Poverty is not necessarily caused by too rapid population growth, although high fertility rates and poverty are related. In fact, many of the world's most populous countries, India and China, for instance, have large segments of their population that are poor, but even with very big populations these countries have begun to reduce poverty levels. Poverty is not caused by people being lazy or uninterested in working. People in extreme poverty work tremendously hard just to survive, and they would work hard at a job if they had one. It is not that they are lazy; it is that there are no jobs for them.

Poverty is a result of a mix of causes. For one, the areas where poverty is increasing have a history of unstable governments or, in some cases, virtually no effective government to coordinate national development or plans that might alleviate extreme poverty and starvation. World relief agencies are reluctant to work in or send food to countries where the national governments cannot guarantee the safety of relief workers or the delivery of food and aid to where it should go. Food convoys may be hijacked or roads blocked by bandits or warlords.

In many countries with high proportions of poverty, the economies have collapsed and the governments have borrowed heavily to remain afloat. As a condition of these international loans, lenders, including the World Bank and the International Monetary Fund, have demanded harsh economic restructuring to increase capital markets and industrial efficiency. These economic reforms may make good sense for some and may lead these countries out of economic ruin over time, but in the short run, these imposed reforms have placed the poor in a precarious position because the reforms also called for drastically reduced government spending on human services.

Poverty is also caused by changes in the world economic system. Although poverty has been a long-term problem and has many causes, increases in poverty and starvation in Africa and Latin America can be attributed in part to the changes in world markets that favored Asia economically but put sub-Saharan Africa and Latin America at a disadvantage. As the price of products declined with more industrialization in places such as India, China, Indonesia, South Korea, Malaysia, and Thailand, commodity-producing nations in Africa and Latin America suffered. In Latin America, the poor have flooded to the cities, hoping to find work, whereas in Africa, they did the opposite, fleeing to the countryside hoping to be able to grow subsistence crops. Governments often had to borrow to provide help to their citizens. Many governments collapsed or found themselves in such great debt that they were unable to help their own people. This has created massive amounts of poverty and starvation.

In sum, poverty has many causes. It is now a major global problem that not only affects the billions of people who are living in poverty, it also affects all people on Earth in one way or another. In some areas, poverty rates are declining as some countries begin to improve their economic situation; however, in other areas of the world, poverty is increasing, and countries are sinking into financial, political, and social chaos.

Globalization and Social Change

As we have seen, many countries in the world are well off, but many countries in the world are poor, some very poor. This global stratification system has been created over several centuries, but the conditions of extreme poverty and even starvation that we see in the world system at this time are relatively new. We must ask, is the world getting better or worse? What will happen in the future?

There is some good news. In some areas of the world, particularly east Asia, but also in Latin America, many countries have shown rapid growth and have emerged as developed countries. These countries are sometimes called the **newly industrializing countries (NICs),** and they include South Korea, Malaysia, Thailand, Taiwan, and Singapore. In these countries, individuals have saved and invested, and the governments have invested in social and economic development. Because some of the NICs have large populations, their success demonstrates that economic development can occur in heavily populated countries. China has embarked on an aggressive policy of industrial growth, and India is also improving economically.

Yet for all the success stories, there are also nations that are not making it. These include nations on all continents. In many cases, governments have collapsed or are functioning only at minimal levels, the economy is bankrupt, the standard of living has plummeted, and people are starving. More disturbing is the situation in many areas of the world where the increase in ethnic hatred has led to mass genocide and forced millions from their homes, creating huge numbers of refugees. These situations have increased poverty and hunger. And these countries, some of which used to be relatively well off, have experienced downward mobility in the global stratification system.

There has been continued growth of capitalism and of capital markets around the world. With the collapse of the Soviet Union and with China moving rapidly in the direction of capitalism, a world capitalistic system has developed. The opening of new markets, the increasing global trade, the growth of multinational corporations, and the development of world financial markets will bring prosperity and wealth to many nations and to many individuals. The growing world market economy may allow some of the emerging countries that were once poor to move into the ranks of the rich nations and share in the newly created wealth. However, whether this wealth will filter down to the people at the lower levels of society is another question.

The new push to further develop the world as one large capital market will also leave some countries behind, and therefore poverty and hunger will continue in many parts of the world. Market economies create opportunities to become wealthy, both for individuals and for nations. For those who can take advantage of these opportunities, the future looks bright, but many nations and individuals do not have this option. Their conditions are so desperate that they do not have a chance to participate in the world market, except at a great disadvantage.

Globalization is a strong force that will continue to shape the future of most nations. Some see globalization simply as the expansion of Western markets and culture into all parts of the world. Western civilization brings new values (including democracy and more equality for women) and certainly new products (movies, clothing styles, and other commercial goods) to other nations, but it also can be a form of imperialism—that is, the domination of Western nations. Resistance to Western globalization and imperialism produces some of the international problems now dominating U.S. and world history, as evidenced in the hostility of militant, fundamentalist Islamic groups against the United States.

Globalization has created great progress in the world—including trade, migration, the spread of diverse cultures, the dissemination and sharing of new knowledge, greater freedom for women, travel, and so forth. Moreover, globalization has not simply extended the values and knowledge of Western culture. Many of the things we now take for granted in our culture originated in non-Western cultures. For example, the decimal system—fundamental to modern math and science—originated in India between the second and sixth centuries and was soon further developed by Arab mathematicians. Western societies certainly get credit for the development of science and technology, but the credit is not theirs alone (Sen 2002).

It is no doubt true that globalization is contributing to the inequality between nations and to the exploitation of some nations and groups by others. Perhaps the solution is not in resisting globalization, but in working so that the benefits of the global economy, global technology, and knowledge reach parts of the world in less exploitative ways. As long as great disparities in standards of living, human rights and basic freedoms, environmental quality, and so forth persist, world conflict is likely to be the result.

Chapter Summary

- ***What is global stratification?***

  Global stratification is a system of unequal distribution of resources and opportunities between countries. A particular country's position is determined by its relationship to other countries in the world. The countries in the global stratification system can be categorized according to their per capita gross national product, or wealth. The world's countries can also be categorized as *first-*, *second-*, or *third-world* countries, which describe their political affiliation and their level of development. The global stratification system can also be described according to the economic power countries have.

- ***How do systems of power affect different countries in the world?***

  The countries of the world can be divided into three levels based on their power in the world economic system. The *core countries* are the countries that control and profit the most from the world system. *Semi-peripheral countries* are semi-industrialized and play a middleman role, extracting profits from the poor countries and passing those profits on to the core countries. At the bottom of the world stratification system are the *peripheral countries,* which are poor and largely agricultural, but with important resources that are exploited by the core countries. Most of these nations are populated by people of color, perpetuating racism as part of the world system.

- ***What are the theories of global stratification?***

  Modernization theory interprets the economic development of a country in terms of the internal attitudes and values. Modernization theory ignores that the development of a country may be due to its economic relationships with other more powerful countries. *Dependency theory* draws on the fact that many of the poorest nations are former colonies of European colonial powers that keep colonies poor and do not allow their industries to develop, thus creating dependency. *World systems theory* argues that no nation can be seen in isolation and that there is a world economic system that must be understood as a single unit.

- ***What are some of the consequences of global stratification?***

  The poorest countries have more than half the world's population and have high birthrates, high mortality rates, poor health and sanitation, low rates of literacy and school attendance, and are largely rural. The richest countries have low birthrates, low mortality rates, better health and sanitation, high literacy rates, high school attendance, and largely urban populations. Although women in the wealthy countries are not completely equal to men, they suffer less inequality than do women in the poor countries.

- ***How do we measure and understand world poverty?***

  Relative poverty means being poor in comparison to others. *Absolute poverty* describes the situation where people do not have enough to survive, measured as having the equivalent of $1.00 per day. *Extreme poverty* is defined as the situation in which people live on less than 75 cents a day. The United Nations has developed a *human poverty index*—a multidimensional measure that accounts for life expectancy, knowledge, economic well-being, and social inclusion. Poverty particularly affects women and children. Children in the very poor countries are forced to work at very early ages and do not have the opportunity for schooling. Street children are a growing problem in many cities of the world. Starvation is also a consequence of the global stratification system.

- ***What is the future of global stratification?***

  The future of global stratification is varied and depends on the country's position within the world economic system. Some countries, particularly those in east Asia—commonly referred to as *newly industrializing countries*—have shown rapid growth and emerged as developed countries. Many nations, though, are not making it. Governments collapse, countries suffer economic bankruptcy, the standard of living plummets, and people starve.

Key Terms

absolute poverty 223
colonialism 216
commodity chain 218
core countries 213
dependency theory 216
extreme poverty 223
first-world countries 213
global stratification 210
gross national income (GNI) 210
human development index 210
human poverty index 223
international division of labor 215
modernization theory 216
multinational corporation 217
neocolonialism 217
newly industrializing countries (NICs) 228
peripheral countries 213
power 213
relative poverty 223
second-world countries 213
semiperipheral countries 213
terrorism 223
third-world countries 213
world cities 219
world systems theory 218

Online Resources

Sociology: The Essentials Companion Website

www.thomsonedu.com/sociology/andersen

Visit your book companion website where you will find more resources to help you study and write your research papers. Resources include Suggested Readings, web links, and a MicroCase Online feature that teaches you how to research society. Other resources include Learning Objectives, Internet exercises, quizzing, and flash cards.

is an easy-to-use online resource that helps you study in less time to get the grade you want NOW.

www.thomsonedu.com/login

Need help studying? This site is your one-stop study shop. Take a Pre-Test and Thomson NOW will generate a Personalized Study Plan based on your test results. The Study Plan will identify the topics you need to review and direct you to online resources to help you master those topics. You can then take a Post-Test to determine the concepts you have mastered and what you still need to work on.

9 Chapter nine

CHAPTER NINE
Chapter

Race and Ethnicity

You might expect a society based on the values of freedom, equality, and brotherhood, such as ours, not to be deeply afflicted by racial conflict, but think of the following situations:

- When Hurricane Katrina struck New Orleans and the Gulf Coast in 2006, hundreds of thousands of people were displaced and billions of dollars of property destroyed. Although the hurricane affected the lives (and deaths) of many, African Americans—many of them poor—were disproportionately killed or left homeless. Millions of Americans were shocked by the images of poor people desperate to survive but left without help for a long time.
- Eight days after the horrific, catastrophic September 11, 2001, terrorist attacks on New York City's now-destroyed World Trade Center and on the Pentagon in Washington, DC, a gunman drove into a Chevron gas station in Mesa, Arizona, and shot to death the owner, a member of the Sikh religious order, who wore a turban on his head. The man who was killed had no known connection whatsoever with suspected Middle Eastern terrorists, but he had dark skin and wore a turban. A few days later, more than two hundred Sikhs had reported instances of harassment.

continued

chapter nine

CHAPTER NI

These ugly incidents have one thing in common—race. Along with gender and social class, race has fundamental importance in human social interaction, and it is an integral part of social institutions. Of course, the races do not always interact as enemies. Nor is interracial tension always obvious. It can be as subtle as a White person who simply does not initiate interactions with African Americans and Latinos or an elderly White man who almost imperceptibly leans backward at a cocktail party as a Japanese American man approaches him.

In everyday human interaction, race still matters and matters a lot. What is race, and what is ethnicity? Why does society treat racial and ethnic groups differently, and why is there social inequality between these groups? How are these divisions and inequalities able to persist so stubbornly, and how extensive are they? These questions fascinate sociologists who do research on racial and ethnic relations and stratification in our society. Just as class stratification differentiates people in society according to class privileges and disadvantages they experience, so does racial and ethnic stratification.

This band, playing for St. Patrick's Day, an Irish holiday, contains people of varied racial-ethnic backgrounds.

Race and Ethnicity

Within sociology, the terms *ethnic, race, minority,* and *dominant group* have very specific meanings, different from the meanings these terms have in common usage. These concepts are important in developing a sociological perspective on race and ethnicity.

Ethnicity

An **ethnic group** is a social category of people who share a common culture, for example, a common language or dialect; a common religion; and common norms, practices, customs, and history. Ethnic groups have a consciousness of their common cultural bond. Italian Americans, Japanese Americans, Arab Americans, Polish Americans, Greek Americans, Mexican Americans, and Irish Americans are examples of ethnic groups in the United States. Ethnic groups are also found in other societies, such as the Pashtuns in Afghanistan or the Shiites and Sunnis in Iraq, whose ethnicity is based on religious differences.

An ethnic group does not exist only because of the common national or cultural origins of a group, however. Ethnic groups develop because of their unique historical and social experiences. These experiences become the basis for the group's *ethnic identity,* meaning the definition the group has of itself as sharing a common cultural bond. Prior to immigration to the United States, Italians, for example, did not necessarily think of themselves as a distinct group with common interests and experiences. Originating from different villages, cities, and regions of Italy, Italian immigrants identified themselves by their family background and community of origin. However, the process of immigration and the experiences Italian Americans faced as a group in the United States, including discrimination, created a new identity for the group (Waters and Levitt 2002; Waters 1990; Alba 1990).

The social and cultural basis of ethnicity allows ethnic groups to develop more or less intense ethnic identification at different points in time. Ethnic identification may grow stronger when groups face prejudice or hostility from other groups. Perceived or real threats and perceived competition from other groups may unite an ethnic group around common political and economic interests. Ethnic unity can develop voluntarily, or it may be involuntarily imposed when ethnic groups are excluded by more powerful groups from certain residential areas, occupations, or social clubs. These exclusionary practices strengthen ethnic identity.

Race

Like ethnicity, race is primarily, though not exclusively, a socially constructed category. A **race** is a group treated as distinct in society based on certain characteristics, some of which are biological, that have been assigned social importance. Because of presumed biologically or culturally inferior characteristics (as defined by powerful groups in society), a race is often singled out for differential and unfair treatment. It is not the biological characteristics per se that define racial groups but *how groups have been treated historically and socially.*

Society assigns people to racial categories, such as Black, White, and so on, not because of science, logic, or fact, but because of opinion and social experience. In other words, how racial groups are defined is a *social* process. This is what is meant when one says that race is "socially constructed." Although the meaning of race begins from perceived biological differences between groups (such as differences in physical characteristics like skin color, lip form, and hair texture), on closer examination, the assumption that racial differences are purely biological breaks down.

Debunking Society's Myths

Myth: Racial differences are fixed, biological categories.

Sociological perspective: Race is a social concept, one in which certain physical or cultural characteristics take on social meanings that become the basis for racism and discrimination. The definition of race varies across cultures within a society and across different societies.

The social categories used to presumably divide groups into races are not fixed, and they vary from society to society (Washington 2006; Morning 2005). Within the United States, laws defining who is Black have historically varied from state to state. North Carolina and Tennessee law historically defined a person as Black if he or she had even one great-grandparent who was Black (thus being one-eighth Black). In other southern states, having any Black ancestry at all defined one as a Black person—the so-called "one drop" (that is, of Black blood) rule (Taylor 2006; Malcomson 2000). This "one drop" rule still applies to a great extent today in the United States.

This is even more complex when we consider the meaning of race in other countries. In Brazil, a light-skinned Black person could well be considered White, especially if the person is of high socioeconomic status; this demonstrates that one's race in Brazil is in part actually *defined* by one's social class. Thus, in parts of Brazil, it is often said that "money lightens" *(o dinheiro embranquence)*. In this sense, a category such as social class can become racialized. In fact, in Brazil people are considered Black only if they are of African descent and have no discernible White ancestry at all. The vast majority of U.S. Blacks would not be considered Black in Brazil (Telles 2004; Surratt and Inciardi 1998).

Racialization is a process whereby some social category, such as a social class or nationality, takes on what *society perceives* to be racial characteristics (Malcomson 2000; Harrison 2000; Omi and Winant 1994). The experiences of Jewish people provide a good example of what it means to say that race is a socially constructed category. Jews are more accurately called an *ethnic group* because of common religious and cultural heritage, but in Nazi Germany Hitler defined Jews as a "race." An ethnic group had thus become *racialized*. Jews were presumed to be biologically inferior to the group Hitler labeled the Aryans—white-skinned, blonde, tall, blue-eyed people. On the basis of this definition—which was supported through Nazi law, taught in Nazi schools, and enforced by the Nazi military—Jewish people were grossly mistreated. They were segregated, persecuted, and systematically murdered in what has come to be called the Holocaust during the Second World War.

Mixed-race people defy the biological categories that are typically used to define race. Is someone who is the child of an Asian mother and an African American father Asian or Black? Reflecting this issue, the U.S. Census's current practice is for a person to check several racial categories, rather than just one (Wright 1994; Waters 1990), although considerable controversy has arisen over this procedure (Harrison 2000). As Table 9.1 shows, the U.S. census has dramatically changed its racial and ethnic classifications since 1890, reflecting the fact that society's thinking about racial and ethnic categorization has not remained constant through time (Rodriquez 2006; Harrison 2000; Mathews 1996; Lee 1993).

The biological characteristics that have been used to define different racial groups vary considerably both within and between groups. Many Asians, for example, are actually lighter skinned than many Europeans and White Americans and, regardless of their skin color, have been defined in racial terms as yellow. Some light-skinned African Americans are

Tiger Woods, considered among the greatest golfers ever, lines up a putt. He is of Asian American and African American parentage. What race is he?

table 9.1 Comparison of U.S. Census Classifications, 1890–2000

Census Date	White	African American	Native American	Asian American	Other Categories
1890	White	Black, Mulatto, Quadroon, Octoroon	Indian	Chinese Japanese	
1900	White	Black	Indian	Chinese Japanese	
1910	White	Black Mulatto	Indian	Chinese Japanese	Other
1990	White	Black or Negro	Indian (American) Eskimo Aleut	Chinese Japanese Filipino Korean Asian Indian Vietnamese	Hawaiian Guamanian Samoan Asian or Pacific Islander Other
2000[a, b]	White	Black or African American	American Indian Alaskan Native	Chinese Japanese Filipino Korean Asian Indian Vietnamese	Native Hawaiian Other Pacific Islander Other

[a]In 2000, for the first time ever, individuals could select more than one racial category. Only 2 percent actually did so.
[b]Hispanics are included under "Other."
Source: Lee, Sharon. 1993. "Racial Classification in the U.S. Census: 1890–1990." *Ethnic and Racial Studies* 16(1): 75–94. U.S. Census Bureau. 2003. "Racial and Ethnic Classification Used in Census 2000 and Beyond." Reprinted by permission of Taylor & Francis Ltd. **www.tandf.co.uk.journals**

also lighter in skin color than some White Americans. Developing racial categories overlooks the fact that human groups defined as races are—biologically speaking—much more alike than they are different.

The biological differences that are presumed to define different racial groups are somewhat arbitrary. Why, for example, do we differentiate people based on skin color and not some other characteristic such as height or hair color? You might ask yourself how a society based on the presumed racial inferiority of red-haired people would compare to other racial inequalities in the United States. The likelihood is that if a powerful group defined another group as inferior because of some biological characteristics, and they used their power to create social institutions that treated this group unfairly, a system of racial inequality would result. In fact, *very few biological differences* exist between racial groups. Most of the variability in almost all biological characteristics, such as blood type and various bodily chemicals, is *within* and not between racial groups (Rodriquez 2006; Malcomson 2000; Lewontin 1996).

Different groups use different criteria to define racial groups. To American Indians, being classified as an American Indian depends upon proving one's ancestry, but this proof varies considerably from tribe to tribe. Among some American Indians, one must be able to demonstrate 75 percent American Indian ancestry to be recognized as such; for other American Indians, demonstrating 50 percent American Indian ancestry is sufficient. It also matters who defines racial group membership. The government makes tribes prove themselves as tribes through a complex set of federal regulations (called the "federal acknowledgment process"); very few are actually given this official status, and the criteria for tribal membership as well as definition as "Indian" or "Native American" have varied considerably throughout American history. Thus, as with African Americans, it has been the state or federal government, and not so much the racial or ethnic group *itself,* that has defined who is a member of the group and who is not!

Official recognition by the government matters. For example, only those groups officially defined as Indian tribes qualify for health, housing, and educational assistance from the Bureau of Indian Affairs (the BIA) or are allowed to manage the natural resources on Indian lands and maintain their own system of governance (Locklear 1999; Brown 1993; Snipp 1989).

This definition of race emphasizes that in addition to physical and cultural differences, race is created and maintained by the most powerful group (or groups) in society. This definition of race also incorporates presumed group differences in the context of social and historical experience. As a result, who is defined as a race is as much a political question as a biological one (Brodkin 2006). For example, although they probably did not think of themselves as a race, Irish Americans in the early twentieth century were defined by more powerful White groups as a "race" that was inferior to White people. At that time, Irish people were not considered by many even to be White (Ignatiev 1995)! In fact, a century ago the Irish were called "Negroes turned inside out," while Negroes (Black people) were called "smoked Irish" (Malcomson 2000).

The social construction of race has been elaborated on in an insightful perspective in sociology known as racial formation theory (Omi and Winant 1994). **Racial formation** is the process by which a group comes to be defined as a race. This definition is supported through official social institutions such as the law and the schools. This concept emphasizes the importance of social institutions in producing and maintaining the meaning of race; it also connects the process of racial formation to the exploitation of so-called racial groups. A good example comes from African American history. During slavery, an African American was defined as being three-fifths of a person for the purposes of deciding how slaves would be counted for state representation in the new federal government and how they would be defined as property in order to be taxed as property. The process of defining slaves in this way served the purposes of White Americans, not slaves themselves, and it linked the definition of slaves as a race to the political and economic needs of the most powerful group in society (Higginbotham 1978).

The process of racial formation also explains how groups such as Asian Americans, American Indians, and Latinos have been defined as races, despite the different experiences and nationalities of the groups composing these three categories. Race, like ethnicity, lumps groups together that may have very different historical and cultural backgrounds, but once they are so labeled, the groups are perceived as a single entity. This reflects a more general principle in the social sciences called **out-group homogeneity effect,** where all members of any out-group are perceived to be similar or even identical to each other and differences among them are perceived to be minor or nonexistent. This has recently been the case in the United States with Middle Easterners: Lebanese, Syrians, Iranians, Iraqis, Jordanians, Egyptians, Afghans, and many others are classified as one group and called Middle Easterners or simply "Arabs."

Minority and Dominant Groups

Minorities are racial or ethnic groups, but not all racial or ethnic groups are always considered minorities. Irish Americans, for instance, are not now thought of as minorities, although they once were in the early part of the twentieth century. A **minority group** is any distinct group in society that shares common group characteristics and is forced to occupy low status in society because of prejudice and discrimination. The group that assigns a racial or ethnic group to subordinate status in society is called the **dominant group.**

A group may be classified as a minority on the basis of ethnicity, race, sexual preference, age, or class status, for example. A minority group is not necessarily a numerical minority but is a group that holds low status in relation to other groups in society, regardless of the size of the group. In South Africa, Blacks outnumber Whites ten to one, but until Nelson Mandela's election as president and the dramatic change of the country's government in 1994, Blacks were an officially oppressed and politically excluded social minority under the infamous *apartheid* (pronounced "aparthate" or "apart-hite") system of government. In general, a racial or ethnic minority group has the following characteristics (Simpson and Yinger 1985):

1. The minority group possesses characteristics (such as race, ethnicity, sexual preference, age, or religion) that are popularly regarded as different from those of the dominant group.
2. The minority group suffers prejudice and discrimination by the dominant group.
3. Membership in the group is frequently ascribed rather than achieved, although either form of status can be the basis for being identified as a minority.
4. Members of a minority group feel a strong sense of group solidarity. There is a "consciousness of kind" or "we" feeling. This bond grows from common cultural heritage and the shared experience of being a recipient of prejudice and discrimination.

Debunking Society's Myths

Myth: Minority groups are those with the least numerical representation in society.

Sociological perspective: A minority group is any group, regardless of size, that is singled out in society for unfair treatment and that generally occupies lower status in the society.

Racial Stereotypes

Racial and ethnic inequality is peculiarly resistant to change. Racial and ethnic inequality in society produces racial stereotypes, and these stereotypes become the lens through which members of different groups perceive one another.

Stereotypes and Salience

In everyday social interaction, people tend to categorize other people. The most common bases for such categorizations are race, gender, and age. A person *immediately* identifies a stranger as Black, Asian, Hispanic, White, and so on; as a man or woman; and as a child, adult, or elderly person. Quick and ready categorizations help people process the huge amounts of information they receive about people with whom they have come into contact. People quickly assign others to a few categories, saving themselves the task of evaluating and remembering every little discernible detail about a person. People are taught to treat each person as a unique individual, but research over the years clearly shows that they do not. Instead, people routinely categorize others in some way or another. This allows thinking about people and processing information about them. Just about everyone does this.

A **stereotype** is an oversimplified set of beliefs about members of a social group or social stratum. It is based on the tendency of humans to categorize a person based on a narrow range of perceived characteristics. Stereotypes are presumed, usually incorrectly, to describe the "typical" member of some social group.

Stereotypes based on race or ethnicity are called *racial–ethnic stereotypes.* Here are some common examples: Asian Americans have been stereotyped as overly ambitious, sneaky, and clannish; African Americans often bear the stereotype of being loud, lazy, naturally musical, and so on. Hispanics are stereotyped as lazy, oversexed, and for Hispanic men, macho; Jews have been perceived as materialistic and unethical. Such stereotypes, presumed to describe the "typical" member of a group, are factually inaccurate for the vast majority of members of a group. *No group in U.S. history has escaped the process of categorization and stereotyping, even White groups.* For example, Italians have been stereotyped as overly emotional and prone to crime, the Irish as heavy drinkers and prone to politics, and so on for virtually any group in U.S. history.

The categorization of people into groups and the subsequent application of stereotypes is based on the **salience principle,** which states that we categorize people on the basis of what appears initially prominent and obvious—that is, salient—about them. Skin color is a salient characteristic; it is one of the first things that we notice about someone. Because skin color is so obvious, it becomes a basis for stereotyping. Gender and age are also salient characteristics of an individual and thus serve as notable bases for group stereotyping.

The choice of salient characteristics is culturally determined. In the United States, skin color, hair texture, nose form and size, and lip form and size have become salient characteristics, and these characteristics determine whether we perceive someone as "intelligent" or "stupid," as "attractive" or "unattractive," or even "trustworthy" or "untrustworthy" (Hunt 2005). We then use these features to categorize people in our minds on the basis of race. In other cultures, religion may be far more salient than skin color. In the Middle East, whether one is Muslim or Christian is far more important than skin color. Religion in the Middle East is a salient characteristic and takes considerable priority over race.

thinking sociologically

Observe several people on the street. What are the first things you notice about them (that is, what is *salient*)? Make a short list of these things. Do these lead you to *stereotype* these people? On what are your stereotypes based?

The Interplay among Race, Gender, and Class Stereotypes

Alongside racial and ethnic stereotypes, gender and social class are among the most prominent features by which people are categorized. In our society, there is a complex interplay among racial or ethnic, gender, and class stereotypes.

Among *gender stereotypes,* those based on a person's gender, the stereotypes about women are more likely to be negative than those about men. The "typical" woman has been traditionally stereotyped as subservient, overly emotional, and talkative, inept at math and science, and so on. Many of these are *cultural stereotypes.* They are conveyed and supported by the cultural media—music, TV, magazines, art, and literature. Men, too, are painted in crude strokes, although usually not as negatively as women. Men in the media are stereotyped as macho, insensitive, and pigheaded and are portrayed in situation comedies as inept. Generally, men are depicted as wanting to have sex with as many women as possible in the shortest time available.

Social class stereotypes are based on assumptions about social class status. Upper-class people are stereotyped (by middle- and lower-class people) as snooty, aloof, condescending, and phony. Some of the stereotypes held about the middle class (by both the upper class and the lower class) are that they are overly ambitious, striving, and obsessed with "keeping up with the Joneses." Finally, stereotypes about lower-class people abound: They are perceived by the upper and middle classes as dirty, lazy, unmotivated, violent, and so on.

The principle of **stereotype interchangeability** holds that stereotypes, especially negative ones, are often interchangeable from one social class to another, from one racial or ethnic group to another, from a racial or ethnic group to a social class, or from a social class to a gender. Stereotype interchangeability is sometimes revealed through humor. Ethnic jokes often interchange different groups as the butt of the humor, stereotyping them as dumb and inept. Take the stereotype of African Americans as inherently lazy. This stereotype has also been applied in recent history to Hispanic, Polish, Irish, and other groups. It has even been applied generally to those people perceived as lower class. In fact, "laziness" is often used to explain *why* someone is lower class or poor.

Middle-class people are more likely to attribute the low status of a lower-class person to something *internal,* such as "inherent" laziness or lack of willpower (Morlan 2005; Krasnodemski 1996; Worchel et al. 2000). Lower-class people are more likely to attribute their status to discrimination or poor opportunities—that is, to an *external* societal factor (Morlan 2005; Krasnodemski 1996; Kluegel and Bobo 1993; Bobo and Kluegel 1991).

The same kinds of stereotypes have historically been applied to women. Many of the stereotypes applied to women in literature and the media—they are childlike, overly emotional, unreasonable, bad at mathematics, and so on—have also been applied to African Americans, lower-class people, the poor, and earlier in the twentieth century, Chinese Americans. A common theme is apparent: Whatever group occupies the lowest social status in society at a given time (whether racial or ethnic minorities, women, or lower-class people) is negatively stereotyped, and often the same negative stereotypes are used between and among these groups, thus demonstrating stereotype interchangeability. The stereotype is then used as an "explanation" for the observed behavior of a stereotyped group's member to justify his or her lower status in society. This in turn subjects the stereotyped group to prejudice, discrimination, and racism, which will now be discussed.

Prejudice, Discrimination, and Racism

Many people use the terms *prejudice, discrimination,* and *racism* loosely, as if they were all the same thing. Typically, in common parlance, people also think of these terms as they apply to individuals, as if the major problems of race were the result of individual people's bad will or biased ideas. Sociologists use more refined concepts to understand race and ethnic relations, distinguishing carefully between prejudice, discrimination, and racism.

Prejudice

Prejudice is the evaluation of a social group and the individuals within it based on conceptions about the social group held despite facts that disprove them; the beliefs involve both prejudgment and misjudgment (Allport 1954; Jones 1997). Prejudices are usually defined as negative predispositions or as evaluations that are rarely positive. Thinking ill of people only because they are members of group X is prejudice.

A negative prejudice against someone not in one's own group is often accompanied by a positive prejudice in favor of someone who *is* in one's own group. Thus, the prejudiced person will have negative attitudes about a member of an *out-group* (any group other than one's own) and positive attitudes about someone simply because he or she is in one's *in-group* (any group a person considers to be one's own).

Most people disavow racial or ethnic prejudice, yet the vast majority of us carry around some prejudices, whether about racial–ethnic groups, men and women, old and young, upper class and lower class, or straight and gay. Virtually no one is free of prejudice. Five decades of research have shown definitively that people who are more prejudiced are also more likely to stereotype and categorize others by race or ethnicity or by gender than those who are less prejudiced (Adorno et al. 1950; Jones 1997; Taylor et al. 2006).

Prejudice based on race or ethnicity is called *racial* or *ethnic prejudice.* If you are a Latino and dislike an Anglo only because he or she is White, then this constitutes prejudice: It is a negative judgment or prejudgment based on race and ethnicity and very little else. If the Latino individual attempts to justify these feelings by arguing that "all Whites have the same bad character," then the Latino is using a stereotype as justification for the prejudice. Note that prejudice can be held by any group against another group.

Prejudice is also revealed in the phenomenon of **ethnocentrism,** which was examined in Chapter 2 on culture. Ethnocentrism is the belief that one's group is superior to all other groups. The ethnocentric person feels that his or her own group is moral, just, and right and that an out group—and thus any member of that out-group—is immoral, unjust, wrong, distrustful, or criminal. The ethnocentric individual uses his or her own in-group as the standard against which all other groups are compared.

Prejudice and Socialization. Where does racial–ethnic prejudice come from? How do moderately or highly prejudiced people end up that way? People are not born with stereotypes and prejudices. Research shows that these attitudes are learned and internalized through the socialization process, including both *primary socialization* (family, peers, teachers) as well as *secondary socialization* (such as the media). Children imitate the attitudes of their parents, peers, and teachers. If the parent complains about "Japs taking away jobs from Americans," then the child grows up thinking negatively about the Japanese, including Japanese Americans. Attitudes about race are formed early in childhood, at about age three or four (Allport 1954; Feagin 2000; Van Ausdale and Feagin 1996). There is a very close correlation between the racial and ethnic attitudes of parents and those of their children. The more ethnically or racially prejudiced the parent, the more ethnically or racially prejudiced the child will be.

Major vehicles for the communication of racial–ethnic attitudes to both young and old are the media, especially television, magazines, newspapers, and books. For many decades, African Americans, Hispanics, Native Americans, and Asians were rarely presented in the media and then only in negatively stereotyped roles. The Chinese were shown in movies, magazines, and early television in the 1950s as bucktoothed buffoons who ran shirt laundries. Japanese Americans were depicted as sneaky and untrustworthy. Hispanics were shown as either ruthless banditos or playful, happy-go-lucky people who took long siestas. American Indians were presented as either villains or subservient characters like the Lone Ranger's famed sidekick, Tonto. Finally, there is the drearily familiar portrayal of the Black person as subservient, lazy, clowning, and bug-eyed, a stereotypical image that persisted from the nineteenth century all the way through the 1950s and early 1960s.

Discrimination

Discrimination is overt negative and unequal treatment of the members of some social group or stratum solely because of their membership in that group or stratum. Prejudice is an attitude; discrimination is overt behavior. *Racial–ethnic discrimination* is unequal treatment of a person on the basis of race or ethnicity.

Discrimination in housing has been a particular burden on minorities. Many studies have been able to reproduce the situation, showing that when two persons identical in nearly all respects (age, education, gender, social class, and other characteristics) present themselves as potential tenants for the same housing, if one is White and the other is a minority, the minority person will often be refused housing by a White landlord and the otherwise identical White applicant will not. A minority landlord who refuses housing to a White person while granting it to a minority person of similar social characteristics is also discriminating, but reverse discrimination of this sort is far less frequent and far less of a problem in society (Feagin 2000; Feagin and Vera 1995; Feagin and Feagin 1993).

Prejudice is an attitude; discrimination is behavior. The discrimination affecting the nation's minorities takes a number of forms—for example, income discrimination and discrimination in housing. Discrimination in employment and promotion and discrimination in education (in Chapter 13) are two other forms of discrimination.

Although the median income of Black and Hispanic families has increased somewhat since 1950, the income gap between these two groups and Whites has remained virtually unchanged since 1967, as can be seen from Figure 9.1. Furthermore, per capita income since 1967 has grown at a somewhat faster rate for Whites. Yet even these median income figures tell only part of the story. Poverty among Blacks decreased from 1960 to 1970, but the poverty level has been somewhat steady since. The current poverty rate (the percent below the poverty level; see Figure 9.2) is highest for African Americans (24 percent) and Hispanics (22 percent) compared with Whites (10 percent) or Asians (11 percent). In all these racial groups, children have the highest rate of poverty.

Discrimination in housing is illegal under U.S. law. Nonetheless, banks and mortgage companies often withhold mortgages from minorities based on "redlining," an illegal practice in which an entire minority neighborhood is designated as "no loan." Racial segregation may also be fostered by *gerrymandering,* the calculated redrawing of election districts, school districts, and similar political boundaries in order to maintain racial segregation. As a result, **residential segregation,** the spatial separation of racial and ethnic groups into different residential areas, called "American apartheid" by Massey and Denton (1993; Frankenberg and Lee 2006), continues to be a reality in this country.

Racism

Racism includes both attitudes and behaviors. A negative attitude taken toward someone simply because he or she belongs to a racial or ethnic group is a prejudice, as has already been discussed. An attitude or prejudice is what you think and feel; a behavior is what you do. **Racism** is the *perception and treatment* of a racial or ethnic group, or member of that group, as intellectually, socially, and culturally inferior to

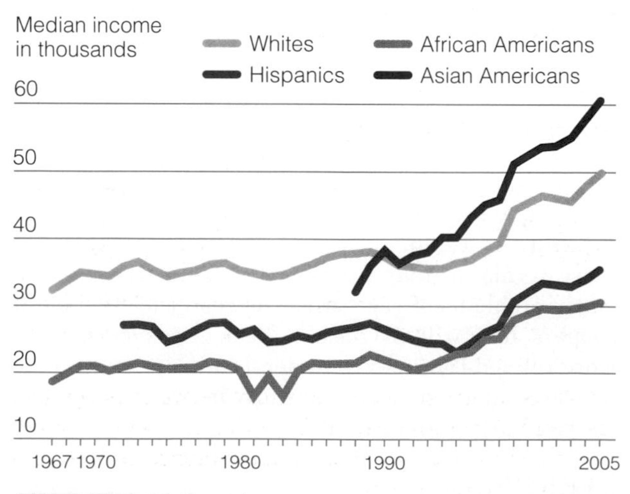

FIGURE 9.1 *The Income Gap*

Source: DeNavas-Walt, Carmen, Bernadette D. Proctor, and Cheryl Hill Lee. 2005. *Income Poverty and Health Insurance Coverage in the United States: 2005.* Washington, DC: U.S. Census Bureau. **www.census gov**

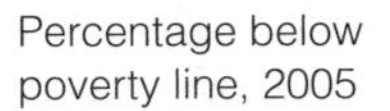

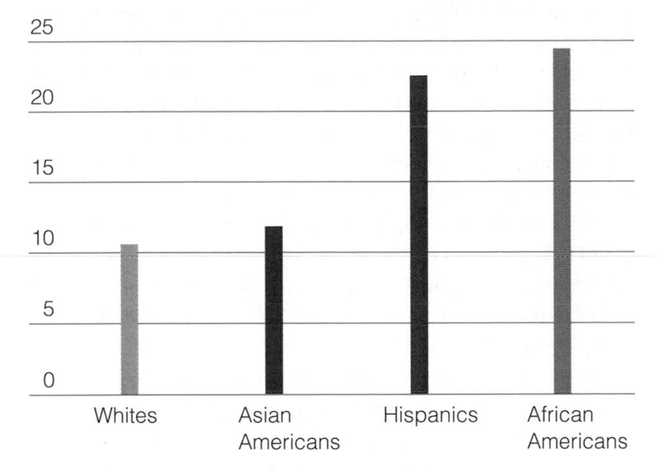

FIGURE 9.2 *Poverty Among Racial Groups*

Source: DeNavas-Walt, Carmen, Bernadette D. Proctor, and Cheryl Hill Lee. 2005. *Income Poverty and Health Insurance Coverage in the United States: 2005*. Washington, DC: U.S. Census Bureau. **www.census gov**

one's own group. It is more than an attitude; it is institutionalized in society. Racism involves negative attitudes that are sometimes linked with negative behavior.

There are different **forms of racism.** Obvious, overt racism, such as physical assaults, from beatings to lynchings, has often been called **old-fashioned racism,** or *traditional racism* (or *Jim Crow racism*), by researchers (for example, Hunter 2002; Bobo 1999). This form of racism has declined somewhat in our society since the 1950s, though it certainly has not disappeared (Schumann et al. 1997). Racism can also be subtle, covert, and nonobvious; this is known as **aversive racism,** another form of racism (Jones 1997). Consistently avoiding interaction with someone of another race or ethnicity is an example of aversive racism. This form of racism is quite common and has remained at roughly the same level for more than thirty years, with perhaps a slight increase (Schumann et al. 1997; Kovel 1970; Dovidio and Gaertner 1986; Katz et al. 1986).

After the Second World War and during the 1950s, a shift to **laissez-faire racism** occurred. This type of racism—also called *symbolic racism* by some (Taylor et al. 2006)—involves several elements:

1. The subtle but persistent negative stereotyping of minorities, particularly Black Americans, especially in the media.
2. A tendency to blame Blacks themselves for the gap between Blacks and Whites in socioeconomic standing, occupational achievement, and educational achievement.
3. Clear resistance to meaningful policy efforts (such as affirmative action, discussed later) designed to ameliorate America's racially oppressive social conditions and practices.

The last element is rooted in perceptions of threat to maintaining the status quo (Bobo 1999; Bobo and Smith 1998).

A close relative of laissez-faire racism is **color-blind racism**—so named because the individual affected by this type of racism prefers to ignore legitimate racial–ethnic, cultural, and other differences and insists that the race problems in America will go away if only race is ignored all together. *Accompanying this belief is the opinion that race differences in America are merely an illusion and that race is not real.* Simply refusing to perceive any differences at all between racial groups (thus, being color-blind) is in itself a form of racism (Bonilla-Silva 2001). This will come as a surprise to many. These types of racism do not necessarily involve explicit or purposeful intent on the part of the nonminority individual to harm the minority person.

Institutional racism as a form of racism is the negative treatment and oppression of one racial or ethnic group *by society's existing institutions* based on the presumed inferiority of the oppressed group. It is a form of racism that exists at the level of social structure and is in Durkheim's sense *external* to the individual—thus institutional. Key to understanding institutional racism is seeing that dominant groups have the economic and political power to subjugate the minority group, *even if they do not have the explicit intent* of being prejudiced or discriminating against others. Power, or lack thereof, accrues to groups because of their position in social institutions, not just because of individual attitudes or behavior. The power that resides in society's institutions can be seen in such patterns as persistent economic inequality between racial groups, which is reflected in high unemployment among minorities, lower wages, and different patterns of job placement (Bobo 1999; Bonilla-Silva 1997).

Racial profiling, already discussed in Chapter 6, is an example of institutional racism in the criminal justice system. African American and Hispanic persons are arrested considerably more often than Whites and Asians. In fact, an African American or Hispanic wrongdoer is more likely to be arrested than a White person who commits the *exact same crime,* even when the White person shares the same age, socioeconomic environment, and prior arrest record as the Black or Hispanic.

Institutional racism is also seen in educational institutions, such as when schools assign Blacks and Latinos to the lower cognitive ability tracks than Whites with the *same* test scores, as will be seen in Chapter 13. In these instances, racism is a characteristic of the institutions and not necessarily of the individuals within the institution. *This is why institutional racism can exist even without prejudice being the cause.*

Consider this: Even if every White person in the country lost all of his or her personal prejudices and even if he or she stopped engaging in individual acts

© Tony Savino/Corbis Sygma

Minorities are more likely to be arrested than Whites for the same offense. Does this reflect institutional racism rather than any individual prejudice of the arresting police officer?

Debunking Society's Myths

Myth: The primary cause of racial inequality in the United States is the persistence of *prejudice.*

Sociological perspective: Prejudice is one dimension of racial problems in the United States, but institutional racism can flourish even while prejudice is on the decline.

of discrimination, institutional racism would still persist for some time. Over the years, it has become so much a part of U.S. institutions (hence, the term *institutional racism*) that discrimination can occur even when no single person is deliberately causing it. Existing at the level of social structure instead of at the level of individual attitude *or* behavior, it is external to the individual personality and is thus a *social fact* of the sort sociological theorist Emile Durkheim observed (Chapter 1).

Theories of Prejudice and Racism

Why do prejudice, discrimination, and racism exist? Two categories of theories have been advanced. The first category consists of psychological theories about prejudice. The second category consists of sociological theories of racism, including institutional racism as well as prejudice and discrimination.

Psychological Theories of Prejudice

Two traditional psychological theories of prejudice are the scapegoat theory and the theory of the authoritarian personality. **Scapegoat theory** argues that, historically, members of the dominant group in the United States have harbored various frustrations in their desire to achieve social and economic success (Feagin and Feagin 1993). As a result of this frustration, they vent their anger in the form of aggression. This aggression is directed toward some substitute that takes the place of the original perception of the frustration. Members of minority groups become these substitutes, that is, the scapegoats. The psychological principle that aggression often follows frustration (originally from the frustration-aggression hypothesis of Dollard et al. [1939]) is central to the scapegoat principle. For example, a White person who perceived that he or she was denied a job because "too many" Mexican immigrants were being permitted to enter the country would be using Mexican Americans as a scapegoat if he or she felt negatively (thus prejudiced) toward a specific Mexican American person, even if that person did not have the job in question and had nothing at all to do with the White person not getting the job.

The second theory, an older one, argues that individuals who possess an authoritarian personality are more likely to be prejudiced against minorities than are nonauthoritarian individuals. The **authoritarian personality** (after Adorno et al. [1950], who coined the term) is characterized by a tendency to rigidly categorize other people, as well as inclinations to submit to authority, strictly conform, be very intolerant of ambiguity, and be inclined toward superstition. The authoritarian person is more likely to stereotype or categorize another and thus readily places members of minority groups into convenient and oversimplified categories or stereotypes. There is some research that links strong authoritarianism with high religious orthodoxy and extreme varieties of political conservatism (Bobo and Kluegel 1991; Altemeyer 1988).

Sociological Theories of Prejudice and Racism

Current sociological theory focuses more on explaining the existence of racism, particularly institutional racism, although speculation about the existence of prejudice is also a component. The three sociological theoretical perspectives considered throughout this text have bearing on the study of racism, discrimination, and prejudice: functionalist theory, symbolic interaction theory, and conflict theory.

Functionalist Theory. Functionalist theory argues that for race and ethnic relations to be functional and thus contribute to the harmonious conduct and stability of society, racial and ethnic minorities and

© Reuters/Jason Reed/Landov

The opening of the National Museum of the American Indian in 2004 (part of the Smithsonian Institution in Washington, DC) was cause for celebration among diverse groups of Native Americans, as well as others.

women must assimilate into that society. **Assimilation** is a process by which a minority becomes socially, economically, and culturally absorbed within the dominant society. The assimilation perspective assumes that to become fully fledged members of society, minority groups must adopt as much of the dominant society's culture as possible, particularly its language, mannerisms, and goals for success, and thus give up much of its own culture. Assimilationism stands in contrast to racial—cultural **pluralism**—the maintenance and persistence of one's culture, language, mannerisms, practices, art, and so on.

Symbolic Interaction Theory. Symbolic interaction theory addresses two issues: first, the role of social interaction in reducing racial and ethnic hostility, and second, how race and ethnicity are socially constructed. Symbolic interactionism asks, What happens when two people of different racial or ethnic origins come into contact with each other, and how can such interracial or interethnic contact reduce hostility and conflict? **Contact theory,** which originated with the psychologist Gordon Allport (Allport 1954; Cook 1988) argues that interaction between Whites and minorities will reduce prejudice within both groups—but only if three conditions are met:

1. The contact must be between individuals of equal status; the parties must interact on equal ground. A Hispanic cleaning woman and the wealthier White woman who employs her may interact, but their interaction will not reduce prejudice. Instead, their interaction is more likely to perpetuate stereotypes and prejudices on the part of both.
2. The contact between equals must be sustained; short-term contact will not decrease prejudice. Bringing Whites together with Hispanics, Blacks, and Native Americans for short weekly meetings will not erase prejudice. Daily contact on the job between individuals of *equal job status* will tend to remove prejudice.
3. Social norms favoring equality must be agreed upon by the participants. Having African Americans and White skinheads interact on a TV talk show, such as the *Jerry Springer Show,* will probably not decrease prejudice; this interaction might well increase it.

Conflict Theory. The basic premise of conflict theory is that class-based conflict is an inherent and fundamental part of social interaction. To the extent that racial and ethnic conflict is tied to class conflict, conflict theorists argue that class inequality must be reduced to lessen racial and ethnic conflict in society.

The current "class versus race" controversy in sociology (reviewed in more detail later in this chapter and in Chapter 7) concerns the question of whether class (namely, economic differences between races) or race (caste differences between races) is more important in explaining inequality and its consequences or whether they are of equal importance. Those focusing primarily upon class conflict, such as sociologist William Julius Wilson (1978, 1987, 1996), argue that class and changes in the economic structure are sometimes more important than race in shaping the life chances of different groups. Wilson argues that being disadvantaged in the United States is more a matter of class, although he sees this clearly linked to race. Sociologists focusing primarily on the role of race (Bonilla-Silva and Baiocchi 2001; Feagin 2000; Bonilla-Silva 1997; Feagin and Feagin 1993; Willie 1979) argue the opposite: They say that race has been and is relatively more important than class—though class is still important—in explaining and accounting for inequality and conflict in society and that directly addressing the question of race forthrightly is the only way to solve the country's race problems (see Table 9.2).

A recent variety of the conflict perspective is the *intersection perspective.* This perspective refers to the interactive or combined effects of racism, classism (elitism), and gender in the oppression of people. An implication of this theory is that the position of women in society should be studied separately within each racial or ethnic group and within each social class group, because the observed differences between women and men are not the same within the different racial or class groups. This perspective notes that not only are the effects of gender and race intertwined, but also both are intertwined with the effects of class. Class conflict is seen as an integral component of gender and race differences in this society, according to the intersection perspective (Andersen and Collins 2004; Collins 1998, 1990).

table 9.2 Comparing Sociological Theories of Race and Ethnicity

	Functionalism	Conflict Theory	Symbolic Interaction
The Racial Order	Has social stability when diverse racial and ethnic groups are assimilated into society	Is intricately intertwined with class stratification	Is based on social construction that assigns groups of people to diverse racial and ethnic categories
Minority Groups	Are assimilated into dominant culture as they adopt cultural practices and beliefs of the dominant group	Have life chances that result from the opportunities formed by the intersection of class, race, and gender	Form identity as the result of sociohistorical change
Social Change	Is a slow and gradual process as groups adapt to the social system	Is the result of organized social movements and other forms of resistance to oppression	Is dependent on the different forms of interaction that characterize intergroup relations

Diverse Groups, Diverse Histories

The different racial and ethnic groups in the United States have arrived at their current social condition through histories that are similar in some ways, yet quite different in other respects. Their histories are related because of a common experience of White supremacy, economic exploitation, and political disenfranchisement.

An historical perspective on each group follows, which will aid in understanding how prejudice, discrimination, and racism has operated throughout the history of U.S. society (see Map 9.1).

© Brown Brothers

Jewish immigrants are questioned at Ellis Island, the point of entry to the United States for many early European immigrants

Native Americans

The exact size of the indigenous population in North America at the time of the Europeans' arrival with Columbus in 1492 has been estimated at anywhere from one million to ten million people. Native Americans were here tens of thousands of years before they were "discovered" by Europeans. Discovery quickly turned to conquest, and in the course of the next three centuries, the Europeans systematically drove the Native Americans from their lands, destroying their ways of life and crushing various tribal cultures. Native Americans were subjected to an onslaught of European diseases. Lacking immunity to these diseases, Native Americans suffered a population decline, considered by some to have been the steepest and most drastic of any people in the history of the world. Native American traditions have survived in many isolated places, but what is left is only a ghostly echo of the original five hundred nations of North America (Nagel 1996; Thornton 1987; Snipp 1989).

At the time of European contact in the 1640s, there was great linguistic, religious, governmental, and economic heterogeneity among Native American tribes. Most historical accounts have underestimated the degree of cultural and social variety, however. Between the arrival of Columbus in the Caribbean in 1492 and the establishment of the first thirteen colonies in North America in the early 1600s, the ravages of disease and the encroachment of Europeans caused a considerable degree of social disorganization. Sketchy accounts of American Indian cultures by early colonists, fur traders, missionaries, and explorers underestimated the great social heterogeneity among the various American Indian tribal groups and the devastating effects of the European arrival on Indian society.

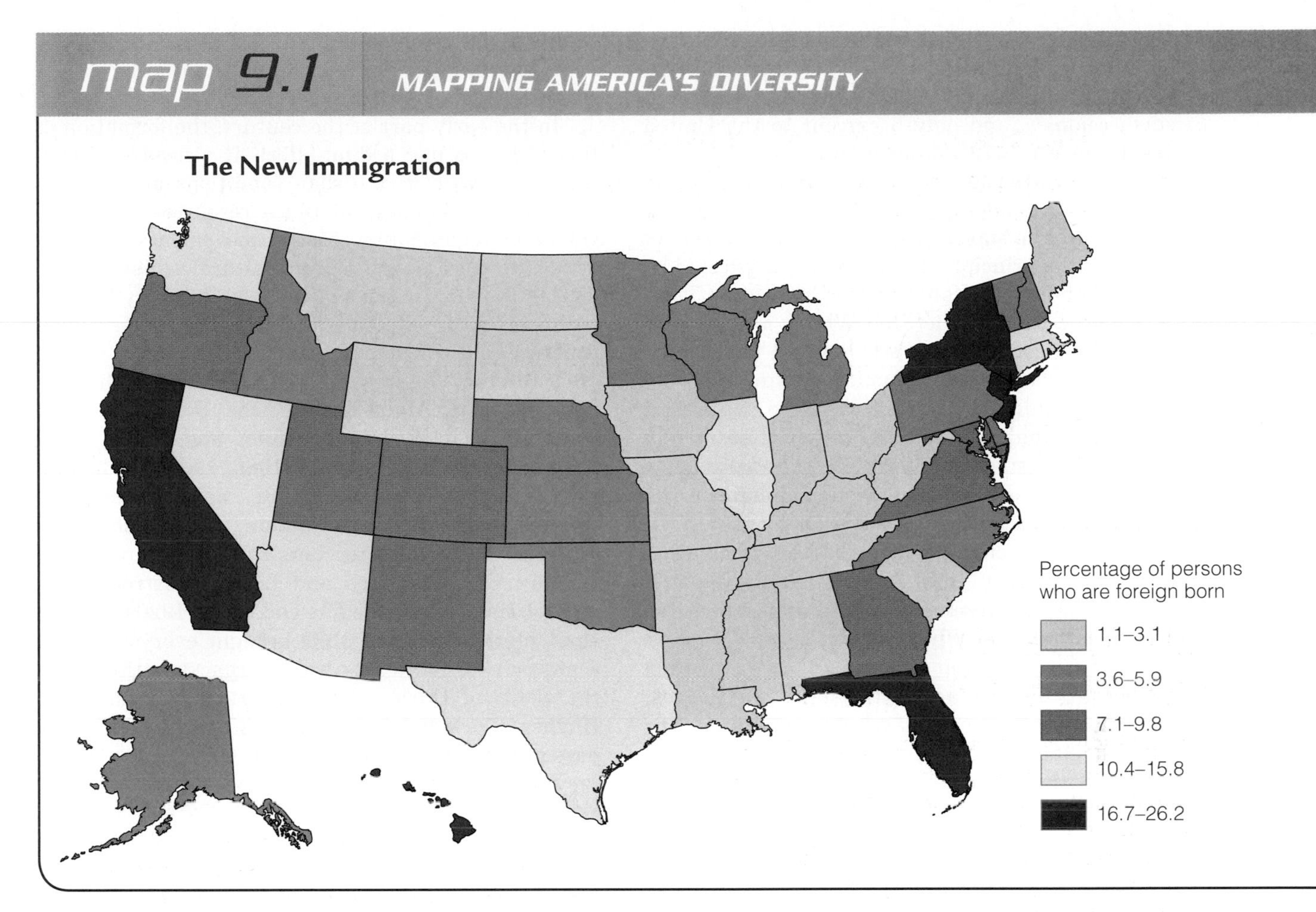

This map shows the total number of immigrants per state. Some states have a high number of immigrants (for example, California, Florida, New York), while other states have fewer immigrants (for example, Wyoming, South Dakota, and Vermont). Where does your state fall in the number of immigrants to the total population?

Data: U.S. Census Bureau. 2003. *American Fact Finder, Foreign-Born Population.* **www.census.gov.**

By 1800, the number of Native Americans had been reduced to a mere 600,000, and wars of extermination against the Indians were being conducted in earnest. Fifty years later, the population had fallen by another half. Indians were killed defending their land, or they died of hunger and disease when taking refuge in inhospitable country. In 1834, 4000 Cherokee died on a forced march from their homeland in Georgia to reservations in Arkansas and Oklahoma, a trip memorialized as the Trail of Tears. The Sioux were forced off their lands by the discovery of gold and the new push of European immigration. Their reservation was established in 1889 (quite recently in history), and they were designated as wards, subjecting them to capricious and humiliating governmental policies. The following year the U.S. Army mistook Sioux ceremonial dances for war dances and moved in to arrest the leaders. A standoff exploded into violence, during which federal troops killed 200 Sioux men, women, and children at the infamous Wounded Knee massacre.

Today about 55 percent of all Native Americans live on or near a reservation, which is land set aside by the U.S. government for their exclusive use. The other 45 percent live in or near urban areas (U.S. Census Bureau 2006; Snipp 1989). The reservation system has served the Indians poorly. Many Native Americans now live in conditions of abject poverty, deprivation, and alcoholism, and they suffer massive unemployment (more than 50 percent among males—extraordinarily high). They are at the lowest rung of the socioeconomic ladder, with the highest poverty rate. The first here in this land are now last in status, a painful irony of U.S. history.

African Americans

The development of slavery in the Americas is related to the development of world markets for sugar and tobacco. Slaves were imported from Africa to provide the labor for sugar and tobacco production and to enhance the profits of the slaveholders. It is estimated that somewhere between twenty and one hundred million

Africans were transported under appalling conditions to the Americas—38 percent went to Brazil; 50 percent to the Caribbean; 6 percent to Dutch, Danish, and Swedish colonies; and only 6 percent to the United States (Genovese 1972; Jordan 1969).

Slavery evolved as a *caste system* in which one caste, the slaveholders, profited from the labor of another caste, the slaves. Central to the operation of slavery was the principle that human beings could be *chattel* (or property). As an economic institution, slavery was based on the belief that Whites are superior to other races, coupled with a belief in a patriarchal social order. The social distinctions maintained between Whites and Blacks were castelike, with rigid categorization and prohibitions, rather than merely classlike, which suggests more pliant social demarcations. Vestiges of this caste system remain in the United States to this day.

The slave system also involved the domination of men over women. In this combination of patriarchy and White supremacy, White males presided over their property of White women as well as their property of Black men and women. This in turn led to gender stratification among the slaves themselves, which reflected the White slaveholder's assumptions about the relative roles of men and women. Black women performed domestic labor for their masters and their own families. White men further exerted their authority in demanding sexual relations with Black women (Blassingame 1973; Raboteau 1978; White 1985; Davis 1981). The predominant attitude of Whites toward Blacks was paternalistic. Whites saw slaves as childlike and incapable of caring for themselves. The more recent stereotypes of African Americans as childlike are directly traceable to the system of slavery.

There exists a widespread belief that slaves passively accepted slavery. Scholarship shows this to be false. Instead, the slaves struggled to preserve both their culture and their sense of humanity and to resist, often by open conflict, the dehumanizing effects of a system that defined human beings as mere property (Myers 1998; Blassingame 1973). Slaves revolted against the conditions of enslavement in a variety of ways, from passive means such as work slowdowns and feigned illness to more aggressive means such as destruction of property, escapes, and outright rebellion.

After slavery was presumably ended by the Civil War (1861–1865) and the Emancipation Proclamation (1863), Black Americans continued to be exploited for their labor. In the South, the system of sharecropping emerged, an exploitative system in which Black families tilled the fields for White landowners in exchange for a share of the crop. With the onset of the First World War and the intensified industrialization of society came the Great Migration of Blacks from the South to the urban North. This massive movement, lasting from the late 1800s through the 1920s, significantly affected the status of Blacks in society because there was now a greater potential for collective action (Marks 1989).

In the early part of the century, the formation of Black ghettos had a dual effect. It victimized Black Americans with grim urban conditions and encouraged the development of Black resources, including volunteer organizations, social movements, political action groups, and artistic and cultural achievements. During the 1920s, Harlem in New York City became an important intellectual and artistic oasis for Black America. The Harlem Renaissance gave the nation great literary figures, such as Langston Hughes, Jesse Fausett, Alain Locke, Arna Bontemps, Zora Neale Hurston, Wallace Thurman, and Nella Larsen (Bontemps 1972; Rampersad 1986, 1988; Marks and Edkins 1999). At the same time, many of America's greatest musicians, entertainers, and artists came to the fore, such as musicians Duke Ellington, Billie Holiday, Cab Calloway, and Louis Armstrong, and painter Hale Woodruff. The end of the 1920s and the stock market crash of 1929 brought everyone down a peg or two, Whites as well as Blacks; although in the words of Harlem Renaissance writer Langston Hughes, Black Americans at the time "had but a few pegs to fall" (Hughes 1967).

Latinos

Latino Americans include Chicanos and Chicanas (Mexican Americans), Puerto Ricans, Cubans, and other recent Latin American immigrants to the United States. It also includes Latin Americans who have lived for generations in the United States; they are not immigrants but very early settlers from Spain and Portugal in the 1400s. The population of Latinos has grown considerably over the past few decades, with the largest increase among Mexican Americans. The terms *Hispanic* and *Latino* or *Latina* mask the great structural and cultural diversity among the various Hispanic groups. The use of such inclusive terms also tends to ignore important differences in their respective entries into U.S. society: Mexican Americans through military conquest of the Mexican War (1846–1848); Puerto Ricans through war with Spain in the Spanish–American War (1898); and Cubans as political refugees fleeing since 1959 from the Communist dictatorship of Fidel Castro, which is opposed by the U.S. government (Glenn 2002; Bean and Tienda 1987).

Mexican Americans. Before the Anglo (White) conquest, Mexican colonists had formed settlements and missions throughout the West and Southwest. In 1834, the U.S. government ordered the dismantling of these missions, bringing them under tight governmental control and creating a period known as the golden age of the ranchos. Land then became concentrated into the hands of a few wealthy Mexican ranchers, who had been given large land grants by the Mexican government. This economy created a class system *within* the Chicano community, consisting of the elite ranchers, mission farmers, and government administrators at the top; *mestizos,* who were small farmers and ranchers, as the middle class; a third class of skilled workers; and a bottom class of manual laborers, who were mostly Indians (Mirandé 1985; Maldonado 1997).

With the Mexican-American War of 1846–1848, Chicanos lost claims to huge land areas that ultimately became Texas; New Mexico; and parts of Colorado, Arizona, Nevada, Utah, and California. White cattle ranchers and sheep ranchers enclosed giant tracts of land, thus cutting off many small ranchers, both Mexican and Anglo. It was at this time that Mexicans, as well as early U.S. settlers of Mexican descent, became defined as an inferior race that did not deserve social, educational, or political equality. This is an example of the *racial formation process,* as noted earlier in this chapter (Omi and Winant 1994). Anglos believed that Mexicans were lazy, corrupt, and cowardly, which launched stereotypes that would further oppress Mexicans; these stereotypes were used to justify the lower status of Mexicans and Anglo control of the land that Mexicans were presumed to be incapable of managing (Moore 1976). As has been noted several times in this chapter, stereotyping has been used in this society as a way of *falsely explaining and justifying* the lower social status of society's minorities.

During the twentieth century, advances in agricultural technology changed the organization of labor in the Southwest and West. Irrigation allowed year-round production of crops and a new need for cheap labor to work in the fields. Migrant workers from Mexico were exploited as a cheap source of labor. Migrant work was characterized by low earnings, poor housing conditions, poor health, and extensive use of child labor. The wide use of Mexican migrant workers as field workers, domestic servants, and other kinds of poorly paid work continues, particularly in the Southwest (Amott and Matthaei 1996).

Puerto Ricans. The island of Puerto Rico was ceded to the United States by Spain in 1899. In 1917, the Jones Act extended U.S. citizenship to Puerto Ricans, although it was not until 1948 that Puerto Ricans were allowed to elect their own governor. In 1952, the United States established the Commonwealth of Puerto Rico, with its own constitution. Following the Second World War, the first elected governor launched a program known as Operation Bootstrap, which was designed to attract large U.S. corporations to the island of Puerto Rico by using tax breaks and other concessions. This program contributed to rapid overall growth in the Puerto Rican economy, although unemployment remained high and wages remained low. Seeking opportunity, unemployed farm workers began migrating to the United States. These migrants were

Activities such as this Puerto Rican Day Parade in New York City reflect pride in one's group culture and result in greater cohesiveness of the group.

interested in seasonal work, and thus a pattern of temporary migration characterized the Puerto Ricans' entrance into the United States (Amott and Matthaei 1996; Rodriguez 1989).

Unemployment in Puerto Rico became so severe that the U.S. government even went so far as to attempt a reduction in the population by some form of population control. Pharmaceutical companies experimented with Puerto Rican women in developing contraceptive pills, and the U.S. government actually encouraged the sterilization of Puerto Rican women. One source notes that by 1974 more than 37 percent of the women of reproductive age in Puerto Rico had been sterilized (Roberts 1997). More than one-third of these women have since indicated that they regret sterilization because they were not made aware at the time that the procedure was irreversible.

Cubans. Cuban migration to the United States is recent in comparison with the other Hispanic groups. The largest migration has occurred since the revolution led by Castro in 1959; between then and 1980, more than 800,000 Cubans—one-tenth of the entire island population—migrated to the United States. The U.S. government defined this as a political exodus, facilitating the early entrance and acceptance of these migrants. Many of the first migrants had been middle- and upper-class professionals and landowners under the prior dictatorship of Fulgencio Batista, but they had lost their land during the Castro revolution. In exile in the United States, some worked to overthrow Castro, often with the support of the federal government. Yet many other Cuban immigrants were of modest means and, like other immigrant groups, came seeking freedom from political and social persecution and escape from poverty.

The most recent wave of Cuban immigration came in 1980, when the Cuban government, still under Castro, opened the port of Mariel to anyone who wanted to leave Cuba. In the five months following this action, 125,000 Cubans came to the United States—more than the combined total for the preceding eight years. The arrival of people from Mariel has produced debate and tension, particularly in Florida, a major center of Cuban migration. The Cuban government had previously labeled the people fleeing from Mariel as "undesirable;" some had been incarcerated in Cuba before leaving. They were actually not much different from previous refugees such as the "golden exiles," who were professional and high-status refugees (Portes and Rumbaut 1996). Because the refugees escaping from Mariel had been labeled as undesirables, and because they were forced to live in primitive camps for long periods after their arrival, they have been unable to achieve much social and economic mobility in the United States. In contrast, the earlier Cuban migrants, who were on average more educated and much more settled, have enjoyed a fair degree of success (Portes and Rumbaut 2001, 1996; Amott and Matthei 1996; Pedraza 1996a).

Asian Americans

Like Hispanic Americans, Asian Americans are from many different countries and diverse cultural backgrounds; they cannot be classified as the single cultural rubric Asians. Asian Americans include migrants from China, Japan, the Philippines, Korea, and Vietnam, as well as more recent immigrants from Cambodia and Laos.

Chinese. Attracted by the U.S. demand for labor, Chinese Americans began migrating to the United States during the mid-nineteenth century. In the early stages of this migration, the Chinese were tolerated because they provided cheap labor. They were initially seen as good, quiet citizens, but racial stereotypes turned hostile when the Chinese came to be seen as competing with White California gold miners for jobs. Thousands of Chinese laborers worked for the Central Pacific Railroad from 1865 to 1868. They were relegated to the most difficult and dangerous work, worked longer hours than the White laborers, and for a long time were paid considerably less than the White workers.

The Chinese were virtually expelled from railroad work near the turn of the twentieth century (in 1890–1900) and settled in rural areas throughout the western states. As a consequence, anti-Chinese sentiment and prejudice ran high in the West. This ethnic antagonism was largely the result of competition between the White and Chinese laborers for scarce jobs. In 1882, the federal government passed the Chinese Exclusion Act, which banned further immigration of unskilled Chinese laborers. Like African Americans, the Chinese and Chinese Americans were legally excluded from intermarriage (Takaki 1989). The passage of this openly racist act, which was preceded by extensive violence toward the Chinese, drove the Chinese populations from the rural areas into the urban areas of the West. It was during this period that several Chinatowns were established by those who had been forcibly uprooted and who found strength and comfort within enclaves of Chinese people and culture (Nee 1973).

Japanese. Japanese immigration to the United States took place mainly between 1890 and 1924, after which passage of the Japanese Immigration Act forbade further immigration. Most of these first-generation immigrants, called *Issei,* were employed in agriculture or in small Japanese businesses. Many Issei were from farming families and wished to acquire their own land, but in 1913 the Alien Land Law of California stipulated that Japanese aliens could lease land for only three years and that lands already owned or leased by them could not be bequeathed to heirs. The second generation of Japanese Americans, or *Nisei,* were born in the United States of Japanese-born parents. They became better educated than their parents, lost their Japanese accents, and in general became more "Americanized," that is, culturally assimilated. The third generation, called *Sansei,* became even better educated and assimilated, yet still met with prejudice and discrimination, particularly where Japanese Americans were present in the highest concentrations, as on the West Coast from Washington to Southern California (Takaki 1989; Glenn 1986).

The Japanese suffered the complete indignity of having their loyalty questioned when the federal government, thinking they would side with Japan after the Japanese attack on Pearl Harbor in December 1941, herded them into concentration camps. By executive order of President Franklin D. Roosevelt, much of the West Coast Japanese American population (many of them loyal second- and third-generation Americans) had their assets frozen and their real estate confiscated by the government. A media campaign immediately followed, labeling Japanese Americans "traitors" and "enemy aliens." Virtually all Japanese Americans in the United States had been removed from their homes by August of 1942, and some were forced to stay in relocation camps until as late as 1946. Relocation destroyed numerous Japanese families and ruined them financially (Glenn 1986; Kitano 1976; Takaki 1989).

In 1986, the U.S. Supreme Court allowed Japanese Americans the right to file suit for monetary reparations. In 1987, legislation was passed, awarding $20,000 to each person who had been relocated and offering an official apology from the U.S. government. One is motivated to contemplate how far this paltry sum and late apology could go in righting what many have argued was the "greatest mistake" the United States has ever made as a government.

Filipinos. The Philippine Islands in the Pacific Ocean fell under U.S. rule in 1899 as a result of the Spanish–American War, and for a while Filipinos could enter the United States freely. By 1934, the islands became a commonwealth of the United States, and immigration quotas were imposed on Filipinos. More than 200,000 Filipinos immigrated to the United States between 1966 and 1980, settling in major urban centers on the West and East Coasts. More than two-thirds of those arriving were professional workers; their high average levels of education and skill have eased their assimilation. By 1985, there were more than one million Filipinos in the United States. Within the next thirty years, demographers project that this population will become the largest group of Asian Americans in the United States, including Chinese Americans and Japanese Americans (Winnick 1990).

Koreans. Many Koreans entered the United States in the late 1960s after amendments to the immigration laws in 1965 raised the limit on immigration from the Eastern Hemisphere. The largest concentration of Koreans is in Los Angeles. As much as half of the adult Korean American population is college educated, an exceptionally high proportion. Many of the immigrants were successful professionals in Korea; upon arrival in the United States, though, they have been forced to take on menial jobs, thus experiencing downward social mobility and status inconsistency. This is especially true of those Koreans who migrated to the East Coast. However, nearly one in eight Koreans in the United States today owns a business; many own small greengrocer businesses. Many of these stores are located in predominantly African American communities and have become one among several sources of ongoing conflict between some African Americans and Koreans. This has fanned negative feeling and prejudice on both sides—among Koreans against African Americans and among African Americans against Koreans (Chen 1991).

Vietnamese. Among the more recent groups of Asians to enter the United States have been the South Vietnamese, who began arriving following the fall of South Vietnam to the Communist North Vietnamese at the end of the Vietnam War in 1975. These immigrants, many of them refugees who fled for their lives, numbered about 650,000 in the United States in 1975. About one-third of the refugees settled in California. Many faced prejudice and hostility, resulting in part from the same perception that has dogged many immigrant groups before them—that they were in competition for scarce jobs. A second wave of Vietnamese immigrants arrived after China attacked Vietnam in 1978. As many as 725,000 arrived in America, only to face discrimination in a variety of locations. Tensions became especially heated when the Vietnamese became a substantial competitive presence in the fishing and shrimping industries in the Gulf of Mexico on the Texas shore. Many communities have welcomed them, however, and most Vietnamese heads of households have become employed full time (Winnick 1990; Kim 1993).

Middle Easterners

Since the mid-1970s, immigrants from the Middle East have been arriving in the United States. They have come from countries such as Syria, Lebanon, Egypt, and Iran, and more recently, especially Iraq. Contrary to popular belief, the immigrants speak no single language and follow no singular religion and thus are ethnically diverse. Some are Catholic, some are Coptic Christian, and many are Muslim. Many are from working-class backgrounds, but many were professionals—teachers, engineers, scientists, and other such positions—in their homelands. About sixty-five percent of those residing in this country were born outside the United States; about half are college-educated (Kohut 2007). Like immigrant populations before them, Middle Easterners have formed their own ethnic enclaves in the cities and suburbs of this country as they pursue the often elusive American dream (Abrahamson 2006).

After the terrorist attacks on the World Trade Center and the Pentagon on September 11, 2001, many male Middle Easterners of several nationalities became unjustly suspect in this country and were subjected to severe harassment; racially motivated physical attacks; and as already noted, out-and-out racial profiling, if only because they had dark skin and—as with some—wore a turban on their heads. Most of these individuals, of course, probably had no discernible connection at all with the terrorists. A recent survey shows that *most* Muslims in this country believe that the September 11, 2001, terrorist attacks were indeed the cause of the increased racial harassment and violence against them (Kohut 2007).

White Ethnic Groups

The story of White ethnic groups in the United States begins during the colonial period. White Anglo-Saxon Protestants (WASPs), who were originally immigrants from England and to some extent Scotland and Wales, settled in the New World (what is now North America). They were the first ethnic group to come into contact on a large scale with those people already here—namely, Native American Indians. WASPs came to dominate the newly emerging society earlier than any other White ethnic group.

In the late 1700s, the WASPs regarded the later immigrants from Germany and France as foreigners with odd languages, accents, and customs. Tension between the "old stock" and the foreigners continued through the Civil War era until around 1860, when the national origins of U.S. immigrants began to change (Handlin 1951). Of all racial and ethnic groups in the United States during that time and

© Bob Daemmrich/PhotoEdit

As illustrated by this photograph of Bangladeshi in Austin, immigrant neighborhoods often exemplify both residential segregation and community.

since, only WASPs do not think of themselves as a nationality. The WASPs came to think of themselves as the "original" Americans despite the prior presence of Native American Indians, who in turn the WASPs described and stereotyped as savages. As immigrants from northern, western, eastern, and southern Europe began to arrive, particularly during the mid- to late-nineteenth century, WASPs began to direct prejudice and discrimination against many of these newer groups. Long discriminated against by the male WASP establishment, women began to assert social and political power, challenging the power of male WASPs in the United States. Much of that WASP dominance remains, however, as is evident in their popular use of the terms "race" and "ethnicity" to describe virtually everyone but themselves (Andersen 2006).

There were two waves of migration of White ethnic groups in the mid- and late-nineteenth century. The first stretched from about 1850 through 1880 and included northern and western Europeans: English, Irish, Germans, French, and Scandinavians. The second wave of immigration occurred from 1890 to 1914 and included eastern and southern European populations: Italians, Greeks, Poles, Russians, and other eastern Europeans. The immigration of Jews to the United States extended for well over a century, but the majority of Jewish immigrants came to the United States during the period from 1880 to 1920.

The Irish arrived in large numbers in the mid-nineteenth century as a consequence of food shortages and massive starvation in Ireland. During the latter half of the nineteenth century and in the early twentieth century, the Irish in the United States were abused, attacked, and viciously stereotyped. It is instructive to remember that the Irish, particularly on the East Coast and especially in Boston, underwent a period of ethnic oppression of extraordinary magnitude. A frequently seen sign posted in Boston saloons of the day proclaimed "No dogs or Irish allowed." German immigrants were similarly stereotyped, as were the French and the Scandinavians. *It is easy to forget that virtually all immigrant groups go through times of oppression and prejudice,* although these periods were considerably longer for some groups than for others. As a rule, where the population density of an ethnic group in a town, city, or region was greatest, so too was the amount of prejudice, negative stereotyping, and discrimination to which that group was subjected.

More than 40 percent of the world's Jewish population lives in the United States, making it the largest community of Jews in the world. Most of the Jews in the United States arrived between 1880 and the First World War, originating from the eastern European countries of Russia, Poland, Lithuania, Hungary, and Romania. Jews from Germany arrived in two phases; the first wave came just prior to the arrival of those from eastern Europe, and the second came as a result of Hitler's ascension to power in Germany during the late 1930s. Because many German Jews were professionals who also spoke English, they assimilated more rapidly than those from the eastern European countries. Jews from both parts of Europe underwent lengthy periods of anti-Jewish prejudice, **anti-Semitism** (defined as the hatred of Jewish people), and discrimination, particularly on Manhattan's Lower East Side. Significant anti-Semitism still exists in the United States (Ferber 1999; Simpson and Yinger 1985; Essed 1991).

In 1924, the National Origins Quota Act was passed, one of the most discriminatory legal actions ever taken by the United States in the area of immigration. By this act, the first real establishment of *ethnic quotas* in the United States, immigrants were permitted to enter the country only in proportion to their numbers already existing in the United States. Thus, ethnic groups who were already here in relatively high proportions (English, Germans, French, Scandinavians, and others, mostly western and northern Europeans) were allowed to immigrate in greater numbers than were those from southern and eastern Europe, such as Italians, Poles, Greeks, and other eastern Europeans. Hence, the act discriminated against southern and eastern Europeans in favor of western and northern Europeans. It has been noted that the European groups who were discriminated against by the National Origins Quota Act tended to be those with darker skins, even though they were White and European.

Immigrants during this period were subject to literacy tests and even IQ tests given in English (Kamin 1974). The act barred anyone who was classified as a convict, lunatic, "idiot," or "imbecile" from immigration. On New York City's Ellis Island, non-English-speaking immigrants, many of them Jews, were given the 1916 version of the Stanford-Binet IQ test in English. Obviously, non-English-speaking persons taking this test were unlikely to score high. On the basis of this grossly biased test, governmental psychologist H.H. Goddard classified fully 83 percent of Jews, 80 percent of Hungarians, and 79 percent of Italians as "feebleminded." It did not dawn on Goddard or the U.S. government that the IQ test, in English, probably did not measure something called intelligence, as intended, but instead simply measured the immigrant's mastery of the English language (Kamin 1994; Gould 1981; Taylor 2002, 1992a, 1980).

Attaining Racial and Ethnic Equality: The Challenge

Race and ethnic relations in the United States have posed a major challenge for the nation, one that is becoming even more complex as the racial–ethnic population becomes more diverse. Intergroup contact has been both negative and positive, obvious and subtle, tragic and helpful. How can the nation respond to its new diversity, as well as to the issues faced by racial and ethnic minorities who have been present since the nation's founding? This question engages significant sociological thought and attention to the nation's record of social change with regard to race and ethnic groups.

The White Immigrants Made It: Why Can't They?

A belief exists for many Americans that with enough hard work and loyalty to the dominant White culture of the country, any minority can make it and thus "assimilate" into American society. It is the often-heard argument that African Americans, Hispanics, and Native Americans need only to pull themselves up "by their own bootstraps" to become a success.

thinking sociologically

Write down your own racial–ethnic background and list one thing that people from this background have positively contributed to U.S. society or culture. Also list one experience (current or historical) in which people from your group have been victimized by society. Discuss how these two things illustrate the fact that racial–ethnic groups have both been *victimized* and have made *positive contributions* to this society. Share your comments with others: What does this reveal to you about the connections between different groups of people and their experiences as racial–ethnic groups in the United States?

This *assimilation perspective* dominated sociological thinking a generation ago and is still prominent in U.S. thought (Alba and Nee 2003; Portes and Rumbaut 2001; Rumbaut 1996a; Glazer, 1970; 1997). The assimilationist believes that to overcome adversity and oppression, the minority person need only imitate the dominant White culture as much as possible. In this sense, minorities must assimilate "into" White culture and White society. The general assumption is that with each new generation, assimilation becomes more and more likely. But one of the questions asked in this perspective is to what extent groups can maintain some of their distinct cultural values and still be incorporated into the society to which they have moved. One could argue for example that the Irish have been able to assimilate quite fully into American culture while still maintaining an ethnic identity—that is particularly salient around St. Patrick's Day!

Many Asian American groups have followed this pattern and have thus been called the "model minority," but this label ignores the fact that Asians are still subject to considerable prejudice, discrimination, racism, and poverty (Takaki 1989; Lee 1996; Woo 1998).

There are problems with the assimilation model. First, it fails to consider the time that it takes certain groups to assimilate. Those from rural backgrounds (Native Americans, Hispanics, African Americans, White Appalachians, and some White ethnic immigrants) typically take much longer to assimilate than those from urban backgrounds.

Second, the histories of Black and White arrivals are very different, with lasting consequences. Whites came voluntarily; Blacks arrived in chains. Whites sought relatives in the New World; Blacks were sold and separated from relatives. For these and other reasons, the experiences of African Americans and Whites as newcomers can hardly be compared, and their assimilation is unlikely to follow the same course.

Third, although White ethnic groups did indeed face prejudice and discrimination when they arrived in America, many entered at a time when the economy was growing rapidly and their labor was in high demand. Thus they were able to attain education and job skills. In contrast, by the time Blacks migrated to northern industrial areas from the rural South, Whites had already established firm control over labor and used this control to exclude Blacks from better-paying jobs and higher education.

Fourth, assimilation is more difficult for people of color because skin color is an especially *salient* characteristic, ascribed and relatively unchangeable. White ethnic group members can change their names, but people of color cannot easily change their skin color.

The assimilation model raises the question of whether it is possible for a society to maintain **cultural pluralism,** which is defined as different groups in society maintaining their distinctive cultures while also coexisting peacefully with the dominant group. Some groups have explicitly practiced cultural pluralism: The Amish people of Lancaster County in Pennsylvania and of north central Ohio—who travel by horse and buggy, use no electricity; and run their own schools, banks, and stores—constitute a good example of a relatively complete degree of cultural pluralism. A somewhat lesser degree of cultural pluralism, but still present, is maintained by "Little Italy" neighborhoods in some U.S. cities and also by certain Black Muslim groups in the United States.

© Bernard Boutrit/Woodfin Camp and Associates

Neighborhoods such as this one in Brooklyn, New York, are indicative of residential segregation.

Segregation and the Urban Underclass

Segregation is the spatial and social separation of racial and ethnic groups. Minorities, who are often believed by the dominant group to be inferior, are compelled to live separately under inferior conditions and are given lower-class educations, jobs, and protections under the law. Although desegregation has been mandated by law (thus eliminating *de jure segregation,* or legal segregation), *de facto segregation*—segregation in fact—still exists, particularly in housing and education.

Segregation has contributed to the creation of an **urban underclass,** a grouping of people, largely minorities and the poor, who live at the absolute bottom of the socioeconomic ladder in urban areas (Massey and Denton 1993; Wilson 1987). Indeed, the level of housing segregation is so high for some groups, especially poor African Americans and Latinos, that it has been termed **hypersegregation,** referring to a pattern of extreme segregation (Massey and Denton 1993). Currently, the rate of segregation of Blacks and Hispanics in U.S. cities is increasing, thus allowing for less and less interaction between White and Black children and White and Hispanic children (Schmitt 2001; Massey and Denton 1993). In education, schools are also becoming more segregated, a phenomenon called *resegregation* since schools are now more segregated than they were even in the 1980s (Frankenberg and Lee 2002).

doing sociological research

American Apartheid

The term *apartheid* (pronounced "apart-hate" or "apart-hite") was used to describe the society of South Africa prior to the election of Nelson Mandela in 1994. It refers to the rigid separation of the Black and White races. Sociological researchers Massey and Denton argue that the United States is now under a system of apartheid and that it is based on a very rigid residential segregation in the country.

Research Question: What is the current state of residential segregation? Massey and Denton note that the terms "segregation" and "residential segregation" practically disappeared from the American vocabulary in the late 1970s and early 1980s. These terms were spoken little by public officials, journalists, and even civil rights officials. This was because the ills of race relations in America were at the time attributed, though erroneously, to other causes such as a "culture of poverty" among minorities, or inadequate family structure among Blacks, or too much welfare for minority groups. The Fair Housing Act was passed in 1968, and the problem of segregation and discrimination in housing was declared solved. Yet, nothing could be farther from the truth.

Research methods and results: Researchers Massey and Denton amassed a large amount of data demonstrating that residential segregation not only has persisted in American society but that it has actually *increased* since the 1960s. Most Americans vaguely realize that urban America is still residentially segregated, but few appreciate the depth of Black and Hispanic segregation or the degree to which it is maintained by ongoing institutional arrangements and contemporary individual actions. Urban society is thus hypersegregated, or characterized by an extreme form of residential and educational segregation.

Conclusions and Implications: Massey and Denton find that most people think of racial segregation as a faded notion from the past, one that is decreasing over time. Today theoretical concepts such as the culture of poverty, institutional racism, and welfare are widely debated, yet rarely is residential segregation considered to be a major contributing cause of urban poverty and the underclass. Massey and Denton argue that their purpose is to redirect the focus of public debate back to race and racial segregation.

Source: Massey, Douglas S., and Nancy A. Denton. 1993. *American Apartheid: Segregation and the Making of the Underclass*. Cambridge, MA: Harvard University Press; Douglas S. Massey. 2005. *Strangers in a Strange Land: Humans in an Urbanizing World*. New York: Norton.

In a seminal study, W. J. Wilson (1987) attributes the causes of the urban underclass to economic and social structural deficits in society. He rejects the "culture of poverty" explanation, an earlier view that attributes the condition of minorities to their own presumed cultural deficiencies, a view attributed to Lewis (1960, 1966) and Moynihan (1965). The problems of the inner city, such as joblessness, crime, teen pregnancy, welfare dependency, and acquired immune deficiency syndrome (AIDS) are seen to arise from social class inequalities, that is, inequalities in the structure of society, and these inequalities have dire behavioral consequences at the individual level, in the form of drug abuse, violence, and lack of education (Wilson 1996, 1987; Sampson 1987). But despite these disadvantages, many individuals nonetheless manage to achieve upward occupational and economic mobility (Newman 1999). Wilson argues that the civil rights agendas need to be enlarged and that the major problem of the underclass, joblessness, needs to be addressed by fundamental changes in the economic institution, such as government-financed jobs and universal health care.

The Relative Importance of Class and Race

The presence of an urban underclass—one composed mostly of African Americans, Latinos, and in some cities, poor Whites and Asian Americans—highlights the significant connection between race and social class. In some regards, class can be seen as having

understanding diversity

Race and Hurricane Katrina

A brutally devastating hurricane, given the name Katrina, hit New Orleans, Louisiana, and other locations along the country's southern gulf coast, such as Biloxi, Mississippi, early in the fall of 2005. Katrina's winds, reaching at times 150 miles per hour, tore apart hundreds of houses, apartment buildings, hospitals, oil wells and derricks, and other structures. Massive flooding devastated New Orleans, with water reaching as high as 20 feet in some locations, stranding people, their pets, and farm animals. Because of the slowness of the federal government's response (it took the president of the United States one full week after the hurricane even to acknowledge the devastation as a national disaster and to visit New Orleans) and because of the extent of the flooding, over one thousand people died from drowning, from direct hits by flying debris, or from lack of medical attention.

The nation was stunned by images of human bodies floating down water-filled streets, the water itself slicked with oil and filled with sewage and other contaminants. In rest homes for the elderly, patients died as a result of lack of electricity and oxygen supplies needed for assisted breathing. In one case, over twenty patients were simply left alone, unattended for one full week by administrators and staff. Each and every patient died from neglect and lack of medical care. Many people died simply waiting to be evacuated from their communities or from temporary transfer locations—another effect of the slowness of the state and federal governments to act. The government's Federal Emergency Management Administration (FEMA) failed miserably in its intended role of organizing and coordinating responses to the devastation and getting people to safe locations quickly. FEMA's response was so anemic that its director was forced to resign within weeks.

The most negatively affected areas of New Orleans were those neighborhoods in the lowest-lying areas of the city, areas up to 20 feet or more below sea level. These neighborhoods were primarily poor and Black or Hispanic; many were *hypersegregated,* that is, almost entirely African American. Clearly, these neighborhoods had the highest potential for flood damage and the aftermath of contamination and disease. Many have argued that if the White and wealthy had been so concentrated in such neighborhoods, the response of the federal government and the president would probably have been much more rapid and perhaps more effective.

An additional manifestation of race and class bias was seen as a result of the evacuation of thousands to the New Orleans Superdome and the New Orleans Convention Center. Within days the Superdome and Convention centers became cauldrons of misery—no water, no food, unbearable heat, grossly inadequate facilities—including few and clogged toilets and a partially collapsed roof—and an utter lack of medical care for the injured, pregnant women, infants, and the elderly. The individuals so evacuated were mainly poor, and they were neglected and ignored for more than a week by the federal government, FEMA, and even the Louisiana state government.

In national polls taken after Hurricanes Katrina, there was a racial divide: Three-quarters of African Americans thought that racism affected the poor response but only one-third of White Americans thought so.

increasing importance in the lives of various groups, including African Americans. Thus, William Julius Wilson (1978, 1996) points out that there has been a simultaneous expansion of both the Black middle class as well as the Black urban underclass. The same pattern is true for Latinos. This does not mean that race is unimportant; but the influence of class is increasing, even though race still remains extremely important. In numerous studies, scholars find that race, in and of itself, still influences such things as income, wealth holdings, occupational prestige, place of residence, educational attainment, and numerous other measures of socioeconomic well-being (Brown et al. 2005; Patillo-McCoy 1999; Oliver and Shapiro 1995).

What is important is understanding the *intersecting* effects of race and class acting together (Andersen and Collins 2007). Racial–ethnic groups live in what has been called a *matrix of domination* (Collins 1990). That is, no single factor alone determines one's location in society. Rather, race—along with class, gender, age, even sexual orientation—together place one in a system of social advantage and disadvantage. Understanding the interrelationship among these social factors is critical to understanding any one of them, including race and ethnicity.

The Civil Rights Strategy

The history of racial and ethnic relations in the United States shows several strategies to achieve greater equality. Political mobilization, legal reform, and social policy have been the basis for much social change in race relations, but there are continuing questions about how best to achieve a greater degree of racial justice in this society. The major force behind most progressive social change in race relations was the civil rights movement. Marked by the strong moral and political commitment and courage of participants, the civil rights movement is probably the single most important source for change in race relations in the twentieth century. The civil rights movement was based on the passive resistance philosophy of Martin Luther King Jr., learned from the philosophy of *satyagraha* ("soul firmness and force") of the Indian Mahatma (meaning "leader") Mohandas Gandhi. This philosophy encouraged resistance to segregation through nonviolent techniques, such as sit-ins, marches, and appealing to human conscience in calls for brotherhood, justice, and equality. Although African Americans had worked for racial justice and civil rights long before this historic movement, the civil rights movement has brought greater civil rights under the law to many groups: women, disabled people, the aged, and gays and lesbians (Andersen 2004).

The major civil rights movement in the United States intensified shortly after the 1954 *Brown v. Board of Education* decision, the famous Supreme Court case that ruled that in education, "separate but equal" was unconstitutional. In 1955, African American seamstress and NAACP secretary Rosa Parks made news in Montgomery, Alabama. By prior arrangement with the NAACP, Parks bravely refused to relinquish her seat in the "White only" section on a segregated bus when asked to do so by the White bus driver. At the time, the majority of Montgomery's bus riders were African American, and the action of Rosa Parks initiated the now-famous Montgomery bus boycott, led by the young Martin Luther King Jr. The boycott, which took place in many cities beyond Montgomery, was successful in desegregating the buses. It got more African American bus drivers hired and catapulted Martin Luther King Jr. to the forefront of the civil rights movement. Impetus was given to the civil rights movement and the boycott by the brutal murder in 1954 of Emmett Till, a Black teenager from Chicago, who was killed in Mississippi merely for whistling at a White woman.

The civil rights movement produced many episodes of both tragedy and heroism. In a landmark 1957 decision, President Dwight D. Eisenhower called out the national guard, after initial delay, to assist the entrance of nine Black students into Little Rock Central High School in Little Rock, Arkansas. Sit-ins followed throughout the South in which White and Black students perched at lunch counters until the Black students were served. Organized bus trips from North to South to promote civil rights, "freedom rides," forged on despite the murders of freedom riders Viola Liuzzo, a White Detroit housewife; Andrew Goodman and Michael Schwerner, two White students; and James Chaney, a Black student. The murders of civil rights workers—especially when they were White—galvanized public support for change.

Radical Social Change

While the civil rights movement developed throughout the late 1950s and 1960s, a more radical philosophy of change also developed, as more militant leaders grew increasingly disenchanted with the limits of the civil rights agenda (which was perceived as moving too slowly). The militant Black power movement, taking its name from the book *Black Power* (published in 1967 by political activist Stokely Carmichael, later Kwame Touré, and Columbia University political science professor Charles V. Hamilton), had a more radical critique of race relations in the United States and saw inequality as stemming not just from moral failures, but from the institutional power that Whites had over Black Americans (Carmichael and Hamilton 1967).

Before breaking with the Black Muslims (the Black Nation of Islam in America) and his religious mentor, Elijah Muhammad, and prior to his assassination in 1965, Malcolm X advocated a form of pluralism demanding separate business establishments, banks, churches, and schools for Black Americans. He echoed an earlier effort of the 1920s led by Marcus Garvey's back-to-Africa movement, the Universal Negro Improvement Association (UNIA).

The Black power movement of the late 1960s rejected assimilation and instead demanded pluralism in the form of self-determination and self-regulation of Black communities. Militant groups such as the Black Panther Party advocated fighting oppression with armed revolution. The U.S. government acted quickly, imprisoning members of the Black Panther Party and members of similar militant revolutionary groups and, in some cases, killing them outright (Brown 1992).

The Black power movement also influenced the development of other groups who were affected by the analysis of institutional racism that the Black power movement developed, as well as by the assertion of strong group identity that this movement encouraged. Groups such as La Raza Unida, a Chicano organization, encouraged "brown power," promoting solidarity and the use of Chicano power to achieve racial justice. Likewise, the American Indian Movement (AIM) used some of the same strategies and tactics that the Black power movement had encouraged, as have Puerto Rican, Asian American, and other racial protest groups. Elements of Black power strategy were also borrowed by the developing women's movement, and Black feminism was developed upon the realization that women, including women of color, shared in the oppressed status fostered by institutions that promoted racism (Collins 1998, 1990). Overall, the Black power movement dramatically altered the nature of political struggle and race and ethnic relations in the United States. It and the other movements it inspired changed the nation's consciousness about race and forced even academic scholars to develop a deeper understanding of how fundamental racism is to U.S. social institutions (Branch 2006, 1998, 1988; Morris 1984).

Affirmative Action

A continuing question from the dialogue between a civil rights strategy and more radical strategies for change is the debate between race-specific versus color-blind programs for change. *Color-blind policies* are those advocating that all groups be treated alike, with no barriers to opportunity posed by race, gender, or other group differences. Equal opportunity is the key concept in color-blind policies.

Race-specific policies are those that recognize the unique status of racial groups because of the long history of discrimination and the continuing influence of institutional racism. Those advocating such policies argue that color-blind strategies will not work because Whites and other racial–ethnic groups do not start from the same position. Even given equal opportunities, continuing disadvantage produces unequal results. The tension between these two strategies for change is a major source for many of the political debates surrounding race relations now.

Affirmative action, a heavily contested program for change, is a race-specific policy for reducing job and educational inequality that has had some success. Affirmative action means two things. First, it means recruiting minorities from a wide base in order to ensure consideration of groups that have been traditionally overlooked, while not using rigid quotas based on race or ethnicity. Second, affirmative action means using admissions slots (in education) or set-aside contracts or jobs (in job hiring) to assure minority representation. The principal objection, heard from both sides of the racial line, is that either interpretation of affirmative action programs is, in effect, use of a quota.

The Legal Defense Fund (LDF), established by the NAACP (National Association for the Advancement of Colored People), has argued forcefully and on legal grounds that the push for affirmative action on the basis of race must continue, though not in the form of rigid quotas. It has noted that in the tradition of its founding lawyer and the future member of the Supreme Court, Thurgood Marshall, the affirmative action policies of the LDF have helped to fundamentally change the composition of the nation's formerly segregated universities and colleges as well as to greatly expand the educational opportunities for African Americans and other minorities. It argues that standardized tests (such as the SATs) are of limited validity and do not adequately predict performance of minorities of color in college; they are thus not a legitimate basis on which to judge Black and White candidates against each other, as has been disputed in a number of the legal cases. (The issue of standardized testing is discussed in Chapter 13.)

Recent data have shown that Blacks admitted to selective colleges and universities under affirmative action programs reveal high rates of social and economic success after graduation. For example, the percentage of Blacks who were admitted to college under affirmative action programs and went on to graduate school and law school was higher than the percentage of Whites from the same schools who did so (Espenshade et al. 2004; Bowen and Bok 1998). This is clearly a benefit of affirmative action in education.

The U.S. Supreme Court decided in 1978 (in *The University of California Regents vs. Bakke*) that race could be used as a criterion for admission to undergraduate, professional, and graduate schools or for job recruitment, as long as race is combined with other criteria and as long as rigid racial quotas are not used. Then twenty-five years later, in 2003, the U.S. Supreme Court decided two cases that modified its 1978 decision. In *Grutter v. Bollinger,* in a five-to-four decision, the high court decided that, as in the 1978 decision, race could indeed be used as a factor in admissions decisions for the University of Michigan Law school as long as race was considered along with other factors and the decision to admit or not admit was made on a case-by-case basis. In a second case (*Gratz v. Bollinger*), a six-to-three decision, the Court threw out as unconstitutional any system of assigning favorable points to minority candidates seeking admission that would increase their chances for admission. This decision thus ruled out the use of any form of minority quotas, interpreting the point system as a type of quota, but it upheld as constitutional a system that considers race among many factors on an individual basis.

The "Tax" on Being a Minority in America: Give Yourself a True-False Test

On the following test, give yourself one point for each statement that is true for you personally. When you are done, total up your points. The higher your score (the more points you have), the less "minority tax" you are paying in your own life.

1. My parents and grandparents were able to purchase a house in any neighborhood they could afford.
2. I can take a job in an organization with an affirmative action policy without people thinking I got my job because of my race.
3. My parents own their own home.
4. I can look at the mainstream media and see people who look like me represented in a wide variety of roles.
5. I can choose from many different student organizations on campus that reflect my interests.
6. I can go shopping most of the time pretty well assured that I will not be followed or harassed when I am in the store.
7. If my car breaks down on a deserted stretch of road, I can trust that the law enforcement officer who shows up will be helpful.
8. I have a wide choice of grooming products that I can buy in places convenient to campus and/or near where I live as a student.
9. I never think twice about calling the police when trouble occurs.
10. The schools I have attended teach about my race and heritage and present it in positive ways.
11. I can be pretty sure that if I go into a business or other organization (such as a university or college) to speak with the "person in charge" that I will be facing a person of my race.

Your total points: ____________________

Your racial identity: ____________________

Your gender: ____________________

How would you describe your social class? ____________________

____

Now gather results from some of your classmates and see if their total points vary according to their own race, gender, and/or social class.

Source: Adapted from the *Discussion Guide for Race: The Power of an Illusion.* **www.pbs.org**

Chapter Summary

- ***How are race and ethnicity defined?***

In virtually every walk of life, race matters. A *race* is a social construction based loosely on physical criteria, whereas an *ethnic group* is a culturally distinct group. A group is *minority* or *dominant* not on the basis of their numbers in a society, but on the basis of which group occupies lower average social status.

- ***What are stereotypes, and how are they important?***

Stereotyping and *stereotype interchangeability* reinforce racial and ethnic prejudices and thus cause them to persist in the maintenance of inequality in society. Racial and gender stereotypes have similar dynamics in society, and both racial and gender stereotypes receive ongoing support in the media. Stereotypes serve to justify and make legitimate the oppression of groups based on race, ethnicity, and gender. Stereotypes such as "lazy" support attributions made to the minority, which attempt to cast blame on the minority in question, thus removing blame from the social structure.

- ***What are the differences between prejudice, discrimination, and racism?***

Prejudice is an attitude usually involving negative prejudgment on the basis of race or ethnicity. *Discrimination* is overt behavior involving unequal treatment. *Racism* involves both attitude and behavior. Racism can take on several forms, such as traditional or *old-fashioned racism, aversive* or subtle racism, *laissez-faire racism, color-blind racism,* and institutional racism. *Institutional racism* is unequal treatment, carrying with it notions of cultural inferiority of a minority, which has become ingrained into the economic, political, and educational institutions of society.

- ***Do all minority groups have different histories, or are they similar?***

Historical experiences show that different groups have unique histories, although they are bound together by some similarities in the prejudice and discrimination they have experienced. Although each group's experience is unique, they are commonly related by a history of prejudice and discrimination.

- ***What are the challenges in attaining racial and ethnic equality?***

Not all immigrant groups and minority groups *assimilate* at the same rate, and some groups (U.S. Black Muslims; the Amish) maintain cultural *pluralism.* An *urban underclass* remains entrenched in the United States, and cities remain *hypersegregated* on the basis of race and ethnicity.

- ***What are some of the approaches to attaining racial and ethnic equality?***

Approaches include Rev. Martin Luther King's nonviolent civil rights strategy, radical social change and movements such as the Black Power movement, La Raza Unida, and the American Indian Movement (AIM), all of which directly addressed *institutional racism. Affirmative action* policies, which are race-specific rather than race-blind programs, continue to be changed and modified through Supreme Court cases. Some policy researchers have argued that only deep-rooted economic changes in the social structure will alleviate persistent economic and racial stratification and inequality.

Key Terms

affirmative action 256
anti-Semitism 251
assimilation 243
authoritarian personality 242
aversive racism 241
color-blind racism 241
contact theory 243
cultural pluralism 252
discrimination 240
dominant group 237
ethnic group 234
ethnocentrism 239
forms of racism 241
hypersegregation 252
institutional racism 241
laissez-faire racism 241
minority group 237
old-fashioned racism 241
out-group homogeneity effect 237
prejudice 239
race 234
racial formation 237
racial profiling 241
racialization 235
racism 240
residential segregation 240
salience principle 238
scapegoat theory 242
segregation 252
stereotype 238
stereotype interchangeability 238
urban underclass 252

Online Resources

Sociology: The Essentials Companion Website

www.thomsonedu.com/sociology/andersen
Visit your book companion website where you will find more resources to help you study and write your research papers. Resources include Suggested Readings, web links, and a MicroCase Online feature that teaches you how to research society. Other resources include Learning Objectives, Internet exercises, quizzing, and flash cards.

is an easy-to-use online resource that helps you study in less time to get the grade you want NOW.

www.thomsonedu.com/login
Need help studying? This site is your one-stop study shop. Take a Pre-Test and Thomson NOW will generate a Personalized Study Plan based on your test results. The Study Plan will identify the topics you need to review and direct you to online resources to help you master those topics. You can then take a Post-Test to determine the concepts you have mastered and what you still need to work on.

10
Chapter ten
CHAPTER TEN

Gender

Imagine suddenly becoming a member of the other sex. What would you have to change? First, you would probably change your appearance—clothing, hairstyle, and any adornments you wear. You would also have to change some of your interpersonal behavior. Contrary to popular belief, men talk more than women, are louder, more likely to interrupt, and less likely to recognize others in conversation. Women are more likely to laugh, express hesitance, and be polite (Robinson and Smith-Lovin 2001; Anderson and Leaper 1998; Crawford 1995; Cameron 1998). Gender differences also appear in nonverbal communication. Women use less personal space, touch less in impersonal settings (but are touched more), and smile more, even when they are not necessarily happy (LaFrance 2002; Lombardo et al. 2001; Basow 1992). Researchers even find that men and women write e-mail in a different style, women writing less opinionated e-mail than men and using it to maintain rapport and intimacy (Sussman and Tyson 2000; Colley and Todd 2002). Finally, you might have to change many of your attitudes because men and women differ significantly on many, if not most, social and political issues (see Figure 10.1).

continued

chapter ten

CHAPTER TE

If you are a woman and became a man, perhaps the change would be worth it. You would probably see your income increase (especially if you became a White man). You would have more power in virtually every social setting. You would be far more likely to head a major corporation, run your own business, or be elected to a political office—again, assuming that you are White. Would it be worth it? As a man, you would be far more likely to die a violent death and would probably not live as long (National Center for Health Statistics 2006).

If you are a man who became a woman, your income would most likely drop significantly. More than forty years after passage of the Equal Pay Act in 1963, men still earn 23 percent more than women; this is counting only those working year-round and full time (DeNavas-Walt et al. 2006). You would probably become resentful of a number of things because poll data indicate that women are more resentful than men about things such as the amount of money available for them to live on, the amount of help they get from their mates around the house, how men share child care, and how they look. Women also report being more fearful on the streets than men. However, women are more satisfied than men with their roles as parents and with their friendships outside of marriage.

For both women and men, there are benefits, costs, and consequences stemming from the social definitions associated with gender. As you imagined this experiment, you may have had difficulty trying to picture the essential change in your *biological* identity: But is this the most significant part of being a man or woman? Nature determines whether you are male or female, but it is society that gives significance to this distinction. Sociologists see *gender* as a social fact, because who we become as men and women is largely shaped by cultural and social expectations.

The Social Construction of Gender

From the moment of birth, gender expectations influence how boys and girls are treated. Now that it is possible to identify the sex of a child in the womb, gender expectations may begin even before birth. Parents and grandparents might select pink clothes and dolls for baby girls, sports clothing and brighter colors for boys. They have little choice to do otherwise because baby products are so typed by gender. Much research shows how parents and others continue to treat children in stereotypical ways throughout their childhood. Girls may be expected to cuddle and be sweet, whereas boys are handled more roughly and given greater independence.

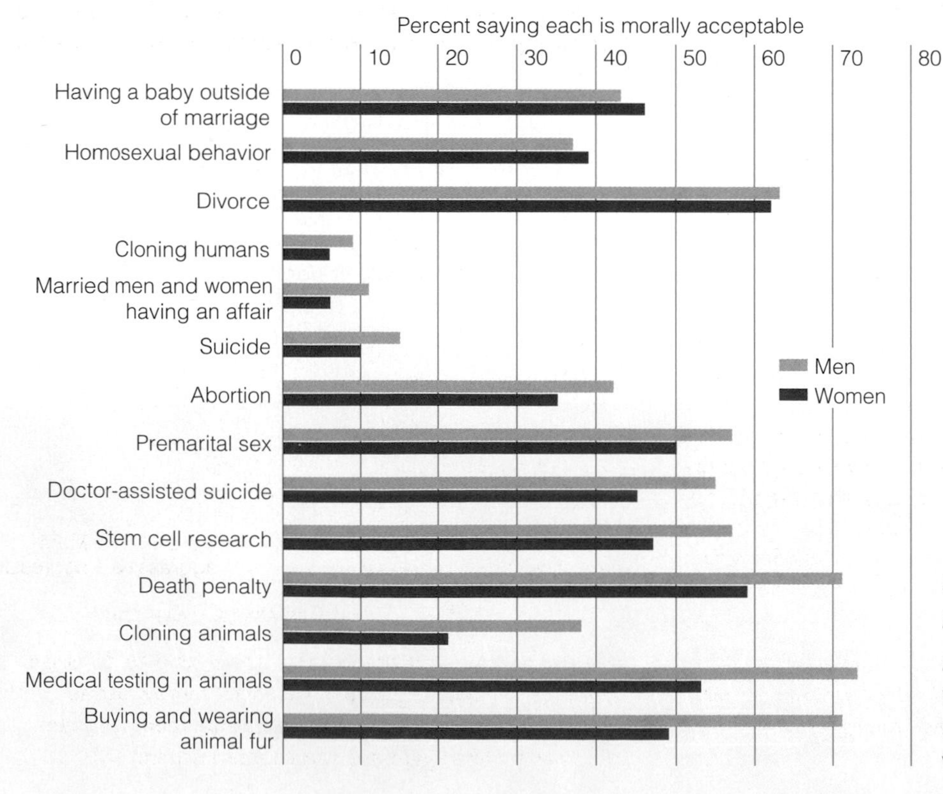

FIGURE 10.1 *The Gender Gap in Attitudes*

Data: Saad, Lydia. 2003. "Pondering 'Women's Issues,' Part II." *The Gallup Poll.* Princeton, NJ: The Gallup Organization. **www.gallup.com**

Defining Sex and Gender

Sociologists use the terms *sex* and *gender* to distinguish biological sex identity from learned gender roles. **Sex** refers to biological identity, being male or female. For sociologists, the more significant concept is **gender**—the socially learned expectations and behaviors associated with members of each sex. This distinction emphasizes that behavior associated with gender is culturally learned.

The cultural basis of gender is especially apparent when we look at other cultures. Across different cultures, the gender roles associated with masculinity and femininity vary considerably. In Western industrialized societies, people tend to think of masculinity and femininity in dichotomous terms, with men and women decisively different, even defined as opposites. The views from other cultures challenge this assumption. The Navajo, for example, offer interesting examples of alternative gender roles. Historically, the berdaches in Navajo society were anatomically normal men who were defined as a third gender considered to fall between male and female. Berdaches married other men who were not considered berdaches and were defined as ordinary men. Moreover, neither the berdaches nor the men they married were considered homosexuals, as they would be considered in many of today's Western cultures (Nanda 1998; Lorber 1994).

We need not look at striking departures from familiar gender categories to see how differently gender can be constructed in other cultures. For example, in China the Chinese Marriage Law of 1950 formally defines marriage as a relationship between equal companions who share responsibility for child care and the family. This is in sharp contrast to age-old gender roles in Chinese tradition. The roles did not change overnight in China. As in many societies backed by strong traditions, contemporary roles can coexist with more traditional ones.

There can also be substantial differences in the construction of gender across social classes or subcultures within a single culture. Within the United States, as we will see, there is considerable variation in the experiences of gender among different racial and ethnic groups (Baca Zinn et al. 2005; Andersen and Collins 2007). In addition, even within a given culture, differences among people of a given gender can be greater than differences across gender (see Figure 10.2). Looking at gender sociologically quickly reveals the social and cultural dimensions of something often popularly defined as biologically fixed.

The hijras *of India are a sexual minority group; this man is considered to be a "third gender"—neither man nor woman. Hijras provide a good illustration of the socially constructed basis of sexuality.*

thinking sociologically

Visit a local toy store and try to purchase a toy for a young child that is not gender-typed. What could you buy? What could you not buy? What does your experiment teach you about gender role socialization? (If you take a child with you, note what toys he or she wants. What does this tell you about how effective gender socialization is?)

Sex Differences: Nature or Nurture?

Despite the known power of social expectations, the belief persists that differences between men and women are biologically determined. Biology is, however, only one component in the differences between men and women. The important question in sociology is not whether biology or culture is more important in forming men and women, but the role of culture in producing a person's gender identity. What is the interplay of nature and nurture?

Biological determinism refers to explanations that attribute complex social phenomena to physical characteristics. The argument that men are more aggressive because of hormonal differences (in

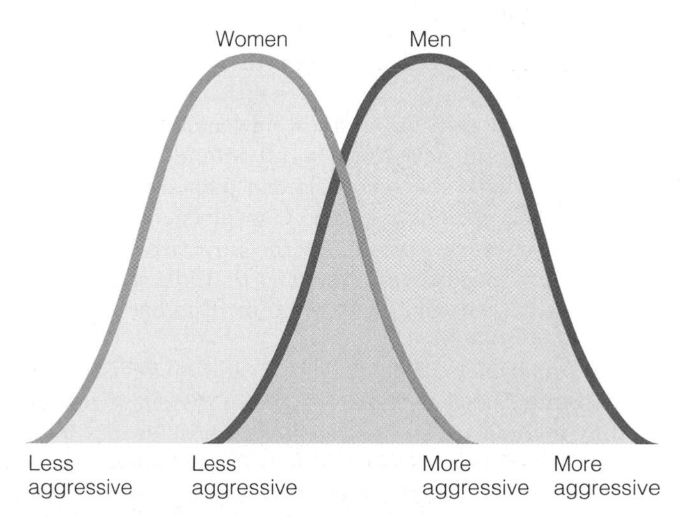

FIGURE 10.2 *Gender Differences: Aggression*

Even when men and women as a whole tend to differ on a given trait, within-gender differences can be just as great as across-gender differences. Some men, for example, may be less aggressive than some women.

particular, the presence of testosterone) is a biologicalally determinist argument. Although people popularly believe that testosterone causes aggressive behavior in men, studies find only a modest correlation between aggressive behavior and testosterone levels. Furthermore, changes in testosterone levels do not predict changes in men's aggression (such as by "chemical castration," the administration of drugs that eliminate the production or circulation of testosterone). What's more, there are minimal differences in the levels of sex hormones between girls and boys during early childhood, yet researchers find considerable differences in the aggression exhibited by boys and girls as children (Fausto-Sterling 1992, 2000).

A person's sex identity is established at the moment of conception when the father's sperm provides either an X or a Y chromosome to the egg at fertilization. The mother contributes an X chromosome to the embryo. The combination of two X chromosomes makes a female, while the combination of an X and a Y makes a male. Under normal conditions, chemical events directed by genes on the sex-linked chromosomes lead to the formation of male or female genitalia.

Hermaphroditism is a condition caused by irregularities in the process of chromosome formation or fetal differentiation that produces persons with mixed biological sex characteristics, also known as *intersexed persons*. In one such form, the child is born with ovaries or testes, but the genitals are ambiguous or mixed. An example would be a child born with female chromosomes but an enlarged clitoris, making the child appear to be male. Sometimes, a child may be a chromosomal male but have an incomplete penis and no urinary canal.

Case studies of intersexed persons reveal the extraordinary influence of social factors in shaping the person's identity (Preves 2003). Parents of intersexed children are usually advised to have their child's genitals surgically assigned to either male or female and also to give the child a new name, a different hairstyle, and new clothes—all intended to provide the child with the social signals judged appropriate to a single gender identity. One physician who has worked on such cases gives the directive to parents that they "need to go home and do their job as child rearers with it very clear whether it's a boy or a girl" (Kessler 1990: 9).

Transgendered people are those who deviate from the binary (that is, male or female) system of gender; they include transsexuals (those whose sexual identity as a man or woman differs from their biological sex identity at birth), cross-dressers, and others who do not fit within the normative expectations of gender. Research on transgendered individuals shows that they experience enormous pressure to fit within the usual expectations. When they were young, for example, they would hide their cross-dressing. Those who change their sex as adults report enormous pressure, particularly during their transition period, because others expect them to be one sex or the other. Most find that whatever their biological sex, they are forced—from fear of rejection and the desire for self-preservation—to manage an identity that would fall into one category or another (Gagné and Tewksbury 1998).

From a sociological perspective, biology alone does not determine gender identity. People must adjust to the expectations of others and the social understanding of what it means to be a man or a woman. A person may remain genetically one sex, while socially being the other—or perhaps something in between. In other words, there is not a fixed relationship between biological and social outcomes. If you only see men and women as biologically "natural" states, you miss some of the fascinating ways that gender is formed in society.

Physical differences between the sexes do, of course, exist. In addition to differences in anatomy, at birth, boys tend to be slightly longer and weigh more than girls. As adults, men tend to have a lower resting heart rate, higher blood pressure, higher muscle mass and muscle density, and more efficient recovery from muscular activity. These physical differences contribute to the tendency for men to be physically stronger than women. However, the public now routinely sees displays of women's athleticism and expects great performances from both men and women in world-class events such as the Olympics. Women can achieve high degrees of muscle mass and muscle density through bodybuilding and can win over men in activities that require endurance, such as the four women who have won the Iditarod—the Alaskan dog sled race considered to be one of the most grueling competitions in the world.

Arguments based on biological determinism assume that differences between women and men are "natural" and, presumably, resistant to change. Like biological explanations of race differences, biological explanations of inequality between women and men tend to flourish during periods of rapid social change. They protect the *status quo* (existing social arrangements) by making it appear that the status of women or people of other races is "natural" and therefore should remain as it is. If social differences between women and men were biologically determined, we would find no variation in gender relations across cultures, but extensive differences are well documented. Moreover, even within the same culture, there can be vast *within-gender* differences. That is, the variation on a given trait, such as aggression or competitiveness, can be as great within a given gender group as the difference across genders. Thus, some women are more aggressive than some men and some men less competitive than some women (see Figure 10.2). We would not exist without our biological makeup, but we would not be who we are without society and culture.

In sum, sociologists emphasize the social basis for both sex and gender. Sociologists see people's sex identity and their gender identity as being influenced by society. Culture defines certain sexual behaviors as appropriate (or not) and establishes particular expectations for women and for men. This is learned through the process of *gender socialization*.

Gender Socialization

As we saw in Chapter 3, socialization is the process by which social expectations are taught and learned. Through **gender socialization,** men and women learn the expectations associated with their sex. The rules of gender extend to all aspects of society and daily life. Gender socialization affects the self-concepts of women and men, their social and political attitudes, their perceptions about other people, and their feelings about relationships with others. Although not everyone is perfectly socialized to conform to gender expectations, socialization is a powerful force directing the behavior of men and women in gender-typical ways.

Even people who set out to challenge traditional expectations often find themselves yielding to the powerful influence of socialization. Women who consciously reject traditional women's roles may still find themselves inclined to act as hostess or secretary in a group setting. Similarly, men may decide to accept equal responsibility for housework, yet they fail to notice when the refrigerator is empty or the child needs a bath—household needs they have been trained to let someone else notice (DeVault 1991). These expectations are so pervasive that it is also difficult to change them on an individual basis. If you doubt this, try buying clothing or toys for a young child without purchasing something that is gender-typed, or talk to parents who have tried to raise their children without conforming to gender stereotypes and see what they report about the influence of such things as children's peers and the media.

The Formation of Gender Identity

One result of gender socialization is the formation of **gender identity,** which is one's definition of oneself as a woman or man. Gender identity is basic to our self-concept and shapes our expectations for ourselves, our abilities and interests, and how we interact with others. Gender identity shapes not only how we think about ourselves and others, but also influences numerous behaviors, including such things as the likelihood of drug and alcohol abuse, violent behavior, depression, or even how aggressive you are in driving (Kulis et al. 2002; Sprock and Yoder 1997; Lupton 2002).

One area where gender identity has an especially strong effect is in how people feel about their appearance. Studies find strong effects of gender identity on body image. Concern with body image begins mostly during adolescence. Thus, studies of young children (that is, preschool age) find no gender differences in how boys and girls feel about their bodies (Hendy et al. 2001), but by early adolescence clear differences emerge. At this age, girls report comparing their bodies to others of their sex more often than boys do. By early adolescence, girls report lower *self-esteem* (that is, how well one thinks of oneself) than boys; they also report more negativity about their body images than do boys. This type of thinking among girls is related to lower *self-esteem* (Jones 2001; Polce-Lynch et al. 2001). Among college students, women are also more dissatisfied with their appearances than are men (Hoyt and Kogan 2001). These studies indicate that idealized images of women's bodies in the media, as well as peer pressures, have a huge impact on young girls' and women's gender identity and feelings about their appearances.

Sources of Gender Socialization

As with other forms of socialization, there are different agents of gender socialization: family, peers, schooling, religious training, mass media, and popular culture, to name a few. Gender socialization is reinforced whenever gender-linked behaviors receive approval or disapproval from these multiple influences.

Parents are one of the most important sources of gender socialization. Parents may discourage children from playing with toys that are identified with the other sex, especially when boys play with toys meant for girls. Research finds that parents are more tolerant of girls not conforming to gender roles than is true for boys (Kane 2006). Direct observations of fathers and mothers interacting with their young children also show that fathers are less likely to provide basic care, although both parents tend to share responsibility for discipline (LaFlamme et al. 2002).

Expectations about gender are changing, although researchers suggest that the cultural expectations about gender may have changed more than people's actual behavior. Thus, mothers and fathers now report that fathers should be equally involved in childrearing, but the reality is different. Mothers still spend more time in child-related activities and have more responsibility for children. Furthermore, the gap that mothers perceive between fathers' ideal and actual involvement in childrearing is a significant source of mothers' stress (Milkie et al. 2002).

Gender socialization patterns also vary within different racial–ethnic families. Latinas, as an example, have generally been thought to be more traditional in their gender roles, although this varies by generation and by the experiences of family members in the labor force. Within families, young women and men learn to formulate identities that stem from both their gender, racial, and ethnic expectations.

seeing society in everyday life

Gender, Bodies, and Beauty

Sociologist Debra Gimlin argues that bodies are "the surface on which prevailing rules of a culture are written" (Gimlin 2002: 6). You see this in many ways, such as how people mark and adorn their bodies in different cultures. This is especially evident with regard to gender. Men and women alike may practice elaborate rituals to achieve particular ideals established by the culture—ideals that vary because of gender roles, as well as class and race relations.

© Mike Watson Images/SuperStock

© Stockbyte/Getty Images

How does gender influence our concepts and practices about the body?

One of the gender differences some have noted is that men generally try to bulk themselves up, sometimes using steroids and other products to do so, while women generally strive to be thin. What are the dominant cultural images of beauty and good looks for women and men? Do they vary because of race?

© Markus Moellenberg/zefa/Corbis

Anorexia is particularly common among White, middle-class women, although the occurrence of anorexia is increasing among women of color. Do you see evidence of this on your campus?

AP Images/Stuart Ramson

© Karl Prouse/Catwalking/Contributor/Getty Images

Beauty ideals in a culture are strongly influenced by popular culture. Fashion models, for example, set some of the standards for women's and men's appearance, but they are often dangerously thin.

Many women—and some men—spend a great deal of money on procedures to alter their appearance. The use of Botox treatments, breast augmentation, liposuction, and other procedures are increasing, even among young women—and solely for cosmetic reasons. How do you explain this?

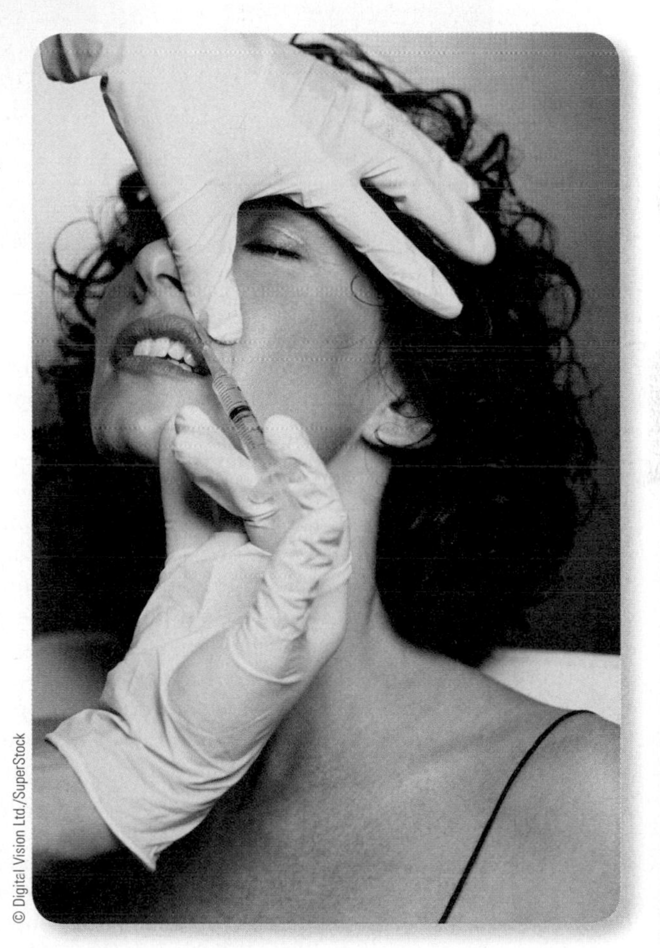
© Digital Vision Ltd./SuperStock

© Think Stock/SuperStock

Although it is now commonplace, popular fashion, including among young girls, tends to sexualize women's bodies.

© David Frazier/The Image Works

Changes in gender roles have involved more men in parenting.

From the time they become aware of their surroundings, children are socialized to adopt behaviors and attitudes judged appropriate for members of their sex. Socialization comes not only from parents and other family members, but also from *peers.* Through play children learn patterns of social interaction, analytical skills, and the values and attitudes of their culture; they also add to their cognitive and physical development. Research shows that girls play more cooperatively when they are in same-sex groups than they do when playing with boys (Neppl and Murray 1997). Boys and girls also organize their play in ways that reinforce not only gender, but also race and age norms (Moore 2001). When they play together, boys exert power over girls and typically establish the conditions of the play activities (Voss 1997).

A variety of other patterns are consistently found in children's play and games. Boys are encouraged to play outside more; girls, inside. Boys' toys are more machine-like and frequently promote the development of militaristic values; they tend to encourage aggression, violence, and the stereotyping of enemies—values rarely associated with girls' toys. Children's books in schools also communicate gender expectations. Even with publishers' guidelines that discourage stereotyping, textbooks still depict men as aggressive, argumentative, and competitive. Men and boys are also more likely to be featured in children's books, although interestingly, systematic analysis of children's books shows that fathers are not very present and, when they are, are most often shown as ineffectual (Anderson and Hamilton 2005).

The gender norms that children learn seem to be more strictly applied to boys than girls. Thus, boys who engage in behavior that is associated with girls are more negatively regarded than girls who play or act like boys (Sandnabba and Ahlberg 1999). While girls may be called "tomboys," boys who are called "sissies" are more harshly judged. Being a tomboy may be acceptable for a girl, but beyond a certain age, the same behavior may result in her being labeled a "dyke." Sociological research shows that girls who become tomboys do so because they recognize the disadvantages of femininity and the privileges of masculinity. They are resisting expected gender roles, even as they also try to conform to existing gender role choices (Carr 1998).

Schools are particularly strong influences on gender socialization because of the amount of time children spend in them. Teachers often have different expectations for boys and girls. In school, boys get more attention, even if it is sometimes negative attention. When teachers of either sex respond more to boys, both positively and negatively, they heighten boys' sense of importance (American Association of University Women 1992, 1998; Sadker and Sadker 1994).

Religion is an often overlooked but significant source of gender socialization. The major Judeo-Christian religions in the United States place strong emphasis on gender differences, with explicit affirmation of the authority of men over women. In Orthodox

© Rob Walls/Alamy

Sports are a source of gender socialization; for young girls, this can boost self-esteem.

Even with social changes in gender roles, boys and girls tend to engage in play activities deemed appropriate for their gender.

Judaism, men offer a prayer to thank God for not having created them as a woman or a slave. The patriarchal language of most Western religions and, in some faiths, the exclusion of women from positions of religious leadership signifies the lesser status of women in religious institutions.

Any religion, interpreted in a fundamentalist way, can be very oppressive to women, as recent events in Islamic societies show. One of the important points to note about any religion and its gender teachings is that religious belief is subject to interpretation and social forces. The most devout believers of any faith tend to hold the most traditional views of women's and men's roles. The influence of religion on gender attitudes cannot be considered separately from other factors. For many, religious faith inspires a belief in egalitarian roles for women and men; both Christian and Islamic women have organized to resist fundamentalist and sexist practices (Gerami and Lehnerer 2001).

The *media* in their various forms (television, film, magazines, music, and so on) communicate strong—some would even say cartoonish—gender stereotypes. Despite some changes in recent years, television and films continue to depict highly stereotyped roles for women and men. Men on television heavily outnumber women, and women are underrepresented in the leading roles in film (Eschholz et al. 2002). Men are not only more visible, but also more formidable, stereotyped in strong, independent roles. Women are more likely now than thirty years ago to be portrayed as employed outside of the home and in professional jobs (Signorielli and Bacue 1999), but it is still more usual to see women depicted as sex objects. In fact, the sexualization of women is so extensive in the media that the American Psychological Association has concluded that there is "massive exposure to portrayals that sexualize women and girls and teach girls that women are sexual objects (American Psychological Association 2007: 5).

Gender stereotypes on television also cross with racial stereotypes. White men are shown as being more prominent and exercising more authority than either White women or African American men and women. African American men, however, are shown as aggressive, and African American women are shown as inconsequential (Coltrane and Messineo 2000).

Social scientists debate the extent to which people actually believe what they see on television, but research with children shows that they identify with the television characters they see. Children report that they want to be like the television characters when they grow up. Boys tend to identify with characters they see based on their physical strength and activity level; girls relate to perceptions of physical attractiveness (Signorielli et al. 1994). Both boys and girls rate the aggressive toys that they see on television commercials as highly desirable. They also judge them as more appropriate for boys' play, suggesting that something as seemingly innocent as a toy commercial reinforces attitudes about gender and violence (Klinger et al. 2001). Even with adults, researchers find that there is a link between viewing sexist images and having attitudes that support sexual aggression, antifeminism, and more traditional views of women (American Psychological Association 2007).

Advertisements are another important outlet for the communication of gender images to the public—one that is especially noted for the communication of idealized, sexist, and racist images of women and men. Women in advertisements are routinely shown in poses that would shock people if the characters were male. Consider how often women are displayed in ads dropping their pants, skirts, or bathrobe, or are shown squirming on beds. How often are men shown in such poses? Men are sometimes displayed as sex objects in advertising, but not nearly as often as women. The demeanor of women in advertising—on the ground, in the background, or looking dreamily into space—makes them appear subordinate and available to men.

Popular culture is also a source of stereotypes that contribute to gender socialization. Greeting cards, CD covers, books, songs, films, and comic strips all communicate images representing the presumed cultural ideals of womanhood and manhood. These popular products have an enormous effect on our ideas and self-concepts. To illustrate this, try to buy a friend a birthday card that does not stereotype women or men. Or notice that Father's Day cards imply that all men want to play golf or fish—activities you will rarely, if ever, find on Mother's Day cards! (As you look at greeting cards, you can also see how much gender stereotypes overlap with stereotypes about aging, family roles, and images of beauty.)

The Price of Conformity

A high degree of conformity to stereotypical gender expectations takes its toll on both men and women. Physical daring and risk-taking associated with men's roles can result in early death from accidents and violence. The strong undercurrent of violence in today's culture of masculinity encourages men to engage in behaviors that put them at risk in a variety of ways.

Violence associated with gender roles also puts women at risk. Men's power is too frequently manifested in physical and emotional violence against

a sociological eye on the media

Cultural Gatekeepers and the Construction of Femininity

Many have noted the distorted images of women that appear in the media. The common argument is that media images present an unrealistic image of women, which shapes women's self-concepts and limits their sense of possibilities for their appearance, their relationships, their careers, and so forth. Femininity is defined in the media by *cultural gatekeepers*—those who make decisions about what images to project. Cultural gatekeepers also have to respond to audience criticism. How they respond is an important part of the institutional process by which media images are sustained.

One sociologist, Melissa Milkie, wanted to explore how images of femininity are constructed in the media, particularly when producers encounter criticism from their audience. As readers of magazines, girls have protested many of the narrow and limiting images in the media, particularly those portraying girl's bodies.

Milkie interviewed ten top editors of leading girls' magazines to find out how they, as cultural gatekeepers, responded to the criticism from girls that images of girls in teen magazines do not reflect what "real girls" are like.

Milkie found that even the top editors think there are institutional limitations on what they can do to respond to girls' criticism. The editors who were very sensitive to the criticisms they received either said there was not much they could do about it or they dismissed the girls' complaints as misguided. They would claim the image was beyond their control, either because of the artistic process, advertisers' needs, or the culture itself. Thus, despite their positions of power, editors believed they could not fully control the images that appear. They pointed to institutional constraints that, in effect, thwarted efforts for change. Some editors simply dismissed the criticisms as girl's misreading the intent or meaning of an image.

Either way, Milkie's research shows how the organizational complexity of media institutions limits how much change is possible in how images of femininity are constructed. Market forces, advertisers, the values of producers, and the values of the public all intertwine in shaping the decisions of cultural gatekeepers. Milkie also shows, however, that people are not passive about what they see in the media, suggesting that how people respond to those images is an important part of the effect of such images in society.

Source: Milkie, Melissa A. 2002. "Contested Images of Femininity: An Analysis of Cultural Gatekeepers' Struggles with the 'Real Girl' Critique." *Gender & Society* 16 (December): 839–859.

women. Violence against women is endemic in the United States, but is also a worldwide problem. Violence takes many forms, including sexual abuse, intimate partner violence, genital mutilation, honor killings, and rape. Around the world, the United Nations is working in various ways to reduce violence against women, including some initiatives to help men examine cultural assumptions about masculinity that promote violence.

Sociologists are also finding that adhering to gender expectations of thinness for women and strength for men is related to a host of negative health behaviors, including eating disorders, smoking, and for men, steroid abuse. The dominant culture promotes a narrow image of beauty for women—one that leads many women, especially young women, to be disturbed about their body image. Striving to be thin, millions of women engage in constant dieting, fearing they are fat even when they are well within or below healthy weight standards. Many develop eating disorders by purging themselves of food or cycling through various fad diets—behaviors that can have serious health consequences. Many young women develop a distorted image of themselves, thinking they are overweight when they may actually be dangerously thin. And, despite the known risks of smoking, increasing numbers of young women smoke and do so not only because they think it "looks cool," but also because they think it will keep them thin. Eating disorders can be related to a woman having a history of sexual abuse, but they also come from the promotion of thinness as an ideal beauty standard for women—a standard that can put girls' and women's health in jeopardy (Hesse-Biber 2007).

Men may also pay the price of conformity if they too thoroughly internalize gender expectations that say they must be independent, self-reliant, and unemotional. Although men are more likely now than in the past to express intimate feelings, gender socialization discourages intimacy among them, affecting the quality of men's friendships. While conformity to traditional gender roles denies women access to power, influence, achievement, and independence in the public world, it denies men the more nurturing and other-oriented worlds that women have customarily inhabited.

Too much conformity to gender roles can be harmful to your health. Such is the case of anorexic *women who starve themselves attempting to meet cultural standards of thinness.*

Race, Gender, and Identity

Because the experiences of race and gender socialization affect each other, men and women from different racial groups have different expectations regarding gender roles. For example, when asked to rate desirable characteristics in men and women, White men and women are more likely than Latinos to select different traits for men and women—counter to the stereotype of Hispanics as holding highly polarized views of manhood and womanhood. African Americans are the group most likely to find value in both sexes displaying traits such as being assertive, athletic, self-reliant, gentle, and eager to soothe hurt feelings (Harris 1994), although African American women tend to be more liberal in this regard than African American men (Hunter and Sellers 1998). Asian American women are more likely than Asian American men to value egalitarian roles for men and women (Chia et al. 1994).

African American women, like White women, are encouraged to become nurturing and other oriented, but African American women are also socialized to become self-sufficient, aspire to an education, desire an occupation, and regard work as an expected part of a woman's role. This is most likely because African American mothers are more likely to have been employed and supporting themselves than has historically been true for White mothers; this encourages more self-sufficient attitudes on the part of daughters (Wharton and Thorne 1997).

Men's gender identity is also affected by race. Latino men, for example, bear the stereotype of *machismo*—exaggerated masculinity. Although machismo is associated with sexist behavior by men, within Latino culture it is also associated with honor, dignity, and respect (Mirandé 1979; Baca Zinn 1995). Researchers find that Latino families are rather egalitarian, with decision making frequently shared by men and women. To the extent that machismo exists, it is not just a cultural holdover from Latino societies, but can be how men defy their racial oppression.

Gender Socialization and Homophobia

Homophobia is the fear and hatred of homosexuals. Homophobia plays an important role in gender socialization because it encourages stricter conformity to traditional expectations, especially for men and young boys. Slurs directed against gays encourage boys to act more masculine as a way of affirming for their peers that they are not gay. As a consequence, homophobia also discourages so-called feminine traits in men, such as caring, nurturing, empathy, emotion, and gentleness. Men who endorse the most traditional male roles also tend to be the most homophobic (Burgess 2001; Alden 2001; Basow and Johnson

doing sociological research

Eating Disorders: Gender, Race, and the Body

Research Question: "A culture of thinness," "the tyranny of slenderness," "the beauty myth": These are terms used to describe the obsession with weight and body image that permeates the dominant culture—especially for girls and women. Just glance at the covers of popular magazines for women and girls and you will very likely find article after article promoting new diet gimmicks, each bundled with a promise that you will lose pounds in a few days if you only have the proper discipline or use the right products. Moreover, the models on the covers of such magazines are likely to be thin, often dangerously so because being too thin causes serious health problems. *Do these body ideals affect all women equally?*

Research Method: Meg Lovejoy wanted to know if the drive for thinness is unique to White women and how gendered images of the body might differ for African American and White women in the United States. Her research is based on reviewing the existing research literature on eating disorders, which has generally concluded that, compared with White women, Black women are less likely to develop eating disorders.

Research Results: Black women are less likely than White women to engage in excessive dieting and are less fearful of fat, although they are more likely to be obese and experience compulsive overeating. White women, on the other hand, tend to be very dissatisfied with their body size and overall appearance, with an increasing number engaging in obsessive dieting. Black and White women also tend to distort their own weight in opposite directions: White women are more likely to overestimate their own weight (that is, saying they are fat when they are not); Black women are more likely to underestimate their weight (saying they are average when they are overweight by medical standards).Why?

Conclusions and Implications: Lovejoy concludes that you cannot understand eating disorders without knowing the different stigmas attached to Black and White women in society. She suggests that Black women develop alternative standards for valuing their appearance as a way of resisting mainstream, Eurocentric standards. Black women who do so are then less susceptible to the controlling and damaging influence of the institutions that promote the ideal of thinness as feminine beauty. On the other hand, the vulnerability that Black women experience in society can foster mental health problems that manifest themselves in overeating. Eating disorders for Black women can also stem from the traumas that result from racism, especially when combined with sexism and other forms of oppression.

Lovejoy, along with others who have examined this issue, concludes that eating disorders must be understood in the context of social structures—gender, race, class, and ethnicity—that affect all women, although in different ways. The cultural meanings associated with bodies differ for different groups in society but are deeply linked to our concepts of ourselves and the basic behaviors—like eating—that we otherwise think of as "natural."

Questions to Consider

1. Pay attention to the music and visual images in popular culture and ask yourself what cultural messages are being sent to different race and gender groups? What messages are being conveyed about appropriate appearance? How do they affect people's body image—and their self-esteem?
2. Lovejoy examines eating disorders in the context of gender, race, class, and ethnicity. What cultural meanings are broadcast with regard to age?
3. Is there a "culture of thinness" among your peers? If so, what impact do you think it has on people's self-concept? If not, are there other cultural meanings associated with weight among people in your social groups?

Source: Lovejoy, Meg. 2001. "Disturbances in the Social Body: Differences in Body Image and Eating Problems among African American and White Women." *Gender & Society* 15 (April): 239–261.

2000). In this way, homophobia is one of the means by which socialization into expected gender roles takes place. The consequence is not only conformity to gender roles, but a learned hostility toward gays and lesbians.

Homophobia is a learned attitude, as are other forms of negative social judgments about particular groups. Homophobia is also deeply embedded in people's definitions of themselves as men and women. Boys are often raised to be manly by repressing so-called feminine characteristics in themselves. Being called a "fag" or a "sissy" is one of the peer sanctions that socializes a child to conform to particular gender roles. Similarly, pressures on adolescent girls to abandon tomboy behavior are a mechanism by which girls are taught to adopt the behaviors and characteristics associated with womanhood. Being labeled a lesbian may cause those with a strong attraction to women to repress this emotion and direct love only toward men. We can see, therefore, how homophobic ridicule, though it may be in the context of play and joking, has serious consequences for both heterosexual and homosexual men and women. Homophobia socializes most people into expected gender roles, and it produces numerous myths about gays and lesbians—examined in more detail in the following chapter.

The Institutional Basis of Gender

The process of gender socialization tells us a lot about how gender identities are formed, but gender is not just a matter of identity: Gender is embedded in social institutions. This means that institutions are patterned by gender, resulting in different experiences and opportunities for men and women. Sociologists analyze gender not just as interpersonal expectations, but as characteristic of institutions. This is what is meant by the term *gendered institution.* This concept means that entire institutions are patterned by gender.

Gendered institutions are the total pattern of gender relations that structure social institutions, including the stereotypical expectations, interpersonal relationships, and the different placement of men and women that are found in institutions. Schools, for example, are not just places where children learn gender roles but are gendered institutions because they are founded on specific gender patterns. Seeing institutions as gendered reveals that gender is not just an attribute of individuals, but is "present in the processes, practices, images and ideologies, and distributions of power in the various sectors of social life" (Acker 1992: 567).

As an example of the concept of gendered institution, think of what it is like to work as a woman in a work organization dominated by men. Women in this situation report that men's importance in the organization is communicated in subtle ways, whereas women are made to feel like outsiders. Important career connections may be made in the context of men's informal interactions with each other—both inside and outside the workplace. Women may be treated as tokens or may think that company policies are ineffective in helping them cope with the particular demands in their lives. These institutional patterns of gender affect men, too, particularly if they try to establish more balance between their personal and work lives. To say that work institutions are gendered institutions means that, taken together, there is a cumulative and systematic effect of gender throughout the institution.

Gender is not just a learned role; it is also part of social structure, just as class and race are structural dimensions of society. Notice that people do not think about the class system or racial inequality in terms of "class roles" or "race roles." It is obvious that race relations and class relations are far more than matters of interpersonal interaction. Race, class, and gender inequalities are experienced within interpersonal relationships, but they extend beyond relationships. Just as it would seem strange to think that race relations in the United States are controlled by race-role socialization, it is also wrong to think that gender relations are the result of gender socialization alone. Like race and class, gender is a system of privilege and inequality in which women are systematically disadvantaged relative to men. There are institutionalized power relations between women and men, and men and women have unequal access to social and economic resources.

Gender Stratification

Gender stratification refers to the hierarchical distribution of social and economic resources according to gender. Most societies have some form of gender stratification, although the specific form varies from country to country. Comparative research finds that women are more nearly equal in societies characterized by the following traits (Chafetz 1984):

- Women's work is central to the economy.
- Women have access to education.
- Ideological or religious support for gender inequality is not strong.
- Men make direct contributions to household responsibilities, such as housework and child care.
- Work is not highly segregated by sex.
- Women have access to formal power and authority in public decision making.

In Sweden, where there is a relatively high degree of gender equality, the participation of both men and women in the labor force and the household (including child care and housework) is promoted by government policies. Women also have a strong role in the political system, although women still earn less than men in Sweden. In many countries, women and girls have less access to education than men and boys, but that gap is closing. Still, in most countries, the illiteracy rate among women is much higher than among men (United Nations 2006b).

As the preceding list suggests, gender stratification is multidimensional. In some societies, women may be free in some areas of life but not in others. In Japan, for example, women tend to be well educated and participate in the labor force in large numbers. Within the family, however, Japanese women have fairly rigid gender roles. Yet the rate of violence against women in Japan (in the form of rape, prostitution, and pornography) is quite low in relation to other nations, even though women are widely employed as "sex workers" in hostess clubs, bars, and sex joints (Allison 1994). Patterns of gender inequality are most reflected in the wage differentials between women and men around the world, as Figure 10.3 shows.

Gender stratification can be extreme, as was the case in Afghanistan under the Taliban regime. The Taliban, an extremist militia group, seized power in Afghanistan in 1996 and stripped women and girls of basic human rights. Women were banished from the labor force, schools were closed to girls, and women who were enrolled in the universities were expelled. In Afghanistan, women were prohibited from leaving their homes unless accompanied by a close male relative. The windows of houses where women lived were painted black to keep women literally invisible to the public. This extreme segregation and exclusion of women from public life has been labeled **gender apartheid.** Gender apartheid is also evident in other nations, although not so extreme as it was under Taliban rule—in Saudi Arabia, women are not allowed to drive and in Kuwait they cannot vote.

Sexism and Patriarchy

Gender stratification tends to be supported by beliefs that treat gender inequality as natural. The term refers to *ideology,* a belief system that tries to explain and justify the status quo. Sexism is an ideology that defines women as different from and inferior to men. Like racism, sexism distorts reality, making behaviors seem natural when they are actually rooted in entrenched systems of power and privilege. The idea that men should be paid more than women because they are the primary breadwinners reflects sexist ideology; however, when this concept becomes embedded in the wage structure, people no longer have to explicitly believe in the original idea for the consequences of sexism to be propagated.

Sexism and racism tend to go hand in hand. Both generate social myths that have no basis in fact, but

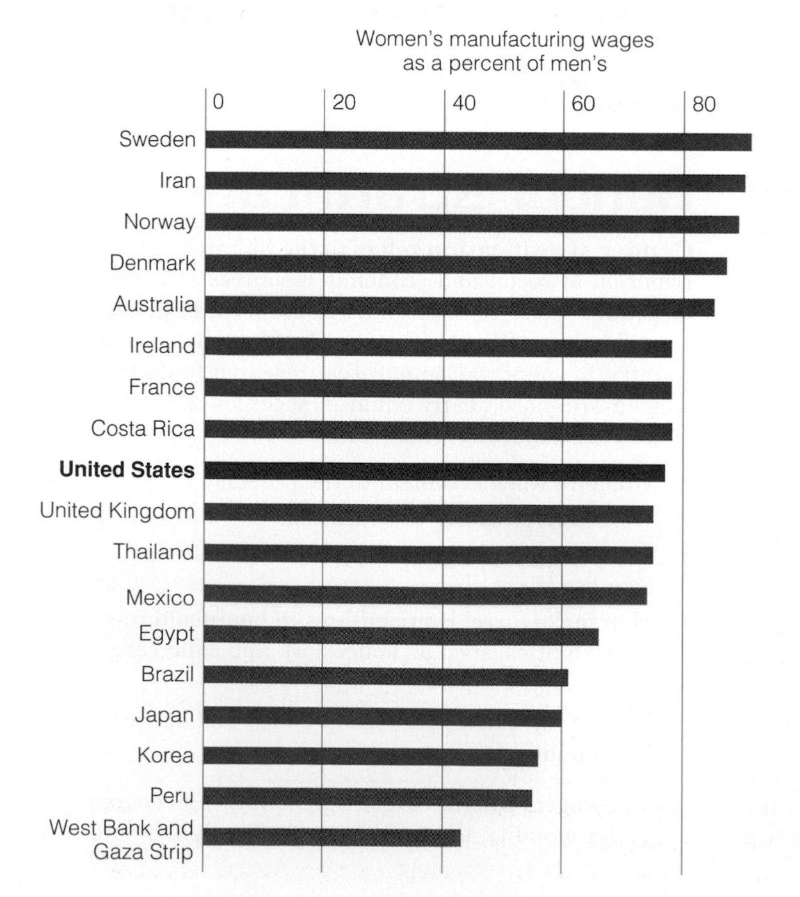

FIGURE 10.3 *The Wage Gap: An International Perspective*

Data: From United Nations. 2006b (August). United Nations Statistics Division: *Statistics and Indicators on Women and Men.* **www.unstats.un.org/demographic/products/default.htm**

they justify the continuing advantage of dominant groups over subordinates. A case in point is the belief that women of color are being hired more often and promoted more rapidly than others. This belief misrepresents the facts. Women rarely take jobs away from men because most women of color work in gender- and race-segregated jobs. The truth is that women, especially women of color, are burdened by obstacles to job mobility that are not present for men, especially White men (Browne 1999; Padavic and Reskin 2002). The myth that women of color get all the jobs makes White men seem to be the victims of race and gender privilege. Although there may be occasional cases where a woman of color (or a man of color, for that matter) gets a job that a White man also applied for, gender and race privilege favors White men.

Debunking Society's Myths

Myth: Because of affirmative action, Black women are taking a lot of jobs away from White men.

Sociological perspective: Sociological research finds no evidence of this claim. Quite the contrary, women of color work in gender- and race-segregated jobs and only rarely in occupations where they compete with White men in the labor market (Browne 1999; Padavic and Reskin 2002).

Patriarchy refers to a society or group in which men have power over women. Patriarchy is common throughout the world. In patriarchal societies, husbands have authority over their wives in the private sphere of their families, and public institutions are also structured around male power. Men hold all or most of the public power positions in patriarchal societies, whether as chief, president, CEO, prime minister, or another leadership position. Forms of patriarchy vary from society to society. In some, it is rigidly upheld in both the public and private spheres, and women may be formally excluded from voting, holding public office, or working outside the home. In societies such as the contemporary United States, patriarchy may be somewhat diminished in the private sphere (at least in some households), but the public sphere continues to be based on patriarchal relations.

Matriarchy has traditionally been defined as a society or group in which women have power over men. Anthropologists have debated the extent to which such societies exist, but new research finds that matriarchies do exist, though not in the form the customary definition implies. Based on her study of the Minangkabau—a matriarchal society in West Sumatra (in Indonesia)—anthropologist Peggy Sanday argues that scholars have used a Western definition of power that does not apply in non-Western societies. The Minangkabau define themselves as a matriarchy, meaning that women hold economic and social power. However, the Minangkabau are not ruled by women. The people believe that rule should be by consensus, including that of men and women. Thus, matriarchy exists but not as a mirror image of patriarchy (Sanday 2002).

In sum, gender stratification is an institutionalized system that rests on specific belief systems supporting the inequality of men and women. Although theoretically one could have a society stratified by gender where women hold power over men, that is not how gender stratification has evolved. In the next sections, different manifestations of gender stratification in the United States will be examined, especially as it involves women's economic position in relation to men.

Women's Worth: Still Unequal

Gender stratification is especially obvious in the persistent earnings gap between women and men (see Figure 10.4). Although the gap has closed somewhat since the 1960s when women earned 59 percent of what men earned, women who work year-round and full time still earn, on average, only 77 percent of what men earn. Women with bachelor's degrees earn the equivalent of men with associate's degrees (see Figure 10.5). In 2005, the median income for women working full time and year-round was $31,858; for men, it was $41,386 (DeNavas-Walt et al. 2006).

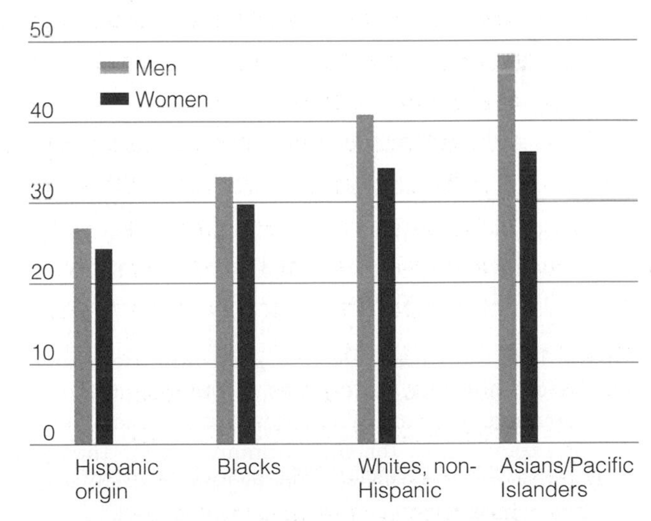

FIGURE 10.4 *Median Income by Race and Gender*

Data include only workers working a full-time job, 50 weeks per year or more.

Source: U.S. Census Bureau. 2007. *Tables of Income, Person Income Table: Work Experience in 2005, People 15 Years and Older by Total Money Earnings, Age, Race, Hispanic Origin, and Sex,* Table PINC -05. Washington, DC: U.S. Department of Commerce. **www.census.gov**

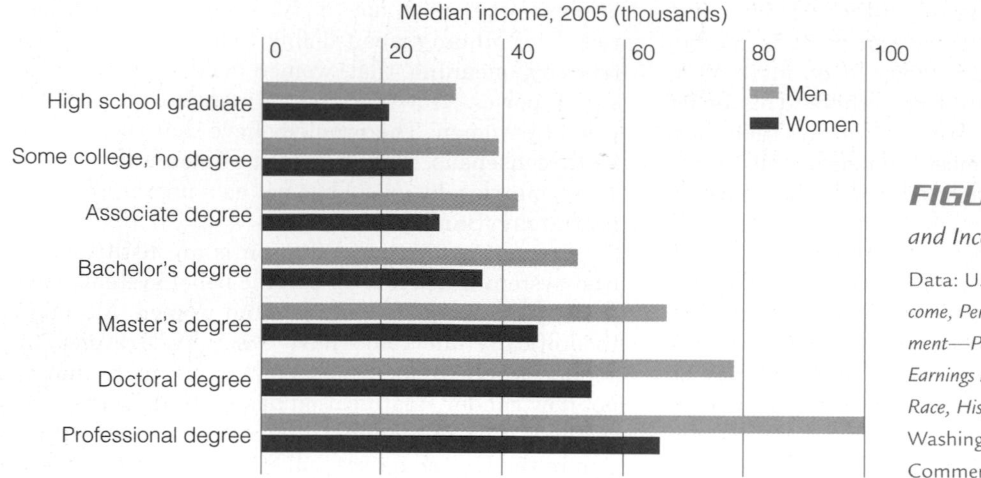

FIGURE 10.5 *Education, Gender, and Income*

Data: U.S. Census Bureau. 2007. *Tables of Income, Person Income Table: Educational Attainment—People 25 Years Old and Over, by Total Money Earnings in 2005, Work Experience in 2005, Age, Race, Hispanic Origin, and Sex,* Table PINC-03. Washington, DC: U.S. Department of Commerce. **www.census.gov**

The income gap between women and men persists despite the increased participation of women in the labor force. The **labor force participation rate** is the percentage of those in a given category who are employed either part time or full time. Sixty percent of all women are in the paid labor force compared with 73 percent of men. Since 1960, married women with children have nearly tripled their participation in the labor force. Two-thirds of mothers are now in the labor force, including more than half of mothers with infants. Current projections indicate that women's labor force participation will continue to rise and men's will decline slightly (U.S. Department of Labor 2005).

This pattern of women being in the labor market has long been true for women of color but now also characterizes the experience of White women; the labor force participation rates of White women and women of color have, in fact, converged. More women in all racial groups are now the sole supporters of their families.

Laws prohibiting gender discrimination have been in place for more than forty years. The Equal Pay Act of 1963 was the first federal law to require that men and women should receive equal pay for equal work, an idea that is supported by the majority of Americans. But wage discrimination is rarely overt. Most employers do not even explicitly set out to pay women less than men. Despite good intentions and legislation, however, differences in men's and women's earnings persist. Why? Research reveals four strong explanations for this: human capital theory, dual labor market theory, gender segregation, and overt discrimination.

Human Capital Theory. Gender differences in wages are explained by **human capital theory** as the result of differences in the individual characteristics that workers bring to jobs. Human capital theory assumes that the economic system is fair and competitive and that wage discrepancies reflect differences in the resources (or human capital) that individuals bring to their jobs. Factors such as age, prior experience, number of hours worked, marital status, and education are human capital variables. Human capital theory asserts that the extent to which human beings vary in these characteristics will influence their worth in the labor market. For example, higher job turnover rates or work records interrupted by childrearing and family responsibilities could negatively influence the earning power of women.

Much evidence supports the human capital explanation for the differences between men's and women's earnings because education, age, and experience do influence earnings. When we compare men and women who have the same level of education, previous experience, and number of hours worked per week, women still earn less than men (see Figure 10.5). Although human capital theory explains some differences between men's and women's earnings, it does not explain it all. Sociologists have looked to other factors to complete the explanation of wage inequality (Browne 1999).

The Dual Labor Market. A second explanation of discrepancies in men's and women's earnings is **dual labor market theory,** which contends that women and men earn different amounts because they tend to work in different segments of the labor market. The dual labor market reflects the devaluation of women's work because it is in low-wage jobs that women are most concentrated. Although it is hard to untangle cause and effect in the relationship between the devaluation of women's work and low wages in certain jobs, once such an earnings structure is established, it is difficult to change. As a result, although equal pay for equal work may hold in principle, it applies to relatively few people because most men and women are not engaged in equal work.

According to dual labor market theory, the labor market is organized in two different sectors: the *primary market* and the *secondary market.* In the primary labor market, jobs are relatively stable, wages are good, opportunities for advancement exist, fringe benefits are likely, and workers are afforded due process. Working for a major corporation in a management job is an example of this. Jobs in the primary labor market are usually in large organizations where there is general stability, steady profits, and a rational system of management. In contrast, the secondary labor market is characterized by high job turnover, low wages, short or nonexistent promotion ladders, few benefits, poor working conditions, arbitrary work rules, and capricious supervision. Many of the jobs students take—such as waiting tables, selling fast food, or cooking and serving fast food—fall into this category. For students, however, these jobs are usually short term.

Within the primary labor market, there are two tiers. The first consists of high-status professional and managerial jobs with potential for upward mobility, room for creativity and initiative, and more autonomy. The second tier comprises working-class jobs, which include clerical work and skilled and semiskilled blue-collar work. Women and minorities in the primary labor market tend to be in the second tier. Although these jobs are secure compared with jobs in the secondary labor market, they are more vulnerable and do not have as much mobility, pay, prestige, or autonomy as jobs in the first tier of the primary labor market.

There is, in addition, an informal sector of the market where there is even greater wage inequality; no benefits; and little, if any, oversight of employment practices. Individuals may hire such workers as private service workers or under the table workers who perform a service for a fee (painting, babysitting, car repairs, and any number of services). Although there are no formal data on the informal sector because much of it tends to be in an underground economy, it is likely that women and minorities form a large segment of this market activity. White men in this sector are also disadvantaged because of the instability and lack of protection in this work.

AP Images/Lincoln Journal Star, Robert Eecker

Data show that occupations where women of color predominate also tend to have the lowest wages.

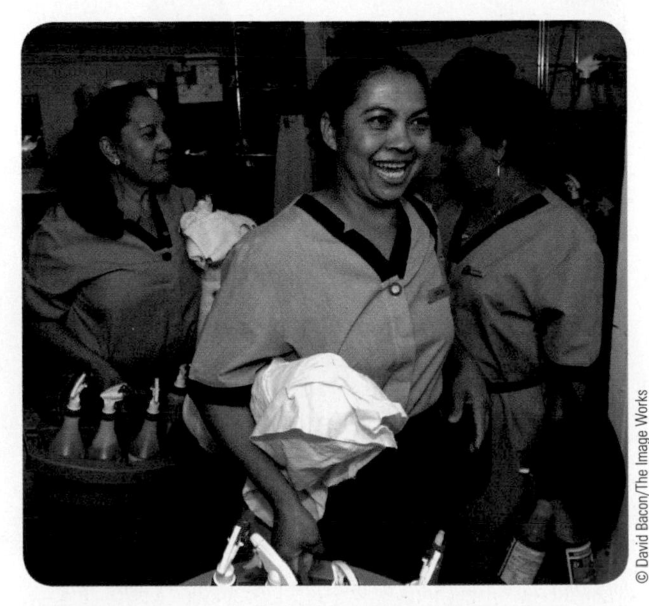
© David Bacon/The Image Works

Women of color, such as these hotel maids, tend to be segregated in low-wage, service work.

Gender Segregation. Dual labor market theory explains wage inequality as a function of the structure of the labor market, not the individual characteristics of workers as suggested by human capital theory. Because of the dual labor market, men and women tend to work in different occupations and, when working in the same occupation, in different jobs. This is referred to as **gender segregation,** a pattern in which different groups of workers are separated into occupational categories based on gender. There is a direct association between the number of women in given occupational categories and the wages paid in those jobs. In other words, the greater the proportion of women in a given occupation, the lower the pay (U.S. Department of Labor 2005). Gender segregation is a specific form of **occupational segregation**; segregation in the labor market can also be based on factors such as race, class, age, or any combination thereof.

Despite several decades of legislation prohibiting discrimination against women in the workplace, most women and men still work in gender-segregated occupations. That is, the majority of women work in occupations where most of the other workers are women, and the majority of men work mostly with men. Women also tend to be concentrated in a smaller range of occupations than men. To this day, almost two-thirds of all employed women work as clerical workers and sales clerks or in service occupations such as food-service workers, maids, health-service workers, hairdressers, and child-care workers. Men are dispersed over a much broader array of occupations. Women make up 81 percent of elementary and middle school teachers, 67 percent of retail sales workers, 97 percent of secretaries, 91 percent of bookkeepers, and 95 percent of child-care workers—stark evidence of the persistence of gender segregation in the labor force (U.S. Department of Labor 2005).

Gender segregation also occurs within occupations. Women usually work in different jobs from men, but when they work within the same occupation, they are segregated into particular fields or job types. For example, in sales work, women tend to do noncommissioned sales or to sell products that are of less value than those men sell. Or, among waiters and waitresses, women often work in lower-priced restaurants where they are likely to be tipped less than men (Hall 1993).

Overt Discrimination. A fourth explanation of the gender wage gap is discrimination. **Discrimination** refers to practices that single out some groups for different and unequal treatment. Despite the progress of recent years, overt discrimination continues to afflict women in the workplace. It is argued that men (especially White men), by virtue of being the dominant group in society, have an incentive to preserve their advantages in the labor market. They do so by establishing rules that distribute rewards unequally. Women pose a threat to traditional White male privileges, and men may organize to preserve their own power and advantage (Reskin 1988).

Within institutions, dominant groups will use their position of power to perpetuate their advantage (Lieberson 1980). Historically, White men used labor unions to exclude women and racial minorities from well-paying, unionized jobs, usually in the blue-collar trades. A more contemporary example is seen in the efforts of some conservative groups, usually led by men, to dilute legislation that has been developed to assist women and racial–ethnic minorities. These efforts can be seen as an attempt to preserve group power.

Another example of overt discrimination is the harassment that women experience at work, including *sexual harassment* and other means of intimidation. Sociologists see such behaviors as ways for men to protect their advantages in the labor force. No wonder that women who enter traditionally male-dominated professions suffer the most sexual harassment; the reverse seldom occurs for men employed in jobs historically filled by women. Although men can be victims of sexual harassment, this is rare. Sexual harassment is a mechanism for preserving men's advantage in the labor force—a device that also buttresses the belief that women are sexual objects for the pleasure of men.

Each of these explanations—human capital theory, dual labor market theory, gender segregation, and overt discrimination—contributes to an understanding of the continuing differences in pay between women and men. Wage inequality between men and women is clearly the result of multiple factors that together operate to systematically place women at a disadvantage in the workplace.

The Devaluation of Women's Work

Across the labor market, women tend to be concentrated in those jobs that are the most devalued, causing some to wonder if the fact that the jobs are held by women leads to devaluation of the jobs. Why, for example, is pediatrics considered a less prestigious specialty than cardiology? Why are elementary school teachers (81 percent of whom are women) paid less than airplane mechanics (97 percent of whom are men)? The association of elementary school teaching with children and its identification as "women's work" lowers its prestige and economic value. Indeed, if measured by the wages attached to an occupation, child care is one of the least prestigious jobs in the nation—paying on average only $334 per week in 2004, which would come out to an income below the federal poverty line if you worked every week of the year (U.S. Department of Labor 2005).

Only a small proportion of women work in occupations traditionally thought to be men's jobs (such as the skilled trades). The representation of women in skilled blue-collar jobs has increased fourfold since 1940 and from 2 to 8 percent in 2004, but that is still less than one in ten blue-collar workers (U.S. Department of Labor 2005). Likewise, very few men work

Jobs that have historically been defined as "women's work" are some of the most devalued in terms of income and prestige, despite their importance for such things as nurturing children.

in occupations historically considered to be women's work, such as nursing, elementary school teaching, and clerical work. Interestingly, men who work in occupations customarily thought of as women's work tend to be more upwardly mobile within these jobs than are women who enter fields traditionally reserved for men (Williams 1992; Budig 2002). Gender segregation in the labor market is so prevalent that most jobs can easily be categorized as men's work or women's work. Occupational segregation reinforces the belief that there are significant differences between the sexes. Think of the characteristics of a soldier. Do you imagine someone who is compassionate, gentle, and demure? Similarly, imagine a secretary. Is this someone who is aggressive, independent, and stalwart? The association of each characteristic with a particular gender makes the occupation itself a gendered occupation.

For all women, perceptions of gender-appropriate behavior influence the likelihood of success within institutions. Even something as simple as wearing makeup has been linked to women's success in professional jobs (Dellinger and Williams 1997). When men or women cross the boundaries established by occupational segregation, they are often considered to be gender deviants. They may be stereotyped as homosexual and have their "true gender identity" questioned. Men who are nurses may be stereotyped as effeminate or gay; women Marines may be stereotyped as "butch." Social practices like these serve to reassert traditional gender identities, perhaps softening the challenge to traditionally male-dominated institutions that women's entry challenges (Williams 1989, 1995).

As a result, many men and women in nontraditional occupations feel pressured to assert gender-appropriate behavior. Men in jobs historically defined as women's work may feel impelled to emphasize their masculinity, or if they are gay, they may feel even more pressure to keep their sexual identity secret. Such social disguises can make them seem unfriendly and distant, characteristics that can have a negative effect on evaluations of their professional performance. Heterosexual women in male-dominated jobs may feel obliged to squash suspicions that they are lesbians or excessively mannish. Lesbian women may be especially wary about having their sexual identity revealed. Studies have found that lesbian women are more likely to be open about their sexual identity at work when they work predominantly with women and have women as bosses (Schneider 1984).

Although they are still a small minority, some women have entered jobs in the skilled trades that were traditionally held only by men.

*Is It True?**

	True	False
1. Men are more aggressive than women.		
2. Parents have the most influence on children's gender identities.		
3. Most women hold feminist values.		
4. In all racial-ethnic groups, women earn less on average than men.		
5. The wage gap between women and men has closed since the 1970s largely as the result of women being more likely to enter the labor force.		
6. In terms of wages, middle-class women have most benefited from antidiscrimination policies.		

*** The answers can be found on page 280.**

Is It True? (Answers)

1. FALSE. Generalizations such as this ignore variation occurring within gender categories; moreover, "aggression" is a broad term that can have multiple meanings.
2. FALSE. There are numerous sources of gender socialization; even parents who try to raise their children not too conform too strictly to gender norms will find that peers, the media, schools, and other socialization agents all push people into the expected behaviors associated with gender.
3. TRUE. Although many women do not use the label "feminist" to define themselves, surveys show that the majority of women agree with basic feminist principles. Self-identification as a feminist is most likely among well-educated, urban women (McCabe 2005).
4. TRUE. However, the gap in median income is not as wide *within* some groups as it is in others. White women, for example, earn 73 percent of what White men earn, but Hispanic women earn 88 percent of what Hispanic men earn (because both have very low earnings on average). Black women earn 83 percent of Black men's earnings and Asian women, 75% of men's earnings. And White and Asian American women, on average, earn *more* than Black and Hispanic men (U.S. Census Bureau 2005).
5. FALSE. The most significant reason for the decline in the wage gap between women and men is the decline in men's wages; a smaller portion of this closing gap is attributed to changes in women's wages (Mishel et al. 2005).
6. FALSE. Although all women do benefit from equal employment legislation, wage data indicate that the group whose wages have increased the most since the 1970s are women in the top 20 percent of earners. Middle- and working-class have seen far lower gains, and poor women's wages have been relatively flat over this period of time (Mishel et al. 2005).

Balancing Work and Family

As the participation of women in the labor force has increased, so have the demands of keeping up with work and home life. Although some changes are evident, women continue to hold primary responsibility for meeting the needs of families, as we will see in more detail in Chapter 12. Many men are now much more involved in housework and child care than has been true in the past, although most of this work still falls to women—a phenomenon that has been labeled "the second shift."

The social speedup that comes from increased hours of employment for both men and women (but especially women), coupled with the demands of maintaining a household, are a source of considerable stress (Hochschild 1989; Jacobs and Gerson 2004). Women continue to provide most of the labor that keeps households running—cleaning, cooking, running errands, driving children around, and managing household affairs. Although more men are engaged in housework and child care, a huge gender gap remains in the amount of such work done by women and men. Women are also much more likely to be providing care, not just for children, but also for their older parents. The strains produced by these demands have made the home seem more and more like work for many; a large number of women and men report that their days at both work and home are harried and that they find work to be the place where they find emotional gratification and social support. In this contest between home and work, simply finding time can be an enormous challenge (Hochschild 1997). It is not surprising then that women report stress as one of their greatest concerns (Newport 2000).

Theories of Gender

Why is there gender inequality? The answer to this question is important, not only because it makes us think about the experiences of women and men, but also because it guides attempts to address the persistence of gender injustice. The major theoretical frameworks in sociology provide some answers, but feminist scholars have also found that traditional perspectives in the discipline are inadequate to address the new issues that have emerged from feminist research.

The Frameworks of Sociology

The major frameworks of sociological theory—functionalism, conflict theory, and symbolic interaction—provide some answers to the question of why gender inequality exists, although, as we will see in the next section, feminist scholars have developed additional new theories to address women's lives directly. Functionalists, for example, have been criticized for interpreting gender as a fixed role in society. *Functionalist theory* purports that men fill instrumental roles in society whereas women fill expressive roles and presumes that this arrangement works to the benefit of society (see Chapter 1). Feminists object to such a characterization, arguing that this presumes sexist arrangements are functional for society. Limiting women's roles to expressive functions and men's to instrumental functions is dysfunctional according to feminists—both for men and women. Although few contemporary functionalist theorists would make such traditionalist arguments, functionalism does emphasize people's socialization into prescribed roles as the major impetus behind gender inequality. Thus, conditions such as wage inequality, a functionalist might argue, are the result of choices women make that may result in their inequality but that nonetheless involve functional adaptation to the competing demands of family and work roles.

Conflict theorists, in contrast, see women as disadvantaged by power inequities between women and men that are built into the social structure. This includes economic inequity, as well as women's disadvantages in political and social systems. Conflict theorists, for example, see wage inequality as produced from men's historic power to devalue women's work and to benefit as a group from the services that women's labor provides. At the same time, conflict theorists have been much more attuned to the interactions of race, class, and gender inequality because they see all forms of inequality as stemming from the differential access to resources that dominant groups in society have.

From an *ethnomethodological* perspective (see Chapters 2 and 4), feminist scholars have developed what is known as "doing gender"—a theoretical perspective that interprets gender as something accomplished through the ongoing social interactions people have with one another (West and Zimmerman 1987; West and Fenstermaker 1995). Seen from this framework, people produce gender through the interaction they have with one another and through the interpretations they have of certain actions and appearances. In other words, gender is not something that is an attribute of different people, as functionalists suggest; rather, it is constantly made up and reproduced through social interaction. When you act like a man or act like a woman, you are confirming gender and reproducing the existing social order (Peralta 2002).

From this point of view, gender is relatively easy to change because all it would take for this to occur is for people to behave differently. This is one reason the theory has been criticized by those with a more macrosociological point of view; they say it ignores the power differences and economic differences that exist based on gender, race, and class. In other words, it does not explain the structural basis of women's oppression (Collins et al. 1995).

SEE FOR YOURSELF

CHANGING GENDER

Try an experiment based on the example of changing *gender* that opens this chapter.

1. First, make a list of everything you think you would have to do to change your behavior if you were a member of a different gender. Separate the things in your list according to whether they are related to such factors as appearance, attitude, or behavior.

2. Second, for a period of twenty-four hours, try your best to change any of these things that you are willing to do. Keep a log that records how others react to you during this period and how the change makes you feel.

3. When your experiment is over, write a report on what your brief experiment tells you about "doing gender" and how gender identities are supported (or not) through social interaction.

Feminist Theory

Feminism is not a single way of thinking and acting; it fundamentally refers to advocating a more just society for women. **Feminist theory** has emerged from the women's movement and refers to analyses that seek to understand the position of women in society for the explicit purpose of improving their position in society. Four major frameworks have developed in feminist theory: liberal feminism, socialist feminism, radical feminism, and multiracial feminism (see Table 10.1).

Liberal feminism emerged from a long tradition that began among British liberals in the nineteenth century. Liberal feminism argues that inequality for women originates in traditions of the past that pose barriers to women's advancement. It emphasizes individual rights and equal opportunity as the basis for social justice and reform. The framework of liberal feminism has been used to support many of the legal changes required to bring about greater equality for women in the United States. Liberal feminists contend that gender socialization contributes to women's inequality because it is through learned customs that inequality is perpetuated. In the interests of social change, liberal feminism advocates the removal of barriers to women's advancement and the development of policies that promote equal rights for women.

Socialist feminism is a more radical perspective that interprets the origins of women's oppression in the system of capitalism. Because women constitute a cheap supply of labor, they are exploited by capitalism in much the same way that the working class is exploited. In the view of socialist feminists, capitalism interacts with patriarchy to make women less powerful, both as women and as laborers. Socialist feminists are critical of liberal feminism for not addressing the fundamental inequalities built into capitalist and patriarchal systems. To these feminists,

table 10.1 Feminist Theory: Comparing Perspectives

	Liberal Feminism	Socialist Feminism	Radical Feminism	Multiracial Feminism
Gender Identity	Identity is learned through patterns of gender role socialization.	Gender division of labor reflects the needs of a capitalist workforce.	Women's identification with men gives men power over women.	Women and men of color form an oppositional consciousness as a reaction against oppression.
Gender Inequality	Inequality is the result of formal barriers to equal opportunity.	Gender inequality stems from class relations.	Patriarchy is the basis for women's powerlessness.	Race, class, and gender intersect to form a matrix of domination.
Social Change	Change is accomplished through legal reform and attitudinal change.	Transformation of the gender division of labor accompanies change in the class division of labor.	Liberation comes as women organize on their own behalf.	Women of color become agents of feminist change through alliances with other groups.

equality for women will come only when the economic and political system is changed.

Radical feminism interprets patriarchy as the primary cause of women's oppression. To radical feminists, the origins of women's oppression lies in men's control over women's bodies; thus, they see violence against women—in the form of rape, sexual harassment, domestic violence, and sexual abuse—as mechanisms that men use to assert their power in society. Radical feminists think that change cannot come about through the existing system because that system is controlled and dominated by men.

Most recently, **multiracial feminism** has developed new avenues of theory for guiding the study of race, class, and gender (Andersen and Collins 2007; Baca Zinn and Dill 1996; Chow et al. 1996; Collins 1990, 1998). Multiracial feminism evolves from studies pointing out that earlier forms of feminist thinking excluded from analysis women of color, which made it impossible for feminists to deliver theories that informed people about the experiences of all women. Multiracial feminism examines the interactive influence of gender, race, and class, showing how together they shape the experiences of all women and men (Baca Zinn and Dill 1996).

These perspectives provide unique ways of looking at the experiences of women and men in society. These theoretical orientations have been the bedrock on which feminists have built their programs of social and political change.

Gender in Global Perspective

Increasingly, the economic condition of women and men in the United States is also linked to the fortunes of people in other parts of the world. The growth of a global economy and the availability of a cheaper industrial labor force outside the United States mean that U.S. workers have become part of an international division of labor. U.S.–based multinational corporations looking around the world for less expensive labor often turn to the developing nations and find that the cheapest laborers are women or children. The global division of labor is thus acquiring a gendered component, with women workers, usually from the poorest countries, providing a cheap supply of labor for manufacturing products that are distributed in the richer industrial nations.

Worldwide, women work as much or more than men. It is difficult to find a single place in the world where the workplace is not segregated by gender. On a worldwide scale, women also do most of the work associated with home, children, and the elderly. While women's paid labor has been increasing, their unpaid labor in virtually every part of the world exceeds that of men. The United Nations estimates that the value of women's unpaid work (both in the home and in the community) amounts to at least $11 trillion (www.un.org).

Despite these general trends, women's situations differ significantly from nation to nation. China is unusual in that there is far greater sharing of household responsibilities than in most other nations. In China, both women and men work long hours in paid employment; a typical work week is six days long, or forty-eight hours. In the paid labor force are 82 percent of women and 83 percent of men, and women are encouraged to stay in the labor force when they have children. There are extensive child-care facilities in China and a fifty-six-day paid maternity leave. Many work organizations have extended this paid leave to six months, although women can lose seniority rights when they are on maternity leave (something that is illegal in the United States).

In contrast, Japan has marked inequality in the domestic sphere. Women are far more likely to leave the labor force after marrying or following childbirth, and Japanese women's identities are more defined by their roles at home, although for many this is changing. Compared with China, Japanese women more closely resemble the pattern that exists in Britain and, to some extent, in the United States, although they are less involved in paid employment than in either of these countries. Ironically, when comparing China, Japan, and Britain, researchers have found that Chinese women are the most discontented with what they perceive as gender injustice, whereas Japanese and British women express greater satisfaction with more limited employment. This may seem surprising, given the greater gender equality of Chinese women with Chinese men. Sociologists explain it as the result of the gap Chinese women see between official ideologies of gender equality and their observations of continuing inequalities in promotions and other benefits of work (Xuewen et al. 1992).

Work is not the only measure by which the status of women throughout the world is inferior to that of men. Women are vastly underrepresented in national parliaments (or other forms of government) everywhere; in only sixteen countries of the entire world is women's representation in national parliaments above 25 percent. Worldwide, women hold only 12 percent of all parliamentary seats. Only twenty-eight nations have ever had a woman as head of state (www.un.org).

The United Nations has also concluded that violence against women and girls is a global epidemic and one of the most pervasive violations of human rights (UNICEF 2000; United Nations 2006b). Violence against women takes many forms, including rape, domestic violence, infanticide, incest, genital mutilation, and murder (including so-called honor killings, where a woman may be killed to uphold the honor of the family if she has been raped or accused of adultery). Although violence is pervasive, some specific groups of women are more vulnerable than others—namely minority groups, refugees, women with disabilities, elderly women, poor and migrant women, and women living in countries where there is armed conflict. Statistics on the extent of violence against women are hard to report with accuracy, both because of the secrecy that surrounds many forms of violence and because of differences in how different nations might report violence. Nonetheless, the United Nations estimates that between 20 and 50 percent of women worldwide have experienced violence from an intimate partner or family member.

As we saw in Chapter 8, many factors put women at risk of violence, including cultural norms, women's economic and social dependence on men, and political practices that either provide inadequate legal protection or provide explicit support for women's subordination (as in the example of the Taliban given earlier).

Gender and Social Change

Few lives have not been touched by the transformations that have occurred in the wake of the feminist movement (see also Chapter 16). The women's movement has changed how women's issues are perceived in the public consciousness. It has created access to work opportunities, generated laws that protect women's rights, and spawned organizations that lobby for public policies on behalf of women. Many young women and men now take for granted freedoms that their generation is among the first to enjoy. These include access to birth control, equal opportunity legislation, and laws protecting against sexual harassment, as well as an increased presence of women on corporate boards, increased athletic opportunities for women, more presence in political life, and greater access to child care, to name a few changes. These impressive changes occurred in a relatively short period. How have they affected public consciousness about gender?

Contemporary Attitudes

In recent years, public attitudes toward gender roles have changed noticeably, especially regarding beliefs about the ideal lifestyle. Only a small minority of people disapprove of women being employed while they have young children, and both women and men say it is not fair for men to be the sole decision maker in the household. Although the majority of women now want to combine work and family, they believe they will be discriminated against in the labor force if they do so. A majority of women (87 percent) say that making laws to establish equal pay should be a legislative priority (Greenhouse 2000).

People's beliefs about appropriate gender roles have evolved as women's and men's lives have changed. Still, nearly half of all men (45 percent) now think one parent should stay at home while the other is employed, compared to 38 percent of all women who think so. Among young people (aged 18–29), 31 percent think one parent should stay home. The majority of men (68 percent) still say they would prefer to work outside the home, rather than staying home to take care of the house and family; 43 percent of women prefer to work outside the home (Moore 2005; McComb 2001). In reality, in more than half (51 percent) of married couple families, both husband and wife are employed (U.S. Bureau of Labor Statistics 2006a). Men's support for women's roles in the family and at work, however, varies across different groups.

Younger men and single men are more egalitarian than older, married men. Among college students, however, women hold more egalitarian views of women's roles than do men; though both become less traditional in their views during college, women change more than men (Bryant 2003). There are also racial and ethnic differences in how different groups view gender roles, with minority men usually being more supportive of egalitarian roles than White men. The mothers of minority men are more likely to have been employed, and these men have different educational and employment backgrounds than White men. Their attitudinal differences reflect the economic necessity that minority men attribute to women working (Wilkie 1993; Blee and Tickamyer 1995).

Debunking Society's Myths

Myth: The men most likely to support equality for women are White, middle-class men with a good education.

Sociological perspective: Although it is true that younger men tend to be more egalitarian than older men, African American men are the most likely to support women's equal rights and the right of women to work outside the home. On most measures of feminist beliefs, African American men tend to be more liberal than White men (Hunter and Sellers 1998).

Old attitudes do not die easily, and changed attitudes do not necessarily mean changed behavior. Gender attitudes change as society changes. However, young people's expectations for being able to "have it all" can be unrealistic. As an example, sociologist Michele Hoffnung surveyed a random sample of college women in their senior year, surveying them again seven years later (in 2001). She found that, as seniors, most of the women wanted careers, marriage, and motherhood, with career development being their top priority in their twenties. But at the seven-year point, those who had become mothers had fewer advanced degrees and lower career status than the nonmothers; marriage was not related to career status (Hoffnung 2004).

© Rick Steele/UPI/Landov

Young women are the group most likely to support feminist goals, although the feminist movement has support across generations, as was apparent at the March for Women's Lives in March 2004 in Washington, DC.

It is likely that further adjustments in the attitudes of men and women are on the way because it seems unlikely that the roles of men and women will return to old patterns in the future. Attitudes, however, are only part of the problem of persistent gender inequality. Social change requires more than changing individual attitudes; it also means changing social institutions.

Legislative Change

Much legislation prohibits overt discrimination against women. In addition to the Equal Pay Act of 1963, the Civil Rights Act of 1964, adopted as the result of political pressure from the civil rights movement, banned discrimination in voting and public accommodations and required fair employment practices. Specifically, Title VII of the Civil Rights Act of 1964 forbids discrimination in employment on the basis of race, color, national origin, religion, or sex.

The passage of the Civil Rights Act, and Title VII in particular, opened up new opportunities to women in employment and education. This was further supported by Title IX, adopted as part of the Educational Amendments of 1972, which forbids gender discrimination in any educational institution receiving federal funds. Title IX prohibits colleges and universities from receiving federal funds if they discriminate against women in any program, including athletics. Adoption of this bill radically altered the opportunities available to women students and laid the foundation for many of the coeducational programs that are now an ordinary part of college life. This law has been particularly effective in opening up athletics to women.

Passage of antidiscrimination policies does not, however, guarantee their universal implementation. Has equality been achieved? In college sports, men still outnumber women athletes by more than two to one, and there is still more scholarship support for male athletes than for women. Title IX allows institutions to spend more money on male athletes if they outnumber women athletes, but it also stipulates that the number of male and female athletes should be roughly proportional to their representation in the student body. Studies of student athletes show that although there has been improvement in support for women's athletics since the implementation of Title IX, there is still a long way to go toward equity in women's sports (Lederman 1992; Sigelman and Wahlbeck 1999). Title IX is being challenged by some who argue that it has reduced opportunities for men in sports. Proponents of maintaining strong enforcement of Title IX counter this, however, by noting that budget reductions in higher education, not Title IX per se, are responsible for any reduction in athletic opportunities for men. Furthermore, they point out that men still greatly predominate in school sports.

In the workplace, a strong legal framework for gender equity is in place, yet equity has not been achieved. Since most women work in different jobs from men, principles of equal pay for equal work do not address all the inequities women experience in the labor market. **Comparable worth** is the principle of paying women and men equivalent wages for jobs involving similar levels of skill. This policy recognizes that men and women tend to work at different jobs. Comparable worth goes beyond the concept of equal pay for equal work by creating job evaluation systems that assess the degree of similarity between different kinds of jobs. Comparable worth schemes have been introduced in only a few places, but where they have been implemented, women's wages have improved (Blum 1991; Steinberg 1992; Jacobs and Steinberg 1990).

Many victories in the fight for gender equity are now also at risk. *Affirmative action* is a method for opening opportunities to women and minorities that specifically redresses past discrimination by taking positive measures to recruit and hire previously disadvantaged groups. Affirmative action has been especially effective in opening new opportunities for women. The national discussion of affirmative action, fueled in part by conservative resistance to these policies, is a good illustration of what many would call the limitations of liberal philosophy. Because affirmative action promotes gender- and race-specific actions to remedy the effects of past discrimination, its opponents have argued that it constitutes reverse discrimination. Those who want social policies that are gender- and race-blind (a classic liberal position, now articulated by conservatives, as well) find it difficult to support policies that are not. The problem is that gender and race inequalities continue even though these inequities may not always result from conscious discrimination. Proponents of affirmative action argue that as long as the structural conditions of gender and race inequality exist, there is still a need for race- and gender-conscious actions designed to address persistent injustices. This viewpoint was upheld by the U.S. Supreme Court in 2003 in the case *Grutter v. Bollinger.*

One solution to the problem of gender inequality is to have more women in positions of public power. Is increasing the representation of women in existing institutions enough? Without reforming the sexism in the institutions, change will be limited and may generate benefits only for groups who are already privileged. Feminists advocate restructuring social institutions to meet the needs of all groups, not just those who already have enough power and privilege to make social institutions work for them. The successes of the women's movement demonstrate that change is possible, but change comes only when people are vigilant about their needs.

AP Images/Fred Beckham

Women are showing that they can compete in sports that have historically been mostly the province of men. Here University of Pittsburgh's Ashleigh Braxton drives to the hoop while being guarded by University of Cincinnati's Carla Jacobs during the Big East women's basketball championship in 2007.

Chapter Summary

- ***How do sociologists distinguish sex and gender?***

  Sociologists use the term *sex* to refer to biological identity and *gender* to refer to the socially learned expectations associated with members of each sex. *Biological determinism* refers to explanations that attribute complex social phenomena entirely to physical or natural characteristics.

- ***How is gender identity learned?***

  Gender socialization is the process by which gender expectations are learned. One result of socialization is the formation of *gender identity*. Overly conforming to gender roles has a number of negative consequences for both women and men, including eating disorders, violence, and poor self-concepts. *Homophobia* plays a role in gender socialization because it encourages strict conformity to gender expectations.

- ***What is a gendered institution?***

  Gendered institutions are those where the entire institution is patterned by gender. Sociologists analyze gender both as a learned attribute and as an institutional structure.

- ***What is gender stratification?***

  Gender stratification refers to the hierarchical distribution of social and economic resources according to gender. Most societies have some form of gender stratification, although they differ in the degree and kind. Gender stratification in the United States is obvious in the differences between men's and women's wages.

- ***How do sociologists explain the continuing earnings gap between men and women?***

  There are multiple ways to explain the pay gap. *Human capital theory* explains wage differences as the result of individual differences between workers. *Dual labor market theory* refers to the tendency for the labor market to be organized in two sectors: the primary and secondary markets. *Gender segregation* persists and results in differential pay and value attached to men's and women's work. *Overt discrimination* against women is another way that men protect their privilege in the labor market.

- ***Are men increasing their efforts in housework and child care?***

  Many men are now more engaged in housework and child care than was true in the past, although women still provide the vast majority of this labor. Balancing work and family has resulted in social speedup, making time a scarce resource for many women and men.

- ***What is feminist theory?***

  Different theoretical perspectives help explain the status of women in society. *Feminist theory* links this explanation to the desire for improving women's lives. The major theoretical perspectives in feminism are *liberal feminism, socialist feminism, radical feminism,* and *multiracial feminism*. Each emphasizes different aspects of women's place in society.

- ***When seen in global perspective, what can be observed about gender?***

  The economic condition of women and men in the United States is increasingly linked to the fortunes of people in other parts of the world. Women provide much of the cheap labor for products made around the world. Worldwide, women work as much or more than men, though they own little of the world's property and are underrepresented in positions of world leadership.

- ***What are the major social changes that have affected women and men in recent years?***

  Public attitudes about gender relations have changed dramatically in recent years. Women and men are now more egalitarian in their attitudes, although women still perceive high degrees of discrimination in the labor force. A legal framework is in place to protect against discrimination, but legal reform is not enough to create gender equity.

Key Terms

biological determinism 263
comparable worth 285
discrimination 278
dual labor market theory 276
feminism 281
feminist theory 281
gender 263
gender apartheid 274
gender identity 265
gender segregation 277
gender socialization 265
gender stratification 273
gendered institution 273
hermaphroditism 264
homophobia 271
human capital theory 276
labor force participation rate 276
liberal feminism 281
matriarchy 275
multiracial feminism 282
occupational segregation 277
patriarchy 275
radical feminism 282
sex 263
socialist feminism 281

Online Resources

Sociology: The Essentials Companion Website

www.thomsonedu.com/sociology/andersen

Visit your book companion website where you will find more resources to help you study and write your research papers. Resources include Suggested Readings, web links, and a MicroCase Online feature that teaches you how to research society. Other resources include Learning Objectives, Internet exercises, quizzing, and flash cards.

is an easy-to-use online resource that helps you study in less time to get the grade you want NOW.

www.thomsonedu.com/login

Need help studying? This site is your one-stop study shop. Take a Pre-Test and Thomson NOW will generate a Personalized Study Plan based on your test results. The Study Plan will identify the topics you need to review and direct you to online resources to help you master those topics. You can then take a Post-Test to determine the concepts you have mastered and what you still need to work on.

11 Chapter eleven

CHAPTER ELEVEN
Chapter

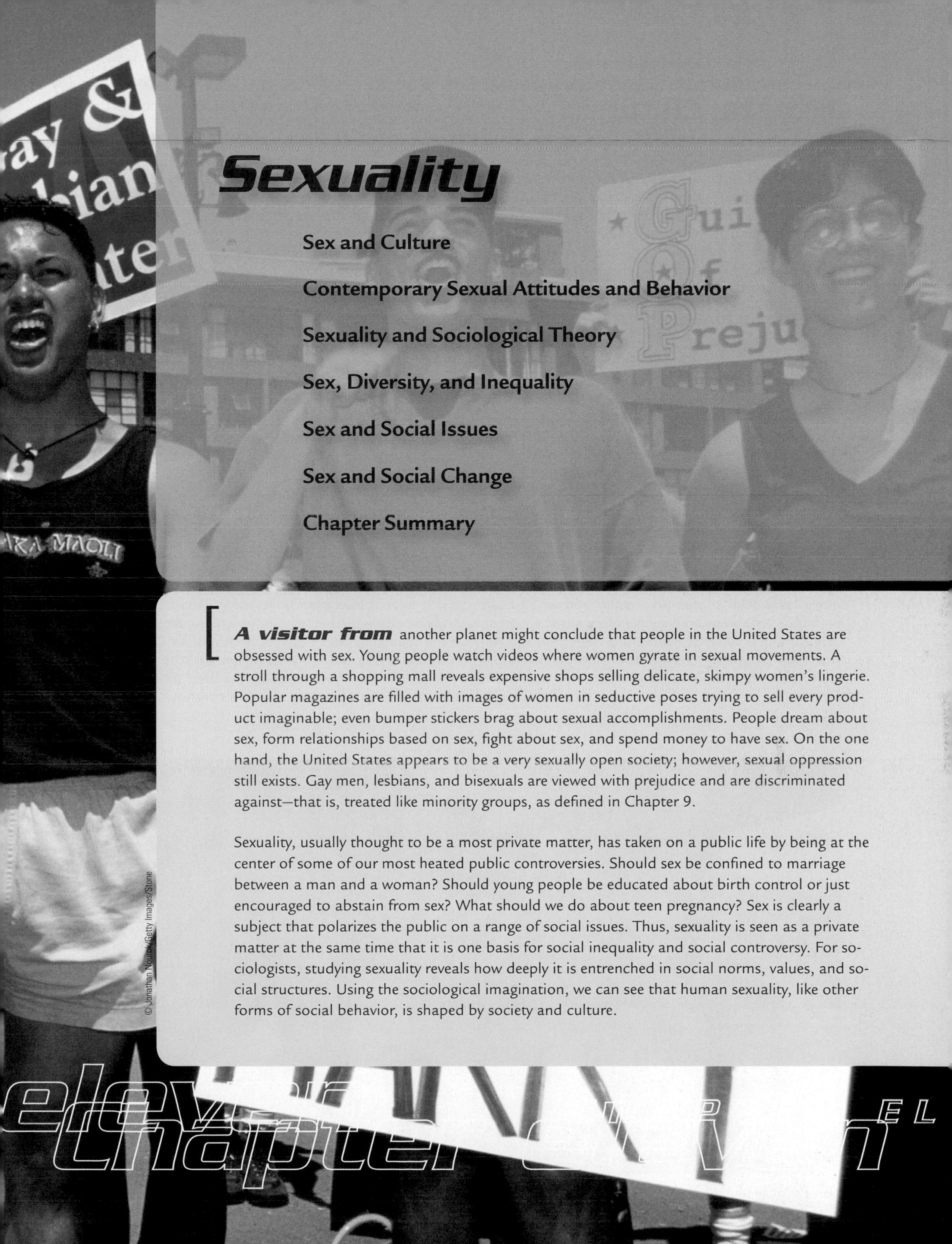

chapter eleven

Sexuality

A visitor from another planet might conclude that people in the United States are obsessed with sex. Young people watch videos where women gyrate in sexual movements. A stroll through a shopping mall reveals expensive shops selling delicate, skimpy women's lingerie. Popular magazines are filled with images of women in seductive poses trying to sell every product imaginable; even bumper stickers brag about sexual accomplishments. People dream about sex, form relationships based on sex, fight about sex, and spend money to have sex. On the one hand, the United States appears to be a very sexually open society; however, sexual oppression still exists. Gay men, lesbians, and bisexuals are viewed with prejudice and are discriminated against—that is, treated like minority groups, as defined in Chapter 9.

Sexuality, usually thought to be a most private matter, has taken on a public life by being at the center of some of our most heated public controversies. Should sex be confined to marriage between a man and a woman? Should young people be educated about birth control or just encouraged to abstain from sex? What should we do about teen pregnancy? Sex is clearly a subject that polarizes the public on a range of social issues. Thus, sexuality is seen as a private matter at the same time that it is one basis for social inequality and social controversy. For sociologists, studying sexuality reveals how deeply it is entrenched in social norms, values, and social structures. Using the sociological imagination, we can see that human sexuality, like other forms of social behavior, is shaped by society and culture.

© Jonathan Nourok/Getty Images/Stone

Sex and Culture

Sexual behavior would seem to be utterly natural. Pleasure and sometimes the desire to reproduce are reasons people have sex, but sexual relationships and identities develop within a social context. It is the social context that establishes what sexual relationships mean and how we define our sexual identities, as well as what social supports are given (or denied) to people based on their sexual identity. *Sexuality is socially defined and patterned.*

Sex: Is It Natural?

From a sociological point of view, little in human behavior is purely natural, as we have learned in previous chapters. Behavior that appears to be natural is the behavior accepted by cultural customs and sanctioned by social institutions. Sexuality creates intimacy between people. People engage in sex not just because it feels good, but also because it is an important part of our social identity. Void of a cultural context and the social meanings attributed to sexual behavior, people might not attribute the emotional commitments, psychological interpretations, spiritual meanings, and social significance to sexuality that it has in different human cultures.

Is there a biological basis to sexual identity? This question is debated in both popular and scientific literature. Gay and lesbian people often say their sexual orientation is natural, something they just are, not just something they choose. Two concepts are important in this discussion: *sexual orientation* and *sexual identity.*

Sexual orientation refers to the attraction that people feel for people of the same or different sex. The term *sexual orientation* implies something deeply rooted in a person. **Sexual identity** is the definition of oneself that is formed around one's sexual relationships. Sexual identity is learned in the context of our social relationships and the social structures in which we live. Although sometimes used interchangeably, sexual orientation and sexual identity are not the same thing, nor is one's sexual identity simply based on one's sexual practices. For example, a man may have sex with other men—perhaps even on a regular basis, but not have a sexual identity as gay. Or, you may have a sexual identity as heterosexual even in the absence of actual sexual relationships. Sexual identity, as gay, lesbian, heterosexual, or bisexual emerges in a social context, as we will see below in discussing the social construction of sexual identity. The point here is to see that sexual identity is not necessarily simply based on biological or "natural" states.

It is important to make the distinction in sexual orientation and sexual identity to untangle the often heated discussion about a presumed biological basis to sexual orientation. Gay, lesbian, and bisexual people often say that they do not choose their sexual orientation and that it is something they just "are," as if it were a biological imperative. Part of the debate about this—heated at times—comes from rejecting the idea that being gay is a choice, as if one could change their sexual orientation at will. There are political reasons for rejecting the idea of homosexuality as a choice because, if it is something inherent in

Sexual relationships, although highly personal, are also shaped by society and culture.

people, then perhaps others will be more accepting of gay, lesbian, and bisexual people.

Perhaps there is some biological basis to sexual orientation, but the evidence is not yet there. Even if a biological influence exists, social experiences are far more significant in shaping sexual identity, even though they are rarely reported in the media with as much acclaim as alleged biological bases to human sexuality (Connell 1992; Lorber 1994; Brookey 2001). Whatever the origins of sexual orientation, there is no doubt that social influences are a very significant part of all people's sexual identity.

Sometimes there is a public claim that scientists have discovered a so-called "gay gene," which presumably directs the sexual orientation of gays and lesbians. Interestingly, there are never claims about a so-called "heterosexual gene" since the implicit assumption seems to be that heterosexuality is the natural state and gay or lesbian behavior is somewhat a mutant form. There is no such scientific evidence. The evidence usually cited to claim a genetic basis to homosexuality was based on a problematic study of gay brothers (identical twins) who were found to have similar DNA markers on one of their X chromosomes. On closer examination, this research was refuted when it was found that the shared DNA markers found in pairs of gay brothers were no more likely than would be expected by chance (Wickelgren 1999). Moreover, the original study did not control for the environment in which the brothers were raised. They grew up in the same family, so obviously they were raised in the same environment. A true scientific test of the hypothesis that homosexuality is genetically based would require a stricter standard of evidence. One could study identical twins raised together who were the offspring of gay parents to see if, controlling for the environments in which they were raised, both turned out to be gay. To date, there are no such studies of biological relatives raised apart (Hamer et al. 1993).

Even if there is some yet undiscovered basis for sexual orientation, there is extensive evidence of the social influences that shape people's sexual identities. Social and cultural environments play a huge part in creating sexual identities. This is what interests sociologists—how sexual identity is constructed through social relationships and in the context of social institutions.

The Social Basis of Sexuality

We can see the social and cultural basis of sexuality in numerous ways:

1. Human sexual attitudes and behavior vary in different cultural contexts. If sex were purely natural behavior, sexual behavior would also be uniform among all societies, but it is not. Sexual behaviors considered normal in one society might be seen as peculiar in another. Think about this: In some cultures women do not believe that orgasm exists, even though biologically it does. In the eighteenth century, European and American writers advised men that masturbation robbed them of their physical powers and that instead they should apply their minds to the study of business. These cultural dictates encouraged men to conserve semen on the presumption that its release would lessen men's intelligence or cause insanity (Freedman and D'Emilio 1988; Schwartz and Rutter 1998).

Culture also affects whether we define certain sexual behaviors as normal or deviant. For example, if the culture allowed men to wear dresses, would there be a category of people known as cross-dressers or transvestites?

2. Sexual attitudes and behavior change over time. Fluctuations in sexual attitudes are easy to document. For example, public opinion polls show that young people today are more permissive in their sexual values than were young people in the past, although tolerance of casual sex has decreased somewhat in recent years (Lyons 2002). Behavioral changes are harder to document in the United States because people in this culture generally consider sex to be a private matter and thus do not always talk frankly about what they do. But, we know that young people are having sex at an earlier age (see Figure 11.1). More people also experience sex before marriage than in the past, and people have more sex partners over their lifetimes (Laumann et al. 1994).

Debunking Society's Myths

Myth: Over time in a society, sexual attitudes become more permissive.

Sociological perspective: All values and attitudes develop in specific social contexts; change is not always in a more permissive direction (Freedman and D'Emilio 1988).

3. Sexual identity is learned. Like other forms of social identity, sexual identity is acquired through socialization and ongoing relationships. Information about sexuality is transmitted culturally and becomes the basis for what we know of ourselves and others. Where did you first learn about sex? For that matter, what did you learn? For some, parents are the source of information about sex and sexual behavior. For many, peers have the strongest influence on sexual attitudes. Long before young people become sexually active, they learn **sexual scripts** that teach us what is appropriate sexual behavior for each gender (Schwartz and Rutter 1998; Thorne and Luria 1986). Children learn sexual scripts by playing roles—playing doctor as a way of exploring their bodies, or hugging and kissing in a way that can mimic heterosexual relationships. Role-playing teaches children social norms about sexuality. The roles learned in youth profoundly influence our sexual attitudes and behavior throughout life.

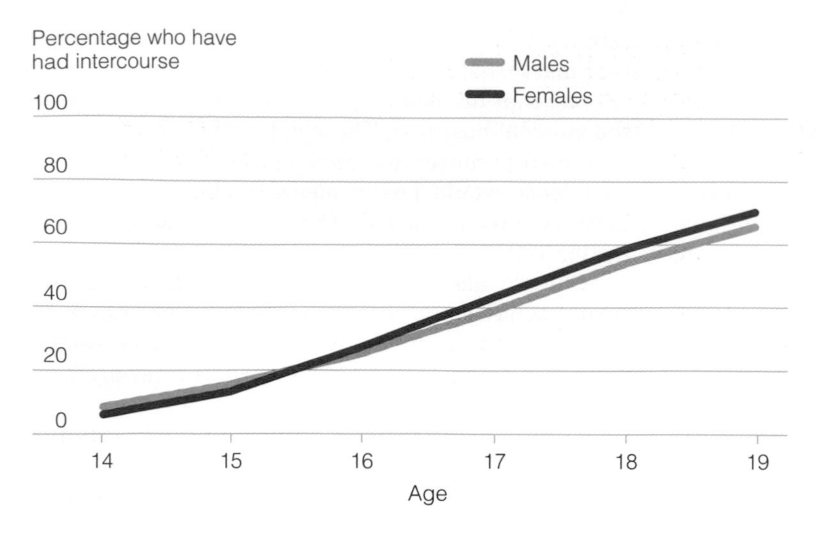

FIGURE 11.1 *Sex Among Teenagers*

Source: Guttmacher Institute. 2006b. "Facts on American Teens' Sexual and Reproductive Health." New York: Guttmacher Institute. **www.guttmacher.org**

4. Social institutions channel and direct human sexuality. Social institutions, such as religion, education, or the family, define some forms of sexual expression as more legitimate than others. For example, being heterosexual is a more privileged status in society than being gay or lesbian. Examples of institutional privileges for heterosexual couples include the right to marry and, as a result, have mutual employee health care benefits and the option to file joint tax returns.

5. Sex is influenced by economic forces in society. Sex sells. In the U.S. capitalist economy, sex appeal is used to hawk everything from cars and personal care products to stocks and bonds. Moreover, sexual acts are bought and sold as some people, particularly poor women, are forced to sell sexual services to earn a living. Sex workers are among some of the most exploited and misunderstood workers. Sometimes the economics of sex and the law intersect as quasi-legal sex trades are regulated in the form of red-light districts, selective enforcement, and even outright state licensing.

6. Public policies regulate sexual and reproductive behaviors. Prohibiting federal spending on abortion, for example, eliminates reproductive choices for women who are dependent on state or federal aid. Government decisions about which reproductive technologies to endorse will influence the choices of birth control technology available to men and to women. Government funding, or lack thereof, for sex education can influence how people understand sexual behavior. In many ways, the government intervenes in people's sexual and reproductive decision making. This fact challenges the idea that sexuality is a private matter and shows how social institutions can direct sexual behavior.

To summarize, human sexual behavior occurs within a cultural and social context. The culture defines certain sexual behaviors as appropriate or inappropriate. Like other forms of social identity, sexual identity is learned.

Contemporary Sexual Attitudes and Behavior

People in the United States today are generally more sexually liberal and have greater tolerance for diverse sexual lifestyles and practices than in the past. The U.S. public has also become much more accepting of gays and lesbians. Today 59 percent of the public think that "homosexual relations between consenting adults should be legal," compared to 43 percent in 1977. At the same time, 89 percent think "homosexuals should have equal rights in terms of job opportunities" (Gallup Organization 2006, 2003; see Figure 11.2). The U.S. Supreme Court ruled in 2003 (in *Lawrence v. Texas*) that private sexual relations are a constitutional liberty, a conclusion widely interpreted as a major victory for gay rights. And, although it will likely be the basis for political, judicial, and legislative action in the years ahead, the Massachusetts Supreme Court ruled in 2003 that same-sex marriages were constitutional. Although the public is divided on this issue (see Chapter 12), there is a changing social climate on issues of gay and lesbian rights.

thinking sociologically

Keep a diary for one week and write down as many examples of *homophobia* and *heterosexism* as you observe in routine social behavior. What do your observations tell you about how *heterosexuality* is enforced?

Changing Sexual Values

Public opinion is now a mix of both liberal and conservative values about sexuality. While adults are much less likely now than in the past to think that premarital sex is wrong (38 percent now versus 68 percent in

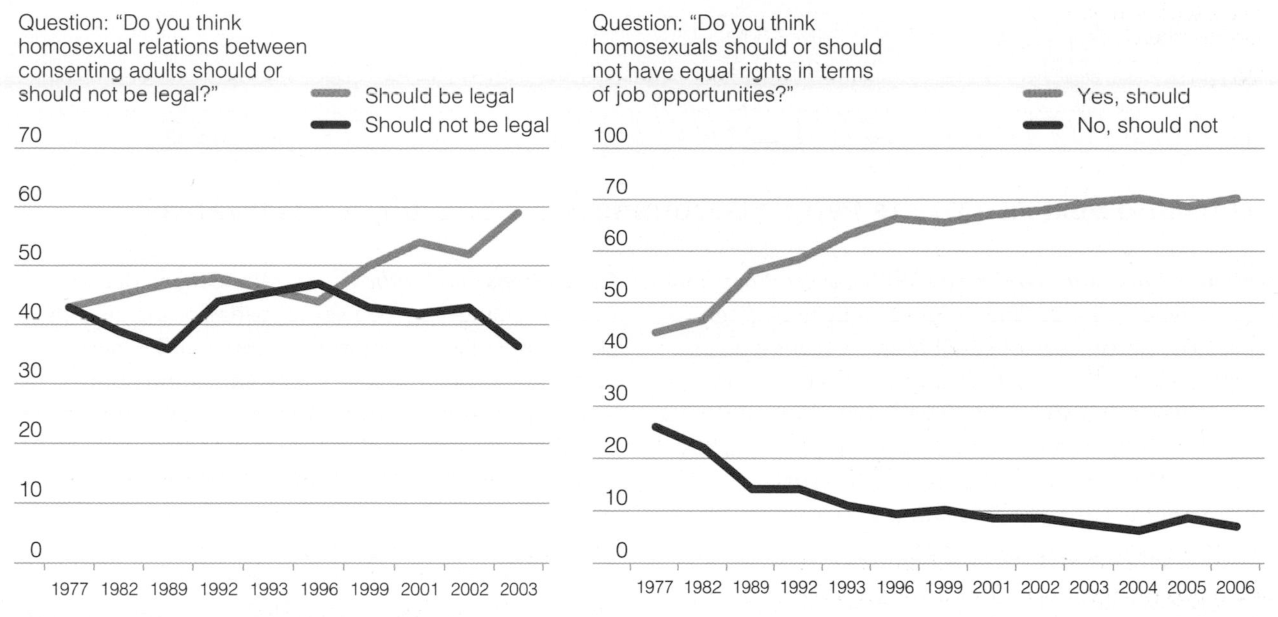

FIGURE 11.2 *Support for Gay Rights*

Data: Saad, Lydia. 2006. "Americans at Odds over Gay Rights." Princeton, NJ: The Gallup Poll, May 31; Newport, Frank. 2003, "Six in Ten Americans Agree that Gay Sex Should Be Legal." Princeton, NJ: The Gallup Poll, June 27. **www.gallup.com**

SEE FOR YOURSELF

CHILDREN'S MEDIA AND SEXUALITY

Watch two movies that are marketed to a specific population of young people (either children, teens, or young adults). Watch the movie very carefully and make a written list of any comments or behaviors you see that suggest *sexual scripts* for men and women. Make note of the assumptions in this script about heterosexual and homosexual behavior. What scripts did you see and who is the intended audience? What do your observations suggest about how sexual scripts are promoted, overtly or not, through popular culture?

1969), among teens the number who think premarital sex is morally wrong is now greater than in the late 1970s (see the box "Doing Sociological Research: Teens and Sex"). More than half of teens now think that they should abstain from sex before marriage, although the odds that they will do so are small (Gallup 2003; Guttmacher Institute 2006b; see also Figure 11.1).

Attitudes about sex vary significantly depending on various social characteristics. For example, men are more likely than women to think that homosexual relations are morally wrong (56 percent of men versus 41 percent of women). Sexual attitudes are also shaped by age. Younger people are more likely than older people to think that homosexuality is acceptable. These differences likely reflect not only the influence of age, but also historical influences on different generations. Religion also matters. Those who attend church weekly are far less likely to support gay rights, compared to those who worship less often (Saad 2007).

Public opinion on controversial matters such as teen pregnancy, AIDS, child care, women's roles in the workplace, and abortion rights tap underlying sexual value systems, often generating public conflicts. In general, sexual liberalism is associated with greater education, youth, urban lifestyle, and political liberalism on other social issues.

Sexual Practices of the U.S. Public

Sexual practices are difficult to document. What we know about sexual behavior is typically drawn from surveys. Most of these surveys ask about sexual attitudes, not actual behavior. What people say they do may differ significantly from what they actually do.

As much as sex is in the news, national surveys of sexual practices are rare. Those that have been conducted tell us the following:

- Young people are becoming sexually active at an earlier age (Guttmacher Institute 2006a).
- Having only one sex partner in one's lifetime is rare (Laumann et al. 1994).
- A significant number of people have extramarital sex (Laumann et al. 1994).
- A significant number of people are lesbian or gay (Laumann et al. 1994; Janus and Janus 1993).

doing sociological research

Teens and Sex: Are Young People Becoming More Sexually Conservative?

Research Question: By the late 1990s, several national studies reported a decline in sexual activity among teens. The percentage of sexually active teens has dropped from the early 1990s, rates of teen pregnancy have fallen, teens are having fewer abortions, and the rate of sexually transmitted diseases among teens has declined. Does this herald a growth in sexual conservatism among young people and the success of policies encouraging sexual abstinence?

Research Method: Sociologists Barbara Risman and Pepper Schwartz based their research on a synthesis of all of the national studies on teen sexuality, as well as data from research organizations on the prevalence of teen sexuality.

Research Findings: Most of the change in teen sex activity is attributable to changes in behavior of boys, not girls. The number of high school boys who are virgins has increased. Girls' behavior has not changed significantly, except among African American girls, whose rates of sexual activity have declined, nearly matching that of White and Hispanic girls. Risman and Schwartz conclude that sexual behavior of boys is then becoming more like girls, the implication being that boys and girls are likely to begin their sexual lives within the context of romantic relationships.

Conclusions and Implications: While many declare that the changes in teen sexual behavior means a decline in the sexual revolution, Risman and Schwartz disagree. Certainly fear of AIDS, education about safe sex, and some growth in conservative values have contributed to changes in teen sexual norms. Risman and Schwartz show that numerous factors influence sexual behavior among teens, just as among adults. They suggest that sexuality is a normal part of adolescent social development and conclude that the sexual revolution—along with the revolution in gender norms—is generating more responsible, not more problematic, sexual behavior among young people.

Questions to Consider

1. Are people in your age group generally sexually conservative or sexually liberal? What factors influence young people's attitudes about sexuality?
2. Following from Question 1, what evidence would you need in order to find out if young people in your community are more liberal than young people in the past? How would you design a study to investigate this question?

Source: Risman, Barbara, and Pepper Schwartz. 2002. "After the Sexual Revolution: Gender Politics in Teen Dating." *Contexts* 1 (Spring): 16–24.

- For those who are sexually active, sex is relatively frequent. Almost two-thirds of Americans report having sex a few times per month or two to three times per week (Laumann et al. 1994).

These facts, however, do not tell us much about the underlying social factors that produce sexual behavior. For that, we turn to sociological theory.

Sexuality and Sociological Theory

Should sex be restricted to marriage? Should prostitution be legal? How are sexual identities formed? These and other questions about sexuality are the subjects of sociological study. Sociological theory puts an analytical framework around the study of sexuality, examining its connection to social institutions and current social issues. How do the major sociological theories frame an understanding of sexuality?

Sex: Functional or Conflict-Based?

The three major sociological frameworks—functionalist theory, conflict theory, and symbolic interaction—take divergent paths in interpreting the social basis of human sexuality (see Table 11.1).

Functionalist theory with its emphasis on the interrelatedness of different parts of society tends to depict sexuality in terms of its contribution to the stability of social institutions. Norms that restrict sex to marriage encourage the formation of families.

table 11.1 Theoretical Perspectives on Sexuality

Interprets:	Functionalism	Conflict Theory	Symbolic Interaction	Queer Theory
Sexual norms	Sexual norms are functional for society because they encourage the formation of stable institutions.	Sexual norms are frequently contested by those who are subordinated by dominant groups.	Sexual norms emerge through social interaction and the construction of beliefs.	Sexual norms are easily contested through play and performances that transgress the dominant sex/gender categories.
Sexual identity	Sexual identity is learned in the family; deviant identities contribute to social disorder.	Sexual identity is regulated by individuals and institutions that enforce compulsory heterosexuality.	Sexual identity is socially constructed as people learn the sexual scripts created in society.	Multiple forms of sexual identity are possible and can be seen in how people cross the ordinarily assumed boundaries.
Changing sexual values	Regulating sexual values and norms is important for maintaining traditional and social stability; too much change results in social disorganization.	Social change comes through the activism of people who challenge dominant belief systems and practices.	Change in sexual value systems evolves as people construct new beliefs and practices over time.	Sexual values can be changed through disrupting taken-for-granted categories of the dominant culture.

Similarly, beliefs that give legitimacy to heterosexual behavior but not homosexual behavior maintain a particular form of social organization where gender roles are easily differentiated and the nuclear family is defined as the dominant social norm. From this point of view, regulating sexual behavior is functional for society because it prevents the instability and conflict that more liberal sexual attitudes supposedly generate. Functionalists would also explain the call for a return to "family values" as producing the uniformity in values necessary for social order.

Conflict theorists see sexuality as part of the power relations and economic inequality in society. Power is the ability of one person or group to influence the behavior of another. Power relations in society influence the power that some sexual groups have over others and influence power within sexual relationships (discussed later in the chapter in "Sexual Politics"). Conflict theorists argue that sexual relations are linked to other forms of stratification, namely, race, class, and gender inequality. According to this perspective, sexual violence such as rape or sexual harassment is the result of power imbalances, specifically between women and men.

At the same time, because conflict theorists see economic inequality as a major basis for social conflict, they tie the study of sexuality to economic institutions. They link the international sex trade to poverty, the status of women in society, and the economics of international development and tourism (Enloe 2001; Altman 2001). Still, conflict theorists do not see all sexual relations as oppressive. Sexuality is an expression of great social intimacy. In connecting sexuality and inequality, conflict theorists are developing a structural analysis of sexuality, not condemning sexual intimacy.

Because both functionalism and conflict theory are macrosociological theories (that is, they take a broad view of society, seeing sexuality in terms of the overall social organization of society), they do not tell us much about the social construction of sexual identities. This is where the sociological framework of symbolic interaction is valuable.

Symbolic Interaction and the Social Construction of Sexual Identity

Symbolic interaction theory uses a **social construction perspective** to interpret sexual identity as learned, not inborn. To symbolic interactionists, culture and society shape sexual experiences. Patterns of social approval and social taboos make some forms of sexuality permissible and others not (Connell 1992; Lorber 1994).

The social construction of sexual identity is revealed by the coming out process. **Coming out**—the process of defining oneself as gay or lesbian—is a series of events and redefinitions in which a person comes to see herself or himself as having a gay identity. In coming out, a person consciously labels that identity either to oneself or others (or both). This is usually not the result of a single homosexual experience. If it were, there would be far more self-identified homosexuals, because researchers find that a substantial portion of both men and women have some form of homosexual experience at some time in their lives.

The development of sexual identity is not necessarily a linear or unidirectional process, with persons moving predictably through a defined sequence of steps or phases. Although they may experience certain milestones in their identity development, some people experience periods of ambivalence about their identity and may switch back and forth between lesbian, heterosexual, and bisexual identity over time (Rust 1995, 1993). Some people may engage in lesbian or gay behavior but not adopt an identity as lesbian or gay. Certainly many gays and lesbians never adopt a public definition of themselves as gay or lesbian, instead remaining "closeted" for long periods, if not entire lifetimes.

One's sexual identity may also change. For example, a person who has always thought of himself or herself as a heterosexual may conclude at a later time that he or she is gay or lesbian. In more unusual cases, people may undergo a sex change operation, perhaps changing their sexual identity in the process. In the case of bisexuals, a person might adopt a dual sexual identity.

Although most people learn stable sexual identities, over the course of one's life, sexual identity evolves. Change is, in fact, a normal outcome of the process of identity formation. Changing social contexts (including dominant group attitudes, laws, and systems of social control), relationships with others, political movements, and even changes in the language used to describe different sexual identities all affect people's self-definition.

AP Images/Peter Lennihan

Gays, lesbians, and their allies have mobilized for social change, fostering pride and celebration as well as a reduction over time in homophobic attitudes.

Queer theory is a perspective that has evolved from recognizing the socially constructed nature of sexual identity. Instead of seeing heterosexual or homosexual attraction as fixed in biology, queer theory interprets society as forcing these sexual boundaries, or dichotomies, on people. By challenging the "either/or" thinking that one is either gay or straight, queer theory challenges the idea that only one form of sexuality is normal and all other forms are deviant or wrong. As a result, queer theory has opened up fascinating new studies of gay, straight, bisexual, and transsexual identities and introduced the idea that sexual identity is a continuum of different possibilities for sexual expression and personal identity (Nardi and Schneider 1997; Seidman 2003; Stein and Plummer 1994).

Queer theory has also linked the study of sexuality to the study of gender, showing how transgressing (or violating) fixed gender categories can reconstruct the possibilities of how all people—men and women, gay, bisexual, or straight—construct their gender and sexual identity. Transgressing gender categories, such as by drag queens or cross-dressers, can show how sex and gender categories are usually constructed in dichotomous categories (that is, opposite or binary types). By violating these constructions, people are liberated from the social constraints that presumably fixed categories of identity create. Thus, queer theory emphasizes how performance and play with gender categories can be a political tool for deconstructing fixed sex and gender identities (Rupp and Taylor 2003).

Sexuality is, of course, also constructed in the context of social institutions. Social institutions tend to define heterosexuality as the only legitimate form of sexual identity and enforce it through social norms and sanctions, including peer pressure, socialization, law and other social policies, and, at the extreme, violence (Rich 1980).

Sex, Diversity, and Inequality

Patterns of sexuality reflect the social organization of society. When you understand this, you also see that sex is related to other social factors—such as race, class, and gender—and you see how sexuality is connected to social institutions and social change.

Sexual Politics

Sexual politics refers to the link between sexuality and power, not just within individual relationships. The feminist movement first linked sexuality to the status of women in society, pointing to the possible exploitation of women within sexual relationships. Sexual politics also refers to the high rates of violence

AP Images/Jennifer Graylock

Having celebrities "come out" about being lesbian or gay has empowered others to be able to do so as well.

against women and sexual minorities and the privilege and power accorded to those presumed to be heterosexual.

The feminist and gay and lesbian liberation movements have put sexual politics at the center of the public's attention by challenging gender role stereotyping and sexual oppression (D'Emilio 1998). Among other things, this has profoundly changed public knowledge of gay and lesbian sexuality. Gay, lesbian, and feminist scholars have argued, and many now concur, that homosexuality is not the result of psychological deviance or personal maladjustment but is one of several alternatives for happy and intimate social relationships. The political mobilization of many lesbian women and gay men and the willingness of many to make their sexual identity public have also raised public awareness of the civil and personal rights of gays and lesbians. These changes make other changes in intimate relations possible.

The Influence of Race, Class, and Gender

Sexual behavior follows gendered patterns, as well as patterns established by race and class relations (Schwartz and Rutter 1998). Gender expectations emphasize passivity for women and assertiveness for men in sexual encounters. The "double standard" is the idea that different standards for sexual behavior apply to men and women. Although this idea is weakening somewhat, men are still stereotyped as sexually overactive; women, less so. Women who openly violate this cultural double standard by being openly sexual are then cast in a negative light as "loose," as if the appropriate role for women is the opposite of "loose," say, "secured" or "caged." The double standard forces women into polarized roles as "good" girls or "bad" girls. The belief that women who are raped must have somehow brought it on themselves rests on the images of women as temptresses. Contrary to popular belief, men do not have a stronger sex drive than women. Men are, however, socialized more often to see sex in terms of performance and achievement, whereas women are more likely socialized to associate sex with intimacy and affection.

Sexual politics are also integrally tied to race and class inequality in society. You can see this in the sexual stereotypes associated with race and class. Latinas are stereotyped as either "hot" or "virgins"; Latin men are stereotyped as "hot lovers"; African American men are stereotyped as overly virile; Asian American women are stereotyped as compliant and submissive, but passionate. Class relations also produce sexual stereotypes of women and men. Working-class and poor men may be stereotyped as dangerous, whereas working-class women may be disproportionately labeled "sluts."

Class, race, and gender hierarchies historically have been justified based on claims that people of color and women are sexually promiscuous and uncontrollable (Nagel 2003). During slavery, for example, the sexual abuse of African American women was one way that slave owners expressed their ownership of African American people. Access to women slaves' sexuality was seen as a right of the slave owner. Under slavery, racist and sexist images of Black men and women were developed to justify the system of slavery. Black men were stereotyped as lustful beasts whose sexuality had to be controlled by the "superior" Whites. Black women were also depicted as sexual animals who were openly available to White men.

A Black man falsely accused of having had sex with a White woman could be murdered (that is, lynched) without penalty to his killers (Genovese 1972; Jordan 1968). Sexual abuse was also part of the White conquest of American Indians. Historical accounts show that the rape of Indian women by White conquerors was common (Freedman and D'Emilio 1988; Tuan 1984), as is the rape of women following war and military conquest.

Poor women and women of color are the groups most vulnerable to sexual exploitation. Becoming a prostitute, or otherwise working in the sex industry (as a topless dancer, striptease artist, pornographic actress, or other sex-based occupation), is often the last resort for women with limited options to support themselves. Women who sell sex are also condemned for their behavior, more so than their male clients—further illustration of how gender stereotypes mix with race and class exploitation. Why, for example, are women, and not their male clients, arrested for prostitution? A sociological perspective on sexuality helps one see how sexuality is linked to other systems of social stratification.

a sociological eye on the media

Publicity Traps: Sex on TV Talk Shows

If you even casually peruse television talk shows, you are likely to see a wide array of sexual nonconformists—people who may be bisexual, transgendered, or in some way sexually nonconforming. Joshua Gamson, a sociologist who specializes in the study of media, sexuality, and social change, has studied television talk shows and used them to think about the cultural visibility of nonconforming sexual groups. Lesbians, gays, transgendered people, bisexuals, and others are usually marginalized in society—treated as invisible, seldom recognized, misunderstood, and usually suppressed in public space. But on television talk shows, the public can witness outrageous, boisterous, even wild guests. In fact, in an attempt to build larger audiences (because, after all, this is about profit), television talk shows have become less likely to feature calm, well-educated, distinguished guests. Instead, they often feature those who least conform to the dominant sexual value system. Gamson has asked, "What is the impact of the visibility of sexual nonconformists for gays and lesbians?"

Gamson sees no simple answer to this question. On the one hand, showcasing sexual nonconformity in a way that may seem freakish, foul-mouthed, and "abnormal" presents a distorted image of gay life and gives legitimacy to those who think that lesbian, gay, bisexual, transgender, and other diverse sexual lifestyles are deviant. The presentation of people on "the fringe," as Gamson puts it, "makes social acceptability harder to gain by overemphasizing difference, often presented as frightening, pathological, pathetic, or silly" (1998: 33).

But Gamson also sees another dimension to this question. He writes that although these images present a distorted image of gay life, the talk shows also make diverse sexual identities public, thereby having the added effect of making sexual nonconformity less shocking and, thus, in the long run, more acceptable. In addition, Gamson argues, these portrayals open up public space, where challenges to sexual conformity can transform the ordinarily fixed boundaries between gay and straight, normal and "queer."

The dilemma comes from the fact that these portrayals are generally not made by or on behalf of gay people. Yet, becoming more visible in a public space such as the media can change people's understandings of who has a right to share public space.

Source: Gamson, Joshua. 1998. "Publicity Traps: Television Talk Shows and Lesbian, Gay, Bisexual, and Transgender Visibility." *Sexualities* 1: 11–41.

A Global Perspective on Sexuality

Cross-cultural studies of sexuality show that sexual norms, like other social norms, develop differently across cultures. Take sexual jealousy. Perhaps you think that seeing your sexual partner becoming sexually involved with another person would naturally evoke jealousy—no matter where it happened. Researchers have found this not to be true. In a study comparing patterns of sexual jealousy in seven different nations (Hungary, Ireland, Mexico, the Netherlands, the United States, Russia, and the former Yugoslavia), researchers found significant cross-national differences in the degree of jealousy when women and men saw their partners kissing, flirting, or being sexually involved with another person (Buunk and Hupka 1987).

Likewise, tolerance for gay and lesbian relationships varies significantly in different societies around the world. Germany has recently legalized gay and lesbian relationships, allowing them to register same-sex partnerships and have the same inheritance rights as heterosexual couples. The new law does not, however, give them the same tax advantages, nor can same-sex couples adopt children. Cross-cultural studies can make someone more sensitive to the varying cultural norms and expectations that apply to sexuality in different contexts. Different cultures simply view sexuality differently. In Islamic culture, for example, women and men are viewed as equally sexual, although women's sexuality is seen as potentially disruptive and needing regulation (Mernissi 1987).

Sex is also big business, and it is deeply tied to the world economic order. As the world has become

Sam Yeh/AFP Photo/Getty Images

The international trafficking of women for sex exploits women—and often children—and puts them at risk for disease and violence.

more globally connected, an international sex trade has flourished—one that is linked to economic development, world poverty, tourism, and the subordinate status of women in many nations.

The *international sex trade,* sometimes also referred to as the "traffic in women" (Rubin 1975), refers to the use of women, worldwide, as sex workers in an institutional context where sex itself is a commodity. Sex is marketed in an international marketplace, and women as sex workers are used to promote tourism, cater to business and military men, and support a huge industry of nightclubs, massage parlors, and teahouses (Enloe 2001).

"Sex capitals" are places where prostitution openly flourishes, such as in Thailand and Amsterdam. Sex is an integral part of the world tourism industry. In Thailand, for example, men as tourists outnumber women by a ratio of three to one. Planeloads of businessmen come to Thailand as tourists, sometimes explicitly to buy sexual companionship. Hostess clubs in Tokyo similarly cater to corporate men. One fascinating study, where the researcher worked as a hostess in a Tokyo club, shows how the men's behavior in these clubs is linked to the expression of their gender identity among other men (Allison 1994). In Amsterdam prostitutes sit in windows in red-light districts to attract customers.

Sociologists see the international sex trade as part of the global economy, contributing to the economic development of many nations and supported by the economic dominance of certain other nations. As with other businesses, the products of the sex industry may be produced in one region and distributed in others. (Think, for example, of the pornographic film industry centered in southern California, but distributed globally). The sex trade is also associated with world poverty; sociologists have found that the weaker the local economy, the more important the sex trade. The international sex trade is also implicated in problems such as the spread of AIDS worldwide, as well as the exploitation of women where women have limited economic opportunities (Altman 2001).

Understanding Gay and Lesbian Experience

Sociological understanding of sexual identity has developed largely through new studies of lesbian and gay experience. Long thought of only in terms of social deviance (see Chapter 6), gays and lesbians have been stereotyped in traditional social science. But the feminist and gay liberation movements have discouraged this approach, arguing that gay and lesbian experience is part of the broad spectrum of human sexuality.

The institutional context for sexuality within the United States, as well as other societies, is one where homophobia permeates the culture. **Homophobia** is the fear and hatred of homosexuality. It is deeply embedded in people's definitions of themselves as men and women; it is manifested in prejudiced attitudes toward gays and lesbians, as well as overt hostility and violence against people suspected of being gay. Homophobia is a learned attitude, as are other forms of negative social judgments about particular groups.

Boys are often raised to be manly by repressing so-called feminine characteristics in themselves. Being called a "fag" or a "sissy" is one peer sanction that socializes a child to conform to particular gender roles. Similarly, verbal attacks on lesbians called "butch" are a mechanism of social control because ridicule can be interpreted as encouraging social conformity.

Homophobia produces many misunderstandings about gay people. One misunderstanding is that gays have a desire to seduce straight people. There is little evidence that this is true. And, if we look at who is most likely to commit sex crimes, we find that heterosexual men are more likely to commit sex crimes than homosexual men. Homosexual men convicted of child molestation are, however, almost seven times more likely to be imprisoned than are heterosexual men who are convicted as child molesters, even when they have the same criminal record (Walsh 1994).

Fears that gay and lesbian parents will have negative effects on their children are also unsubstantiated by research. The ability of parents to form good relationships with their children is far more significant in children's social development than is their parents' sexual orientation (Stacey and Biblarz 2001).

[understanding diversity

Gay-Bashing and Gender Betrayal

Carmen Vázquez has related gay-bashing to the phenomenon she labels *gender betrayal.* She explains this by the following:

> *At the simplest level, looking or behaving like the stereotypical gay man or lesbian is reason enough to provoke a homophobic assault. Beneath the veneer of the effeminate gay male or the butch dyke, however, is a more basic trigger for homophobic violence. I call it gender betrayal.*
>
> *The clearest expression I have heard of this sense of gender betrayal comes from Doug Barr, who was acquitted of murder in an incident of gay bashing in San Francisco that resulted in the death of John O'Connell, a gay man. Barr is currently serving a prison sentence for related assaults on the same night that O'Connell was killed. . . . When asked what he and his friends thought of gay men, he said, "We hate homosexuals. They degrade our manhood. We was brought up in a high school where guys are football players, mean and macho. Homosexuals are sissies who wear dresses. I'd rather be seen as a football player."*
>
> *Doug Barr's perspective is one shared by many young men. I have made about 300 presentations to high school students in San Francisco, to boards of directors and staff of nonprofit organizations, and at conferences and workshops on the topic of homophobia or "being lesbian and gay." Over and over again, I have asked, "Why do gay men and lesbians bother you?" The most popular response to the question is, "Because they act like girls," or "Because they think they're men." I have even been told, quite explicitly, "I don't care what they do in bed, but they shouldn't act like that."*
>
> *They shouldn't act like that. Women who are not identified by the relationship to a man, who value their female friendship, who like and are knowledgeable about sports, or work as blue-collar laborers and wear what they wish are very likely to be "lesbian-baited" at some point in their lives. Men who are not pursuing sexual conquests of women at every available opportunity, who disdain sports, who choose to stay at home and be a house-husband, who are employed as hairdressers, designers, or housecleaners, or who dress in any way remotely resembling traditional female attire (an earring will do) are very likely to experience the taunts and sometimes the brutality of "fag bashing."*
>
> *The straitjacket of gender roles suffocates many lesbians, gay men, and bisexuals, forcing them into closets without an exit and threatening our very existence when we tear the closet open.*

Source: Vazquez, Carmen. 1992. "Appearances." Pp. 157–166 in *Homophobia: How We All Pay the Price,* edited by Warren J. Blumenfeld. Boston, MA: Beacon Press.

Other myths about gay people are that they are mostly White men with large discretionary incomes who work primarily in artistic areas and personal service jobs (such as hairdressing). This stereotype prevents people from recognizing that gays and lesbians come from all racial–ethnic groups, may be working class or poor, and are employed in a wide range of occupations (Gluckman and Reed 1997). Some lesbians and gays are also elderly, though the stereotype defines gay people as primarily young or middle-aged (Smith 1983). These different misunderstandings reveal just a few of the many unfounded myths about gays and lesbians. Support for these attitudes comes from homophobia, not from empirical truth.

Heterosexism refers to the institutionalization of heterosexuality as the only socially legitimate sexual orientation. Heterosexism is rooted in the belief that heterosexual behavior is the only natural form of sexual expression and that homosexuality is a perversion of "normal" sexual identity. Heterosexism is reinforced through institutional mechanisms that project the idea that only heterosexuality is normal. Institutions also provide different benefits to people presumed to be heterosexual. Businesses and communities, for example, rarely recognize the legal rights of people in homosexual relationships, although this is changing. Within an institution, individual beliefs can reflect heterosexist assumptions. Thus, a person may be accepting of gay and lesbian people (that is, not be homophobic) but still benefit from heterosexual privileges. At the behavioral level, heterosexist practices can exclude lesbians and gays, such as when coworkers talk about dating the other sex, assuming that everyone is interested in a heterosexual partner.

In the absence of institutional supports from the dominant culture, lesbians and gays have invented their own institutional support systems. Gay communities and gay rituals, such as gay pride marches, affirm gay and lesbian identities and provide a support system that is counter to the dominant heterosexual culture. Those who remain "in the closet" deny themselves this support system.

The absence of institutionalized roles for lesbians and gays affects the roles they adopt within relationships. Despite popular stereotypes, gay partners typically do not assume roles as the dominant or submissive sexual partner. They are more likely to adopt roles as equals. Gay couples and lesbian couples are also more likely than heterosexual couples to both be employed, another source of greater equality within the relationship. Researchers have also found that the quality of relationships among gay men is positively correlated with the social support the couple receives from others (Smith and Brown 1997; Metz et al. 1994; Harry 1979).

Lesbians and gays are a minority group in our society, denied equal rights and singled out for negative treatment in society. *Minority groups* are not necessarily numerical minorities; they are groups with similar characteristics (or at least perceived similar characteristics) who are treated with prejudice and discrimination (see Chapter 9). As a minority group, gays and lesbians have organized to advocate for their civil rights and to be recognized as socially legitimate citizens. Some organizations and municipalities have enacted civil rights protections on behalf of gays and lesbians, typically prohibiting discrimination in hiring. The Supreme Court ruled in 1996 *(Romer v. Evans)* that gays and lesbians cannot be denied equal protection under the law. This ruling overturned ordinances that had specifically denied civil rights protections to lesbians and gays (Greenhouse 1996). The implications of this case for domestic partner benefits, child custody and adoption by gay and lesbian parents, and gay marriage are not yet known, but will likely be determined in the courts in the years ahead.

© Terry Schmitt/UP/Landov

Many gay and lesbian couples jubilantly took formal vows of marriage when San Francisco legally recognized the right to gay unions.

Sex and Social Issues

In studying sexuality, sociologists tap into some highly contested social issues of the time. Birth control, reproductive technology, abortion, teen pregnancy, pornography, and sexual violence are all subjects of public concern and are important in the formation of social policy. Debates about these issues hinge in part on attitudes about sexuality and are shaped by race, class, and gender relations. These social issues can generate personal troubles that have their origins in the structure of society—recall the distinction made by C. Wright Mills between personal troubles and social issues (see Chapter 1).

Birth Control

The availability of birth control is now less debated than it was in the not too distant past, but this important reproductive technology has been strongly related to the status of women in society (Gordon 1977). Reproduction has been controlled by men; to this day, it is mostly men who define the laws and make the scientific decisions about what types of birth control will be available. Women are also most likely seen as responsible for reproduction because it is a presumed part of their traditional role. At the same time, changes in birth control technology have also made it possible for women to change their roles in society, given that breaking the link between sex and reproduction has freed women from some traditional constraints.

The right to birth control is a recently won freedom. It was not until 1965 that the Supreme Court, in *Griswold v. Connecticut,* defined the use of birth control as a right, not a crime. This ruling originally applied only to married people; unmarried people were not extended the same right until 1972 in the Supreme Court decision, *Eisenstadt v. Baird.* Today birth control is routinely available by prescription, but there is heated debate about whether access to birth control should be curtailed for the young—at a time when youths are experimenting with sex at younger ages and risking teenage pregnancy and AIDS. Some argue that increasing access to birth control will only encourage more sexual activity among the young.

Class and race relations have also had a role in shaping birth control policy. In the mid-nineteenth century, increased urbanization and industrialization ended the necessity for large families, especially in the middle class, because fewer laborers were needed to support the family. Early feminist activists such as Emma Goldman and Margaret Sanger also saw birth control as a way of freeing women from unwanted pregnancies and allowing them to work outside the home if they chose. As the birthrate fell among White upper- and middle-class families during this period, these classes feared that immigrants, the poor, and racial minorities would soon outnumber them.

The *eugenics* movement of the early twentieth century grew from the fear of domination by immigrant groups. **Eugenics** sought to apply scientific principles of genetic selection to "improve" the offspring of the human race. It was explicitly racist and class-based, calling for, among other things, the compulsory sterilization of those who eugenicists thought were unfit. Eugenicist arguments appeal to a public that fears the social problems that emerge from race and class inequality: Instead of attributing these problems (such as crime) to the structure of society, blame the genetic composition of the least powerful groups in society.

New Reproductive Technologies

Practices such as surrogate mothering, in vitro fertilization, and new biotechnologies of gene splicing, cloning, and genetic engineering mean that reproduction is no longer inextricably linked to biological parents (Rifkin 1998). A child may be conceived through means other than sexual relations between one man and one woman. One woman may carry the child of another. Offspring may be planned through genetic engineering. A sheep can be cloned (that is, genetically duplicated). So can monkeys. Are humans next? With such developments, those who could not otherwise conceive children (infertile couples, single women, or lesbian couples) are now able to do so, thereby raising new questions. To whom are such new technologies available? Which groups are most likely to sell reproductive services, which groups to buy? What are the social implications of such changes?

There are no simple answers to such questions, but sociologists would point first to the class, race, and gender dimensions of these issues (Roberts 1997). Poor women, for example, are far more likely than middle-class or elite women to sell their eggs or offer their bodies as biological incubators. Groups that can afford new, costly methods of reproduction may do so at the expense of women whose economic need places them in the position of selling themselves for financial necessity.

Breakthroughs in reproductive technology raise especially difficult questions for makers of social policy. Developments in the technology of reproduction have ushered in new possibilities and freedoms but also raise questions for social policy. With new reproductive technologies, there is potential for a new eugenics movement. Sophisticated prenatal screenings make it possible to identify fetuses with presumed defects. Might society then try to weed out those perceived as undesirables—the disabled, certain racial groups, certain sexes? Will parents try to produce "designer children"? If boys and girls are differently valued, one sex may be more often aborted, a frequent practice in India and China—two of the most populous nations on Earth. Because of population pressures, state policy in China, for example, encourages families to have only one child. Because girls are less valued than boys, the aborting and selling of girls is common and has created a U.S. market for the adoption of Chinese baby girls.

There are no traditions to guide us on such questions. Public thinking about sexuality and reproduction will have to evolve. Although the concept of reproductive choice is important to most people, choice is conditioned by the constraints of race, class, and gender inequalities in society. Like other social phenomena, sexuality and reproduction are shaped by their social context.

Abortion

Abortion is one of the most seriously contended political issues. Of the U.S. public, 30 percent think abortion should be legal in any circumstances; 53 percent think it should be legal only under certain circumstances; and 15 percent think it should be illegal in all circumstances (see Figure 11.3). Public support for abortion has remained relatively constant over the last twenty years.

The right to abortion was first established in constitutional law by the *Roe v. Wade* decision in 1973. In *Roe v. Wade,* the Supreme Court ruled that at different points during a pregnancy, separate but legitimate rights collide—the right to privacy, the right of the state to protect maternal health, and the right of the state to protect developing life. To resolve this conflict of rights, the Supreme Court ruled that pregnancy occurred in trimesters. In the first, women's right to privacy without interference from the state prevails; in the second, the state's right to protect maternal health takes precedence; in the third, the state's right to protect developing life prevails. In the second trimester, the government cannot deny the right to abortion, but it can insist on reasonable standards of medical procedure. In the third, abortion may be performed only to save the life or health of the mother. More recently, the Supreme Court has allowed states to impose restrictions on abortion, but it has not, to date, overturned the legal framework of *Roe v. Wade.*

Data on abortion show that it occurs across social groups, although certain patterns do emerge. The

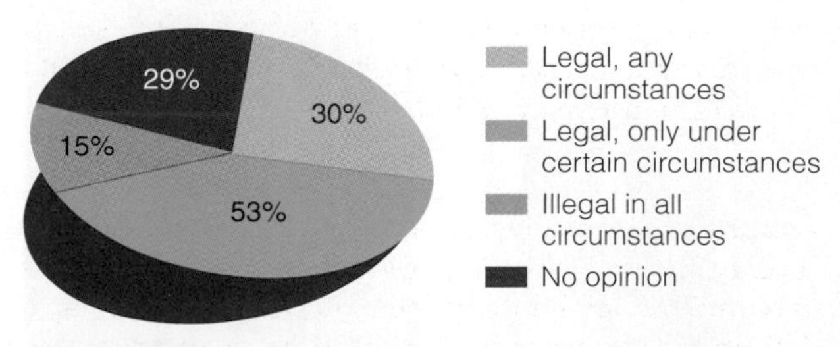

FIGURE 11.3 *Attitudes toward Abortion*

Data: Gallup Poll. 2006. "Abortion." Princeton, NJ: The Gallup Organization, **www.gallup.com**

FIGURE 11.4 *Deaths from Abortion: Before and After* Roe v. Wade

Source: Alan Guttmacher Institute. 1999. *Trends in Abortion 1973–2000*. New York: The Alan Guttmacher Institute, **www.agiusa.org**. Reproduced with the permission of the Alan Guttmacher Institute.

abortion rate has declined since 1980 from a rate of 29.3 per 1000 women to 20.8 now (among women age 15 to 44). And, as you can see in Figure 11.4, the number of deaths from illegal abortions plummeted in the years following the *Roe v. Wade* decision. Now, young women (age 15 to 19) are the most likely group to get abortions, although the second most likely group is women over 40. Women of color are three times more likely to have abortions than White women; poor and low-income women account for half of all abortions. The majority of women having abortions are already mothers (U.S. Census Bureau 2006; Finer and Henshaw 2003; National Center for Health Statistics 2006).

The abortion issue provides a good illustration of how sexuality has entered the political realm. Abortion rights activists and antiabortion activists hold very different views about sexuality and the roles of women. Antiabortion activists tend to believe that giving women control over their fertility breaks up the stable relationships in traditional families. They tend to view sex as something that is sacred, and they are disturbed by changes that make sex less restrictive. Abortion rights activists see women's control over reproduction as essential for their independence. They also tend to see sex as an experience that develops intimacy and communication between people who love each other. The abortion debate can thus be interpreted as a struggle over the right to terminate a pregnancy, as well as a battle over differing sexual values and a referendum on the nature of men's and women's relationships (Luker 1984).

Pornography

Little social consensus has emerged about the acceptability and effects of pornography. Part of this debate is about defining what is obscene. The legal definition of obscenity is one that changes over time and in different political contexts. Public agitation over pornography has divided people into those who think it is solidly protected by the First Amendment, those who want it strictly controlled, those who think it should be banned for moral reasons, and those who think it must be banned because it harms women. Who is right?

Research shows that exposure to violent pornography in a laboratory setting affects sexual attitudes. After exposure to violent pornography, men are more likely to see victims of rape as responsible for their assault, less likely to regard them as injured, and more likely to accept the idea that women enjoy rape. These effects are found, however, only after exposure to pornography that is both sexually explicit and violent (Donnerstein et al. 1987). Little evidence supports the assertion that exposure to pornography in general increases sexual promiscuity or sexual deviance.

Despite public concerns about pornography, a two-thirds majority think pornography should be protected by the constitutional guarantees of free speech and a free press. At the same time, people believe that pornography dehumanizes women. Women tend to be more negative than men about pornography, perhaps because pornography generally portrays women in demeaning ways. The controversy about pornography is not likely to go away, because it taps so many different sexual values among the public.

Teen Pregnancy

Each year about 349,000 teenage girls (under age 19) have babies in the United States; 78 percent of these births are unplanned. The United States has the highest rate of teen pregnancy among developed nations, even though levels of teen sexual activity around the world are roughly comparable. Teens account for about 10 percent of all births. Teen pregnancy has declined since 1990, a decline caused almost entirely from the increased use of birth control. Analysts find that abstinence accounts for a very small portion decline—probably only about 10 percent of the difference (Santelli et al. 2007; U.S. Census Bureau 2006; Guttmacher Institute 2006a).

Although the rate of pregnancy among teens has dropped, so has the rate of marriage for teens who become pregnant. Thus, most babies born to teens will be raised by single mothers—a departure from the past when teen mothers often got married (U.S. Census Bureau 2006). What concerns people about teen parents is that teens are more likely to be poor than other mothers, although sociologists have cautioned that this is because teen mothers are more likely poor *before* getting pregnant (Luker 1996). Teen parents are among the most vulnerable of all social groups.

Teenage pregnancy correlates strongly with poverty, lower educational attainment, joblessness, and health problems. Teen mothers have a greater incidence of problem pregnancies and are most likely to deliver low-birth-weight babies, a condition associated with a myriad of other health problems. Teen parents face chronic unemployment and are less likely to complete high school than those who delay childbearing. Many continue to live with their parents, although this is more likely among Black teens than among Whites. The costs of adolescent pregnancy then fall most heavily on those families least able to help financially (Kaplan 1996).

Although teen mothers feel less pressure to marry now than in the past, if they raise their children alone, they suffer the economic consequences of raising children in female-headed households—the poorest of all income groups. Teen mothers report that they do not marry because they do not think the fathers are ready for marriage. Sometimes their families also counsel them against marrying precipitously. These young women are often doubtful about men's ability to support them. They want men to be committed to them and their child, but they do not expect their hopes to be fulfilled (Farber 1990; Edin and Kefalas 2005). Research shows that low-income single mothers are distrustful of men, especially after an unplanned pregnancy. They think they will have greater control of their household if they remain unmarried. Many teen mothers also express fear of domestic violence as a reason for not marrying (Edin 2000). Both Black and White teen parents see marriage as an impediment to their career ambitions, although Black teen mothers are more likely than Whites to continue attending school (Farber 1990; Trent and Harlan 1990).

Why do so many teens become pregnant given the widespread availability of birth control? Teens typically delaying the use of contraceptives until several months after they become sexually active. Teens who do not use contraceptives when they first have sex are twice as likely to get pregnant as those who use contraception (Guttmacher Institute 2006a). In recent years, the percentage of teens using birth control (especially condoms) has increased, although the pill is still the most widely used method (see Figure 11.5; Guttmacher Institute 2006a).

Debunking Society's Myths

Myth: Providing sex education to teens only encourages them to become sexually active.

Sociological perspective: Comprehensive sex education actually delays the age of first intercourse; abstinence-only education has not been shown to be similarly effective in delaying intercourse (Kirby 1997; Risman and Schwartz 2002).

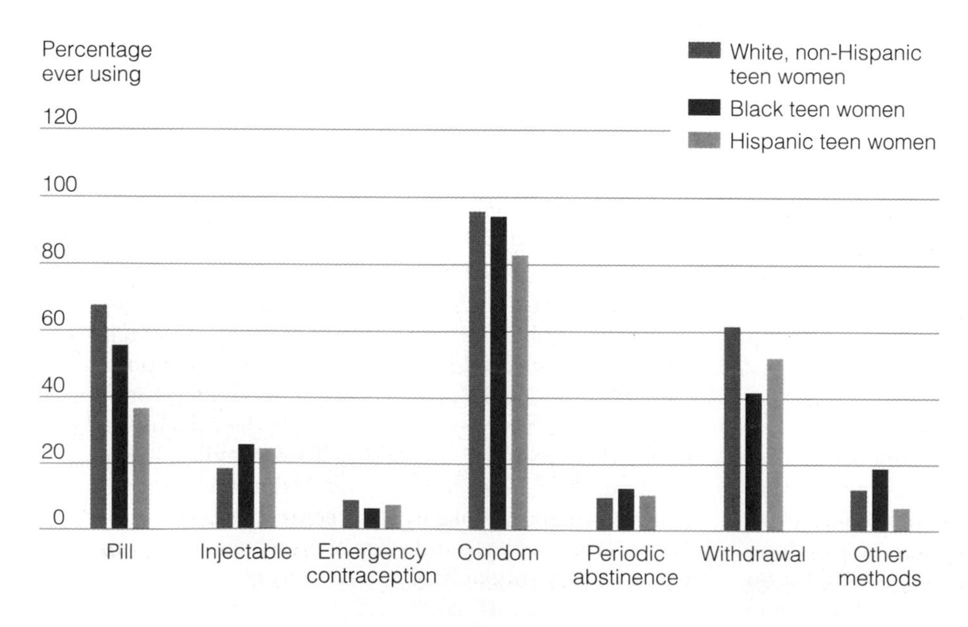

FIGURE 11.5 *Contraceptive Use Among American Teen Women*

These data include teen (age 15–19) women's contraceptive use, including only those who have had sexual intercourse. The data indicate contraceptive methods *ever used,* not necessarily those used regularly. What do you think explains the use of different methods by different groups? How do social factors influence the use of contraception among different groups?

Source: Abma, J.C. et al. 2002. *Teenagers in the United States: Sexual Activity, Contraceptive Use, and Childbearing, 2002.* Washington, DC: U.S. Department of Health and Human Services.

© Topham/The Image Works

Despite public concerns about teen pregnancy, rates of teen pregnancy have actually declined in recent years.

Sociologists have argued that the effective use of birth control requires a person to identify himself or herself as sexually active (Luker 1975). Teen sex, however, tends to be episodic. Teens who have sex on a couple of special occasions may not identify themselves as sexually active and may not feel obliged to take responsibility for birth control. Despite many teens initiating sex at an earlier age, social pressure continues to discourage them from defining themselves openly, or even privately, as sexually active.

Teen pregnancy is integrally linked to the gender expectations of men and women in society. Some teen men consciously avoid birth control, thinking it takes away from their manhood. Teen women often romanticize motherhood, thinking that becoming a mother will give them social value they do not otherwise have. For teens in disadvantaged groups, motherhood confers a legitimate social identity on those otherwise devalued by society (Horowitz 1995). Although their hopes about motherhood are not realistic, they indicate how pessimistic the teenagers feel about their lives that are often marked by poverty, a lack of education, and few good job possibilities (Ladner 1986). For young women to romanticize motherhood is not surprising in a culture where motherhood is defined as a cultural ideal for women, but the ideal can seldom be realized when society gives mothers little institutional or economic support.

Sexual Violence

Before the development of the feminist movement, sexual violence was largely hidden from public view. One great success of the women's movement has been to identify, study, and advocate better social policies to address the problems of rape, sexual harassment, domestic violence, incest, and other forms of sexual coercion. Sexual coercion is not just a matter of sexuality, but also a form of power relations shaped by the social inequality between women and men. In the thirty years or so that these issues have been identified as serious social problems, volumes of research have been published on these different subjects and numerous organizations and agencies have been established to serve victims of sexual abuse and to advocate reforms in social policy.

Rape and sexual violence were covered in Chapter 6 on deviance and crime, in keeping with the argument that these are forms of deviant and criminal behavior, not expressions of human sexuality. Here we point out that various forms of sexual coercion (rape, domestic violence, and sexual harassment) can best be understood (and therefore changed) by understanding how social institutions shape human behavior and how social interactions are influenced by social factors such as gender, race, age, and class.

Take, for example, the phenomenon now known as *acquaintance rape* (sometimes also called *date rape*). Acquaintance rape is forced and unwanted sexual relations by someone who knows the victim (even if only a brief acquaintance). This kind of rape is common on college campuses, although it is also the most underreported form of rape. Researchers estimate, based on

© Bob Daemmrich/PhotoEdit

Domestic violence survivors Angela De Hoyos, Marylu De Hoyos, and Venus De Hoyos speak at a rally for Texas Council on Family Violence outside the State Capitol, Austin, Texas.

surveys, that 15 to 25 percent of college women will experience some form of acquaintance rape (Fisher et al. 2000; see also Chapter 6).

Studies show that, although rape is an abuse of power, it is related to people's gender attitudes. Holding stereotypical attitudes about women is strongly related to adversarial sexual beliefs, accepting rape myths, and tolerating violence against women (American Psychological Association 2007).

Violence against women is also more likely to occur in some contexts than others, especially in organizations that are set up around a definition of masculinity as competitive, where alcohol abuse occurs, and where women are defined as sexual prey. This is one explanation given for the high incidence of rape in some college fraternities (Martin and Hummer 1989; Armstrong et al. 2006; Stombler and Padavic 1997).

Research on violence against women also finds that Black, Hispanic, and poor White women are more likely to be victimized by various forms of violence, including rape (U.S. Bureau of Justice Statistics 2005). African American and American Indian women and men report the highest incidence of intimate partner violence; Asian Americans and Pacific Islanders have the lowest incidence (Tjaden and Thoennes 2000). Studies also find that Black women are more aware of their vulnerability to rape than are White women and are more likely to organize themselves to resist rape collectively (Stombler and Padavic 1997).

In sum, sociological research on sexual violence shows how strongly sexual coercion is tied to the status of diverse groups of women in society. Rather than explaining sexual coercion as the result of maladjusted men or the behavior of victims, feminists have encouraged a view of sexual coercion that links it to an understanding of dominant beliefs about the sexual dominance of men and the sexual passivity of women. Researchers have shown that those holding the most traditional gender role stereotypes are most tolerant of rapists and least likely to give credibility to victims of rape (Marciniak 1998; Varelas and Foley 1998). Understanding sexual violence requires an understanding of the sociology of sexuality, gender, race, and class in society.

Sex and Social Change

As with other forms of social behavior, sexual behavior is not static. Sexual norms, beliefs, and practices emerge as society changes. As we saw in Chapter 10 on gender, some major changes affecting sexual relations come from changes in gender roles. But technological change, as well as the emphasis on consumerism in the United States, also affects sexuality. As you think about sex and social change, you might try to imagine what other social factors influence human sexual behavior.

The Sexual Revolution: Is It Over?

The **sexual revolution** refers to the widespread changes in men's and women's roles and a greater public acceptance of sexuality as a normal part of social development. Many changes associated with the sexual revolution have been changes in women's behaviors. Essentially, the sexual revolution has narrowed the differences in the sexual experiences of men and women. The feminist and the gay and lesbian movements have put the sexual revolution at the center of public attention by challenging gender role stereotyping and sexual oppression, profoundly changing our understanding of gay and lesbian sexuality. The sexual revolution has meant greater sexual freedom, especially for women, but it has not eliminated the influence of gender in sexual relationships.

Technology, Sex, and Cybersex

Technological change has also brought new possibilities for sexual freedom. One significant change is the widespread availability of the birth control pill. Sex is no longer necessarily linked with reproduction; new sexual norms associate sex with intimacy, emotional ties, and physical pleasure (Freedman and D'Emilio 1988). These sexual freedoms are not equally distributed among all groups, however. For women, sex is still more closely tied to reproduction than it is for men because women are still more likely to take the responsibility for birth control.

Contraceptives are not the only technology influencing sexual values and practices. Now the Internet has introduced new forms of sexual relations as many people seek sexual stimulation from pornographic websites or online sexual chat rooms. Cybersex, as sex via the Internet has come to be known, can transform sex from a personal, face-to-face encounter to a seemingly anonymous relationship with mutual online sex. This introduces new risks; for example, two-thirds of those visiting chat rooms are adults masquerading as children (Lamb 1998; Wysocki 1998; Cooper and Scherer 1999). The Internet has introduced new forms of deviance that are difficult to regulate.

Commercializing Sex

At the same time that there are new sexual freedoms for women, many worry that this will only increase the sexual objectification of women. Furthermore, sexuality is becoming more and more of a commodity in our highly consumer-based society. Thus, girls are being sexualized at younger ages, evidenced by the marketing of thongs to very young girls, the promotion of "sexy" dolls sold to young girls, and the highly sexualized content of media images that young girls consume (Levy 2005).

© RGraeme Robartson/Getty Images

The popular Bratz dolls are being marketed to young girls, selling an image of women as sexual objects.

Definitions of sexuality in the culture are heavily influenced by the advertising industry, which narrowly defines what is considered "sexy." Thin women, White women, and rich women are all depicted as more sexually appealing in the mainstream media. Images defining "sexy" are also explicitly heterosexual. The commercialization of sex uses women and, increasingly, men in demeaning ways. Poor women are also most likely to have to sell their sexuality for economic survival by working in the sex trade. Thus, although the sexual revolution has removed sexuality from many of its traditional constraints, the inequalities of race, class, and gender still shape sexual relationships and values.

The combination of sexualization and commodification means that people become "made" into things for others' use. When people are held to narrow definitions of sexual attractiveness or seen as valuable solely for their sexual appeal, you have social conditions that are ripe for exploitation and damage to people's sense of self-worth and value (American Psychological Association 2007). Thus, even with a seemingly increasingly "free" sexual society, sexuality is still nested in American culture within a system of power relations—power relations that, despite the sexual revolution, continue to influence how different groups are valued and defined.

11 Chapter Summary

- ***In what sense is sexuality, seemingly so personal an experience, a part of social structure?***

Sexual relationships develop within a social and cultural context. Sexuality is learned through socialization, is channeled and directed by social institutions, and reflects the race, class, and gender relations in society.

- ***What evidence is there of contemporary sexual attitudes and behavior?***

Contemporary sexual attitudes vary considerably by social factors such as age, gender, race, and religion. Sexual behavior has also changed in recent years. Young people initiate sex at an earlier age. Although the public believes in marital fidelity, many people have extramarital affairs. The number of gays and lesbians in the population, though debated, is significant. Cultural expectations about sexuality are also deeply influenced by gender expectations in the society.

- ***What does sociological theory have to say about sexual behavior?***

Functionalist theory depicts sexuality in terms of its contribution to the stability of social institutions. *Conflict theorists* see sexuality as part of the power relations and economic inequality in society. *Symbolic interaction* focuses on the social construction of sexual identity.

- ***What is meant by the social construction of sexual identity?***
  Sociologists see *sexual identity,* whatever its form, as constructed through socialization experiences. They have moved away from studying gay and lesbian experience as deviant behavior. Debate continues about the underlying origins of homosexuality, although evidence for a biological basis of homosexuality is not as strong as the evidence for social causes. *Homophobia* is the fear and hatred of homosexuals. *Heterosexism* is the institutionalization of heterosexuality as the only socially legitimate sexual orientation.

- ***How is sexuality related to contemporary social issues?***
  Sexuality is related to some of the most difficult social problems—including birth control, abortion, reproductive technologies, teen pregnancy, pornography, and sexual violence. Such social problems can be understood by analyzing the sexual, gender, class, and racial politics of society.

- ***How is sex related to social change?***
  The *sexual revolution* refers to widespread changes in the roles of men and women and a greater acceptance of sexuality as a normal part of social development. The sexual revolution has been fueled by social movements, such as the feminist movement and the gay and lesbian rights movement. Technological changes, such as the development of the pill, have also created new sexual freedoms. Now, sexuality is also influenced by the growth of cyberspace and its impact on personal and sexual interactions. At the same time, sex is also treated as a commodity in this society—it is bought and sold and used to sell various products.

Key Terms

coming out 295
eugenics 302
heterosexism 300
homophobia 299
queer theory 296
sexual identity 290
sexual orientation 290
sexual politics 296
sexual revolution 306
sexual scripts 291
social construction perspective 295

Online Resources

Sociology: The Essentials Companion Website

www.thomsonedu.com/sociology/andersen

Visit your book companion website where you will find more resources to help you study and write your research papers. Resources include Suggested Readings, web links, and a MicroCase Online feature that teaches you how to research society. Other resources include Learning Objectives, Internet exercises, quizzing, and flash cards.

is an easy-to-use online resource that helps you study in less time to get the grade you want NOW.

www.thomsonedu.com/login

Need help studying? This site is your one-stop study shop. Take a Pre-Test and Thomson NOW will generate a Personalized Study Plan based on your test results. The Study Plan will identify the topics you need to review and direct you to online resources to help you master those topics. You can then take a Post-Test to determine the concepts you have mastered and what you still need to work on.

12
Chapter twelve
CHAPTER TWELVE

Families and Religion

[***Suppose you were*** to ask a large group of people in the United States to describe their families. Many would describe divorced families. Some would describe single-parent families. Some would describe stepfamilies with new siblings and a new parent stemming from remarriage. Others would describe gay or lesbian households, perhaps with children present. Also included would be adoptive families and families with foster children. Others would describe the so-called traditional family with two parents living as husband and wife in the same residence as their biological children. Families have become so diverse that it is no longer possible to speak of "the family" as if it were a single thing.

continued

The traditional *family ideal*—a father employed as the breadwinner and a mother at home raising children—has long been the dominant cultural norm, communicated through a variety of sources, including the media, religion, and the law. Few families now conform to this ideal, and the number of families that ever did is probably fewer than generally imagined (Coontz 1992). Regardless of their form, families now face new challenges—possibly living on one income or managing family affairs when both parents are employed. Many families also feel that they are under siege by changes in society that are dramatically altering all family experiences.

Many view the changes taking place in families as positive. Women have new options and greater independence. Fathers are discovering that there can be great pleasure in domestic and child-care responsibilities. Change, however, also brings difficulties: balancing the demands of family and employment, coping with the interpersonal conflicts caused by changing expectations, and striving to make ends meet in families without sufficient financial resources. These changes bring new questions to the sociological study of families.

Family affairs are believed to be private, but as an institution, the family is very much part of the public agenda. Many people believe that "family breakdown" causes society's greatest problems. Public policies shape family life directly and indirectly, and family life is now being openly negotiated in political arenas, corporate boardrooms, and courtrooms, as well as in the bedrooms, kitchens, and "family" rooms of individual households.

As a social institution, the family is intertwined with other social institutions, such as religion. Religious values and customs tend to be learned first within families, and in the United States, religion influences beliefs about how families should be organized. Many major family events, such as weddings, christenings, and funerals, are observed with religious ceremonies, and many family behaviors, such as reproduction, marriage, divorce, and sexual behavior, are affected by religious values.

Religion thus has a profound effect on society and human behavior. This is also easily observed in daily life outside the family. Church steeples dot the landscape. Invocations to a religious deity occur at the beginning of many public gatherings. The news frequently reports on events generated by religious conflict.

Religious beliefs have led to conflict, but religious beliefs have also been the soul of some of the most liberating social movements, including the civil rights movement and other human rights movements around the world. Some of life's sweetest moments are marked by religious celebration, and some of its most bitter conflicts persist because of unshakable religious conviction. Religion is both an integrative force in society and the basis for many of our most deeply rooted social conflicts. The family and religion are, for most people, the first institutions encountered in life. In this chapter we examine both as social institutions.

Family diversity is the norm in American society, with no one type of family shaping people's experience.

Defining the Family

As a *social institution*, the family is an established social system that emerges, changes, and persists over time. Studying the family as an institution simply recognizes that families are organized in socially patterned ways. Institutions are "there;" we do not reinvent them every day, although people adapt in ways that make institutions constantly evolve. Institutions shape and direct our actions and make it appear that the only options available to us are those deemed acceptable by society. Perhaps you find it hard to imagine living in a family where the husband is expected to have both a wife and a concubine or where children do not live with their parents. The institutional structure of the family in this society does not support such family practices. Institutions shape both the form of families and the expectations that we have for family life.

Like other institutions, families are also shaped by their relationship to systems of inequality in society. Race, class, gender, and age stratification affect how society values certain families, and they influence the resources available to different families. In addition, these systems of stratification influence the power that individual members have within families. Children, for example, have fewer rights than adults, and often the very old are relatively powerless within families. Heterosexual privileges also shape the resources available to certain families since our social institutions presume that families will be heterosexual. Those living in gay and lesbian families then have to invent new practices that are not typically supported by the institutional fabric of society.

Given the diversity among families, how do we define the family? The family has traditionally been defined as a social unit of those related through marriage, birth, or adoption who reside together in officially sanctioned relationships and who engage in economic cooperation, socially approved sexual relations, reproduction, and child rearing (Gough 1984). Changes in contemporary family life require some flexibility in this definition because not all families fit these conditions. Following divorce, for example, the family does not typically share a common residence. Or some working parents leave their children in the care of others as they pursue seasonal or regional employment, sometimes far from their families (including across national borders).

As social scientists have become more aware of the diversity in family life, they have refined their analyses to better show the different realities of family situations. Now, we can define the **family** broadly to refer to a primary group of people—usually related by ancestry, marriage, or adoption—who form a cooperative economic unit to care for any offspring (and each other) and who are committed to maintaining the group over time (adapted from Lamanna and Riedmann 2003: 10).

thinking sociologically

Identify two popular "family" shows on television. Devise a systematic way to conduct a content analysis of the family ideal portrayed by these shows. What do your observations reveal about how the *family ideal* is communicated through the popular media?

Comparing Kinship Systems

Families are part of what are more broadly considered to be kinship systems. A **kinship system** is the pattern of relationships that define people's family relationships to one another. Kinship systems vary enormously across cultures and during different times. In some societies, marriage is seen as a union of individuals; in others, marriage is seen as creating alliances between groups. The form families take in different societies varies enormously. In some societies, such as India and China, many still follow traditional practices of arranged marriage, some even involving a broker whose job it is to conduct the financial transactions and arrange the marriage ceremonies (Croll 1995). In still other societies, maintaining multiple marriage partners may be the norm. At the heart of all these diverse family patterns are the social norms and structures associated with kinship systems. Kinship systems can generally be categorized by the following features:

- How many marriage partners are permitted at one time
- Who is permitted to marry whom
- How descent is determined
- How property is passed on
- Where the family resides
- How power is distributed

Polygamy is the practice of men or women having multiple marriage partners. Polygamy usually involves one man having more than one wife, technically referred to as *polygyny; polyandry* is the practice of a woman having more than one husband, an extremely rare custom. Within the United States, polygamy is commonly associated with Mormons, even though polygamy is practiced now by only a few Mormon fundamentalists (about 2 percent of the state population of Utah) who do so without official church sanction (Bachman and Esplin 1992; Driggs 1990; Brooke 1998).

Monogamy is the practice of a sexually exclusive marriage with one spouse at a time. It is the most common form of marriage in the United States and other Western industrialized nations. In the United States, monogamy is a cultural ideal that is prescribed through law and promoted through religious teachings. Lifelong monogamy is not always realized, however, as evidenced by the high rate of divorce and extramarital affairs. Many sociologists characterize modern marriage as *serial monogamy* in which individuals may, over a lifetime, have more than one marriage, but only one spouse at a time (Lamanna and Riedmann 2003).

Kinship systems determine whom one can marry. In contemporary U.S. society, people are expected to marry outside their own kin group; in other societies, people may be expected to marry within the kinship network. **Exogamy** is the practice of selecting mates from outside one's group; the group may be based on religion, territory, racial identity, and so forth. **Endogamy** is the practice of selecting mates from within one's group. In the United States, neither exogamy nor endogamy is mandated by law, although some religious doctrines condemn marriage outside the faith. Even if certain forms of marriage are not explicitly outlawed, society establishes normative expectations about who is an appropriate marriage partner. In general, people in the United States marry people with very similar social characteristics. For example, there is a clear tendency for people to select mates from similar class, race, religion, and educational backgrounds (Kalmijn 1991). Interracial marriage, although increasing, is still infrequent—only about 2 percent of married couples (U.S. Census Bureau 2007). The increase in interracial marriage contributes to a more multiracial and multicultural society with an increase in those who have a mixed racial identity (Root 2001, 1996).

Although interracial marriages are not common, historically a tremendous amount of energy has been put into preventing them. Laws have prohibited marriage between various groups, including between Whites and African Americans and between Whites and Chinese, Japanese, Filipinos, Hawaiians, Hindus, and Native Americans. Not until 1967 were laws prohibiting interracial marriage declared unconstitutional (Kennedy 2003; Takaki 1989).

Kinship systems shape the distribution of property in society by determining descent. **Patrilineal kinship** systems trace descent through the father; **matrilineal kinship** systems, through the mother. **Bilateral kinship** (or bilineal) traces descent through both.

Extended and Nuclear Families

Extended families are the whole network of parents, children, and other relatives who form a family unit. Sometimes extended families, or parts thereof, live together, sharing their labor and economic resources. In some contexts, "kin" may refer to those who are not related by blood or marriage but who are intimately involved in the family support system and are considered part of the family (Stack 1974). As an example, among African Americans *othermothers* are "women who assist bloodmothers by sharing responsibilities" (Collins 1990: 119). This is an adaptation to the demands of motherhood and work that characterizes African American women's experience. Their dual responsibility in the family and work has meant that African American women have often created alternative means of providing family care for children—a situation that the majority of all women now face. An othermother may be a grandmother, sister, aunt, cousin, or a member of the local community, but she is someone who provides extensive child care and receives recognition and support from the community around her.

The system of *compadrazgo* among Chicanos is another example of an extended kinship system. In this system, the family is enlarged by the inclusion of godparents, to whom the family feels a connection that is the equivalent of kinship. The result is an extended system of connections between "fictive kin" (those who are not related by birth but are considered part of the family) and actual kin that deeply affects family relationships among Chicanos (Baca Zinn and Eitzen 2005).

In the **nuclear family,** a married couple resides together with their children. Like extended families, nuclear families develop in response to economic and social conditions. The origin of the nuclear family in Western society is tied to industrialization. Before industrialization, families were the basic economic unit of society. Large household units produced and distributed goods, whether in small communities or large plantation and feudal systems where slaves and peasants provided most of the labor. Production took place primarily in households, and all family members were seen as economically vital. There was no sharp distinction between economic and domestic life because household and production were one. Women's role in the preindustrial family, although still marked by patriarchal relations, was publicly visible and economically valued. Women performed and supervised much of the household work, engaged in agricultural labor, and produced cloth and food. The work of women, men, and children was also highly interdependent. Although the tasks each performed might differ, together they were a unit of economic production.

With industrialization, paid labor was performed mostly away from the home in factories and public marketplaces. The transition to wages for labor created an economy based on cash rather than domestic production. Families became dependent on the wages that workers brought home. The shift to wage labor was accompanied by a patriarchal assumption that men should earn the "family wage" (that is, be the breadwinner). Thus, men who worked as paid laborers were paid more than women, and women became more economically dependent on men. At the same time, men's status was enhanced by having a wife who could afford to stay at home—a privilege seldom

understanding diversity

Interracial Dating and Marriage

Picture this: A young couple, stars in their eyes, holding hands, intimacy in their demeanor. Newly in love, the couple imagines a long and happy life together. When you visualize this couple, who do you see? If your imagination reflects the sociological facts, odds are that you did not imagine this to be an interracial couple. Although interracial couples are increasingly common (and have long existed), people are more likely to form relationships with those of their same race—as well as social class, for that matter. What do sociologists know about interracial dating and marriage?

First, patterns of interracial dating are influenced by both race and gender. Black men are those most likely to say they would date a person of another race (81 percent say they would), although, along with White women, they are the least likely to say they would marry someone of another race. White women are least likely to say they would date someone of another race. Both Blacks and Whites in interracial relationships report negative reactions from their families; the majority of these were not extremely hostile, but strong enough to put pressure on the interracial couple (Majete 1999). And although most Blacks and Whites profess to have a color-blind stance toward interracial marriages, when pressed, they raise numerous qualifications and concerns about such pairings (Bonilla-Silva and Hovespan 2000). Among college students, approval of family and friends is the strongest indicator of one's attitude toward interracial dating. Racial minorities are more accepting of interracial dating than are White students; students in the Greek system on college campuses are less accepting than non-Greeks (Khanna et al. 1999).

© Robin Nelson/PhotoEdit

Interracial marriage is increasingly common, although still a small proportion of all marriages.

Regardless of these attitudes, interracial marriage is on the rise, although they are still a small percentage of marriages formed. The most likely interracial marriages are between Black men and White women and between Hispanics and non-Hispanics. National data on Asian American marriage is limited, but studies find that Asian Americans are increasingly likely to marry other Asian Americans, though often someone of a different Asian heritage (Shinagawa and Pang 1996). The data on interracial dating and marriage show how something seemingly "uncontrollable," such as love, is indeed shaped by many sociological factors.

Sources: Khanna, Nikki D., Cherise Harris, and Rana Cullers. 1999. "Attitudes Toward Interracial Dating." Paper presented at the Annual Meetings of the Society for the Study of Social Problems; Bonilla-Silva, Eduardo, and Mary Hovespan. 2000. "If Two People Are in Love: Deconstructing Whites' Views on Interracial Marriage with Blacks." Paper presented at the annual meetings of the Southern Sociological Society; Majete, Clayton A. 1999. "Family Relationships and the Interracial Marriage." Paper presented at the Annual Meetings of the American Sociological Association; Shinagawa, Larry, and Gin Yong Pang. 1996. "Asian American Panethnicity and Intermarriage." *Amerasia Journal* 22 (Spring): 127–152.

accorded to working-class or poor families. The *family wage system* has persisted and is reflected in the unequal wages of men and women today.

The unique social conditions that racial–ethnic families have experienced also affect the development of family systems. Disruptions posed by the experiences of slavery, migration, and urban poverty affect how families are formed, their ability to stay together, the resources they have, and the problems they face. For example, historically, Chinese American laborers were explicitly forbidden to form families by state laws designed to regulate the flow of labor. Only a small number of merchant families were exempt from the law. Under slavery, African American

SEE FOR YOURSELF

ANALYZING FAMILY STRUCTURES

What is the kinship structure of your family? Make a list of all the people whom you consider to be part of your family. List them by name and then by your relationship (i.e., father, sister, partner, othermother, and so forth). Then using the descriptive concepts that describe *kinship systems,* identify the specific form of kinship that describes your family (nuclear, extended, matrilineal, patriarchal, and so forth). Are there characteristics of your family that suggest any need to revise these concepts? How does your family structure compare to that of other students in the class? Are there any differences that appear because of the different social statuses (ethnicity, regional origin, etc.)? What does this tell you about the social structure of families?

families faced a constant threat of disruption. Marriages among slaves were not officially recognized by law since slaves were not considered to be citizens. Nonetheless, slaves formed families, although the children of slave parents were the legal property of the slave owners who maintained strict control over slave life and, when it served their needs, would sell or separate family members.

During the westward expansion of the United States, many Mexicans who had settled in the Southwest were displaced. The loss of their land disrupted their families and kinship systems. In the rapidly industrializing United States, many Mexican Americans were able to find work in the mines opening in the new territories or by building the railroads spreading from the East toward the Pacific. Employers apparently thought that they had better control over laborers if their families were not there to distract them, so families were typically prohibited from being with the worker. One result was the development of prostitution camps, which followed workers from place to place (Dill 1988).

Families continue to be influenced by social structural forces. Some families, particularly those with marginal incomes, find it necessary for the entire family to work to meet the economic needs of the household. Migration to a new land and exposure to new customs also disrupts traditional family values. The ability to form and sustain nuclear families is directly linked to the economic, political, and racial organization of society.

Sociological Theory and Families

Is the family a source of stability or change in society? Are families organized around harmonious interests, or are they sources of conflict and differential power? How do new family forms emerge, and how do people negotiate the changes that affect families? These questions and others guide sociological theories of the family (see Table 12.1).

table 12.1 Theoretical Perspectives on Families

	Functionalism	Conflict Theory	Feminist Theory	Symbolic Interactionism
Families	Meet the needs of society to socialize children and reproduce new members	Reinforce and support power relations in society	Are gendered institutions that reflect the gender hierarchies in society	Emerge as people interact to meet basic needs and develop meaningful relationships
	Teach people the norms and values of society	Inculcate values consistent with the needs of dominant institutions	Are a primary agent of gender socialization	Are where people learn social identities through their interactions with others
	Are organized around a harmony of interests	Are sites for conflict and diverse interests of different family members	Involve a power imbalance between men and women	Are places where people negotiate their roles and relationships with each other
	Experience social disorganization ("breakdown") when society undergoes rapid social changes	Change as the economic organization of society change	Evolve in new forms as the society becomes more or less egalitarian	Change as people develop new understandings of family life

Functionalist Theory and Families

Functionalist theorists interpret the family as filling particular societal needs, including socializing the young, regulating sexual activity and procreation, providing physical care for family members, and giving psychological support and emotional security to individuals. According to functionalism, families exist to meet these needs and to ensure a consensus of values in society. In the functionalist framework, the family is conceptualized as a mutually beneficial exchange, wherein women receive protection, economic support, and status in return for emotional and sexual support, household maintenance, and the production of offspring (Glenn 1987). At the same time, in traditional marriages, men get the services that women provide—housework, nurturing, food service, and sexual partnership. Functionalists also see families as providing care for children, who are taught the values that society and the family support. In addition, functionalists see the family as regulating reproductive activity, including cultural sanctions about sexuality.

According to functionalist theory, when societies experience disruption and change, institutions such as the family become disorganized, weakening social cohesion. Currently, some analysts interpret the family as "breaking down" under societal strains. Functionalist theory suggests that this breakdown is the result of the disorganizing forces that rapid social change has fostered.

Functionalists also note that, over time, other institutions have begun to take on some functions originally performed solely by the family. For example, as children now attend school earlier in life and stay in school for longer periods of the day, schools (and other caregivers) have taken on some functions of physical care and socialization originally reserved for the family. Functionalists would say that the diminishment of the family's functions produces further social disorganization, because the family no longer carefully integrates its members into society. To functionalists, the family is shaped by the template of society, and such things as the high rate of divorce and the rising numbers of female-headed and single-parent households are the result of social disorganization.

© Ariel Skelley/Jupiter Images

The family is the major institution where socialization of children occurs.

Conflict Theory and Families

Conflict theory interprets the family as a system of power relations that reinforces and reflects the inequalities in society. Conflict theorists are also interested in how families are affected by class, race, and gender inequality. This perspective sees families as the units through which the advantages, as well as the disadvantages, of race, class, and gender are acquired. Conflict theorists view families as essential to maintaining inequality in society, because they are the vehicles through which property and social status are acquired (Eitzen and Baca Zinn 2007).

The conflict perspective also emphasizes that families in the United States are shaped by capitalism. The family is vital to capitalism because it produces the workers that capitalism requires. Accordingly, within families, personalities are shaped to the needs of a capitalist system; thus families socialize children to become obedient, subordinate to authority, and good consumers. Those who learn these traits become the kinds of workers and consumers that capitalism needs. Families also serve capitalism in other ways; giving a child an allowance teaches the child capitalist habits involving money.

Whereas functionalist theory conceptualizes the family as an integrative institution (meaning it has the function of maintaining social stability), conflict theorists depict the family as an institution subject to the same conflicts and tensions that characterize the rest of society. Families are not isolated from the problems facing society as a whole. The struggles brought on by racism, class inequality, sexism, homophobia, and other social conflicts are played out within family life.

Feminist Theory and Families

Feminist theory has contributed new ways of conceptualizing the family by focusing sociological analyses on women's experiences in the family and by making gender a central concept in analyzing the family as a social institution. Feminist theories of the family emerged initially as a criticism of functionalist theory. Feminist scholars argued that functionalist theory assumed that the gender division of labor in the household is functional for society. Feminists have also been critical of functional theory for assuming an inevitable gender division of labor within the family. Feminist critics argue that, although functionalists may see the gender division of labor as functional, it is based on stereotypes about men's and women's roles.

Influenced by the assumptions of conflict theory, feminist scholars do not see the family as serving the needs of all members equally. Quite the contrary, feminists have noted that the family is one of the primary institutions producing the gender relations found in society. Feminist theory conceptualizes the family as a system of power relations and social conflict (Thorne and Yalom 1992). In this sense, it emerges from conflict theory, but adds that the family is a gendered institution (see Chapter 10).

Symbolic Interaction Theory and Families

Symbolic interaction emphasizes that meanings people give to their behavior and that of others is the basis of social interaction. Those who study families from this perspective tend to take a more microscopic view of families. A symbolic interactionist might ask how different people define and understand their family experience. Symbolic interactionists also study how people negotiate family relationships, such as deciding who does what housework, how they will arrange child care, and how they will balance the demands of work and family life.

To illustrate, when people get married, they form a new relationship and new identities with specific meanings within society. Some changes may seem very abrupt—a change of name certainly requires adjustment, as does being called a husband or wife. Some changes are more subtle—how one is treated by others and the privileges couples enjoy (such as being a recognized legal unit). Symbolic interactionists see the married relationship as socially constructed; that is, it evolves through the definitions that others in society give it, as well as through the evolving definition of self that married partners make for themselves.

The symbolic interactionist perspective understands that roles within families are not fixed, but rather evolve as participants define and redefine their behavior toward each other. Symbolic interaction is especially helpful in understanding changes in the family because it supplies a basis for analyzing new meaning systems and the evolution of new family forms over time. Each theoretical perspective used to analyze families illuminates different features of family experiences.

Diversity Among Contemporary American Families

Today the family is one of the most rapidly changing of all of society's institutions. Families are systems of social relationships that emerge in response to social conditions that, in turn, shape the future direction of society. There is no static or natural form for the family. Change and variation in families are social facts.

Among other changes, families today are smaller than in the past. There are fewer births, and they are more closely spaced, although these characteristics of families vary by social class, region of residence, race, and other factors. Because of longer life expectancy, childbearing and child rearing now occupy a smaller fraction of parents' adult life. During earlier periods, death (often from childbirth) was more likely to claim the mother than the father of small children; thus, men in the past would have been more likely than now to raise children on their own after the death of a spouse. That trend is now reversed; it is now women who are more likely to be widowed with children, and death, once the major cause of early family disruption, has been replaced by divorce (Rossi and Rossi 1990).

Demographic and structural changes have resulted in great diversity in family forms. Compared to thirty years ago, married couples now make up a smaller proportion of households; single-parent households have increased dramatically, and divorced and never-married people make up a larger proportion of the population (Fields 2004; see Figure 12.1). Overall, married-couple families are three-quarters of household types, but this varies significantly by race. Single-parent households (typically headed by women), post-childbearing couples, gay and lesbian couples, childless households, and single people are increasingly common. Now people may also spend more years caring for elderly parents than they did raising their children.

Female-Headed Households

One of the greatest changes in family life is the increase in the number of families headed by women. One-quarter of all children live with one parent. Most single-parent households (88 percent) are headed by women, although the number of households headed by single fathers has also increased. The odds of living in a single-parent household are even greater for African American and Latino children (Fields 2004).

The two primary causes for the growing number of women heading their own households are the high rate of pregnancy among unmarried teens and the high divorce rate, with death of a spouse also contributing. Although as we discussed in Chapter 10, the rate of pregnancy among teenagers (of all races), married and unmarried, has declined, the proportion of teen births that occur outside marriage has increased. As you can see in Map 12.1, the rate of teen pregnancy also varies from region to region.

Divorce, the second reason for the increase in the number of female-headed households, also contributes to the growing rate of poverty among women, as we saw in Chapter 7. Most women see a substantial decline in their income in the year following divorce (Hoffman and Duncan 1988; Peterson 1996a, 1996b;

[a sociological eye on the media

Idealizing Family Life

Cultural norms about motherhood and fatherhood come from many places, but the media is certainly a strong influence on how family ideals—and the ideals for mothers and fathers—are created in society. Media images of the family have certainly changed since the inception of television. In the 1930s, Hollywood Codes—that is, official rules in Hollywood about what could and could not be seen in movies—and later, on television—meant that families were always shown with two parents, marriage intact, father working, and mom staying at home. Parents never talked about sex; indeed, it appeared as though they never had it because the Codes forbade any nudity and required that scenes of passion not excite the audience. As a result, if bedroom scenes between married couples were shown, there were typically only twin beds.

Now, family images on television are more diverse. Fathers are shown in parenting roles, and some families are divorced or blended following divorce; working mothers are typical, not exceptional; and children may be stepchildren. Not all TV families are heterosexual, though most on television are. Still, the media continues to construct an ideal for family life—one that continues to stereotype men and women in family roles. What does research show about these social constructions?

- Most family characters are middle class.
- Men appearing with children are most likely to be shown outside; they are also more likely to be seen with boys, not girls.
- Fathers are infrequently seen with infants.
- Fathers are shown playing with, reading to, talking with, and eating with children, but not preparing meals, cleaning house, changing diapers, and so forth.
- Women are disproportionately shown in family settings in the media.

What gender stereotypes do such images project? How do they influence people's views of ideal family roles? How realistically do they portray the actual gender division of labor in families? What race and class images confound these results?

Sources: Kaufman, Gayle. 1999."The Portrayal of Men's Family Roles in Television Commercials." *Sex Roles* 41 (September): 439–458. Coltrane, Scott, and Melinda Messineo. 2000. "The Perpetuation of Subtle Prejudice: Race and Gender Imagery in 1990s Television Advertising." *Sex Roles* 42 (March): 363–389.

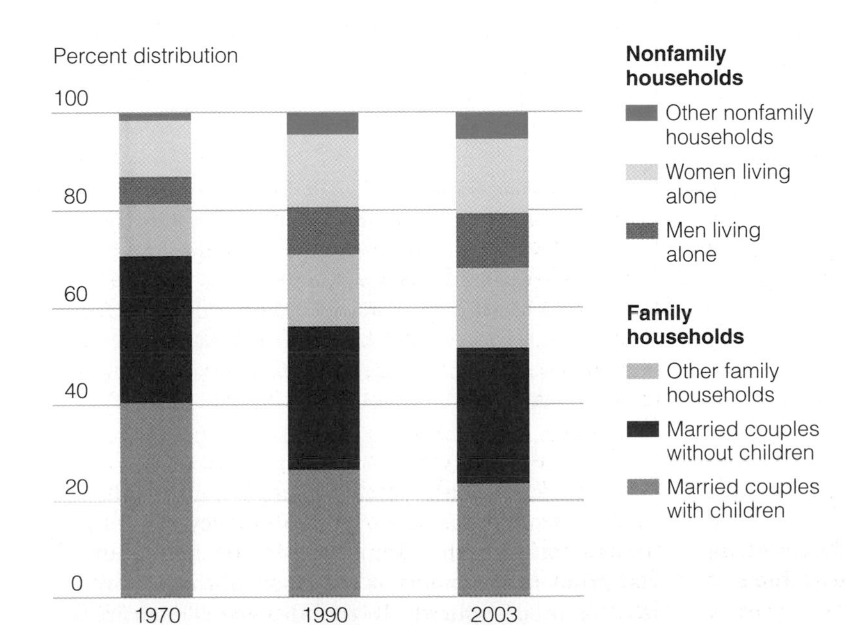

FIGURE 12.1 *Diversity in U.S. Families*

Source: Fields, Jason. 2004. *America's Families and Living Arrangements: 2003.* Current Population Reports, P20-553. Washington, DC: U.S. Bureau of the Census, p. 4.

map 12.1 MAPPING AMERICA'S DIVERSITY

Births to Teenage Mothers

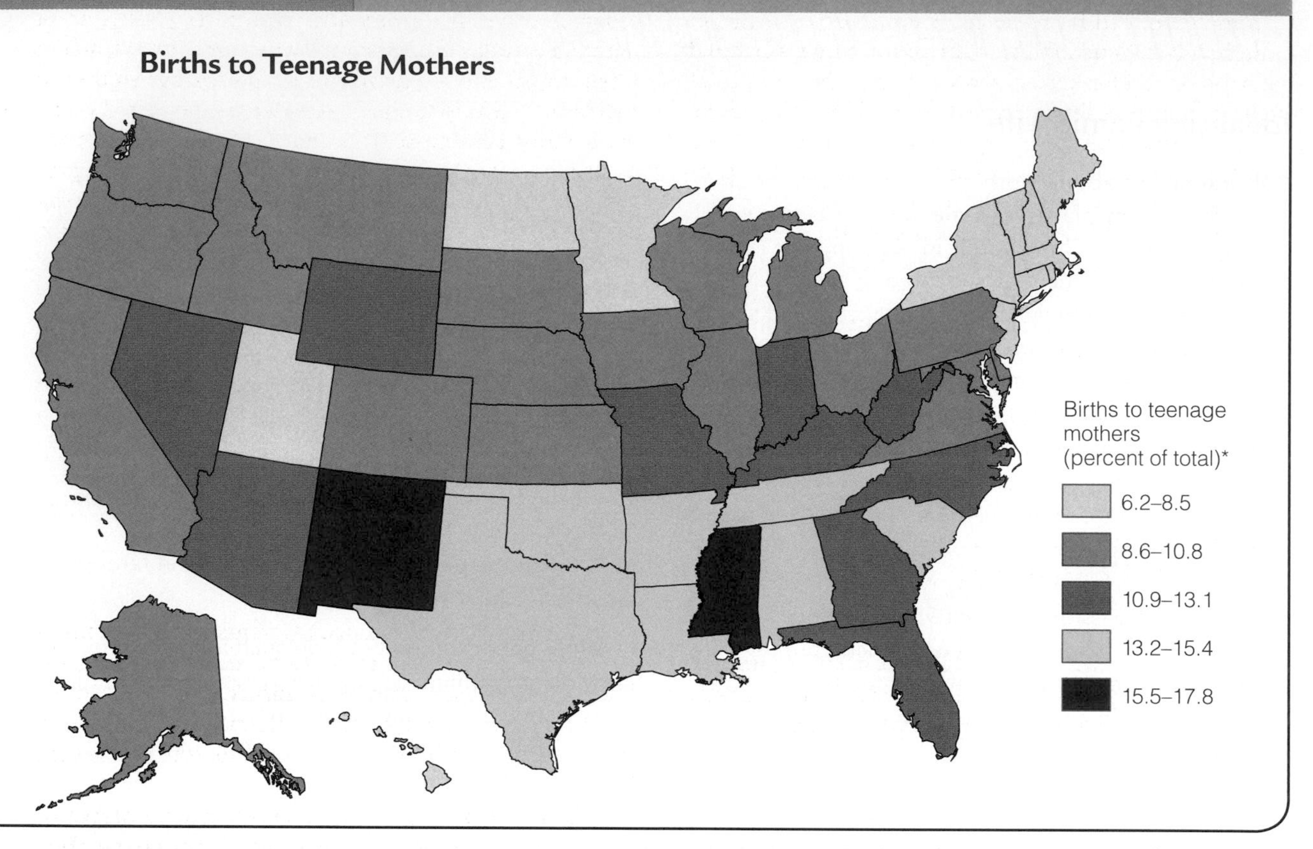

This map shows the percentage of births to teenage mothers as a percentage of all births. Although you cannot see the historic change from this map, it is the case that the rate of teen pregnancy has been declining in the United States, but so has the likelihood of marriage for teen mothers who give birth. These data include births to both married and unmarried teen mothers. If you were trying to explain why the rates of teen pregnancy vary from state-to-state, what factors would you want to examine and why?

Data: From the U.S. Census Bureau. 2007. *Statistical Abstract of the United States, 2007.* Washington, DC: U.S. Government Printing Office, p. 68.

Weitzman 1985). Contrary to popular belief, most men also experience a decline in their standard of living following divorce, mostly because of the loss of wives' income. However, men's loss is not as great as women's, and some men actually benefit economically from divorce (McManus and Diprete 2001). Following divorce, of those supposed to receive child support payments (almost all of whom are mothers), less than half receive all that is due. The average amount actually paid is only $4274 per year (U.S. Census Bureau 2007).

Many people see the increase of female-headed households as representing a breakdown of the family and a weakening of social values. An alternate view, however, is that the rise of female-headed households reflects the growing independence of women, some of whom are making decisions to raise children on their own. Not all female-headed households are women who have never married; many are divorced and widowed women, whose circumstances may be quite different from those of a younger, never-married woman.

Debunking Society's Myths

Myth: Absence of fathers is the cause of numerous social problems; if these fathers would just adopt "family values," families would be stronger and children wouldn't get into so much trouble.

Sociological perspective: This idea usually targets African American and poor men, but research on never-married, poor, noncustodial African American fathers finds that they want to be able to provide for their children; they may spend a lot of time with their children; and they want to be good fathers. Understanding what fatherhood means to them requires understanding the context in which they live (Hamer 2001).

Some claim that female-headed households are linked to problems such as delinquency, the school dropout rate, children's poor self-image, and other social problems. Sometimes the cause of these troubles is attributed explicitly to the absence of men in the family. Sociologists, however, have not found the absence of men as the sole basis for such problems; rather, it is the presence of economic pressure faced by female-headed households, compared with that of male-headed households, that puts female-headed households under great strain—with the threat of poverty being by far the greatest problem they face. Among households headed by women with children, one-third live below the poverty line, with the rates of poverty highest among Black and Hispanic female-headed households (DeNavas-Walt et al. 2006). It is not the makeup of households headed by women that is a problem but the fact that they are most likely to be poor (Brewer 1988). This phenomenon is discussed in Chapter 7 as the *feminization of poverty.*

Although the majority of single-parent families are headed by women, families headed by a single father are also increasing. Today, 69 percent of all children live with both parents; 23 only with their mother; 5 percent only with their fathers; and 4 percent in households with neither parent present (Fields 2003). Male-headed households are less likely than female-headed households to experience severe economic problems. Unlike female-headed households where a man is not present to help with housework and children, single fathers commonly get domestic help from women—either girlfriends, daughters, or mothers (Popenoe 2001).

Married-Couple Families

Among married-couple families, one of the greatest changes in recent years has been the increased participation of women in the paid labor force. Indeed, as we saw in Chapter 7, families sustain a median income level by having both husband and wife in the paid labor force—a practice historically common among most African American families, many Latino and Asian American families, and many White, working-class families. Even though increased labor force participation is most dramatic among White families, most families are experiencing substantial speedup. This is reflected in longer hours worked—especially for women (Mishel et al. 2005).

Women's labor force participation has created other changes in family life. One is the number of married couples who have *commuter marriages,* when work requires one partner in a dual-career couple to reside in a different city, separated by jobs too distant for a daily commute. The common image of a commuter marriage is one consisting of a prosperous professional couple, each holding important jobs, flashing credit cards, and using airplanes like taxis. However, working-class and poor couples do their share of long-distance commuting: Agricultural workers follow seasonal work; skilled laborers sometimes have to leave their families to find jobs; many families cross national borders in search of work, often separating one or both parents and children. Although their commute may be less glamorous than that of professional spouses, they are commuting nonetheless. When all types of commuter marriages are included, this form of marriage is more prevalent than is typically imagined.

Stepfamilies

Stepfamilies are becoming more common in the United States, matching the rise in divorce and remarriage. Stepfamilies take numerous forms, including married adults with stepchildren, cohabiting stepparents, and stepparents who do not reside together (Stewart 2001).

About 40 percent of marriages involve stepchildren. These families face a difficult period as they deal with the troubles that arise when two family systems blend or when new people are introduced into an existing family system. Parents and children discover that they must learn new roles when they become part of a stepfamily. Children accustomed to being the oldest child in the family, or the youngest, may find that their status in the family group is suddenly transformed. New living arrangements may require children to share rooms, toys, and time with people they perceive as strangers.

*Is It True?**

	True	False
1. Half of all marriages end up in divorce.		
2. Children who grow up in gay or lesbian families are likely to become gay.		
3. Single people are a larger proportion of the population than was true in the past.		
4. Children are better off growing up in a home where mothers are not employed.		
5. Women and men find great satisfaction in trying to balance family and work.		
6. Fathers' involvement in child care leads to more stable marriages.		

*** The answers can be found on page 322.**

Is It True? [Answers]

1. FALSE. The divorce rate is based on the number of divorces in one year per 1,000 people in the population; the marriage rate, the number of marriages in one year per 1,000 people. This does NOT mean that half of all marriages end in divorce because marriages made in one year can last many years beyond.
2. FALSE. There is little difference in outcomes for children growing up in gay or lesbian households relative to those living in heterosexual households, including their later sexual orientation (Stacey and Biblarz 2001).
3. TRUE. The number of never-married people has increased substantially since the 1970s, as have the number of divorced people in the population (Fields 2003).
4. FALSE. Researchers find no negative influence of parental employment per se on children; far more significant are other conditions, especially the economic stability of the family and other stresses for children that emerge from conflict and violence (Perry-Jenkins et al. 2000).
5. TRUE. Although balancing the demands of both work and family is stressful for both women and men, both report that being able to do so produces satisfaction. Stress results when women and men feel they have to make trade-offs at work in order to meet family demands (Milkie and Peltola 1999).
6. TRUE. Researchers find that fathers' increased involvement in child care is linked to more stable marriages, in part because mothers are then happier (Kalmijn 1999).

In stepfamilies, the parenting roles of mothers and fathers suddenly expand to include more children, each with his or her needs. Jealousy, competition, and demands for time and attention can make the relationships within stepfamilies tense. The problems are compounded by the absence of norms and institutional support systems for stepfamilies. Without norms to follow, people have to adapt by creating new language to refer to family members and new relationships. Many develop strong relationships within this new kinship system; others find the adjustment extremely difficult, resulting in a high probability of divorce among remarried couples with children (Baca Zinn and Eitzen 2005).

Gay and Lesbian Households

The increased visibility of gay and lesbian households challenges the traditional understanding of families as only heterosexual and has led to a greater acceptance of gays and lesbians in society. Although they have neither the official blessings nor supports of social institutions, many gay and lesbian couples form long-term, primary relationships that they define as marriage. Like other families, gay and lesbian couples share living arrangements and household expenses, make decisions as partners, and in many cases, raise children (Dalton and Bielby 2000; Dunne 2000).

Researchers have found that gay and lesbian couples tend to be more flexible and less gender-stereotyped in their household roles than heterosexual couples. Lesbian households, in particular, are more egalitarian than are either heterosexual or gay male couples. Money also has less effect on the balance of power in lesbian relationships than is true for heterosexual couples. However, where one partner is the primary breadwinner and the other the primary caregiver for children, the partner staying at home becomes economically vulnerable and less able to negotiate her needs, just as in heterosexual relationships (Sullivan 1996).

Gay and lesbian marriages are producing new social forms and new social debates. Should gay marriage be recognized in law? The Supreme Court of Massachusetts has recognized gay marriage as legal, setting off a public debate and prompting conservative groups to propose that there should be an amendment to the U.S. Constitution defining marriage as only between a man and a woman. Thirty-nine percent of the American public believes that gay marriages should be recognized as valid and given the same rights as heterosexual couples; 58 percent disagree. A larger number (49 percent) believe that civil unions between gays should be legally valid, suggesting that the word "marriage" *per se* is imbued with meaning that many do not want to extend to gay couples (Moore and Carroll 2004). Gay marriage is also more acceptable in the eyes of younger people than to older groups, raising the question of whether over time social support for gay marriages will increase or whether, as young people age, they will shift their values (see Figure 12.2). In the meantime, in most states and municipalities, gays and lesbians who form strong and lasting relationships do so without formal institutional support and, as a result, have had to be innovative in producing new support systems.

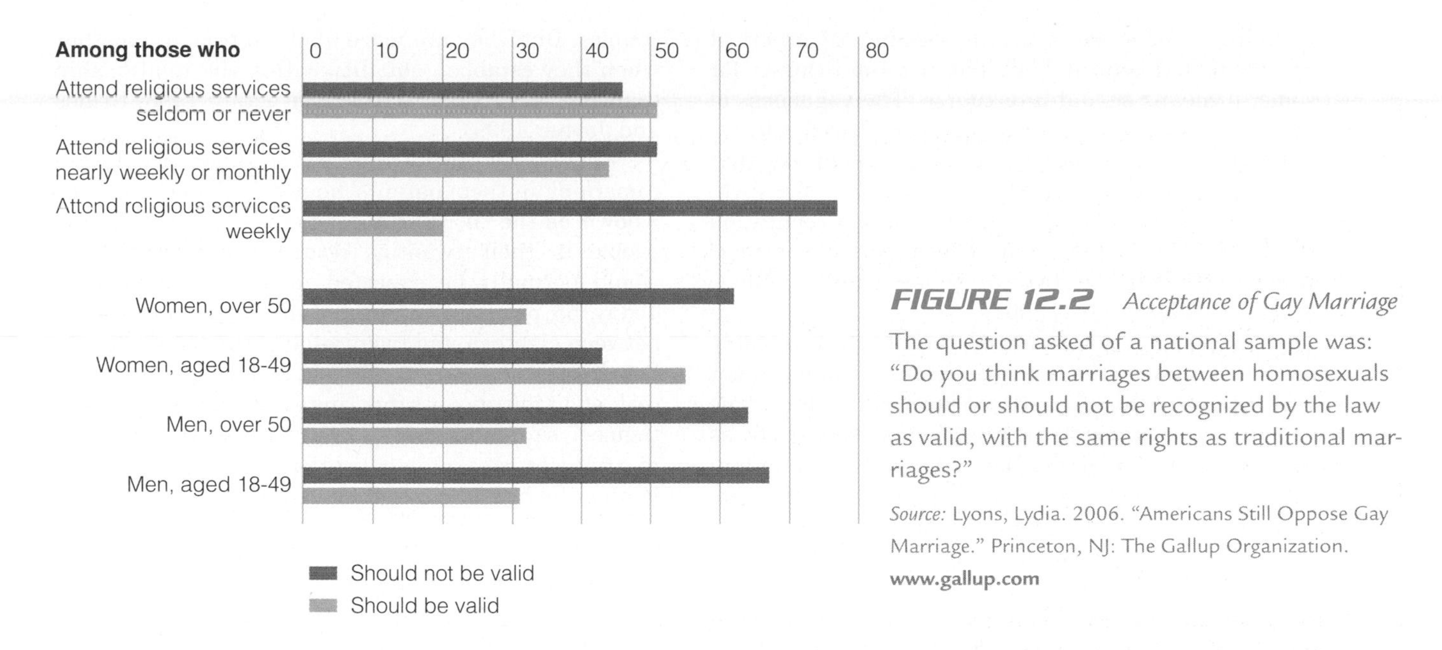

FIGURE 12.2 *Acceptance of Gay Marriage*

The question asked of a national sample was: "Do you think marriages between homosexuals should or should not be recognized by the law as valid, with the same rights as traditional marriages?"

Source: Lyons, Lydia. 2006. "Americans Still Oppose Gay Marriage." Princeton, NJ: The Gallup Organization. **www.gallup.com**

The new family forms that lesbian and gay couples are creating mean that they have to actively construct new meanings of such things as motherhood. Researchers find that they do so in ways that are collaborative, including elaborate networks of family and friends (Dunne 2000; Dalton and Bielby 2000). To date, most gay fathers are those who have children from a previous heterosexual marriage, although the number of gay men adopting children is increasing.

Public debate about gay marriage often centers on the implications for children raised in gay and lesbian families. Research on children in gay and lesbian households finds that for the most part there is little difference in outcomes for children raised in gay and lesbian households compared to those raised in heterosexual households. What differences are found are the result of other factors—not just the sexual orientation of parents. The greatest differences are the result of the homophobia that is directed against children in lesbian and gay families, who are very likely to be stigmatized by others. But, such children are also less likely to develop stereotypical gender roles and are more open-minded about sexual matters, although they are no more likely to become gay themselves (Allen 1997; Stacey and Biblarz 2001). If we lived in a society more tolerant of diversity, the differences that emerge might be viewed as strengths, not deficits.

Single People

Single people, including those never married, widowed, or divorced, today constitute 45 percent of the population (over age 15)—an increase from 29 percent in 1970. Some of the increase is a result of the rising number of divorced people, but there has also been an increase in the number of those never married. Men and women are marrying at a later age—25.3 years on average for women, 27.1 for men, compared with 20.8 for women and 23.2 for men in 1970 (Fields 2004).

Among singles, patterns of establishing intimate relationships have changed significantly. "Hooking up"—a phrase referring to a casual, sexual alliance between two people—has supplanted dating as the pattern by which young people get to know each other. Courtship no longer follows preestablished norms. Hooking up is widespread on college campuses and influences the campus culture, although only a minority of students engage in it (about 40 percent of college women). Hooking up carries multiple

Popular shows like Friends *gain some of their appeal by the growing proportion of the population who are single.*

meanings. For some it means kissing; for others it means sexual-genital play, but not intercourse; for some, it means sexual intercourse. The vagueness of the term contributes to its becoming a shared cultural phenomenon. A majority of college women say that hooking up makes them feel desirable, but also awkward, and they are wary of getting a bad reputation from hooking up too often. The majority of college women still want to meet a spouse while at college (Glenn and Marquardt 2001).

The path to a committed relationship, possibly marriage, involves increasing phases of commitment and sexual exclusivity. Many people find the same sexual and emotional gratification in single life as they would in marriage. Being single is also no longer the stigma it once was, especially for women. Nonetheless, studies find that married people are generally happier than singles; this changes in different historical periods, however, and depends a lot on other factors, such as financial resources, marital conflict and violence, the extent of other personal ties, and the presence of stressful life problems (Aldous and Ganey 1999; Keith 1997; Lee et al. 1991; Waite 2000).

Cohabitation (living together) has become increasingly common. Some of the increase is the result of better census taking, but the increase is also real. Estimates are that one-quarter of all children will at some time during their childhood live in a family headed by a cohabiting couple (Graefe and Lichter 1999), indicating how common cohabitation has become. Cohabitation is more common among the young and the less educated and among Black and Native American couples (Cohen 1999; Marquis 2003). Some cohabit because they are critical of the existing norms surrounding marriage (Elizabeth 2000). Cohabitants tend to have more egalitarian attitudes toward gender roles, although they tend to develop a household division of labor similar to that of married heterosexual couples. But, they are more likely to remain together when they establish equality within the relationship (Sanchez et al. 1998; Barber and Axinn 1998; Brines and Joyner 1999).

Finally, a growing number of single people are remaining in their parents' homes for longer periods. Known as the "boomerang generation," these young people in their twenties return home when they would normally be expected to live independently. Today, 55 percent of men and 45 percent of women between eighteen and twenty-four years of age live at home with their parents (Fields 2004). The increased cost of living means that many young people find themselves unable to pay their own way, even after marrying or getting an education. They economize by joining the household of their parents.

Marriage and Divorce

Even with the extraordinary diversity of family forms in the United States, the majority of people will still marry at some point in their lives (see Figure 12.3). Indeed, the United States has the highest rate of marriage of any Western industrialized nation, as well as a high divorce rate.

Marriage

The picture of marriage as a consensual unit based on intimacy, economic cooperation, and mutual goals is widely shared, although marital relationships also involve a complex set of social dynamics, including cooperation and conflict, different patterns of resource allocation, and a division of labor. Sociologists must be careful not to romanticize marriage to the point that they miss other significant social patterns within marriage.

Gender roles are a significant reality of family life, shaping power dynamics within marriage, as

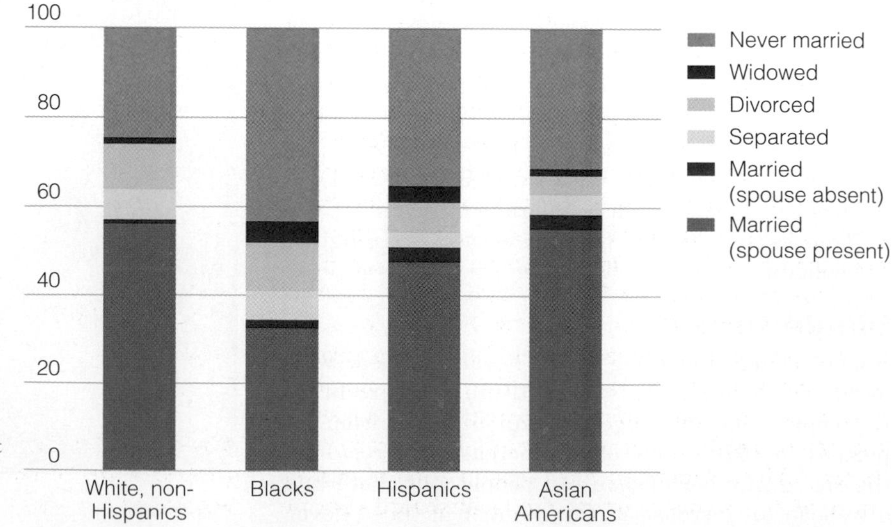

FIGURE 12.3 *Marital Status of the U.S. Population by Race*

Data: U.S. Census Bureau. 2003. *America's Families and Living Arrangements: 2003, Detailed Tables.* Current Population Reports. Washington, DC: U.S. Bureau of the Census. **www.census.gov**

well as the allocation of work, the degree of marital happiness, the likelihood of marital violence, and even the leisure time that each partner has. Although people do not like to think of marriage as a power relationship, gender shapes the power that men and women have within marriage, as it does in other relationships. For one thing, sociologists have long found that the amount of money a person earns establishes that person's relative power within the marriage, including the ability to influence decisions, the degree of autonomy and independence held by each partner, and the control of expectations about family life. Despite changes in women wanting to work outside the home, in most marriages (64 percent) men are the sole or major earners (Raley et al. 2006). But studies also find that even when wives earn substantially more than their husbands, rare as that is, couples tend to negotiate marital power within the confines of traditional gender expectations (Tichenor 2005).

Within marriage, gender also shapes the division of household labor. Women do far more work in the home and have less leisure time (Baca Zinn and Eitzen 2005). Most employed mothers do two jobs—the so-called *second shift* of housework after working all day in a paid job (Hochschild 1989). With more women in the labor market, wives with children have increased their working hours substantially more than men have (Mishel et al. 2005). Little wonder that people are feeling that they have less and less time.

Are men more involved in housework than in the past? Yes and no. Men report that they do more housework, but they devote only slightly more of their time to housework than in the past. Estimates vary regarding the amount of housework that men do, but studies generally find a large gap between the number of hours women give to housework and child care and the hours men give. Among couples where both partners are employed, only 28 percent equally share the housework. Fathers do more when there is a child in the house under two years of age, but the increase is mostly accounted for by the amount of child care men provide, not the housework they do. The end result is that men have about eleven more hours of leisure per week than women do (Press and Townsley 1998). Interestingly, sociologists have found that the allocation of housework is greatly affected by men's and women's experience in their own families of origin; those from households with a more egalitarian division of labor are likely to carry this into their own relationships (Cunningham 2001).

Despite a widespread belief that young professional couples are the most egalitarian, studies find that there is little difference across social class in the amount of housework that men do (Wright et al. 1992). African American husbands provide a greater share of housework than do White husbands. Latino households have more diversity in gender roles than stereotypes about machismo would lead us to believe (McLoyd 2000).

Although marriage can be seen as a romantic and intimate relationship between two people, it can also then be seen within a sociological context. Marriage relationships are shaped by a vast array of social factors, not just the commitment of two people to each other. You see this especially when examining marital conflicts. Life events, such as the birth of a child, job loss, retirement, and other family commitments, such as elder care or caring for a child with special needs, all influence the degree of marital conflict and stability (Moen et al. 2001; Crowley 1998). As conditions in society change, people make adjustments within their relationships, but how well they can cope within a marriage depends on a large array of sociological—not just individual—factors.

Divorce

The United States leads the world not only in the number of people who marry, but also in the number of people who divorce. More than sixteen million people have divorced but not remarried in the population today; more women are in this group than men since women are less likely to remarry following a divorce. Since 1960, the rate of divorce has more than doubled, although it has declined recently since its all-time high in 1980 (see Figure 12.4).

You will often hear that one in every two marriages ends in divorce, but this is a misleading statistic. The marriage rate is 7.8 marriages per 1000 people and the divorce rate, 3.7 per 1000 people (U.S. Census Bureau 2007). At first glance, it appears that

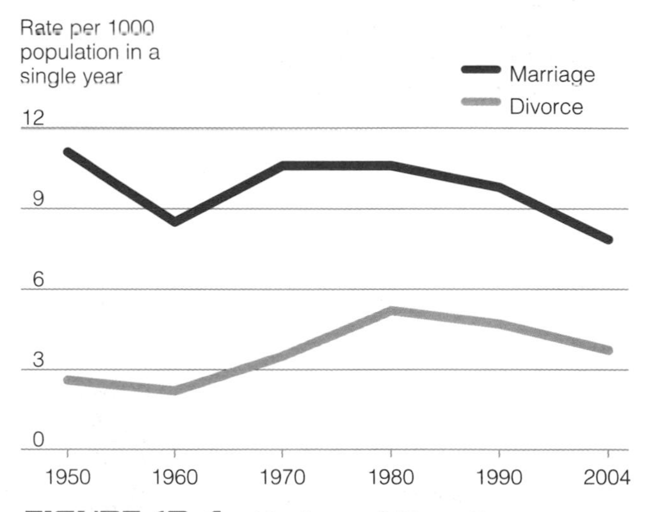

FIGURE 12.4 *Marriage and Divorce Rates, 1950–2004*

Data: U.S. Bureau of the Census. 2007. *Statistical Abstract of the United States, 2007.* Washington, DC: U.S. Department of Commerce, p. 63.

doing sociological research

Men's Caregiving

Research Question: Much research has documented the fact that women do the majority of the housework and child care within families. Why? Many have explained it as the result of gender socialization—women learn early on to be nurturing and responsible for others, while men are less likely to do so. Yet, things are changing, and some men are more involved in the "care work" of family life. What explains whether men will be more engaged in family care work?

Research Method: Sociologists Naomi Gerstel and Sally Gallagher studied a sample of 188 married people. They interviewed ninety-four husbands and ninety-four wives, married to each other; the sample was 86 percent White and 14 percent African American but was too small to examine similarities or differences by race.

Research Results: You might expect that men who had attitudes expressing support for men's family responsibilities would be more involved in family care (defined by Gerstel and Gallagher to include elder care, child care, and various household tasks). But, this is not what Gerstel and Gallagher found. Gender attitudes did *not* influence men's involvement in caregiving. Rather, the characteristics of the men's families were the most influential determinant of their engagement in housework and child care. Men whose wives spent the most time helping kin and men who had daughters were more likely to help kin. Having sons had no influence. But, in addition, men with more sisters spend less time helping with elder parents than men with fewer sisters. Furthermore, men's employment (neither hours employed, job flexibility, and job stability) did not affect their involvement in care work.

Conclusions and Implications: It is the social structure of the family, not gender beliefs, that shape men's involvement in family work. As they put it, "It is primarily the women in men's lives who shape the amount and types of care men provide" (2001: 211). This study shows a most important sociological point: social structure, not just individual attitudes, is the most significant determinant of social behavior.

Questions to Consider

1. Who does the work in your family? Is it related to the social organization of your family, as Gerstel and Gallagher find in other families?
2. Do you think that men's gender identity changes when they become more involved in care work? What hinders and/or facilitates men's engagement in this kind of work?

Source: Gerstel, Naomi, and Sally Gallagher. 2001. "Men's Caregiving: Gender and the Contingent Character of Care." *Gender & Society* 15 (April): 197–217.

there are half as many divorces as marriages. But these are divorces out of *all married couples, not just those formed in one year,* so divorce is not as widespread as occurring in one in two marriages.

Still, the rate of divorce is high, and it has risen since 1950, though it has fallen again in recent years. The likelihood of divorce is not equally distributed across all social groups, however. Divorce is more likely for couples who marry young, while in their teens or early twenties. Second marriages are more likely than first marriages to end in divorce. Divorce is somewhat higher among low-income couples, a fact reflecting the strains that financial problems create. Divorce is also somewhat higher among African Americans than among Whites, partially because African Americans make up a disproportionate part of lower-income groups. Hispanics have a lower rate of divorce than either Whites or Blacks, probably the result of religious influence. Recently, the divorce rate among Asian Americans has also risen, interpreted as the possible shedding of cultural taboos (Armas 2003). This explanation seems supported by the fact that Asian Americans born in the United States are more likely to be divorced than Asians who immigrated (McLoyd et al. 2000).

A number of factors contribute to the current high rate of divorce in the United States. Demographic changes (shifts in the composition of the population) are part of the explanation. The rise in life expectancy, for example, has an effect on the length of marriages.

In earlier eras, people died younger, and thus the average length of marriages was shorter. Some marriages that earlier would have ended with the death of a spouse may be now be dissolved by divorce. Still, cultural factors also contribute to divorce.

In the United States, individualism is a cultural norm, placing a high value on a person's satisfaction within marriage. The cultural orientation toward individualism may predispose people to terminate a marriage in which they are personally unhappy. In other cultural contexts (including this society years ago), marriage, no matter how difficult, may have been seen as an unbreakable bond, regardless of whether one was unhappy. As we have seen in the previous data, for racial–ethnic groups in the United States, cultural factors can also influence the likelihood of divorce.

Changes in women's roles also are related to the rate of divorce. Women today are now less financially dependent on husbands than in the past, even though they still earn less. As a result, the economic interdependence that bound women and men as a marital unit is no longer as strong. Although most married women would be less well off without access to their husband's income, they could probably still support themselves. This can make it possible for people to end marriages that they find unsatisfactory.

To people in unhappy marriages, divorce, though painful and financially risky, can be a positive option (Kurz 1995). The belief that couples should stay together for the sake of the children is now giving way to a belief, supported by research, that a marriage with protracted conflict is more detrimental to children than divorce. Although there are periodic public outcries about the negative effect of divorce on children, many other factors influence their long-term psychological and social adjustment. Few children feel relieved or pleased by divorce; feelings of sadness, fear, loss, and anger are common, along with desires for reconciliation and feelings of conflicting loyalties. But most children adjust reasonably well after a year or so. Moreover, children's adjustment is influenced most by factors that precede the divorce. The single most important factor influencing children's poor adjustment is marital violence and prolonged discord (Stewart et al. 1997; Arendell 1998; Furstenberg 1998; Cherlin et al. 1998; Amato and Booth 1997). The emotional strain on children is significantly reduced if the couple remains amicable. If both parents remain active in the upbringing of the children, the evidence shows that children do not suffer from divorce; especially important is the ability of the mother to be an effective parent after a divorce. Her ability to be effective can be influenced by the resources she has and her ongoing relationship with the father (Buchanan et al. 1996; Simons et al. 1996; Furstenberg and Nord 1985).

In the aftermath of divorce, many fathers become distant from their children. Sociologists have argued that the tradition of defining men in terms of their role as breadwinners minimizes the attachment they feel for their children. If the family is then disrupted, they may feel that their primary responsibility, as financial provider, is lessened, leaving them with a diminished sense of obligation to their children.

Family Violence

Generally speaking, the family is depicted as a private sphere where members are nurtured and protected, existing away from the influences of the outside world. Although this is the experience of many, families can also be locales for violence, disruption, and conflict. Family violence, hidden for many years, is a phenomenon that has recently been the subject of much sociological research.

Partner Violence. Estimates of the extent of domestic violence are hard to come by and notoriously unreliable because the majority of cases of domestic violence go unreported. The National Violence Against Women Office estimates that 25 percent of women will be raped, physically assaulted, or stalked by an intimate partner in their lifetime. Men also experience partner violence, although far less frequently. Women who experience violence are also twice as likely as men to be injured. Violence also occurs in gay and lesbian relationships, although silence around the issue may be even more pervasive given the marginalized status of gays and lesbians. Men living with male partners are just as likely to be raped, assaulted, or stalked as are women living with men, but the incidence of violence against women by women partners is about half as likely as heterosexual violence. Researchers conclude that this is because most domestic violence is committed by men. Violence is usually accompanied by emotionally abusive and controlling behavior. Jealous and dominating partners are the most likely perpetrators of domestic violence (Tjaden and Thoennes 2000; West 1998; Renzetti 1992).

One of the most common questions asked about domestic violence is why victims stay with their abuser. The answers are complex and stem from sociological, psychological, and economic problems. A victim tends to believe that the batterer will change, but they also find they have few options; they may perceive that leaving will be more dangerous, because violence can escalate when the abuser thinks he (or she) has lost control. Many women are also unable to support their children and meet their living expenses without a husband's income. Mandatory arrest laws in cases of domestic violence can exacerbate this problem because they may, despite their intentions, discourage a woman from reporting violence for fear her batterer will lose his job (Miller 1997). Despite the belief that battered women do not leave their abusers, however, the majority do leave and seek ways to prevent further victimization (Gelles 1999).

Sociological analyses of violence in the family have led to the conclusion that women's relative powerlessness in the family is at the root of high rates of violence against women. Because most violence in the family is directed against women, the imbalance of power between men and women in the family is the source of most domestic violence. Since women are relatively powerless within the society, they may not have the resources to leave their marriage (Kurz 1989; McCloskey 1996).

Child Abuse. Violence within families also victimizes many children who experience child abuse. Not all forms of child abuse are alike. Some people consider repeated spanking to be abusive; others think of this as legitimate behavior. Child abuse, however, is behavior that puts children at risk and may include physical violence and neglect. As with battering, the exact incidence of child abuse is difficult to know, but 872,000 children are known to be victims of abuse or neglect (Child Welfare Information Gateway 2007). Whereas men are the most likely perpetrators of violence in other forms of domestic abuse and in sexual abuse, with child abuse women are just as likely to be the perpetrators as are men.

Research on child abuse finds a number of factors associated with abuse, including chronic alcohol use by a parent, unemployment, and isolation of the family. Sociologists point to the absence of social supports—in the form of social services, community assistance, and cultural norms about the primacy of motherhood—as related to child abuse, because most abusers are those with weak community ties and little contact with friends and relatives (Baca Zinn and Eitzen 2005).

Incest. Incest is a particular form of child abuse involving sexual relations between persons who are closely related. A history of incest has been related to a variety of other problems, such as drug and alcohol abuse, runaways, delinquency, and various psychological problems, including the potential for violent partnerships in adult life. Studies find that fathers and uncles are the most frequent incestuous abusers and that incest is most likely in families where mothers are debilitated (such as by mental illness or alcoholism). In such families, daughters often take on the mothering role, being taught to comply with men's demands to hold the family together. Scholars have linked women's powerlessness within families to the dynamics surrounding incest (Herman 1981).

Elder Abuse. The National Center on Elder Abuse estimates that between one and two million elders are abused in the United States, but it is difficult to gauge the true extent of the problem. Elder abuse tends to be hidden in the privacy of families, and victims are reluctant to talk about their situations, so estimates are only approximations. What is known is that reports of elder abuse have increased. Whether this reflects an actual increase or more reporting is open to speculation (Teaster 2000; National Center on Elder Abuse 2007).

Why are the elderly abused? One explanation is that caring for the elderly is very stressful for the caregiver—usually a daughter who may be employed in addition to caring for the elderly person. Research finds that abusers are most likely to be middle-aged women and (sadly) the daughter of the victim—the person most likely to be caring for the older person. Sons, however, are most likely to be engaged in direct physical abuse, accounting for almost half of the known physical abusers. Sometimes the physical abuser is a husband, where the abuse is a continuation of abusive behavior in the marriage. The same factors that affect family life in any generation contribute to the problem of elder abuse (Teaster 2000).

Changing Families in a Changing Society

Like other social institutions, the family is in a constant state of change, particularly as new social conditions arise and as people in families adapt to the changed conditions of their lives. Some changes affect only a given family—the individual changes that come from the birth of a new child, the loss of a partner, divorce, migration, and other life events. These changes are what C. Wright Mills referred to as "troubles" (see Chapter 1). Some may even be happy events; the point is that they are changes that happen at the individual level, as people adjust to the presence of a new child, adjust to a breakup with a long-term partner, or grieve the loss of a spouse.

As Mills would have pointed out, many microsociological events that people experience in families have their origins in the broader *macrosociological* changes affecting society as a whole.

Global Changes in Family Life

Changes in the institutional structure of families are also being affected by the process of globalization. The increasing global basis of the economy means that people often work long distances from other family members—a phenomenon that occurs at all points on the social class spectrum, although the experience of such global mobility varies significantly by social class. A corporate executive may accumulate thousands—even millions—of first-class flight miles, crossing the globe to conduct business. A regional sales manager may spend most nights away from a family, likely staying in modestly priced motels and eating in fast-food franchises along the way. Truckers may sleep in the cabs of their tractor trailers after logging extraordinary numbers of hours of driving in a given week. Laborers may move from one state to the next, following the pattern of the harvest, living in camps away from families and being paid by the amount they pick.

These patterns of work and migration have created a new family form, the **transnational family,** defined as families where one parent (or both) lives and works in one country while their children remain in the country of origin. A good example is found in Hong Kong, where most domestic labor is performed by Filipina women who work on multiple-year contracts managed by the government, typically on a live-in basis. They leave their children in the Philippines, usually cared for by a relative, and send money home; the meager wages they earn in Hong Kong far exceed the average income of workers in the Philippines. This pattern is so common that the average Filipino migrant worker supports five people at home; one in five Filipinos directly depends on migrant workers' earnings (Constable 1997; Parreñas 2001).

One need not go to other nations to see such transnational patterns in family life. In the United States, Caribbean women and African American women have had a long history of having to leave their children with others while they sought employment in different regions of the country. Central American and Mexican women may come to work in the United States, while their children stay behind. Mothers may return to see their children whenever they can, or alternatively, children may spend part of the year with their mothers, part with other relatives.

Mothers in transnational families have to develop new concepts of their maternal role, because their situation means giving up the idea that biological mothers should raise their own children. Many have expanded their definition of motherhood to include breadwinning, traditionally defined as the role of fathers. Transnational women also create a new sense of home, one not limited to the traditional understanding of "home" as a single place where mothers, fathers, and their children reside (Hondagneu-Sotelo 2001; Hondagneu-Sotelo and Avila 1997; Das Gupta 1997; Alicea 1997).

Families and Social Policy

Family social policies are the subject of intense national debate. Should gay marriages be recognized by the state? What responsibility does society have to help parents balance the demands of work and family? Many issues on the front lines of national social policy engage intense discussions of families. Some claim the family is breaking down. Others celebrate the increased diversity among families. Many blame the family for the social problems our society faces. Drugs, low educational achievement, crime, and violence are often attributed to a crisis in "family values," as if rectifying these attitudes is all it will take to solve our nation's difficulties. The family is the only social institution that typically takes the blame for all of society's problems. Is it reasonable to expect families to solve social problems? Families are afflicted by most of the structural problems that are generated by racism, poverty, gender inequality, and class inequality. Expecting families to solve the problems that are the basis for their own difficulties is like asking a poor person to save us from the national debt.

Balancing Work and Family. Balancing the multiple demands of work and family is one of the biggest challenges for most families. With more parents employed, it is difficult to take time from one's paid job to care for newborn or newly adopted children, tend to sick children, or care for elderly parents or other family members. As more families include two earners, more people feel pulled in multiple directions, always strategizing to find the time to get everything done. Work institutions are structured on a gendered model of the male breadwinner, where family and work are assumed to be separate, nonintersecting spheres. But now there is significant "spillover" between family and work—work seeping into the home and home also affecting people's work (Moen 2003).

The **Family and Medical Leave Act (FMLA),** adopted by Congress in 1993, is meant to provide help for these conflicts. It requires employers to grant employees a total of twelve weeks in unpaid leave to care for newborns, adopted children, or other family members with a serious health condition. The FMLA is the first law to recognize the need of families to care for children and other dependents. A number of conditions, however, limit the effectiveness of the FMLA, not the least of which is that the leave is unpaid, making it impossible for many employed parents. Many workers in firms where there are family-friendly policies worry that taking advantage of these policies will harm their prospects for career advancement (Blair-Loy and Wharton 2002). Currently, only 15 percent of workers have child-care benefits available to them from employers (U.S. Bureau of Labor Statistics 2006b). Among industrialized nations, the United States provides the least in support for maternity and child-care policies (see Table 12.2).

Child Care. Family leave polices, much as they are needed, also do not address the ongoing needs for child care. Almost one-half of families with children under age thirteen have child-care expenses, typically taking 9 percent of their earnings. Single-parent families pay an even larger percentage of earnings on child care (Giannarelli and Barsimantov 2000).

Many parents struggle to find good and affordable child care for their children, some relying on relatives for care; others, on paid providers; and some, a combination of both. In the United States, one-half of three-year olds and two-thirds of four-year-olds now spend

table 12.2 Maternity Leave Benefits: A Comparative Perspective

Country	Length of Maternity Leave	Percentage of Wages Paid in Covered Period	Provider of Coverage
Zimbabwe	90 days	60–75%	Employer
Cuba	18 weeks	100%	Social Security
Iran	90 days	66.7% for 16 weeks	Social Security
China	90 days	100%	Employer
Saudi Arabia	10 weeks	50 or 100%	Employer
Canada	17–18 weeks	55% for 15 weeks	Unemployment insurance
Germany	14 weeks	100%	Social Security to a ceiling; employer pays difference
France	16–26 weeks	100%	Social Security
Italy	5 months	80%	Social Security
Japan	14 weeks	60%	Social Security or health insurance
Russian Federation	140 days	100%	Social Security
Sweden	14 weeks	450 days, 100% paid	Social Security
United Kingdom	14–18 weeks	90% for 6 weeks; flat rate thereafter	Social Security
United States	12 weeks	n/a	n/a

Source: United Nations. 2000. *The World's Women 2000: Trends and Statistics.* New York: United Nations, pp. 140–143.

much of their time in child-care centers. But the national approach is one of patching together different programs and primarily relying on private initiatives for care. Compare this with France. Although participation is voluntary, in France almost all parents enroll young children in the *école maternelle* system, where a place is guaranteed to every child aged three to six. These child-care centers are integrated with the school system and are seen as a form of early education. Moreover, in the United States child-care costs match tuition costs at public universities, but in France child care is seen as a social responsibility and is paid by the government. National norms about whether families are a private or public responsibility clearly shape social policy (Clawson and Gerstel 2002; Folbre 2001).

Care work—work that sustains life, including child care, elder care, housework, and other forms of household labor—is increasingly provided to middle- and upper-class families by women of color and immigrant women. Nannies, cleaners, and personal attendants now do much of the domestic work that was once provided by wives and mothers. These trends raise many new questions for sociologists, such as how work is negotiated outside of the public labor market, how "mothering" is defined when it is provided by multiple people, how domestic workers care for their own families, and what work conditions exist in the lives of domestic laborers (Hondagneu-Sotelo 2001). Clearly, new social policies are needed to address the needs of diverse families (Moen 2004).

Elder Care. The shrinking size of families means that the proportion of elderly people is growing faster than the number of younger potential caretakers. As life expectancy has increased and people live longer, elder care becomes a greater and greater need. Family members provide almost all long-term care for the elderly—work that is often taken for granted (Glazer 1990; Meyer 1994).

Women, who shoulder the work of elder care, can now expect to spend more years as the child of an elderly parent than as the mother of children under eighteen. Indeed, young people now can expect to spend more years caring for an elderly parent than raising their own children. The effects of the burden of care are apparent in the stress that women report from this role. Women also believe they are better at elder care than their husbands and brothers, but with the rapid increase in the older population that lies ahead, these social norms may have to change. As the U.S. population ages, social policies will likely need to respond to this growing need.

Because families are so diverse, different families need different social supports. Family leave policies that give parents time off to care for their children or sick relatives are helpful, but of little use to people who cannot afford to take time off work without pay. Strong enforcement of laws to protect victims of domestic violence can help, but not if removing partners only further reduces the income support for families (Miller 1997). Greater employer support for child care can help men and women meet family needs. Some policies will benefit some groups more than others—one reason why policy makers need to be sensitive to the diversity of family experiences. Social policies cannot solve all the problems that families face, but they can go a long way toward creating the conditions under which diverse family units can thrive.

Defining Religion

Sociologists study religion as both a belief system and a social institution. The belief systems of religion have a powerful hold on what people think and how they see the world. The patterns and practices of religious institutions are among the most important influences on people's lives. Sociologists are interested in several questions about religion: How are religious belief and practice related to other social factors, such as social class, race, age, gender, and level of education? How are religious institutions organized? How does religion influence social change? In using sociology to understand religion, what is important is not what one believes about religion, but one's ability to examine religion objectively in its social and cultural context.

What is religion? Most people think of it as a category of experience separate from the mundane acts of everyday life, perhaps involving communication with a deity or communion with the supernatural (Johnstone 1992). Sociologists define **religion** as an institutionalized system of symbols, beliefs, values, and practices by which a group of people interprets and responds to what they feel is sacred and that provides answers to questions of ultimate meaning (Glock and Stark 1965: Johnstone 1992). The elements of this definition bear closer examination:

1. **Religion is institutionalized.** Religion is more than just beliefs: It is a pattern of social action organized around the beliefs, practices, and symbols that people develop to answer questions about the meaning of existence. As an institution, religion presents itself as larger than any single individual; it persists over time and has an organizational structure into which members are socialized.
2. **Religion is a feature of groups.** Religion is built around a community of people with similar beliefs. It is a cohesive force among believers because it is a basis for group identity and gives people a sense of belonging to a community or organization. Religious groups can be formally organized, as in the case of bureaucratic churches, or they may be more informally organized, ranging from prayer groups to cults. Some religious communities are extremely close-knit, as in convents; other communities are more diffuse, such as people who identify themselves as Protestant but attend church only on Easter.
3. **Religions are based on beliefs that are considered sacred.** The **sacred** is that which is set apart from ordinary activity for worship, seen as holy, and protected by special rites and rituals. The sacred is distinguished from the **profane,** which is of the everyday world and specifically not religious (Durkheim 1947/1912; Chalfant et al. 1987). Each religion defines what is to be considered sacred; most religions have sacred objects and sacred symbols. The holy symbols are infused with special religious meaning and inspire awe.

   A **totem** is an object or living thing that a religious group regards with special reverence. A statue of Buddha is a totem and so is a crucifix hanging on a wall. Among the Zuni (a Native American group), fetishes are totems; these are small, intricately carved animal objects representing different dimensions of Zuni spirituality. A totem is important not for what it is, but for what it represents. To a Christian taking communion, a piece of bread is defined as the flesh of Jesus; eating the bread unites the communicant mystically with Christ. To a nonbeliever, the

© Bob Daemmrich/PhotoEdit

Religious spirituality takes many forms but produces feelings of awe and reverence among believers.

bread is simply that—a piece of bread (McGuire 1996). Likewise, Native Americans hold certain ground to be sacred and are deeply offended when the holy ground is disturbed by industrial or commercial developers who see only potential profit.

4. **Religion establishes values and moral proscriptions for behavior.** A proscription is a constraint imposed by external forces. Religion typically establishes proscriptions for the behavior of believers, some of them quite strict. For example, the Catholic Church defines living together as sexual partners outside marriage as a sin. Often religious believers come to see such moral proscriptions as simply "right" and behave accordingly. At other times, individuals may consciously reject moral proscriptions, although they may still feel guilty when they engage in a forbidden practice. Of course, what people believe and what they do can be very contradictory, perhaps best exemplified by the various scandals recently reported involving sexual abuse of young boys by Catholic priests.
5. **Religion establishes norms for behavior.** Religious belief systems establish social norms about how the faithful should behave in certain situations. Worshippers may be expected to cover their heads in a temple, mosque, or cathedral, or wear certain clothes. Such behavioral expectations may be quite strong. The next time you are at a gathering where a prayer is said before a meal, note how many people bow their heads, even though some of those present may not believe in the deity being invoked.
6. **Religion provides answers to questions of ultimate meaning.** The ordinary beliefs of daily life are **secular** beliefs and may be institutionalized, but they are specifically not religious. Science, for example, generates secular beliefs based on particular ways of thinking—logic and empirical observation are at the root of scientific beliefs. Religious beliefs, in contrast, often have a supernatural element. They emerge from spiritual needs and may provide answers to questions that cannot be probed with the profane tools of science and reason. Think of the difference in how religion and science explain the origins of life. Whereas science explains this as the result of biochemical and physical processes, different religions have other accounts of the origin of life.

thinking sociologically

For one week keep a daily log, noting every time you see an explicit or implicit reference to religion. At the end of the week, review your notes and ask yourself how religion is connected to other social institutions. Based on your observations, how do you interpret the relationship between the *sacred* and the *secular* in this society.

The Significance of Religion in the United States

The United States is one of the most religious societies in the world. Two-thirds of Americans think religion can solve all or most of society's problems. Forty-two percent of the population describe themselves as "born again" (Winseman 2004). Religion is, for millions of people, the strongest component of their individual and group identity. Much of the world's most celebrated art, architecture, and music has its origins in religion, whether in the classical art of western Europe, the Buddhist temples of the East, or the gospel rhythms of contemporary rock.

Religion is also strongly related to a number of social and political attitudes. Religious identification is a good predictor of how traditional a person's beliefs will be. People who belong to religious organizations that encourage intolerance of any form are most likely to be racially prejudiced. However, there is not a simple relationship between religious belief and prejudice, because religious principles are also often the basis for lessening racial prejudice. Those with deeper religious involvement tend to have more traditional gender attitudes, and *homophobia* has also been linked to religious belief, although some religious congregations have actively worked to encourage the participation of gays and lesbians.

The Dominance of Christianity

Despite the U.S. Constitution's principle of the separation of church and state, Christian religious beliefs and practices dominate U.S. culture. Indeed, Christianity is often treated as if it were the national religion. It is commonly said that the United States is based on a Judeo-Christian heritage, meaning that our basic cultural beliefs stem from the traditions of the pre-Christian Old Testament of the Bible (the Judaic tradition) and the Gospels of the New Testament. The dominance of Christianity is visible everywhere. State-sponsored colleges and universities typically close for Christmas break, not Yom Kippur. Christmas is a national holiday, but not Ramadan, the most sacred holiday among Muslims. Despite the dominance of Christianity, however, the pattern of religion in the United States is a mosaic one.

Measuring Religious Faith

Religiosity is the intensity and consistency of practice of a person's (or group's) faith. Sociologists measure religiosity both by asking people about their religious beliefs and by measuring membership in religious organizations and attendance at religious services. The majority of people in the United States identify themselves as Protestant or Catholic, though there is great religious diversity within the United States (see Figure 12.5).

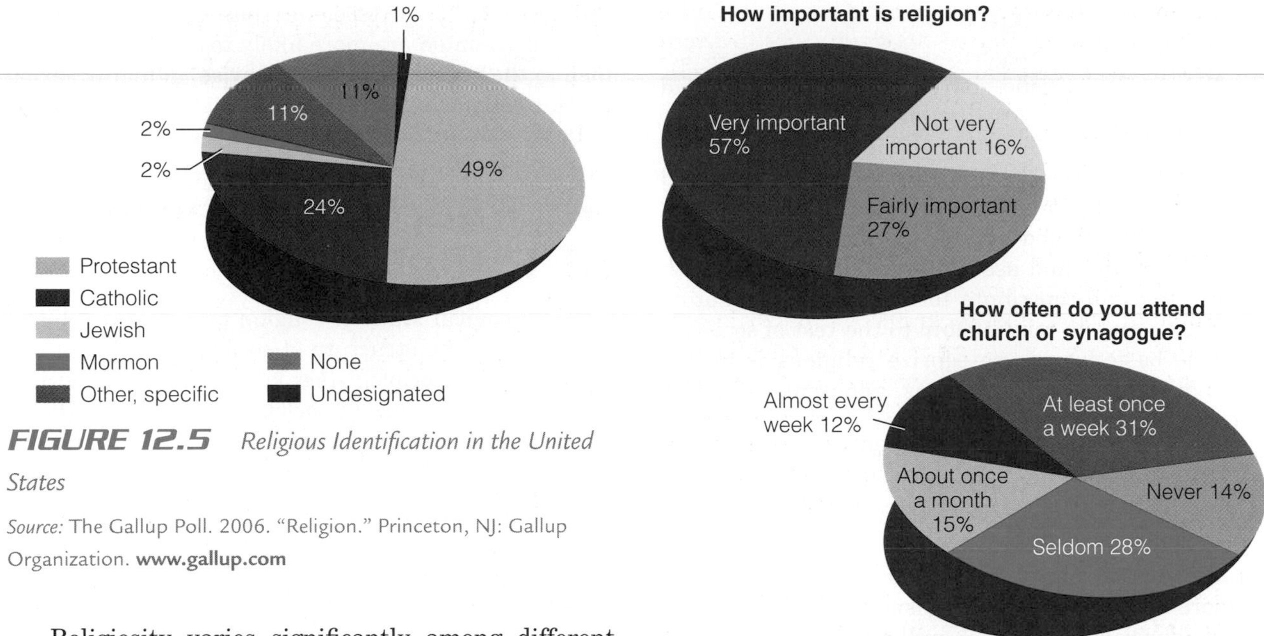

FIGURE 12.5 *Religious Identification in the United States*

Source: The Gallup Poll. 2006. "Religion." Princeton, NJ: Gallup Organization. **www.gallup.com**

FIGURE 12.6 *Measuring Religiosity*

Source: The Gallup Poll. 2006. "Religion." Princeton, NJ: Gallup Organization. **www.gallup.com**

Religiosity varies significantly among different groups in society. Church membership and attendance is higher among women than men and more prevalent among older than younger people. African Americans are more likely than Whites to belong to and attend church. On the whole, church membership and attendance fluctuate over time; membership has decreased slightly since 1940, but attendance has remained largely the same since (see Figure 12.6). Large, national religious organizations, such as the mainline Protestant denominations, have lost many members, whereas smaller, local congregations have increased membership.

In recent years, there has been a decrease in the number of people who think that religion can answer all or almost all of today's problems. Changes in immigration patterns have also affected religious patterns in the United States, with Muslims, Buddhists, and Hindus now accounting for several million believers (see Figure 12.7; Niebuhr 1998; Haddad et al. 2003). One of the greatest changes has been a tremendous increase in the number identifying as evangelical Protestants, but Islam has also been one of the fastest growing religions in the United States in recent years (Dudley and Roozen 2001; Gallagher 2003).

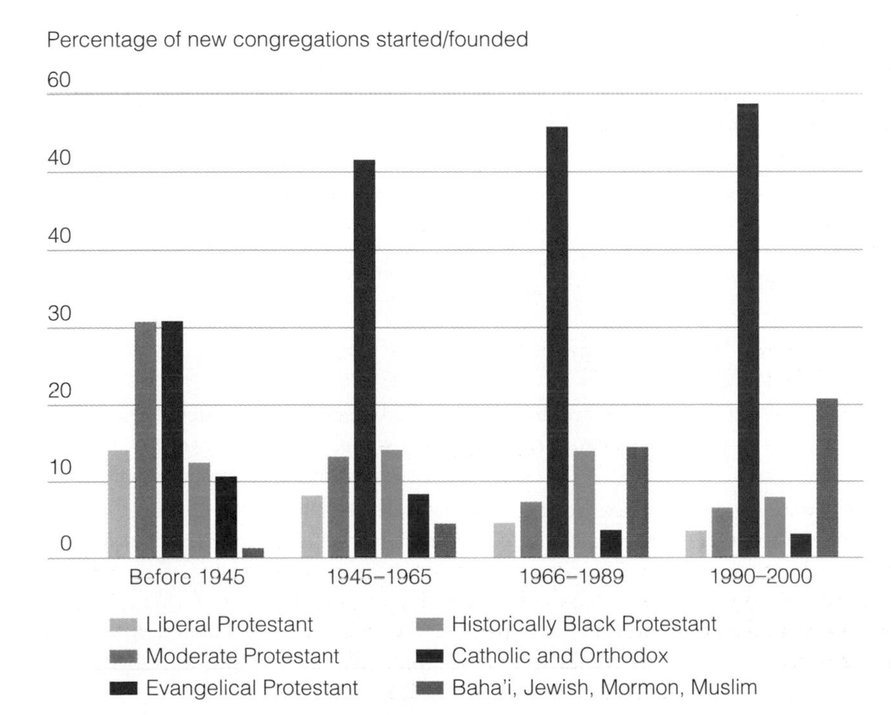

FIGURE 12.7 *The New Face of Religion*

Source: Dudley, Carl S., and Roozen, David A. 2001. *Faith Communities Today: A Report on Religion in the United States Today.* Hartford, CT: Hartford Institute for Religion Research, Hartford Seminary.

Forms of Religion

Religions can be categorized in different ways according to the specific characteristics of faiths and how religious groups are organized. In different societies and among different religious groups, the form religion takes reflects differing belief systems and reflects and supports other features of the society. Believing in one god or many, worshiping in small or large groups, and associating religious faith with gender roles all contribute to the social organization of religion and its relationship to the rest of society.

One basic way to categorize religions is by the number of gods or goddesses adherents worship. **Monotheism** is the worship of a single god. Christianity and Judaism are monotheistic in that both Christians and Jews believe in a single god who created the universe. Monotheistic religions typically define god as omnipotent (all-powerful) and omniscient (all-knowing). **Polytheism** is the worship of more than one deity. Hinduism, for example, is extraordinarily complex, with millions of gods, demons, sages, and heroes—all overlapping and entangled in religious mythology; within Hinduism, the universe is seen as so vast that it is believed to be beyond the grasp of a single individual, even a powerful god (Grimal 1963).

Religions may also be patriarchal or matriarchal. **Patriarchal religions** are those in which the beliefs and practices of the religion are based on male power and authority. Christianity is a patriarchal religion; the ascendancy of men is emphasized by the role of women in the church, the instruction given on relations between the sexes, and even the language of worship itself. **Matriarchal religions** are based on the centrality of female goddesses, who may be seen as the source of food, nurturance, and love, or who may serve as emblems of the power of women (McGuire 1997). In societies based on matriarchal religions, women are more likely to share power with men in the society at large. Likewise, in highly sexist, patriarchal societies, religious beliefs are also likely to be patriarchal.

Sociological Theories of Religion

The sociological study of religion probes how religion is related to the structure of society. Recall that one basic question sociologists ask is, "What holds society together?" Coherence in society comes from both the social institutions that characterize society and the beliefs that hold society together. In both instances, religion plays a key role. From the functionalist perspective of sociological theory, religion is an integrative force in society because it has the power to shape collective beliefs. In a somewhat different vein, the sociologist Max Weber saw religion in terms of how it supported other social institutions. Weber thought that religious belief systems provided a cultural framework that supported the development of specific social institutions in other realms, such as the economy. From yet a third point of view, based on the work of Karl Marx and conflict theory, religion is related to social inequality in society (see Table 12.3).

Emile Durkheim: The Functions of Religion

Emile Durkheim argued that religion is functional for society because it reaffirms the social bonds that people have with each other, creating social cohesion and integration. Durkheim believed that the cohesiveness of society depends on the organization of its belief system. Societies with a unified belief system

table 12.3 Theoretical Perspectives on Religion

	Functionalism	Feminist Theory	Symbolic Interaction
Religion and the social order	Is an integrative force in society	Is the basis for intergroup conflict; inequality in society is reflected in religious organizations, which are stratified by factors such as race, class, or gender.	Is socially constructed and emerges with social and historical change
Religious beliefs	Provide cohesion in the social order by promoting a sense of collective consciousness	Can provide legitimation for oppressive social conditions	Are socially constructed and subject to interpretations; they can also be learned through religious conversion
Religious practices and rituals	Reinforce a sense of social belonging	Define in-groups and out-groups, thereby defining group boundaries	Are symbolic activities that provide definitions of group and individual identity

© AP Images/Gregorio Borgia

Emile Durkheim theorized that public rituals, such as the ordination of new priests before Pope Benedict shown here, provide cohesion in society.

are highly cohesive; those with a more diffuse or competing belief systems are less cohesive.

Religious **rituals** are symbolic activities that express a group's spiritual convictions. Making a pilgrimage to Mecca, for example, is an expression of religious faith and a reminder of religious belonging. In Durkheim's view, religious rituals are vehicles for the creation, expression, and reinforcement of social cohesion. Groups performing a ritual are expressing their identity as a group. Whether the rituals of a group are highly elaborated or casually informal, they are symbolic behaviors that sustain group awareness of unifying beliefs. Lighting candles, chanting, or receiving a sacrament are behaviors that reunite the faithful and help them identify with the religious group, its goals, and its beliefs (McGuire 1996). Durkheim believed that religion binds individuals to the society in which they live, by establishing what he called a **collective consciousness,** the body of beliefs common to a community or society that gives people a sense of belonging. In many societies, religion establishes the collective consciousness and creates in people the feeling that they are part of a common whole.

Durkheim's analysis of religion suggests some of the key ideas in symbolic interaction theory, particularly in the significance he gave to symbols in religious behavior. Symbolic interaction theory sees religion as a socially constructed belief system, one that emerges in different social conditions. From the perspective of symbolic interaction, religion is a meaning system that gives people a sense of identity, defines one's network of social belonging, and confers one's attachment to particular social groups and ways of thinking.

Max Weber: The Protestant Ethic and the Spirit of Capitalism

Theorist Max Weber also saw a fit between the religious principles of society and other institutional needs. In his classic work *The Protestant Ethic and the Spirit of Capitalism,* Weber argued that the Protestant faith supported the development of capitalism in the Western world. He began by noting a seeming contradiction: How could a religion that supposedly condemns extensive material consumption coexist in a society such as the United States with an economic system based on the pursuit of profit and material success?

Weber argued that these ideals were not as contradictory as they seemed. As the Protestant faith developed, it included a belief in predestination—one's salvation is predetermined and a gift from God, not something earned. This state of affairs created doubt and anxiety among believers, who searched for clues in the here and now about whether they were among the chosen—called the "elect." According to Weber, material success was taken to be one clue that a person was among the elect and thus favored by God, which drove early Protestants to relentless work as a means of confirming (and demonstrating) their salvation. As it happens, hard work and self-denial—the key features of the **Protestant ethic**—lead not only to salvation but also to the accumulation of capital. The religious ideas supported by the Protestant Ethic therefore fit nicely with the needs of capitalism. According to Weber, these austere religionists stockpiled wealth, had an irresistible motive to earn more (that is, eternal salvation), and were inclined to spend little on themselves, leaving a larger share for investment and driving the growth of capitalism (Weber 1958/1904).

© Ethel Wolvovitz/The Image Works

Religion can be interpreted both as producing social conflict and as promoting social justice.

Karl Marx: Religion, Social Conflict, and Oppression

Durkheim and Weber concentrated on how religion contributes to the cohesion of society. Religion can also be the basis for conflict, as we see in the daily headlines of newspapers. In the Middle East, differences between Muslims and Jews have caused decades of political instability. These conflicts are not solely religious, but religion plays an inextricable part. Certainly religious wars, religious terrorism, and religious genocide have contributed some of the most violent and tragic episodes of world history. The image of religion in history has two incompatible sides: piety and contemplation on the one hand, battle flags on the other. Domestic conflicts over ethical issues such as abortion, assisted suicide, and school prayer evolve from religious values even though they are played out in the secular world of politics and public opinion. Conflict theory illuminates many of the social and political conflicts that engage religious values.

The link between religion and social inequality is also key to the theories of Karl Marx. Marx saw religion as a tool for class oppression. According to Marx, oppressed people develop religion, with the urging of the upper classes, to soothe their distress (Marx, 1843/1972). The promise of a better life hereafter makes the present life more bearable, and the belief that "God's will" steers the present life makes it easier for people to accept their lot. To Marx, religion is a form of *false consciousness* (see Chapter 7) because it prevents people from rising up against oppression. He called religion the "opiate of the people" because it encourages passivity and acceptance.

Marx saw religion as supporting the status quo and being inherently conservative (that is, resisting change and preserving the existing social order). To Marx, religion promotes stratification because it supports a hierarchy of people on Earth and the subordination of humankind to divine authority. Christianity, for example, supported the system of slavery. When European explorers first encountered African people, they regarded them as godless savages, and they justified the slave trade by arguing that slaves were being converted to the Christian way of life. Principles of Christianity thus legitimated the system of slavery in the eyes of the slave owners and allowed them to see themselves as good people, despite their enslavement of other human beings.

At the same time, religion can be the basis for liberating social change. In the civil rights movement in the United States and in Latin American liberation movements, the words and actions of religious organizations have been central in mobilizing people for change. This does not undermine Marx's main point, however, because there remains ample evidence of the role of religion in generating social conflict and resisting social change.

Symbolic Interaction: Becoming Religious

Symbolic interaction theory emphasizes the process by which people become religious. The process can be slow and gradual, as when someone switches to a new religious faith (such as a Christian person converting to Judaism), or it can be more dramatic, as when a person joins a cult or some other extreme religious group. Different religious beliefs and practices emerge in different social and historical contexts because context frames the meaning of religious belief.

The emphasis on meaning that is typical of symbolic interaction helps explain how the same religion can be interpreted differently by different groups or in different times. From this perspective, religious texts are not simply "truth" but have been interpreted by people. Thus, while the Bible may be seen as the literal word of God to some, others may interpret the same document in a different way.

Diversity and Religious Belief

The world is marked by diverse religious beliefs. Christianity has the largest membership, followed by Islam. But Hindus, Jews, Confucianists, Buddhists, and other folk religions also comprise the world's religions (see Figure 12.8). In the United States, religious identification varies with a number of social factors, including age, income level, education, and political affiliation. Younger people are more likely than older people to express no religious preference. Those in higher income brackets are more likely to identify as Catholic or Jewish than those in lower income brackets, who are more likely to identify as Protestant, although these trends vary among Protestants by denomination. Fundamentalist Protestants, for example, are most likely to come from lower-income groups.

The Influence of Race and Ethnicity

Race is one of the most significant indicators of religious orientation. African Americans are more likely to identify as Protestant than are Whites or Hispanics, although the number of African American Catholics is increasing (Bezilla 1990). Many urban African Americans have also become committed Black Muslims, which involves strict regulation of dietary habits and prohibition of many activities, such as alcohol use, drug use, gambling, use of cosmetics, and hair straightening. The emphasis among Black Muslims on self-reliance and traditional African identity has earned it a fervent following, although the actual number of Black Muslims in the United States is relatively small. For many African Americans, religion has been a defense against the damage caused by racism. Churches have served as communal centers,

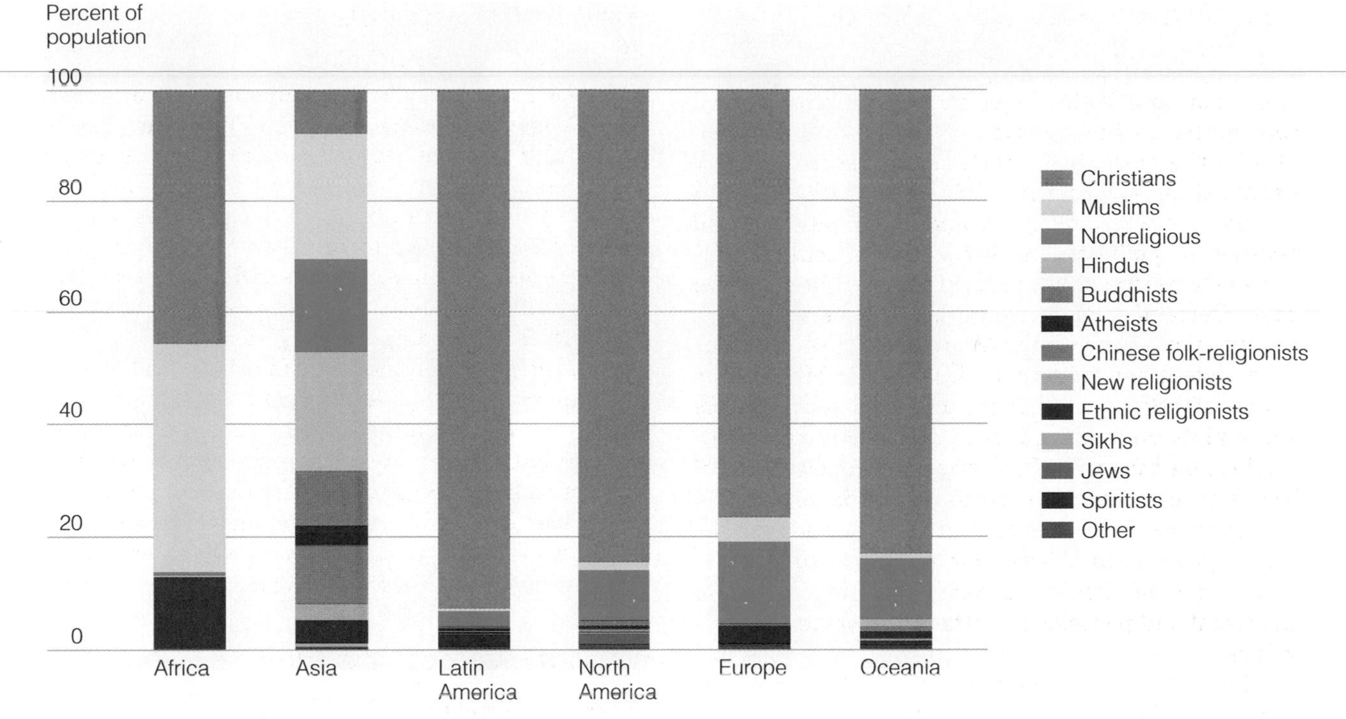

FIGURE 12.8 *Viewing Society in Global Perspective: The World's Religions*

Data: From the U.S. Census Bureau. 1999. *Statistical Abstract of the United States, 1999.* Washington, DC: U.S. Government Printing Office, p. 831.

political units, and sources of social and community support, making churches among the most important institutions within the African American community (Gilkes 2000).

Religion has also been a strong force in Latino communities, with the largest number identifying as Catholic. However, there are a growing number of Latino Protestants, both in mainstream Protestant denominations and in fundamentalist groups.

Asian Americans have a great variety of religious orientations, in part because the category "Asian American" is constructed from so many different Asian cultures. Hinduism and Buddhism are common among Asians, but so is Christianity. As with all groups whose family histories include immigration, religious belief and practice among Asian Americans frequently changes between generations. The youngest generation may not worship as their parents and grandparents did, although some aspects of the inherited faith may be retained. Within families, the discontinuity with a religious past brought on by cultural assimilation can be a source of tension between grandparents, parents, and children. Within the United States, Asian Americans often mix Christian and traditionally Buddhist, Confucian, or Hindu beliefs, resulting in new religious practices.

Religious Extremism

Religious extremism refers to actions and beliefs that are driven by high levels of religious intolerance. Religious extremists tend to see the world in simplistic either/or terms—dividing people into either good or evil, godly or demonic. Such divisive imagery reduces the complexity of human life into simplistic categories—categories that fuel hate and conflict. Thus, religious extremism usually becomes the basis for extremely violent behavior. And it produces martyrs and enemies, as if the world's people were divided along some "axis of evil," good people on one side and everyone else on the other (Anthony et al. 2002). When such religious fanaticism is intertwined with the power of a state government, religiously inspired leaders can use the power of the military and government propaganda to wield extraordinary power.

Religious extremism is now associated with terrorism in the Middle East, but it has also fueled other horrendous acts, including mass executions and genocide, enslavement, and other heinous crimes against humanity. All religions, taken to an extreme, are dangerous social forces because they can drive adherents to think they are doing sacred work even when they are engaging in violent, murderous behavior.

It is easy to see the acts of religious extremists as the work of misled individuals, but those who study religious extremism know that it has social origins. Religious extremism is learned—usually within a narrowly circumscribed social world, such as the *madrassas*—religious camps in Pakistan (and other areas) where young boys are taught a strict interpretation of Islam. For young boys uprooted from families by war, detached from other social contacts, and

© JOHN GRESS/Reuters /Landov

Megachurches—*those with memberships numbering into the thousands—are becoming increasingly common. As just one example of a growing trend, the Creek Community Church in South Barrington, Illinois (shown here), can attract as many as 20,000 people on a weekend.*

with no other education, it is easy to be socialized into a narrow worldview that gives them a cause to fight for (Rashid 2000).

Knowing this helps explain how young men in Pakistan and other places could celebrate the death of Americans by cheering and waving guns in the streets—behavior that can easily emerge when all you are taught is to perceive some other group as your enemy.

Religious extremism typically emerges in countries (or subcultures) where people are very poor and without access to education. Young men without an education and with little life opportunity can be attracted to violence as a way of asserting a collective identity (Khashan and Kreidie 2001; Stern 2003). Many of the young suicide bombers who have committed horrendous acts of violence come from needy families who gain status and money from having their children do the footwork of terrorists.

Also, where there is a lack of modernism—and where people perceive that their traditional way of life is being overtaken by Western influences—religious extremism can come from trying to defend a traditional way of life (Pain 2002). Where most people are poor, frustrations can be channeled into extremist movements fueled by highly traditional religious beliefs (Heilman and Kaiser 2002).

Finally, religious extremist movements tend to be highly patriarchal—that is, based on the power of men and the subordination of women. This is true not only in the extremist factions of contemporary Islamic movements, but also in extremist segments of the Christian right in the United States (Antrobus 2002; Ferber 1998). When religious extremism links with militaristic and patriarchal values, it becomes extremely dangerous.

Religious Organizations

Sociologists have organized their understanding of the various religious organizations into three types: churches, sects, and cults. These are *ideal types* in the sense that Max Weber used the term. That is, the ideal types convey the essential characteristics of some social entity or phenomenon, even though they do not explain every feature of each entity included in the generic category.

Churches are formal organizations that tend to see themselves, and are seen by society, as the primary and legitimate religious institutions. They tend to be integrated into the secular world to a degree that sects and cults are not. They are sometimes closely tied to the state. Churches are organized as complex bureaucracies with division of labor and different roles for groups within. Generally, churches employ professional, full-time clergy who have been formally ordained following a specialized education. Church membership is renewed as the children of existing members are brought up in the church; churches may also actively proselytize. Churches are less exclusive than cults and sects, because they see all of society as potential members of the fold (Johnstone 1992).

Sects are groups that have broken off from an established church. They emerge when a faction within an established religion questions the legitimacy or purity of the group from which they are separating. Many sects form as offshoots of an existing religious organization. Sects tend to place less emphasis on organization (as in churches) and more emphasis on the purity of members' faith. The Shakers, for example, were formed by departing from the Society of Friends (the Quakers). They retained some Quaker practices, such as simplicity of dress and a belief in pacifism, but departed from Quaker religious philosophy. The Shakers believed that the second coming of Christ was imminent, but that Christ would appear in the form of a woman (Kephart 1993). Sects tend to admit only truly committed members, refusing to compromise their beliefs. Some sects hold emotionally charged worship services, although others, like the Amish, are more stoical. The Shakers, for example, had such emotional services that they shook, shouted, and quivered while "talking with the Lord," earning them their name. The only bodily contact permitted among the Shakers was during the unrestrained religious rituals; they were celibate (did not have sexual relations) and gained new members only through adoption of children or recruitment of newcomers (Kephart 1993).

Cults, which are like sects in their intensity, are religious groups devoted to a specific cause or charismatic leader. Many cults arise within established religions and sometimes continue to peaceably reside within the parent religion as simply a fellowship of people with a particular, often mystical, dogma. As they are developing, it is common for tension to exist between cults and the society around them. Cults tend to exist outside the mainstream of society, arising when believers think that society is not satisfying their spiritual needs and attracting those who feel a longing for meaningful attachments. Internally, cults seldom develop an elaborate organizational structure but are instead close-knit communities held together by personal attachment and loyalty to the cult leader.

Cults form around leaders with great **charisma,** a quality attributed to individuals believed by their followers to have special powers (Johnstone 1992). Typically, followers are convinced that the charismatic leader has received a unique revelation or possesses supernatural gifts. Although there are exceptions, cult leaders are usually men, probably because men are more likely to be seen as having the characteristics associated with charismatic leadership.

Debunking Society's Myths

Myth: People who join extreme religious cults are maladjusted and have typically been brainwashed by cult leaders.

Sociological perspective: Conversion to a religious cult is usually a gradual process wherein the convert voluntarily develops new associations with others and through these relationships develops a new worldview.

Religion and Social Change

What is the role of religion in social change? Durkheim saw religion as promoting social cohesion; Weber saw it as culturally linked to other social institutions; Marx assessed religion in terms of its contribution to social oppression. Is religion a source of oppression, or is it a source of personal and collective liberation from worldly problems? There is no simple answer to this question. Religion has had a persistent conservative influence on society, but it has also been an important part of movements for social justice and human emancipation.

The role of religious organizations in social change has of late become a question of public policy. Should faith-based organizations receive government support for work they do in helping people, or does this violate the constitutional separation of church and state? The constitutional issues will ultimately be settled by law, but sociological research sheds light on the implications of such organizations. Though liberals fear that faith-based organizations will infuse religion into government too much and conservatives hope that faith-based initiatives will support their political agenda, research shows that faith-based initiatives enhance the participation of traditionally disadvantaged groups in the democratic process (Wood 2002; Kniss 2003).

The public debate about faith-based initiatives is occurring at a time when evangelical groups have increased membership and influence and have affiliated with conservative political causes, dramatically increasing the influence of religion on politics. At the same time, there has been a decrease in the importance of religion to many people. As a social institution, religion is in transition. Religion, like other aspects of society, is also becoming more commercialized. A large self-help industry has developed in religious publication; religious music is increasingly successful as a form of enterprise. All sorts of religious products are bought and sold in what sociologists now call a "spiritual marketplace" (Roof 1999; Wuthnow 1998). Clearly, religion influences social change, but it is also influenced by the same changes that affect other social institutions.

At the same time, religion continues to have an important role in liberation movements around the world. Throughout the world, liberation theologians have used the prestige and organizational resources of the Catholic Church to develop a consciousness of oppression among poor peasants and working-class people. Likewise, in the United States, churches have had a prominent role in the civil rights movement (Marx 1967/1867; Morris 1984). Churches supplied the infrastructure of the developing Black protest movements of the 1950s and 1960s, and the moral authority of the church was used to reinforce the appeal to Christian values as the basis for racial justice. Now, they continue to be important places for the mobilization of Black politics and provide an important source of community support—often when other institutions have abandoned the Black community (Zuckerman 2002).

The role of women is also changing in most religious organizations. Women have long been denied the right to full participation in many faiths. Some religions still refuse to ordain women as clergy, but the public generally supports the ordination of women. Women now make up a large portion of divinity students. Whereas traditional religious images of women have provided the basis for the subordination of women, those stereotypes are eroding. In sum, religion is a force of both social change and social stability.

Chapter Summary

- ***How are different kinship systems defined?***
  All societies are organized around a *kinship system,* varying in how many marriage partners are allowed, who can marry whom, how descent is determined, family residence, and power relations within the family. *Extended family* systems develop when there is a need for extensive economic and social cooperation. The *nuclear family* is the result of the rise of Western industrialization that separated production from the home.

- ***What does sociological theory contribute to our understanding of families?***
  Functionalism emphasizes that families have the function of integrating members to support society's needs. *Conflict theorists* see the family as a power relationship, related to other systems of inequality. *Feminist theory* emphasizes the family as a gendered institution and is critical of perspectives that take women's place in the family for granted. *Symbolic interaction* takes a more microscopic look at families, emphasizing how different family members experience and define their family experience.

- ***What changes characterize the diversity in contemporary families?***
  One of the greatest changes in families has been the increase in female-headed households, which are most likely to live in poverty. The increase in women's labor force participation has also affected families, resulting in dual roles for women. Stepfamilies face unique problems stemming from the blending of two households. Gay and lesbian households are also more common and challenge traditional heterosexual definitions of the family. Single people make up an increasing portion of the population, due in part to the later age when people marry.

- ***Is marriage declining?***
  The United States has both the highest marriage rate and the highest divorce rate of any industrialized nation. The high divorce rate is explained as the result of a cultural orientation toward individualism and personal gratification, as well as structural changes that make women less dependent on men within the family.

- ***Why is family violence such a problem?***
  Family violence takes several forms, including partner violence, child abuse, incest, and elder abuse. Power relationships within families, as well as gender differences in the division of labor, help explain domestic violence.

- ***What major changes are affecting contemporary families?***
  Changes at the global level are producing new forms of families—*transnational families*—where at least one parent lives and works in a different nation from the children. Social policies designed to assist families should recognize the diversity of family forms and the interdependence of the family with other social conditions and social institutions.

- ***What are the elements of a religion?***
  Sociologists are interested in religion because of the strong influence it has in society. *Religion* is an institutionalized system of symbols, beliefs, values, and practices by which a group of people interprets and responds to what they feel is sacred and that provides answers to questions of ultimate meaning.

- ***How do sociologists measure the significance of religion for people, and what forms does religion take?***
  The United States is a deeply religious society. Christianity dominates the national culture, even though the U.S. Constitution specifies a separation between church and state. *Religiosity* is the measure of the intensity and practice of religious commitment.

- ***How do the different sociological theories analyze religion?***
  Durkheim understood religions and religious rituals as creating social cohesion. Weber saw a fit between the ideology of the *Protestant ethic* and the needs of a capitalistic economy. Religion is also related to social conflict. Marx saw religion as supporting societal oppression and encouraging people to accept their lot in life. *Symbolic interaction theory* focuses on the process by which people become religious. Religious conversion involves a dramatic transformation of religious identity and involves several phases through which individuals learn to identify with a new group and lose other existing social ties.

- ***What diversity exists in religious faith and practice?***
  The United States is a diverse religious society. Protestants, Catholics, Jews, and, increasingly, Muslims, make up the major religious faiths in the United States. *Religious extremism* can emerge in any religion and is generated by certain societal characteristics.

- ***How is a religion organized?***
  Churches are formal religious organizations. They are distinct from *sects,* which are religious groups that have withdrawn from an established religion. *Cults* are groups that have also rejected a dominant religious faith, but they tend to exist outside the mainstream of society.

- ***How has religion been affected by social change?***
  In recent years, there has been an enormous growth in conservative religious groups. Religion is a conservative influence in society, but religion also has an important part in movements for human liberation, including the civil rights movement and the move to ordain women in the church.

Key Terms

Online Resources

Sociology: The Essentials Companion Website

www.thomsonedu.com/sociology/andersen
Visit your book companion website where you will find more resources to help you study and write your research papers. Resources include Suggested Readings, web links, and a MicroCase Online feature that teaches you how to research society. Other resources include Learning Objectives, Internet exercises, quizzing, and flash cards.

is an easy-to-use online resource that helps you study in less time to get the grade you want NOW.

www.thomsonedu.com/login
Need help studying? This site is your one-stop study shop. Take a Pre-Test and Thomson NOW will generate a Personalized Study Plan based on your test results. The Study Plan will identify the topics you need to review and direct you to online resources to help you master those topics. You can then take a Post-Test to determine the concepts you have mastered and what you still need to work on.

13 Chapter thirteen

CHAPTER THIRTEEN

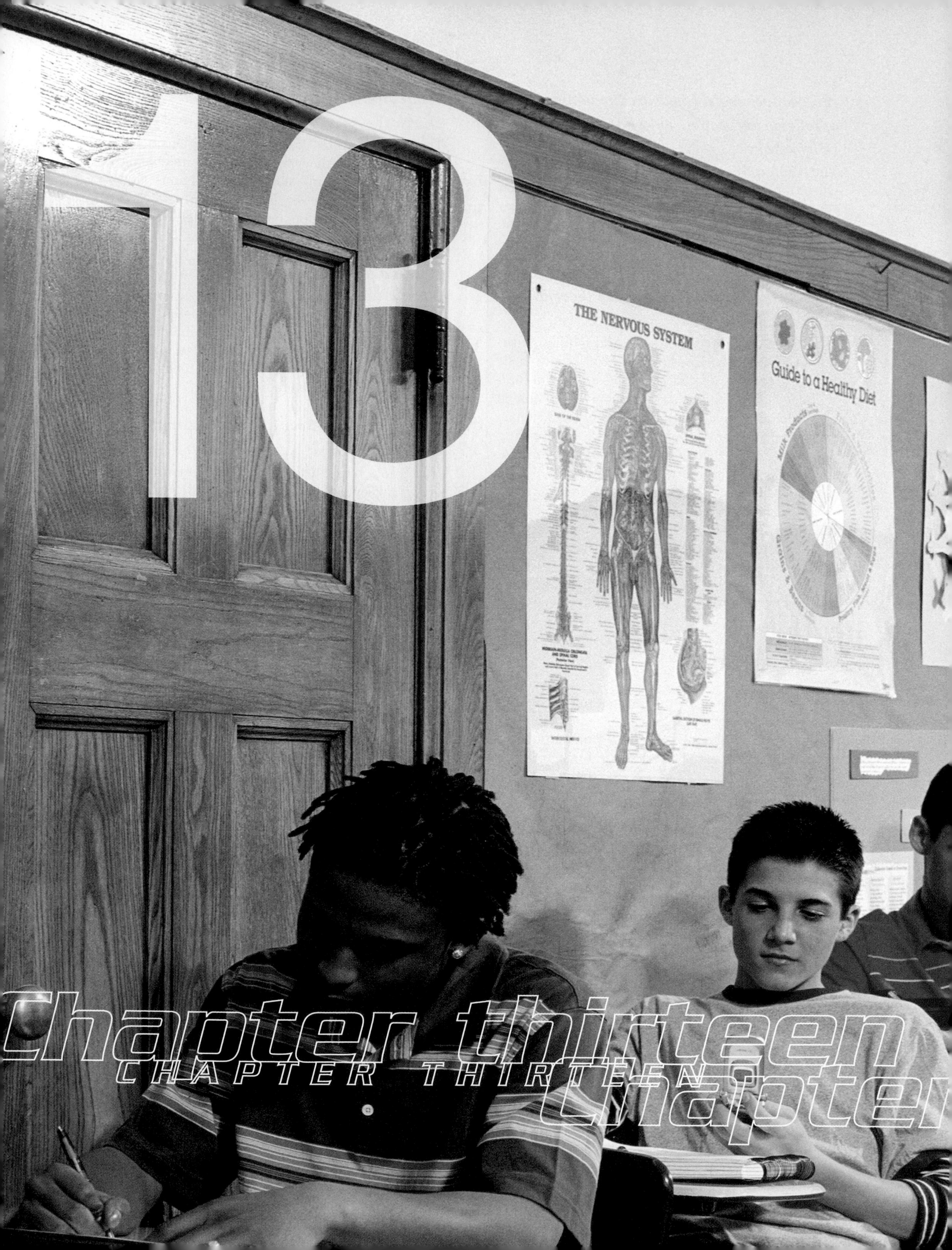

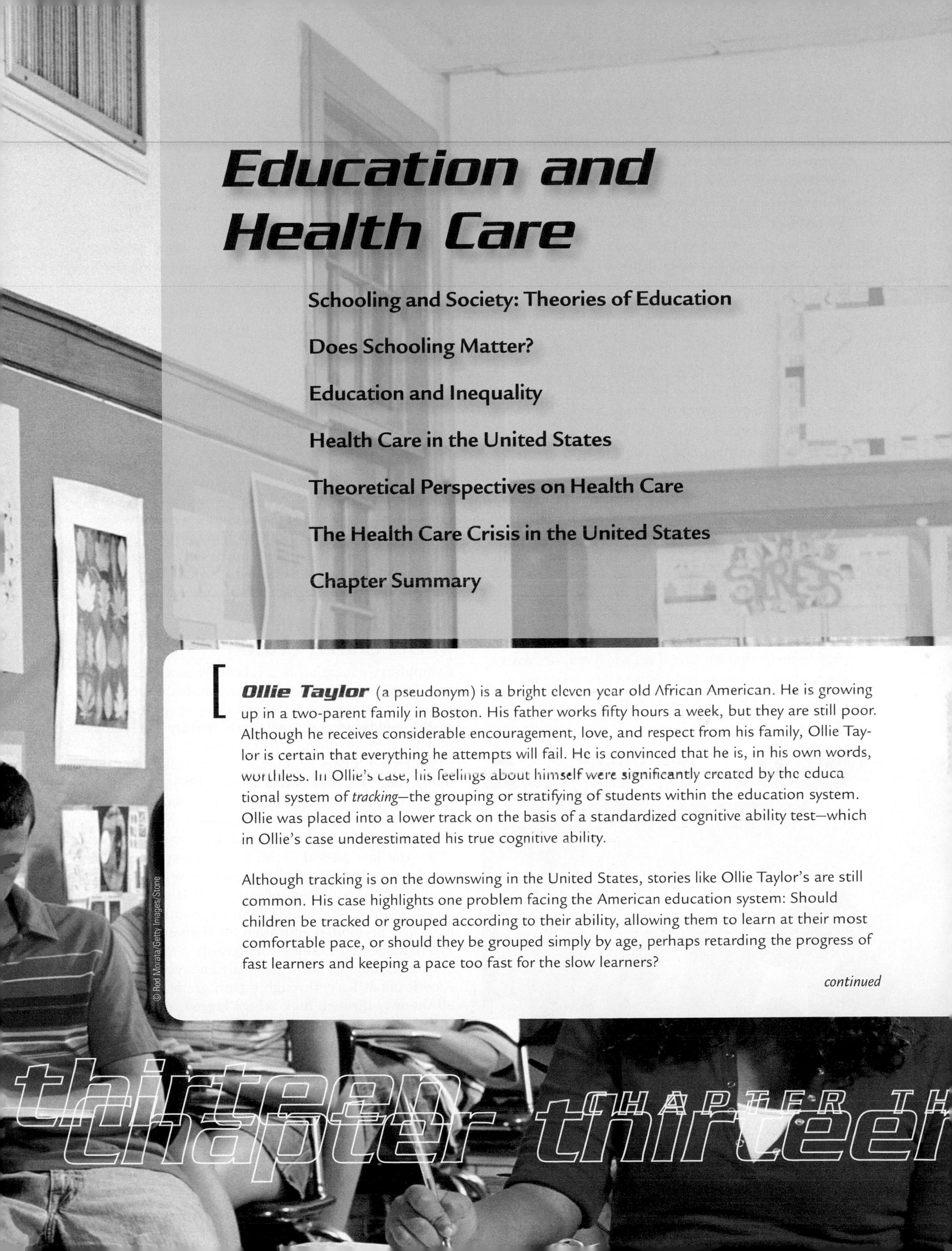

Education and Health Care

Schooling and Society: Theories of Education

Does Schooling Matter?

Education and Inequality

Health Care in the United States

Theoretical Perspectives on Health Care

The Health Care Crisis in the United States

Chapter Summary

Ollie Taylor (a pseudonym) is a bright eleven year old African American. He is growing up in a two-parent family in Boston. His father works fifty hours a week, but they are still poor. Although he receives considerable encouragement, love, and respect from his family, Ollie Taylor is certain that everything he attempts will fail. He is convinced that he is, in his own words, worthless. In Ollie's case, his feelings about himself were significantly created by the educational system of *tracking*—the grouping or stratifying of students within the education system. Ollie was placed into a lower track on the basis of a standardized cognitive ability test—which in Ollie's case underestimated his true cognitive ability.

Although tracking is on the downswing in the United States, stories like Ollie Taylor's are still common. His case highlights one problem facing the American education system: Should children be tracked or grouped according to their ability, allowing them to learn at their most comfortable pace, or should they be grouped simply by age, perhaps retarding the progress of fast learners and keeping a pace too fast for the slow learners?

continued

chapter thirteen

There are both positive and negative consequences to early tracking of students (a system discussed in detail later in this chapter), but the negatives outweigh the positives. Ollie Taylor's case exemplifies the negative consequences, yet he may also benefit from the positive aspects of getting an education in the United States, even with tracking in place: As an African American, he is six times more likely now than in 1940 to graduate from high school.

His chances of dropping out prior to graduation are, unfortunately, still fairly high. They are considerably higher than for a White youngster of his age, but significantly lower than for a Hispanic young man. If he graduates, he will stand a better chance of avoiding the grim path of many urban Black and Latino male dropouts and many poor White youth—a path leading to unemployment and possibly drug abuse, poor health, incarceration, and death at an early age.

Like education, health care in the United States is an institution. And like education, health care as an institution is in crisis. The public wonders: Has the quality of health care in the United States declined over the last two decades? Are doctors and nurses as good as they used to be? Are doctors overpaid? Are hospitals overcharging? Are lawsuits against the medical profession increasing, and if so, why? Why are costs for medical insurance so high, and why do so many people not have health insurance?

Ethnic and cultural diversity in the public high school classroom are increasingly common.

These public issues may seem to be simple matters of personal responsibility or individual rights, but sociologists show that illness in our society is strongly influenced by social factors. How we treat illnesses is highly influenced by race, social class, gender, and age; the same can be said of how we treat ill people. In different societies, the *same* physical state may be regarded as either sickness or wellness, demonstrating that the definition of illness is at least partly socially constructed. Physical conditions regarded as illness might be treated by prayer or witchcraft, surgery or radiation therapy, depending on the culture or society in question.

Schooling and Society: Theories of Education

Education in a society is concerned with the systematic transmission of the society's knowledge. This includes teaching formal knowledge such as the "three R's," reading, writing, and arithmetic, as well as morals, values, and ethics. Education prepares the young for entry into society and is thus a form of socialization. Sociologists refer to the more formal, institutionalized aspects of education as **schooling.**

The Rise of Education in the United States

Compulsory education is a relatively new idea. During the nineteenth century, many states did not yet have laws requiring education for everyone. Most jobs in the middle of the nineteenth century demanded no education or literacy whatsoever. Education was considered a luxury, available only to children of the upper classes (Cookson and Persell 1985). Education for slaves was prohibited by law until 1900, long after the Emancipation Proclamation passed in 1863.

When compulsory education was established in 1900, the law passed in all states except a few in the South, where Black Americans were still largely denied formal education of any kind (Higginbotham 1978). In the past, state laws in the South and West have also prohibited education for Hispanics, American Indians, and Chinese immigrants. State laws requiring attendance were generally enforced for White Americans at least through eighth grade. Education all the way through high school lagged considerably. In 1910, less than 10 percent of White eighteen-year-olds in the United States had graduated from high school.

Attendance in both high school and college has expanded considerably, such that in the United States today almost 90 percent of those under thirty-five have received at least a high school diploma (see Figure 13.1). High school and college graduation

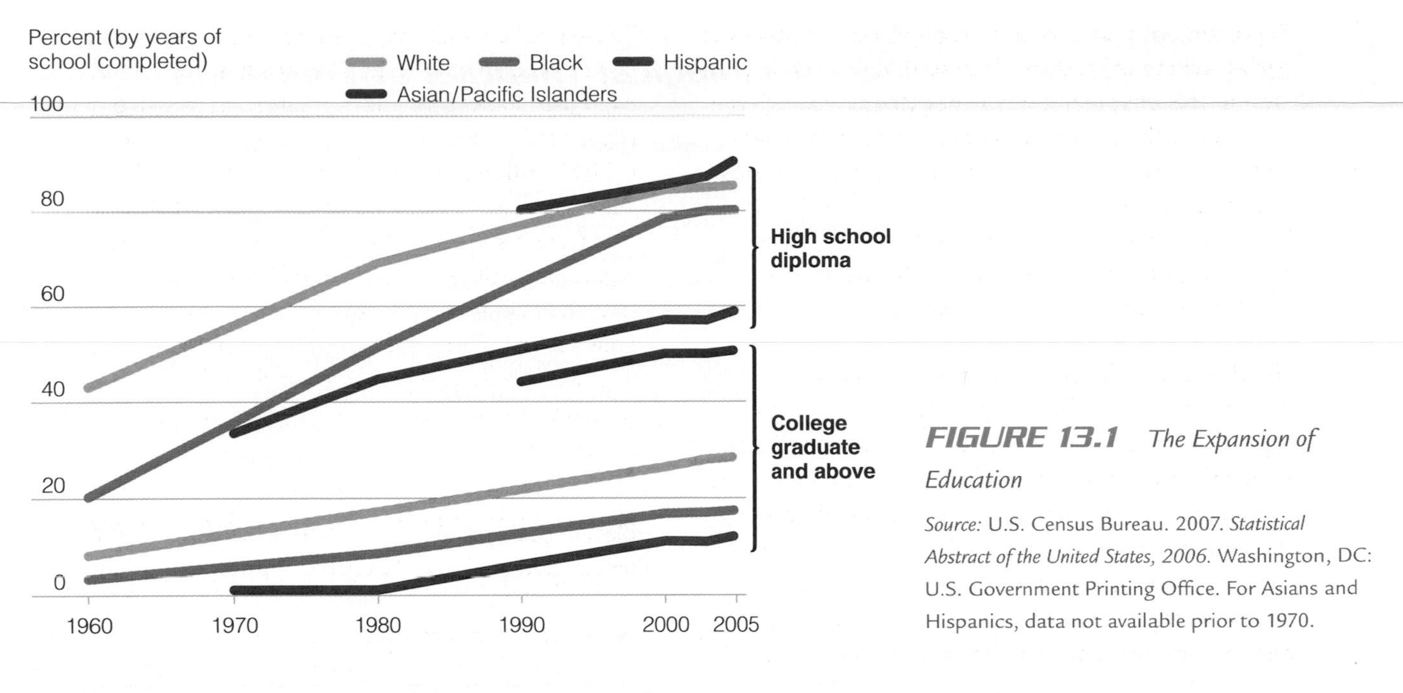

FIGURE 13.1 The Expansion of Education

Source: U.S. Census Bureau. 2007. *Statistical Abstract of the United States, 2006.* Washington, DC: U.S. Government Printing Office. For Asians and Hispanics, data not available prior to 1970.

rates however are not equal across racial groups. High school completion for Blacks lags behind that of Whites, and that of Hispanics lags significantly behind that of Blacks. As Figure 13.1 shows, similar trends exist for college graduations.

The Functionalist View of Education

All known societies have an education system of some sort. In the United States, as in other industrialized societies, the education system is large and highly formalized. In other societies, such as pastoral societies, it may consist only of parents teaching their children how to till land and gather food. Under these circumstances, the family is both the education system and the kinship institution.

Why does an education institution exist? Functionalist theory in sociology argues that education accomplishes the following consequences, or "functions," for a society.

Socialization is brought about as the cultural heritage is passed on from one generation to the next. This heritage includes a lot more than "book knowledge." It includes moral values, ethics, politics, religious beliefs, habits, and norms—in short, culture.

Occupational training is another function of education, especially in an industrialized society such as the United States. In less complex societies, as was the United States prior to the nineteenth century, jobs and training were passed from father to son, or more rarely, from father or mother to daughter. A significant number of occupations and professions today are still passed on from parent to offspring, particularly among the upper classes (a father passing on a law practice to his son). It also occurs among certain highly skilled occupations (plumbers, ironworkers, and electricians), where both training and union memberships may be passed on. Modern industrialized societies need a system that trains people for jobs. Most jobs today require at least a high school education, and many professions require a college or post-graduate degree.

Social control, or the regulation of deviant behavior, is also a function of education, although a less obvious one. Such indirect, subtle consequences emerging from the activities of institutions are called **latent functions** of the institution. Increased urbanization and immigration beginning in the late nineteenth century were accompanied by rises in crime, overcrowding, homelessness, and other urban ills. One perceived benefit of compulsory education (that is, one latent function) was that it kept young people off the streets and out of trouble (see Table 13.1).

Debunking Society's Myths

Myth: To get ahead in society, all you need is an education.

Sociological perspective: Education is necessary, but not sufficient, for getting ahead in society. Success depends significantly on one's class origins; the formal education of one's parent or parents; and, one's race, ethnicity, and gender.

The Conflict View of Education

In contrast to functionalist theory, which emphasizes how education unifies and stabilizes society, conflict theory emphasizes the disintegrative and disruptive aspects of education. Conflict theory focuses on the competition between groups for power, income, and table 13.1 **Sociological Theories of Education**

	Functionalism	Conflict Theory	Symbolic Interaction
Education in society	Fulfills certain societal needs for socialization and training; "sorts" people in society according to their abilities	Reflects other inequities in society, including race, class, and gender inequality, and perpetuates such inequalities by tracking practices, for example	Emerges depending on the character of social interaction between groups in schools
Schools	Inculcate values needed by the society	Are hierarchical institutions reflecting conflict and power relations in society	Are sites where social interaction between groups (such as teachers and students) influences chances for individual and group success
Social change	Means that schools take on functions that other institutions, such as the family, originally fulfilled	Threatens to put some groups at continuing disadvantage in the quality of education	Can be positive as people develop new perceptions of formerly stereotyped groups

social status, emphasizing the prevailing importance of institutions in the conflict. One intersection of education between group and class competition is embodied in the significant correlations that exist between education and class, race, and gender. The unequal distribution of education allows it to be used to separate groups. The higher the educational attainment of a person, the more likely that person will be middle to upper class, White, and male. Conflict theorists argue that educational level is a mechanism for producing and reproducing inequality in our society.

According to conflict theorists, educational level can be a tool for discrimination by using the mechanism of **credentialism**—the insistence upon educational credentials for their own sake, even if the credentials bear little relationship to the intended job (Collins 1979; Marshall 1997). This device can be used by potential employers to discriminate against minorities, working-class people, or women—that is, those who are often less educated and least likely to be credentialed because discriminatory practices within the education system limited their opportunities for educational achievement.

Although functionalists argue that jobs are becoming more technical and thus require workers with higher education, conflict theorists argue that the reverse is true—most new opportunities appearing today are in categories such as assembly-line work, jobs that are becoming less complex and less technical and therefore require less traditional education or training. Nonetheless, potential employers will insist on a particular degree for the job even though there should be little expectation that educational level will affect job performance. In this case, education serves as a discriminatory barrier.

The Symbolic Interaction View of Education

Symbolic interaction focuses on what arises from the operation of the interaction process during the schooling experience. Through interaction between student and teacher, certain expectations arise on the part of both. As a result, the teacher begins to expect or anticipate certain behaviors, good or bad, from students. Through the operation of the **teacher expectancy effect,** the expectations a teacher has for a student can actually create the very behavior in question. Thus fulfilled, the behavior is actually caused by the expectation rather than the other way around. For example, if a White teacher expects Latino boys to perform below average on a math test, relative to White students, over time the teacher may act in ways that encourage the Latino boys to get below average math test scores. Later we will examine just how the expectancy effect works.

Does Schooling Matter?

How much does schooling really matter? Does more schooling actually lead to a better job, more annual income, and greater happiness? Is the effect of education great or small?

Effects of Education on Occupation and Income

One way that sociologists measure a person's social class or socioeconomic status (SES) is to determine the person's amount of schooling, income, and type of occupation (see Chapter 7 on class stratification). Sociologists call these the *indicators* of SES. In the general population there is a strong relationship between formal education and occupation, although the relationship is not perfect. Measuring occupations in terms of social status or prestige, we find that the higher a person's occupational status, the more formal education he or she is likely to have received. Overall, we know that on average, doctors, lawyers, professors, and nuclear physicists spend many more years in school than unskilled laborers, such as garbage collectors and shoe shiners. This relationship is strong enough that we can often, although not always, guess a person's educational attainment just by knowing their occupation. There are indeed instances of semiskilled laborers, such as taxi drivers, who have Ph.D.s, but they are relatively rare. Also exceedingly rare is the reverse: the self-educated, self-made individual who completed only the fourth grade and is now the CEO of a major corporation. Finally, at least one researcher (Kasarda 1999) argues that there is a serious mismatch between the skills youths learn today and the skills required to enter the job market.

The connection between income (and jobs) and education is not independent of gender. Gender heavily influences the relationship between income and education. Note from Table 13.2 that although the higher one's education, the higher one's (average) income, it is nonetheless true that the average income for women is less than the average income for men at *each* education level. This is because, in general, throughout our society the average woman earns less than a man of the same *or even less* education. Men with professional degrees (law, medicine, and so forth) earn a median annual income of above $90,000, whereas a woman with that same education earns only about $60,000, about two-thirds of what a man earns. A man with no graduate education but a college-only education earns more than a woman with a master's degree. And men with some college, but no bachelor's degree, earn more than women with a bachelor's degree.

Effects of Social Class Background on Education and Social Mobility

Education has traditionally been viewed in the United States as the way out of poverty and low social standing—the main route to upward social mobility. The assumption has been that a person can overcome modest beginnings, starting by staying in school.

Much sociological research has demonstrated that the effect of education on a person's eventual job and income greatly depends on the social class that the person was born into. Among White people of the upper classes, including those who inherited wealth, as well as professionals and high-level managers, social class origin is *more important than education* in determining occupation and income (Blau and Duncan 1967; Jencks et al. 1972; Taylor 1973b; Jencks et al. 1979; Bielby 1981; Cookson and Persell 1985; Persell 1990; Jencks 1993).

Class and race work together to "protect" the upper classes from downward social mobility—and to block the lower classes from too much upward mobility. Education is used by the upper classes to avoid downward mobility by means such as sending their children to elite private secondary schools. A disproportionate number of upper-class children attend

table 13.2 *Median Income, by Education and Gender (in dollars)*

Level of schooling	Men	Women
Less than 9th grade	$16,321	$ 9,496
9th–12th grade (no diploma)	20,934	11,136
High school graduate	30,134	16,695
Some college, no degree	36,930	21,545
Associate degree	41,903	26,074
Bachelor's degree	51,700	32,668
Master's degree	64,468	44,385
Professional degree	90,878	59,934
Doctorate degree	76,937	56,820

Source: U.S. Census Bureau. 2005. Historical Income Tables—People, Table, p-16. Washington, DC: U.S. Department of Commerce. **www.census.gov**

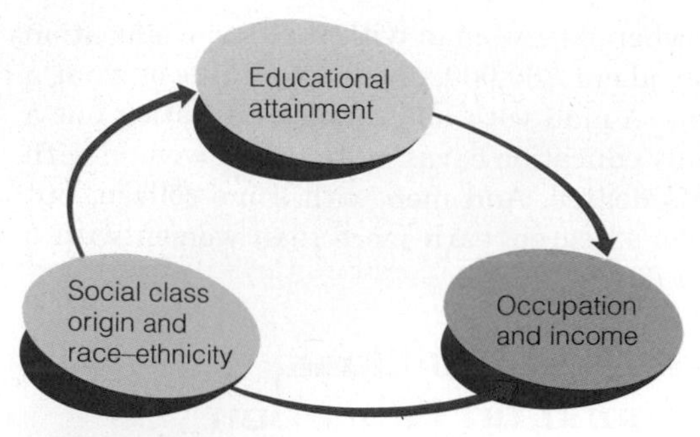

FIGURE 13.2 *Relationship of Social Class, Race–Ethnicity, Education, Occupation, and Income*

elite boarding schools and day schools, compared to working-class children who are considerably underrepresented in such schools (Rendon and Hope 1996; Cookson and Persell 1985).

Among middle-class Whites, education considerably improves the chances of getting middle-class jobs; yet access to upper-class positions is limited. Among those of lower-class origins, such as unskilled laborers or the chronically unemployed, chances of getting a good education as well as a prestigious job are poor. In sum, education is affected by social-class origins, and occupation (and income) is heavily influenced by social class and education. Individuals with lower-class origins are less likely to get a college education and thus are less likely to get a prestigious job. These interrelationships are summarized in Figure 13.2, which shows that social class origin affects occupation and income both directly and indirectly by way of education.

Debunking Society's Myths

Myth: Education is more important than social class in determining one's job and income.

Sociological perspective: Although education has an effect on the job one gets and the income one earns, overall, social class origin is more important than acquired education in determining one's job and earned income.

Education, Social Class, and Mobility Seen Globally

It is sometimes argued that because of our education system, there is more occupational and income mobility in the United States than in other industrialized nations, particularly England, Germany, and Japan. In general this is true, but not by much. Until a few years ago, students in England were required to take an examination, called the Eleven Plus, at age eleven. A person's score on this examination determined whether he or she was put on a track to prestigious universities such as Oxford or Cambridge or went directly into the labor force from high school. Children of the upper classes stood a *far* better chance of scoring high on this examination than did middle- or working-class children, and the average scores of women and minorities, especially Africans and East Indians, were considerably lower than those of upper-class White males.

A similar situation exists in the United States, as illustrated in this chapter's opening vignette on Ollie Taylor. Students from lower-class families have lower average scores on exams such as the Scholastic Assessment Test (SAT) and the American College Testing Program (ACT). As shown in Table 13.3, there is a smooth and dramatic increase in average (mean) SAT score as family income is higher, for both SAT verbal as well as math scores. In this sense, one's SAT score is a "proxy" measure of one's social class: *Within a certain range, you can guess one's likely SAT score from knowing only the income and social class of one's parents!* As you can see from Table 13.3, each additional $10,000 in family income is worth about 10 to 15 more points on either the SAT verbal or the SAT math tests (thus 20 or 30 points more for combined score)! This is truly ironic, since the multiple-choice SAT was originally designed in the 1940s as an "objective" test to combat the pattern of children from wealthy families having an advantage for admission to college.

With lower scores comes a diminished chance of getting into the best colleges or universities. African Americans, Latinos, and American Indians score on the average less than Whites, and women tend to score lower than men on the quantitative (mathematical) sections of the SAT (Table 13.4). Asian Americans as a group have scored higher than Whites in recent years on the quantitative sections of the SAT, but lower on the verbal sections. Women of any ethnic group score lower on the quantitative sections than men of the same ethnic group. These patterns indicate that the SAT has an effect in the United States similar to that of the Eleven Plus test in England, directing the futures of young people according to the results of widely administered exams.

In Germany, an examination called the *Abitur* is taken during the equivalent of the junior year in high school. A high score on the *Abitur* facilitates admission to a university; a low score inhibits getting into a university. Low-scoring students must take two or three more years of courses and then reapply to a university if they wish to attend.

In Japan, a similar examination given at age twelve determines even more rigidly a child's subsequent educational opportunities. Students who wish to continue their education at a college or university must score high enough to gain admission to prep schools. Especially high scores guarantee admission to prestigious prep schools, which is necessary for later admission to the best universities. Low scorers are virtually shut out from prep school admission,

table 13.3 Average SAT Scores by Family Income

Family Income	SAT Verbal Average Scores[a]	SAT Math Average Scores[b]
<$10,000	429	457
$10,000–$20,000	445	465
$20,000–$30,000	462	474
$30,000–$40,000	478	488
$40,000–$50,000	493	501
$50,000–$60,000	500	509
$60,000–$70,000	505	515
$70,000–$80,000	511	521
$80,000–$90,000	523	534
>$100,000	549	564

[a,b]SAT scores, on either the verbal or the math sections, range from 200 to 800, with 800 being a perfect score.
Source: College Board, 2006. *College Bound Seniors 2006: A Profile of SAT Program Test Takers.* New York: The College Board. Used with permission.

and these students become ineligible for a university education. In recent years, many parents have begun to send their children to weekend "cram" seminars called *jukos* to prepare for this examination, adding a grueling additional regimen to the already stiff requirements of the Japanese school system. Many Japanese, including educators, are becoming concerned that the extreme competitiveness of this system and the great burden of work placed on students are brutalizing the youngsters.

Overall, the educational system in the United States appears to allow for a bit more social mobility

table 13.4 Average SAT Scores, by Ethnicity and Gender

	SAT Verbal Average Scores		SAT Math Average Scores	
SAT test takers who described themselves as:	*Male*	*Female*	*Male*	*Female*
American Indian or Alaskan Native	489	485	512	478
Asian, Asian American, or Pacific Islander	511	509	594	562
African American or Black	430	437	438	423
Hispanic or Latino:				
• Mexican or Mexican American	459	451	485	450
• Puerto Rican	462	458	472	443
• Latin American, South American, Central American, or other Hispanic or Latino	464	454	484	448
White	529	526	555	520
Other	494	494	534	497

Source: College Board, 2006. *College Bound Seniors 2006: A Profile of SAT Program Test Takers.* New York: The College Board. **www.collegeboard.com/about/news_info/ebsenior/yr2006/html/links.html**

Japanese students and their parents at this juko, *or "cram" seminar, vow academic success at a meeting in a hotel in Tokyo over the New Year's holiday.*

than can be achieved in Germany, possibly England, and certainly Japan. The stratification system in those countries and others with similar systems are more rigid, or *castelike,* than the United States. However, there is a danger in concluding that the U.S. educational system permits *much* more social mobility. In general, the similarities tend to be more prominent than the differences.

Education and Inequality

In its original nineteenth-century conception, the education system was to serve as a leveling force in American society—the road to full equality for all citizens regardless of race, social-class origin, nationality, religion, or gender. Jew and gentile, Irish and Polish, Black and White, rich and poor, male and female would learn together side by side. Through education, each student would learn the ways of others and thus come to understand and respect them. Full equality for humankind was to follow.

Education has indeed reduced many inequalities in society since compulsory education began at the turn of the twentieth century. The percentage of high school graduates has risen among Whites and minorities, both male and female, as have certain types of social mobility. Despite continuing inequalities in college enrollments comparing African Americans, Hispanics, and Whites, the enrollment of minorities has risen overall. Furthermore, as more minorities and women attend and graduate from two- and four-year colleges, the result has been more employment in mid-level and high-level jobs. Nonetheless, many inequalities still exist in U.S. education.

Cognitive Ability and Its Measurement

Since as long ago as classical Greece, humans have sought to measure a "mental faculty" or "intelligence." It is now called **cognitive ability**—the capacity for abstract thinking. Since early in the twentieth century, educators in our society, from preschools to universities, have attempted to measure intelligence by means of **standardized ability tests,** such as the SAT or IQ tests, which are intended to measure ability or potential. They are not the same as **achievement tests,** which are intended to measure what has actually been learned, in addition to ability or potential. Advanced Placement (AP) exams are achievement tests taken before entering college. Students who score high demonstrate that they have already mastered certain material and can skip those courses in college.

The education system in the United States has relied heavily upon the idea that intelligence, or ability, or potential is a single unitary trait. Cognitive ability has been gauged according to the numerical results of the standardized tests. There has been a will to reduce measurements of cognitive ability to a single number, or perhaps two numbers, such as language and math scores of SAT tests and IQ tests in the recent past.

There are three major criticisms made regarding using standardized tests such as the SAT or ACT as measures of cognitive ability. First, the tests tend to measure only limited ranges of abilities (such as quantitative aptitude or verbal aptitude) while ignoring other cognitive endowments such as creativity, musical ability, spatial perception, or even political skill and athletic ability (Zwick 2004; Freedle 2003; Gardner 1999; Lehmann 1999; Sternberg 1988).

Second, the tests possess at least some degree of *cultural bias* and *gender bias.* As a result, they may perpetuate inequality between different cultural or racial groups and social classes, as well as perpetuate social, economic, and educational inequality between men and women. The tests were designed primarily by middle-class White males, and the "standardization" they strive to achieve mirrors middle-class White male populations. Many studies show that although standardized ability tests are somewhat capable of predicting future school performance for White males, a significant number of studies show less accurate forecasts for the success of minorities, especially Hispanics, African Americans, and American Indians, and they sometimes predict school performance less accurately for women than for men (Zwick 2004; Epps 2002; Taylor 2002, 1992a, 1980; Jencks and Phillips 1998; Fleming and Garcia 1998; Pennock-Roman 1994; Young 1994; Crouse and Trusheim 1988). In other words, the **predictive validity** of the tests, which is the extent to which the tests accurately predict later college grades, is compromised for minorities, women, and persons of working class origins.

© Lara Jo Regan/Getty Images

Contrary to the impression given in this photo, girls are frequently underrepresented in scientific and technical classes in school, particularly in upper grades and college.

The third criticism of the SATs is that they do not predict school performance very well, even for Whites. For example, SAT scores are only modestly accurate predictors of college grades even for White persons (Zwick 2004; Fleming and Garcia 1998; Manning and Jackson 1984). This fact is not well known. Grade point average in high school (and school class rank as well) is also only a modestly accurate predictor of success in college. High school grades are about as accurate as the SATs in predicting college grades—which is to say, not very accurate.

Ability and Diversity

As already noted, average scores on IQ tests and cognitive ability tests such as the SAT differ by racial–ethnic group, social class, and gender. Overall, Whites score higher on average than minorities, and in general the higher a person's social class, the higher his or her test scores. The differences between groups are regarded by experts as primarily environmental in origin, reflecting group differences in years of parental education, social class status, childhood socialization, language, nutrition, and cultural advantages received in the home and during youth. There is no evidence that *between-group* differences by gender, race, or social class are in any way genetically inherited. Certain *within-group* differences may reflect genetic differences among individuals within the *same* racial or ethnic group, social class, or gender. But even the within-group effect of genes is estimated to be much smaller than the within-group effect of social environment. That is, the effects of social environment are greater than the effect of genes, even though genes have some effect (Kamin and Goldberger 2002; Taylor 2007, 2002, 1980; Jang et al. 1996; Chipuer et al. 1990; Gould 1981; Goldberger 1979; Kamin 1974; Jencks et al. 1972).

Both women and minorities have been catching up somewhat with White men in SAT math scores, at least up until about 1995 (College Board, 2000 and 2006), after which the differences have remained roughly constant. The change, coinciding with some social gains on the part of minorities and women over the last decade or so, tends to discredit the traditional belief that women, Blacks, and Latinos have less mathematical ability than White men. This belief has been used in the past to support the argument that women and minorities are less fit to perform high-level executive jobs that require number crunching and abstract analytical reasoning. This belief tends to perpetuate both occupational and educational inequality between women and minorities on the one hand and White males on the other.

The "Cognitive Elite" and *The Bell Curve* Debate

The book *The Bell Curve* caused a major stir, still ongoing, among educators, lawmakers, teachers, public officials, policy makers, and the general public. In this book, which contains analyses of great masses of data, authors Herrnstein and Murray (1994) and others (Bouchard and McGue 2003) argue that not only does the distribution of intelligence in the general population closely approximate a bell-shaped curve (called the *normal distribution*). They also argue there is one basic, fundamental kind of intelligence, not several independent kinds of intelligences.

Herrnstein and Murray and subsequent researchers estimate that this fundamental intelligence is about 70 percent genetically heritable and only 30 percent determined by environment (Bouchrd and McGue 2003; Bouchard et al. 1990). Therefore, they argue, intelligence is inherited; that is, it is determined primarily by one's genes rather than by one's social and educational environment. How do they arrive at such a figure? To do this, they reviewed data on identical twins who have been separated early in life and then raised apart. The idea is that because identical twins (as opposed to fraternal twins) are genetic *clones* (exact genetic duplicates), any similarities that remain between them after their separation (such as having the same or similar intelligence) must necessarily be caused by their identical genes rather than by similarities in their social or educational environments. The authors argue that the similarity in intelligence between separated twins (the heritability) is about 70 percent.

Critics, however, point out that some of the identical twins in the studies cited by Herrnstein and Murray and other researchers were actually more separated than others. That is, some were not really very separated at all, whereas some were separated for longer periods during their lives, and had fewer similarities in their social and educational environments. When this is taken into account, it is seen that twins who were more separated (attended different schools or were raised in different families with different socioeconomic circumstances) were also less similar in intelligence. In general, *the more separated the identical twins were, the less similar they were in intelligence.* This shows the effect of their differing social environments over the effect of their identical genes, suggesting a heritability of about 40 percent, not 70 percent (Kamin and Goldberger 2002; Taylor 2007, 1980; Chipuer et al. 1990; Kamin 1974; Jencks et al. 1972).

Debunking Society's Myths

Myth: Intelligence is mostly determined by genetic inheritance.

Sociological perspective: Intelligence is a complex concept not easily measured by one thing and is likely shaped as much by environmental factors as by genetic endowment.

Another point made by *The Bell Curve* authors is that if intelligence is primarily inherited, and if, on average, social classes differ in intelligence, it follows that the lower classes are less endowed with genes for high intelligence than the upper classes. The authors reason that the upper- and upper-middle classes constitute a *genetically based* **cognitive elite** in the United States, those with high IQs, high incomes, and prestigious jobs. They strongly imply, but do not state outright, that any two groups presumed to differ in average intelligence (such as Blacks versus Whites or Latinos versus Whites, and any other minority versus the dominant comparison—and women versus men, as well) may very well differ in genes for intelligence.

The main problem with the cognitive elite argument is that the authors base a between-group conclusion on a within-group estimate of genetic heritability. Thus they base their conclusions about women versus men, minority versus White, and lower class versus upper class on heritability results attained on White men only (Fischer et al. 1996; Lewontin 1996, 1970; Hauser et al. 1995; Kamin 1995; Gould 1994, 1981; Taylor 1995, 1980).

Tracking and Labeling Effects

Over half of America's secondary schools and elementary schools currently use some kind of **tracking** (also called *ability grouping*), which is the separating of students according to some measure of cognitive ability (Lucas 1999; Oakes and Lipton 1996; Maldonado and Willie 1996; Oakes 1985). Tracking has taken place for more than seventy years. Starting as early as first grade, children are divided into high-track, middle-track, and lower-track groups. In high school, the high-track students take college preparatory courses in math and science and read Shakespeare. The middle-track students take courses in business administration and typing. The lower-track students like little Ollie Taylor take vocational courses in auto mechanics, metal shop, and cooking. While this kind of tracking is now on the decline in the United States, it is still with us in many schools (Hallinan 2003; Lucas 1999; Oakes 1990, 1985).

"When you lie about yourself, is it to appear closer to or farther away from the middle of the bell curve?"

The original idea behind tracking is that students will get a better education and be better prepared for life after high school if they are grouped early according to cognitive ability. Tracking is supposed to benefit the gifted, the slow learners, and everyone in the middle, but is now severely criticized. Theoretically, students in all tracks learn faster because the curriculum is tailored to their ability level, and the teacher can concentrate on smaller, more homogenous groups.

The opposite argument is given by advocates of *detracking*. The detracking movement is based on the belief that combining students of varying cognitive abilities benefits the students more than tracking, especially by the time students get to junior high and high school. Students of high and low ability can thus learn from each other; the high-ability students are not seen to be "held back" by students with less ability, but are enriched by their presence. Finally, advocates of detracking point out that students in the lower tracks get less teacher attention and simply learn less, in effect penalized for being in the lower tracks. The idea is mix, don't match.

Which approach is better? Most researchers and educators who have studied tracking agree that not all students should be mixed together in the same classes. The differences between students can be too great, and their needs too dissimilar. Some degree of tracking has always had advocates based on its presumed benefits for all students. This presumption is under attack. One of the most consistent research findings on tracking is that students in the higher tracks receive positive effects, but that the lower-track students suffer negative effects (Lucas 1999; Owens 1998; Rendon and Hope 1996; Cardenas 1996; Perez 1996; Oakes and Lipton 1996; Oakes 1990, 1985; Gamoran 1992; Gamoran and Mare 1989; Braddock 1988).

thinking sociologically

Were you in a tracked elementary school? What were the tracks? Describe them. Did you get the impression that teachers devoted different amounts of actual time to students in different tracks? Did teachers "look down" on those in the lower tracks? What about the students—did they treat some tracks as "better" or "worse" than others (were they perceived as differing in *prestige*)? Based on your recollections, what does this tell you about *tracking* and *social class*?

To begin with, students in the lower tracks learn less because they are, quite simply, taught less. They are asked to read less and do less homework. High-track students are taught more; furthermore, they are consistently rewarded by teachers and administrators for their academic abilities. Less is expected of lower-track students, and as a result, their academic performance is lower (Hallinan 2003: Owens 1998; Slavin 1993; Gamoran and Mare 1989; Braddock 1988).

Who gets assigned to which tracks? Research shows that track assignment is not solely based on the performance in cognitive ability tests. Social class and race are involved. Students with the *same* test scores often get assigned to different tracks because of differences in their social class and race. Few administrators or teachers consciously and deliberately assign students to tracks based on these criteria, but it occurs nevertheless. Researchers have consistently found that when following two students with identical scores on cognitive ability tests, the student of higher social class is more likely than the student of lower social class status to get assigned to the higher track.

Teacher Expectancy Effect

Similar to the labeling effect of tracking is the teacher expectancy effect, which is the effect of teacher expectations on a student's actual performance—*independent* of the student's actual ability. What the teacher expects students to do affects what they will do. The expectations a teacher has for a student's performance can dramatically influence how much the student learns.

Insights into the teacher expectancy effect come from symbolic interactionist theory. In a classic study, Rosenthal and Jacobson (1968) told teachers of several grades in an elementary school that certain children in their class were academic "spurters," who would increase their performance that year. The rest of the students were called "nonspurters." The researchers selected the "spurters" list *completely at random,* unbeknown to the teachers. The distinction had no relation to an ability test the children took early in the school year, although the teachers were told (falsely) that it did. At the end of the school year, it was found that although all students improved somewhat on the achievement test, those labeled spurters made greater gains than those designated nonspurters, especially among first and second graders. This experiment isolates the effect of the label (hence showing a *labeling effect*), because it is the only difference that distinguishes the two groups. Variations of this clever and revealing study have been conducted many times over and the results are generally similar.

How are expectations converted into performance? By the powerful mechanism of the **self-fulfilling prophecy,** in which merely applying a label has the effect of justifying the label (Taylor et al. 2006; Cardenas 1996; Darley and Fazio 1980). Recall the quote from early sociologist W.I. Thomas in Chapters 1 and 4: "If men [sic] define situations as real, they are real in their consequences" (Thomas 1928: 572). If a student is defined (labeled) as a certain type, the student becomes that type. The process unfolds in stages (see Figure 13.3). First, a teacher is told that a student merits a label such as spurter; perhaps the designation originates with administrators or comes from the scoring key of a standardized exam. The teacher's perception of the student is then colored by the label; a student labeled a spurter may be coaxed and praised more often than nonspurters. The student then reacts to the teacher's behavior; students expected to perform well and encouraged to excel actually perform better in class and on exams than other students. Finally, the original prophecy fulfills itself: The teacher observes the behavior of the student, notes the increase in performance, and concludes that the designation spurter is affirmed because the so-called spurters actually perform better by objective measures than the nonspurters (Hallinan 2003; Lucas 1999; Rendon and Hope 1996; Cardenas 1996; Gamoran 1972).

Schooling and Gender

Teachers hold different expectations for girls and boys in school, and it comes as no surprise that gender affects teacher expectations and therefore the actual performance of these children. Tracking, particularly in high school, is significantly dependent on gender; for example, a far greater proportion of young men are in the science and math tracks than are young women. Throughout schooling, from preschool through graduate or professional school, women and men are treated differently, with women discriminated

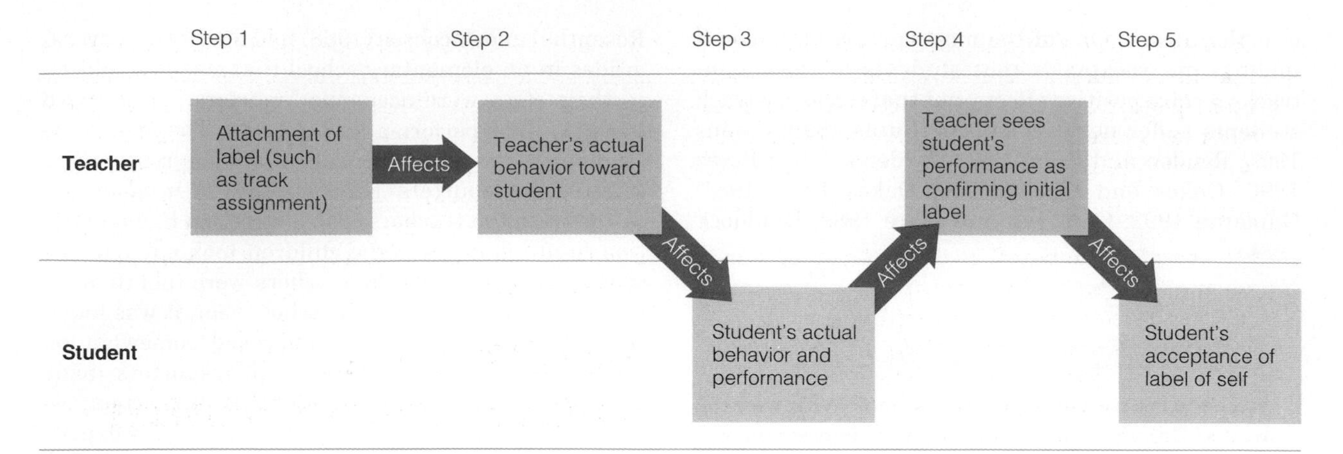

FIGURE 13.3 *The Self-Fulfilling Prophecy*

Source: Adapted from Taylor, Shelley E., Letitia Anne Peplau, and David O. Sears, 2006. *Social Psychology,* 12th ed. Upper Saddle River, NJ: Prentice Hall.

against. This consistent and long-term differential treatment has had profound consequences. *Girls and boys start out in school roughly equal in skills and confidence,* a fact that contradicts the often-heard assertion that girls are "inherently" worse than boys in math, spatial, and other kinds of skills

By the end of high school, however, women trail men, especially in mathematics and the sciences. What happens in between has been documented in a comprehensive report commissioned by the American Association of University Women (AAUW) that summarizes the results of more than 1000 publications and studies. This extensive research shows the following with regard to schooling and gender (Zwick 2004; American Association of University Women 1998).

1. In general, teachers pay less attention to girls and women. In elementary school as well as high school, teachers direct more interaction to boys than to girls. As a result, boys tend to talk more in class, and thus teachers tend to interact with them more. The difference is particularly notable in math and science classes.
2. On national tests of reading and writing, girls perform equally to boys. But on advanced mathematics and some science tests, differences emerge. The most dramatic gender differences appear on tests with the highest stakes—the quantitative sections of the Preliminary Scholastic Assessment Tests (PSAT), Scholastic Assessment Tests (SAT), and Advanced Placement (AP) tests. Researchers explain this as the result of several factors, including course-taking patterns, differences in family socialization, and, as we have noted, bias in the standardized tests themselves. With regard to course-taking, young men and women tend to take the same number of math and science courses, but what they take differs substantially. In science, girls are more likely to take biology and chemistry, and boys are more likely to take physics. Girls also tend to end their math studies after a second algebra course; boys are more likely to take trigonometry and calculus, thus giving them a relative advantage on higher-level math skills.
3. Some standardized math and science tests still retain gender bias, despite ten years or more of effort to weed out bias on the part of education specialists and testing organizations such as the Educational Testing Service, manufacturer of the SAT. Bias is especially prevalent in mathematical word problems. Certain mathematics word problems are gender-typed because they employ words and concepts more familiar to men than women (Chipman 1991; American Association of University Women 1998).
4. Standardized tests in math tend to underpredict women's actual grades in mathematics. Women tend to do somewhat better in math courses than their test scores predict (Zwick 2004).
5. Teachers tend to treat Black women and White women differently by rebuffing Black girls and interacting more with White girls. This is particularly true of Black and White girls during the elementary school years. The trend appears to be independent of the teacher's own race; namely, Black teachers tend to interact more with White girls, just as do White teachers.
6. Textbooks still tend to either ignore women or stereotype them. In this respect, textbooks are gender-role socializers. In elementary school texts, boys are portrayed as building things, being clever, and leading others. Girls tend to be shown performing dull tasks and following the leadership of boys.
7 As girls and boys approach adolescence, their self-esteem tends to drop, with the erosion of self-esteem occurring more quickly among girls

than boys. The trend is further exacerbated by the discrimination against women in the classroom, gender bias in standardized tests, and stereotyped presentations of women in presumably authoritative textbooks.

Stereotype Threat

As has already been noted, racial and gender stereotypes can affect actual behavior. To what extent can a negative stereotype about *oneself* affect one's *own* behavior and academic performance? As with the self-fulfilling prophecy, to what extent do minorities and women internalize negative stereotypes about themselves that show such effects?

An answer has been provided recently by the research of Claude M. Steele and associates (Steele 1999, 1997, 1996, 1992; Steele and Aronson 1995; Massey and Fisher 2007; Woods et al. 2005; McIntyre et al. 2003; Aronson et al. 2002; Brown and Josephs 1999). They note that two common stereotypes exist in the United States and other countries as well. First, because on average Blacks perform less well than Whites on math and verbal ability tests, Blacks must have, or so it is believed, some "inherent" deficiency in math and verbal abilities relative to Whites. Second, because women perform less well than men on math ability tests, women must therefore have some "inherent" deficiency in math ability.

To the extent that Black students in high school or college may believe (internalize) such stereotypes, they may perform less well on a test if they are told that "this is a genuine test of your true ability," because telling them that would tend to increase their anxiety about the test. The same should apply to women on a math ability test. Either Whites, or males, who are told this should be *less* affected, since these societal stereotypes are not *about* Whites or males.

Results show that this is just what happens. Black college students who are told that the test is a "genuine" test of their true verbal ability perform less well than Whites who are also told the same thing—even though the Whites and Blacks start out equal in their average test scores. This is the **stereotype threat effect.** If the groups of Blacks and Whites are told nothing, then they perform about the same on the test. Note that nothing is said to the students specifically about Black and White test performance, only that the test was designed to be a "genuine" test of verbal ability. Perhaps Ollie Taylor was told this when he was subjected to a cognitive ability test and thus suffered some stereotype threat effect, which then resulted in his assignment to a low track, even though he was quite bright.

Stereotype threat appears to operate in the same way with regard to the presumed female–male difference in math ability test performance. When both men and women are told that a math test given to them is a "genuine" test of their true math ability, the women do worse than the men. If men and women are told nothing, then women and men perform about the same on the test (Smith 2006; McIntyre et al. 2003; Gonzales et al. 2002; Lau 2002; Brown and Josephs 1999). This suggests that the long-believed female deficit in math ability may stem at least in part simply from what they are told before they take the test, and less from "inherent" differences between men and women in math ability.

Health Care in the United States

Like education, the health care system is one of America's major institutions. Generally speaking, the citizens of the United States are quite healthy in relation to the rest of the world. However, there are very great discrepancies among people within the United States in terms of longevity, general health, and access to health care, with the least advantaged part of the curve consisting primarily of minorities, the lower classes, and for a number of ailments, women. But the fact remains that we are a robust nation.

The Emergence of Modern Medicine

The highly technological, scientific, corporate-based health care that now characterizes modern medicine in the United States is not how health care was delivered in the past. In Colonial times, American physicians received their training in Europe. Their competitors in the healing arts included alchemists, herbalists, ministers, faith healers, and even barbers. Treatments were a combination of folk wisdom, superstition, tried-and-true regimens, and often, dangerous quackery. A simple scratch, once infected, could easily cost a limb or a life. Patients were often "bled" as a cure, which involved the removal of "bad" blood from the patient, sometimes by means of drilling holes in the skull. Needless to say, the cure was often worse than the disease.

By the start of the nineteenth century, advances in biology and chemistry ignited a century of explosive growth in medical knowledge. One fruit of the scientific revolution of the mid-1800s was the *germ theory,* the idea that many illnesses were caused by microscopic organisms, or germs. Now considered scientific fact, the claim that something called *germs* caused illness was then hotly debated. Shortly thereafter, germ theory established itself as a foundation of medicine. Doctors were able to show that isolating infected people and sealing infected wounds could stop the spread of illness by stopping the spread of germs. Relentless study and research transformed medicine into a science. Coincidentally, the social prestige of medicine greatly increased, contributing mightily to the status of physicians, who had formerly enjoyed more modest social standing.

The development of new technologies has transformed the delivery of health care. The "house call" by a physician is now a rarity, although it is being renewed in a few communities. Many New technologies, such as this heart bypass procedure assisted by robots, would have been unimaginable not many years ago.

The American Medical Association (AMA) was founded in 1847, after half a century of sweeping away rivals in the healing arts many of whom were women. The AMA successfully campaigned to outlaw or delegitimize alternative therapies and emerged as the most powerful organization in U.S. health care.

In the late 1800s the image of medicine as an upper-class profession took hold. A medical education was expensive, and medical schools drew on White, male, urban populations for their students. Those trained as physicians took their place in the upper social strata. Herbalists and faith healers came more frequently from the rural lower class and generally remained there. As the ranks of the medical profession swelled with wealthy Whites, African Americans and Hispanics became proportionately more affiliated with older folk practices and midwifery. This overall trend continued through the early part of the twentieth century, although today folk practices continue to have adherents among rural lower-class Whites and rural and urban lower-class Blacks, Hispanics, and Native Americans (Starr 1982).

Health, Diversity, and Social Inequality

Prominent problem areas in the U.S. health care system include the following.

- **Unequal distribution of health care by race–ethnicity, social class, or gender.** Health care is more readily available and more readily delivered to White or middle-class individuals in urban and suburban areas than to minorities. The lack of health care delivery to Native American populations is particularly serious. Likewise, men and women receive unequal treatment for certain types of medical conditions, with women more likely than men to receive truncated treatment.
- **Unequal distribution of health care by region.** Each year, many in the United States die because they live too far away from a doctor, hospital, or emergency room. Doctors and hospitals are concentrated in cities and suburbs; they are much less likely to be situated in isolated rural areas. Rural people in Appalachia and some parts of the South and Midwest may have to travel 100 miles or more to get to the doctor or emergency room.
- **Inadequate health education of inner-city and rural parents.** Many inner-city and rural parents do not understand the importance of immunizing their children against smallpox, tuberculosis, and other illnesses, and they are often suspicious of immunization programs. This hesitancy is reinforced by the depersonalized and inadequate health care ghetto residents often encounter when care is available at all.

The definitions of *sick* and *well* have varied greatly over time in the United States. For example, from the early 1900s until the mid-1940s, thinness was associated with poverty and hunger. If you were skinny, that meant you were in bad health. From the late 1950s through the present, a positive value has been placed on being thin. Female role models in our society, such as movie stars and fashion models, have firmly established that "thin is in." Millions of young women have tried to copy this look, and one result has been an increased incidence of anorexia nervosa (Weitz 2001; Williams and Collins 1995; Taylor et al. 2006).

Anorexia nervosa (*anorexia* for short) is an eating disorder characterized by compulsive dieting.

understanding diversity

The Americans with Disabilities Act

The Americans with Disabilities Act (ADA), passed by Congress in 1990, prohibits discrimination against disabled persons. In 1999, the U.S. Supreme Court restricted the definition of disability to exclude disabilities that can be corrected with devices such as eyeglasses or with medication. The decision was based on cases in which people had been denied employment because they did not meet a health standard required by the employer, even though, with correction, the standard was met. One case involved two nearsighted women denied jobs as airline pilots because, without glasses, their vision did not meet the airline company's standard of 20/40. With glasses, their vision was 20/20. Another case involved a man whose high blood pressure was above the federal standard for driving trucks. He was denied a job as a trucker—even though with medication, his pressure was regulated to an acceptable level. The Court's decision denied these people the right to claim discrimination.

The decision raises several interesting sociological questions about disabilities and civil rights. Writing for the majority, then-Justice Sandra Day O'Connor argued that the law requires people to be assessed based on each individual's condition, not as members of groups affected in a particular way. The disability rights movement would argue that disabled people are a minority group with certain civil rights. The issue being mediated by the courts involves competing definitions of what constitutes a disability. This point intrigues symbolic interactionists who would note the role of socially constructed meanings in legal negotiations. Furthermore, as one lawyer who drafted the original ADA law has pointed out, according to the original decision, someone may be disabled enough to be excluded from a job, but not disabled enough to claim discrimination. This raises interesting questions about the rights of employers to establish physical and medical standards for certain jobs, even if those standards result in the exclusion from employment for some groups of people.

The disability rights movement has opened up new opportunities to those who face the challenge of disability.

Additional Resources: Smith, Bonnie G. and Beth Hutchison, eds. 2004. *Gendering Disability.* Piscataway, NJ: Rutgers University Press.

Victims of this illness starve themselves, sometimes to death, even though sufferers do not typically define themselves as ill; they tend to see themselves as overweight, even though they are dangerously thin. A related malady, *bulimia,* is an eating disorder characterized by alternate binge eating and then purging or inducing vomiting to lose weight.

Like many other diseases, anorexia has social as well as biological causes. A majority of people suffering from the disease are young White women, from well-to-do families, most often two-parent families. Many behavioral scientists have noted that generally anorexics have been pressured excessively by their parents to be high achievers. Others have detected a link between anorexia and the socially constructed ideals of beauty in our society, which are fixated on thinness and so-called ideal body types. Images of physical "perfection" are emblazoned across television, magazines, and billboards, with slenderness displayed as the ideal of femininity. Researchers note that these social values, which encourage compulsive dieting, are comparable to the foot-binding once practiced in China and other forms of female mutilation found in some foreign cultures (Wolf 1991; Chernin 1991).

Anorexia is least likely to afflict African American women, Latinos, and lesbians. According to researchers (Logio 1998; Thompson 1994), many in these groups overeat, rather than self-starve, thus producing a more familiar disorder: *obesity.* Thompson interprets this behavior as a reaction to oppressive life experiences associated with racism, sexism, and homophobia.

Men have not been exempt from the pressure of such values (Logio 1998). Since the 1940s, and especially from the mid-1970s, one of the most persistent male physical ideals has been the rippling physique of the bodybuilder or weight lifter. Young men have been urged by the media and their peers to "pump iron" for the perfect body. Many athletes, professional and amateur, have been goaded by athletic ambitions to use *anabolic steroids,* powerful hormones that stimulate the growth of muscle. Used widely (despite dire warnings by physicians), steroids not only build muscle as advertised, they can also shrivel the testicles and cause impotence, hair loss, heart arrhythmia, liver damage, strokes, and very possibly some forms of cancer.

Obesity has recently become defined as a public health problem. Obesity has traditionally been considered a matter of individual habit, but in 2004 officials in the Medicare program changed their policy to include obesity as a disease (and thus various weight loss programs are now eligible for Medicare funding). As obesity has come to be defined as a medical problem, analysts tend to emphasize the physical causes of being overweight. But a sociological perspective on obesity would also look to cultural and social structural factors as possible sources of obesity (Czerniawski 2007).

Debunking Society's Myths

Myth: The health care system works with the best interests of clients in mind.

Sociological perspective: The health care system is structured along the same lines as other social institutions, thus reflecting similar patterns of inequality in society.

Race and Health Care

Health and sickness in the United States is influenced by social factors. **Epidemiology** is the study of all the factors—biological, social, economic, and cultural—associated with disease in society. **Social epidemiology** is the study of the effects of social, cultural, temporal, and regional factors in disease and health. Among the more important social factors that affect disease and health in the United States are race, ethnicity, social class, gender, and age.

Among women, African Americans are more likely than Whites to fall victim to diseases such as cancer, heart disease, stroke, and diabetes. Death of the mother during childbirth is three times higher among African American women than among White women, and African American women are three times more likely to die while pregnant (Williams and Collins 1995; Livingston 1994). In the 45–64 age group, African American women die at twice the rate of White women. The occurrence of breast cancer is lower among African American women than White women, yet the *mortality rate* (death rate) for breast cancer for African American women is considerably greater than for White women (Weitz 2001; National Cancer Institute 2002). This reflects the fact that White women are more likely to get high-quality care, and get it more rapidly, than African American women. African American women also develop cervical cancer at twice the rate of White women. Hypertension (high blood pressure) afflicts 25 percent of all African American women and only 11 percent of White women. Although differences in culture, diet, and lifestyle account for some of these differences, it is nevertheless clear that African American women and men do not receive medical attention as early as Whites, and when they do eventually get treatment, the stage of their illness is more advanced and the treatment they receive is not of the same quality (National Center for Health Statistics 2006; National Cancer Institute 2002).

Hispanics, like African Americans, Native Americans, and other minorities, are significantly less healthy than Whites (Office of Minority Health 2004). Hispanics contract tuberculosis at a rate seven times that of Whites. Other indicators of health, such as infant mortality, reveal a picture for Hispanics similar to that of African Americans and Native Americans. Hispanics are less likely than Whites to have a regular source of medical care, and when they do, it is likely to be a public health facility or an outpatient clinic. Because of language barriers as well as other cultural differences, Hispanics are less likely than other minority groups to use available health services, such as hospitals, doctors' offices, and clinics (National Center for Health Statistics 2006; Weitz 2001).

One of the challenges for the health care system in a society with increasing diversity is responding to the different cultural orientations of various groups in society. Immigrants, for example, who may have limited English language skills and may come from cultures with very different health care practices, may be especially confused by the practices within the U.S. health care system (Suro 2000). Developing the ability for greater cross-cultural administration of health care will likely continue to be a challenge in the future. Also, because the majority of physicians are White, patients of a different racial or ethnic background will likely feel some social distance between themselves and the health care provider, contributing to a reluctance to seek care (Malat 2001). At the same time, researchers find that patients are much more satisfied with their care when their physician is of the same race as they are (Laveist and Nuru-Jeter 2002). Such findings indicate the continuing significance that race and color have in shaping people's health.

Social Class and Health Care

In the United States, social class has a pronounced effect on health and the availability of health services. The lower the social class status of the person or family, the less access they have to adequate health care (Jacobs and Morone 2004). Consequently, the lower one's social class, the less long one will live. People with higher incomes who are asked to rate their own health tend to rate themselves higher than people with lower incomes. Almost 50 percent of those in the highest income bracket rate their health as excellent, whereas only 25 percent in the lowest income bracket do so. In fact, the effects of social class are nowhere more evident than in the distribution of health and disease, showing up dramatically in the rates of infant mortality, stillbirths, tuberculosis, heart disease, cancer, arthritis, diabetes, and a variety of other illnesses. The reasons lie partly in personal habits that are themselves partly dependent on one's social class. For example, those with lower socioeconomic status smoke more often, and smoking is the major cause of lung cancer and a significant contributor to cardiovascular disease.

© Tracy Frankel/Getty Images;Photographer's Choice

As many people have become aware of the importance of good health, a major industry has developed to provide assorted health services, such as this yoga class.

Social circumstances also have an effect on health. Poor living conditions, elevated levels of pollution in low-income neighborhoods, and lack of access to health care facilities all contribute to the high rate of disease among lower classes. Another contributing factor is the stress caused by financial troubles. Research has consistently shown correlations between psychological stress and physical illness (Taylor et al. 2006; Worchel et al. 2000; Jackson 1992; Thoits 1991; House 1980). The poor are more subject to psychological stress than the middle and upper classes, and it shows up in their comparatively high level of illness.

With the exception of the elderly, now subsidized by Medicare, low-income people remain largely outside the mainstream of private health care. Nearly 41 million Americans—14.5 percent of the population—have no health insurance at all (see Map 13.1). The main sources for medical care for many of them are hospital emergency rooms in the inner cities, often called the "doctor's office of the poor." Even then, emergency room service is lacking. Treatment is given only for specific critical ailments, and rarely is there any follow-up care or comprehensive treatment.

Sociologists have found that during interactions between health care providers and poor patients, the poor, particularly Black and Hispanic poor, are most likely to be *infantilized* and to receive health counseling that is incorrect, incomplete, or delivered in inappropriate language not easily understood by the patient (Weitz 2001; Gonzalez 1996). The symbolic interaction perspective in sociology has noted the tendency for inadequate care. It has also been attributed to an attitude among health counselors that the poor are charity cases who should be satisfied with whatever they get, because they are probably not paying for their own care (Diaz-Duque 1989).

Gender and Health Care

Although women live longer on average than men, older women are more likely than older men to suffer from stress, obesity, hypertension, and chronic illness. National health statistics show that hypertension is more common among men than women until age 55, when the pattern reverses. This may reflect differences in the social environment experienced by men and women, with women finding their situation to be more stressful as they advance toward old age. Under the age of 35, men are more likely to be overweight than women; after that, women are more likely to be overweight. Women have a higher likelihood of contracting chronic disease than men, although men are more likely to be disabled by disease (National Center for Health Statistics 2006; Atchley 2000).

The rate of death caused by infectious diseases is higher for men than women and has declined significantly for both since the early 1900s. Researchers cite differences in male and female roles and cultural practices to explain the differences between the genders. In particular, male occupational roles

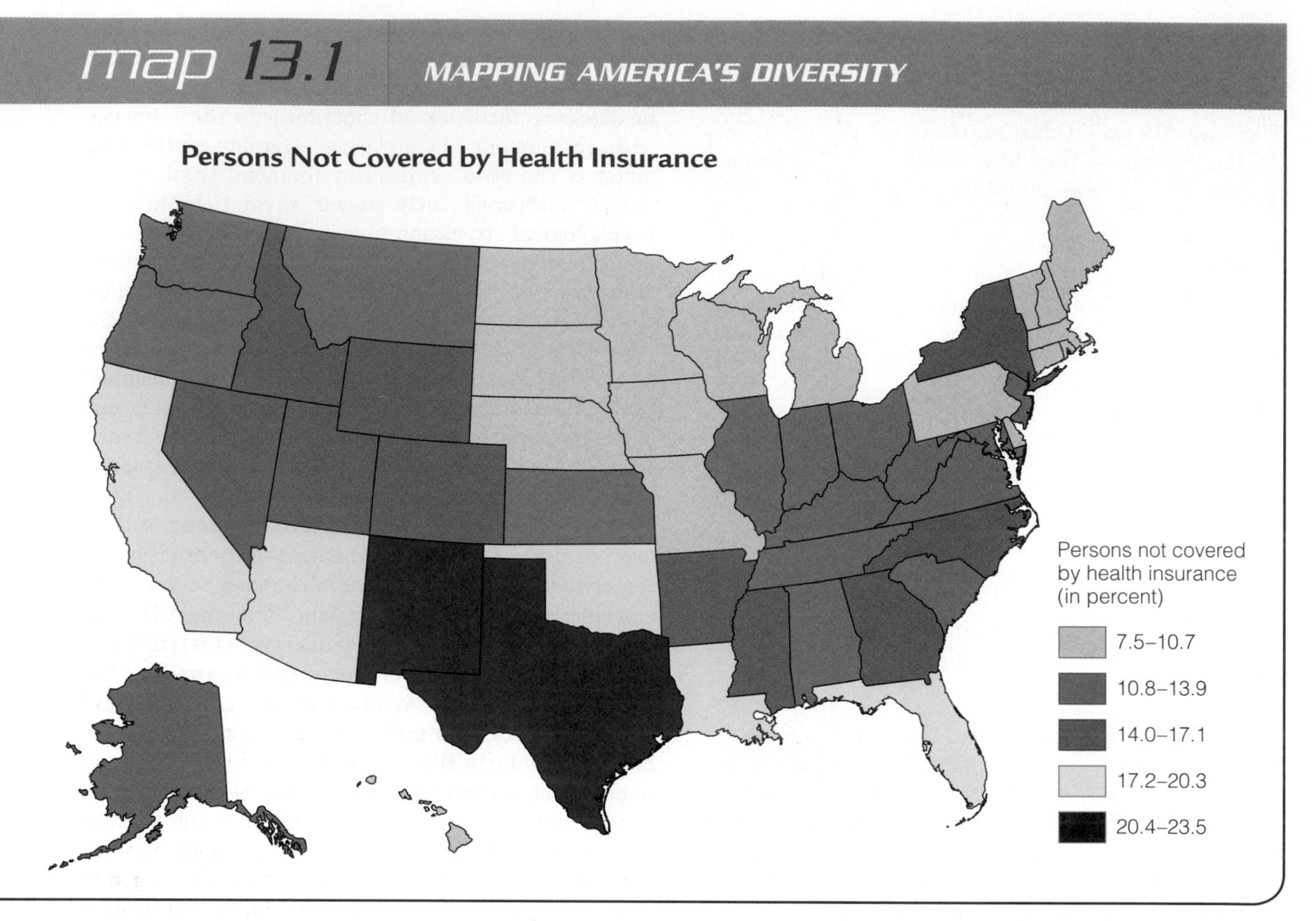

This map shows the percentage of the population in each state that was not covered by health insurance in 2005. Texas has the highest percentage of people not covered (23.9%), followed by New Mexico (20.1%), then California (19%). Iowa has the lowest percentage of those not covered by health insurance (2.5%), followed by Rhode Island (10.9%). Were you to design a study to explain state-by-state variation in coverage, what explanatory factors would you want to examine?

Source: From the U.S. Census Bureau, 2007. *Statistical Abstract of the United States, 2006.* Washington, DC: U.S. Department of Commerce.

call for more travel and more exposure to other people, the major sources of infection. In addition, men smoke more, and the overall deleterious effect of smoking on health can account for some of the differences in infectious diseases between men and women (National Cancer Institute 2002; Williams and Collins 1995).

It is typically assumed that the work-oriented, hard-driving lifestyle associated with the traditional role of men tends to produce elevated levels of heart disease and other health problems. In general, this is true. However, the role of women in society is changing, with associated changes in women's health. The health of women and men varies with their social circumstances. People who are "tokens" in the workplace—meaning mainly women and Blacks, especially Black women—suffer more stress in the form of depression and anxiety than nontokens, or women and Blacks who work in places where there is nothing exceptional about their presence (Jackson et al. 1995). Research has also found that perceiving that there is discrimination against you is significantly related to both physical and mental health (Pavalko et al. 2003). Those who experience discrimination by both gender and race are therefore the most vulnerable.

Housewives have higher rates of illness than women who work outside the home, indicating that employment can have positive effects on women's health. Moreover, housewives who never worked outside the home are healthier than those who once worked outside the home but then returned to homemaking. Employed women are more likely than housewives to get well quickly when sick, and they tend to return to their normal activities sooner after illness than do housewives (Andersen 2006).

AIDS and Sexually Transmitted Diseases (STDs)

Approximately fifty sexually transmitted diseases have been medically diagnosed. The four major STDs are syphilis, gonorrhea, genital herpes, and AIDS (acquired immune deficiency syndrome). Less frequent are the more esoteric diseases such as lymphogranuloma venereum (LGV), which untreated devastates the body with open sores. The incidence of all STDs increased during the sexual revolution of the sixties and seventies, when sex was thought of as simply a pleasant way to express affection. Little fear was harbored about contracting a venereal disease because most, such a syphilis or gonorrhea, were known to be medically curable, and the remaining diseases were thought of a too rare to sorry about. With the subsequent dramatic rise in STDs, especially AIDS, the late 1980s and early 1990s witnessed a new revolution in contrast to the sexual revolution. The new counterrevolution caused people to reevaluate the nature of sex, the possibly fatal risks involved in unprotected sexual activity, and sexual behavior in general (Laumann et al. 1994; Alan Guttmacher Institute 1994).

Syphilis and gonorrhea have been around for a long time. Both are caused by microorganisms transmitted through sexual contact involving the mucous membranes of the body. It is virtually impossible to get either syphilis or gonorrhea any other way. If untreated, syphilis causes damage to major body organs, blindness, mental deterioration, and death. Each result was seen all too dramatically in the infamous *Tuskegee Experiment,* beginning in 1932 and ending in 1972, which was done on a group of Black men who were never told they were infected with syphilis and who were never treated. Untreated gonorrhea can cause sterility in both women and men. Both diseases are quickly curable with penicillin or other appropriate antibiotic medication—which incidentally was not made available to those Black men suffering from the Tuskegee study, even after it had been invented and though it had become available to the general public (from the early 1950s on).

Genital herpes (Herpes Simplex II) is more widespread than either syphilis or gonorrhea and affects roughly 30 million people in the United States alone. That represents one person in seven. Although genital herpes can remain dormant for the life of the infected individual, it is nonetheless to date incurable. Genital herpes began to receive attention in the early 1980s, when concern about the risks of STDs generally was on the rise. A person with genital herpes may have no symptoms or may experience blisters in the genital area and a fever as well. Genital herpes is not fatal to adults, but it can be fatal to infants born through vaginal delivery, but not via cesarean section.

People tend to regard sexually transmitted diseases as not merely diseases but as punishment for being immoral. As a result, people who contract an STD become negatively stigmatized, sometimes resulting in their not seeking treatment. This shows the power that social influence can have on the treatment of disease.

A **stigma** occurs when an individual is socially devalued because of some malady, illness, misfortune, or similar attribute. A stigma is viewed as a relatively permanent characteristic of the stigmatized individual; the negative attribute of the stigmatized person is expected to persist, with no cure in sight (Goffman 1963b; Jones et al. 1986; Epstein 1996). When the disease AIDS first appeared in the early 1980s, it was mostly associated with gay men and was heavily stigmatized. The result was that the federal government (during the Reagan administration) largely ignored this spreading problem and devoted little in research funds to identify its causes. The stigma associated with AIDS with gay men, and the resulting delay by the medical and scientific community in researching treatment likely cost many lives (Shilts 1988).

AIDS is the term for a category of disorders that result from a breakdown of the body's immune system. HIV (human immuno deficiency virus), the virus that causes AIDS, was first identified in 1981. The incubation period between infection with HIV and the development of AIDS can stretch longer than ten years. Thus, one can be infected with HIV and yet not have full-blown AIDS for up to ten years. It is not the actual HIV infection that causes death; rather, it is caused by a complex of severe illnesses that thrive in the absence of a working immune system, such as pneumonia, certain cancers, and a number of other illnesses rare enough that their presence is judged to be diagnostic of AIDS. Since the 1980s, the disease has spread rapidly, with over 830,000 cases reported in the United States since 1981.

Over 34 million adults and children worldwide are infected with HIV, 14.8 million of whom are

In recent years, the public has mobilized in the fight against HIV/AIDS.

women. The global AIDS epidemic among women is overwhelmingly the result of heterosexual contact, almost entirely so in Africa and South and Southeast Asia. Analysts have argued that the high rate of transmission to women worldwide results from women's financial dependence on men. Because of such dependence, women may have little control over when and with whom they have sex. Many women have to exchange sex for financial support, and in highly patriarchal cultures, women are not expected, nor allowed, to make decisions about sex. If they refuse sex or request condom use, they risk abuse and violence or may be suspected of infidelity, which can also put them at great risk, sometimes the risk of death. These facts mean that treating the worldwide AIDS epidemic requires that an analysis of gender relations in different cultures might be a part of the solution to this health epidemic. World leaders have also severely criticized the United States for limiting its international funding of AIDS-prevention programs to only those that promote abstinence (Associated Press 2004).

The AIDS disease is transmitted through the exchange of bodily fluids, particularly blood and semen. A little more than one-third of all new cases of AIDS are the result of male-to-male sexual contact. Although AIDS initially affect primarily White, gay men, now one-third of new AIDS cases are women, the largest share of whom acquire AIDS through heterosexual contact. The second most frequent cause of transmission for women and men is intravenous drug use (U.S. Census Bureau 2004). In the United States, AIDS has hit inner-city minority communities disproportionately hard. The contextual problems of poverty, poor health, inadequate health care, drug and alcohol abuse, and violence must be taken into account in explaining the high rate of AIDS in these communities (see Figure 13.4). Because drug use and other causes vary by race, social class, and gender, this points to different interventions for different communities.

AIDS can be understood in medical terms, and new treatment protocols promise to bring more effective treatment for those who contract the disease. But to reduce the incidence of AIDS requires a sociological perspective that takes into account the *social networks* and *social norms* that contribute to the transmission of this disease. Social norms affecting the age of first intercourse, number of sexual partners, drug use, and homosexual and bisexual sexual practices influence the spread of AIDS.

Theoretical Perspectives on Health Care

A deeper understanding of the nation's health care system and its problems can be achieved by applying the three major theoretical paradigms of sociology: functionalism, conflict theory, and symbolic interaction theory (see Table 13.5).

The Functionalist View of Health Care

Functionalism argues that any institution, group, or organization can be interpreted by looking at its positive and negative functions in society. Positive functions contribute to the harmony and stability

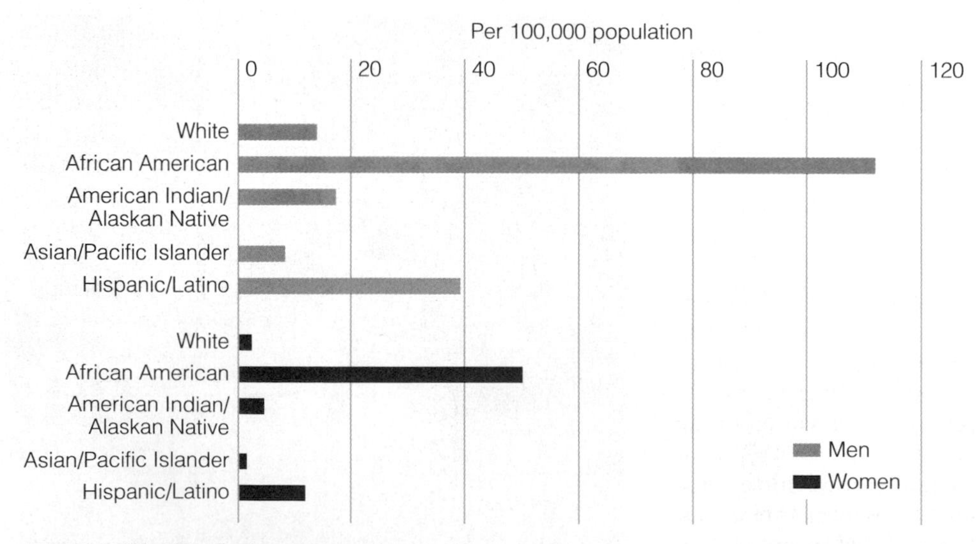

FIGURE 13.4 *AIDS Cases by Race and Gender*

Data: National Center for Health Statistics. 2004. *Health United States 2004*. Washington, DC: U.S. Department of Health and Human Services.

table 13.5 **Theoretical Perspectives on the Sociology of Health**

	Functionalism	Conflict Theory	Symbolic Interaction
Central point	The health care system has certain functions, both positive and negative.	Health care reflects the inequalities in society.	Illness is partly socially constructed.
Fundamental problem uncovered	The health care system produces some negative functions.	Excessive bureaucratization of the health care system and privatization lead to excess cost.	Patients are patronized and infantilized.
Policy implications	Policy should decrease negative functions of health care system for minority groups, the poor, and women.	Policy should improve access to health care for minority racial-ethnic groups, the poor, and women.	Doctors, nurses, and other medical personnel should periodically take the sick role of the patient, as an instructional device.

of society. The positive functions of the health care system are the prevention and treatment of disease. Ideally, this would mean the delivery of health care to the entire population without regard to race, ethnicity, social class, gender, age, or any other characteristic. At the same time, the health care system is notable for a number of negative functions, those that contribute to disharmony and instability of society (some to be reviewed in the section titled "The Health Care Crisis in the United States").

Functionalism also emphasizes the systematic way that various social institutions are related to each other, together forming the relatively stable character of society. You can see this with regard to how the health care system is entangled with government through such things as federal regulation of new drugs and procedures. The government is also deeply involved in health care through scientific institutions such as the National Institutes of Health, a huge government agency that funds new research on various matters of health and health care policy. As a social institution, health care is also one of the nation's largest employers and, thus, is integrally tied to systems of work and the economy.

The Conflict Theory View of Health Care

Conflict theory stresses the importance of social structural inequality in society. From the conflict perspective, the inequality inherent in our society is responsible for the unequal access to medical care. Minorities, the lower classes, and the elderly, particularly elderly women, have less access to the health care system in the United States than Whites, the middle and upper classes, and the middle-aged. (To the contrary, functionalists argue that relatively greater access of the middle and upper classes to medical care is good for society because the upper classes are more beneficial—"functional"—to society.) Restricted access is further exacerbated by the high costs of medical care, stemming from high fees and the abuses of the fee-for-service and third-party-payment systems (explained later in the chapter). The exceptionally high incomes of medical professionals amplify the social chasm between medical practitioners and an increasingly resentful public.

Excessive bureaucratization is another affliction of the health care system that adds to the alienation of patients. The U.S. health care system is burdened by endless forms for both physician and patient, including paperwork to enter individuals into the system, authorize procedures, dispense medicines, monitor progress, and process payments. Long waits for medical attention are normal, even in the emergency room. Prolonged waits have reached alarming proportions in the emergency rooms of many urban hospitals in the United States and can only deepen the alienation of patients. One study showed that up to 15 percent of emergency room patients give up and leave before receiving care because of long waits—some stretching to an abysmal fifteen hours.

Symbolic Interactionism and Health Care

Symbolic interactionists hold that illness is partly (although obviously not totally) socially constructed. The definitions of illness and wellness are culturally relative—sickness in one culture may be wellness in another. It is time dependent as well. A condition considered optimal in one era (such as being thin) may be defined as sickness at another time in the same culture (at the turn of the twentieth century in this country, a healthy woman was supposed to be plump). Similarly, the health care system itself has a socially constructed aspect. The ways we behave toward the ill, toward doctors, and toward innovative ventures such as HMOs are all social creations.

The symbolic interaction perspective highlights a number of socially constructed problems in the health care system. Medical practitioners frequently subject patients to *infantilization* (treat them like children, even if adult). The patient is assigned a role that depends heavily on the physician and the health care system, much as an infant is dependent on its parents. Doctors and nurses may begin patronizing the patient from the initial greeting, a condescending "How are *we* today?" Physicians commonly address patients by their first names, yet patients virtually always address physicians as "Doctor." Women patients are far more likely than men to be addressed by their first names. Such patronizing and infantilizing of patients is common in emergency rooms, where minority patients are infantilized most often (Weitz 2001; Gonzalez 1996).

The symbolic interactionist analysis of the health care system allows us to see these problems more clearly. One solution is to give health care professionals training about such matters in medical school. This is just starting to happen in a few U.S. medical schools. For example, a social issue addressed in some medical school courses on gynecology is how to manage the interaction when a male gynecologist treats his woman patient. Women patients feel uncomfortable and vulnerable when they lie partially naked on an examination table with their heels in elevated stirrups and their legs apart. Furthermore, strongly fixed social norms say that when a man touches a woman's genitals, it is an act of intimacy. Yet the gynecological examination is supposed to be completely impersonal. Male gynecologists are notorious for their failure to appreciate the discomfort of their female patients (Scully 1994; Emerson 1970).

The Health Care Crisis in the United States

Currently, the cost of medical care in the United States is approximately 14 percent of our gross domestic product, making health care the nation's

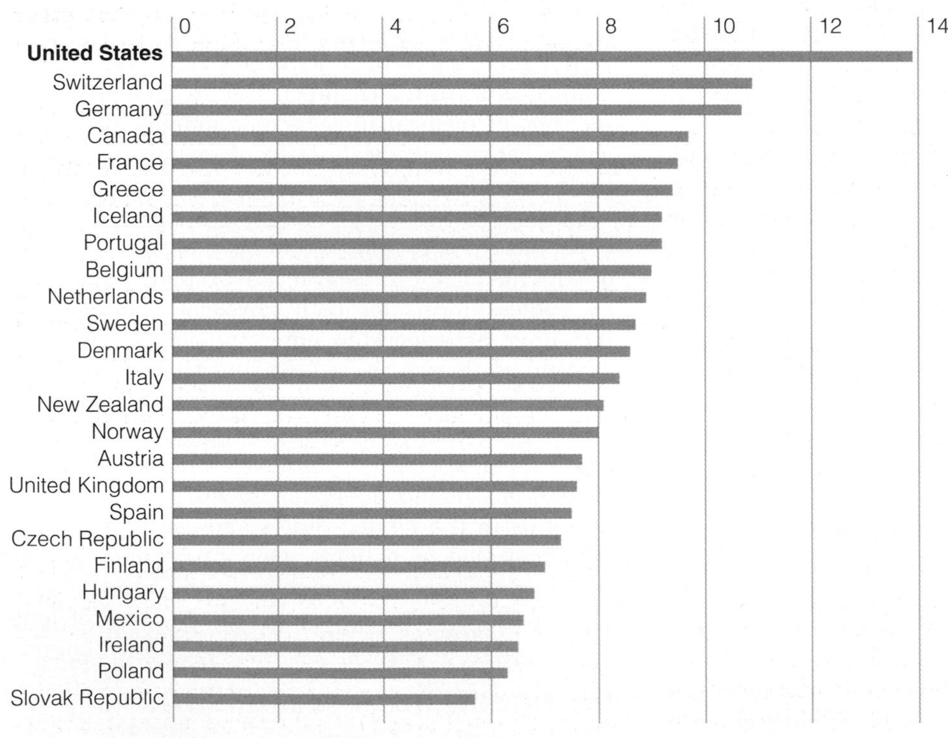

FIGURE 13.5 *Health Care Expenditures as a Percentage of Gross National Product (GNP)*

Source: U.S. Census Bureau. 2007. *Statistical Abstract of the United States, 2006*. Washington, DC: U.S. Printing Office.

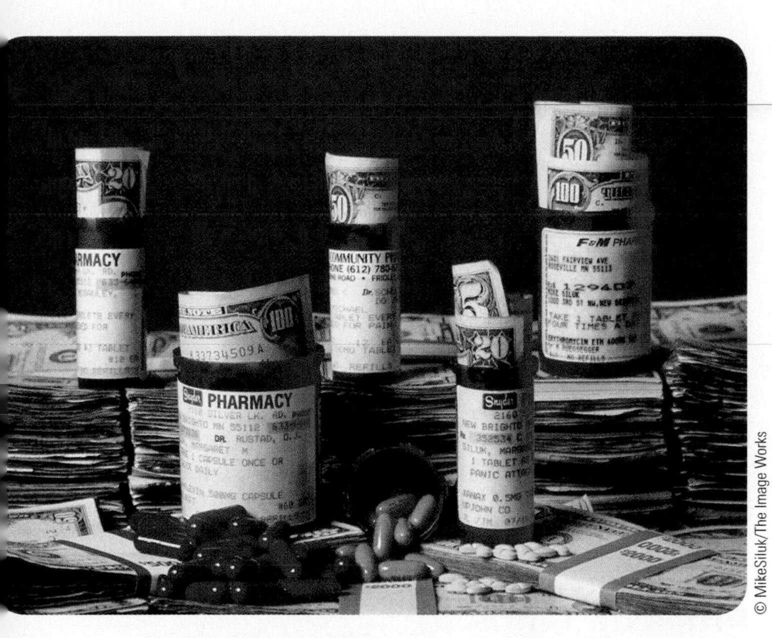

The high cost of prescription drugs is indicative of the problems generated in a profit-based health care system.

third leading industry. The cost of health care has at times risen at twice the annual rate of inflation. The United States tops the list of all countries in per person expenditures for health care (see Figure 13.5). Despite all that spending, the U.S. health care system is not the best in the world. Other countries spend considerably less money and deliver a level of health care at least as good. For example, Sweden and France spend roughly one-half as much per capita as the United States, and Great Britain spends a bit more than one-third as much, yet these countries have national health insurance programs that cover virtually their entire populations and deliver a level of care not much different overall from that achieved in the United States.

The battlefronts of the health care crisis extend to the cost of health care, the epidemic of malpractice suits, the development of HMOs, and the debate over universal health care (Starr 1982, 1995; Scarpitti et al. 1997). Here we examine each.

The Cost of Health Care

Health care must be paid for, and in the United States the structures in place for paying the doctor's bill are in a state of chaos. The central element in the system is the fee-for-service principle. Under this arrangement, the patient is responsible for paying the fees charged by the physician or hospital. Fortunate patients will be adequately insured and able to pass on their health expenses to the insurance company. Many people are not insured, however, especially people with lower incomes. This group must reckon with large bills by drawing on their own limited resources. Overall, 31 million people in the United States are without health insurance. Most severely hit are the southern and southwestern states, where people without health insurance reach more than 15 percent of the population (see Map 13.1). A third source of payment is the government, which is increasingly the ultimate payer of health care expenses.

The greatest contributors to skyrocketing health care costs are the soaring costs of hospital care and the rise in fees for the services of physicians. The services of specialists in particular are hugely expensive. Another culprit is the third-party payment system. With hospitals interested in making a profit and patients simply passing bills on to insurance providers, there is little pressure to keep down the cost of care. Patients may be treated on an in-patient basis when a hospital stay is not actually necessary, unnecessary treatments may be given, and some diagnostic procedures may be ordered simply because expensive equipment is available that needs to be used to pay for itself.

The U.S. government has for some time sought to have some form of widespread guaranteed health service, at least for certain categories of people, such as veterans, the poor, and the elderly. The **Medicare** program, begun in 1965 under the administration of President Lyndon Johnson, provides medical insurance covering hospital costs for all individuals age 65 or older. The Medicare program does not cover physician costs incurred outside the hospital, but other programs do, although the patient must pay a portion of the cost.

Medicaid is a governmental program that provides medical care in the form of health insurance for the poor, welfare recipients, and the disabled. The program is funded through tax revenues. The costs covered per individual vary from state to state because the state must provide funds to the individual in addition to the funds that are provided by the federal government. The Medicare and Medicaid programs together are as close as the United States has come to the ideal of universal health insurance. More recent attempts to establish universal coverage have not been successful (Weitz 2001; Starr 1995).

Medical Malpractice

A rising problem is malpractice insurance, a problem forcing many physicians to abandon some services. Annual insurance premiums (costs) for physicians have risen tenfold in the past decade and range now from about $5000 per year for physicians in family practice to well over $150,000 per year for physicians in specialties such as radiology, anesthesiology, and surgery. The cost of these insurance premiums is simply passed along to consumers (patients) and has contributed to the rise in the overall cost of health care.

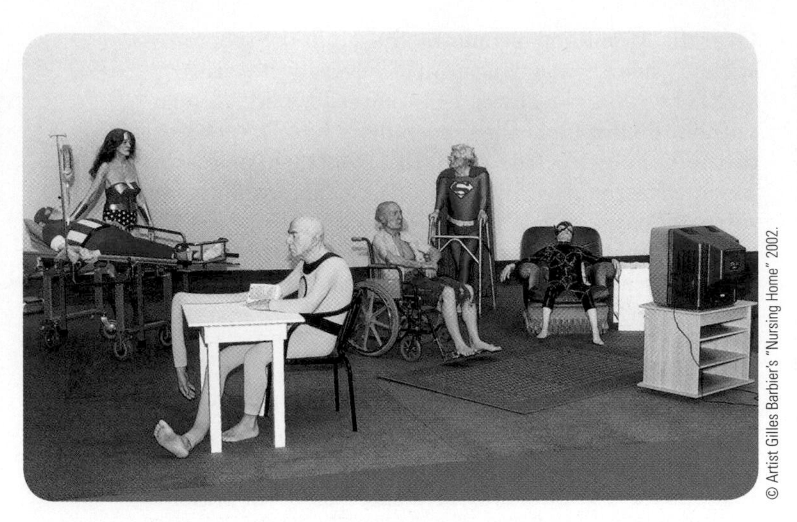

Elder care for Wonder Woman, Superman, Captain America—can you name the rest?

The U.S. public has traditionally accorded physicians a high social status and high incomes. The recent increase in malpractice suits suggests that the public is beginning to question the privileged status of doctors (Pescosolido et al. 2001). Several specific reasons have been put forward for client revolts against the medical profession. Attorneys claim that the primary cause of the malpractice dilemma is a declining standard of medical care and a rising incidence of medical negligence. Another argument is that, as a result of specialization, doctors now fail to establish old-fashioned rapport with patients, and thus patients are less attached to their physicians and more likely to turn hostile. A third notion is that the high cost of medical care and resentment about the outsized incomes of physicians has generated animosity in patients that is quick to show itself.

Contributing to the public's increasing mistrust of the medical profession is the incidence of *medical errors* and the medical profession's response to them. A doctor's actions often mean the difference between life and death for patients. On occasion, a doctor's own actions can become the direct cause of death or further disability of the patient. Removal of the wrong lung or amputation of the wrong leg are very rare accidents, but such things do indeed happen. When a suspected medical error occurs, a staff meeting at the involved hospital is called, but a number of researchers have noted that such meetings seem designed more to downplay or cover up medical errors than to account for them or prevent further errors (Weitz 2001; Bosk 1979).

Adding to the crisis is the eagerness with which lawyers take on malpractice suits. About one doctor in five is sued each year; awards from medical malpractice suits can be in the millions of dollars. This has dramatically increased the malpractice insurance costs for doctors. One unfortunate result of all this is simply that physicians are increasingly leaving the medical field and going into other lines of work (Eisenberg and Siegler 2003).

To protect themselves against potential malpractice suits, doctors increasingly practice **defensive medicine,** which entails ordering excessively thorough tests, X-rays, and so on at the least indication that something might be amiss. This is done partly to ensure that nothing is missed and partly to document that the highest possible level of care was given. Although this extra attentiveness may contribute marginally to favorable medical outcomes, it contributes mightily to the overall cost of health care.

A Response to the Problem: HMOs

Health maintenance organizations (HMOs) are an innovation in health care first begun shortly after the Second World War but becoming widespread only within the last twenty years. The staffs in HMOs can include anywhere from several doctors to several hundred doctors. People who join HMOs are assigned to a physician who administers care and when necessary gives referrals to specialists affiliated with the HMO. The doctors in an HMO earn salaries rather than fees. All services are ultimately paid for by the membership fee paid by subscribers to the HMO. Eliminating both the fee-for-service system and the third-party insurer drives costs down in several ways. Physicians have an incentive to give the most economical levels of care, while presumably retaining their motive to maintain a high standard of care because they want to secure the reputation of the HMO and avoid malpractice suits. The opportunity for fraud, which costs third-party insurers so dearly, is eliminated. Finally, the corporate structure of HMOs presumably offers the economies of scale and economies of organization enjoyed by other profit-oriented corporations, yet often found lacking in the U.S. hospital system.

The American Medical Association (AMA) has a history of opposing HMOs on the grounds that they decrease the rights of physicians to determine treatments and limit the rights of patients to choose their own doctors or seek out treatments from specialists. The AMA has also argued in a number of forums that HMOs are inclined to pay too much attention to the bottom line and not enough to patient welfare (Scheid 2003). The problems with HMOs—and health care, more generally—has created a call for major reform in the health care system, but to date there is no new national plan, even though public concern is high.

SEE FOR YOURSELF

YOUTH AND HEALTH INSURANCE

Young people are known to be the group least likely to be covered by health insurance. Identify a group of young people you know and ask them if they are covered by health insurance. If they are insured, where does their insurance come from? Who pays? If they are not insured, ask them why not and whether they think this is important. Would they support a national health insurance program?

Having conducted your interviews, write a short paper reporting on the information you have learned. Include such social factors as the age, race, ethnicity, gender, and educational/occupational status of those you interviewed. Then discuss whether you think any or all of these social characteristics are related to the likelihood that these people are covered by health insurance and whether these characteristics are related to their attitudes about coverage. What are the implications of your results for public support for new health care policies?

The Universal Health Care Debate

Periodically, in the United States plans for the reorganization of the health care system are suggested in order to solve problems of high costs and other problems. Universal health care for all Americans is suggested as part of such intended programs.

The core of such programs is called **managed care.** Under this plan, all individuals in the United States would belong to a complex of managed care organizations, rather like HMOs, that would use their collective bargaining force to drive down the cost of health insurance, while accepting the responsibility for operating their own facilities in an economical manner and continuing to provide high-quality care. Everyone would join the managed care complex. However, individuals would still be free to retain their personal physicians if they so choose, as long as their physicians meet government-stipulated criteria. The plan was intended to achieve the advantages of socialized medicine systems as administered in Great Britain, Canada, and other places where everyone is entitled to see a doctor when they need one, while retaining elements of the profit motive in the system.

13 Chapter Summary

- ***What is the importance of the education institution?***

  Education is the social institution that is concerned with the formal transmission of society's knowledge. It is therefore part of the socialization process. More money is spent in the United States on education than on any other activity except health care.

- ***How does education affect, or not affect, the likelihood of social mobility?***

  The number of years of formal education for individuals has important, but in many ways modest, effects on their ultimate occupation and income. Social class origin affects the extent of educational attainment (the higher the social class origins, the more education is ultimately attained) as well as on occupation and income (higher social class origin likely means a more prestigious occupation and more income).

- ***How does the educational system perpetuate inequality, instead of reduce it?***

  Although the education system in the United States has traditionally been a major means for reducing racial, gender, and class inequalities among people, it is also true that the educational institution has perpetuated these inequalities. Test biases based on culture, language, race, gender, and class have only been slightly reduced in standardized tests. *Tracking* continues to disproportionately and negatively affect minorities, women, and the working classes. Students are subject to *teacher expectancy effect,* which affects both teacher expectations and student performance. The *self-fulfilling prophecy* and *stereotype threat effect* work to the detriment of minorities, working-class people, and women.

- ***What are the social factors in illness?***
  Race–ethnicity, social class, and gender are major factors in the state of health care in the United States: Whites live longer than Blacks and Hispanics, due partly to worse health care, and less access to health care, for minorities. Black women are more likely than White women to fall victim to cancer, heart disease, strokes, and diabetes. The mortality rate among Native Americans is one and one-half times that of the general population.

- ***How serious are the effects of these social factors?***
  The lower one's social class status, the greater are one's chances of tuberculosis, heart disease, cancer, and arthritis. The poor are more likely to be depersonalized and treated inadequately by the health care system than those of higher social class status. Men and women tend to be treated differently in the health care system because there is still a tendency for the male-dominated profession to regard the problems of women as non-mainstream and "special."

- ***How have AIDS and other sexually transmitted diseases developed?***
  A person who is known to be HIV positive, or who has full-blown AIDS, is partly blamed, negatively *stigmatized,* and treated as a social outcast. Although once thought to be primarily a disease among gay men, AIDS has increased among intravenous drug users and among women who are infected from contact with heterosexual men. Among new cases of AIDS worldwide, half are women.

- ***What is the health care crisis in the United States?***
  High costs and malpractice suits have resulted in a policy crisis today in the U.S. health care system. The growth of *health maintenance organizations (HMOs)* shows promise for easing the crisis, although programs for large HMOs and national universal health insurance (including programs for managed care) have had only limited success.

Key Terms

Online Resources

Sociology: The Essentials Companion Website

www.thomsonedu.com/sociology/andersen

Visit your book companion website where you will find more resources to help you study and write your research papers. Resources include Suggested Readings, web links, and a MicroCase Online feature that teaches you how to research society. Other resources include Learning Objectives, Internet exercises, quizzing, and flash cards.

is an easy-to-use online resource that helps you study in less time to get the grade you want NOW.

www.thomsonedu.com/login

Need help studying? This site is your one-stop study shop. Take a Pre-Test and Thomson NOW will generate a Personalized Study Plan based on your test results. The Study Plan will identify the topics you need to review and direct you to online resources to help you master those topics. You can then take a Post-Test to determine the concepts you have mastered and what you still need to work on.

14

Chapter fourteen

CHAPTER FOURTEEN

Chapter

Politics and the Economy

Defining the State

Power and Authority

Theories of Power

Government: Power and Politics in a Diverse Society

Economy and Society

The Changing Global Economy

Theoretical Perspectives on Work

Characteristics of the Labor Force

Power in the Workplace

Chapter Summary

[***Picture a couple*** lounging together on a public beach on a national holiday. They imagine themselves married someday, perhaps having children they will send to public school. Does the government have anything to say about the couple's children? They may decide to use birth control until they are ready for children; if so, their choice of birth control methods will be limited to those the government allows. What if they cannot have children? They might consider adopting—if state agencies judge them acceptable as parents.

As you imagined this couple, whom did you picture? Were they the same race? Not so long ago, the law would have forbidden the marriage if it were interracial. Is this a man and a woman, or is this a same-sex couple? If they are lesbian or gay, despite being in love and wanting to form a lifelong relationship, in most states they are prevented by law from marrying.

continued

© Andy Sacks/Getty Images/Stone

chapter fourteen CHAPTER FOURTEEN

This couple is probably giving no thought to how much their life is influenced by the state at that moment. Permission to walk on the public beach comes from the state; the day off to celebrate a holiday is sanctioned by the state; the drive to the beach took place in a car registered with the state and inspected by the state, driven by a motorist licensed by the state. The range of things regulated by the state is simply enormous, and yet many are never noticed.

Every day millions of people within this state get up, get dressed, and go to work. Another several million wish they could find work. Many who work wish they earned more or had jobs with better working conditions. Work has an enormous effect on most aspects of your life, including relationships with your family, how much power you have, and what resources are available.

When thinking about work, most people think in terms of what kind of work they might do, how much they can earn, whether their work will be satisfying, and whether their job will give them opportunities for promotion and advancement. Students, for example, may see their education as a way to get work opportunities or promotions, but they may also worry about what work they will find and whether their work will be meaningful, use their talents, and provide a lifetime of rewards. These concerns can be understood in a sociological framework even though people tend to think each is an individual problem. Sociologists study the social forces that shape people's experiences and understand how downsizing, globalization of the economy, corporate restructuring, and technological change affect work and the politics of the state.

Defining the State

The sociological definition of the **state** is the organized system of power and authority in society. The term *state* does not refer to the state in the United States where someone lives. It is an abstract concept that includes the institutions that represent official power in society, such as the government and its legal system (with the courts and the prison system), the police, and the military.

Theoretically, the state exists to regulate social order, ranging from individual behavior and interpersonal conflicts to international affairs, although it does not always do so fairly or equitably. The guarantee of life, liberty, and the pursuit of happiness promised in the Declaration of Independence, when examined carefully, is not evenly distributed by the state. Less powerful groups in the society may see the state more as an oppressive force than a protector of individual rights. However, they may still turn to the state to rectify injustice. For example, when African Americans sought to end segregation, they sought legal reform through the state, that is, through the courts.

The state has a central role in shaping class, race, and gender relations in society. It determines the rights and privileges of many groups—and defines who is a member in each group. The involvement of the state may include the resolution of management and labor conflicts (such as in airline strikes), congressional legislation determining the benefits of different groups (such as the Americans with Disabilities Act of 1990), or Supreme Court decisions interpreting the U.S. Constitution. The state also provides the basic institutional structure of society by determining through state laws what institutional forms will receive societal support. Laws regulating family relationships also emanate from the state. Thus, families defined as illegitimate by law (or other mechanisms through which the state exerts its will, such as tax policy) do not receive the same benefits and rights as families that the state defines as legitimate, even if someone in the family is quite ill.

Sociological analyses of the state focus on several different questions. One important issue is the relationship between the state and inequality in society. State policies can have very different impacts on different groups, as we will see later in this chapter. Another issue explored by sociological theory is the connection between the state and other social institutions—the state and religion, the state and the family, and so on.

The Institutions of the State

Numerous institutions make up the state, including the government, the legal system, the police, and the military. The *government* creates laws and procedures that regulate and guide a society. The *military* is the branch of government responsible for defending the nation against domestic and foreign conflicts. The *court system* is designed to punish wrongdoers and adjudicate disputes. Court decisions also determine the guiding principles or laws of human interaction. Law is a fundamental type of formal social control that outlines what is permissible and what is forbidden. The *police* are responsible for enforcing law in the community and for maintaining public order. The *prison system* is the institution responsible for punishing those who have broken the law, although as we have already seen in some detail (particularly in Chapter 6), prisons often punish differently (even the innocent) depending on a person's race, gender, social class, and other factors.

The State and Social Order

Throughout this book we have seen that a variety of social processes contribute to order in society, including the learning of cultural norms (that is, socialization), peer pressure, and the social control of deviance. Each plays a part in producing social order, but none so explicitly and unambiguously as the official system of power and authority in society. In making laws, the state clearly decrees which actions are legitimate and which are not. Punishments for illegitimate actions are enforced, and systems for administering punishment are maintained. The state also influences public opinion through its power to regulate the media. In some cases the state circulates **propaganda,** which is disseminated with the intention to justify the state's power. Any group or organization may circulate propaganda. Censorship is another means by which the state can direct public opinion. The movement to censor sexually explicit materials on the Internet is an example of state-based censorship.

The state's role in maintaining public order is also apparent in how it manages dissent. Protest movements perceived by those in power to challenge state authority or threaten the disruption of society may be repressed through state action. Options available to the state range from surveillance, through imprisonment, and all the way to military force. Throughout the late 1950s and 1960s, for example, the FBI conducted illegal surveillance operations targeting Martin Luther King Jr. and other civil rights leaders (Garrow 1981). Also, witness the use of increased surveillance that has occurred in the aftermath of 9/11 through security screenings at airports, increased power to intercept e-mail and voice mail via the Patriot Act of 2001, and even more cameras at traffic intersections.

Global Interdependence and the State

Internationally, there are increasingly strong ties between the state and the global economy. The interdependence of national economies means that political systems are also elaborately entangled—a phenomenon that can be observed daily in the newspaper.

The process of globalization profoundly affects the character of states and their relationships. One example is the World Trade Organization (WTO), created in 1994 to monitor and resolve trade disputes. Member nations may challenge the decisions made by local and national governments, with the WTO Council in Geneva resolving disagreements. Another is the European Union (EU), an alliance of separate nations established to promote a common economic market and develop political unions with Western Europe. The creation of these organizations has produced a system that transcends individual nation-states by formalizing and strengthening the rules that govern many aspects of world trade. This represents the trend, in an increasingly international economy, toward the creation of a single state.

Power and Authority

The concepts of power and authority are central to sociological analyses of the state. **Power** is the ability of one person or group to exercise influence and control over others. The exercise of power can be seen in relationships ranging from the interaction of two people (husband and wife, police officer and suspect) to a nation threatening or dominating other nations. Sociologists are most interested in how power is structured in society: who has it, how it is used, and how it is built into institutions such as the state. Power can be structured, such as when men as a group have power over women as a group. In the United States, a society that is heavily stratified by race, class, and gender, power is structured into basic social institutions in ways that reflect these inequalities. Sociologists also understand that institutionalized power in society influences the social dynamics within individual and group relationships.

The exercise of power may take the form of persuasion or coercion. For example, a group may be encouraged to act a certain way based on a persuasive argument. A strong political leader may persuade the nation to support a military invasion or a social policy. Or, power may be exerted by sheer force. Between persuasion and coercion are many gradations. Generally speaking, groups with the greatest material resources most likely have the advantage in transactions involving power, but this is not always the case. A group may by sheer size be able to exercise power, or groups may use other means, such as armed uprisings or organized social protests, to exert power.

Power can be legitimate—accepted by the members of society as right and just—or it can be illegitimate. **Authority** is power perceived by others as legitimate, and emerges not only from the exercise of power, but from the belief of constituents that the power is legitimate. In the United States, the source of the president's domestic power is not just his status as commander of the armed forces but also the belief by most people that his power is legitimate. The law is also a source of authority in the United States. Those who accept the status quo as a legitimate system of authority perceive the guardians of law to be exercising *legitimate power.* In contrast, *coercive power* is achieved through force, often against the will of the people being forced. Those people may be only a few dissidents or most of the citizenry of an entire nation. A dictatorship often relies on its ability to exercise coercive power through its control of the military or state police.

Types of Authority

Max Weber (1864–1920), the German classical sociologist, postulated that three types of authority exist in society: traditional, charismatic, and rational–legal (Weber 1978/1921). **Traditional authority** stems from long-established patterns that give certain people or groups legitimate power in society. A monarchy is an example of a traditional system of authority. Within a monarchy, kings and queens rule, not necessarily because of their appeal, or because they have won elections, but because of long-standing traditions within the society.

Charismatic authority is derived from the personal appeal of a leader. Charismatic leaders are often believed to have special gifts, even magical powers, and their presumed personal attributes inspire devotion and obedience. Charismatic leaders often emerge from religious movements, but they come from other realms also. President John F. Kennedy was for many a charismatic leader, admired for the vigorous image he projected of himself and the nation. Arnold Schwarzenegger, a movie star, was elected governor of California partially as the result of his charismatic authority.

Rational–legal authority stems from rules and regulations, typically written down as laws, procedures, or codes of conduct. This is the most common form of authority in the contemporary United States. People obey not because national leaders are charismatic or because of social traditions, but because there is an unquestioned legal system of authority established by formalized rules and regulations. Rulers gain legitimate authority by election or appointment in accordance with society's rules. In systems based on rational–legal authority, the rules are upheld by state agents such as the police, judges, social workers, and other state functionaries to whom power is delegated.

AP Images

Bureaucracies such as the Department of Motor Vehicles are organized according to hierarchical and rule-driven forms of social organization.

The Growth of Bureaucratic Government

According to Weber, rational–legal authority leads inevitably to the formation of bureaucracies. As we noted in Chapter 5, a **bureaucracy** is a type of formal organization characterized by an authority hierarchy, a clear division of labor, explicit rules, and impersonality. Bureaucratic power comes from the accepted legitimacy of the rules, not personal ties to individuals. The rules may change, but they do so through formal, bureaucratic procedures. People who work within bureaucracies are selected, trained, and promoted based on how well they apply the rules. Those who establish the rules are unlikely to be the same people who administer them. Bureaucracies are hierarchical, and the bureaucratic leadership may be quite remote. Power in bureaucracies is dispersed downward through the system to those who actually carry out the bureaucratic functions. It is an odd feature of bureaucracy that those with the least power to influence how the rules are formulated—those at the bottom of the hierarchy—are very often the most adamant about strict adherence to the rules.

Is the bureaucratic growth of government in the United States becoming unwieldy? Many would say yes, judging from the sheer size of bureaucratic governmental institutions. Bureaucracy has become the modern system of administration. In principle, bureaucracies are highly efficient modes of organization, but the reality is often very different. Bureaucracies tend to proliferate rules, often causing the organization to become ensnared in its own bureaucratic requirements. For example, government procurement of desktop computers has become so overburdened by bureaucratic government rules for review, comparison, bidding, and approval that by the time the federal government approves a purchase, the approved models may be obsolete.

Within bureaucracies, personal temperament and individual discretion are not supposed to influence the application of rules. Of course, we know that the face of bureaucracy is not always perfectly stony. As we saw in Chapter 5, bureaucracy has another face. Bureaucratic workers frequently exercise discretion in applying rules and procedures. Some people who encounter bureaucracies know how to "work the system," perhaps by personalizing the interaction and making a willing accomplice of the bureaucrat in dodging bureaucratic stipulations, or perhaps by using knowledge of some rules to evade other rules. People who lack privileged relationships or privileged information are continually at a disadvantage because of their inability to negotiate within the system. Despite the supposed impersonal administration in bureaucracies, people may receive widely varied treatment from bureaucratic workers who may favor some people while discriminating against others based on race, gender, age, or other characteristics.

thinking sociologically

Observe the national evening news for one week, noting the people featured who have some kind of *authority*. List each of them and note their area of influence. What form of authority would you say each represents: *traditional, charismatic,* or *rational–legal*? How is the kind of authority a person has reflected in his or her position in society (that is, race, class, gender, occupation, education, and so on)?

Our picture of the state so far—bureaucratic, powerful, omnipresent—presents an important question to be addressed by the sociological imagination: Does the state act in the interests of its different constituencies, or does it merely reflect the needs of a select group positioned at the top of the pyramid of state power? This question has spawned much sociological study and debate and has resulted in several theoretical models of state power.

Theories of Power

How is power exercised in society? Four theoretical models have been developed by sociologists to answer this question: the *pluralist model,* the *power-elite model,* the *autonomous state model,* and *feminist theories of the state.* Each begins with a different set of assumptions and arrives at different conclusions (see Table 14.1).

The Pluralist Model

The **pluralist model** interprets power in society as derived from the representation of diverse interests of different groups in society. This model assumes that in democratic societies, the system of government works to balance the different interests of groups in society. An **interest group** can be any constituency in society organized to promote its own agenda, including large, nationally based groups such as the American Association of Retired Persons (AARP), the National Gay and Lesbian Task Force, the National Rifle Association (NRA), the National Organization for Women (NOW), and the National Urban League; groups organized around professional and business interests, such as the American Medical Association (AMA), the National Association of Manufacturers, and the Tobacco Institute; or groups that concentrate on one political or social goal, such as Greenpeace and Mothers Against Drunk Driving (MADD). According to the pluralist model, interest groups achieve power and influence through their organized mobilization of concerned people and groups.

The pluralist model has its origins in functionalist theory. This model sees the state as benign and representative of the whole society. No particular group is seen as politically dominant; rather, the pluralist model sees power as broadly diffused across the public. Groups that want to effect a change or express their point of view need only mobilize to do so. The pluralist model also suggests that members of diverse ethnic, racial, and social groups can

table 14.1 Theories of Power in Society

	Pluralism	Power Elite	Autonomous State	Feminist Theory
Interprets the state	As representing diverse and multiple groups in society	As representing the interests of a small, but economically dominant, class	As taking on a life of its own, perpetuating its own form and interests	As masculine in its organization and values (that is, based on rational principles and a patriarchal structure)
Interprets political power	As derived from the activities of interest groups and as broadly diffused throughout the public	As held by the ruling class	As residing in the organizational structure of state institutions	As emerging from the dominance of men over women
Interprets social conflict	As the competition between diverse groups that mobilize to promote their interests	As stemming from the domination of elites over less powerful groups	As developing between states, as each vies to uphold its own interests	As resulting from the power men have over women
Interprets social order	As the result of the equilibrium created by multiple groups balancing their interests	As coming from the interlocking directorates created by the linkages among those few people who control institutions	As the result of administrative systems that work to maintain the status quo	As resulting from the patriarchal control that men have over social institutions

The pluralist model of the state sees diverse interest groups as mobilizing to gain political power.

participate equally in a representative and democratic government assuming that power does not depend on social status or wealth. Various special interest groups compete for government attention and action with equal political opportunity available for any group that organizes to pursue its interests. The pluralist model sees special interest groups as an integral part of the political system, even though they are not an official part of government. In the pluralist view, special interest groups make government more responsive to the needs and interests of different people, an especially important function in a highly diverse society.

The pluralist model helps explain the importance of **political action committees (PACs),** groups of people who organize to support candidates they feel will represent their views. In 1974, Congress passed legislation enabling employees of companies, members of unions, professional groups, and trade associations to support political candidates with money they raise collectively. The number of political action committees has now grown to almost 4000. PACs have enormous influence on the political process. One measure of their growing influence is the tremendous increase in financial contributions that PACs made to political campaigns over the last two decades—in excess of $300 million in the year 2004 (Federal Election Commission 2004).

The Power Elite Model

The **power elite model** originated in the work of **Karl Marx** (1818–1883) and developed from the framework of conflict theory. According to Marx, the dominant or ruling class controls all the major institutions in society; the state itself is simply an instrument by which the ruling class exercises its power. The Marxist view of the state emphasizes the power of the upper class over the lower classes, the small group of elites over the rest of the population. The state, according to Marx, is not a representative, rational institution, but an expression of the will of the ruling class (Marx 1972/1845).

Marx's theory was elaborated much later by **C. Wright Mills** (1956), who popularized the term *power elite.* Mills attacked the pluralist model, arguing that the true power structure consists of people well positioned in three areas: the economy, the government, and the military. These three institutions are considered the bastions of the power elite, although some have argued that Mills overemphasized the role of the military (Domhoff 2002). While sharing common beliefs and goals, the power elite shape political agendas and outcomes in the society along the narrow lines of their particular collective interests. This small and influential group at the top of the pyramid of power holds the key positions in the major state institutions, giving them extensive power and control over the rest of society.

The power elite model posits a strong link between government and business, a view supported by the strong hand government takes in directing the economy and by the role of military spending as a principal component of U.S. economic affairs. The power elite model also emphasizes how power overlaps between influential groups. **Interlocking directorates** are organizational linkages created when the same people sit on the board of directors for numerous corporations. People in elite circles may serve on the boards of major companies, universities, and foundations at the same time. People drawn from the same elite group receive most of the major government appointments; thus, the same relatively small group of people tends to the interests of all these organizations and the interests of the government. These interests naturally overlap and reinforce each other.

The power elite model sees the state as part of the structure of domination in society, one in which the state is simply a piece of the whole. Members of the upper class do not need to occupy high office themselves to exert their will, as long as they are in a position to influence people who are in power (Domhoff 2002). The majority of the power elite are White men, which means that the interests and outlooks of White men dominate the national agenda.

The Autonomous State Model

A third view of power developed by sociologists, the **autonomous state model,** interprets the state as its own major constituent. From this perspective, the state develops interests of its own, which it seeks to promote independent of other interests and the public that it allegedly serves. The state does not reflect the needs of the dominant groups, as Marx and power elite theorists would contend. It is an administrative organization with its own needs, such as maintenance of its complex bureaucracies and protection of its special privileges (Evans et al. 1985; Skocpol 1992; Rueschmeyer and Skocpol 1996).

Autonomous state theorists note that states tend to grow over time, including the possible expansion beyond their original boundaries by military incursions. Another example of expansion is the North American Free Trade Agreement (NAFTA), in which the United States is expanding its state interests by regulating business not only in the United States, but also in Mexico and Canada.

The huge government apparatus now in place in the United States is a good illustration of autonomous state theory. The government provides a huge array of social support programs, including Social Security, unemployment benefits, agricultural subsidies, public assistance, and other economic interventions intended to protect citizens from the vagaries of a capitalist market system (Collins 1988). The purpose of these programs is to serve people in need. Autonomous state theory argues that the government has grown into a massive, elaborate bureaucracy, run by bureaucrats more absorbed in their own interests than in meeting the needs of the people. As a consequence, government can become paralyzed in conflicts between revenue-seeking state bureaucrats and those who must fund them. This can lead to revolt against the state, as in the tax revolts appearing sporadically throughout the country (Lo 1990; Collins 1988).

Feminist Theories of the State

Feminist theorists diverge from the preceding theoretical models by seeing men as having the most important power in society. The pluralists see power as widely dispersed through the class system, power elite theorists see political power directly linked to upper-class interests, and autonomous state theorists see the state as relatively independent of class interests.

Some feminist theorists argue that all state institutions reflect men's interests; they see the state as fundamentally patriarchal, its organization embodying the fixed principle that men are more powerful than women. Feminist theories of the state conclude that despite the presence of a few powerful women, the state is devoted primarily to men's interests, and moreover, the actions of the state will tend to support gender inequality (Haney 1996; Blankenship 1993). One historical example would be laws denying women the right to own property once they married. Such laws protected men's interests at the expense of women.

Evidence that "the state is male" (MacKinnon 2006, 1983) is easy to observe by looking at powerful political circles. Despite the recent inclusion of more women in powerful circles and the presence of some notable women as major national figures, most of the powerful are men. The U.S. Senate is 86 percent men; groups that exercise state power, such as the police and military, are predominantly men. Moreover, these institutions are structured by values and systems that can be described as culturally masculine—that is, based on hierarchical relationships, aggression, and

The world's leadership: Showing far too few women.

understanding diversity

Diversity in the Power Elite

As society has become more diverse, has it made a difference in the makeup of the power elite? Various groups—women, racial-ethnic groups, lesbians, and gays—have vied for more representation in the halls of power, but have their efforts succeeded? If they make it to power, does this change the corporations, military, or government—the major institutions composing the power elite?

Sociologists Richard L. Zweigenhaft and G. William Domhoff examined these questions by analyzing the composition of boards of directors and chief executive officers (CEOs) of the largest banks and corporations in the United States, as well as analyzing Congress, presidential cabinets, and the generals and admirals who form the military elite. In addition, they examined the political party preferences and the political positions of people found among the power elite. Do women and minorities bring new values into power, thereby changing society as they move into powerful positions, or do their values match those of the traditional power elite or become absorbed by a system more powerful than they are? Zweigenhaft and Domhoff's study looks as well at whether those who do make it into the power elite are within the innermost circles or whether they are marginalized.

They find that women, Jews, gays, lesbians, Black Americans, and Hispanics have become more numerous within the power elite, but only to a small degree. The power elite is still overwhelmingly White, wealthy, Christian, and male. Women and other minorities who make it into the power elite also tend to come from already privileged backgrounds as measured by their social class and education. Among African Americans and Latinos, skin color continues to make a difference, with darker-skinned Blacks and Hispanics less likely to achieve prominence compared with lighter-skinned people. Furthermore, Zweigenhaft and Domhoff find that the perspectives and values of women and minorities who rise to the top do not differ substantially from their White male counterparts. Some of this is explained by the common class origins of those in the power elite. The researchers also attribute the managing of one's identity to avoid challenging the system as a sorting factor that perpetuates the dominant worldview and practices of the most powerful.

The authors of this study conclude that "the irony of diversity" is that greater diversity may have strengthened the position of the power elite, because its members appear to be more legitimate through their inclusion of those previously left out. But, by including only those who share the perspectives and values of those already in power, little is actually changed. Condoleeza Rice and Clarence Thomas may make it seem that women and African Americans have it made, but as long as they support the positions of White male elites, the power elite goes undisturbed.

Sources: Zweigenhaft, Richard L., and G. William Domhoff. 2006. *Diversity in the Power Elite: Have Women and Minorities Reached the Top?* New Haven: Yale University Press; G. William Domhoff. 2002. *Who Rules America?* New York: McGraw-Hill.

force. Feminist theory begins with the premise that an understanding of power cannot be sound without a strong analysis of gender (Haney 1996).

Government: Power and Politics in a Diverse Society

The terms *government* and *state* are often used interchangeably. More precisely, the government is one of several institutions that make up the state. The **government** includes those institutions that represent the population, making rules that govern the society. The government of the United States is a **democracy;** therefore it is based on the principle of representing all people through the right to vote.

The actual makeup of the government, however, is far from representative of society. Not all people participate equally in the workings of government, either as elected officials or as voters, nor do their interests receive equal attention. Women, the poor and working class, and racial–ethnic minorities are less likely to be represented by government than are White middle- and upper-class men. Sociological research on political power has concentrated on inequality in government affairs and demonstrated large, persistent differences in the political participation and representation of various groups in society.

Diverse Patterns of Political Participation

One would hope that in a democratic society all people would be equally eager to exercise their right to vote and be heard. That is far from the case. Among democratic nations, the United States has one of the *lowest* voter turnouts (see Figure 14.1). In the 2004 presidential election, the percentage of eligible voters who went to the polls was only 61 percent of the population. A turnout of 50 percent or less is typical of national elections; voter turnout in congressional and local elections is even lower.

The group most likely to vote is older, better educated, and financially better off than the average citizen. In sociological terms, age, income, and education are the strongest predictors of whether someone will vote. The higher a person's social class, the higher the likelihood that she or he will be a voter. As a result, the upper social classes are far more likely to vote and have their voices heard in the political process. Despite the fact that almost all young people say it is important for people in a democracy to vote (97 percent say so in polls), only about one-third of young people (those between the ages of 18 and 29) actually vote in presidential elections. Analysts interpret this as result of cynicism because large numbers of young people think their vote won't matter (Rosenberg 2004).

There is also significant variation in voting patterns by race. Overall, African Americans are less likely to vote than Whites, a fact reflecting the disproportionate numbers of African Americans in the poor and working classes—groups that typically are less likely to vote than middle- and upper-class people. During the late 1960s, however, controlling for factors such as age, education, and income, African Americans were actually more likely to vote than White Americans, probably because of the heightened sense of the importance of the franchise generated at the time by the civil rights movement and voter registration drives. Beginning in the 1970s, there was a convergence in the likelihood of voting among Blacks and Whites of similar social class, education, and age (Fletcher 2000; Ellison and Gay 1989).

A subjective factor influencing the voting behavior of African Americans is that they are generally less trusting of the political system than Whites and are more alienated from politics. Sociological research shows that when Black Americans are approached directly by party representatives, the likelihood they will vote increases substantially—more than it does when Whites are canvassed. In addition, the more that Black Americans and Latinos identify themselves as groups with distinct needs and interests, the more likely they will participate in the political process, reflecting the long-standing sociological finding that a strong sense of ethnic identity is linked to increased political participation (Pantoja and Segura 2003; Welch and Sigelman 1993).

Not only do social factors influence the likelihood of voting, they also influence for whom people vote. Race is one major social factor that influences voting. African Americans and Latinos, with the exception of Cuban Americans, tend to be markedly Democratic and voted overwhelmingly for John Kerry,

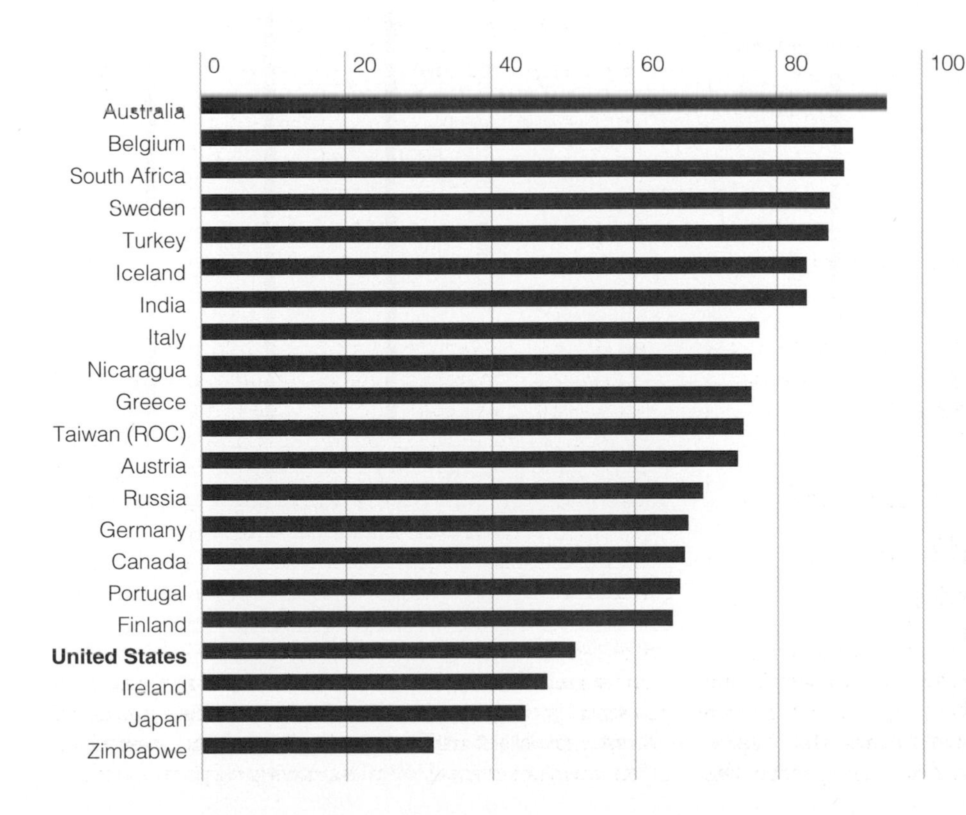

FIGURE 14.1 *Voter Participation in Democratic Nations, Average Participation, 1945–1989.*

Source: Federal Election Commission. 2006. "International Voter Turnout." Washington, DC: Federal Election Commission. **www.fec.gov/ votregis/InternatTO.htm**

the 2004 Democratic presidential candidate. Cuban Americans are disproportionately Republican. One reason both Latinos and African Americans have been more attracted to the Democratic party and its candidates is that Democrats historically have been more committed to government action that promotes social and economic equality. The allegiance of Latino and Black voters to Democratic candidates is also explained in part by socioeconomic factors such as age, gender, income, education, and religion; all these factors, regardless of racial–ethnic identity, are linked to political behavior. Even without controlling for these factors, however, Latinos are more liberal than Anglos. Latinos are generally also more likely than other groups to support government spending (Edison/Mikofsky 2004).

Ethnicity of the candidates also influences voting patterns. People tend to vote for those of their own ethnic group. Sociologists have also found that ethnicity can outweigh socioeconomic factors such as occupation, income, and education in predicting voting patterns. White Catholics, for example, have traditionally been aligned with the Democratic Party, although that link has been weakened somewhat by economic changes that have had a strong influence on voting patterns (Legge 1993; R. Smith 1993).

Gender is also a major social factor influencing political attitudes and behavior. **Gender gap** refers to the differences between men and women in political attitudes and behavior. One aspect of the gender gap is that women are more likely than men to identify and vote as Democrats and to have liberal views on a variety of social and political issues. For many decades, men were more likely to vote than women, and, when asked, women reported voting the same as their husbands and fathers. In recent years, women have been as likely to vote as men, but there are significant differences in their political outlooks (Saad 2002; Wilcox et al. 1996; Eliason et al. 1996; Jackson et al. 1996).

Political Power: Who's in Charge?

Although the government in a democracy is supposed to be representative, the class, race, and gender composition of the ruling bodies in this country indicate that this is hardly the case (see Figure 14.2). Most members of Congress are White, well–educated men, from upper middle- or upper-class backgrounds with an Anglo-Saxon Protestant heritage. One-third of the people in Congress are lawyers; another third are businesspeople and bankers. Most other members come from professional occupations; very few senators or representatives were blue-collar workers before coming to Congress (Freedman 1997; Ornstein et al. 1996; Saunders 1990).

Many of the members of Congress are millionaires, and many in Congress have large financial interests in the industries that they regulate. Simply getting into politics requires a substantial investment of money. The average cost of a Senate campaign is

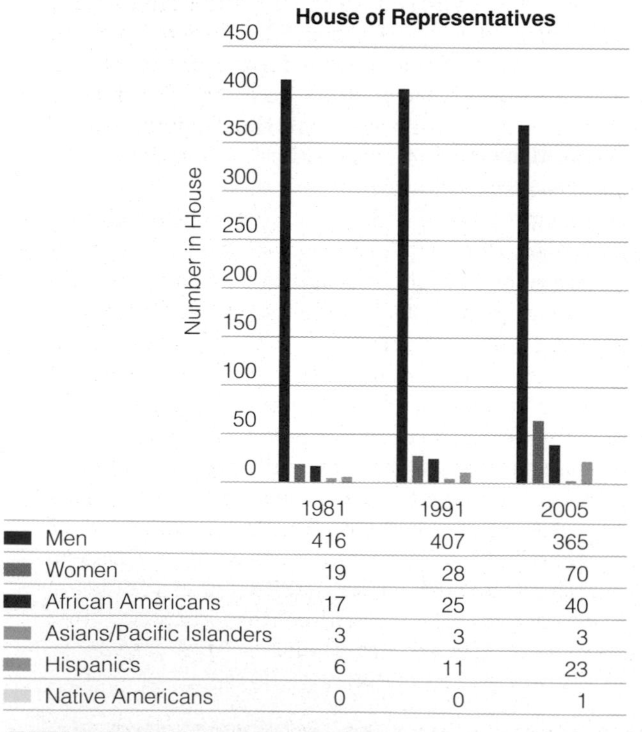

	1981	1991	2005
Men	416	407	365
Women	19	28	70
African Americans	17	25	40
Asians/Pacific Islanders	3	3	3
Hispanics	6	11	23
Native Americans	0	0	1

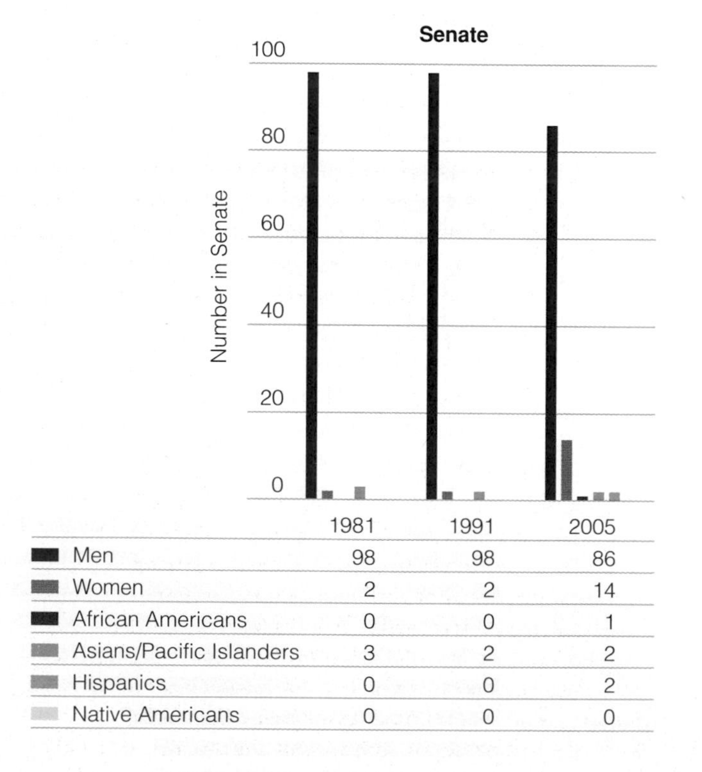

	1981	1991	2005
Men	98	98	86
Women	2	2	14
African Americans	0	0	1
Asians/Pacific Islanders	3	2	2
Hispanics	0	0	2
Native Americans	0	0	0

FIGURE 14.2 *A Representative Government?*

Source: U.S. Census Bureau, 2007. *Statistical Abstract of the United States, 2006.* Washington, DC: U.S. Government Printing Office. **http://www.census.gov/compendia/statab/**

nearly $4 million; a successful run for the House of Representatives costs nearly $700,000. In the 2004 presidential election, President George W. Bush spent a record $339 million and John Kerry, the Democratic candidate, spent $299 million, much of it coming from PACs (Political Action Committees) associated with business and other organizations.

All candidates depend on contributions from individuals and groups to finance their election campaigns. Wealthy families and individuals are among the largest campaign contributors to presidential elections (Allen and Broyles 1991). The largest individual contributors typically have interests in the same industries that fund political action committees. Much of the money given by individuals and PACs goes to incumbents, who in the past had an overwhelming edge in elections and who already sit on the committees where public policy is hammered out. This picture of elites and business interests funneling money to candidates, who return to the same donors for more money when the next campaign rolls around, has shaken the faith of many Americans in their political system. There is little belief that the political process is a democratic and populist mechanism by which the "little people" can select political leaders to represent them. In fact, national surveys show that only 30 percent of U.S. citizens have a great deal of confidence in Congress, and only 50 percent have some confidence in the Supreme Court (Saad 2004; Gallup Organization 2002).

Women and Minorities in Government

Although there have been some gains in the number of women and minorities in government, they are still underrepresented—both at the federal and state level. The 110th Congress, convening in 2007, had the first-ever woman Speaker of the House, Nancy Pelosi.

There are 14 women in the Senate (out of 100) and 70 in the House of Representatives (out of 435), an all-time high in both houses of government. Among those in the House of Representatives, 40 are African Americans, 23 are Latino/a, only 3 are Asian-American, and only one is Native American.

Note that only 14 out of the 100 senators are women, although women make up 51 percent of the U.S. population. And, although 13 percent of the population is African American, there is only one Black senator, Barack Obama, a major contender for the Democratic presidential nomination when this book went to press. Hispanics are 13 percent of the U.S. population but only 2 out of 100 (thus 2 percent) of the Senate. There are only 2 Asian-Americans in the Senate, and no Native Americans (Ben Nighthorse Campbell, elected in 1992, was for awhile the only Native American in the Senate). Clearly, there is a long way to go before we have a representative government in this country.

Researchers offer several explanations for why women and racial–ethnic minorities continue to be underrepresented in government. Certainly prejudice plays a role. It was not long ago, in the 1960 Kennedy–Nixon election, that Kennedy became the first Catholic president elected. In the 2000 presidential election, Joseph Lieberman was the first Jewish candidate to appear on a major national ticket. Gender and racial prejudice run just as deep in the public mind. Although the percentage of Americans who say they would vote for a woman for president has climbed to 92 percent (53 percent in 1969), a substantial number (42 percent) also say they think a man would make a better president than a woman (Simmons 2001).

Individual prejudice alone, however, cannot account for the lack of representation. Societal causes are a major factor in the successful elections of women and people who are better represented in local political office. Women and minority candidates receive a great deal of political support from local groups, but at the national level, they do not fare as well. The power of incumbents, most of whom are White men, is a disadvantage to any new office seeker.

The Military

The military arm of the state is among the most powerful and influential social institutions in almost all societies. In the United States, the military is the largest single employer. Approximately 3 million men and women serve in the U.S. military, 1.4 million on active duty, including those in Iraq and Afghanistan, and the rest are in the reserves. This does not include the many hundreds of thousands who are employed in industries that support the military, nor does it include civilians who work for the Department of Defense and other military-affiliated agencies (U.S. Department of Defense 2003).

Highly esteemed in the value system of the United States is *militarism,* the pervasive influence of military goals and values throughout the culture. The militaristic bent in our culture is evident in many ways. The toys children play with often inculcate militaristic values. Military garb falls in and out of fashion. The plots of each summer's biggest blockbuster movies almost always revolve around a burly character hugely gifted in the military arts, often a veteran of the armed forces, who generally stocks up on the latest weaponry. Fighter jets flash over the Super Bowl. Military brass sparkles in parades. Being the strongest nation, militarily, in the world is frequently in the national political rhetoric. Most people in the United States place a lot of confidence in the military (Saad 2004).

The Military as a Social Institution. Institutions are stable systems of norms and values that fulfill certain functions in society. The military is a social institution whose function is to defend the

AP Images

The men and women who serve in the Armed Forces, such as this young man in Iraq, are often separated from families and loved ones for long periods of time.

nation against external (and sometimes internal) threats. A strong military is often considered an essential tool for maintaining peace, although the values that promote preparedness in the armed forces are perilously close to the warlike values that lead to military aggression against others (see Map 14.1).

The military is one of the most hierarchical social institutions, and its hierarchy is extremely formalized. People who join the military are explicitly labeled with rank, and if promoted, they pass through a series of well-defined levels, each with clearly demarcated sets of rights and responsibilities. An explicit line exists between officers and enlisted personnel, and officers have many privileges that others do not. Higher ranks are also entitled to absolute obedience from the ranks below them, with elaborate rituals created to remind both dominants and subordinates of their status. As in other social institutions, military enlistees are carefully socialized to learn the norms of the culture they have joined. Military socialization places a high premium on conformity and eliminates individuality. All new recruits have their heads shaved, are issued identical uniforms, and are allowed to retain very few of their personal possessions. They must quickly learn new, strictly enforced codes of behavior.

Race and the Military. The greatest change in the military as a social institution is the representation of racial minority groups and women within the armed forces. Picture a U.S. soldier. Whom do you see? At one time you would have almost certainly pictured a young, White male, possibly wearing army green camouflage and carrying a weapon. Today the image of the military is much more diverse. Drawing on the cultural images you have stored in your mind, you are just as likely to picture a young African American man in military blues with a stiffly starched shirt and a neat and trim appearance, or perhaps a woman wearing a flight helmet in the cockpit of a fighter plane.

African Americans have served in the military for almost as long as the U.S. armed forces have been in existence. Except for the Marines, which desegregated in 1942, the armed forces were officially segregated until 1948, when President Harry Truman signed an executive order banning discrimination in the armed services. Although much segregation continued after this order, the desegregation of the armed forces is often credited with promoting more positive interracial relationships and increased awareness among Black Americans of their right to equal opportunities. Until that time the widespread opinion among Whites was that to allow Black and White soldiers to serve side by side would destroy soldiers' morale.

Currently 34 percent of military personnel are racial minorities—20 percent African American, 8 percent Hispanic, and 6 percent other racial minorities (U.S. Department of Defense 2002; Moskos and Butler 1996). For groups with limited opportunities in civilian society, joining the military seems to promise an educational and economic boost. Is this realized? Within the military today, there is equal pay for equal rank. African Americans and Latinos, however, are overrepresented in lower-ranking support positions. Often they are excluded from the higher-status, technologically based positions—those most likely to bring advancement and higher earnings both in the military and beyond. Most minorities remain in positions with little supervisory responsibility. At the highest ranks in the military, there are few African Americans, Latinos, Asian Americans, or Native Americans (U.S. Department of Defense 2002; Moskos and Butler 1996; Moskos 1988).

For both Whites and racial minorities, serving in the military leads to higher earnings relative to one's nonmilitary peers. The economic payoff is greatest for African Americans and Latinos relative to those of comparable background, education, experience, and ability. The earnings difference between military and civilian groups is substantial. Young Black and Latino men earn $3000 to $5000 more per year in the military than do their civilian peers.

Women in the Military. The military academies did not open their enrollment to women until 1976. Since then, the armed services have profoundly changed their admission policies, although the hostile reaction to women who have entered these academies

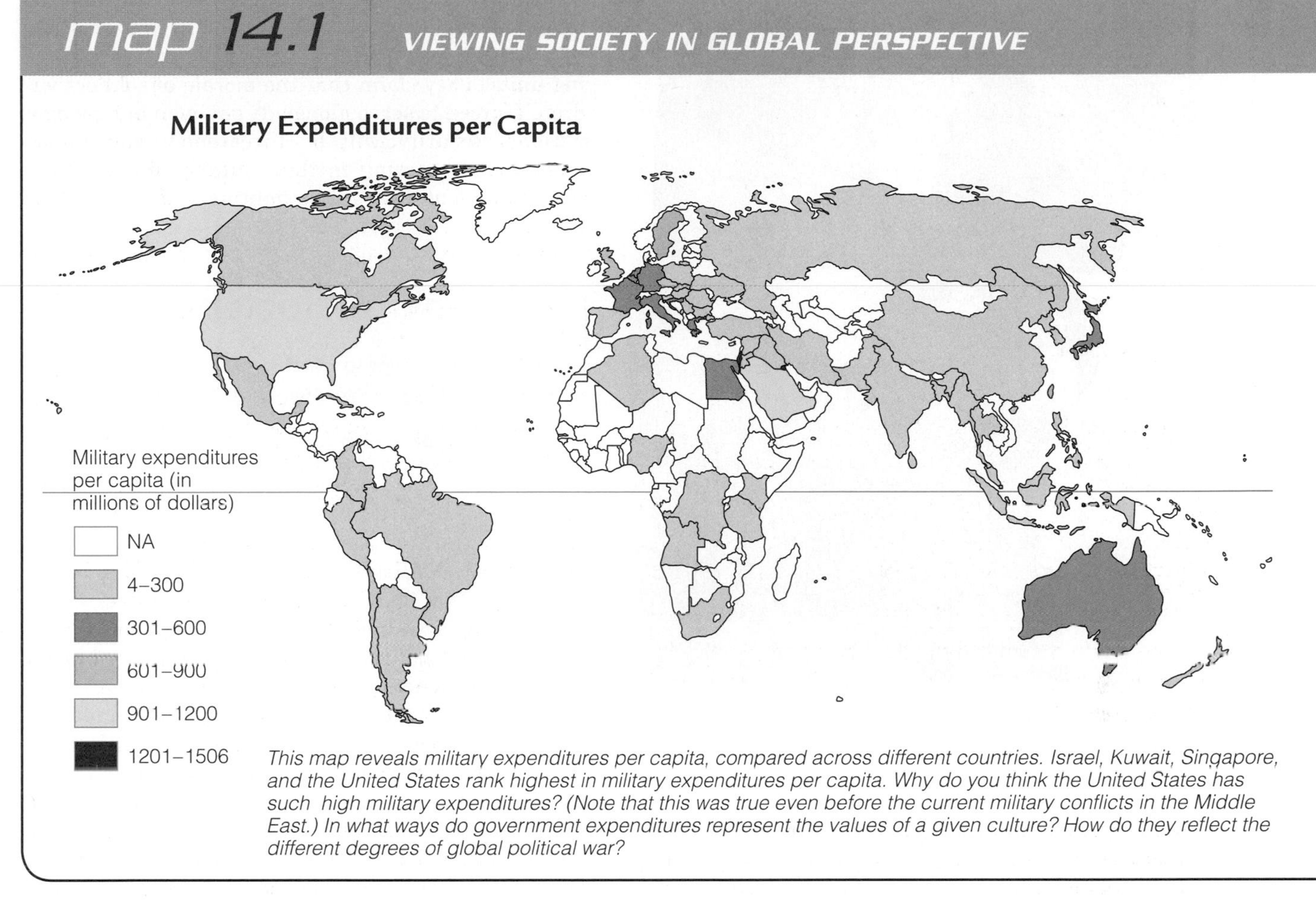

This map reveals military expenditures per capita, compared across different countries. Israel, Kuwait, Singapore, and the United States rank highest in military expenditures per capita. Why do you think the United States has such high military expenditures? (Note that this was true even before the current military conflicts in the Middle East.) In what ways do government expenditures represent the values of a given culture? How do they reflect the different degrees of global political war?

Data: U.S. Census Bureau. 2007. *A Statistical Abstract of the United States, 2006.* Washington, DC: U.S. Government Printing Office. **www.census.gov/statab/www/.**

shows how fierce the resistance to including women in the military can be. Shannon Faulkner, a young woman who won the right to enter the Citadel, an elite military academy, had to embark on a two-and-a-half-year legal struggle to gain admission. Once admitted, she was harassed and ridiculed. When she succumbed to exhaustion during the first week of intense physical training and had to leave, many of the cadets whooped and shouted in glee, leaving many to argue that none of the men had been forced to endure such tribulations. Later, two other women left the Citadel after serious hazing in which, among other things, the women were sprinkled with nail polish remover and their clothes set on fire.

The Supreme Court ruled in 1996 (in *United States* v. *Virginia*) that women cannot be excluded from state-supported military academies such as the Citadel and the Virginia Military Institute (VMI). This was a landmark decision that opened new opportunities for women who want the rigorous physical and academic training that military academies provide (Kimmel 2000).

The marginalization of women in the military has been rationalized by the popular conviction that women should not serve in combat. The traditional belief that men are protectors and women are dependents has led many to believe that women, especially mothers and wives, should stay on home soil where they can safely carry out their "womanly" duties. Despite these beliefs, women have fought in combat to defend this country. For example, 4100 women served in combat in the Persian Gulf War of 1991; 7.2 percent of the active combat forces in that war were women (Holm 1992; Becraft 1992a). Women are about 10 percent of the forces serving in the Afghanistan and Iraq wars (The Woman's Research and Education Institute 2004).

The involvement of women in the military has reached an all-time high in recent years. Now, almost 200,000 women are on active duty, with an additional 151,000 in the reserves, not including the Coast Guard and its reserves. The Air Force has the highest proportion of women (18 percent), followed by the Army (15 percent), the Navy (13 percent), and the Marines (6 percent). Minority women are overrepresented in the military, relative to their presence in the general population.

Gender relations in the military extend beyond just women who serve. The experience of military

© Bob Daemmrich/PhotoEdit

The men and women who serve in the Armed Forces, such as this young woman returning to Iraq, are often separated from families and loved ones for long periods of time.

wives, for example, is greatly affected by their husbands' employment as soldiers. Frequent relocation means that military wives are less competitive in the labor market. They have lower labor force participation than their women peers and earn less than their peers at the same educational level (Enloe 1993; Moskos 1992; Payne et al. 1992). Sociologists have found that the presence of a military base depresses wages and employment for all women in the surrounding community (Booth et al. 2000).

Gays and Lesbians in the Military. According to the "don't ask, don't tell" policy, recruiting officers cannot ask about sexual preference, and individuals in the military who keep their sexual preference a private matter shall not be discriminated against. Those who publicly reveal that they are gay or lesbian can be expelled from the military based on their sexual orientation. In essence, this policy is simply an extension of the already existing military attitude toward gays and lesbians. It does not explicitly permit the military to seek out and discharge service people on the grounds of homosexuality, nor can recruits be screened for the same reason; nevertheless, the policy does not clearly assert the civil rights of gays and lesbians. Gay soldiers and potential recruits have filed several lawsuits against the U.S. military. It remains unclear whether gays and lesbians will ultimately be permitted to live openly as homosexuals while also pursuing careers in the armed services.

Supporters of the ban on gays in the military often use arguments similar to the arguments used before 1948 to defend the racial segregation of fighting units. They claim that the morale of soldiers will drop if forced to serve alongside gay men and women, national security will be threatened, and known homosexuals serving in the military will upset the status quo and destroy the fighting spirit of military units.

Debunking Society's Myths

Myth: Letting lesbians and gays serve in the military erodes morale and weakens the armed services.

Sociological perspective: Similar arguments were made in the past to exclude Black Americans from military service; this belief is an ideology that serves to support the status quo.

The long-standing policy against gays in the military has kept homosexuality in the military hidden, but not nonexistent. Even the military has had to admit that there have always been some gays and lesbians in all branches of the U.S. armed forces. Because their presence clearly did not cause the armed forces to disintegrate, the military has been forced to adapt their argument against gays and lesbians (Enloe 1993). Nonetheless, despite some increased tolerance of gays and lesbians in the military, homophobia is a pervasive part of military culture (Myers 2000; Becker 2000).

Economy and Society

Politics, then, is part of the political institution of society. We move next to an examination of work and the economy in society. To understand *work,* you first must see it in the context of the broad social institution known as the economy. All societies are organized around an economic base. The **economy** of a society is the system by which goods and services are produced, distributed, and consumed. We first look at the economic significance of the historic transformation from agriculturally based societies to industrial and now postindustrial societies.

The Industrial Revolution

In Chapter 4, we discussed the evolution of different types of societies. One of the most significant of these changes was, first, the development of agricultural societies and, later, the far-ranging impact of the Industrial Revolution. Now, the Industrial Revolution is giving way to the growth of postindustrial societies—a development in the economic system with far-reaching consequences for how society is organized.

The Industrial Revolution transformed labor, moving it from the household to the factory—or other sites where work was mechanized and oriented to mass production.

The Industrial Revolution is usually pinpointed as beginning in mid-eighteenth–century Europe, soon thereafter spreading throughout other parts of the world. The Industrial Revolution led to numerous social changes since Western economies were organized around the mass production of goods. The Industrial Revolution led to the creation of factories, which, as we briefly saw in Chapter 4, separated work and family by relocating the place where most people were employed.

We still live in a society that is largely industrial, but that is quickly giving way to a new kind of social organization: postindustrial society. Whereas industrial societies are primarily organized around the production of goods, **postindustrial societies** are organized around the provision of services. Thus, in the United States, we have moved from being a manufacturing-based economy to an economy centered on the provision of services. *Service industry* is a broad term meant to encompass a wide range of economic activities now common in the labor market. It includes banking and finance, retail sales, hotel and restaurant work, and health care; it also includes parts of the information technology industry—not electronics assembly, but areas such as software design and the exchange of information (through publishing, video production, and the like).

Comparing Economic Systems

The three major economic systems found in the world today are *capitalism, socialism,* and *communism.* These are not totally distinct, that is, many societies have a mix of these economic systems. **Capitalism** is an economic system based on the principles of market competition, private property, and the pursuit of profit. Within capitalist societies, stockholders own corporations—or a share of the corporation's wealth. Under capitalism, owners keep a surplus of what is generated by the economy; this is their *profit,* which may be in the form of money, financial assets, and other commodities.

Socialism is an economic institution characterized by state ownership and management of the basic industries; that is, the means of production are the property of the state not of individuals. Modern socialism emerged from the writings of Karl Marx, who predicted that capitalism would give way to egalitarian, state-dominated socialism, followed by a transition to stateless, classless communism.

Many European nations, for example, have strong elements of socialism that mix with the global forces of capitalism. Sweden supports an extensive array of state-run social services, such as health care, education, and social welfare programs, but Swedish industry is capitalist. Other world nations are more strongly socialist although they are not immune from the penetrating influence of capitalism. The People's Republic of China was formerly a strongly socialist society that is currently undergoing transformation to a mix of socialist and capitalist principles, including state encouragement of a market-based economy, the introduction of privately owned industries, and increased engagement in the international capitalist economy.

Communism is sometimes described as socialism in its purest form. In pure communism, industry is not the private property of owners. Instead, the state is the sole owner of the systems of production. Communist philosophy argues that capitalism is fundamentally unjust because powerful owners take more from laborers (and society) than they give and use their power to maintain the inequalities between the worker and owner classes. Communist theorists in the nineteenth-century declared that capitalism would inevitably be overthrown as workers worldwide united against owners and the system that exploited them. Class divisions were supposed to be erased at that time, along with private property and all forms of inequality. History has not borne out these predictions.

The Changing Global Economy

As we start a new century, one of the most significant developments is the creation of a global economy, which affects work in the United States and worldwide. The concept of the **global economy** acknowledges that all dimensions of the economy now cross national borders, including investment, production, management, markets, labor, information, and technology (Altman 2001; Carnoy et al. 1993). Economic events in one nation now can have major reverberations throughout the world. When the economies of Brazil, Japan, China, or Russia are unstable, the effects are felt worldwide.

Multinational corporations—those that draw a large share of their revenues from foreign investments and conduct business across national borders—have become increasingly powerful, spreading their influence around the globe. The global economy links the lives of millions of Americans to the experiences of other people throughout the world. You can see the internationalization of the economy in everyday life: Status symbols such as high-priced sneakers are manufactured for just a few cents in China. The Barbie dolls that young girls accumulate are inexpensive by U.S. standards, yet it would require one month's wages for the Indonesian worker who makes the doll to buy it for her child.

In the global economy, the most developed countries control research and management and assembly-line work is performed in nations with less privileged positions in the global economy. A single product, such as an automobile, may be assembled from parts made all over the world—the engine assembled in Mexico, tires manufactured in Malaysia, electronic parts constructed in China. The relocation of manufacturing to wherever labor is cheap has led to the emergence of the *global assembly line,* a new international division of labor in which research and development is conducted in the United States, Japan, Germany, and other major world powers, and the assembly of goods is done primarily in underdeveloped and poor nations—mostly by women and children.

Within the United States, the development of a global economy has also created anxieties about foreign workers, particularly among the working class. Because it is easier to blame foreign workers for unemployment in the United States than it is to understand the complex processes that have produced this phenomenon, U.S. workers have been prone to **xenophobia,** the fear and hatred of foreigners. Campaigns to "buy American" reflect this trend, although the concept of buying American is increasingly antiquated in a global economy.

When buying a product from a U.S. company, it is likely that the parts, if not the product itself, were built overseas. In a global economy, distinctions between U.S. and foreign businesses blur. Moreover, the label "Made in U.S.A." does not necessarily mean that the product was made by well-paid workers in the United States. In the garment industry, sweatshop workers, many of whom are recent immigrants and primarily women, are likely to have stitched the clothing that bears such a label. Moreover, these workers are likely to be working under exploitative conditions. For example, recent surveys of the garment industry in Los Angeles, where 25 percent of all women's outerwear is made, find that 96 percent of the garment firms are in violation of health and safety regulations for workers, and 61 percent are violating wage regulations (Bonacich and Applebaum 2000; Louie 2001).

The development of a global economy is part of the broad process of **economic restructuring,** which refers to the contemporary transformations in the basic structure of work that are permanently altering the workplace. This process includes the changing composition of the workplace, deindustrialization, and use of enhanced technology. Some changes are *demographic*—that is, resulting from changes in the population. The labor force is becoming more diverse, with women and people of color becoming the majority of those employed. Other changes are driven by *technological developments.* For example, the economy is based less on its earlier manufacturing base and more on service industries—those in which the primary business is not the production of goods, but the delivery of services (banking, health care, provision of food, or the like). All these developments are happening within a global context.

thinking sociologically

Identify a job you once held (or currently hold) and make a list of all the ways that workers in this segment of the labor market are being affected by the various dimensions of *economic restructuring: Demographic changes, globalization of the economy,* and *technological change.* What does your list tell you about how people's individual work experiences are shaped by social structure?

A More Diverse Workplace

A more diverse workplace is becoming a common result of economic restructuring. In a few years, women and racial minorities are expected to increase to more than one-half of the labor force. Another upcoming change in the labor market will be an increase in the number of older people (those over age fifty-five) as the population bulge of the Baby Boomer generation passes through late middle age (Dohm 2000; Bowman 1999; Fullerton 1999).

These changes in the social organization of work and the economy are creating a more diverse labor force, but much of the growth in the economy is projected to be in service industries, where, for the better jobs, education and training are required. People without these skills will not be well positioned for success. Manufacturing industries, where racial minorities have in the past maintained a foothold on employment, are now in decline. New technologies and corporate layoffs have reduced the number of entry-level corporate jobs, which recent college graduates have always used as a starting point for career mobility. Many college graduates are employed in jobs that do not require a college degree. College graduates, however, do still have higher earnings than those with less education.

Deindustrialization

Deindustrialization refers to the transition from a predominantly goods-producing economy to one based on the provision of services (Harrison and Bluestone 1982). This does not mean that goods are no longer produced, but that fewer workers in the United States are required to produce goods because machines can do the work people once did and many goods-producing jobs have moved overseas. Different from traditional manufacturing jobs, such as the production of cloth or automobiles, service-based industries provide the delivery of a product or provision of a service.

Deindustrialization is most easily observed by looking at the decline in the number of jobs in the manufacturing sector of the U.S. economy since the Second World War. At the end of the war in 1945, the majority of workers (51 percent) in the United States were employed in manufacturing-based jobs. Now the majority (at least 70 percent) are employed in the *service sector* (U.S. Department of Labor 2006; Wilson 1978). The service sector includes two segments: service delivery (such as food preparation, cleaning, or child care) and information processing (such as banking and finance, computer operation, or clerical work). Service delivery consists of many low-wage, semiskilled, and unskilled jobs and employs many women and people of color.

The human cost of deindustrialization can be severe. Deindustrialization has led to **job displacement,** the permanent loss of certain job types that occurs when employment patterns shift. When a manufacturing plant shuts down, many people may lose their jobs at the same time, and whole communities can be affected. Among the areas hardest hit by deindustrialization are communities that were heavily dependent on a single industry, such as steel towns or automobile-manufacturing cities such as Detroit, Flint, Lansing, Cleveland, and Akron. Some of these communities have rebounded by investing in new, high-tech industries, but this has still left many groups disadvantaged. Groups concentrated in the inner cities at these locations have suffered the most, because they have been the most dependent on manufacturing jobs for economic survival. They are not located in emerging suburban areas to take advantage of new industries.

In the early twentieth century, many African Americans migrated from the South to the North in pursuit of manufacturing jobs; Latino groups have more recently flowed into cities for the same reason. Both groups have been hard hit by the declines in urban manufacturing jobs that have characterized the late twentieth century. As a result, there are now extremely high rates of joblessness and dim prospects for economic recovery in these urban communities (Wilson 1996). African American and Latino youth have very high rates of unemployment; in some inner cities, teen unemployment rates exceed 40 percent (U.S. Department of Labor 2006; Wilson 1996).

Technological Change

Coupled with deindustrialization, rapidly changing and developing technologies are bringing major changes in work, including how it is organized, who does it, and how much it pays. One of the most influential technological developments of the twentieth century has been the invention of the semiconductor. Some have argued that the computer chip has as much significance for social change as the earlier inventions of the wheel and the steam engine. Computer technology has made possible workplace transactions that would have seemed like science fiction just a few years ago. Electronic information can be transferred around the world in less than a second. Employees can provide work for corporations located on another continent; thus, a woman in Southeast Asia or the Caribbean can type a book manuscript for a publishing house in New York. Corporations in the United States are lured overseas by the extremely low wages of these workers.

Increasing reliance on the rapid transmission of electronic data has produced *electronic sweatshops,* a term referring to the back offices found in many industries, such as airlines, insurance firms, mail-order houses, and telephone companies, where workers at computer terminals process hundreds or thousands of transactions in a day. Increasingly, workers' performance is monitored by their computer, thus conjuring images of "Big Brother" invisibly watching. For example, telephone operators are typically given twenty-five seconds to root out a number; the speed of the transaction is recorded by computer. Computers can also measure how fast cashiers ring up groceries and how fast ticket agents book reservations. Records derived from computer monitoring then become the basis for job performance evaluation.

Technological innovation in the workplace is a mixed blessing. **Automation**—the process by which human labor is replaced by machines—eliminates many repetitive and tiresome tasks, and it makes rapid communication and access to information possible. But critics worry that our increasing dependence on technology also makes workers subservient

Automation means that machines can now supply the labor originally provided by human workers, such as these soccer-playing robots.

to machines. For example, robots now perform 98 percent of the spot-welding on Ford automobiles. Some sophisticated robots are capable of highly complex tasks, enabling them to assemble finished products or flip burgers in fast-food restaurants, replacing the human employees. Robots are expensive to buy, but employers can calculate their advantage given that "the robot hamburger-flipper would need no lunch or bathroom breaks, would not take sick days, and most certainly would neither strike nor quit" (Rosengarten 2000: 4).

Deskilling is a side effect of automation and technological progress in which the level of skill required to perform certain jobs declines over time. This may result when a job is automated or when a more complex job is divided into a sequence of easily performed units. As work roles become deskilled, employees are paid less and have less control over their tasks. Less mental labor is required of workers, and jobs become routine and boring. Observers of the workplace note that deskilling contributes to polarization of the labor force. The best jobs require increasing levels of skill and technological knowledge, whereas at the bottom of the occupational hierarchy, people stuck in dead-end positions become alienated from their work (Apple 1991).

Along with deskilling has come an increasing reliance on temporary or **contingent workers** who do not hold regular jobs, but their employment is dependent on labor demand—contract workers, temporary workers, on-call workers (those called only when needed), the self-employed, and day laborers. Women are more likely than men to be employed in these jobs, and women are concentrated in the least desirable jobs—those with the lowest pay and least likelihood of providing benefits. Considering race, Whites are more likely to be independent contractors or self-employed, whereas Blacks and Hispanics are more likely to be found in temporary and part-time work (Kalleberg et al. 2000; Cook 2000; Hudson 1999).

Theoretical Perspectives on Work

Sociologists have devised a number of theoretical perspectives on the nature of work. These perspectives are explored in this section.

Defining Work

Most people think of work as an activity for which a person gets paid. This definition is not adequate for a sociological definition of work because it devalues work that people do without pay. Unpaid jobs such as housework, child care, and volunteer activities make up much of the work done in the world. Sociologists define **work** as productive human activity that creates something of value, either goods or services. Work takes many forms. It may be paid or unpaid. It may be performed inside or outside the home. It may involve physical or mental labor, or both.

Under the sociological definition, housework, though unpaid, is defined as work, even though it is not included in the official measures of productivity that economists use to indicate the work output of the United States. Influenced by feminist studies, sociologists now recognize that housework and other forms of unpaid labor are an important part of the productivity of a society.

Some sociologists argue that work has been too narrowly defined as referring to only physical and mental labor. Arlie Hochschild (1983) has introduced the concept of emotional labor to address some forms of work that are common in a service-based economy. **Emotional labor** is work specifically intended to produce a desired state of mind in a client and often involves putting on a false front before clients. Many jobs require some handling of other people's feelings, and emotional labor is performed where inducing or suppressing a feeling in the client is one of the primary work tasks. Airline flight attendants perform emotional labor—their job is to please the passenger and, as Hochschild suggests, to make passengers feel as though they are guests in someone's living room.

The Division of Labor

The **division of labor** is the systematic *interrelatedness* of different tasks that develop in complex societies. When different groups engage in different economic activities, a division of labor is said to exist. In a relatively simple division of labor, one group

Fritz Hoffman/The Image Works

Race and gender segregation in the labor market mean that women of color are concentrated in occupations where most other workers are also women of color.

may be responsible for planting and harvesting crops, whereas another group is responsible for hunting game. As the economic system becomes more complex, the division of labor becomes more elaborate.

In the United States, the division of labor is affected by gender, race, class, and age—the major axes of stratification. The *class division of labor* can be observed by looking at the work done by people with different educational backgrounds, because education is a fairly reliable indicator of class. People with more education tend to work in higher-paid, higher-prestige occupations. Class also leads to perceived distinctions in the value of manual labor versus mental labor. Those presumed to be doing mental labor (management and professional positions) tend to be paid more and have more job prestige than those presumed to be doing manual labor. Class thus produces stereotypes about the working class; manual labor is presumed to be the inverse of mental labor, meaning it is presumed to require no thinking. By extension, workers who do manual labor may be assumed not to be very smart, regardless of their intelligence.

Think about the labor market in the region where you live. What racial and ethnic groups have historically worked in various segments of this labor market and how would you now describe the *racial and ethnic division of labor*?

The *gender division of labor* refers to the different work that women and men do in society. In societies with a strong gender division of labor, the belief that some activities are women's work (for example, secretarial work) and other activities men's work (for example, construction) contributes greatly to the propagation of inequality between women and men, especially because cultural expectations usually place more value (both social and economic) on men's work. This helps explain why librarians and social workers are typically paid less than electricians despite the likelihood the women have higher education.

Similarly, the *racial division of labor* is seen in the pattern of workers from different racial groups working, in general, in different jobs. The labor performed by racial minority groups has often been the lowest paid, least prestigious, and most arduous work. The racial division of labor found today is rooted in the past.

Functionalism, Conflict Theory, and Symbolic Interaction

The major theoretical perspectives identified in this book also provide the frameworks for understanding the social structural forces that are transforming work. Each viewpoint—conflict theory, functionalist theory, and symbolic interaction—offers a unique analysis of work and the economic institution of which it is a part (see Table 14.2).

Conflict theorists view the transformations taking place in the workplace as the result of inherent tensions in the social systems, tensions that arise

table 14.2 Theoretical Perspectives on Work

	Functionalism	Conflict Theory	Symbolic Interaction
Defines work	As functional for society because work teaches people the values of society and integrates people within the social order; more "talented" people rank higher	As generating class conflict because of the unequal rewards associated with different jobs	As organizing social bonds between people who interact within work settings
Views work organizations	As functionally integrated with other social institutions	As producing alienation, especially among those who perform repetitive tasks	As interactive systems within which people form relationships and create beliefs that define their relationships to others
Interprets changing work systems	As an adaptation to social change	As based in tensions arising from power differences between different class, race, and gender groups	As the result of the changing meanings of work resulting from changed social conditions
Explains wage inequality	As motivating people to work harder	As reflecting the devaluation of different classes of workers	As producing different perceptions of the value of different occupations

from the power differences between groups vying for social and economic resources. Class conflict is then a major element of the social structure of work, and conflict theorists would look to the class division of labor as the source of unequal rewards that workers receive for work and the unequal way the value of their work is perceived. Conflict theorists analyze the fact that some forms of work are more highly valued than others, both in how the work is perceived by society and how it is rewarded. As noted long ago by social-economic theorist **Thorstein Veblen** (1931), mental labor has always been more highly valued than manual labor.

Functionalist theorists, in contrast, interpret the work and the economy as a functional necessity for society. Certain tasks must be done to sustain society, and how work is organized reflects the values and other characteristics of a given social order. Functionalists argue that society "sorts" people into occupations, with the more able sorted into the more prestigious occupations that pay more because, so the functionalist argues, they are more valuable—more "functional"—for society. When society changes too rapidly, as you could argue is the case with new technological and global developments in the world, then work institutions generate social disorganization—perhaps creating **alienation,** a feeling of powerlessness and separation from society.

Symbolic interaction brings a different perspective to the sociology of work. Less interested in the workings of the whole society, *symbolic interaction theorists* might study what work means to those who do it and how social interactions in the workplace form social bonds between people. Some classic symbolic interactionist studies have examined how new workers learn their new roles and how workers' identities are shaped by the social interactions in the workplace. Some symbolic interactionist studies also look at creative ways that people deal with routinized jobs; they sometimes develop elaborate and exaggerated displays of routine tasks to bring a human dimension to otherwise dehumanizing work (Leidner 1993).

Characteristics of the Labor Force

Data on characteristics of the U.S. labor force typically are drawn from official statistics reported by the U.S. Department of Labor. In 2006, the labor force included approximately 150 million people (U.S. Department of Labor 2006). This is 68 percent of the working-age population (those sixteen years old and older)—an increase from 59 percent at the end of the Second World War. The specific characteristics of those in the labor force have also changed considerably.

Who Works?

Employment varies significantly for different groups in the population. Looking at all groups, Hispanic men are the most likely group to be employed, Hispanic women, the least (see Figure 14.3).

Several trends regarding who works can be discerned. One of the most dramatic changes in the labor force since the Second World War has been the increase in the number of women employed. Since 1948, the employment of women has increased from 35 to 60 percent of all women. The calculation is weighted toward White women, because women of color historically have always had a high rate of employment. In fact, the percentage of employed Black and White women has converged in recent years—both are now equally likely to be employed (U.S. Department of Labor 2006). However, Black women are generally employed in lower-paying and less prestigious jobs.

The increase in employment of women with children has also created new demands in the workplace. Some organizations have responded by creating on-site child-care facilities or facilities to care for children who are sick to allow their parents to come to work. The Family and Medical Leave Act of 1993 has also forced companies to write more progressive policies for medical, maternity, and paternity leave for their employees.

Changes in the employment status of diverse groups reflect the economic restructuring of the workplace and, as conflict theorists would point out, produce strife between groups competing for the jobs that are left. Jobs where White men have predominated are declining, especially in the manufacturing

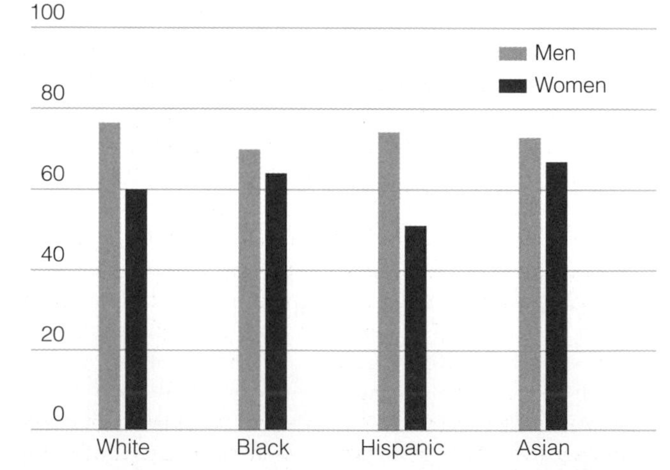

FIGURE 14.3 *Employment Patterns by Race and Gender*

Data: From the U.S. Department of Labor, 2007. *Employment Statistics of the Civilian Population, by Race, Sex, and Age.* Washington, DC: U.S. Department of Labor.

SEE FOR YOURSELF

ANATOMY OF WORKING CLASS JOBS

Identify an occupation that you think of as a working-class job. Then find someone who does this kind of work and who would be willing to talk with you. Ask them to tell you exactly what they do and what they find most difficult and most rewarding about their work. Where does this job fit in the *division of labor*? Who does it and how are they rewarded (or not)? How would each of the three major theoretical perspectives in sociology (functionalism, conflict theory, and symbolic interaction) interpret this work in the context of the broader society?

sector, whereas jobs in race- and gender-segregated niches of the labor market are those most likely to increase. The popular belief that women and minorities are taking jobs from White men derives from this fact. Note, however, that women are not taking the jobs where White men have predominated in the past; these jobs simply are not as numerous as before, and growth in the labor market is in areas that have traditionally been considered women's work or other menial jobs where minorities have been likely to be employed. The conflicts that one sees over labor market issues—beliefs that foreign workers are taking U.S. jobs, competition between groups for scarce positions, and debates about the role of immigrant labor in the U.S. economy—are the result of social structural transformations, that is, sociological factors that lie behind statistical patterns.

Unemployment and Joblessness

The U.S. Department of Labor regularly reports data on the **unemployment rate,** the percentage of those not working but officially defined as looking for work. Currently more than nine million people are officially unemployed (U.S. Department of Labor 2006). The official number does not include all people who are jobless. It includes only those who meet the official definition of a job seeker—someone who has actively sought to obtain a job during the previous four weeks and who is registered with the unemployment office. Several categories of job seekers are excluded from the list, including those who earned money at any job during the week prior to the data being collected, even one day of work with no prospect of more employment. People are excluded who have given up looking for full-time work, because they are discouraged, have settled for part-time work, are ill or disabled, or cannot afford the child care needed to allow them to go to work, among other reasons. Because the number of job-seeking people excluded from the official definition of unemployment can be quite large, the official reported rates of unemployment seriously underestimate the extent of actual joblessness in the United States.

People most likely to be undercounted in unemployment statistics are those for whom unemployment runs the highest—the youngest and oldest workers, women, and members of racial minority groups. These groups are least likely to meet the official criteria of unemployment, because they are most likely to have left jobs that do not qualify them for unemployment insurance. Migrant workers and other transient or mobile populations are also undercounted in the official statistics. People who work only a few hours a week are excluded even though their economic position may differ little from that of someone with no work at all. Workers on strike are also counted as employed, even if they receive no income while they are on strike.

Official unemployment rates also take no notice of **underemployment**—the condition of being employed at a skill level below what would be expected given a person's training, experience, or education. This condition can also include working fewer hours than desired. A laid-off autoworker flipping hamburgers at a fast-food restaurant is underemployed and so is a person with a law degree who drives a taxi.

The group experiencing the greatest unemployment in the United States at this time is Native Americans. Almost one-half of all Native Americans are unemployed. The greatest barrier to employment is the simple absence of jobs, especially for those Native Americans who live on reservations (about one quarter of the population). Few jobs are on or near reservations; the few that do exist are often held by Whites. Reservation land is typically cut up so the fertile areas are owned by Whites. Tribal crafts and small-scale farming are usually all that is left to Native Americans. For most jobs, one must leave the community to find work. Lack of appropriate training in schools and job-training programs means that many American Indians are not prepared for the few available jobs (Snipp 1996). Finally, on measures of health, Native Americans fare worse than any other group, making work impossible or difficult for many.

Unemployment among African Americans, Puerto Ricans, and Mexican Americans is currently at a level considered to be a major economic depression (between 8 and 10 percent). Contributing to the problem are some of the trends already discussed in this chapter—globalization, economic restructuring, and the move to a service economy—that fall especially hard on minority workers. In addition to being more likely to be unemployed, African Americans are more likely than Whites to experience the negative effects of job displacement, and the reemployment of African American men after losing a job is significantly lower

doing sociological research

All in the Family: Children of Immigrants and Family Businesses

A relatively common pattern among Asian American immigrants is to establish a small family business, using the labor of family members to establish and run the business. Owning a family business is part of the American dream, thus idealized as promoting entrepreneurial and family values, including the value of hard work, devotion to family, and self-sacrifice.

Research Question: What kind of life does this produce for the children of immigrant families? This is what sociologist Lisa Park wanted to know in her research on the children of Asian American entrepreneurs.

Research Methods: Park, a child of such entrepreneurs herself, conducted her study by interviewing Korean American and Chinese American high school and college students who grew up in small family businesses. She utilized several qualitative research methods, including (1) in-depth individual interviews, (2) focus group interviews, and (3) participant observation in family stores.

Research Results: Although many teens hold jobs—often to help support their families, sometimes for their own spending money—Park found that the life of Asian American teenagers who are sons and daughters of entrepreneurs was difficult. Her research subjects reported wanting to have the freedom to be bored that they associated with White, middle-class teens. They were critical of the popular images of teens in families as only worrying about their hair, their clothes, and their friends.

In contrast to the stereotype of Asian American families as close-knit and conflict-free, Park found that teens in these families report feeling overworked and with little family "quality time." Some of these young people have worked since they were six years old. Often parents set up a small "home away from home" in the family restaurant or business—a small table in the corner where children can read, color, or watch television. But her subjects also reported being burdened and not able to have a normal childhood. They often had to do all the household chores while parents worked, as well as serving customers or helping in other ways.

Conclusions and Implications: Park's research shows immigration shapes family structures and affects children's lives. She challenges simplistic views of the family business as an ideal way of life and helps us understand some of the social forces that are part of the immigrant experience and that shape the life of some Asian American youth.

Questions to Consider

1. Are there businesses in your community run by immigrant families? If so, what evidence of children's involvement in the business do you see and how do you think this influences the children?
2. In addition to immigrant businesses, how are children involved in the work that their parents do? In what ways does this influence the children's identities, values about work, future opportunities?

Source: Park, Lisa Sun-Hee. 2002. "A Life of One's Own." *Contexts* 1 (Summer): 56–57.

than the reemployment of White men (U.S. Department of Labor 2006).

Popular explanations often attribute unemployment to the individual failings of workers. Many often claim that unemployed people simply do not try hard enough to find jobs or prefer a welfare check to hard work and a paycheck. This viewpoint reflects the common myth that anyone who works hard enough can succeed. This individualistic perspective traces the cause of unemployment to personal motivation. Rather than blame the victim for joblessness, sociologists examine how changes in the social organization of work contribute to unemployment by looking at structural problems in the economy—rapidly changing technology (which reduces the need for human labor), discriminatory employment practices, deindustrialization, corporate downsizing, and the exportation of jobs overseas, where cheap labor is abundant.

Debunking Society's Myths

Myth: Because of programs such as affirmative action, women and minorities are taking jobs away from White men.

Sociological perspective: Although women and minorities have made many gains as a result of programs such as affirmative action, women and minority workers still are clustered in occupations that are segregated by race and gender. There is little actual evidence that the progress of these groups has been at the expense of White men.

Some also argue that unemployment for White men is the result of hiring more women and minorities, as if men's unemployment were the result of affirmative action programs. Although on the surface this may seem reasonable because there are a finite number of jobs, the truth is that the employment of women and minorities does not cause White male unemployment, if for no other reason than race and gender segregation. Women are typically not employed in the same jobs as men; race segregation also means that racial minorities are concentrated in certain segments of the labor force.

Popular wisdom holds that the bulk of new immigrants are illegal, poor, and desperate, but the data show otherwise. The contemporary labor force is also being shaped by the employment of recent immigrants (Rumbaut 1996a, 1996b; Pedraza and Rumbaut 1996; Lamphere et al. 1994). In fact, the proportion of professionals and technicians among legal immigrants exceeds the proportion of professionals in the labor force as a whole. Note, however, that this figure is based on formal immigration data that exclude undocumented immigrants, most of whom are working class. Still, those who migrate are usually not the most downtrodden in their home country; seldom are the poorest able to migrate. Even undocumented immigrants tend to have higher levels of education and occupational skills than the typical workers in their homeland. Among immigrants can be found both the most educated and the least educated segments of the population.

Diversity in the U.S. Occupational System

From a sociological perspective, jobs are organized into an *occupational system.* This system is the array of jobs that together constitute the labor market. Within the occupational system, people are distributed in patterns that reflect the race, class, and gender organization of society. Jobs vary in their economic rewards, their perceived value and prestige, and the opportunities they hold for advancement, a fact that sociologists explain through a variety of perspectives. Fundamental to sociological analyses of the occupational system is the idea that individual and group experiences at work are the result of social structures, *not just individual attributes.* There is, for example, a rough correlation between the desirability of given jobs in the occupational system and the social status of the group most likely to fill those jobs. This tells us that there is a relationship between work and social inequality. When sociologists study the occupational system, they acknowledge the importance of individual attributes such as motivation, training, and previous work experience in predicting the place of a worker in the occupational system, but they see a greater importance in the societal patterns that structure the experience of workers.

The Dual Labor Market. A branch of conflict theory is the **dual labor market theory,** which views the labor market as comprising two major segments: the *primary labor market* and the *secondary labor market,* as we saw in Chapter 10 (see Table 14.3). The primary labor market offers jobs with relatively high wages, benefits, stability, good working conditions, opportunities for promotion, job protection, and due process for workers (meaning workers are treated according to established rules and procedures that are allegedly fairly administered). Blue-collar and service workers in the primary labor market are often unionized, which leads to better wages and job benefits. High-level corporate jobs and unionized occupations fall into this segment of the labor market.

The secondary labor market is characterized by low wages, few benefits, high turnover, poor working conditions, little opportunity for advancement, no job protection, and arbitrary treatment of workers. Many service jobs such as waiting tables, nonunionized assembly work, and domestic work are in the secondary labor market. Women and minority workers are the most likely groups to be employed in the secondary labor market. Dual labor market theory traces some of the causes of race and gender inequality to this divided structure within the labor market (U.S. Department of Labor 2006; McCall 2001; Padavic and Reskin 2002; Bonacich 1972).

Occupational Distribution. Occupational distribution describes the pattern by which workers are located in the labor force. Workers are dispersed throughout the occupational system in patterns that vary greatly by race, class, and gender, revealing a certain **occupational segregation** on the basis of such characteristics. Women are most likely to work in technical, sales, and administrative support, primarily because of their heavy concentration in clerical work. This is now true for both White women and women of color. White men are most likely found in managerial and professional jobs, whereas African American and Hispanic men are most likely employed

table 14.3 The Segmented Labor Market: Examples of Occupations

Primary Labor Market	Secondary Labor Market	Underground Economy
Upper tier:	*Upper tier:*	
High-status professional specialties (physicians, lawyers, professors, engineers, and so on)	Food servers, bartenders	Prostitution
Business executives	Hairdressers, cosmetologists	Con artists
Supervisors	Sales workers, cashiers	Unreported domestic labor
Farm managers	Clerical workers, clerks Machine operators	Undocumented workers Sweatshop labor Thieves Drug dealers
Lower tier:	*Lower tier:*	
Lower-status professional specialties (teachers, librarians, social workers, actors)	Service work (cooks, maids, janitors, child care)	
Middle managers	Private household workers	
Protective service (police and fire)	Farm labor	
Truck driving	Food counter workers	
Mechanics		
Precision craft, repair, production (electricians, plumbers, welders, construction, and so on)		
Technicians		
Health assessment and training		

Adapted from: Ammot, Teresa, and Julie Matthaei. 1996. *Race, Gender and Work: A Multicultural History of Women in the United States.* Boston: South End Press, p. 344; U.S. Department of Labor. 2006. *Employment and Earnings.* Washington, DC: U.S. Department of Labor.

as operators and laborers—among the least well paid and least prestigious in the occupational system (U.S. Department of Labor 2006).

When looking at occupational distribution over time, changes are noticeable. One decline has been the number of Black women employed in private domestic work. In 1960, 38 percent of all Black women in the labor force were employed as private domestic workers; by the 1990s, this had declined to less than 2 percent. Over time, there has also been some increase in the number of women employed in working-class jobs traditionally held only by men. Today, women are 12 percent of precision production, craft, and repair workers, compared with 1 percent in 1970 (U.S. Department of Labor 2006). This is a significant increase in the number of women in these jobs, although women are still a small proportion of all such workers. The women most likely to be in such jobs are African American women, low-income mothers, women who previously held lower-rank jobs and disliked their previous job, and women who can work hours other than during the day. From this information, we can conclude that women are attracted to these jobs not just because they prefer them, but because they offer better opportunities (Landry 2000).

Occupational Prestige and Earnings. *Occupational prestige* is the perceived social value of an occupation in the eyes of the general public (see Chapter 7). Sociologists have found a strong correlation between occupational prestige and the race and gender of people employed in given jobs. The influence of race on occupational prestige is stronger than that of gender. Thus, African American and Latino men are disproportionately found in jobs that have the lowest occupational prestige scores; White and Asian American men hold the jobs with the highest occupational prestige, followed by White women, Asian American women, African American women, and Latinas. The higher the socioeconomic status of the occupational group, the smaller the proportion of employed African American men (Xu and Leffler 1992; Stearns and Coleman 1990).

Gender also has an effect on occupational prestige. Three patterns are apparent. First, women receive less prestige for the same work as men. Second, the gender composition of the job and its occupational prestige are linked: Jobs that employ mostly women are lower in prestige than those that employ more men. Indeed, jobs often lose their prestige as many women enter a given profession. Likewise, the prestige of jobs increases as more men enter the field. Finally, men and women assign occupational prestige differently; women give occupations where most of the incumbents are women a higher prestige ranking than men do (Bose and Rossi 1983; Tyree and Hicks 1988).

An important question for sociologists, as well as for policy makers and individuals, for that matter, is why different groups earn different incomes. Sociologists have extensively documented that earnings from work are highly dependent on race, gender, and class, as shown in Figure 14.4. White men earn the most, with a gap between men's and women's earnings among all groups. African American women and Hispanic men and women earn the least. Occupations in which White men are the numeric majority tend to pay more than occupations in which women and minorities are a majority of the workers. Not all men benefit equally, however, from the earnings hierarchy. CEOs have the highest pay of all—a whopping 150 times (and more) than what workers earn.

Why are there such disparities? Again, we have to turn to theory to put the facts into perspective. According to functionalist theories, workers are paid according to their value—derived from the characteristics they bring to the job: education, experience, training, and motivation to work. As we saw in Chapter 7, when we reviewed functionalist perspectives on inequality, the functionalist theorists see inequality as what motivates people to work. From this point of view, the high wages and other rewards associated with some jobs are the incentive for people to spend long years in training and garnering experience; otherwise, the jobs would go unfilled. To functionalists, then, differential wages are a source of motivation and a means to ensure that the most talented workers fill jobs essential to society and that different wages reflect the differently valued characteristics (education, years experience, training, and so forth) that workers bring to a job.

Conflict theorists strongly disagree with the functionalist point of view, arguing that many talented people are thwarted by the systems of inequality they encounter in society. Thus, far from ensuring that the most talented will fill the most important jobs, conflict theorists note that some of the most essential jobs are, in fact, the most devalued and underrewarded. From a conflict perspective, wage inequality is one way that systems of race, class, and gender inequality are maintained.

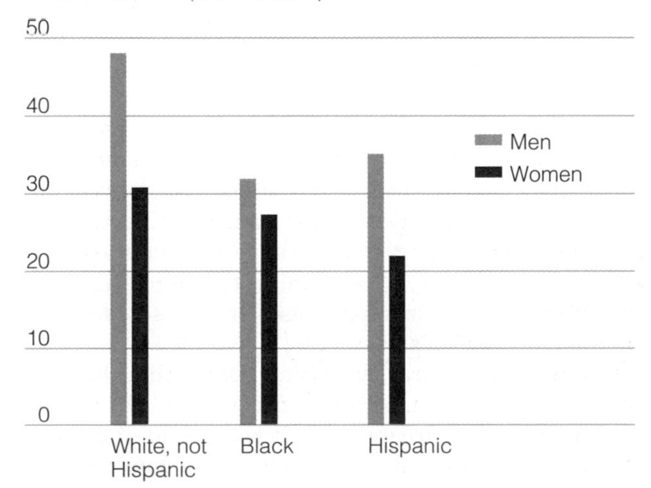

FIGURE 14.4 *The Income Gap*

Data: DeNavas-Walt, Carmen, Bernadette D. Proctor, and Jessica Smith. 2007. *Income, Poverty, and Health Insurance Coverage in the United States 2006.* Washington, DC: U.S. Census Bureau.

Power in the Workplace

Obvious tangible factors that influence the degree of job satisfaction are the employee's level of pay and benefits package. Also very important but less quantifiable is the value accorded to one's job and the opportunity for advancement.

As we saw in Chapter 5, the **glass ceiling** is to the term used to describe the limits that women and minorities experience in job mobility. Over the last few decades, many barriers to the advancement of women and minorities have been removed, yet as they try to advance, invisible barriers still prevent further advancement. The existence of the glass ceiling is well documented. Although there has been an increase in the number of managers who are women and minorities, most top management jobs are still held by White men, who are far more likely than other groups to control a budget, participate in hiring and promotion, and have subordinates who report to them. Women and racial minorities remain clustered at the bottom of managerial hierarchies.

Sexual Harassment

In addition to having less income and authority on the job than men, women workers are also more likely to experience **sexual harassment,** defined as unwanted physical or verbal sexual behavior that occurs in the context of a relationship of unequal power and that is experienced as a threat to the victim's job or educational activities (Saguy 2003; Martin 1989). Two primary forms of sexual harassment are recognized in the law. *Quid pro quo sexual harassment* forces sexual compliance in exchange for an employment or educational benefit. A professor who suggests

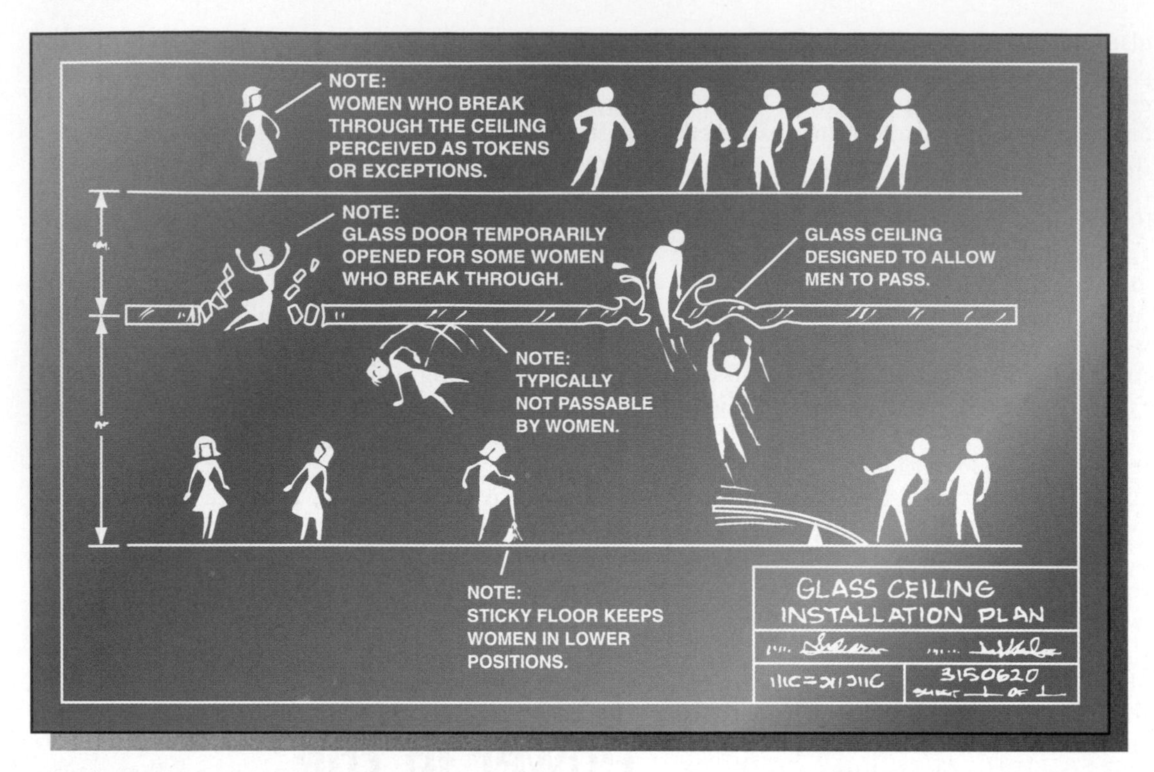

The Glass Ceiling

Whimsically depicted as an architectural drawing by Norman Andersen, the glass ceiling refers to the structural obstacles still inhibiting upward mobility for women workers. Most employed women remain clustered in low status, low-wage jobs that hold little chance for mobility. Those who make it to the top often report being blocked and frustrated by patterns of exclusion and gender stereotyping.

to a student that going out on a date or having sex would improve the student's grade is engaging in quid pro quo sexual harassment. The other form of sexual harassment recognized by law is the creation of a *hostile working environment* in which unwanted sexual behaviors are a continuing condition of work. This kind of sexual harassment may not involve outright sexual demands, but includes unwanted behaviors such as touching, teasing, sexual joking, and other kinds of sexual comments.

Sexual harassment was first made illegal by Title VII of the 1964 Civil Rights Act, which identifies sexual harassment as a form of sex discrimination. In 1986, the U.S. Supreme Court upheld the principle that sexual harassment violates federal laws against sex discrimination (*Meritor Savings Bank* v. *Vinson*). The law defines sexual harassment as discriminatory because it places workers at a disadvantage on the basis of their sex. By creating a hostile working environment, sexual harassment makes productive work difficult and discourages educational and work advancement.

More recent Supreme Court cases have also upheld that same-sex harassment also falls under this law, as does harassment directed by women against men. Fundamentally, sexual harassment is an abuse of power by which perpetrators use their position to exploit subordinates. The law also makes employers liable for financial damages if they do not have policies appropriate for handling complaints or have not educated employees about their paths of redress.

The true extent of sexual harassment is difficult to estimate because it tends to be underreported. Most studies find that typically neither women nor men are aware of the proper channels for reporting sexual harassment. Women are less likely than men to report sexual harassment primarily because they believe that nothing will be done to stop the behavior (VanRoosmalen and McDaniel 1998; Cortina et al. 1998). Most surveys indicate that as many as one-half of all employed women experience some form of sexual harassment at some time in their working lives. Men are sometimes the victims of sexual harassment, although far less frequently than women; most studies find that less than 3 percent of all sexual harassment cases involve women supervisors harassing male employees.

The typical harasser is male, older than his victim, and of the same race and ethnicity. He is also likely to have harassed other women in the past. There is some evidence that women of color are more likely to be harassed than White women (Gruber 1982). Same-gender harassment also occurs; when men are harassed by other men, they react with more severe consequences than do men who have been harassed by women (DuBois et al. 1998).

Gays and Lesbians in the Workplace

The increased willingness of lesbians and gay men to be open about their sexual identity has resulted in more attention paid to their experience in the workplace. Surveys find that a large majority of the U.S. population (85 percent) endorse the general concept that lesbians and gay men should have equal rights in job opportunities, but when asked about specific occupations, many people think gays should not be hired in some occupations. Thus 78 percent of Americans now say that gays should be hired as doctors, compared with only 44 percent who thought so in the 1970s; 72 percent also think that gays should be employed in the armed forces and be included in the president's cabinet; 63 percent think they should be employed as high school teachers—up from only 36 percent in 1989 (Mazzuca 2002; Newport 2001).

There is increasing acceptance of lesbians and gays generally, but negative experiences in the workplace still affect self-esteem, productivity, and economic and social well-being for gay and lesbian workers. Although there are few empirical studies of gay and lesbian work experiences, those that do exist find that gays fear they will suffer adverse career consequences if heterosexual coworkers know they are gay. This leads many to "pass" as heterosexual at work or keep their private lives secret, which puts gay workers at a disadvantage. Shielding themselves from antagonism or rejection may make them appear distant and isolate them from social networks. These behaviors can actually have a negative effect on their performance reviews since fellow workers may find them unfriendly and withdrawn. Research finds that the relationships of gay employees with their coworkers are less stressful when the employees are open about their identity (Schneider 1984). The work organization is also improved for lesbian and gay employees when employee assistance programs encourage open communication, policies are sensitive to lesbian and gay needs, and discriminatory practices can be identified and stopped (Sussal 1994). Nonetheless, one-third of gay employees keep their social life separate from work.

Disability and Work

Not too many years ago, people with disabilities were not thought of as a social group; rather, disability was thought of as an individual frailty or perhaps a stigma. Sociologist **Irving Zola** (1935–1994) was one of the first to suggest that people with disabilities face issues similar to those of minority groups. Instead of using a medical model that treats disability like a disease and sees individuals as impaired, conceptualizing disabled people as a minority group enabled people to think about the social, economic, and political environment that disabled people face. Instead of seeing disabled people as pitiful victims, this approach emphasizes the group rights of the disabled, illuminating things such as access to employment and education (Zola 1989, 1993).

The Americans with Disabilities Act provides legal protection for disabled workers, giving them rights to reasonable accommodations by employers and access to education and jobs.

As we discussed in Chapter 13, now those with disabilities have the same legal protections afforded to other minority groups. Key to these rights is the Americans with Disabilities Act (ADA), adopted by Congress in 1990. Building on the Civil Rights Act of 1964 and earlier rehabilitation law, in particular the Rehabilitation Act of 1973, the Americans with Disabilities Act protects disabled people from discrimination in employment and stipulates that employers and other providers (such as schools and public transportation systems) must provide "reasonable accommodation." Disabled people must be qualified for the jobs or activities for which they seek access, meaning that they must be able to perform the essential requirements of the job or program without accommodation to the disability. For students, reasonable accommodation includes provision of adaptive technology, exam assistants, and accessible buildings.

The law prohibits employers with fifteen or more employees from discriminating against job applicants who are disabled or current employees who become disabled, including job application, wages and benefits, advancement, and employer-sponsored social activities. The law applies to state and local governments, as well as employers. ADA also legislates that public buses, trains, and light rail systems must be accessible to disabled riders; airlines are excluded from this requirement. The law also requires businesses and public accommodations to be accessible and requires telephone companies to provide services that allow hearing- and speech-impaired people to communicate by telephone.

Not every disabled person is covered by ADA. To be considered disabled, a person must have a condition or the history of a condition that impairs a major life activity. The law specifically excludes the use of illegal drugs as a cause of disability and excludes pregnancy as a disability—which is ironic because pregnancy is treated as a disability by other federal laws for purposes of maternity leave.

What effect does pregnancy have on women's work patterns? The common myths that women who become pregnant drop out of the labor force are simply untrue. Because most women work out of economic necessity, maternity leaves are generally short. Women who work the most hours and women in higher-status jobs are the least likely to leave a job following the birth of a child. Family circumstances are also a strong predictor of whether a woman will return to work following a birth and how soon. Contrary to what one might expect, mothers with a spouse or another adult in the household return to work more quickly than those without other adults in their household. The greater the proportion of the family income a mother provides, the sooner she returns to work. There are no differences by race in these patterns (Wenk and Garrett 1992).

Chapter Summary

- ***What is the state?***

  The *state* is the organized system of power and authority in society. It comprises different institutions, including the government, the military, the police, the law and the courts, and the prison system. The state is supposed to protect its citizens and preserve society, but it often protects the status quo, sometimes to the disadvantage of less powerful groups in the society.

- ***How do sociologists define power and authority?***

  Power is the ability of a person or group to influence another. *Authority* is power perceived to be legitimate. There are three kinds of authority—*traditional authority,* based on long-established patterns; *charismatic authority,* based on an individual's personal appeal or charm; and *rational–legal authority,* based on the authority of rules and regulations (such as law).

- ***What theories explain how power operates in the state?***

  Sociologists have developed four theories of power. The *pluralist model* sees power as operating through the influence of diverse interest groups in society. The *power elite model* sees power as based on the interconnections between the state, industry, and the military. *Autonomous state theory* sees the state as an entity in itself that operates to protect its own interests. *Feminist theorists* argue that the state is patriarchal, representing primarily men's interests.

- ***How well does the government represent the diversity of the U.S. population?***

  An ideal democratic government would reflect and equally represent all members of society. The makeup of the U.S. government does not reflect the diversity of the general population. African Americans, Latinos, Native Americans, Asians, and women are underrepresented within the government. Political participation also varies by a number of social factors, including income, education, race, gender, and age. African Americans and Latinos, however, are overrepresented in the military, in part because of the opportunity the military purports to offer groups otherwise disadvantaged in education and the labor market; however, both are underrepresented at the levels of high-level commissioned officers. There is an increased presence of women in the military. However, prejudice and discrimination continue against lesbians and gays in the military.

- ***How are societies economically organized?***

  Societies are organized around an economic base. The *economy* is the system on which the production, distribution, and consumption of goods and services are based. *Capitalism* is an economic system based on the pursuit of profit, market competition, and private property. *Socialism* is characterized by state ownership of industry; *communism* is the purest form of socialism.

- ***How do sociologists define work?***

  Sociologists define *work* as human activity that produces something of value. Some work is judged to be more valuable than other work. *Emotional labor* is work that is intended to produce a desired state of mind in a client. The *division of labor* is the differentiation of work roles in a social system. In the United States there is a class, gender, and racial division of labor.

- ***How has the global economy changed?***
  As capitalism has spread throughout the world, *multinational corporations* conduct business across national borders. A number of countries have undergone *deindustrialization,* or changeover from a goods-producing economy to a services-producing one, causing many heavy-industry jobs (such as those in steel production) in U.S. cities to vanish, thus increasing the unemployment rate in those cities. Changes in information technology, plus increased *automation,* have resulted in the further elimination of jobs in both the United States and abroad.

- ***What are some of the characteristics of the contemporary labor force?***
  In recent years, the employment of women has been increasing whereas that of men has been decreasing. Official *unemployment rates* underestimate the actual extent of joblessness. Women and racial minorities are most likely to be unemployed. Recent immigrants also have a unique status in the labor market, often in the lowest paying and least prestigious jobs. The U.S. occupational system is characterized by a *dual labor market.* Jobs in the primary sector of the labor market carry better wages and working conditions, whereas those in the secondary labor market pay less and have fewer job benefits. Women and minorities are disproportionately employed in the secondary labor market. Patterns of occupational distribution also show tremendous segregation by race and gender in the labor market. Race and gender also affect the occupational prestige, as well as the earnings, of given jobs.

- ***How does power in the workplace affect women, racial–ethnic groups, and gays and lesbians?***
  Women and minorities often encounter the *glass ceiling*—a term used to describe the limited mobility of women and minority workers in male-dominated organizations. In addition, women more often than men face *sexual harassment* at work—defined as the unequal imposition of sexual requirements in the context of a power relationship. Homophobia in the workplace also negatively affects the working experience of gays and lesbians.

Key Terms

alienation 390
authority 373
automation 387
autonomous state model 377
bureaucracy 374
capitalism 385
charismatic authority 374
communism 385
contingent worker 388
deindustrialization 386
democracy 378
division of labor 388
dual labor market theory 393
economic restructuring 386
economy 384
emotional labor 388
gender gap 380
glass ceiling 395
global economy 385
government 378
interest group 375
interlocking directorate 376
job displacement 387
multinational corporations 385
occupational segregation 393
pluralist model 375
political action committee (PAC) 376
postindustrial society 385
power 373
power elite model 376
propaganda 373
rational–legal authority 374
sexual harassment 395
socialism 385
state 372
traditional authority 373
underemployment 391
unemployment rate 391
work 388
xenophobia 386

Online Resources

Sociology: The Essentials Companion Website

www.thomsonedu.com/sociology/andersen

Visit your book companion website where you will find more resources to help you study and write your research papers. Resources include Suggested Readings, web links, and a MicroCase Online feature that teaches you how to research society. Other resources include Learning Objectives, Internet exercises, quizzing, and flash cards.

is an easy-to-use online resource that helps you study in less time to get the grade you want NOW.

www.thomsonedu.com/login

Need help studying? This site is your one-stop study shop. Take a Pre-Test and Thomson NOW will generate a Personalized Study Plan based on your test results. The Study Plan will identify the topics you need to review and direct you to online resources to help you master those topics. You can then take a Post-Test to determine the concepts you have mastered and what you still need to work on.

15

Chapter fifteen

CHAPTER FIFTEEN

Chapter

Population, Urbanization, and the Environment

The study of population and environment is a most important topic as we begin the new millennium. A great many issues of central concern to the entire country are driven by population. Population growth and density are responsible for the major policy issues of the day, such as overcrowding in cities, traffic jams, long lines at markets and other stores, environmental pollution, family planning, diminishing resources, and food shortages. Population issues head the federal government's list of problems to be solved, and the U.S. Census Bureau debates with Congress the methods for accurately counting the number of people who constitute the population of the United States. The federal as well as the state governments endlessly

continued

fifteen chapter fifteen

debate who should or should not be permitted to immigrate into the United States. Finally, events such as the terrorist attacks on September 11, 2001, presumably by foreign persons residing in the United States, have—rightly or wrongly—resulted in stricter government immigration policies.

The estimated population of the United States is presently more than 280 million. At the current rate of growth, the country will reach almost 300 million by the year 2025. The population has more than doubled since 1946, when it stood at approximately 132 million. In that year, a spike in the number of births began that lasted until 1964—from the end of the Second World War to the beginning of the Vietnam War. The Baby Boom is the name given to that crop of 75 million babies, who are now middle-aged and older adults constituting nearly one-third of all the people in the United States. The generation that in the tumultuous 1960s warned, "Don't trust anyone over thirty," is now hovering in its mid-fifties and early sixties and starting to have grandchildren.

Babies are born every year to about 70 in 1000 American women aged 15 to 45, averaged across all social classes and racial-ethnic groups. Planning for children greatly affects the goals that adults set for themselves. For example, if you want to go directly into a career, you may choose to postpone having a child. Having a career and a child at the same time is a heavy strain—but many people do it nonetheless. College or professional school may cause you to postpone your decision to have a child. That first job or overseas assignment may cause you to put off children yet again. The decision to have or to adopt children is among the most important that you will ever make, affecting not only your own life, but that of your child, the rest of your family, and ultimately, society.

The education and occupation structure of the country has been greatly affected by the decisions of so many people to have children during the previous years. Today, as globalization, economic restructuring, terrorism, and war cause young people to feel insecure about their future, many twenty-something couples are deciding, at least for the present, not to have children. Their decisions will ripple forward in the years to come, as fewer youngsters are enrolled in grade school and then in college two decades down the road. Inevitably, there will follow a decline in the number of middle-aged people to take care of others in their old age.

In this chapter, we examine the nature of human populations. How do births, deaths, and migrations affect society? Who in our society is most likely to die young or die old? How do demographic hazards such as accidental death or exposure to toxic waste differ across racial and ethnic groups? How does the increasingly urban basis of society affect us? Finally, we will address a question that many people find most chilling of all: Are there too many people on this planet?

Demography and the U.S. Census

The scientific study of population is called **demography,** which is the study of the current state and changes over time in the size, distribution, and composition of human populations. This field of sociology draws on huge bodies of data generated by a variety of sources. One source is the U.S. Census Bureau.

A **census** is a head count of the entire population of a country, usually done at regular intervals. The census conducted every ten years by the U.S. Census Bureau, as required by the U.S. Constitution, attempts not only to enumerate every individual, but also to obtain information such as gender, race, ethnicity, age, education, occupation, and other social factors.

The 2000 U.S. Census is estimated to have undercounted a small percentage of the country's population. Among those most likely to have been undercounted by the census are the homeless, immigrants, minorities living in ghetto neighborhoods, and other residents of low social status. The constitutional requirement for a census was included to ensure fair apportionment of representatives in the federal government. Undercounting minorities and the underclass tends to leave these groups underrepresented in government. The estimated undercount for the entire U.S. population overall is only about 2 percent, yet the undercount for African Americans nationally has been estimated to be as high as 20 percent, and for Hispanics, as high as 25 percent (NAACP Legal Defense and Education Fund 2006; Harrison 2000).

The U.S. Congress and the U.S. Census Bureau have fiercely debated the following issue: Should people be allowed to select multiracial (or "mixed race") responses on the census questionnaire, which it now does, by allowing someone to check more than one racial–ethnic category from among several, such as African American, Hispanic, non-Hispanic White, Native American, Asian, Eskimo, and Aleut; or, should individuals be limited to checking one category only. The use of the multiracial response option gives individuals an opportunity to define themselves as mixed race. One argument against this option is that it subtracts from the number of people who would have otherwise indicated only one category, especially a racial–ethnic minority category, further undercounting African Americans, Hispanics, and Native Americans. A recent study estimates, however, that only about 2 percent of the persons responding actually indicate a multiracial response (Harrison 2000).

Another body of data used in demography is **vital statistics,** which includes information about births, marriages, deaths, migrations in and out of the country, and other fundamental quantities related to population. From the census and vital statistics gathered using a wide variety of sources, we can create a picture of the U.S. population—who we are, how we are changing, and even who we will be in the future.

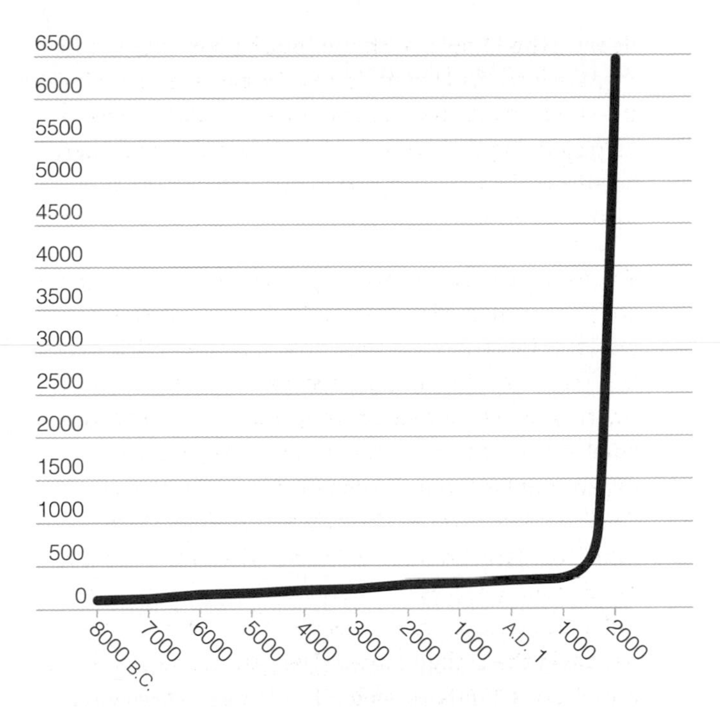

FIGURE 15.1 *World Population Growth (in Millions)*

Data: From the Population Reference Bureau. 2007. Washington, DC. **www.prb.org**. Reprinted with permission.

Diversity and the Three Basic Demographic Processes

The total number of people in a society at any given moment is determined by only three variables: births, deaths, and migrations. These three variables show different patterns for different racial and ethnic groups, different social strata, and both genders. Births add to the total population, and deaths subtract from it. Migration into a society from outside, called **immigration,** adds to the population, whereas **emigration,** the departure of people from a society (also called "out-migration"), subtracts from the population.

The total population of the world is increasing at a rate of approximately 270,000 people per day, or just less than 200 people per minute. The world's population does not increase in a linear fashion—the line on a graph of population does not rise in a straight line, with the same number of people added each year. Instead, the population grows exponentially, with an upward accelerating curve, as shown in Figure 15.1. An ever increasing number of people are added each year. At the present rate of growth, the world's population will double in forty years, barring some major catastrophe such as world war, an international epidemic of an untreatable disease, or some such global calamity.

Birthrate

The **crude birthrate** of a population is the number of babies born each year for every 1000 members of the population or, alternatively, the number of births divided by the total population, multiplied by 1000:

$$\text{Crude birthrate (CBR)} = \frac{\text{number of births}}{\text{total population}} \times 1000$$

The crude birthrate for the entire world population is approximately 27.1 births per 1000 people. Various countries and different subgroups within a country can have dramatically different birthrates. For example, the country with the highest birthrate in the world is Niger, with 50.4 births per 1000 people. The lowest birthrate is found in Belarus, with only 9 births per 1000 people (U.S. Census Bureau 2006).

The crude birthrate reflects what is called the *fertility* of a population, which is live births per number of women in the population. Fertility is different from *fecundity,* which is the potential number of children in a population that could be born (per 1000 women) if every woman reproduced at her maximum biological capacity during the childbearing years. If a majority of women had on the average around twenty or more children, fecundity is what current demographic estimates suggest as a maximum estimate. Fertility rates are lower than fecundity rates because few women in any given population reproduce at anywhere near the theoretical biological maximum.

The overall birthrate for the United States is approximately 16.5 births per 1000 people, compared with the all-time high rate of 27 births per 1000 people in 1946, the start of the so-called Baby Boom that followed the Second World War. The rate varies according to racial–ethnic group, region, socioeconomic status, religion, and other factors. Overall, we find that for different racial–ethnic groups, the birthrates in the United States are as follows (National Center for Health Statistics 2006):

Group	Birthrates per Thousand
Whites	14.1
African Americans	17.6
Latinos (Mexican Americans, Puerto Ricans, and other Hispanic groups combined)	24.0
Native Americans	17.1
Asian Americans	17.8

In general, minority groups tend to have somewhat higher birthrates than White nonminority groups, and lower socioeconomic groups tend to have higher birthrates than those higher on the socioeconomic scale. Native Americans have a birthrate higher than the national average, even though the Native American birthrate has *declined* steadily over the past fifty years.

The effects of birthrates are somewhat cumulative. For example, minorities tend to be overrepresented at the lower end of the socioeconomic scale, compounding the likelihood of a high birthrate. Similarly, religious and cultural differences can make themselves felt. Catholics, for example, have a higher birthrate than non-Catholics of the same socioeconomic status. Hispanic Americans have a high likelihood of being Catholic, another factor that contributes to the higher birthrate among Hispanic Americans. Because minorities tend to have higher birthrates, assuming that present migration rates continue and assuming that death rates do not outstrip the birthrates, in the coming years the United States will have a significantly greater proportion of minorities, thus a relatively lower proportion of Whites.

Given current birthrates, the population of Latinos in the United States will double by the year 2030, and the population of Asian Americans will double by the year 2035. This increase has great implications for local and national policy. Local, state, and federal governments must increasingly take these projected increases into account for future legislation and programs.

Death Rate

The **crude death rate** of a population is the number of deaths each year per 1000 people, or the number of deaths divided by the total population, times 1000:

$$\text{Crude death rate (CDR)} = \frac{\text{number of deaths}}{\text{total population}} \times 1000$$

The crude death rate can be an important measure of the overall standard of living for a population. In general, the higher the standard of living enjoyed by a country, or a group within the country, the lower the death rate. The death rate of a population also reflects the quality of medicine and health care. Poor medical care, which goes along with a low standard of living, will correlate with a high death rate.

Another measure that can reflect the standard of living in a population is the **infant mortality rate,** which is the number of deaths per year of infants less than one year old for every 1000 live births. In the United States, the overall infant mortality rate is approximately seven infant deaths for every 1000 live births. The highest infant mortality rates among seventy-seven countries throughout the world are Angola, with a rate of 200, and Afghanistan, with a rate of 150 (U.S. Census Bureau 2006).

Infant mortality rates, a measure of the chances of the very survival of members of the population, are important to compare across racial–ethnic groups and across social class strata. They are a good indicator of the overall quality of life, as well as the survival chances for members of that racial or class group. Inadequate health care and health facilities cause higher infant mortality rates. Consequently, the higher infant mortality among minorities and those in lower socioeconomic strata in the United States suggests the lack of adequate health care and adequate access to health facilities. There are also many other causes of higher infant mortality, such as presence of toxic wastes, malnutrition of the mother, inadequate food, and outright starvation.

The **life expectancy** of a population or group is defined as the average number of years a member of the group can expect to live. In the United States, life expectancy has gone from forty years of age in 1900 to over seventy-seven years of age for people born now. Although a life expectancy of seventy-seven years

As shown in this painting by African American artist Jacob Lawrence, in the 1920s from virtually every southern town, African Americans left by the thousands to go north to enter northern industry. This has come to be known as the Great Black Migration in America.

table 15.1 Life Expectancy and Infant Mortality (by country, industrialized countries only)

Country	Life Expectancy	Infant Mortality Rate[a]
Japan	80.8	3.9
Australia	79.8	5.0
Canada	79.4	5.0
Italy	79.1	5.8
France	78.9	4.5
Spain	78.9	4.9
Netherlands	78.4	4.4
United Kingdom	77.8	5.5
Germany	77.6	4.7
United States	76.6	6.8
China	71.6	28.1
Russia	67.3	20.1

[a]Per 1000 live births
Source: U.S. Census Bureau. 2007. *Statistical Abstract of the United States, 2006*. Washington, DC: U.S. Census Bureau.

might seem high, the truth is that the United States does not compare very well to other developed nations in either life expectancy or infant mortality. Table 15.1 shows that the United States ranks near the *bottom* among *industrialized nations* in life expectancy, behind Japan, the Netherlands, Canada, and several other countries. The picture is similar regarding infant mortality. Interestingly, Russia, with whom the United States was engaged in a long-running, expensive Cold War, also has a low life expectancy and an exceptionally high infant mortality rate.

We saw in Chapter 13 that life expectancy varies with gender, race–ethnicity, and social class. White women live on average five years longer than African American women, and White men on average live seven years longer than African American men. African Americans are twice as likely as Whites of comparable age to die from diseases such as hypertension or the flu, as are Hispanics. Both groups are also considerably more likely than Whites to die of acquired immunodeficiency syndrome (AIDS).

Sadly, racial and ethnic differences are also striking when it comes to infant survival: African American babies are almost *thirty* times more likely to contract AIDS than White babies, and Hispanic babies are twenty-five times more likely. Native Americans are ten times more likely than Whites to get tuberculosis, and nearly seventy times more likely to get dysentery (National Urban League 2006). Once again, we see the dramatic effects of gender and race–ethnicity, as well as class, this time on the odds of avoiding the devastation of chronic disease and surviving to a relatively old age. Those odds are lowered considerably in the United States if you are poor and minority.

Debunking Society's Myths

Myth: High infant mortality and a high death rate for young adults are problems in underdeveloped countries, but not in the United States.

Sociological perspective: Both infant mortality and adult death rates are significantly greater among lower-class than among middle-class persons in the United States, greater among people of color (Latinos, Native Americans, and African Americans) than among Whites of the same social class, and greater among men than women.

Migration

Joining the birthrate and death rate as factors in determining the size of a population is the migration of people into and out of the country. Migration affects society in many ways. Israel, since its establishment in 1948, has experienced considerable growth in its population, primarily due to a tremendous migration of Jews from Europe and the United States. These migrants tend to be younger on average than the rest of the Israeli population, therefore their arrival has certain direct consequences, such as increasing the birthrate, which is higher among the young than the older Israelis.

Migration can also occur within the boundaries of a country. In the 1980s, internal migration by African Americans, Hispanics, Asians, and Pacific Islanders within the borders of the United States has occurred at a rate unmatched since the First World War and the great Black migration from the South to the North in the United States early in the twentieth century. In that era, Blacks migrated from the South to major industrial urban areas in the North, such as Chicago, New York, Detroit, and Cleveland, where jobs were available. The recent pattern of migration for African Americans has been not only from the South, but also from the major northern urban centers to the West, the Southwest, and New England. This migration has carried many African Americans into areas that were previously all White and has frequently been associated with the presence of the migrants at institutions such as military bases or universities.

Among Hispanics, migration patterns have traditionally been loosely linked to the agriculture industry. More recently, these patterns are linked to other industries such as meatpacking and textiles and to industries centered in urban areas (Portes and Rumbaut 1996; Edmondson 2000). Although Mexican Americans have traditionally settled in the West and Southwest, recent movement has taken them to Michigan, Washington State, and New England. Farm workers from Mexico have settled near the beet fields of northern Minnesota, tripling the Hispanic population there since 1980. Puerto Rican Americans have migrated northward in increasing numbers to rural, urban, and suburban areas. In New England towns that were once predominantly White, such as Lynn and Lowell, Massachusetts, the Hispanic population has increased 180 percent since 1980 (Rumbaut and Portes 2001; Portes and Rumbaut 2001).

As noted in Chapter 14, the structure of the U.S. labor force is being shaped by the employment of recent Hispanic immigrants. Contrary to present stereotypes, these new immigrants are not poor and unmotivated, but in contrast show a proportion of professionals and technicians higher than in the labor force as a whole (Portes and Rumbaut 2001; Pedraza and Rumbaut 2001).

Asians in the last decade have migrated to a variety of destinations within the country. Many Vietnamese joined the fishing industry on Louisiana's Gulf Coast. The Lutheran Church resettled hundreds of Laotians in Wisconsin. In Smyrna, Tennessee, where Nissan has built a truck plant, the Asian population has more than doubled. Finally, Asians, African Americans, and Hispanics have joined Whites in migrating from urban to suburban areas.

Population Characteristics

The composition of a society's population can reveal a tremendous amount about the society's past, present, and future. The populations of many nations show a striking imbalance in the number of men and women of certain ages, with many fewer men than would be expected. Quite often, the explanation can be found by looking back in time to when men in this category were the right age for military service—the demographic vacancy usually coincides with a major war from which many young men failed to return. The demographic data are thus a record of national history. The Second World War was responsible for killing many men, and some women, in the early 1940s—people in their twenties. As a result, this age category in the United States revealed a shortage of people, and considerably fewer men than women, as the entire population of the country aged.

Sociologists put together data about population characteristics to develop pictures of the population in slices or as a whole. Important characteristics of the population are sex ratio, age composition, the age–sex population pyramid, and age cohorts. We examine these main approaches to describing the population in the next section.

Sex Ratio and the Population Pyramid

Two factors that affect the composition of a population are its sex ratio and its age–sex pyramid. The **sex ratio** (also called *gender ratio*) is the number of males per 100 females, or the number of males divided by the number of females, times 100.

$$\text{Sex ratio} = \frac{\text{number of males}}{\text{number of females}} \times 100$$

A sex ratio above 100 indicates there are more males than females in the population; below 100 indicates there are more females than males. The sex ratio could just as easily have been defined as the number of females per 100 males. A ratio of exactly 100 indicates the number of males equals the number of females. In almost all societies, there are more boys born than girls, but because males have a higher infant mortality rate and a higher death rate after infancy, there are usually more females in the overall population. In the United States, approximately 105 males are born for every 100 females, thus giving a sex ratio for live births of 105. After factoring in male mortality, the sex ratio for all ages for the entire country ends up being 94; there are 94 males for every 100 females.

The *age composition* of the U.S. population is presently undergoing major changes. More and more people are entering the sixty-five and older age bracket. This trend is known as the *graying of America.* The elderly will soon become the largest population category in our society. As our society gets grayer, its older members will have more influence on national policy and a greater say in matters such as health care, housing, and other areas where the elderly traditionally experience age discrimination.

Gender and age data are often combined into a graphic format called an **age–sex pyramid** (or age–gender pyramid) that represents the age and gender structure of a society (see Figure 15.2). The left side of the figure shows the age–sex pyramid for the United States. Note that there are slightly more males than females in the younger age ranges (due to the higher birthrate for males), a trend that reverses in the middle-age brackets. At the upper range of the age scale, women outnumber men. The pronounced bulge around the middle of the pyramid represents the Baby Boom generation, people now in their late fifties and sixties. As these Baby Boomers age, the bulge will rise toward the top of the pyramid, to be replaced underneath by whatever birth trends occur in the coming years. This restructuring of the population pyramid will necessitate restructuring of society's institutions. Marketing will have to be directed more often toward those over age 65, considerably more funds will be needed for their health care, and recreational facilities for the elderly will have to be greatly expanded.

The birthrate of a society or group tends to rise and fall in line with the shape of the age–sex pyramid. The right side of Figure 15.2 shows the age–sex pyramid for Mexico. It is quite wide at the bottom and very narrow at the top—exactly pyramidal in shape. This suggests a high birthrate and a death rate that increases rapidly with age. Relatively few males or females survive to fill the elderly age categories. Countries with a high birthrate tend to have a high proportion of women in their childbearing years. A pyramid of this shape is more characteristic of developing nations than the major industrial powers. The birthrate is obviously a statistic of great consequence for the future: Since the children being born will themselves grow up and have children, a high birthrate in the present can be projected to indicate many births a generation further into the future. Within the United States, the population pyramid for African Americans tends to be shaped somewhat like the one in Figure 15.2. The same pyramid shape applies to Mexican Americans and Puerto Rican Americans.

Cohorts

A birth cohort, or more simply, a **cohort,** consists of all the persons born within a given period. A cohort can include all persons born within the same year, decade, or other time period. Over time, cohorts either stay the same in size or get smaller owing to deaths, but can never grow larger. If we have knowledge of the death rates for this population, we can predict quite accurately the size of the cohort as it passes through the stages of life from infancy to old age. This enables us to predict things such as how many people will enter the first grade at age 6 between the years 2010 and 2015, how many will enroll in college, and how many will arrive at retirement decades down the road. Administrators of social entities such as schools and pension funds can make preparations on the basis of cohort predictions.

To see how dramatically a single cohort can affect a society over time, consider the effect of the

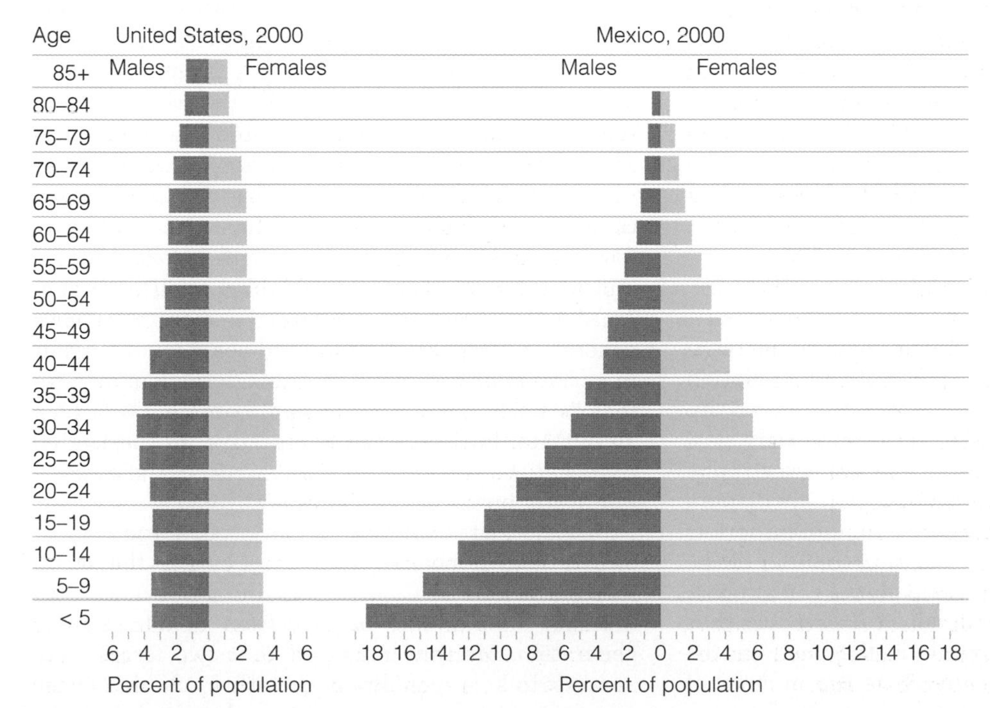

FIGURE 15.2 *A Comparison of Two Age–Sex Pyramids*

Source: John R. Weeks, 2004. *Population: An Introduction to Concepts and Issues,* 9th ed. Belmont, CA: Wadsworth Publishing Co. **www.thomsonrights.com**

cohort born in the United States between 1946 and 1964—the Baby Boom cohort. Many parents of students reading this book are Baby Boomers. The Baby Boom has been one of the most significant demographic events in all of U.S. history, along with the Great Black Migration from the South to the North in the late 1880s through the 1920s, the simultaneous massive immigration of Europeans and other groups, and current migration of Hispanic groups (Portes and Rumbaut 2001; Weeks 2004; Kennedy 1989).

The Baby Boom cohort, now comprising nearly one-third of the entire U.S. population, has had a major impact on the practices, politics, habits, preferences, and culture of our society. Raised in the relatively permissive late 1950s and 1960s, watching the likes of *Howdy Doody* and the *Mickey Mouse Club* on television, they became a large part of the "greed generation" of the 1990s. Dr. Benjamin Spock's book of down-to-earth advice about how to raise healthy children, *Baby and Child Care,* became the greatest bestseller of the twentieth century, with more than 30 billion copies sold. Among its principles was the suggestion that parents should communicate with their children and forego physical punishments when possible. For encouraging this brand of "permissiveness," Dr. Spock has often been blamed at one time or another for nearly every social problem at the time, including the youth rebellions on campuses of the 1960s, protests against the Vietnam War, and experimentation with drugs such as marijuana, LSD, and cocaine.

Once they start to pass age 65 beginning in 2010, the Baby Boom cohort will greatly increase the ranks of the elderly. One effect will be an increase in political power for those over age 65. A heavy burden will befall those born between 1965 and 1975, because they will be the main contributors to the Social Security fund just as the fund is required to meet the needs of the giant Boomer population bulge as it comes to rest in its retirement years (Weitz 2004; Robey 1982).

Theories of Population Growth Locally and Globally

Among the major problems facing modern-day civilization is the specter of uncontrolled population growth. As noted earlier, the world population increases by approximately 270,000 people every day. Some view overpopulation as an epochal catastrophe about to roll over us like a tidal wave. Others dispute whether the problem exists at all, explaining there is no scientific consensus on the *carrying capacity* of the planet. The number of people the planet can support on a sustained basis, and technological advances that have dependably met our needs in the past, can be counted on to do so in the future as the number of mouths to feed continues to grow.

Is the world overpopulated? Is the United States? What can we expect from the future? If the less optimistic scenarios turn out to be accurate, these could be the most important questions facing humankind.

Malthusian Theory

Humans, like animals, can survive and reproduce only when they have access to the means of *subsistence,* meaning the necessities of life, such as food and shelter. Human populations and animal populations are both prone to decline and die off when encountering checks on population growth such as famine, disease, and war. In the face of a daunting environment, humans have managed to thrive, and their population has doubled many times over. The period of doubling gets shorter and shorter.

Thomas R. Malthus, a clergyman born in 1766 in Scotland and educated in England, pondered the realities of life on earth and assembled his observations into a chilling depiction of disastrous population growth called **Malthusian theory,** the idea that a population tends to grow faster than the subsistence needed to sustain it. Malthus propounded his gloomy views in his *First Essay on Population,* published in 1798 (Malthus 1798), and continued to revise and extend his theory until his death in 1834. In sum, Malthus declared that the earth must be near the limits of its ability to support so many humans and that the future must inevitably hold catastrophe and famine.

Malthus noted that populations tend to grow not by *arithmetic increase,* adding the same number of new individuals each year, but by *exponential increase,* in which the number of individuals added each year grows, with the larger population generating an even larger number of births with each passing year. Arithmetic increase would cause a population to double in size at a decelerating rate. Exponential increase, in contrast, causes a population to double ever faster. It took 200 years for the world population to double from .5 billion in 1650 to 1 billion in 1850. From 2 billion to 4 billion took only forty-five years during the twentieth century. The mathematical power of doubling is awesome when applied to population. If we were to start with just one couple and imagine a lineage that doubled itself each generation by having four children, a mere thirty-two generations, roughly six hundred years, would result in a population of 8.4 billion—considerably more than today's world population of approximately 6 billion. To put Malthus's fears in perspective, in the United States at the time Malthus was writing, the average number of births per couple was seven!

Unchecked doubling of any population must swiftly spawn enough people to carpet the earth many times over. Clearly something prevents populations, human and others, from doubling every generation. Malthus reasoned that two forces were at work to keep population growth in check: (1) the growth in the amount of food produced tends to be

Bubonic plague (called the black death) severely decreased the population of medieval Europe.

only arithmetic and not exponential; and, (2) Malthus surmised that there were three major *"positive" checks* on population growth—famine, disease, and war. In Malthus's time, disease could reach apocalyptic scales. The outbreak of bubonic plague in Europe from 1334 to 1354 eliminated one-third of the population; a smallpox epidemic in 1707 wiped out *three-fourths* of the populations of Mexico and the West Indies. Wars took a toll on European men, with deaths in battle causing semipermanent gaps in the population pyramids of European populations. Along with positive checks on population growth, Malthus acknowledged what he called *preventive checks,* such as sexual abstinence, but he knew that sexual abstinence was unlikely to be the behavior change that halted uncontrolled population growth.

Malthusian theory actually predicted rather well the population of many agrarian societies such as Egypt from A.D. 500 through Malthus's own lifetime. However, Malthus failed to foresee three revolutionary developments that derailed his predictions of growth and catastrophe. In agriculture, technological advances have permitted farmers to work larger plots of land and grow more food per acre, resulting in subsistence levels higher than Malthus would have predicted. In medicine, science has fought off diseases that Malthus expected to periodically wipe out entire nations, such as the bubonic plague. Finally, the development and widespread use of contraceptives in many countries have kept the birthrate at a level lower than Malthus would have thought possible.

The technological victories of this century have not completely erased the specter of Malthus. The worldwide AIDS epidemic warns us that nature can still hurl catastrophes our way. Heartrending pictures of swollen, starving babies remind us that famine continues to destroy human populations in some parts of the world just as it has for thousands of years. Overall, Malthus's theory has served as a warning that subsistence and natural resources are limited. The Malthusian doomsday has not yet occurred, but some believe that Malthus's warning was not in error, just premature.

Debunking Society's Myths

Myth: The average number of children per family in the United States may have decreased a little bit since Malthus's time (around 1800), but not by much.

Sociological perspective: The average number of children has been about 2.1 per family in the United States for the last twenty years or so (which is close to zero population growth level), but the average was *seven* children per family in Malthus's time! This may have seemed high even for Malthus—which is one reason he became so concerned with the effects of overpopulation.

Demographic Transition Theory

An alternative to Malthusian theory is demographic transition theory, developed initially in the early 1940s by Kingsley Davis (Davis 1945) and extended by Ansley Coale (1974, 1986) and others (Weeks 2004). **Demographic transition theory** proposes that countries pass through a consistent sequence of population patterns linked to the degree of development in the society and ending with a situation in which the birthrates and death rates are both relatively low. Overall, the population level is predicted to eventually stabilize, with little subsequent increase or decrease over the long term.

There are three main stages to population change, according to demographic transition theory (see Figure 15.3). Stage 1 is characterized by a high birthrate and high death rate. The United States during its colonial period was in this stage. Women began bearing children at younger ages, and it was not uncommon for a woman to have twelve or thirteen children—a very high birthrate. However, infant mortality was also high, as was the overall death rate owing to primitive medical techniques and unhealthy sanitary conditions.

Stage 2 in the demographic transition is characterized by a high birthrate but a declining death rate. Hence, the overall level of the population increases.

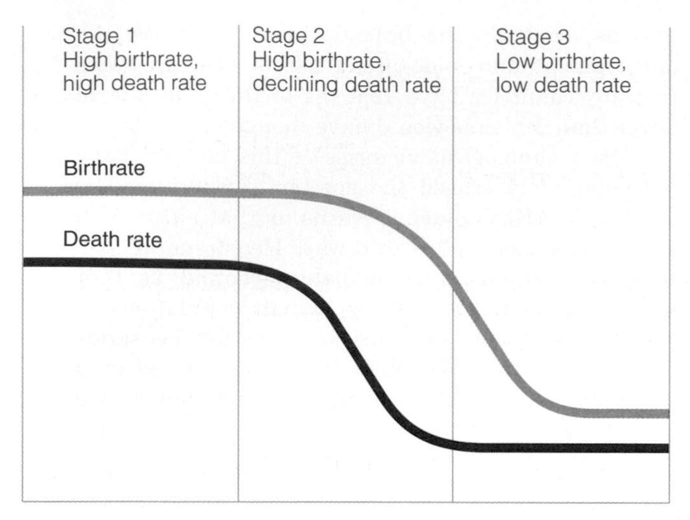

FIGURE 15.3 *Demographic Transition Theory*

The United States entered stage 2 in the second half of the nineteenth century as industrialization took hold in earnest. The norms of the day continued to encourage large families, thereby causing high birthrates, while advances in medicine and public sanitation whittled away at the infant mortality rate and the overall death rate. Life expectancy increased, and the population grew in size.

The characteristics of stage 2 did not apply across all social groups or social classes, however. Minorities at the time (Blacks, and especially Native Americans and Chinese in the Midwest and West) were less likely to benefit from medical advances. The infant mortality and overall mortality rates for Blacks, Native Americans, and Chinese remained high, while their life expectancy remained considerably shorter than life expectancy for the White population. Lack of access to quality medical care was particularly devastating to Native Americans, who had very high death rates for all ages during this period. Demographic transition theory is not completely accurate for different racial–ethnic groups within the same society.

Stage 3 of the demographic transition is characterized by a low birthrate and low death rate. The overall level of the population tends to stabilize in stage 3. Medical advances continue, the general prosperity of the society is reflected in lowered death rates, and cultural changes take place, such as a lowering of the family size. The United States entered this stage prior to the Second World War, and with the notable exception of the Baby Boom, has exhibited stage 3 demographics since then. Again, however, the trend does not apply equally to all groups in U.S. society. Some groups, such as African Americans and Hispanics, continue to have high infant mortality rates and higher death rates owing to the disproportionate impact of disease on these groups, as well as disproportionate lack of access to health resources. Whites in the United States are further along into stage 3 than the rest of U.S. society.

Demographic transition theory has received some criticism primarily because it tends to be based on heavily industrialized countries with a White majority (Coale 1986). In this respect the theory is *ethnocentric* (Weeks 2004). It argues that industrialized countries in the past have gone from high birthrates and death rates to low birthrates and death rates, with the demographic "transition" phase in between. Therefore all countries, and ethnic groups within these countries, should go through these three phases and thus experience a decline in birthrate as well. The theory carried with it the presupposition that nations and ethnic groups of color were considerably less industrialized and were thus at stage 1 with high birthrates and death rates. Yet it is not true that such groups are necessarily nonindustrialized. Thus, such nations and groups can be industrialized yet still be at stage 1.

The "Population Bomb" and Zero Population Growth

In 1968, a modern Malthus appeared in the form of Paul Ehrlich, a biology professor at Stanford University. His book, *The Population Bomb* (Ehrlich 1968), was the first in a series of writings in which Ehrlich argued that many of the dire earlier predictions of Malthus were not far from wrong. The growth in world population, according to Ehrlich, was a time bomb ready to go off in the near future, with dismal consequences. Ehrlich supported his argument with scientific research. This commodity was used rather loosely in the work of Malthus, whose reasoning depended heavily on speculative guesswork about how many people the planet could and should bear. Ehrlich stated that the sheer mathematics of population growth worldwide were sufficient to demonstrate that world population could not possibly continue to expand at its present rates.

Ehrlich pointed out that worldwide population growth has outgrown food production and that massive starvation must inevitably follow. He went beyond Malthusian theory, however, to state that the problem transcended food production and was in fact a problem of the environment. Ehrlich was among the earliest modern thinkers to argue that the quality of the environment, especially the availability of clean air and water, was a critical factor in the growth and health of populations.

Many of the dire predictions Ehrlich made back in 1968 have come true. In a later book Ehrlich wrote with his wife, Anne (*The Population Explosion,* 1991), they pointed to a variety of disasters that confirmed the predictions of the 1968 work and even of Malthus—mass starvation in parts of Africa, starvation in the United States among Black and Hispanic populations, increased homelessness in cities among minorities in particular, acid rain, rampant extinctions of plant and animal species, the irrecoverable

table 15.2 A Comparison of Demographic Theories

	Malthusian Theory	Demographic Transition Theory	Zero Population Growth (ZPG) Theory
Main point	A population grows faster than the subsistence (food supply) needed to sustain it.	Populations go through predictable stages ("transitions") from high birth and death rates to a stable population with low birth and death rates.	Achievement of zero population growth solves the Malthusian problem of unchecked population growth.
"Positive" checks on population growth	Famine, disease, and war are likely.	Famine, disease, and war are moderately likely.	Famine, disease, and war are unlikely.
"Preventive" checks on population growth	Sexual abstinence.	Sexual abstinence, birth control, and contraceptive methods.	Sexual abstinence, birth control, and contraceptive methods.
Predictions for the future	Pessimistic, despite positive and preventive checks, a population will ultimately outstrip its food supply.	Optimistic, given technology and medical advances in a population.	Very optimistic; zero population growth has already been achieved in the United States and other countries.

destruction of environments such as the rain forests, and particularly **global warming,** the systematic increase in world-wide surface temperatures—the latter underscored in former Vice President Al Gore's book and movie, *An Inconvenient Truth* (2006).

According to Ehrlich, only limits on population growth can avert disaster. He has advocated the implementation of Malthusian preventive checks by the government by supporting organizations such as Zero Population Growth (ZPG), dedicated to reaching the **population replacement level,** (the combined birthrate and death rate of a population simply sustains the population at a steady level) called the *equilibrium level.* To achieve zero population growth, couples would have to limit themselves to approximately two children per family. By 1980, the United States had indeed reached the replacement level of reproduction with an average of 2.1 children per family, ignoring race–ethnicity and class differences.

Despite these advances, worldwide population growth is still out of control. Population-related maladies such as urban crime, unemployment, and increasing homelessness remain a scourge of worldwide populations, adding an updated twist to the positive checks of the past, such as disease and war, which are still with us. The Ehrlichs deliver a final message: The population bomb must inevitably explode. What remains in doubt is whether we will arrive at that historic moment having made humane preparations, perhaps mitigating the problem with preventive checks, or whether the population crisis will end tragically with a realization of Malthus's vision of people starving worldwide and disease rampaging through a weakened population (Ehrlich and Ehrlich 1991).

thinking sociologically

What is the one major demographic change that would have to take place for a society to achieve *zero population growth* (ZPG)? Can you think of the advantages of ZPG? Any disadvantages? What effect would ZPG likely have on the *demographic transition* process? What is a major sociological barrier that would cause a population to resist adopting a governmental policy of birth control?

Checking Population Growth

As early as the 1950s, most countries had accepted population growth as a problem that must be addressed. By the 1980s, countries representing 95 percent of the world's population had formulated some policies aimed at stemming population growth; however, no consensus exists on how to control population. For example, many religious and political authorities have argued against the use of birth control. Efforts to encourage the use of contraceptives among rapidly growing populations have therefore had mixed support, whereas other attempts to curb population growth, such as encouraging changes in social habits, generally meet with little success.

Family Planning and Diversity

Many governments, including that of the United States, make contraceptives available to individuals and families. Doing so, however, is not always

© Dilip Mehta/Contact Press Images

A field worker in Bangladesh explains birth control pills and their use to village women.

consistent with the beliefs and cultural practices of all groups in the society. Catholics are taught that it is acceptable to use natural means of birth control (such as the "rhythm method") but are forbidden to use contraceptive devices such as the "pill." Many Catholics choose to use contraceptives anyway; some studies have shown that the majority of Catholics in the United States practice forms of birth control forbidden by their church.

Governmental programs that advocate contraception can be successful only if couples themselves choose smaller families over larger ones. This is most likely to occur if the wider culture supports that decision. Creating a new image of the ideal family size is a central concern of those involved in the family planning movement.

Birthrate and family size are known to be related to the overall level of economic development of a country, or even the economic status of certain ethnic groups within a country. As countries in general become more economically developed, their birthrates and average family size generally drop, as predicted by demographic transition theory. Industrialized countries generally prefer smaller families. Developing countries tend to have higher birthrates and give greater cultural value to large families. The same is true of ethnic and racial minorities living in developed nations but not entirely assimilated, such as Hispanics in the United States. Large families are seen as a demonstration of potency, a source of needed labor, and as a preparation for old age when offspring are expected to care for their elders. Before contraceptive devices such as condoms, and the pill will be widely adopted, cultural values in favor of undisturbed fertility must be countered (Vanlandingham 1993; Westoff et al. 1990), although some studies have shown that certain countries, such as Bangladesh, in stage 2 of demographic transition, have become very receptive to birth control programs and contraceptive devices (Stevens 1994).

Population Policy and Diversity

Findings such as those in Bangladesh have an important bearing on population policy elsewhere, including in the United States. Family planning programs offer great potential for achieving large declines in birthrates. In fact, in underdeveloped, overpopulated countries where family planning programs can have the most effect, the demand for resources surpasses the supply. Some cultural resistance is evident from some U.S. racial and ethnic groups to government-sponsored contraceptive programs, including some Hispanic groups and some African American groups. They argue that decreasing the birthrate by means of contraceptive methods borders on genocide and threatens the survival of members of these groups.

If contraceptive programs are sponsored by the federal government, they are perceived as direct governmental attempts to reduce the number of minority people in the country. To the extent that the government disproportionately promotes the programs in Hispanic or African American communities and promotes them less aggressively in traditionally White or upper-class areas, the programs are perceived to be racist. As a consequence, individuals are less likely to be convinced to adopt some sort of contraceptive method or device.

At the borderline of policy and social custom are still other cultural barriers against the use of contraceptive methods, including the belief among some young urban men, for instance, that condoms are unmasculine. Still, the popularity of condoms and other means of contraception is increasing, with sterilization (including vasectomies) leading the list, followed by the pill and the condom. Because of controversy over the side effects of the pill, its use has not increased as rapidly as sterilization and the condom.

Urbanization

The growth and development of cities—centers of human activity with high degrees of population density—is a relatively recent occurrence in the course of human history. Scholars locate the development of the first city around 3500 B.C. (Flanagan 1995). The study of the urban, the rural, and the suburban is the task of *urban sociology,* a subfield of sociology that examines the social structure and cultural aspects of the city compared with rural and suburban centers. These comparisons involve what urban sociologist Gideon Sjøberg (1965) calls the *rural–urban continuum,* those structural and cultural differences that exist as a consequence of differing degrees of urbanization. **Urbanization** is the process by which a community acquires the characteristics of city life and the "urban" end of the rural–urban continuum.

Urbanization as a Lifestyle

Early German sociological theorist **Georg Simmel** (1905/1950) argued that urban living had profound social psychological effects on the individual. Thus he was among the early theorists who argued that social structure could affect the individual. He argued that urban life has a quick pace and is stimulating, but as a consequence of this intense style of life, individuals become insensitive to people and events that surround them. The urban dweller tends to avoid the emotional involvement that, according to Simmel, was more likely found in rural communities. Interaction tends to be characterized as economic rather than social, and close, personal interaction is frowned on and discouraged. Yet urban dwelling can increase the likelihood of other ills: Early theorist **Emile Durkheim** noted that the suicide rate per 10,000 people was greater in more urbanized areas than in rural areas (Durkheim 1897/1951).

The sociologist **Louis Wirth** (1938), focusing on Chicago in the 1930s, also argued that the city was a center of distant, cold interpersonal interaction, and as a result the urban dweller experienced alienation, loneliness, and powerlessness. One positive consequence of all this, according to Wirth and Simmel, was the liberating effect that arises from the relative absence of close, restrictive ties and interactions. Thus, city life offered individuals a certain feeling of freedom.

A contrasting view of urban life is offered by Herbert Gans (1962/1982), who studied people in Boston in the late 1950s and concluded that many city residents develop strong loyalties to others and are characterized by a sense of community. Such subgroupings he referred to as the *urban village,* which is characterized by several "modes of adaptation," among them *cosmopolites*—typically students, artists, writers, and musicians, who together form a tightly knit community and choose urban living to be near the city's cultural facilities. A second category are the *ethnic villagers,* people who live in ethnically and racially segregated neighborhoods. Such "urban enclaves," as they are called by researcher Mark Abrahamson (2006), tend to develop their own unique identities, such as San Francisco's "Chinatown" or Miami's "Little Havana." Today's urban underclass would encompass what Gans called the *trapped* of late 1950s Boston, individuals unable to escape from the city because of extreme poverty, homelessness, unemployment, and other familiar urban ills.

Race, Class, and the Suburbs

The impact of race and class can clearly be seen in the distinction between city and suburb. Today, only approximately one-fourth of African Americans live in suburban areas. Echoing the earlier *urban villagers* principle of Gans, as well as Abrahamson's (2006) principle of urban enclave, closely knit communal subgroups form in the suburbs, but the subgroupings tend to form on the basis of class and race, with class according to some (such as DeWitt 1995) being more important than race. In the suburbs, people choose neighbors and friends on the basis of educational and occupational similarity in addition to race.

Pleasant though the suburbs can be, it is also true that people of color, particularly African Americans, often become as segregated there as they do in cities (Feagin and Feagin 1993; Massey and Denton 1993). Racial segregation persists not only in suburban neighborhoods, but in schools as well. Although some may argue that moving to the suburbs and taking up residence is a matter of personal choice, there is ample evidence that residential segregation is maintained by landlords, homeowners, and White realtors who steer people of color to segregated neighborhoods (Feagin and Feagin 1993). The practice of *redlining* by banks, which renders it impossible for a person of color to get a mortgage loan for a specific property, further intensifies residential segregation. Finally, these practices serve only to encourage further segregation in the realm of interpersonal interaction.

The New Suburbanites

The United States ended the twentieth century as it began it—in a great wave of immigration. The 1924 National Origins Quota Law encouraged immigration from northern and western Europe (England, France, Germany, Switzerland, and the Scandinavian countries) but discouraged immigration from eastern and southern Europe (Greece, Italy, Poland, Turkey, and eastern European Jews generally, among others) (see Chapter 9). Despite this openly discriminatory law, millions of eastern Europeans successfully made the journey to Ellis Island and thence the U.S. mainland, only to face prejudice, discrimination, and the accusation that they were taking jobs that would have otherwise gone to the already-present White majority.

At the end of the twentieth century, once again American shores received millions from abroad, and once again prejudice and discrimination were part of the picture, but often in more subtle forms and at times in clearly and fortunately decreased forms. Neighborhoods are now invigorated and culturally enriched by mosques or Buddhist temples; by whole neighborhoods of Vietnamese Catholics, Koreans, or Asian Indians; or by war refugees from Somalia and Bosnia. The most prominent immigrants in suburban neighborhoods are Hispanic Americans and Asian Americans, the groups that presently make up most of the country's foreign-born population (Edmondson 2000).

One long-term consequence of the current immigration settlement is what at least one demographer (William Frey of the University of Michigan) calls the new *demographic divide.* Because today's immigrants settle in a relatively small number of big cities and their suburbs, mostly located on either the east or west coast, many parts of the United States are still not feeling the effects of these new immigrants. This distinction has a number of consequences. For example, for immigrant groups who are youthful and have a relatively high birthrate, schooling has become a major political issue. But in other areas with a higher proportion of older immigrants, tax cuts for the elderly and Medicare are the major issues.

Although flows of immigrants can change from year to year, most immigrants to the United States recently have been from Mexico, the Philippines, China, and countries of the former Soviet Union. Each group brings its own culture, politics, and differing ages. A consequence of these new suburbanites present across the country is that suburban Whites now often are outnumbered by "minorities" such as African Americans, Asians, and Hispanics. As a result, many Whites now refer to themselves as the "minority majority." One consequence is this "minority majority" may perceive itself to be in competition with the new immigrants for jobs and other resources. In this respect, relations between Whites and the new immigrants may resemble the history of race and ethnic relations of the 1950s and the early 1960s in the United States (Edmondson 2000).

Ecology and the Environment

It should be apparent by now that population size has an important social dimension. Social forces can cause changes in the size of the population, and population changes can transform society. Values are acquired from one's culture, including values about what family size is considered ideal, or what degree of crowding in a city is considered tolerable. The values people hold will affect the number of children they want, the place they want to live, and the degree to which they consider population control an urgent problem.

Population density is the total number of people per unit of area, usually per square mile. As population density rises to high levels, as it has in today's cities, the familiar problems of urban living appear, including high rates of crime and homelessness. Interacting with these problems are crises of the physical environment, such as air and water pollution, acid rain, and the growing output of hazardous wastes. Humans sometimes forget that, like all other creatures, we are intimately dependent on the physical environment.

Human ecology is the scientific study of the interdependencies that exist between humans and our physical environment. A **human ecosystem** is any system of interdependent parts that involves human beings in interaction with one another and the physical environment. A city is a human ecosystem; so is a rural farmland community. In fact, the entire world is a human ecosystem. Two fundamental and closely related problems confront our present ecosystems: overpopulation and the destruction or exhaustion of natural resources (Weeks 2004; Hawley 1986).

Vanishing Resources

In all ecosystems, whether human, animal, or plant, organisms depend for their survival on one another as well as on the physical environment. Plants use carbon dioxide and give off oxygen, which all animals need to survive. Terrestrial creatures metabolize oxygen and produce carbon dioxide, which is then used by plants, completing a cycle. Humans and other animals require nutrients they can get only by eating plants. When they die, they decompose and provide nutrients to the soil that are taken up by plants, completing another cycle.

The examination of ecosystems has demonstrated two things: The supply of many natural resources is finite, and if one element of an ecosystem is disturbed, the entire system is affected. For much of the history of humankind, the natural resources of the earth were so abundant compared with the amounts used by humans that they may as well have been infinite. No more. Some resources, such as certain fossil fuels, are simply nonrenewable and will be gone soon. Other resources, such as timber or seafood, are renewable as long as we do not plunder the sources of supply so recklessly that they disappear. This is an ecological blunder we have made many times before. Finally, some natural resources are so abundant that they still seem infinite, such as the planet's stock of air and water; but at this stage of our technological development, we are learning that our powers extend to such heights and depths that we can even destroy the near-infinite resources (Gore 2006).

On the planetary scale, our gaseous wastes are gnawing away at the ozone layer, and our buried chemical wastes are trickling into the water table creating underground pools of poison. One study (Barlow and Clarke 2002) actually notes that pollution has damaged the earth's surface water so badly that worldwide underground water reserves are being mined faster than nature can replenish them. This has led to attempts to privatize water, as in Alltandra Township, South Africa, thereby cutting off water supply to those in this poverty-stricken community who cannot afford it.

One of the best demonstrations of how each part of the ecosystem affects all the other parts was seen in the use of DDT in the United States during the 1940s and 1950s. Dichlorodiphenyltrichloroethane (DDT)

© Joel Sartore/National Geographic Collection

This woman was told by a gas company in Powder River Basin, Wyoming, that the drilling for gas near her house "would never cause you to lose your water." Shortly after drilling began near her, her well water turned into a muddy methane slurry, which she unhappily holds in a glass in this photo.

was sprayed on plants by farmers and suburbanites to kill a variety of pests. After just a few years, DDT had seeped through the soil, into the groundwater, on into the seas, then into fish, and finally into birds that ate the fish. Birds that accumulated DDT in their systems appeared healthy enough, but their population plummeted. It was soon discovered that DDT caused a disastrous brittleness in the eggs laid by the birds, decimating succeeding generations. The chemical also found its way into the human food supply, with dangerous consequences such as the contamination of human breast milk in the United States (Carson 1962; Ehrlich and Ehrlich 1991). By the time DDT was identified as a major environmental hazard, tremendous damage had already been done.

If growing population is a problem of the developing world, shrinking resources are a problem of the industrialized world. The United States consumes more than 40 percent of the world's aluminum and coal as well as about 30 percent of its platinum and copper (Ehrlich et al. 1977; Ehrlich and Ehrlich 1991). According to some observers, within forty years we will reach the end of the world's supply of lead, silver, tungsten, and mercury—mainstays of heavy and high-tech manufacturing, including the computer hardware industry (see Map 15.1).

In the United States alone, real estate development takes over millions of acres of farmland each year. In the western and southwestern United States, the groundwater supply is being depleted at a rapid pace. We are racing through our nonrenewable natural resources and destroying much that should be renewable; however, some activists have begun to claim that the environmental situation, though perilous, is improving (Simon 2001). There is evidence to support this view, but even optimists who wish to show that things are getting better are quick to point out that "better" is not sufficient if the situation is bad enough to begin with (for example, Montagne 1995).

SEE FOR YOURSELF

THE WASTEFUL SOCIETY

For just one day (a full 24-hour period), make a list of everything that you use up or discard. Include everything that you throw away, including garbage, waste from cooking and eating, gasoline in your car, and so forth. At the end of the day, list the things you discarded. Indicate whether there were any alternatives to discarding these things. How might one reduce the amount of waste produced in society generally?

Environmental Pollution

It would be a bitter irony if we managed to avoid exhausting the resources of the planet only to find that what we had conserved was too degraded by pollution to be useful. The most threatening forms of pollution are the poisoning of the planet's air and water. Air pollution is not only ugly and uncomfortable; it is deadly. The skies of all major cities around the world are stained with pollution hazes, and in cities that rest within geological basins, such as Mexico City and Los Angeles, the concentrations of pollutants can rise so high that pollution-sensitive individuals cannot leave their homes. The occurrence of respiratory cases in the hospitals rises and falls with the passing of weather systems that cause the pollutants to concentrate or disperse.

Water pollution has an especially insidious side to it—most people never see how much is dumped into the world's waterways. Watchers are not present far offshore when the interior walls of vast tankers—acres on acres of fouled surface area—are rinsed with hot seawater and the waste flushed into the ocean. Small factories dump invisibly into canals that feed streams that lead to the sea.

The leading air and water polluters are the United States, Japan, Russia, China, and Poland. When the reign of secrecy ended in the former Soviet Union, bloodcurdling stories of environmental vandalism emerged. Nuclear-powered ships at the ends of their useful lives were blithely sunk—contaminated reactors, spent fuel, and all—in Antarctic seas. In the farther reaches of the giant Russian territory (which

map 15.1 VIEWING SOCIETY IN GLOBAL PERSPECTIVE

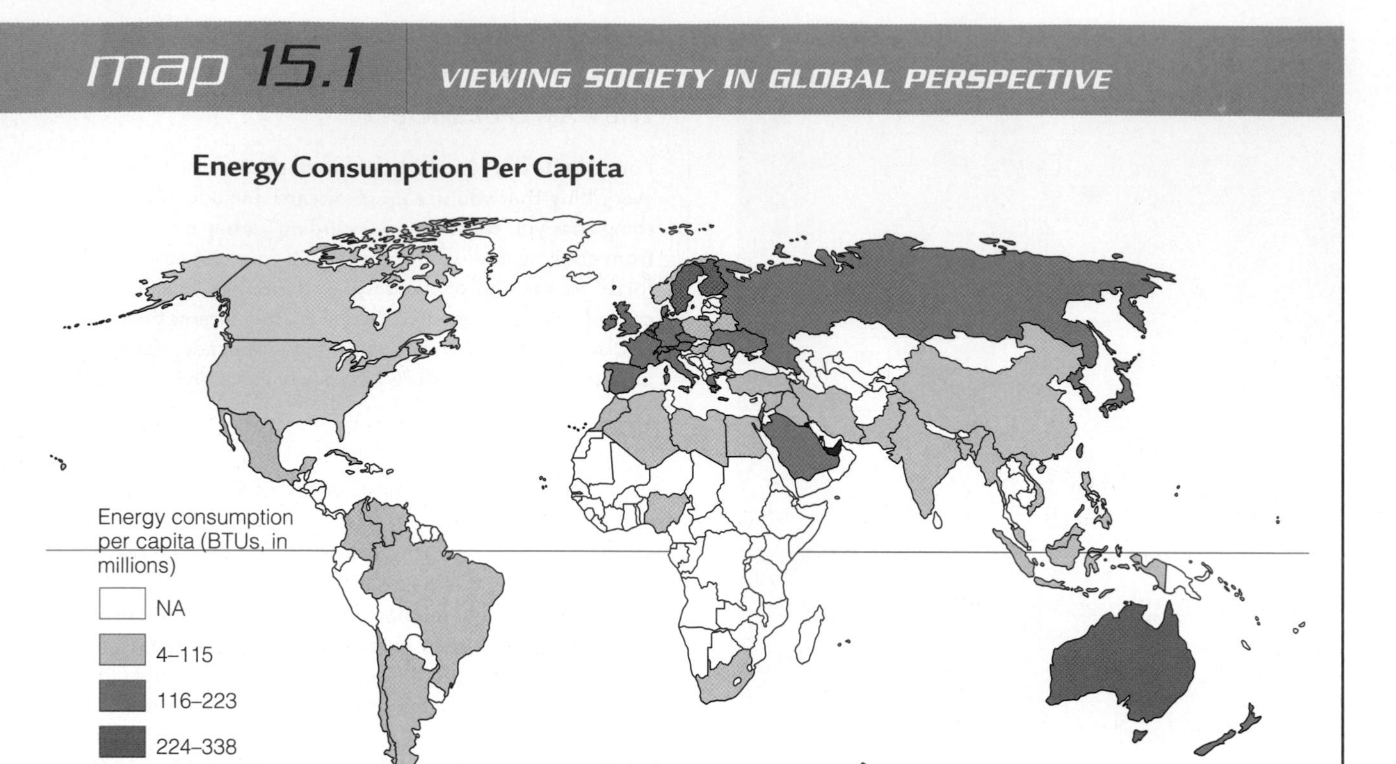

As you can see from this map, the United States and Canada are among the highest consumers of energy in the world. For the United States to reduce its energy consumption and thus become less dependent on foreign energy sources, what lifestyle changes would Americans need to make?

Data: From the U.S. Census Bureau, 2007. *Statistical Abstract of the United States, 2006.* Washington, D.C.: U.S. Census Bureau, **www.census.gov.**

crosses eleven time zones), areas were designated as open dumping sites for toxic wastes and then sealed off. Similarly, a toxic quarantine area exists downwind of the Chernobyl nuclear power plant that blew up in 1986, spewing radioactive poisons over Europe in amounts never accurately determined, but now appearing to be *far in excess* of what was once thought to be the worst case imaginable. For comparison, the Three Mile Island nuclear accident in Pennsylvania in 1979 released 15–20 curies of radioactive iodine-131 into the environment; Chernobyl is now estimated to have released as much as 50 *million* curies of the same dangerous isotope. Incredibly, the Chernobyl nuclear complex was shut down only recently, in November 2000.

A huge portion of the pollutants released into the air come from the exhaust pipes of motor vehicles. The major component of this exhaust is carbon monoxide, a highly toxic substance. Also found in exhaust fumes are nitrogen oxides, the substances that give smog its brownish-yellow tinge. The action of sunlight causes these oxides to combine with hydrocarbons also emitted from exhausts, forming a host of health-threatening substances (see Figure 15.4).

On the industrial side, the Environmental Protection Agency (EPA) has estimated that hazardous and cancer-causing pollutants released into the air by industry are responsible for more than 12,000 deaths a year. Electric utility companies and other industries often burn low-grade fossil fuels that emit harmful sulfur dioxide. When mixed with other chemicals normally present in the air, sulfur dioxide turns into sulfuric acid, which gets carried back to the earth in droplets of *acid rain.* Acid rain can change the acidity of lakes, soil, and forests so severely they can no longer support life.

A daunting international issue has grown around a group of chemicals called chlorofluorocarbons (CFCs), used as a coolant in refrigerators, in the manufacture of plastics, and as an aerosol can propellant. CFCs released into the air find their way to the ozone layer in the upper atmosphere, where they eliminate the highly reactive ozone. The ozone layer is a shield that blocks dangerous ultraviolet light, and

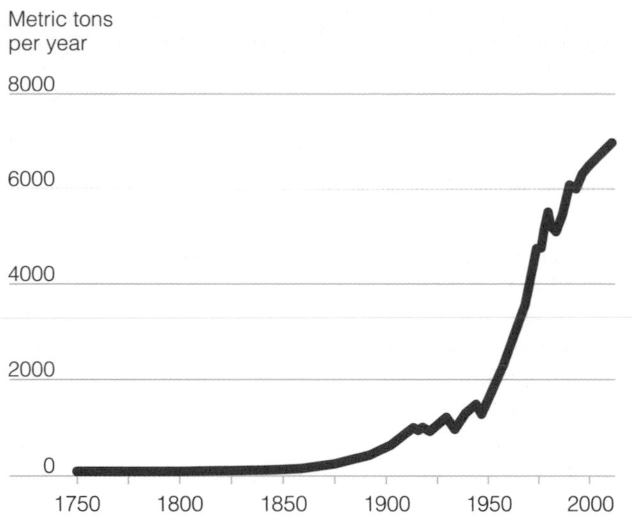

FIGURE 15.4 *Worldwide Carbon Dioxide (CO_2) Emissions from Burning Fossil Fuels, 1751–2000 (Metric Tons Per Year)*

Source: U.S. Census Bureau 2007. *Statistical Abstract of the United States, 2006*. Washington, D.C.: U.S. Department of Commerce.

AP Images/EyePress

A power plant in Moscow, Russia, spews pollution into the environment, a problem affecting nearly all major industrial cities throughout the world.

as this shield is destroyed, more ultraviolet light gets through, causing an increase in sunburn, skin cancer, and other illnesses. In 1987, a treaty named the Montreal Protocol on Substances that Deplete the Ozone Layer called for drastic reduction in the use of CFCs, but the alternatives are expensive, and developing nations are being called on to forgo refrigeration.

Related to the problem of ozone depletion is the **greenhouse effect.** As the sun's energy pours onto the earth, some is reflected from the earth's surface to the earth's atmosphere. Of the reflected energy, a portion is captured by carbon dioxide in the earth's atmosphere, while the rest radiates into space. If the amount of solar energy trapped by carbon dioxide rises, the temperature increases, producing global warming. Rather small changes in the average temperature of the earth can have dramatic consequences. A few degrees of difference can cause greater melting in the arctic regions, which raises the level of the sea, affecting water, land, and weather systems worldwide. Some researchers have noted that public opinion in countries outside the U.S., including less-developed nations, reveals more concern about global warming than that in the U.S. (Brechen 2003).

In the face of warning about water pollution, many people take comfort in the vastness of the ocean—three-quarters of the earth's surface is covered with water. However, only a tiny portion of the planet's water is usable by humans. Nearly all the water on earth is seawater—too salty to drink. Of the freshwater on the earth, most is locked in the polar ice caps. Of the remainder, most is inaccessible groundwater. All that is left and available to us is the trickle of the planet's river systems and its lakes, and this fragile supply we are polluting. Since about 1955, it has been clear that water is not the infinite resource we once thought it to be. The nation's rivers and lakes have long been dumping grounds for heavy industry. Yet these same industries—paper, steel, automobile, and chemical—depend on clean water for their production processes, during which they take water from the rivers and lakes and return it heated and polluted. The difference in temperature can alter aquatic habitats and kill aquatic life, earning this change the name of **thermal pollution.** The chemical pollutants that industries discharge into rivers, lakes, and the oceans include solid wastes, sewage, nondegradable by-products, synthetic materials, toxic chemicals, and radioactive substances. Add the polluting effects of sewage systems of towns and large cities, detergents, oil spills, pesticide runoff, and runoff from mines, and the enormity of the problem is clear.

The EPA estimates that 63 percent of rural Americans may be drinking water contaminated by agricultural runoff and the improper disposal of toxic substances in landfills. Thousands of rural water wells have been abandoned due to contamination. Households served by municipal water systems are also endangered; fully 20 percent of the country's public water systems do not meet the minimum toxicity standards set by the government (Weeks 2004; Shaberoff 1988).

Federal and state statutes now prohibit industry from polluting the nation's water, but the pollution continues. Why? The answer is economic, political, and sociological. Industries that contribute to a vigorous economy have traditionally met little interference from the government. Public awareness and outrage can force the government to crack down on major polluters. Nevertheless, the federal government often chooses a look-the-other-way attitude.

Many argue that of all environmental problems facing the United States today, the most urgent is the dumping of hazardous wastes, if only for the sheer noxiousness of the materials being dumped (see Map 15.2). It is estimated that since 1970, the production of toxic wastes increased fivefold (Weeks 2004). This dramatic increase in the amount and variety of hazardous waste production is traceable to new and profitable industrial technologies. The public creates great demand for products that inevitably produce hazardous wastes—insecticides and other useful poisons; products requiring mercury or lead (both acutely toxic when released into the environment); a variety of dyes, pigments, and paints; and an endless list of specialty plastics whose manufacture produces dangerous by-products.

Environmental Racism and Classism

As already noted earlier in this book, the brutally devastating hurricanes Katrina and Rita hit New Orleans, Louisiana, as well as Beaumont, Texas, and other locations along the country's southern gulf coast early in the fall of 2005. Because of slowness of response by the federal government, by FEMA (the Federal Emergency Management Administration), and by the president himself, over 1000 people died from drowning, from direct hits from flying debris, and from lack of medical attention. The most negatively affected neighborhoods were the lowest-lying ones, which were primarily poor and Black or Hispanic, thus *hypersegregated.* News commentators and some social scientists argued that if the White and wealthy had been so concentrated in such neighborhoods, the response of the federal government and the president would probably have been much more rapid and perhaps more effective (Dyson 2006).

Adding to the problem is the fact that toxic wastes are dumped with disproportionate frequency in areas that have high concentrations of minorities, particularly American Indians, Hispanics, and African Americans, as well as people of lower socioeconomic status (Downy 2005; Pellow 2003; Holmes 2000; Boer et al. 1997; Pollock and Vittas 1995; Bullard 1994a, 1994b). One study determined that it was "virtually impossible" that dumps were being placed so often in minority and lower socioeconomic status

map 15.2 *MAPPING AMERICA'S DIVERSITY*

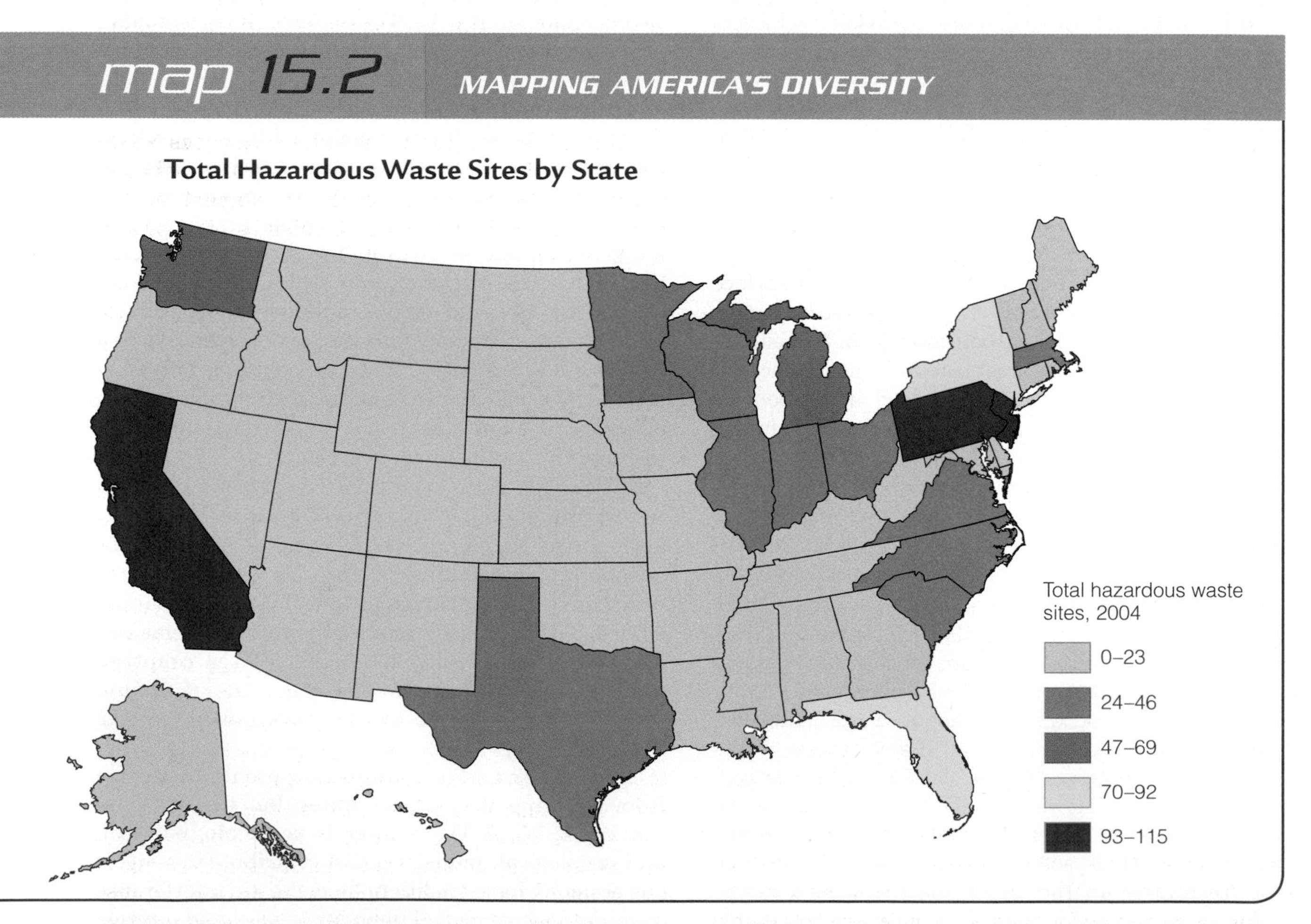

The total number of waste sites in a state varies greatly from state to state. Pick out your home state on this map. What is the approximate number of hazardous waste sites in it?

Data: From the U.S. Census Bureau, 2007. *Statistical Abstract of the United States, 2006.* Washington, D.C.: U.S. Census Bureau, **www.census.gov.**

Debunking Society's Myths

Myth: Environmental pollution is in fact more common in or near economically poor areas, but race has nothing to do with it.

Sociological perspective: Even when comparing areas of the same low economic status but different racial compositions, the areas with a greater percent of minorities are on average closer to polluted areas than those with a lower percentage of minorities.

© Jim West/PhotoEdit

This man is protesting a plan by Michigan Waste Services to burn mercury-producing medical and bio-level radioactive waste in incinerators.

(SES) communities by chance alone (Bullard 1994a, 1994b). The same study found that communities with the greatest number of toxic dumps had the highest concentration of non-White residents. Such communities also tended to fall below the national average economically and educationally. **Environmental racism** consists of the dumping of toxic wastes with disproportionate frequency at or very near areas with high concentrations of minorities.

One study of households throughout Florida found that Native American, Hispanic, and particularly African American populations reside disproportionately closer to toxic sources than do Whites. This pattern is *not* explainable by social class differences alone. That is, when communities of the same socioeconomic characteristics but different racial–ethnic compositions are compared, Native Americans, Hispanics, and African Americans of a given socioeconomic level live still closer to toxic dumps than do Whites of the *same* socioeconomic level (Holmes 2000; Pollock and Vitas 1995).

Visual concept by Norman Andersen

Environmental racism refers to the pattern whereby people living in predominantly minority communities are more likely exposed to toxic dumping and other forms of pollution. Nuclear waste and testing in the American Southwest, for example, have been located in areas predominantly inhabited by Native Americans. In other areas, African Americans and Latinos are exposed to the effects of industrial waste.

One of the largest commercial hazardous waste landfills in the nation is located in Emelle, Alabama, where Blacks comprise nearly 80 percent of the population. In Scotlandville, Louisiana, the site of the fourth-largest toxic landfill in the country, Blacks comprise 93 percent of the population. In Kettleman City, California, the site of the fifth-largest toxic landfill, more than 78 percent of the residents are Hispanic. In many cases, the siting of the landfills was linked to the economic interests of both residents and corporations. Companies with profit in mind negotiated favorable deals with the residents of these communities to permit the dumping of wastes, often misrepresented as nontoxic, in exchange for jobs and other economic incentives, which were often slow in coming (Bullard 1994b).

From the 1950s through the 1970s, the Navajo population of Shiprock, New Mexico, was exposed to waste from uranium mining and dumping that was 90 to 100 times more radioactive than the level permissible by law. Kerr-McGee, the corporation responsible, was forced out of the area in the early 1970s. When it left, it simply abandoned the site, leaving 70 acres of radioactive mine tailings (the residue from the separation of ores). An even worse situation developed during the same time in Laguna, New Mexico, involving the Pueblo Indians. Anaconda Copper, a subsidiary of the Atlantic-Richfield Corporation, virtually wrecked the traditional Pueblo economy by recruiting the community's youth for hazardous jobs even as it contaminated their environment with the

thinking sociologically

The *human ecosystem* can be adversely affected by environmental hazards such as toxic wastes. Have you ever witnessed the effects on a population of a major environmental hazard disaster such as Love Canal, Three Mile Island, or the toxic waste left over from Hurricane Katrina? Did you grow up in or near such an area, or know someone who did? From your observations, are areas containing large amounts of toxic wastes more likely to be in or near areas that are working class or heavily occupied by people of color, as shown by research?

wastes from uranium mining. A high rate of cancer deaths in these communities serves as testimony to the horrible consequences of carelessly discarding toxic wastes (Churchill 1992).

Feminism and the Environment

Women and men do not regard environmental issues equally. In general, women tend to be more concerned with issues of environmental risk, and this has important policy implications. Lack of attention on the part of local and federal governments to environmental issues can be interpreted as lack of attention to policy that differentially affects women. In this respect, it is a feminist issue.

In one recent study (Bord and O'Connor 1997), women and men were both asked a set of detailed questions pertaining to their perceptions of risk to themselves from environmental hazards. Women consistently showed more concern than men for environmental issues and perceived themselves to be at considerably more risk from environmental hazards than did men. For example, women were more likely than men to believe that abandoned waste sites cause cancer and produce miscarriages and other health problems. They were also more likely than men to feel that waste sites posed dangers to trees, fish, and other wildlife. Women were also more likely than men to perceive dangers in global warming; they were more likely to predict, as a result, coastal flooding from polar ice meltdown; loss of forests; increased killing off of certain animal species; increases in hurricanes and tornadoes; and increased air and water pollution.

The issue is not, for example, whether global warming will definitely result in such calamities, because there is room for debate on the issue. The issue is that women feel more vulnerable than men to the risks posed by such environmental problems (Bord and O'Connor 1997; Blocker and Eckberg 1997), and as a consequence, women are more concerned that policy makers act to reduce these risks.

Environmental Policy

Environmental policy of the U.S. government over the last thirty years or so has been affected, to greater or lesser degrees, by what has come to be called the environmental movement, a social movement (see Chapter 16) consisting of various loosely organized groups such as the Earth Day groups, NIMBY ("Not in my backyard"), the Sierra Club, and other similar organizations. Biologist Rachel Carson's seminal book, *The Silent Spring* (1962), served as a significant impetus to the environmental movement. The book was a scathing indictment concerning the pesticide chemical DDT and its far-reaching polluting effects from toxic runoff, including the severe reduction of many bird populations and the contamination of human breast milk. The Earth Day mobilization in 1970, now noted every year, received tremendous support from the U.S. public (nearly 70 percent according to one survey; Dunlop and Mertig 1992).

The infamous Love Canal debacle is often cited as one of the historical spurs to the environmental movement. During the 1980s, in an area of upstate New York near Niagara Falls called Love Canal, new homes were discovered to have been built on land previously used as large toxic waste dumps and landfills. The pollution was so severe that many of the homeowners discovered a dark, gray chemical ooze issuing from their backyard lawns (Levine 1982). The homeowners abandoned their properties, which remain unoccupied and unsellable to this day.

In the past three decades, federal and local agencies have made concerted efforts to bring the problems of environmental pollution under control, although in most cases environmental pollution is seen by these agencies as affecting women and men equally. The main lines of attack have been stiffer antipollution laws and the encouragement of alternative technologies.

Antipollution laws have been resisted by industry because the laws require expensive adaptations of manufacturing processes. They have also been resisted by unions that fear the added expense to industry would cost jobs. Despite these points of opposition, many antipollution laws have been passed since the late 1960s and have won great public support. However, in the early 1980s, President Ronald Reagan's administration relaxed antipollution standards based on the claim that they were too costly for industry, and these relaxed policies were continued under the administration of President George H.W. Bush. Questions still persist involving the balance of interests of big business and environmental protection.

The development of new technologies in the last twenty years has played a major role in the reduction of certain kinds of pollution. Since the early 1970s, emissions controls for automobile exhaust have been widely installed. Electric cars are once again enjoying a vogue, as is the search for alternative fuels to replace gasoline, such as methanol or methane from the fermentation of human and animal waste products.

Globalization: Population and Environment in the Twenty-First Century

The U.S. Census Bureau predicts that the world's population will increase from the 6 billion it is now to 7.9 billion by the year 2020. Even more upsetting is the United Nation's revised predictions concerning stabilization of the world's population. A few years ago, the United Nations Division on Population estimated that the world population would stabilize at around 9 billion. That estimate has been revised to 10 billion, with a high estimate of as much as 14 billion.

Sociologists predict that the United States will continue to experience increasing suburban development, with accompanying increases in heavy industry, and thus additional pollution. The rate at which people are leaving the centers of today's cities will slow somewhat. A major concern that today's sociologists have for the future is the effect that a changing planet will have on our lifestyle, and the reverse, the effect our lifestyle will have on the planet (Weitz 2004; Brown et al. 2000; Logan and Molotch 1987). For the first time in our history, our lifestyle may threaten our existence. Perhaps, as philosopher Matthew Arnold has said, humankind "carries with it the seeds of [its] own destruction."

One analyst, writing even before the devastations of Hurricane Katrina in 2005, foresaw a depressing outcome to our abuse of the environment:

> The year—2035. In an effort to hold back the rising seawater, massive dikes have been built around . . . New York and Miami. Phoenix is baking in its third week of temperatures of 155 degrees. Decades of drought have laid waste to the once fertile Midwest farm belt. Hurricanes batter the Gulf Coast and forest fires continue to blacken thousands of acres across the country (Rifkin 1989).

Ecological concern has stimulated developments in the field of **ecological demography,** which combines the studies of demography and ecology (Namboodiri 1988). This field would monitor experimentation with alternative fuels, fertilizers, and pesticides; efforts at the recycling of toxic wastes; and protection for the ozone layer. It would also observe the use and success of alternative technologies, such as solar, wind, and geothermal power. The development of such alternative technologies offers some hope for deflecting Rifken's dire predictions.

As the twenty-first century dawns, earth seems to be shrinking, giving ecology a worldwide, or global, dimension. **Ecological globalization** is now upon us—the worldwide dispersion of problems and issues involving the relationships between humans and the physical and social global environment. Goods, money, people, ideas, and pollution are traveling around the world on an unprecedented scale. With the upward soaring of international trade in fish and timber comes the internationalization of environmental issues. Environmental problems are escalating on the international politics agenda, at times occupying the attention of diplomats as much as the issues of arms control and war. Fears arising from the introduction of *genetically modified organisms* (GMOs) permeate the governments of many nations. Many of these governments are ill-suited for managing environmental problems that transcend their borders, whether via air, water, or international commerce. Yet global environmental governance is still in its infancy, and nations and economies are realizing that international coordination in ecological globalization has become necessary for any significant likelihood of global human survival.

- ***What is demography?***

  Demography is the study of population, a field that focuses on three fundamental processes, all of which determine the level of population at a given moment: births, deaths, and migrations. We noted that the United States ranks very low in *life expectancy,* and high in *infant mortality,* relative to other Western countries.

- ***How is diversity relevant?***

  Diversity is of great significance since both *infant mortality* and *life expectancy* are not equal across all races, social classes, or for men and women. Women have a greater life expectancy than men in virtually all countries and at all social class levels. However, the lower one's social class, the lower one's life expectancy, regardless of gender, and the greater the infant mortality. Minority group individuals, especially African Americans, Hispanics, and American Indians, all have lower life expectancies and higher infant mortality, than Whites.

- ***What about current migration patterns?***
Current migration patterns show that in addition to movement from city to suburb, not only by Whites but by people of color as well, including the "new suburbanite" populations from other countries and cultures, large portions of populations, such as the underclass, remain stuck in central cities. Yet recent populations of Hispanics have become upwardly occupationally mobile and have become new suburbanites. The study of a *cohort,* the Baby Boomers, showed how a demographic cohort can both use and produce cultural change.

- ***What is the Malthusian problem, and why is it important?***
Malthusian theory, still relevant today, warns us about the dangers of exponential population growth along with only arithmetic increases in food and natural resources. To avoid the so-called Malthusian positive checks of famine and war, population control can be, and is being, instituted by programs for family planning and birth control. The level of zero population growth and replacement-level birthrates can be achieved and already has been accomplished in many parts of the United States.

- ***How can population growth be checked?***
Whether methods of contraception advocated by the government are adopted, either in the United States or in other countries, depends largely on the culture of the group in question. Some Latino groups and some African Americans consider advocacy of contraception for people of color, without equally rigorous contraception advocacy for middle-class Whites, to border on genocide.

- ***What are the current problems pertaining to urbanization, human ecology, and the environment?***
Any society is a *human ecosystem* with interacting and interdependent forces, consisting of human populations, natural resources, and the state of the environment. *Urbanization* has caused such forces to become more prominent. Depletion of one natural resource affects many other things in the ecosystem. The dumping of toxic wastes is a very major problem in our society, especially when toxic dumps are found more frequently, as they are, in or very near African American, Hispanic, and Native American communities. Such practices constitute what some researchers call *environmental racism.* Surveys have shown that environmental risks are of more concern to women than to men. As a consequence, environmental policy has more impact on women than on men. *Ecological globalization* and *global warming* show that environmental issues are now international in scope and interconnected among countries around the globe.

Key Terms

age–sex pyramid 407
census 402
cohort 407
crude birthrate 403
crude death rate 404
demographic transition theory 409
demography 402
ecological demography 421
ecological globalization 421
emigration 403
environmental racism 419
global warming 411
greenhouse effect 417
human ecology 414
human ecosystem 414
immigration 403
infant mortality rate 404
life expectancy404
Malthusian theory 408
population density 414
population replacement level 411
thermal pollution 418
urbanization 413
vital statistics 403

Online Resources

Sociology: The Essentials Companion Website

www.thomsonedu.com/sociology/andersen

Visit your book companion website where you will find more resources to help you study and write your research papers. Resources include Suggested Readings, web links, and a MicroCase Online feature that teaches you how to research society. Other resources include Learning Objectives, Internet exercises, quizzing, and flash cards.

is an easy-to-use online resource that helps you study in less time to get the grade you want NOW.

www.thomsonedu.com/login

Need help studying? This site is your one-stop study shop. Take a Pre-Test and Thomson NOW will generate a Personalized Study Plan based on your test results. The Study Plan will identify the topics you need to review and direct you to online resources to help you master those topics. You can then take a Post-Test to determine the concepts you have mastered and what you still need to work on.

16

Chapter sixteen

Social Change and Social Movements

Technological innovation can transform an entire society. Consider the following: A gigabyte of information can travel from China to the United States in less time than it takes you to read this paragraph. At the same time, someone can send a fax in the opposite direction while also sipping coffee grown in Colombia, roasted in Seattle, distributed across the nation, and delivered via express mail. Was this imaginable fifty years ago? Even thirty years ago? Not really. What will the next thirty to fifty years bring?

We are quick to think of change as progress, but sometimes people in a changing society wonder if the things gained from progress are not overbalanced by the things lost. The Alaskan pipeline and related innovations have brought considerable wealth to the communities of northern Alaska, and many Eskimo entrepreneurs have prospered from the changes. Others have fared less well. Some researchers have attributed increases in suicide and alcohol-related deaths to changes that simply came too rapidly in the social structure and culture of the Eskimos (Klausner and Foulks 1982; Simons 1989).

continued

chapter sixteen

Social movements often produce major social change in a society. Fads and fashions often serve as a basis for relatively permanent social change. For example, blue jeans have been around since the late 1800s and show no signs of becoming less popular. Skateboarding, begun in the early 1960s, has endured for nearly fifty years. Other social movements have resulted in deeper and more far-reaching cultural and structural changes in society, such as the civil rights movement, the social movement to protect the environment, or the animal rights movement—examples of true social movements. Finally, changes in society resulting from the September 11, 2001, terrorist attacks are sure to be significant—in fact, already have been.

What is social change? What causes it? What has the power to launch deep changes in society in norms, habits, practices, beliefs, gender roles, racial and ethnic relations, and class distinctions? This chapter examines the causes and consequences of societies in change.

What Is Social Change?

Social change is the alteration of social interactions, institutions, stratification systems, and elements of culture over time. Societies are in a constant state of flux. Some changes are rapid, such as those brought about by desktop computers in little more than ten years. Other changes are more gradual, such as the increasing urbanization that characterizes the contemporary world. Sometimes people adapt quickly to change, in response to the development of electronic communication. Other times people resist change or are slow to adapt to new possibilities. Despite decades of effort to promote contraceptive methods, in some developing nations there are only the most sluggish gains (Goldman et al. 1989). The speed of social change varies from society to society and from time to time within the same society.

As societies become more complex, the pace of change increases. In U.S. society, this truism can be seen by comparing the rate of change now compared to earlier periods of time. Most people would agree that change is becoming more rapid, especially given the effects that rapidly emerging new technologies have on various forms of social life.

Microchanges are subtle alterations in the day-to-day interactions between people. A fad "catching on" is an example of a microchange. Take the popularity of bungee jumping. Although not as widespread as some previous fads, this highly dangerous recreation is one of a group of "extreme sports" that have become popular across the country. Bungee jumping has caused quite a few serious injuries and deaths, but has also provided thrilling footage for soft-drink commercials. This may account for why a large number of youths have suddenly developed a taste for putting themselves in bone-smashing danger. Although the overall change in the structure of society caused by fads is quite small, some minor effects may persist. Skateboarding has had several rises and falls as a fad, starting in the early 1960s, and in that time has never completely faded from the repertoire of youthful recreations—an example of how a microchange can persist.

Macrochanges are gradual transformations that occur on a broad scale and affect many aspects of society. In the process of *modernization,* societies absorb the changes that come with new times and shed old ways. One frequently noted trend accompanying

Social norms about dress and human activity are sometimes made evident through historical contrast.

modernization is that societies develop greater differentiation in social rank, divisions of labor, and so on. The effects of the fast-food industry and its impact on social structure exemplify a macrochange (see the discussion of "McDonaldization" in Chapter 5). In the United States, the rise of the computer through all its generations from vacuum tube to microchip has dramatically changed society, another example of a macrochange. Not many years ago, who would have imagined that you could "surf the Web" to the extent possible today? As recently as 1990, few people had heard of the Internet, much less used it (Friedman 1999). Now it is a daily presence in the lives of millions. Although this macrochange to a digital culture was swift, some macrochanges can take generations. Whatever time they require, macrochanges represent deep and pervasive changes in social structure and culture.

Large or small, fast or slow, social change generally has in common the following characteristics:

1. **Social change is uneven.** The various parts of a society do not all change at the same rate; some parts lag behind others. This is the principle of *culture lag,* a term coined by sociological theorist **William F. Ogburn** (1922) and first described in Chapter 2. Recall that culture lag refers to the delay between when social conditions change and when cultural adjustments are made. Often the first change is a development in material culture (such as a technological change in computer hardware), which is followed by a change in nonmaterial culture (meaning the habits and mores of the culture). The symptoms of culture lag can be seen in the uneven dissemination of computer power. Some organizations and bureaucracies adopt state-of-the-art hardware and software more quickly than others, leaping ahead of their colleagues and competitors. Even within single organizations, change occurs unequally, with older employees tending to adapt to new technology more slowly than younger employees.
2. **The onset and consequences of social change are often unforeseen.** The inventors of the atomic bomb in the early 1940s could not predict the vast changes in the character of international relations, including a Cold War that lasted until the demise of the Soviet Union in the early 1990s. Television pioneers, who envisioned a mode of mass communication more compelling than radio, could not predict television would become such a dominant force in determining the interests and habits of youth and the activities and structure of the family. Culture lag is present in both examples: A change in material culture (invention of the atomic bomb, invention of television) precedes later changes in nonmaterial culture (international relations, youth culture, and family structure).
3. **Social change often creates conflict.** Change often triggers conflicts along racial–ethnic lines, social class lines, and gender lines. Terrorism—both in the United States and abroad—focuses attention on the deep conflicts that exist worldwide in political, ethnic, and religious division. These conflicts not only produce international tension, they often drive the world events that generate social change.
4. **The direction of social change is not random.** Change has "direction" relative to a society's history. A populace may want to make a good society better, or it may rebel against a status quo regarded as unendurable. Whether change is wanted or resisted, when it occurs, it takes place within a specific social and cultural context.

Social change cannot erase the past. As a society moves toward the future, it carries along its past, its traditions, and its institutions (Lenski et al. 1998; McCord 1991; McCord and McCord 1986). A generally satisfied populace that strives to make a good society better obviously wishes to preserve its past, but even when a society is in revolt against a status quo that is intolerable, the social change that occurs must be understood in the context of the past as much as the future.

Theories of Social Change

As we have seen, social change may occur for different reasons. It may occur quickly or slowly, may be planned or unplanned, and may represent microchange or macrochange. Different theories of social change emphasize different aspects of the change process. The three main lines of contention in social change theory are functionalist theories, conflict theories, and cyclical theories (see Table 16.1). Later in this chapter we consider three additional global theories: modernization theory, world systems theory, and dependency theory. We will consider each individually.

Functionalist and Evolutionary Theories

Recall from previous chapters that functionalist theory builds on the postulate that all societies, past and present, possess basic elements and institutions that perform certain functions permitting a society to survive and persist. A *function* is a consequence of a social element that contributes to the continuance of a society. For example, the function of an institution such as the family is to provide the society with sufficient population to assure its continuance.

The early theorists **Herbert Spencer** (1882) and **Emile Durkheim** (1964/1895) both argued that as societies move through history, they become more complex. Spencer argued that societies moved from "homogeneity to heterogeneity." Durkheim similarly argued that societies moved from a state of *mechanical solidarity,* a cohesiveness based on the similarity among its members, to *organic solidarity* (also called *contractual solidarity*), a cohesiveness based on difference; a division of labor that exists among its members joins them together, because each depends on the others to perform specialized tasks (see Chapter 4). Through the creation of specialized roles, structures, and institutions, societies thus move from a condition of relative undifferentiation to higher social differentiation.

According to functional theorists, societies that are structurally simple and homogeneous, such as foraging or pastoral societies, where all members engage in similar tasks, move to societies more structurally complex and heterogeneous, such as agricultural, industrial, and postindustrial societies, where great social differentiation exists in the division of labor among people who perform many specialized tasks. The consequence (or function) of increased differentiation and division of labor is a higher degree of stability and cohesiveness in the society, brought about by mutual dependence (Parsons 1951a, 1966).

Evolutionary theories of social change are a branch of functionalist theory. One variety, called **unidimensional evolutionary theory,** argued that societies follow a single evolutionary path from simple, undifferentiated societies to more complex and highly differentiated societies. The more differentiated societies are then perceived as more "civilized." Early theorists such as Lewis Morgan (1877) labeled the distinctions between societies as "primitive" and "civilized," an antiquated notion that has been severely criticized; there is no reason to suppose that an undifferentiated society is necessarily more "primitive" than a more differentiated one. Furthermore, these earlier theories offered no firm definitions for the terms *primitive* or *civilized.* Nevertheless, the notion that some societies are primitive continues to persist today.

Unidimensional theories of social change fell out of favor because social change occurs in several dimensions and affects a variety of institutions and cultural elements. Meeting the need for a theory that better matches what is actually observed, **multidimensional evolutionary theory** (also called *neoevolutionary* theory) argues that the structural, institutional, and cultural development of a society can simultaneously follow many evolutionary paths, with the different paths all emerging from the circumstances of the society in question.

One formulation of multidimensional evolutionary theory is that of Lenski and his associates (Lenski et al. 1998). Lenski gives a central role to technology, arguing that technological advances are significantly (though not wholly) responsible for other changes, such as alterations in religious preference, the nature of law, the form of government, and relations between races and genders. Although the role of technology is presented as central, other relationships among institutions continue to be important. For example, advances in computer hardware and software can produce changes in the legal system by creating a need for new laws to deal with computer crimes—such as identity theft.

In support of the overall argument that social change is in fact evolutionary—cumulative and not easily reversible—Lenski and his associates point out that many agricultural societies have transformed into industrial societies throughout history, but few have made the reverse trip from industrial to agricultural, although certain countercultural groups have tried, such as hippie communes of the 1960s and 1970s. Lenski also argues that social advances can be reversed. For example, a major cataclysm such as an earthquake, flood, a tsunami, or a major hurricane such as Hurricane Katrina that hit New Orleans in 2005 (see Chapter 9) can humble a technologically advanced society. Following a natural disaster, especially in the developing world, city dwellers may find themselves foraging for food if the elaborate infrastructure supporting urban life is destroyed. The September 11, 2001, terrorist attack that destroyed the twin towers of the World Trade Center in New York City has already resulted in major economic and social changes—dramatically increased airport security and changes in U.S. immigration policy.

Unlike the early theories of Spencer, Durkheim, and Parsons, newer functionalist theories emphasize the role of racial–ethnic, social class, and gender differences in the process of social change (Lenski et al. 1998; Alexander and Colomy 1990; McCord 1991; McCord and McCord 1986). The same early theories made the implicit assumption that European and American societies, predominantly White, were more evolved or advanced. Societies largely comprising people of color were usually assumed to be less evolved and more primitive. This bias was often projected onto analyses *within* a society. For example, in the United States, Native American and Latino cultures were often seen as less advanced, and therefore less intelligent. The new functionalists, as well as social and behavioral researchers, reject the "primitive" versus "civilized" dichotomy and its implicit commentary on racial groups.

Debunking Society's Myths

Myth: Societies change in linear, directed fashion from primitive to civilized.

Sociological perspective: Social change can occur in several directions at roughly the same time. Furthermore, the terms *primitive* and *civilized* are out of favor as concepts, in that they imply an ethnocentric value judgment about the relative sophistication of diverse cultures.

Conflict Theories

Karl Marx, the founder of conflict theory (Marx 1967/ 1867), was influenced by the early functionalist and evolutionary theories of Herbert Spencer. Marx agreed that societies change and social change has direction, the central principle in Spencer's social evolutionary theory, but Marx placed greater emphasis on the role of economics. He argued that societies could indeed "advance" and that advancement was to be measured by the movement from a class society to a society with no class structure. Marx believed that, along the way, class conflict was inevitable.

As noted earlier in this book, the central notion of conflict theory is that conflict is inherently built into social relations (Dahrendorf 1959). For Marx, social conflict, particularly between the two major social classes—working class versus upper class, proletariat versus bourgeoisie—was not only inherent in social relations but was indeed the driving force behind all social change. Marx believed that the most important causes of social change were the tensions between social groups, especially those defined along social class lines. Different classes have different access to power, with the relatively lower class carrying less power. Although the groups Marx originally referred to were indeed social classes, subsequent interpretations include conflict between any socially distinct groups that receive unequal privileges and opportunities (G. Marx 1967; Rodney 1974). However, be aware that the distinction between class and other social variables is necessarily murky. For example, conflict between Whites and minorities is at least partly (but not wholly) class conflict, because minorities are disproportionately represented among the less well-off classes.

Racial and ethnic conflict in the United States involves far more than class differences alone: Many *cultural* differences exist between Whites and Native Americans, Latinos, Blacks, and Asians. Furthermore, cultural differences exist *within* broadly defined ethnic groups as well. We have pointed out earlier in this book that there are broad differences in norms and heritage among Chinese Americans, Japanese Americans, Vietnamese Americans, and so on—all often grouped rather coarsely as Asian Americans. The central idea of conflict theory is the notion that social groups will have competing interests regardless of how they are defined. Conflict is an inherent part of the social scene in any society.

A central theme in Marx's writing is that revolution and dramatic social change would come about when class conflict inevitably led to a decisive social rupture. Marx predicted that the capitalist class would progressively eliminate or absorb competitors and relentlessly pursue profits while squeezing the wages of the working class, thereby crushing dissent. Among the working classes discontent was supposed to blossom into a recognition that the common enemy of the worker was the capitalist class. The workers would then join in revolution, overthrow the system of capitalism, eliminate privately owned property, and establish a new economic system that would exist for the good of all.

Although the worldwide revolution predicted by Marx has never occurred, his highly refined analyses of class-related conflict have advanced our understanding of social change, and his work continues to be of interest. However, Marx seems to overemphasize the role of economics in the network of social tensions he observed, while ignoring the importance of other relevant factors related to class.

Sociologist Theda Skocpol (1979) has noted that in France, Russia, and China—countries where major revolutions have occurred—serious internal conflicts between social classes were combined with major international crises that the elite social classes proved unable to resolve before they were overthrown. The French Revolution, begun in 1789, erupted during a period when the newly arisen capitalist class was asserting itself worldwide against the old monarchies. In addition, while France was bankrupt from the many wars of the seventeenth and eighteenth century, the country intervened in the American Revolution. The Russian Revolution occurred while Russia was flattened from its disastrous defeat in the First World War, and the communist revolution in China occurred as the world was still putting out the flames of the Second World War. In each case, internal social change was linked to relations between entire societies as well as relations among social classes within each society.

The collapse of the Soviet Union (now Russia) offers some support for Skocpol's hypothesis. Years of international trade sanctions and rejection of its currency by foreign traders had eroded the economic foundations of the Soviet Union, and headlong military competition with the United States helped drive the Soviet Union into bankruptcy. Although economic considerations were paramount in the fall of the Soviet Union, the relationships *between* societies, and not just among classes within a society, were crucial in undoing the former superpower. Nevertheless, conflicts among Russians along religious, social class, ethnic, and regional lines also contributed to the social and cultural disintegration of the Soviet state.

Cyclical Theories

Cyclical theories of social change invoke patterns of social structure and culture that are believed to recur at regular intervals. Cyclical theories build on the idea that societies have a life cycle, like seasonal plants, or at least a life span, like humans. **Arnold J. Toynbee,** a social historian and a principal theorist of cyclical social change, argues that societies are born, mature, decay, and sometimes die (Toynbee and Caplan 1972). For at least part of his life, Toynbee believed that Western society was fated to self-destruct as energetic social builders were replaced by entrenched elite minorities who ruled by force, and society would wither under the sterile regimes. Some believe that societies become more decrepit, only to be replaced by more youthful societies. This belief is typified in Oswald Spengler's famous work, *The Decline of the West* (1932), which held that Western European culture was already deeply in decline, following a path Spengler believed was observable in all cultures.

Sociological theorists **Pitrim Sorokin** (1941) and, more recently, Theodore Caplow (1991), have argued that societies proceed through three phases or cycles. In the first phase, the *idealistic culture,* the society wrestles with the tension between the ideal and the practical. An example would be the situation captured in **Gunnar Myrdal's** classic work, *An American Dilemma* (1944), in which our nation declared a belief in equality for all, despite intractable racial, class, and gender stratification.

The second phase, *ideational culture,* emphasizes faith and new forms of spirituality as a phase in social change. The current New Age spirituality movement stresses nontraditional techniques of meditation and the use of such things as crystals, yoga, and chanting in a journey toward self-fulfillment and spiritual peace (Wuthnow 1994).

The third phase is *sensate culture,* which stresses practical approaches to reality, and involves the hedonistic and the sensual ("sex, drugs, and rock and roll"). Sorokin may have foreseen the hedonistic elements of popular culture in the 1960s and 1970s as indicative of sensate culture. According to the theory, when a society tires of the sensate, the cyclical process begins again with the society seeking refuge in idealistic culture. The emphasis beginning in the late 1980s and continuing now for a return to "family values," meaning older and more traditional values, is an example of a return to idealistic culture, presumably as a response to a perceived sensate culture.

Global Theories of Social Change

Globalization is the increased interconnectedness and interdependence of numerous societies around the world. No longer can the nations of the world be viewed as separate and independent societies. The irresistible current trend has been for societies to develop deep dependencies on each other, with interlocking economies and social customs. In Europe, this trend proceeded as far as developing a common currency, the *euro,* for all nations participating in the newly constructed common economy.

If the world is becoming increasingly interconnected, does this mean we are moving toward a single, homogeneous culture—the culture that futurist Marshall McLuhan once called the "global village" (Griswold 1994)? Are cities such as New York, London, and Tokyo, becoming *world cities*—cities that connect entire societies, as some (Sassen 1991) have argued? In such formulations, electronic communications, computers, and other developments would erase the geographic distances between cultures and, eventually, the cultural differences. However, greater interconnectedness among societies may *magnify* the cultural differences between interacting groups by emphasizing their incompatibilities.

In fact, both processes take place. As societies become more interconnected, cultural diffusion between them creates common ground, while cultural differences may become more important as the relationships among nations becomes more intimate. The different perspectives on globalization are represented by three main theories that we will review: modernization theory, world systems theory, and dependency theory, which are included in Table 16.1.

Modernization Theory

Modernization theory states that global development is a worldwide process including nearly all societies affected by technological change. As a result, societies are more homogeneous in terms of differentiation and complexity. Modernization theory traces the beginnings of globalization to technological advances in western Europe and the United States that propelled them ahead of the less developed nations of the world, which were left to adopt the new technologies years later. Homogenization resulted, with developing nations being shaped in the mold of the Western nations that had modernized first.

Proponents of modernization theory, such as William McCord and Arline McCord, reject the assumption that only western European countries and the United States have led the technological globalization (McCord 1991; McCord and McCord 1986). The McCords argue that non-Western societies, most notably Japan, have also been leaders in modernization. As a result, Japanese culture has profoundly influenced other countries and cultures with its emphasis on the importance of small friendship groups in the workplace and a traditional work ethic. According to the McCords, Japan and other technological leaders, such as Taiwan and South Korea, have added to the impetus of global economic growth.

World Systems Theory

Formulated by theorist Immanuel Wallerstein (1989, 1979, 1974), **world systems theory** argues that all nations are members of a worldwide system of unequal political and economic relationships that benefit the developed and technologically advanced countries at the expense of the less technologically advanced and less developed. Less developed nations are thus shortchanged in the world system. As discussed in Chapter 8, this has resulted in a worldwide

table 16.1 Theories of Social Change

	Global Theories			General Theories		
	Modernization Theory	**World Systems Theory**	**Dependency Theory**	**Functionalist/ Evolutionary**	**Theory Conflict Theory**	**Cyclical Theory**
How do societies change?	Societies become more homogenous as a result of technological change.	Unequal political and economic relationships between nations result in some ("core nations") becoming more advanced than others ("peripheral nations").	The most successful nations control the development of less powerful nations, which become dependent on them.	Societies change from simple to complex and from an undifferentiated to a highly differentiated division of labor.	Conflict is inherent in social relations, and society changes from a class-based to a classless society.	Societies develop in cycles from idealistic to sensate culture.
What is the primary cause of social change?	Technology and global development	Growth of international capitalism	Economic inequality in the global economy	Technology	Economic conflict between social classes.	Necessary for growth

(global) system of stratification—the stratification of entire countries.

Wallerstein divides the world system into two camps. *Core nations,* such as the United States, England, and Japan, produce goods and services both for their own consumption and for export. The core nations import raw materials and cheap labor from the *noncore nations* (or *peripheral nations*), of Africa, Latin America, South America, and parts of Asia. These nations occupy lower positions in the global economy, thus showing its stratification. Certain populations in the noncore nations suffer as a result. Children are used as low-paid laborers in parts of Malaysia, Singapore, and Latin America, for the manufacturing of shirts, soccer balls, and blankets. By manufacturing these goods, the noncore nations end up contributing to the wealth of the core nations.

Debunking Society's Myths

Myth: Different countries, especially smaller ones, exist largely to themselves, and what they do economically or socially has little effect on other countries in the world.

Sociological perspective: Countries are part of a worldwide network of interdependencies. World systems theory notes that this interdependence tends to benefit the developed and technologically advanced countries at the expense of the less technologically advanced and less developed.

Dependency Theory

Closely allied with Wallerstein's world systems theory is **dependency theory,** which maintains that highly industrialized nations tend to imprison developing nations in dependent relationships rather than spurring the upward mobility of developing nations with transfers of technology and business acumen (Rodney 1981, 1974; Reich 1991). Dependency theory sees the highly industrialized core nations as transferring only those narrow capabilities that it serves them to deliver. Once these unequal relationships are forged, core nations seek to preserve the status quo because they benefit from the cheap raw materials and cheap labor from the noncore nations. In this sense the core nations actively *prevent* upward mobility within the developing noncore nations. In the meantime, the developing nations remain dependent on the core nations for markets and support in maintaining what industry they have acquired, while they experience minimal social development, limited economic growth, and increased income stratification among their own population. Rodney (1981) has argued that this pattern of dependency is to blame for the exceptional underdevelopment of a number of African countries.

Borrowing dependency is a form of a dependent relationship. Former Secretary of Labor Robert Reich (1991) has noted that core nations have been willing to lend money to noncore nations, but often at high interest rates that put severe economic strain on these nations, sometimes requiring interventions such as wage and price freezes in the developing societies to maintain solvency. The hardship produced falls disproportionately on the lower social classes; the upper classes are less affected, and occasionally they benefit extravagantly.

As economist Reich (1991) and sociologists McCord and McCord (1986; also McCord 1991) have noted, the network of dependency is complicated by the fact that in today's global economic system, it is often difficult to determine just who owns what. For example, in the early 1990s, General Motors in the United States owned almost half of the stock of Isuzu in Japan. Japanese companies own a large number of American enterprises in the automotive and entertainment business. Such takeovers can also spawn humorous incidents. In the early 1990s, a construction company in California turned down a low bid from a Japanese company (Haimatsu) to spite them for a previous deal and accepted a higher bid from John Deere, the tractor maker. As it turned out, Haimatsu was a U.S.-owned company, and John Deere was Japanese owned!

Modernization

As societies grow and change, in a general sense they become more modern. As already noted, sociologists use the term *modernization* in a specific sense: **Modernization** is a process of social and cultural change initiated by industrialization and followed by increased social differentiation and division of labor. Societies can, of course, experience social change without industrialization. Modernization is a specific type of social change that industrialization tends to bring about. The change toward an industrialized society can have positive consequences, such as improved transportation and a higher gross national product, or negative consequences, such as pollution, elevated stress, and increases in certain job discrimination.

Modernization has three general characteristics (Berger et al. 1974). First, *modernization is typified by the decline of small, traditional communities.* The individuals in foraging or agrarian societies live in small-scale settlements with their extended families and neighbors. The primary group is prominent in social interaction. Industrialization causes an overall decline in the importance of primary group interactions and an increase in the importance of secondary groups, such as colleagues at work. Second, *with increasing modernization, a society becomes more bureaucratized;* interactions come to be shaped by formal organizations. Traditional ties of kinship and neighborhood feeling decrease; members of the society tend to experience feelings of uncertainty and powerlessness. Third, *there is a decline in the importance of religious institutions,* and with the mechanization of daily life, people begin to feel that they have lost control of their own lives; people may respond by building new religious groups and communities (Wuthnow 1994).

From Community to Society

The German sociologist **Ferdinand Tönnies,** who died in 1936, formulated a theory of modernization that still applies to today's societies (Tönnies 1963/1887). Tönnies viewed the process of modernization as a progressive loss of **gemeinschaft** (German for "community"), a state characterized by a sense of common feeling, strong personal ties, and sturdy primary group memberships, along with a sense of personal loyalty to one another. Tönnies argued the Industrial Revolution, which emphasized efficiency and task-oriented behavior, destroyed the sense of community and personal ties associated with an earlier rural life. At the crux of this was a society organized on the basis of self-interest, which caused the condition of **gesellschaft** (German for "society"), a kind of social organization characterized by a high division of labor, less prominence of personal ties, the lack of a sense of community among the members of society, and the absence of a feeling of belonging—maladies often associated with modern urban life.

According to Tönnies, the United States was characterized by gemeinschaft through the year 1900. Life was mainly rural, characterized by families that had lived for generations in villages, and one's work was closely tied to the family. In terms of gender roles, patriarchy was prominent, and most women centered their lives on the home, with very few women holding jobs outside the home, at least among the white middle class. At the time, there were no radios, no televisions, and few telephones. As a result, family members depended on each other for entertainment, information, and support. Despite the relative intimacy of the gemeinschaft, social interaction tended to remain within racial–ethnic and social class boundaries. Mass transportation was not yet developed, and people tended to base their lives in their own town. These characteristics of the United States at the turn of the century are preserved today in the communities of the Amish people in parts of Pennsylvania and Ohio. The Amish are a classic example of a present-day gemeinschaft.

The United States since the 1940s has become a gesellschaft, with social interaction less intimate and less emotional, although certain primary groups such as the family and the friendship group still permit strong emotional ties. However, Tönnies noted that the role of the family is considerably less prominent in a gesellschaft than in a gemeinschaft. Patriarchy is less prominent, yet more public, and more women are employed outside the home. In the large cities that characterize the gesellschaft, people live among strangers and pass people on the street who are unfamiliar. In a gemeinschaft, most of the people one encounters one has seen before. The level of interpersonal trust is considerably less in a gesellschaft. Social interaction tends to be even more confined within ethnic, racial, and social class groups. To find personal contact and to satisfy the need for intimate interaction, individuals often join small church groups, training groups, or personal awareness groups (Wuthnow 1994).

Mass Society and Bureaucracy

According to sociologists Rolf Dahrendorf (1959) and Peter Berger and his associates (1974), modernization has produced what they call a *mass society,* one in which industrialization and bureaucracy reach exceedingly high levels. In the mass society, the change from gemeinschaft to gesellschaft is accelerated, and the breakup of primary, family, and kinship ties is particularly pronounced. The government and its functions expand to the extent that much of someone's personal life falls under government management, including tasks previously performed by family. Care for the elderly, for example, may be placed in the hands of unfamiliar, faceless bureaucrats who run elder-care facilities and administer financial benefits for the aged.

Dahrendorf, Berger, and other mass society theorists argue that not only have we moved from gemeinschaft to gesellschaft, with all the attendant negatives described by Tönnies, but that bureaucracies have obtained virtually complete control of an individual's life.

As people moved from town to city during the twentieth century, divisions of labor became more pronounced, and social differentiation increased in the workplace, education, government, and other institutions. It became more common to identify people by secondary attributes, such as their job ("He's John, a lawyer") or their gender ("She's Ms. Blackburn, a woman judge"), rather than by their kinship ("She's a Smith") or their hometown ("He's from Mantua, Ohio"), which is more commonly done in the gemeinschaft. The importance of newspapers, television, magazines, radio, and movies took on more prominent roles in society. People became more mobile geographically, and less dependent on neighbors and kin. All these changes worked together to increase the feeling that most people in someone's immediate environment are strangers.

The rise of large government is a major part of the overall increased bureaucratization of social life. In the preindustrial societies of both the United States and Europe, government may have been only a clergyman, a nobleman, a justice of the peace, or a sheriff. Industrialization allowed government to expand at the national, state, and community levels, therefore becoming more complex and bureaucratized. Government demonstrates an eagerness to involve itself in many aspects of life formerly left to community standards or private resolution—regulating working conditions, setting wages and salaries, establishing standards for products and medicines, health care, and the care of the poor, as well as all sorts of intimate behaviors. Most political and social power today resides in such large bureaucracies, thus leaving individuals a diminishing degree of control over their lives.

Social Inequality, Powerlessness, and the Individual

Another product of modernization, along with mass society, is pronounced social stratification, according to theorists such as **Karl Marx** (1967/1867) and **Jurgen Habermas** (1970). In their view, the personal feelings of powerlessness that accompany modernization are due to social inequalities related to race, ethnicity, class, and gender stratification. Marx argued that inequalities are the inevitable product of the capitalist system. Habermas argued that inequalities are the cause of social conflict.

The social structural conditions that arise from modernization, such as increased social stratification, are influenced by the individuals. Building a stable personal identity is difficult in a highly modernized society that presents the individual with complex and conflicting choices about how to live. Many individuals flounder among lifestyles while searching for personal stability and a sense of self. According to Habermas, individuals in highly modernized environments are more likely than their less modernized peers to experiment with new religions, social movements, and lifestyles in search of a fit with their conception of their own "true self." These individual responses to social structural conditions reveal how the social structure can affect personality.

Social theorist David Riesman (1970, 1950) argued that there are three main orientations of personality that can be traced to social structural conditions:

- **other-directedness,** wherein the behavior of the individual is guided by the observed behavior of others and is characterized by rigid conformity and attempts to "keep up with the Joneses"
- **inner-directedness,** wherein the individual is guided by internal principles and morals and is relatively impervious to the superficialities of other people
- **tradition-directedness,** or strong conformity to long-standing and time-honored norms, practices, and styles of life

According to Riesman, modernization tends to produce other-directedness, whereas less modernized gemeinschafts, such as horticultural or agricultural societies, tend to produce tradition-directedness. The inner-directed individuals, because they are guided by internal rather than external forces, are less likely to sway with the presence or absence of modernization.

thinking sociologically

Did you grow up in what was primarily a rural community *(gemeinschaft)* or in what was primarily an urban or suburban environment *(gesellschaft)*? Do you remember how your community or neighborhood changed over time? What were some of those changes—in population, in ethnic composition, in the nature of interpersonal relationships? Would you say that your community encouraged *inner-directedness, other-directedness,* or *tradition-directedness?*

If modernization tends to produce other-directedness, then anyone who happens to be inner-directed or tradition-directed in a highly modernized and rapidly changing society, such as the United States, is most likely seen as deviant. The other-directed person, in contrast to the inner-directed or tradition-directed person, is highly flexible, capable of rapid personal change, and more open to the influences of group pressures, changing styles, and shifting interests. These qualities can leave the other-directed individual in the highly modernized society stranded and searching for his or her "true self." The poignant question "Who am I?" can give rise to feelings of individual powerlessness. The influential social theorist Herbert Marcuse (1964) has argued that modernized society fails to meet the basic needs of people, among them the need for a fulfilling identity. In this respect, modern society and its attendant technological advances are not stable and rational, as is often argued, but unstable and irrational. The technological advances of modern society do not increase the feeling of having control over one's life, but instead reduce that control and lead to feelings of powerlessness.

This powerlessness leads to the *alienation* of the individual from society—the individual experiences feelings of separation from the group or society. This alienation is most likely to affect people traditionally denied access to power, such as racial minorities, women, and the working class. This alienation from the highly modernized, technological society is, in Marcuse's view, one of the most pressing problems of civilization today. Marcuse argues that, despite the popular view that technology is supposed to yield efficient solutions to the world's problems, it may be more accurate to say that technology is a primary cause of many problems in modern society.

The Causes of Social Change

The causes of social change are many and varied but fall into several broad areas, including cultural diffusion, inequality, changes in population, war, technological innovation, and the mobilization of people through collective behavior and social movements. We examine each of these topics in the following sections.

Revolution

A **revolution** is the overthrow of state or the total transformation of central state institutions. A revolution thus results in far-reaching social change. Numerous sociologists have studied revolutions and identified the conditions under which revolutions are likely to occur. Revolutions can sometimes break down a state and various disenfranchised groups. An array of groups in a society may be dissatisfied with the status quo and organize to replace established institutions. Dissatisfaction alone is not enough to produce a revolution, however. The opportunity must exist for the group to mobilize en masse. Thus, revolutions can result when structured opportunities are created, such as through war or an economic crisis, or mobilization through a *social movement* as we will see later in this chapter.

Social structural conditions that often lead to revolution can include a highly repressive state—so repressed that a strong political culture develops out of resistance to state oppression. A major economic crisis can also produce revolution—as can the development of a new economic system, such as capitalism—that transforms the world economy.

Cultural Diffusion

Cultural diffusion (as noted in Chapter 2) is the transmission of cultural elements from one society or cultural group to another. Cultural diffusion can occur by trade, migration, mass communications media, and social interaction. Anthropologist **Ralph Linton** (1937) long ago alerted us to the fact that much of what many people regard as "American" originally came from other lands—cloth (developed in Asia), clocks (invented in Europe), coins (developed in Turkey), and much more.

Expressions and cultural elements found in the English-speaking United States have been harvested from all over the world. Barbecued ribs, originally eaten by Black slaves in the South after the ribs were discarded by White slave owners who preferred meatier parts of the pig, are now a delicacy enjoyed throughout the United States by virtually all ethnic and racial groups. One theorist, Robert Ferris Thompson (1993), points out that an exceptionally large range of elements in material and nonmaterial culture that originated in Africa have diffused throughout virtually all groups and subcultures in the United States, including aspects of language, dance, art, dress, decorative styles, and even forms of greeting. These examples illustrate cultural diffusion not only from one place to another (such as from West Africa to the United States), but also diffusion across

© Terry Wild Studio

Here members of a predominantly African American fraternity, Kappa Alpha Psi (KAΨ) put on a step show, a highly rhythmic and energetic dance form with roots in slave society as well as in West Africa.

time from a community in the past to many diverse ethnic groups in the present.

Similarly, the immigration of Latino groups into the United States over time has dramatically altered U.S. culture by introducing new food, music, language, slang, and many other cultural elements (Muller and Espenshade 1985). By a similar token, popular culture in the United States has diffused into many other countries and cultures: Witness the adoption of American clothing styles, rock, rap, and Big Macs in countries such as Japan, Germany, Russia, and China. In grocery shops worldwide, from the rain forests of Brazil to the ice floes of Norway, can be found the Coca-Cola logo.

At one time, it was thought that slavery killed off most of the institutions and cultural elements that the slaves brought to the Americas (Herskovits 1941). Extensive research over the past three decades now demonstrates that elements of culture carried from Africa by Black slaves continue to survive among African Americans and, thanks to cultural diffusion, among many other groups as well. A *step show* is an energetic, highly rhythmic, group choreography performed as a special event by predominantly Black fraternities and sororities. Researchers have traced these displays to traditional West African and central African group dances (Thompson 1993; Gates 1988, 1992). The step show has recently been noticed by non-Black students at universities and colleges all over the country, with a few White groups, fraternal, sororal, and otherwise, taking up "steppin'."

Many religious practices among African Americans are also traceable to Africa. As noted by religious historian Albert J. Raboteau (1978), the religious singing styles of the slaves, influenced by their African heritage, had four attributes: polyrhythms, syncopation, glissandos (slides) from one musical note to another, and repetition. African lineage is clearly detectable in jazz, rock, rhythm and blues, and rap music, diffusing far beyond the African American community today.

Inequality and Change

Inequalities between people on the basis of class, ethnicity, gender, or other social structural characteristics can be a powerful spur toward social change. Social movements may blossom into full-blown revolution if the underlying tension is great enough. An example of the mechanism of change can be seen when inequalities between the middle class and the urban underclass produce governmental initiatives, such as increased education for the poor, which are designed to reduce this inequality.

Culture itself can sometimes contribute to the persistence of social inequality and thus becomes a source of discontent among the individuals in the society. Inequalities within the education system often have a cultural basis. For example, a poor child in the United States, who adopts a language useful in the ghetto, is at a disadvantage in the classroom where standard English is used. Culturally specific linguistic systems such as urban Black English, or "Ebonics," which is not merely a different dialect (as many people assume) but a distinct linguistic system with its own grammar and syntax (Dillard 1972; Harrison and Trabasso 1976; DiAcosta 1998), are generally not adopted by schools. As a result, this may serve to strengthen the inequalities between the poor and the privileged—unless the child is bicultural and can speak both standard English and Ebonics, which many African Americans can do.

Compounding the problem, a female student may shrink from studying mathematics because she has received the cultural message that to be adept at mathematics is not feminine (Lau and Taylor 2003; Lau 2002; Shih et al. 1999; Sadker and Sadker 1994). The perpetuation of the inequalities of class, race, and gender can stoke the desire for social change on the part of the disadvantaged groups.

Population and Change

Changes in the population, that is, demographic change (see Chapter 15), can greatly influence the nature of interpersonal relations in society. Limitations placed on the population by the natural environment can shape social relations. In Japan, a small country with a large population, crowding is a fact of life that affects how people interact with one another. In Japanese cities, bus drivers negotiate streets that U.S. bus drivers would consider far too narrow for even a small bus. Japanese subways are packed so tightly that white-gloved "pushers" must squeeze commuters bodily into subway cars. Riders on the subway are so tightly packed that their entire bodies are in constant contact, a situation in the United States that would be considered taboo, because close physical contact has sexual overtones. In Japan, the contact is not considered sexual (although there are a few exceptions), nor is it considered a violation of the other's personal space, as it would be in the United States. This is one example of how population density can affect the nature of interpersonal relations among people and thus affect social change.

Immigration has profound effects on the overall ethnic and racial composition of the United States. Each year roughly 750,000 immigrants take up permanent residence in the United States (Edmondson 2000; Espenshade 1995; U.S. Census Bureau 2006). Approximately one-half are from Mexico or Latin America, and about one-fourth are from Asia. The United States is presently about 12.5 percent Hispanic; by the year 2025, that figure will be about 18 percent. Currently, approximately 4 percent of the U.S. population is of Asian ancestry. By the year 2025, that figure will be approximately 7 percent. By the year 2050, it is expected that Hispanics will be 25 percent of the population and Asians will be 9 percent. These population shifts will cause major changes in society's institutions. The structure of the economy will change, as will the ethnic mix in education, the ethnic complexion of jobs, and the strong influence of Hispanic culture, including styles of dance, language, and other behaviors.

SEE FOR YOURSELF

SOCIAL CHANGE AND TECHNOLOGICAL INNOVATION

Identify a technological device that you use in your daily life. For a 24-hour period, write down all the ways you use this device. As you do so, also write about whether this is something you use on your own or in relationship to other people. What forms of social interaction are involved in your use of this device? What forms of social interaction are excluded?

Now imagine that you have moved back in time and that this device is not available. How might the absence of this device change the social interactions you have observed? What might change in your life today if you did not have this device? Are there others in society without access to this technology? If so, how does this affect their lives?

© Figaro Magahn/Photo Researchers

Population density can affect social interaction and cultural norms, as illustrated here in Japan, where a subway worker (a "pusher") causes close physical contact among subway riders.

War, Terrorism, and Social Change

War and severe political conflict result in large and far-reaching changes for both the conquering society, or a region within a society (as in civil war), and for the conquered. The conquerors can impose their will on the conquered and restructure many of their institutions, or the conquerors can exercise only minimal changes.

The U.S. victory over Japan and Germany in the Second World War resulted in societal changes in each country. The war transformed the United States into a mass-production economy and affected family structure (father's absence increasing and women not previously employed joining the labor force) and education (men of college age went off to war in large numbers). Many in the armed forces who returned from the war were educated under a scholarship plan called the GI Bill.

The war also transformed Germany in countless ways, given the vast physical destruction brought on by U.S. bombs and the worldwide attention brought to anti-Semitism and the Nazi Holocaust. The cultural and structural changes in Japan were extensive, as well. The decimation of the Jewish population in Germany, and other nations throughout Europe, resulted in the massive migration of Jews to the United States. More recently, the Vietnam War also resulted in many social changes, including the migration of Vietnamese to the United States.

As we noted in Chapter 6, terrorism is a type of crime—the use of force or violence to coerce a government or population in the furtherance of political or social objectives. Threats of bioterrorism have resulted in recent microchanges in the United States, where some families wrapped their houses in plastic and duct tape. As we have already noted in Chapter 6, Osama Bin Laden had a role in the international opium trade. His al Qaeda organization was headquartered in Afghanistan, the world's largest grower of opium-producing poppies. The profits from this trade may well have financed the September 11th terrorist attacks in the United States. Terrorism has global significance, and social changes resulting from terrorism occur not only in that society, but in other societies as well.

Technological Innovation and the Cyberspace Revolution

Technological innovations can be strong catalysts of social change. The historical movement from agrarian societies to industrialized societies has been tightly linked to the emergence of technological innovations and inventions (see Chapter 4). Inventions often come about because they answer a need in the society that promises great rewards. The waterwheel promised agrarian societies greater power to raise crops despite dry weather, while also saving large amounts of time and labor. It is possible to trace a timeline from the use of the waterwheel to the use of the large hydroelectric dams that power industrialized societies and along the way find evidence of how each major advance changed society.

In today's world, the most obvious technological change transforming society is the rise of the digital computer and the subsequent development of desktop computing since the 1980s. The invention and development of the Internet and the resulting communication is now called *cyberspace,* which includes the use of computers for communication between persons and communication between persons and computers. Unique in its vastness and lack of a required central location, the Internet has very rapidly become so much a part of human communication and social

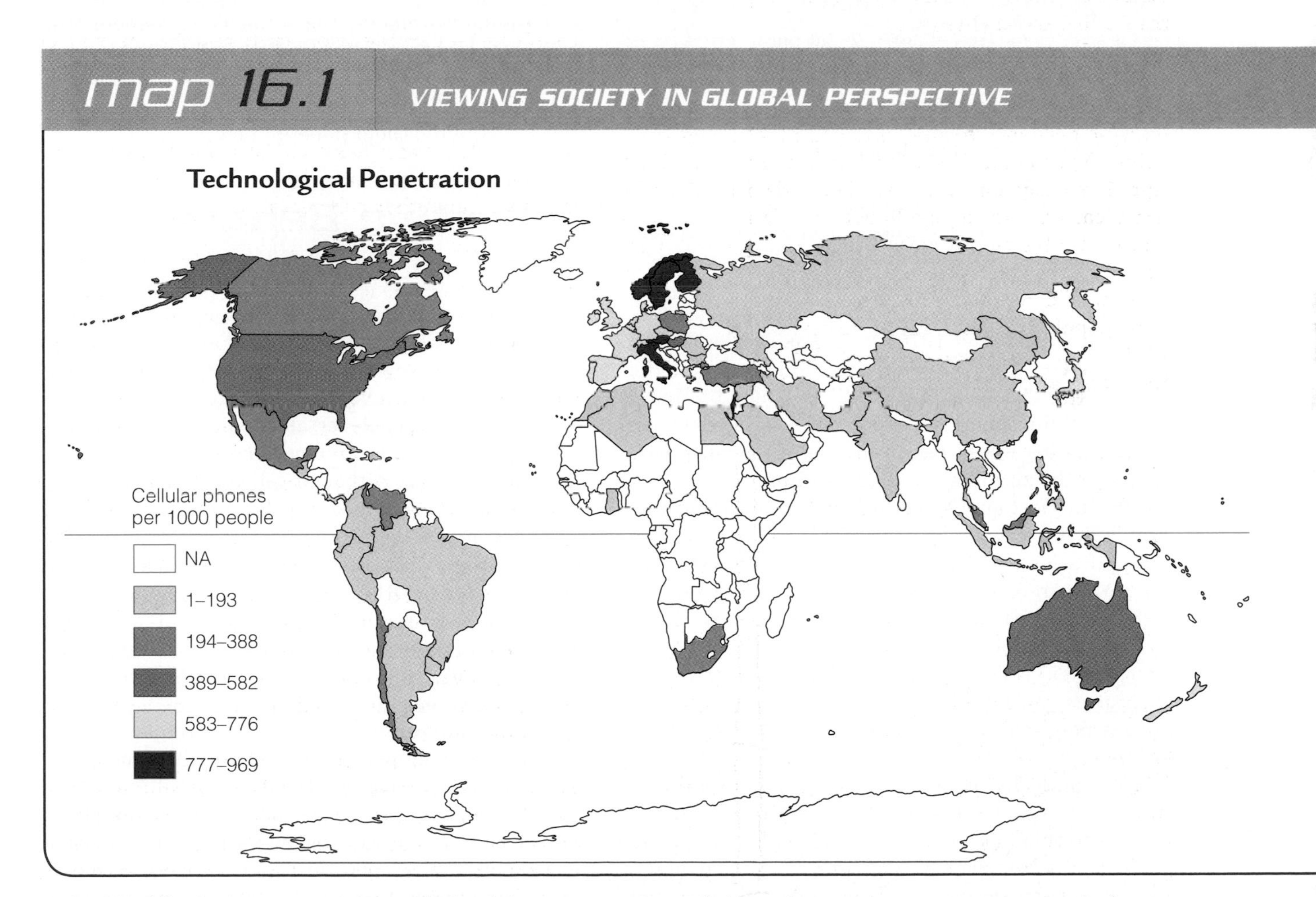

Technological development is a major source of cultural change in any society. Cell phones, for example, are now commonplace in the United States and other nations. What cultural changes inspire the use of cell phones? And what cultural changes does the increased use of cell phones then create?

Data: U.S. Census Bureau. 2007. *Statistical Abstract of the United States, 2006.* Washington, DC: U.S. Government Printing Office, **www.census.gov**

The Kaipo people of Brazil wear colorful formal dress. Technology from outside this society (TV, guns) presently threatens the persistence of such cultural practices. Recently, the Kaipo have mobilized to oppose outside intervention such as oil drilling.

reality that it pervades and has transformed literally every social institution—educational, economic, political, familial, and religious.

The cyberspace revolution began with vacuum tube mainframe computers in the 1950s and early 1960s, followed by the transistorized computer of the mid-1960s and the integrated circuit computers of the late 1960s and 1970s, and was accelerated by the advent of the PC (personal computer) and the invention of the microchip. What can now be stored in a microchip memory the size of a wristwatch, in the late 1960s would have required a transistorized computer the size of a small auditorium.

Few institutional structures have not been transformed by the cyberspace revolution—which includes its applications such as e-mail, My Space, and YouTube. Most transactions that are part of everyday life, such as cooking and banking, have been changed by the permeation of computers into everyday life. Even interpersonal relationships are being affected by the cyberspace revolution. Computer users can now develop a new self, a new identity with the use of the Internet.

The path by which technology is introduced into society often reflects the predominant cultural values in that society. Some cultural values may prevent a technological innovation from changing a society. For example, anthropologists have noted that new technologies introduced into an agrarian society very often meet with resistance even though the new technology might greatly benefit the society. The Yanomami, an agrarian society existing deep in the rain forests of South America, live without electricity, automobiles, guns, and other items of material culture associated with industrialized societies. The Yanomami place great positive cultural value on their way of hunting and engaging in war. The recent introduction of steel into Yanomami culture, however, may have introduced major social changes and changed them into a more warlike society (Tierney 2000).

Mobilizing People for Change

Social change does not develop in the abstract. Change comes from the actions of human beings. Collective behavior and social movements are ways that people organize to promote, or in some cases, to resist change. *Collective behavior* occurs when normal conventions cease to guide people's behavior, and people establish new patterns of interaction and social structure. *Social movements* are organized and persistent forms of collective behavior. The purpose of a social movement is often to initiate or vigorously resist social change. Examples abound: the civil rights movement, the women's movement, the environmental movement, and the militia movement, just to name a few. So significant are the changes that result from collective behavior and social movements that we examine them in detail now.

Collective Behavior and Social Movements

Social change is propelled by the actions of people who engage in collective behavior and social movements. As we have seen, **collective behavior** occurs when the usual conventions to guide behavior are suspended and people establish new norms of behavior in response to an emerging situation (Turner and Killian 1988). Although collective behavior may emerge spontaneously, it can be predicted. Some of the phenomena defined as collective behavior are whimsical and fun, such as fads and fashions and certain crowds. Other collective behaviors can be terrifying, as in panics or riots. Whether whimsical or awesome, collective behavior is innovative, sometimes revolutionary, and it is this feature that links collective behavior to social change.

Social movements are led by groups that act with some continuity and organization to promote or resist change in society (Turner and Killian 1988). Social movements tend to persist over time more than other forms of collective behavior. Evidence of social movements can be seen throughout society, in movements to protect the environment, promote racial justice, defend the rights of diverse groups (including animal rights), attack the government (such as militia groups), or advocate particular beliefs (such as the Christian Coalition).

Debunking Society's Myths

Myth: Rebellious social movements such as the Black Power movement or the gay rights movements are simply people blowing off steam. They do not result in long-lasting changes in society.

Sociological perspective: Social movements are one of several major causes of long-lasting social change, resulting in enduring structural and cultural changes in society.

Characteristics of Collective Behavior

Collective behavior exhibits certain common characteristics.

1. **Collective behavior always represents the actions of groups of people, not individuals.** The action of a lone gunman who opens fire in a post office or a high school cafeteria is not collective behavior because it is one person acting alone. However, groups that gather at the scene of the incident to observe the emergency response are engaged in collective behavior.
2. **Collective behavior involves new or emergent relationships in groups that arise in unusual or unexpected circumstances.** The behavior of people who commute to work together is not considered collective behavior because commuting is an ordinary part of their everyday life. But when a community is struck by a natural disaster, such as a flood, earthquake, or hurricane such as Hurricane Katrina, suddenly nothing can be taken for granted—not food, water, transportation, electricity, nor shelter. Collective behavior emerges to meet the new needs that people in the community face. Following the collapse of the World Trade Center, there were numerous reports of people helping each other to the exits, as well as heroic efforts of firefighters and other rescue workers who quickly organized—even at risk of death—to save as many people as they could.
3. **Collective behavior captures the more novel, dynamic, and changing elements of society.** An example is fads that introduce something new into everyday social life. Collective behavior may mark the beginnings of more organized social behavior and often precedes the establishment of formal social movements. People who spontaneously organize to protest something may develop structured ways of sustaining their protest. The environmental justice movement includes a wide array of Native American, African American, Latino, and other communities that have organized to protest the dumping and pollution that imperil their neighborhoods (Pellow 2003; Bullard 1994a).
4. **Collective behavior is patterned and not the irrational, overly emotional behavior of crazed individuals.** People may follow new guidelines of social behavior, but they do follow guidelines. Even episodes of panic, which appear to be asocial and disorganized, follow a relatively orderly pattern. Episodes of collective behavior often exhibit the more emotional side of life.
5. **During collective behavior, people may communicate extensively through rumors.** Lacking communication channels or distrusting the ones available, people use rumors to help define an otherwise ambiguous situation (Turner et al. 1986). Rumors are transmitted by people who are piecing together information about a story whose facts are partly obscured. Rumoring is most common when there is inadequate information to interpret a problematic situation or event (Dahlhamer and Nigg 1993; Shibutani 1966).

Debunking Society's Myths

Myth: In the face of disasters and other unexpected events, people do not behave under the normal social influences.

Sociological perspective: When faced with unexpected events, people develop norms to guide their behavior, often drawing on previous social behavior and knowledge to guide new interactions.

Collective behavior is spontaneous, as when people created memorials to those lost on 9/11/01.

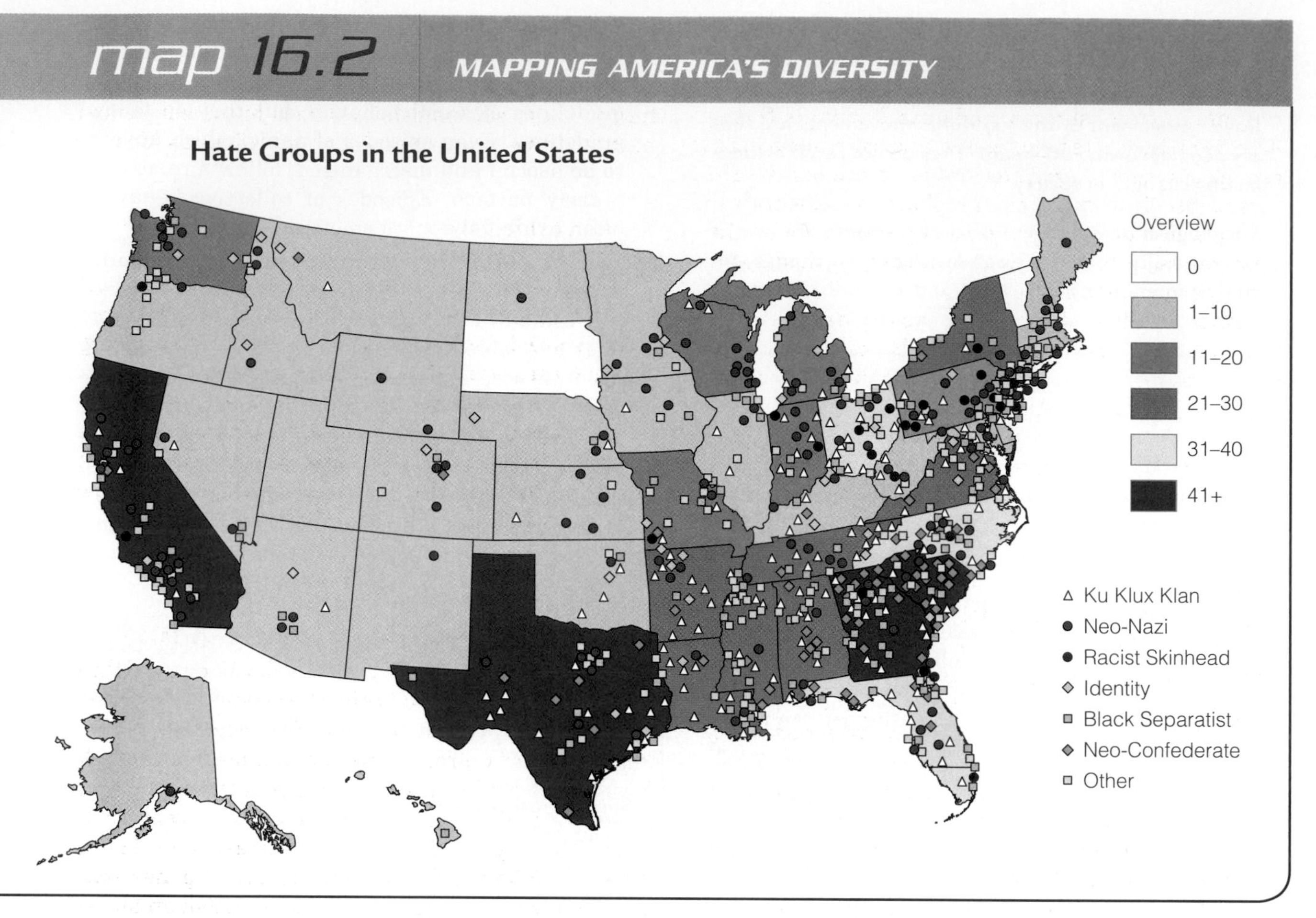

The Southern Poverty Law Center, originally founded to protest and monitor the activities of the Ku Klux Klan, now monitors numerous hate groups throughout the country, estimating that there are over eight hundred such groups currently active. The Center also provides extensive teaching materials to promote tolerance and interracial understanding. You can see that different kinds of hate groups populate various regions of the United States. Many of them involve interracial hatred, such as the Ku Klux Klan, Neo-Nazis, and Skinheads. But they also include other extremist groups. The website of the Southern Poverty Law Center has an interactive version of this map, where you can click in a given location and learn more about the hate groups and statewide legislation to outlaw hate crimes. Based on the map presented here, what part of the county has the highest concentration of hate groups? What factors do you think lead to the presence of hate groups?

Source: Taken from Southern Poverty Law Center 2006. Used by permission, **www.splcenter.org**

The Organization of Social Movements

Social movements are both spontaneous and structured, although within movements, tension typically exists between spontaneity and structure (Freeman 1983a). Unlike everyday organizations, they thrive on spontaneity and often must swiftly develop new strategies and tactics in the quest for change. During the civil rights movement, students improvised the technique of sit-ins, which quickly spread because they succeeded in gaining attention for the activists' concerns. Social movements differ from other forms of collective behavior, however, by containing routine elements of organization and lasting for longer periods.

There are three broad types of social movements: personal transformation movements, political or social change movements, and reactionary movements. **Personal transformation movements** aim to change the individual. Rather than pursuing social change, participants adopt a new identity—one they use to redefine their life, both in its current and former state. They focus on the development of new meaning within individual lives (Klapp 1972). The contemporary New Age movement defines mainstream life as stressful and overly rational and promotes relaxation and spiritualism as an emotional release and route to expanded perceptions. New Age music, crystals, massage therapy, and meditation are intended to restore the New Age person to a state of unstressed wholeness. Like many other social movements, New

Ageism is supported by an array of dues-charging organizations and commercial products, from tapes and crystals to sessions with the spirits of the dead, retreats in yoga ashrams, and guided tours of Native American holy sites. Religious and cult movements are also personal transformation movements. The recent rise in evangelical religious movements can be explained in part by the need people have to give clear meaning to their lives in a complex and sometimes perplexing society.

Social/political change movements aim to change some aspect of society, such as the environmental movement, the gay and lesbian movement, the civil rights movement, the animal rights movement, and the religious right movement. All seek social change, although in distinct and sometimes oppositional ways. Some movements want radical change in existing social institutions; others want a retreat to a former way of life or even a move to an imagined past (or future) that does not exist. Social movements use a variety of tactics, strategies, and organizational forms to achieve their goals. The civil rights movement, starting in the early 1950s, used collective action during sit-ins and organizational activity to overturn statutes that supported the "separate but equal" principle of segregation. These efforts culminated in the U.S. Supreme Court decision *Brown v. Board of Education* of Topeka, Kansas, which declared the "separate but equal" doctrine unconstitutional (Morris 1999).

Social change movements may be either reformist or radical (Turner and Killian 1988). **Reform movements** seek change through legal or other mainstream political means, typically working *within* existing institutions. **Radical movements** seek broader-based fundamental change in the basic institutions of society. Instead of working within society's institutions, radical social movements tend to work outside of society's institutions. Although most movements are primarily reformist or radical, within a given movement both factions may exist. In the environmental movement, for example, the Sierra Club is a classic reform movement that lobbies within the existing political system to promote legislation protecting the environment. Greenpeace, in contrast, is a more radical group that sometimes uses dramatic tactics to disrupt activities the group finds objectionable, such as the killing of whales.

Reactionary movements are organized to resist change or to reinstate an earlier social order that participants perceive to be better, and they are reacting against contemporary changes in society. The Aryan Nation, a White supremacist coalition, is a reactionary movement that wants to suppress Jews and minorities and institute a "racially pure" decentralized state (Ferber 1998). The militia movement is also a reactionary movement, seeking to resist government authority and reinstate the perceived lost power of White people (Esterberg 2003). Reactionary movements represent an organized backlash against social values and societal changes that participants find deeply objectionable. Rather than adjust to the future, they may instead hark back to some mythical past that their movement ideology defines as ideal.

thinking sociologically

Identify a *social movement* in your community or school that you find interesting. By talking to the movement leaders and participants, and if possible, by examining any written material from the movement, how would you describe this movement in sociological terms? Is it *reform, radical,* or *reactionary*? How have the available resources helped or hindered the movement's development? What tactics does the movement use to achieve its goals? How is it connected to other social movements?

Origins of Social Movements

Social movements do not typically develop out of thin air. For a movement to begin, there must be a preexisting communication network (Freeman 1983a). The importance of a preexisting communication network is well illustrated by the beginning of the civil rights movement, which most people date to December 1, 1955, the day Rosa Parks was arrested in Montgomery, Alabama, for refusing to give up her seat on a municipal bus to a White man. Although she is typically understood as simply too tired to give up her seat that day, Rosa Parks had been an active member of the movement against segregation in Montgomery. Parks was secretary of the local NAACP chapter that had already been organizing to boycott the Montgomery buses. She was the one chosen to be the plaintiff in a test case against the bus company because she was soft-spoken, middle aged, and a model citizen. NAACP leaders believed her to make a credible and sympathetic plaintiff in what they had long hoped would be a significant legal challenge to segregation in public transportation.

When, according to plan, Rosa Parks refused to give up her seat, the movement stood ready to mobilize. News of her arrest spread quickly via networks of friends, kin, church, and school organizations. A small group, the Women's Political Council, had previously discussed plans to announce a boycott of the bus system. The arrest of Rosa Parks presented an opportunity for implementation. One member of the Women's Political Council, Jo Ann Gibson Robinson, called a friend who had access to the mimeograph machine at Alabama State College. In the middle of the night, Robinson and two of her students duplicated 52,500 leaflets calling for the boycott to be distributed among Black neighborhoods (Robinson 1987).

Movements are fueled by the news media, which bring public attention to the movement's cause. With

The 9/11 Commission Report on Terrorist Attacks Against the U.S.

Late in 2004, the government released its report on the September 11, 2001, terrorist attacks on the World Trade Center and the Pentagon. The full report is entitled: *The 9/11 Commission Report: Final Report of the National Commission on Terrorist Attacks Upon the United States* (2004).

The report itself, despite its flaws, was a piece of painstaking research that interviewed over 1200 individuals in ten different countries, took public testimony from 160 witnesses, and examined more than 2.5 million pages of documents. The researchers found that the United States was extremely unprepared for the attacks. While the report correctly notes the heroic intervention of the "first responders" (ordinary people in the neighborhood vicinities), the New York City Fire Department, public officials, and the passengers on United Airlines Flight 93 who overpowered the plane's hijackers, the report also notes *numerous* government failures such as communication breakdowns between governmental offices; communication breakdowns among the airlines (airline companies) whose planes were involved; failure to arm the fighter jets that were scrambled to shoot down hostile aircraft; errors in intelligence-gathering concerning al Qaeda, the terrorist organization presumed to have engineered and carried out the attacks; and problems with U.S. governmental culture such as the tendency to initially underestimate the seriousness of early threats and warnings. Such unfortunate tendencies tend to be based on sociological phenomena such as "groupthink" (recall the discussion from Chapter 6), oversimplification, and premature pressure toward consensus in group meetings.

The report itself, however, contains numerous flaws and misjudgments. The report glosses over or ignores factors that are key to the public's understanding of what happened on 9/11/01. For example, little information is given on New York City's lack of preparedness for attacks on its World Trade Center towers, despite the fact that the towers had been partially bombed in 1993. The report fails to detail comparisons between New York City and other cities, such as Los Angeles, which is considerably better prepared to deal with such emergencies, given its history of earthquakes, fires, major riots, and other community disasters. The report blithely singles out "communication problems" as in need of repair, but fails to note that to date no system of communications and information-sharing has yet been devised that can address fundamental problems of large-scale disaster responses. The report de-emphasizes the critical role in emergency evacuation played by the people occupying the towers when they were hit, even well before official rescue efforts were underway. Thus, the working public itself was a major contributor toward dealing with the many problems of the attack. The report is painfully brief in giving recommendations for solutions to the problems of preparedness for attack, although this was supposed to have been one of its major goals. In short, the report says nothing about the pressing need ignored by the present U.S. administration: the need to actively engage the public itself in preparedness and response efforts for all types of disastrous events.

Sources: Perrow, Charles, David M. Mednicoff, and Kathleen Tierney. 2005. "A Symposium on the 9/11 Commission Report," *Contemporary Sociology* 34 (March), pp. 99–120; *The 9/11 Commission Report: Final Report of the National Commission on Terrorist Attacks Upon the United States*. 2004. Baton Rouge, LA: Claitor's Publishing Division.

AP Images/Susan Walsh

Social movements often use highly visible tactics to promote their causes, as in the March for Women's Lives in Washington, DC.

the extraordinarily widespread availability of television and now the Internet, cellular phones, and standard radio transmissions, news of movements around the world spreads quickly. The Internet in particular has become increasingly important for the mobilization of social movements. Communication via the Internet was an important organizing tactic for the diverse groups that participated in the Seattle demonstrations against the World Trade Organization in 1999 and the more recent anti-war demonstrations. The Internet has also been a tool of White supremacist groups for recruiting new members and promoting their views. This is particularly effective because of the appeal of the Internet to young people and the anonymity it can provide.

Celebrities can also advance the cause of social movements by bringing visibility to the movement, although sociologists have found that celebrities typically involve themselves in less controversial movements that already have widespread popular support. The entry of celebrities into social movements can unintentionally hurt a movement's development, because celebrities tend to depoliticize it (Meyer and Gamson 1995). Their endorsement and presence, nonetheless, brings visibility to a movement.

Another cause for a movement to begin must be a perceived sense of injustice among the potential participants. The environmental justice movement has emerged largely from grassroots organizations formed in communities where residents have organized to clean up a toxic waste site, close a polluting industry, or protect children from a perceived threat to their health and safety.

As movements develop, they quickly establish an organizational structure. The shape of the movement's organization may range from formal bureaucratic structures to decentralized, interpersonal, and egalitarian arrangements. Many movements combine both. Examples of large bureaucracies are the National Organization for Women, the National Association for the Advancement of Colored People (NAACP), Amnesty International, Greenpeace, the Jewish Defense League, and the National Rifle Association. As social movements become institutionalized, they are most likely to take on bureaucratic form. Bureaucratically organized movements affirm Max Weber's prediction that social movements gradually rationalize their structures to efficiently meet the needs of members and the goals of the organization.

Debunking Society's Myths

Myth: Social movements develop usually as a result of extremists who are single-minded in their interests.

Sociological perspective: Social movements often develop from the everyday concerns of ordinary people who mobilize to address conditions in their lives that they find unacceptable.

Theories of Social Movements

Sociologists have developed several theories to explain the development of social movements, including resource mobilization theory, political process theory, and new social movements theory.

Resource mobilization theory is an explanation of the development of social movements that focuses on how movements gain momentum by successfully garnering resources, competing with other movements, and mobilizing the resources available to them (Marx and McAdam 1994; McCarthy and Zald 1973). Money, communication technology, special technical or legal knowledge, and people with organizational and leadership skills are all examples of resources that can be put to use in organizing a social

© Thomas Frey/EPA/Landov

Social movements such as the disability rights movement can raise public awareness and result in new forms of social behavior.

movement (Zald and McCarthy 1975). Interpersonal contacts are one of the most important resources a group can mobilize because the contacts provide a continuous supply of new recruits, as well as money, knowledge, skills, and other assistance (Snow et al. 1986). Sometimes social movement organizations acquire the resources of other organizations.

Aldon Morris, an African American sociologist, has used resource mobilization theory to explain the development of the civil rights movement. That nationwide movement relied heavily on Black churches and colleges for resources such as money, leadership, meeting space, and administrative support (Morris 1984). In other words, it mobilized existing resources on behalf of its own cause.

Resource mobilization theory, among other things, notes how movements are connected to each other. The gay and lesbian movement has used many strategies used by the anti-violence against women movement in developing its own campaign to halt hate crimes against gay men and lesbian women (Jenness and Broad 1997; Jenness 1995).

Political process theory posits that movements achieve success by exploiting a combination of internal factors, such as the ability of organizations to mobilize resources, and external factors, such as changes occurring in the society (McAdam 1982). Some structural conditions provide opportunities for collective action. A war, pressure from international parties, demographic shifts, or an economic crisis may create the possibility for those who challenge the social order to mobilize a movement (McAdam 1999). Political process theory explicitly acknowledges the effects that larger social and political processes have on the mobilization of social movements. Large-scale changes such as industrialization, urbanization, or ending repression may provide opportunities for the mobilization of social movements not present previously. Charles Tilly argued that as capitalism developed after popular uprisings in seventeenth-century France, markets emerged in cities for grain from the countryside. To meet the needs of the new markets, grain had to be stored and transported. Rural peasants, angry that their food supply was being transported away, saw the opportunity for collective action and attacked and sabotaged storage and transportation facilities (Tilly 1986).

Political process theory also stresses the vulnerability of the political system to social protest. For example, several groups in Watsonville, California, used the Loma Prieta earthquake in 1989 as an opportunity to protest. Organized Latino groups demanded better low-income housing in that community for approximately a decade (Simile 1995). Among Mexican farm laborers in the valley, there was already a long tradition of labor organizing, including a cannery workers' strike in 1986 (Bardacke 1988; Amott and Matthaei 1996). After the 1989 earthquake, these groups mobilized around the issues of temporary and permanent housing and the way that aid had been distributed to Latinos. In other words, groups already organized for protest on other issues were ready to squeeze an opportunity out of disaster.

Resource mobilization theory and political process theory are social structural as opposed to individual-level explanations. Other explanations of social movements are more cultural in focus (see Table 16.2). People will not organize within social movements unless they develop a shared definition of the situation that gives meaning to their action, in other words, a culture.

Sociologists use the concept of *framing* to explain this process. **Frames** are specific schemes of interpretation that allow people to perceive, identify, and label events within their lives that can become the basis for collective action (Snow et al. 1986; Snow and Benford 1988; Goffman 1974). The framing process emphasizes that social movements do not emerge unless

© Mystic Seaport, Rosenfeld Collection, Mystic, CT © Ezra Shaw/Getty Images © Nick Latham/Getty Images

Social change is reflected not only in the methods of play of tennis today, but in the presence of more minorities in professional tennis.

table 16.2 Sociological Theories of Social Movements

	Resource Mobilization Theory	Political Process Theory	New Social Movement Theory
How do social movements start?	People garner resources and organize movements by utilizing such things as money, knowledge, and skills	Movements exploit social structural opportunities, such as economic crises and wars	New forms of identity are created as people participate in social movements
What does the theory emphasize?	Linkages among groups within a movement	Vulnerability of political system to social protest	Interconnection between social structural and cultural perspectives

people have a shared understanding of the causes of a perceived injustice or a shared definition of their opposition. As an example, framing can help you understand the emergence of Ralph Nader and the Green Party. Here the frame is a shared understanding of corporate domination as the major source of political grievance, providing the lens through which people have been mobilized to join this movement.

Frames can also help us understand how movements change or break down. During the civil rights movement, participants shared a frame, or definition of the situation, that Jim Crow laws were the main obstacle to Black social progress. When this singular frame broke down as more radical forms of protest emerged in the late 1960s and as laws were passed to eliminate old forms of segregation, the movement subsided. Cultural and social structural explanations of social movements come together in what is called **new social movement theory,** which conceptually links culture, ideology, and identity to explain how new identities are forged within social movements. Whereas resource mobilization theory emphasizes the rational basis for social movement organization, new social movement theorists are especially interested in how identity is socially constructed through participation in social movements (Gamson 1995; Larana et al. 1994; Calhoun 1994). This new development in social movement theory links structural explanations with cultural and social-psychological theories, investigating how social movements provide foundations for people to construct new identities (Gamson 1992; Gamson 1995; Morris and Mueller 1992).

Each theoretical viewpoint is helpful in explaining different aspects of how movements organize and how, in some cases, they fail. One team of sociologists has examined the demise of the American Indian Movement (AIM)—a specific organization that was prominent in the Native American liberation movement from about 1968 to 1990. These sociologists argue that AIM ultimately failed to reach its goal of building a pantribal sense of identity among American Indians because they were unable to mobilize resources in the face of strong government repression. AIM was a threat to both government and corporate interests because of its focus on reclaiming energy resources on lands Native Americans argued were rightfully theirs. The authors conclude that resource mobilization theory, political process theory, and new social movement theory each explain different dimensions of this movement (Stotik et al. 1994). This is a good illustration of how sociological theory can be used to explain the contemporary social movement activity.

Dramatic tactics are often used by social movements to bring attention to their causes. Here, La Tigresa protests the logging of redwood forests, calling attention to the environmental movement.

Diversity, Globalization, and Social Change

Social movements are a major source of social change. Around the world, as people have organized to protest what they perceive to be oppressive forms of government, the absence of civil rights, or economic injustices, social change often results. What would the United States be like had the civil rights movement not been inspired by Mahatma Gandhi's liberation movement in India? How would contemporary politics be different had the African National Congress and other movements for the liberation of Black South Africans not dismantled apartheid? How is the world currently affected by the development of a more fundamental Islamic religious movement in the Middle East?

These and countless other examples show the significance of collective behavior and social movements for the many changes affecting our world. Sometimes collective behavior and social movements can be the basis for revolutionary events—those that change the course of world history. At other times, the persistence of social movements more slowly change the world—or a particular society. Persistence has been the case for the civil rights movement in the United States, a movement that not only has transformed American society, but has also inspired similar movements throughout the world. Likewise, the women's movement is now a global movement, although the particular issues for women vary from place to place.

Another way to think about globalization and social movements is through the concept of a **transnational social movement,** in which an organization crosses national borders, such as the reactionary terrorist group al Qaeda. Although al Qaeda is not typically thought of as a social movement, analyzing it as such helps explain how al Qaeda works and the impact it has on world affairs. For example, al Qaeda is organized in small "cells" that are networks of people affiliated with this reactionary movement. Concepts from the sociology of social movements help explain how this terrorist organization works.

In the United States, the most significant social movements are those associated with the nation's diverse population. The women's movement, the civil rights movement, the gay and lesbian movement are all major sources of activism in contemporary society, and these movements have generated some of the most transformative changes in the nation's social institutions. Despite the persistence of class, race, and gender inequality, no longer is segregation legally mandated. Equal opportunity laws, at least in theory, protect diverse minority groups from discriminatory treatment. Large segments of the public have become more conscious of the effects of racism, sexism, and homophobia in society. All these changes can be attributed to the successful mobilization of diverse social movements.

16 Chapter Summary

- ***What is social change and what are its types?***
  Social change is that process by which social interaction, the social stratification system, and entire institutions in a society change over time. Social change can take place quickly or it can take longer, sometimes involving *microchanges* (such as a fad that "catches on"), and at other times, *macrochanges* (such as a technological innovation).

- ***How do sociologists explain social change?***
  Functionalist theories and *evolutionary theories* predict that societies move or evolve from the structurally simple to the structurally complex. *Unidimensional evolutionary theory* predicts that societies evolve along a path from simpler socially undifferentiated societies to more complex highly differentiated ones. *Multidimensional evolutionary theory* predicts that societies follow not one but several different paths in the process of social change. *Conflict theories* predict that social conflict is an inherent part of any social structure and that conflict between social class strata or racial–ethnic groups can bring about social change. *Cyclical theories,* such as those of Arnold J. Toynbee and Pitrim Sorokin, predict that certain patterns of social structure and culture recur in a society at different times.

  Modernization theory states that global development among societies is a worldwide process not confined to any one society. *World systems theory,* formulated by Immanuel Wallerstein, predicts that societies are members of a worldwide system of inequality that benefits the more technologically developed societies (core nations) at the expense of the less technologically developed countries (noncore or peripheral nations). *Dependency theory* notes that industrialized core nations tend to keep noncore nations economically dependent, and this retards the economic development and upward mobility of the noncore nations.

- ***What is modernization and what are its aspects?***
  Modernization is a complex set of processes initiated by industrialization. It results from the decline of small, traditional communities (*gemeinschafts*) and the change to generally larger, more differentiated, and impersonal societies (*gesellschafts*). With modernization comes increased bureaucratization of a society. Modernization tends to produce other-directedness, where one person's behavior is strongly guided by the perceived behavior of others; it also tends to produce individual feelings of powerlessness and an increased search for individual identity and one's "true self."

- ***What are the causes of social change?***
  Sources of social change include revolution, cultural diffusion, social inequalities, demographic (population) changes, war and terrorism, technological innovation, collective behavior, and social movements.

- ***What are collective behavior and social movements?***
  Collective behavior occurs when the usual conventions to guide behavior are suspended and people establish new norms of behavior. *Social movements* are organized social groups that have some continuity and promote or resist social change. Social movements start when there is a preexisting communication network, a collective sense of grievance, a precipitating factor initiating the movement, and mobilization of a group of people. *Resource mobilization theory* suggests that social movements develop when people can compete for and gain resources needed for mobilization. *Political process theory* suggests that large-scale social changes, such as industrialization or urbanization, provide the conditions that spawn social movements. *New social movement theory* adds that social movements are places where people construct their identities.

- ***What effects do globalization and diversity have on social change?***
  Social movements can be the basis of revolutionary change. Some movements originating in one nation can also spill over to affect movements in another. *Transnational social movements* are those whose organizational structures cross national borders. Some of the most profound changes in the United States have come as a result of social movements from the nation's diverse population.

Key Terms

Online Resources

Sociology: The Essentials Companion Website

www.thomsonedu.com/sociology/andersen

Visit your book companion website where you will find more resources to help you study and write your research papers. Resources include Suggested Readings, web links, and a MicroCase Online feature that teaches you how to research society. Other resources include Learning Objectives, Internet exercises, quizzing, and flash cards.

is an easy-to-use online resource that helps you study in less time to get the grade you want NOW.

www.thomsonedu.com/login

Need help studying? This site is your one-stop study shop. Take a Pre-Test and Thomson NOW will generate a Personalized Study Plan based on your test results. The Study Plan will identify the topics you need to review and direct you to online resources to help you master those topics. You can then take a Post-Test to determine the concepts you have mastered and what you still need to work on.

Glossary

absolute poverty: the situation in which individuals live on less than $365 a year, or $1.00 a day

achieved status: a status attained by effort

achievement test: test intended to measure what is actually learned rather than potential

adult socialization: the process of learning new roles and expectations in adult life

affirmative action: programs in education and job hiring that recruit minorities over a wide range without using rigid quotas, but that may consider race as a factor

age cohort: an aggregate group of people born during the same time period

age discrimination: different and unequal treatment of people based solely on their age

age prejudice: a negative attitude about an age group that is generalized to all people in that group

age stereotype: preconceived judgments about what different age groups are like

age stratification: the hierarchical ranking of age groups in society

ageism: the institutionalized practice of age prejudice and discrimination

age–sex (or age–gender) pyramid; also population pyramid: a graphic representation of the age and gender structure of a society

alienation: the feeling of powerlessness and separation from one's group or society

altruistic suicide: the type of suicide that can occur when there is excessive regulation of individuals by social forces

anomic suicide: the type of suicide occurring when there are disintegrating forces in the society that make individuals feel lost or alone

anomie: the condition existing when social regulations (norms) in a society break down

anorexia nervosa: a condition characterized by compulsive dieting resulting in self-starvation

anticipatory socialization: the process of learning the expectations associated with a role one expects to enter in the future

anti-Semitism: the belief or behavior that defines Jewish people as inferior and that targets them for stereotyping, mistreatment, and acts of hatred

ascribed status: a status determined at birth

assimilation: process by which a minority becomes socially, economically, and culturally absorbed within the dominant society

attribution error: error made in attributing the causes for someone's behavior to their membership in a particular group, such as a racial group

attribution theory: the principle that dispositional attributions are made about others (what the other is "really like") under certain conditions, such as outgroup membership

authoritarian personality: a personality characterized by a tendency to rigidly categorize people and to submit to authority, rigidly conform, and be intolerant of ambiguity

authority: power that is perceived by others as legitimate

automation: the process by which human labor is replaced by machines

autonomous state model: a theoretical model of the state that interprets the state as developing interests of its own, independent of other interests

aversive racism: subtle, nonovert, and nonobvious racism

beliefs: shared ideas held collectively by people within a given culture

bilateral kinship: a kinship system where descent is traced through the father and the mother

biological determinism: explanations that attribute complex social phenomena to physical characteristics

bioterrorism: a form of terrorism involving the dispersion of chemical or biological substances intended to cause widespread disease and death

bureaucracy: a type of formal organization characterized by an authority hierarchy, a clear division of labor, explicit rules, and impersonality

capitalism: an economic system based on the principles of market competition, private property, and the pursuit of profit

caste system: a system of stratification (characterized by low social mobility) in which one's place in the stratification system is determined by birth

census: a count of the entire population of a country

charisma: a quality attributed to individuals believed by their followers to have special powers

charismatic authority: authority derived from the personal appeal of a leader

church: a formal organization that sees itself and is seen by society as a primary and legitimate religious institution.

class: see social class

class consciousness: the awareness that a class structure exists and the feeling of shared identification with others in one's class with whom one perceives common life chances

cognitive ability: the capacity for abstract thinking

cognitive elite: the idea that there is a presumably genetically based elite class in the United States containing those with high IQs, high incomes, and prestigious jobs

cohort (birth cohort): see age cohort

collective behavior (action): behavior that occurs when the usual conventions are suspended and people collectively establish new norms of behavior in response to an emerging situation

collective consciousness: the body of beliefs that are common to a community or society and that give people a sense of belonging

colonialism: system by which Western nations became wealthy by taking raw materials from other societies (the colonized) and reaping profits from products finished in the homeland

color-blind racism: ignoring legitimate racial, ethnic, and cultural differences between groups, thus denying the reality of such differences

coming out: the process of defining oneself as gay or lesbian

commodity chain: the network of production and labor processes by which a product becomes a finished commodity. By following the commodity chain, it is evident which countries gain profits and which ones are being exploited

communism: an economic system where the state is the sole owner of the systems of production

comparable worth: the principle of paying women and men equivalent wages for jobs involving similar levels of skill

conflict theory: a theoretical perspective that emphasizes the role of power and coercion in producing social order

contact theory: the theory that prejudice will be reduced through social interaction with those of different race or ethnicity but of equal status

content analysis: the analysis of meanings in cultural artifacts like books, songs, and other forms of cultural communication

contingent worker: those who do not hold regular jobs, but whose employment is dependent upon demand

controlled experiment: a method of collecting data that can determine whether something actually causes something else

core countries (core nations): within world systems theory, those nations that are more technologically advanced

correlation: a statistical technique that analyzes patterns of association between pairs of sociological variables

counterculture: subculture created as a reaction against the values of the dominant culture

credentialism: the insistence upon educational credentials only for their own sake

crime: one form of deviance; specifically, behavior that violates criminal laws

criminology: the study of crime from a scientific perspective

crude birthrate: the number of babies born each year for every 1000 members of the population

crude death rate: the number of deaths each year per 1000 members of the population

cult: a religious group devoted to a specific cause or charismatic leader

cultural capital: (also known as *social capital*) cultural resources that are socially designated as being worthy (such as knowledge of elite culture) and that give advantages to groups possessing such capital

cultural diffusion: the transmission of cultural elements from one society or cultural group to another

cultural hegemony: the pervasive and excessive influence of one culture throughout society

cultural pluralism: pattern whereby groups maintain their distinctive culture and history

cultural relativism: the idea that something can be understood and judged only in relationship to the cultural context in which it appears

culture: the complex system of meaning and behavior that defines the way of life for a given group or society

culture lag: the delay in cultural adjustments to changing social conditions

culture of poverty: the argument that poverty is a way of life and, like other cultures, is passed on from generation to generation

culture shock: the feeling of disorientation that can come when one encounters a new or rapidly changed cultural situation

cyberspace interaction: interaction occurring when two or more persons share a virtual reality experience via electronic communication and interaction with each other

cyberterrorism: the use of the computer to commit one or more terrorist acts

cyclical theory: the idea that societies go through a "life cycle," for example, from idealistic through hedonistic culture and back to idealistic

data: the systematic information that sociologists use to investigate research questions

data analysis: the process by which sociologists organize collected data to discover what patterns and uniformities are revealed

debunking: looking behind the facades of everyday life

deductive reasoning: the process of creating a specific research question about a focused point, based on a more general or universal principle

defensive medicine: practiced by physicians who order extra tests on a patient in an effort to fend off a lawsuit by the patient

deindividuation: the feeling that one's self has merged with a group

deindustrialization: the transition from a predominantly goods-producing economy to one based on the provision of services

democracy: system of government based on the principle of representing all people through the right to vote

demographic transition theory: the theory that societies pass through phases based on economic development, which affects the birth and death rates

demography: the scientific study of population

dependency theory: the global theory maintaining that industrialized nations hold less-industrialized nations in a dependent, thus exploitative, relationship that benefits the industrialized nations at the expense of the less-industrialized ones

dependent variable: the variable that is a presumed effect (*see* independent variable)

deviance: behavior that is recognized as violating expected rules and norms

deviant career: continuing to be labeled as deviant even after the initial (primary) deviance may have ceased

deviant community: groups that are organized around particular forms of social deviance

deviant identity: the definition a person has of himself or herself as a deviant

differential association theory: theory that interprets deviance as behavior one learns through interaction with others

discrimination: overt negative and unequal treatment of the members of some social group or stratum solely because of their membership in that group or stratum

disengagement theory: theory predicting that as people age, they gradually withdraw from participation in society and are simultaneously relieved of responsibilities

diversity: the variety of group experiences that result from the social structure of society

division of labor: the systematic interrelation of different tasks that develops in complex societies

dominant culture: the culture of the most powerful group in society

dominant group: the group that assigns a racial or ethnic group to subordinate status in society

dual labor market theory: the idea that women and men have different earnings because they tend to work in different segments of the labor market

dyad: a group consisting of two people

ecological demography: a field that combines the studies of demography and human ecology

ecological globalization: the worldwide dispersion of problems involving relationships between humans and the physical environment

economic restructuring: contemporary transformations in the basic structure of work that are permanently altering the workplace, including the changing composition of the workplace, deindustrialization, the use of enhanced technology, and the development of a global economy

economy: the system on which the production, distribution, and consumption of goods and services is based

educational attainment: the total years of formal education

ego: the part of the self representing reason and common sense

egoistic suicide: the type of suicide that occurs when people feel totally detached from society

elite deviance: the wrongdoing of powerful individuals and organizations

emigration (vs. immigration): migration of people from one society to another (also called out-migration)

emotional labor: work that is explicitly intended to produce a desired state of mind in a client

empirical: refers to something that is based on careful and systematic observation

endogamy: the practice of selecting mates from within one's group

Enlightenment: the period in eighteenth- and nineteenth-century Europe characterized by faith in the ability of human reason to solve society's problems

environmental racism: the dumping of toxic wastes with disproportionate frequency at or very near areas with high concentrations of minorities

epidemiology: the study of all factors—biological, social, economic, and cultural—that are associated with disease and health

estate system: a system of stratification in which the ownership of property and the exercise of power is monopolized by an elite or noble class, which has total control over societal resources

ethnic group: a social category of people who share a common culture, such as a common language or dialect, a common religion, or common norms, practices, and customs

ethnocentrism: the belief that one's in-group is superior to all out-groups

ethnomethodology: a technique for studying human interaction by deliberately disrupting social norms and observing how individuals attempt to restore normalcy

eugenics: a social movement in the early twentieth century that sought to apply scientific principles of genetic selection to "improve" the offspring of the human race

evaluation research: research assessing the effect of policies and programs

evolutionary (social) theory: the idea that societies go through stages of advancement, as from structurally undifferentiated to more complex and structually differentiated

exogamy: the practice of selecting mates from outside one's group

expressive needs: needs for intimacy, companionship, and emotional support

extended families: the whole network of parents, children, and other relatives who form a family unit and often reside together

extreme poverty: the situation in which people live on less than $275 a year, or 75 cents a day

false consciousness: the thought resulting from subordinate classes internalizing the view of the dominant class

family: a primary group of people—usually related by ancestry, marriage, or adoption—who form a cooperative economic unit to care for any offspring (and each other) and who are committed to maintaining the group over time

Family and Medical Leave Act (FMLA): federal law requiring employers of a certain size to grant leave to employees for purposes of family care

feminism: a way of thinking and acting that advocates a more just society for women

feminist theory: analyses of women and men in society intended to improve women's lives

feminization of poverty: the process whereby a growing proportion of the poor are women and children

first-world countries: industrialized nations based on a market economy and with democratically elected governments

folkways: the general standards of behavior adhered to by a group

formal organization: a large secondary group organized to accomplish a complex task or set of tasks

frames: specific schemes of interpretation that allow people to perceive, identify, and label events within their lives that can become the basis for collective action

functionalism: a theoretical perspective that interprets each part of society in terms of how it contributes to the stability of the whole society

game stage: the stage in childhood when children become capable of taking a multitude of roles at the same time

gemeinschaft: German for *community,* a state characterized by a sense of common feeling among the members of a society, including strong personal ties, sturdy primary group memberships, and a sense of personal loyalty to one another; associated with rural life

gender: socially learned expectations and behaviors associated with members of each sex

gender apartheid: the extreme segregation and exclusion of women from public life

gender gap: gender differences in behavior, such as voting behavior

gender identity: one's definition of self as a woman or man

gender segregation: the distribution of men and women in different jobs in the labor force

gender socialization: the process by which men and women learn the expectations associated with their sex

gender stratification: the hierarchical distribution of social and economic resources according to gender

gendered institutions: the total pattern of gender relations that structure social institutions, including the stereotypical expectations, interpersonal relationships, and the different placement of men and women that are found in institutions

generalization: the ability to make claims that a finding represents something greater than the specific observations on which the finding is based

generalized other: an abstract composite of social roles and social expectations

gesellschaft: a type of society in which increasing importance is placed on the secondary relationships people have—that is, less intimate and more instrumental relationships

glass ceiling: popular concept referring to the limits that women and minorities experience in job mobility

global culture: the diffusion of a single culture throughout the world

global economy: term used to refer to the fact that all dimensions of the economy now cross national borders

global stratification: the systematic inequalities between and among different groups within nations that result from the differences in wealth, power, and prestige of different societies relative to their position in the international economy

global warming: the systematic increase in world-wide surface temperatures

globalization: increased economic, political, and social interconnectedness and interdependence among societies in the world

government: those state institutions that represent the population and make rules that govern the society

greenhouse effect: a rise in the earth's surface temperature caused by heat trapped by excess carbon dioxide in the atmosphere; global warming

gross national income (GNI): the total output of goods and services produced by residents of a country each year plus the income from nonresident sources, divided by the size of the population

group: a collection of individuals who interact and communicate, share goals and norms, and who have a subjective awareness as "we"

group size effect: the effect upon the person of groups of varying sizes

groupthink: the tendency for group members to reach a consensus at all costs

hate crime: assaults and other malicious acts (including crimes against property) motivated by various forms of bias, including that based on race, religion, sexual orientation, ethnic and national origin, or disability

health maintenance organization (HMO): a cooperative of doctors and other medical personnel who provide medical services in exchange for a set membership fee

hermaphroditism: a condition produced when irregularities in chromosome formation or fetal differentiation produces persons with biologically mixed sex characteristics

heterosexism: the institutionalization of heterosexuality as the only socially legitimate sexual orientation

HMO: see Health Maintenance Organization

homophobia: the fear and hatred of homosexuality

human capital theory: a theory that explains differences in wages as the result of differences in the individual characteristics of the workers

human development index: a compilation of data indicating various levels of national well-being

human ecology: the study of the interdependence between humans and their physical environment

human ecosystem: any system of interdependent parts that involves human beings in interaction with one another and with the physical environment

human poverty index: a multidimensional measure of poverty meant to indicate the degree of deprivation in four dimensions: life expectancy, education, economic well-being, and social inclusion

hypersegregation: a pattern of extreme racial, ethnic, and/or social class residential segregation, such that nearly all individuals in an area are of one such group

hypothesis: a statement about what one expects to find in research

id: the part of the personality that includes various impulses and drives, including sexual passions and desires, biological urges, and human instincts

identity: how one defines oneself

ideology: a belief system that tries to explain and justify the status quo

imitation stage: the stage in childhood when children copy the behavior of those around them

immigration (vs. emigration): the migration of people into a society from outside it (also called in-migration)

impression management: a process by which people attempt to control how others perceive them

income: the amount of money brought into a household from various sources during a given year (wages, investment income, dividends, etc.)

independent variable: a variable that is the presumed cause of a particular result (*see* dependent variable)

index crimes: the FBI's tallying of violent crimes of murder, manslaughter, rape, robbery, and aggravated assault, plus property crimes

indicator: something that points to or reflects an abstract concept

inductive reasoning: the process of arriving at general conclusions from specific observations

infant mortality rate: the number of deaths per year of infants under age 1 for every 1000 live births

inner-directedness: a condition wherein the individual's behavior is guided by internal principles and morals

institutional racism: racism involving notions of racial or ethnic inferiority that have become ingrained into society's institutions

instrumental needs: emotionally neutral, task-oriented (goal-oriented) needs

interest group: a constituency in society organized to promote its own agenda

interlocking directorate: organizational linkages created when the same people sit on the boards of directors of a number of different corporations

international division of labor: system of labor whereby products are produced globally, while profits accrue only to a few

issues: problems that affect large numbers of people and have their origins in the institutional arrangements and history of a society

job displacement: the permanent loss of certain job types when employment patterns shift, as when a manufacturing plant shuts down

kinship system: the pattern of relationships that define people's family relationships to one another

labeling effect: the effect of track (role) assignment as distinct from the effect of cognitive ability

labeling theory: a theory that interprets the responses of others as most significant in understanding deviant behavior

labor force participation rate: the percentage of those in a given category who are employed

laissez-faire racism: maintaining the status quo of racial groups by persistent stereotyping and blaming of minorities themselves for achievement and socioeconomic gaps between groups

language: a set of symbols and rules that, when put together in a meaningful way, provides a complex communication system

latent functions: subtle, unintended consequences of an institutional element of which the participants (the people) are usually unaware

law: the written set of guidelines that define what is right and wrong in society

liberal feminism: a feminist theoretical perspective asserting that the origin of women's inequality is in traditions of the past that pose barriers to women's advancement

life chances: the opportunities that people have in common by virtue of belonging to a particular class

life course: the connection between people's personal attributes, the roles they occupy, the life events they experience, and the social and historical context of these events

life expectancy: the average number of years individuals and the particular groups can expect to live

looking glass self: the idea that people's conception of self arises through reflection about their relationship to others

macroanalysis: analysis of the whole of society, how it is organized and how it changes

macrochange: gradual transformations in a society that occur on a broad scale and affect many aspects of a society

Malthusian theory: after T.R. Malthus, the principle that a population tends to grow faster than the subsistence (food) level needed to sustain it

managed care: use of collective bargaining as part of a large collection of HMOs

mass media: channels of communication that are available to very wide segments of the population

master status: some characteristic of a person that overrides all other features of the person's identity

material culture: the objects created in a given society

matriarchal religion: a religion based on the centrality of a female goddess or goddesses

matriarchy: a society or group in which women have power over men

matrilineal kinship: kinship systems in which family lineage (or ancestry) is traced through the mother

mean: the sum of a set of values divided by the number of cases from which the values are obtained; an average

mechanical solidarity: unity based on similarity, not difference, of roles

median: the midpoint in a series of values that are arranged in numerical order

median income: the midpoint of all household incomes

Medicaid: a governmental assistance program that provides health care assistance for the poor, including the elderly

medicalization of deviance: explanations of deviant behavior that interpret deviance as the result of individual pathology or sickness

Medicare: a governmental assistance program established in the 1960s to provide health services for older Americans

meritocracy: a system in which one's status is based on merit or accomplishments

microanalysis: analysis of the smallest, most immediately visible parts of social life, such as people interacting

microchange: subtle alterations in the day-to-day interaction between people, such as a fad "catching on"

minority group: any distinct group in society that shares common group characteristics and is forced to occupy low status in society because of prejudice and discrimination

mobilization: the process by which social movements and their leaders secure people and resources for their movement

mode: the value (or score) that appears most frequently in a set of data

modernization: a process of social and cultural change that is initiated by industrialization and followed by increased social differentiation and division of labor

modernization theory: a view of globalization in which global development is a worldwide process affecting nearly all societies that have been touched by technological change

monogamy: the marriage practice of a sexually exclusive marriage with one spouse at a time

monotheism: the worship of a single god

mores: strict norms that control moral and ethical behavior

multidimensional evolutionary theory: a theory predicting that over time societies follow not one but several evolutionary paths

multinational corporation: corporations that conduct business across national borders

multiracial feminism: form of feminist theory noting the exclusion of women of color from other forms of theory and centering its analysis in the experiences of all women

neocolonialism: a form of control of poor countries by rich countries, but without direct political or military involvement

new social movement theory: a theory about social movements linking culture, ideology, and identity conceptually to explain how new identities are forged within social movements

newly industrializing countries (NICs): countries that have shown rapid growth and have emerged as developed countries

nonmaterial culture: the norms, laws, customs, ideas, and beliefs of a group of people

norms: the specific cultural expectations for how to act in a given situation

nuclear family: family in which a married couple resides together with their children

occupational prestige: the subjective evaluation people give to jobs as better or worse than others

occupational segregation: a pattern in which different groups of workers are separated into different occupations

old-fashioned racism: overt and obvious expressions of racism, such as physical assaults, lynchings, and other such acts against a minority

organic metaphor: refers to the similarity early sociologists saw between society and other organic systems

organic (contractual) solidarity: unity based on role differentiation, not similarity

organized crime: crime committed by organized groups, typically involving the illegal provision of goods and services to others

other-directedness: a condition wherein the individual's behavior is guided by the behavior of others

out-group homogeneity effect: the tendency for an in-group member to perceive members of any out-group as similar or identical to each other

participant observation: a method whereby the sociologist becomes both a participant in the group being studied and a scientific observer of the group

patriarchal religion: religion in which the beliefs and practices of the religion are based on male power and authority

patriarchy: a society or group where men have power over women

patrilineal kinship: a kinship system that traces descent through the father

peers: those of similar status

percentage: the number of parts per hundred

peripheral countries (nations): poor countries, largely agricultural, having little power or influence in the world system

personal crimes: violent or nonviolent crimes directed against people

personal transformation movements: social movements that aim to change the individual

personality: the cluster of needs, drives, attitudes, predispositions, feelings, and beliefs that characterize a given person

play stage: the stage in childhood when children begin to take on the roles of significant people in their environment

pluralist model: a theoretical model of power in society as coming from the representation of diverse interests of different groups in society

political action committees (PACs): groups of people who organize to support candidates they feel will represent their views

political process theory: explanation of social movements positing that movements achieve success by exploiting a combination of internal organizational factors as well as external changes in society

polygamy: a marriage practice in which either men or women can have multiple marriage partners

polytheism: the worship of more than one deity

popular culture: the beliefs, practices, and objects that are part of everyday traditions

population: a relatively large collection of people (or other unit) that a researcher studies and about which generalizations are made

population density: the number of people per square mile

population replacement level: a situation in which the birthrate and death rate of a population sustains the population at a steady level

positivism: a system of thought that regards scientific observation to be the highest form of knowledge

postindustrial society: a society economically dependent upon the production and distribution of services, information, and knowledge

postmodernism: a theoretical perspective based in the idea that communication in all its forms is reality, such that understanding society requires studying all forms of communication, including cultural ideas, language, texts, and self-conceptions

poverty line: the figure established by the government to indicate the amount of money needed to support the basic needs of a household

power: a person or group's ability to exercise influence and control over others

power elite model: a theoretical model of power positing a strong link between government and business

predictive validity: the extent to which a test accurately predicts later college grades, or some other criterion such as likelihood of graduating

preindustrial society: one that directly uses, modifies, and/or tills the land as a major means of survival

prejudice: the negative evaluation of a social group, and individuals within that group, based upon conceptions about that social group that are held despite facts that contradict it

prestige: the value with which different groups or people are judged

primary group: a group characterized by intimate, face-to-face interaction and relatively long-lasting relationships

profane: that which is of the everyday, secular world and is specifically not religious

propaganda: information disseminated by a group or organization (such as the state) intended to justify its own power

property crimes: crimes involving theft of or harm to property without bodily harm to the victim(s)

Protestant ethic: belief that hard work and self-denial lead to salvation

proxemic communication: meaning conveyed by the amount of space between interacting individuals

psychoanalytic theory: a theory of socialization positing that the unconscious mind shapes human behavior

qualitative research: research that is somewhat less structured than quantitative research but that allows more depth of interpretation and nuance in what people say and do

quantitative research: research that uses numerical analysis

queer theory: a theoretical perspective that recognizes the socially constructed nature of sexual identity

race: a social category, or social construction, that we treat as distinct on the basis of certain characteristics, some biological, that have been assigned social importance in the society

racial formation: process by which groups come to be defined as a "race" through social institutions such as the law and the schools

racial profiling: the use of race alone as the criterion for deciding whether to stop and detain someone on suspicion of their having committed a crime

racialization: a process whereby some social category, such as a social class or nationality, is assigned what are perceived to be race characteristics

racism: the perception and treatment of a racial or ethnic group, or member of that group, as intellectually, socially, and culturally inferior to one's own group

radical feminism: feminist theoretical perspective that interprets patriarchy as the primary cause of women's oppression

radical movements: social movements that seek fundamental change in the structure of society

random sample: a sample that gives everyone in the population an equal chance of being selected

rate: parts per some number (e.g., per 10,000; per 100,000)

rational–legal authority: authority stemming from rules and regulations, typically written down as laws, procedures, or codes of conduct

reactionary movements: social movements organized to resist change or to reinstate an earlier social order that participants perceive to be better

reference group: any group (to which one may or may not belong) used by the individual as a standard for evaluating her or his attitudes, values, and behaviors

reflection hypothesis: the idea that the mass media reflect the values of the general population

reform movements: social movements that seek change through legal or other mainstream political means, by working within existing institutions

relative poverty: a definition of poverty that is set in comparison to a set standard

reliability: the likelihood that a particular measure would produce the same results if the measure were repeated

religion: an institutionalized system of symbols, beliefs, values, and practices by which a group of people interprets and responds to what they feel is sacred and that provides answers to questions of ultimate meaning

religiosity: the intensity and consistency of practice of a person's (or group's) faith

religious extremism: actions and beliefs that are driven by high levels of religious intolerance

replication study: research that is repeated exactly, but on a different group of people at a different point in time

research design: the overall logic and strategy underlying a research project

residential segregation: the spatial separation of racial and ethnic groups in different residential areas

resocialization: the process by which existing social roles are radically altered or replaced

resource mobilization theory: theory of how social movements develop that focuses on how movements gain momentum by successfully garnering organizational resources

revolution: the overthrow of a state or the total transformation of central state institutions

risky shift (also polarization shift): the tendency for group members, after discussion and interaction, to engage in riskier behavior than they would while alone

rite of passage: ceremony or ritual that symbolizes the passage of an individual from one role to another

ritual: symbolic activities that express a group's spiritual convictions

role: behavior others expect from a person associated with a particular status

role conflict: two or more roles associated with contradictory expectations

role modeling: imitation of the behavior of an admired other

role set: all roles occupied by a person at a given time

role strain: conflicting expectations within the same role

sacred: that which is set apart from ordinary activity, seen as holy, and protected by special rites and rituals

salience principle: categorizing people on the basis of what initially appears prominent about them

sample: any subset of units from a population that a researcher studies

Sapir–Whorf hypothesis: a theory that language determines other aspects of culture since language provides the categories through which social reality is defined and perceived

scapegoat theory: argument that dominant group aggression is directed toward a minority as a substitute for frustration with some other problem

scientific method: the steps in a research process, including observation, hypothesis testing, analysis of data, and generalization

secondary group: a group that is relatively large in number and not as intimate or long in duration as a primary group

second-world countries: socialist countries with state-managed economies and typically without a democratically elected government

sect: groups that have broken off from an established church

secular: the ordinary beliefs of daily life that are specifically not religious

segregation: the spatial and social separation of racial and ethnic groups

self: our concept of who we are, as formed in relationship to others

self-fulfilling prophecy: the process by which merely applying a label changes behavior and thus tends to justify the label

semiperipheral countries: semi-industrialized countries that represent a kind of middle class within the world system

sex: used to refer to biological identity as male or female

sex ratio (gender ratio): the number of males per 100 females

sexual harassment: unwanted physical or verbal sexual behavior that occurs in the context of a relationship of unequal power and that is experienced as a threat to the victim's job or educational activities

sexual identity: the definition of oneself that is formed around one's sexual relationships

sexual orientation: the attraction that people feel for people of the same or different sex

sexual politics: the link feminists argue exists between sexuality and power and between sexuality and race, class, and gender oppression

sexual revolution: the widespread changes in men's and women's roles and a greater public acceptance of sexuality as a normal part of social development

sexual scripts: the ideas taught to us about what is appropriate sexual behavior for a person of our gender

significant others: those with whom we have a close affiliation

social capital: see cultural capital

social change: the alteration of social interaction, social institutions, stratification systems, and elements of culture over time

social change movements: movements that aim to change some aspect of society

social class: the social structural hierarchical position groups hold relative to the economic, social, political, and cultural resources of society

social construction perspective: a theoretical perspective that explains sexual identity (or any other identity) as created and learned within a cultural, social, and historical context

social control: the process by which groups and individuals within those groups are brought into conformity with dominant social exceptions

social control agents: those who regulate and administer the response to deviance, such as the police, or mental health workers

social control theory: theory that explains deviance as the result of the weakening of social bonds

social Darwinism: the idea that society evolves to allow the survival of the fittest

social differentiation: the process by which different statuses in any group, organization, or society develop

social epidemiology: the study of the effects of social and cultural factors upon disease and health

social facts: social patterns that are external to individuals

social institution: an established and organized system of social behavior with a recognized purpose

social interaction: behavior between two or more people that is given meaning

social learning theory: a theory of socialization positing that the formation of identity is a learned response to social stimuli

social mobility: a person's movement over time from one class to another

social movement: a group that acts with some continuity and organization to promote or resist social change in society

social network: a set of links between individuals or other social units such as groups or organizations

social organization: the order established in social groups

social/political change movement: a type of social movement that intends to change some status quo aspect of society, such as the civil rights movement or the environmental movement

social sanctions: mechanisms of social control that enforce norms

social stratification: a relatively fixed hierarchical arrangement in society by which groups have different access to resources, power, and perceived social worth; a system of structured social inequality

social structure: the patterns of social relationships and social institutions that make up society

socialism: an economic institution characterized by state ownership and management of the basic industries

socialist feminism: a feminist theoretical perspective that interprets the origins of women's oppression as lying in the system of capitalism

socialization: the process through which people learn the expectations of society

socialization agents: those who pass on social expectations

society: a system of social interaction that includes both culture and social organization

socioeconomic status (SES): a measure of class standing, typically indicated by income, occupational prestige, and educational attainment

sociological imagination: the ability to see the societal patterns that influence individual and group life

sociology: the study of human behavior in society

standardized ability test: tests given to large populations and scored with respect to population averages

state: the organized system of power and authority in society

status: an established position in a social structure that carries with it a degree of prestige

status attainment: the process by which people end up in a given position in the stratification system

status inconsistency: exists when the different statuses occupied by the individual bring with them significantly different amounts of prestige

status set: the complete set of statuses occupied by a person at a given time

stereotype: an oversimplified set of beliefs about the members of a social group or social stratum that is used to categorize individuals of that group

stereotype interchangeability: the principle that negative stereotypes are often interchangeable from one racial group (or gender or social class) to another

stereotype threat effect: the effect of a negative stereotype about one's self upon one's own test performance

stigma: an attribute that is socially devalued and discredited

structural strain theory: a theory that interprets deviance as originating in the tensions that exist in society between cultural goals and the means people have to achieve those goals

subculture: the culture of groups whose values and norms of behavior are somewhat different from those of the dominant culture

superego: the dimension of the self representing the cultural standards of society

symbolic interaction theory: a theoretical perspective claiming that people act toward things because of the meaning things have for them

symbols: things or behavior to which people give meaning

taboos: those behaviors that bring the most serious sanctions

taking the role of the other: the process of imagining oneself from the point of view of another

teacher expectancy effect: the effect of the teacher's expectations on the student's actual performance, independent of the student's ability

Temporary Assistance to Needy Families: federal program by which grants are given to states to fund welfare

terrorism: the unlawful use of force or violence against persons or property to intimidate or coerce a government or population in furtherance of political or social objectives

thermal pollution: the heating up of the earth's rivers and lakes as a result of the chemical discharges of heavy industry

third-world countries: countries that are poor, underdeveloped, largely rural, and with high levels of poverty; typically governments in such countries are autocratic dictatorships and wealth is concentrated in the hands of a small elite

total institution: an organization cut off from the rest of society in which individuals are subject to strict social control

totem: an object or living thing that a religious group regards with special awe and reverence

tracking: grouping, or stratifying, students in school on the basis of ability test scores

traditional authority: authority stemming from long-established patterns that give certain people or groups legitimate power in society

tradition-directedness: conformity to long-standing and time-honored norms and practices

transnational family: families where one parent (or both) lives and works in one country while the children remain in their countries of origin

transnational social movement: a social movement that crosses national borders, such as al Qaeda

triad: a group consisting of three people

triadic segregation: the tendency for a triad to separate into a dyad and an isolate

troubles: privately felt problems that come from events or feelings in one individual's life

underemployment: the condition of being employed at a skill level below what would be expected given a person's training, experience, or education

unemployment rate: the percentage of those not working, but officially defined as looking for work

unidimensional evolutionary theory: a theory that predicts that societies over time follow a single path from simple and structurally undifferentiated to more complex and structurally differentiated

urban underclass: a grouping of people, largely minority and poor, who live at the absolute bottom of the socioeconomic ladder in urban areas

urbanization: the process by which a community acquires the characteristics of city life

validity: the degree to which an indicator accurately measures or reflects a concept

values: the abstract standards in a society or group that define ideal principles

variable: something that can have more than one value or score

verstehen: the process of understanding social behavior from the point of view of those engaged in it

victimless crimes: violations of law not listed in the FBI's serious crime index, such as gambling or prostitution

vital statistics: information about births, deaths, marriages, and migration

wealth: the monetary value of what someone actually owns, calculated by adding all financial assets (stocks, bonds, property, insurance, value of investments, etc.) and subtracting debts; also called net worth

work: productive human activity that produces something of value, either goods or services

world cities: cities that are closely linked through the system of international commerce

world systems theory: theory that capitalism is a single world economy and that there is a worldwide system of unequal political and economic relationships that benefit the technologically advanced countries at the expense of the less technologically advanced

xenophobia: the fear and hatred of foreigners

References

Aberle, David F., Albert K. Cohen, A. Kingsley Davis, Marion J. Levy, Jr., and Francis X. Sutton. 1950. "The Functional Prerequisites of a Society." *Ethics* 60 (January): 100–111.

Abrahamson, Mark. 2006. *Urban Enclaves: Identity and Place in the World,* 2nd ed. New York: Worth.

Acker, Joan. 1992. "Gendered Institutions: From Sex Roles to Gendered Institutions." *Contemporary Sociology* 21 (September): 565–569.

Acker, Joan, Sandra Morgen, and Lisa Gonzales. 2002. "Welfare Restructuring, Work & Poverty: Policy from Oregon." Working Paper, Eugene, OR: Center for the Study of Women in Society.

Adler, Patricia, and Peter Adler. 1998. *Peer Power: Preadolescent Culture and Identity.* New Brunswick, NJ: Rutgers University Press.

Adorno, T. W., Else Frenkel-Brunswik, D. J. Levinson, and R. N. Sanford. 1950. *The Authoritarian Personality.* New York: Harper and Row.

AFL-CIO. 2002. "Executive Pay Watch," www.afl-cio.org

Alan Guttmacher Institute. 1994. *Sex and America's Teenagers.* New York: Alan Guttmacher Institute.

Alba, Richard. 1990. *Ethnic Identity: The Transformation of Ethnicity in the Lives of Americans of European Ancestry.* New Haven, CT: Yale University Press.

Alba, Richard, and Gwen Moore. 1982. "Ethnicity in the American Elite." *American Sociological Review* 47 (June): 373–383.

Alba, Richard D., and Victor Nee. 2003. *Remaking the American Mainstream: Assimilation and Contemporary Immigration.* Cambridge, MA: Harvard University Press.

Albas, Daniel, and Cheryl Albas. 1988. "Aces and Bombers: The Post-Exam Impression Management Strategies of Students." *Symbolic Interaction* 11: 289–302.

Albelda, Randy, and Chris Tilly. 1996. "It's a Family Affair: Women, Poverty, and Welfare." Pp. 79–86 in *For Crying Out Loud: Women's Poverty in the United States,* edited by Diane Dujon and Ann Withorn. Boston, MA: South End Press.

Albelda, Randy, and Ann Withorn (eds.). 2002. *Lost Ground: Welfare Reform, Poverty, and Beyond.* Boston, MA: South End Press.

Aldag, Ramon J,. and Sally R. Fuller. 1993. "Beyond Fiasco: A Reappraisal of the Groupthink Phenomenon and a New Model of Group Decision Processes." *Psychological Bulletin* 113: 533–552.

Alden, Helena L. 2001. "Gender Role Ideology and Homophobia." Paper presented at the Annual Meeting of the Southern Sociological Society.

Aldous, Joan and Rodney F. Ganey. 1999. "Family Life and the Pursuit of Happiness: The Influence of Gender and Race." *Journal of Family Issues* 20(2): 155–180.

Aldrich, Howard, and Martin Ruef. 2006. *Organizations Evolving.* Thousand Oaks, CA: Sage.

Alexander, Jeffrey C., and Paul Colomy (eds.). 1990. *Differentiation Theory and Social Change: Comparative and Historical Perspectives.* New York: Columbia University Press.

Alicea, Marixsa. 1997. "'What Is Indian About You?' A Gendered, Transnational Approach to Ethnicity." *Gender & Society* 11 (October): 597–626.

Allen, Katherine R. 1997. "Lesbian and Gay Families." Pp. 196–218 in *Contemporary Parenting: Challenges and Issues,* edited by Terry Arendell. Thousand Oaks, CA: Sage.

Allen, Michael Patrick, and Philip Broyles. 1991. "Campaign Finance Reforms and the Presidential Campaign Contributions of Wealthy Capitalist Families." *Social Science Quarterly* 72 (December): 738–750.

Allison, Anne. 1994. *Nightwork: Sexuality, Pleasure, and Corporate Masculinity in a Tokyo Hostess Club.* Chicago, IL: University of Chicago Press.

Allport, Gordon W. 1954. *The Nature of Prejudice.* Reading, MA: Addison-Wesley.

Altemeyer, Bob. 1988. *Enemies of Freedom: Understanding Right-Wing Authoritarianism.* San Francisco, CA: Jossey-Bass.

Altman, Dennis. 2001. *Global Sex.* Chicago, IL: University of Chicago Press.

Amato, Paul R., and Alan Booth. 1996. "A Prospective Study of Divorce and Parent-Child Relationships." *Journal of Marriage and the Family* 58(2): 356–365.

American Association of University Women. 1992. *How Schools Shortchange Girls.* Washington, DC: American Association of University Women.

American Association of University Women. 1998. *Gender Gaps: Where Schools Still Fail Our Children.* Washington, DC: American Association of University Women.

American Psychological Association. 2007. *Report of the APA Task Force on the Sexualization of Girls.* Washington, DC: American Psychological Association.

Amott, Teresa L., and Julie A. Matthaei, 1996. *Race, Gender, and Work: A Multicultural History of Women in the United States,* 2nd ed. Boston, MA: South End Press.

Andersen, Margaret L. 2000. *Thinking About Women: Sociological Perspectives on Sex and Gender,* 5th ed. Boston, MA: Allyn and Bacon.

Andersen, Margaret L. 2003. *Thinking About Women: Sociological Perspectives on Sex and Gender,* 6th ed. Boston, MA: Allyn and Bacon.

Andersen, Margaret L. 2004. "From Brown to Grutter: The Diverse Beneficiaries of Brown v. Board of Education." *University of Illinois Law Review* 2004 (5): 1073–1097.

Andersen, Margaret L. 2006. *Thinking About Women: Sociological Perspectives on Sex and Gender.* 7th ed. Boston, MA: Allyn and Bacon.

Andersen, Margaret L., and Patricia Hill Collins. 2007. *Race, Class, and Gender: An Anthology,* 5th ed. Belmont, CA: Wadsworth.

Anderson, David A., and Mykol Hamilton. 2005. "Gender Role Stereotyping of Parents in Children's Picture Books: The Invisible Father." *Sex Roles: A Journal of Research* 52(3–4): 145–151.

Anderson, Elijah. 1976. *A Place on the Corner.* Chicago, IL: University of Chicago Press.

Anderson, Elijah. 1990. *Streetwise: Race, Class, and Change in an Urban Community.* Chicago, IL: University of Chicago Press.

Anderson, Elijah. 1999. *Code of the Street: Decency. Violence, and the Moral Life of the Inner City.* New York: W.W. Norton.

Anderson, Kristin J., and Campbell Leaper. 1998. "Meta-Analyses of Gender Effects of Conversational Interruption: Who, What, When, Where, and How." *Sex Roles* 39 (August): 225–252.

Andreasse, Arthur. 1997. "Evaluating the 1995 Industry Employment Projections." *Monthly Labor Review* 120 (September): 9–31.

Angotti, Joseph. 1997. "Content Analysis of Local News Programs in Eight U.S. Television Markets." Miami, FL: University of Miami Center for the Advancement of Modern Media.

Anthony, Dick, Thomas Robbins, and Steven Barrie-Anthony. 2002. "Cult and Anticult Totalism: Reciprocal Escalation and Violence." *Terrorism and Political Violence* 14 (Spring): 211–239.

Antrobus, Peggy. 2002. "Feminism as Transformational Politics: Towards Possibilities for Another World." *Development* 45 (June): 46–52.

Apple, Michael W. 1991. "The New Technology: Is It Part of the Solution or Part of the Problem in Education?" *Computers in the Schools* 8 (April-October): 59–81.

Arendell, Terry. 1998. "Divorce American Style." *Contemporary Sociology* 27 (May): 226–228.

Argyris, Chris. 1990. *Overcoming Organizational Defenses: Facilitating Organizational Learning.* Boston, MA: Allyn and Bacon.

Armas, Genaro C. 2003. "Asian American Divorce Rate Up." *San Francisco Chronicle:* A4.

Armstrong, E.A., L. Hamilton, and B. Sweeney. 2006. "Sexual Assault on Campus: A Multilevel, Integrative Approach to Party Rape." *Social Problems* 53: 483–499.

Aronson, Joshua, Carrie B. Fried, and Catherine Good. 2002. "Reducing the Effects of Stereotype Threat on African American College Students by Shaping Theories of Intelligence." *Journal of Experimental Social Psychology* 38: 113–125.

Asch, Solomon. 1951. "Effects of Group Pressure upon the Modification and Distortion of Judgments." In *Groups, Leadership, and Men,* edited by H. Guetzkow. Pittsburgh, PA: Carnegie Press.

Asch, Solomon. 1955. "Opinions and Social Pressure." *Scientific American* 19 (July): 31–35.

Associated Press. 2004. "AIDS Meeting Warns of Global Dangers." Associated Press, July 16, 2004.

Atchley, Robert C. 2000. *Social Forces and Aging,* 9th ed. Belmont, CA: Wadsworth.

Avakame, Edem F., James J. Fyfe, and Candace McCoy. 1999. "'Did You Call the Police? What Did They Do?' An Empirical Assessment of Black's Theory of Mobilization of Law." *Justice Quarterly* 16 (December): 765–792.

Baca Zinn, Maxine. 1995. "Chicano Men and Masculinity." Pp. 33–41 in *Men's Lives,* 3rd ed., edited by Michael S. Kimmel and Michael A. Messner. Boston, MA: Allyn and Bacon.

Baca Zinn, Maxine, and D. Stanley Eitzen. 2005. *Diversity in Families,* 7th ed., Boston, MA: Allyn and Bacon.

Baca Zinn, Maxine, and Bonnie Thornton Dill. 1996. "Theorizing Difference from Multiracial Feminism." *Feminist Studies* 22 (Summer): 321–331.

Baca Zinn, Maxine, Pierrette Hondagneu-Sotelo, and Michael Messner, eds. 2005. *Through the Prism of Difference.* New York: Oxford University Press.

Bachman, Daniel, and Ronald K. Esplin. 1992. "Plural Marriage." *Encyclopedia of Mormonism,* Vol. 3. New York: Macmillan.

Bailey, Garrick. 2003. *Humanity: An Introduction to Cultural Anthropology,* 6th ed. Belmont, CA: Wadsworth.

Bandura, Albert, and R. H. Walters. 1963. *Social Learning and Personality Development.* New York: Holt, Reinhart, and Winston.

Barber, Benjamin R. 1995. *Jihad vs. McWorld: How Globalism and Tribalism Are Reshaping the World.* New York: Random House.

Barber, Jennifer S., and William G. Axinn. 1998. "Gender Role Attitudes and Marriage among Young Women." *Sociological Quarterly* 39 (Winter): 11–31.

Bardacke, Frank. 1988. "Watsonville: A Mexican Community on Strike." Pp. 149–182 in *Reshaping the US Left: Popular Struggles in the 1980s,* edited by Mike Davis and Michael Sprinker. London: Verso.

Barlow, Maude and Tony Clarke. 2002. "Who Owns Water?" *The Nation* (Sept. 2/9): 11–14.

Basow, Susan A. 1992. *Gender: Stereotypes and Roles,* 3rd ed. Pacific Grove, CA: Brooks/Cole.

Basow, Susan A., and Kelly Johnson. 2000. "Predictors of Homophobia in Female College Students." *Sex Roles* 42 (March): 391–404.

Bean, Frank D., and Marta Tienda. 1987. *The Hispanic Population of the United States.* New York: Russell Sage Foundation.

Becker, Elizabeth. 2000. "Harassment in the Military Is Said To Rise." *The New York Times* (March 10): 14.

Becker, Howard S. 1963. *Outsiders: Studies in the Sociology of Deviance.* New York: Free Press.

Becraft, Carolyn H. 1992a. Pp. 8–17 in *Women in the Military,* edited by E. A. Blacksmith. New York: H. W. Wilson.

Belknap, Joanne. 2001. *The Invisible Woman.* Belmont, CA: Wadsworth.

Oxford University Press/McGraw-Hill.

Belknap, Joanne, Bonnie S. Fisher, and Francis T. Cullen. 1999. "The Development of a Comprehensive Measure of the Sexual Victimization of College Women." *Violence Against Women* 5 (February): 185–214.

Bell, Daniel. 1973. *The Coming Crisis of Postindustrial Society.* New York: Basic Books.

Bellah, Robert (ed.). 1973. *Emile Durkheim on Morality and Society: Selected Writings.* Chicago, IL: University of Chicago Press.

Beller, Emily, and Michael Hout. 2006. "Intergenerational Social Mobility: The United States in Comparative Perspective." *The Future of Children* 16 (Fall): 19–36.

Benedict, Ruth. 1934. *Patterns of Culture.* Boston, MA: Houghton Mifflin.

Ben-Yehuda, Nachman. 1986. "The European Witchcraze of the Fourteenth-Seventeenth Centuries: A Sociologist's Perspective." *American Journal of Sociology* 86: 1–31.

Berger, Joseph, and Morris Zelditch, Jr. (eds.). 1985. *Status, Rewards, and Influence.* San Francisco, CA: Jossey-Bass.

Berger, Peter L. 1963. *Invitation to Sociology: A Humanistic Perspective.* Garden City, NY: Doubleday Anchor.

Berger, Peter L., and Thomas Luckmann. 1967. *The Social Construction of Reality: A Treatise in the Sociology of Knowledge.* Garden City, NY: Anchor Books.

Berger, Peter L., Brigitte Berger, and Hansfried Kellner. 1974. *The Homeless Mind: Modernization and Consciousness.* New York: Vintage Books.

Berkman, Lisa F., Thomas Glass, Ian Brissette, and Teresa E. Seeman. 2000. "From Social Integration to Health: Durkheim in the New Millenium." *Social Science and Medicine* 51 (September): 843–857.

Bernstein, Nina. 2002. "Side Effect of Welfare Law: The No-Parent Family." *The New York Times,* July 29: A1.

Berscheid, Ellen, and Harry R. Reis. 1998. "Attraction and Class Relationships." Pp. 193–281 in *The Handbook of Social Psychology,* 4th ed., edited by Daniel T. Gilbert, Susan T. Fiske, and Gardner Lindzey. New York:

Best, Joel. 1999. *Random Violence: How We Talk About New Crimes and New Victims.* Berkeley, CA: University of California Press.

Best, Joel. 2001. *Damned Lies and Statistics: Untangling Numbers from the Media, Politicians, and Activists.* Berkeley, CA: University of California Press.

Bezilla, Robert. 1990. *Religion in America 1990: Approaching the Year 2000.* Princeton, NJ: Princeton Religious Research Center.

Bielby, William T. 1981."Models of Status Attainment." Pp. 3–26 in *Research in Social Stratification and Mobility,* edited by Donald Treiman and Robert Robinson. Greenwich, CT: JAI Press.

Binns, Allison. 2003. *White Gold, Weed, and Blow: The Drug Trades of Afghanistan, Columbia, and Mexico in Comparative Historical Perspective.* Senior Thesis, Princeton University, Princeton, NJ.

Blair-Loy, Mary, and Amy S. Wharton. 2002. "Employees' Use of Work-Family Policies and the Workplace Social Context." *Social Forces* 80(3): 813–845.

Blake, C. Fred. 1994. "Footbinding in Neo-Confucian China and the Appropriation of Female Labor." *Signs* 19 (Spring): 676–712.

Blankenship, Kim. 1993. "Bringing Gender and Race In: U.S. Employment Discrimination Policy." *Gender & Society* 7 (June): 204–226.

Blaskovitch, J. 1973. "Blackjack and the Risky Shift." *Sociometry* 36 (March): 42ff.

Blassingame, John. 1973. *The Slave Community: Plantation Life in the Antebellum South.* New York: Oxford University Press.

Blau, Francine, and Lawrence Kahn. 2004. "The U.S. Gender Pay Gap in the 1990s: Slowing Convergence." *National Bureau of Economics Research Working Papers,* #10853. www.nber.org.

Blau, Peter M. 1986. *Exchange and Power in Social Life,* revised ed. New Brunswick, NJ: Transaction.

Blau, Peter M., and Otis Dudley Duncan. 1967. *The American Occupational Structure.* New York: Wiley.

Blau, Peter M., and W. Richard Scott. 1974. *On the Nature of Organizations.* New York: Wiley.

Blee, Kathleen M., and Ann R. Tickamyer. 1995. "Racial Differences in Men's Attitudes about Women's Gender Roles." *Journal of Marriage and the Family* 57 (February): 21–30.

Blinde, Elaine M., and Diane E. Taub. 1992a. "Homophobia and Women's Sport: The Disempowerment of Athletes." *Sociological Focus* 25 (May): 151–166.

Blinde, Elaine M., and Diane E. Taub. 1992b. "Women Athletes as Falsely Accused Deviants: Managing the Lesbian Stigma." *Sociological Quarterly* 33 (Winter): 521–533.

Blinde, Elaine M., Diane E. Taub, and Lingling Han. 1993. "Sport Participation and Women's Personal Empowerment: Experiences of the College Athlete." *Journal of Sport and Social Issues* 17 (April): 47–60.

Blinde, Elaine M., Diane E. Taub, and Lingling Han. 1994. "Sport as a Site for Women's Group and Social Empowerment: Perspectives from the College Athlete." *Sociology of Sport Journal* 11 (March): 51–59.

Block, Alan, and Frank R. Scarpitti. 1993. *Poisoning for Profit: The Mafia and Toxic Waste in America.* New York: Morrow.

Blocker, T. Jean, and Douglas Lee Eckberg. 1997. "Gender and Environmentalism: Results from the 1993 General Social Survey." *Social Science Quarterly* 78 (December): 641–650.

Blum, Linda. 1991. *Between Feminism and Labor: The Significance of the Comparable Worth Movement.* Berkeley, CA: University of California Press.

Blumer, Herbert, 1969. *Studies in Symbolic Interaction.* Englewood Cliffs, NJ: Prentice Hall.

Bobo, Lawrence D. 1999. "Prejudice as Group Position: Microfoundations of a Sociological Approach to Racism and Race Relations." *Journal of Social Issues* 55 (3): 445–492.

Bobo, Lawrence, and James R. Kluegel. 1991. "Modern American Prejudice: Stereotypes, Social Distance, and Perceptions of Discrimination Toward Blacks, Hispanics, and Asians." Paper presented before the American Sociological Association, Cincinnati, OH.

Bobo, L., and R. A. Smith. 1998. "From Jim Crow Racism to Laissez-Faire Racism: An Essay on the Transformation or Racial Attitudes in America." Pp. 182–220 in *Beyond Pluralism: Essays on the Conception of Groups and Group Identities in America,* edited by W. Katkin, N. Landsmand, and A. Tyree. Urbana, IL: University of Illinois Press.

Bobo, Lawrence D., and Michael Dawson. 2006."Katrina: Unmasking Race, Poverty, and Politics in the 21st Century." *DuBois Review* 3 (Spring): 1–6.

Boer, J. Tom, Manuel Pastor, Jr., James L. Sadd, and Lori D. Snyder. 1997. "Is There Environmental Racism? The Demographics of Hazardous Waste in Los Angeles County." *Social Science Quarterly* 78 (December): 793–810.

Boeringer, Scott B. 1999. "Associations of Rape Supportive Attitudes with Fraternal and Athletic Participation." *Violence against Women* 5 (January): 81–90.

Boggs, Vernon W. 1992. *Salsiology: Afro-Cuban Music and the Evolution of Salsa in New York City.* Westport, CT: Greenwood Press.

Bonacich, Edna. 1972. "A Theory of Ethnic Antagonism: The Split Labor Market." *American Sociological Review* 37 (October): 547–559.

Bonacich, Edna, and Richard P. Applebaum. 2000. *Behind the Label: Inequality in the Los Angeles Apparel Industry.* Berkeley, CA: University of California Press.

Bonilla-Silva, Eduardo. 1997. "Rethinking Racism: Toward a Structural Interpretation." *American Sociological Review* 62 (3): 465–480.

Bonilla-Silva, Eduardo. 2001. *White Supremacy and Racism in the Post-Civil Rights Era.* Boulder, CO: Lynne Rienner.

Bonilla-Silva, Eduardo, and Gianpaolo Baiocchi. 2001. "Anything But Racism: How Sociologists Limit the Significance of Racism." *Race and Society* 4 (No. 2): 117–131.

Bontemps, Arna (ed.). 1972. *The Harlem Renaissance Remembered.* New York: Dodd, Mead.

Booth, Bradford, William W. Falk, David R. Segal, and Mady Wechsler Segal. 2000. "The Impact of Military Presence in Local Labor Markets and the Employment of Women." *Gender & Society* 14 (April): 318–332.

Bord, Richard J., and Robert E. O'Connor. 1997. "The Gender Gap in Environmental Attitudes: The Case of Perceived Vulnerability to Risk." *Social Science Quarterly* 78 (December): 831–840.

Bose, Christine E., and Peter H. Rossi. 1983. "Gender and Jobs: Prestige Standings of Occupations as Affected by Gender." *American Sociological Review* 48 (June): 316–330.

Bosk, Charles L. 1979. *Forgive and Remember: Managing Medical Failure.* Chicago, IL: University of Chicago Press.

Bouchard, Thomas J. Jr., and Matt McGue. 2003. "Genetic and Environmental Influences on Human Psychological Differences." *Journal of Neurobiology* 54(1): 4–45.

Bouchard, Thomas J. Jr., David T. Lykken, Matthew McGue, Nancy Segal, and Auke Tellegen. 1990. "Sources of Human Psychological Differences: The Minnesota Study of Twins Reared Apart." *Science* 250 (October 12): 223–228.

Bourdieu, Pierre. 1984. *Distinction: A Social Critique of the Judgement of Taste,* translated by Richard Nice. Cambridge, MA: Harvard University Press.

Bowditch, Christine. 1993. "Getting Rid of Troublemakers: High School Disciplinary Procedures and the Production of Dropouts." *Social Problems* 40 (November): 493–510.

Bowen, William G., and Derek Bok. 1998. *The Shape of the River: Long-Term Consequences of Considering Race in College and University Admissions.* Princeton, NJ: Princeton University Press.

Bowles, Samuel, and Herbert Gintis. 1976. *Schooling in Capitalist America.* New York: Basic Books.

Bowman, Charles. 1999. "BLS Projections to 2008: A Summary." *Monthly Labor Review* 122: 3–4.

Braddock, Jomills H. 1988. *National Education Longitudinal Study of 1988.* Washington, DC: U.S. Department of Education, National Center for Educational Statistics.

Branch, Taylor. 1988. *Parting the Waters: America in the King Years, 1954–1963.* New York: Simon & Schuster.

Branch, Taylor. 1998. *Pillar of Fire: America in the King Years, 1963–1965.* New York: Simon and Shuster.

Branch, Taylor. 2006. *At Canaan's Edge: America in the King Years, 1965–1968.* New York: Simon and Shuster.

Brechin, Steven R. 2003. "Comparative Public Opinion and Global Climatic Change and the Kyoto Protocol: The U.S. Versus the World?" *International Journal of Sociology and Social Policy* 23: 106–134.

Brehm, S. S., R. S. Miller, D. Perlman, and S. M. Campbell. 2002. *Intimate Relationships.* 3rd ed. Boston, MA: McGraw-Hill.

Bridges, George, and Robert Crutchfield. 1988. "Law, Social Standing and Racial Disparities in Imprisonment." *Social Forces* 66 (June): 699–724.

Brignall, Thomas Wells III, and Thomas Van Valey. 2005. "The Impact of Internet Communications on Social Interaction." *Sociological Spectrum* 25: 335–348.

Brines, Julie, and Kara Joyner. 1999. "The Ties That Bind: Principles of Cohesion in Cohabitation and Marriage." *American Sociological Review* 64 (June): 333–355.

Britt, Chester L. 1994. "Crime and Unemployment among Youths in the United States, 1958–1990: A Time Series Analysis." *American Journal of Economics and Sociology* 53 (January): 99–109.

Brodkin, Karen. 2006. "How Did Jews Become White Folks?" Pp. 59–66 in Elizabeth Higginbotham and Margaret L. Andersen, eds., *Race and Ethnicity in Society: The Changing Landscape.* Belmont, CA: Wadsworth.

Brooke, James. 1998. "Utah Struggles with a Revival of Polygamy." *The New York Times* (August 23): A12.

Brookey, Robert Alan. 2001. "Bio-Rhetoric, Background Beliefs, and the Biology of Homosexuality." *Argumentation and Advocacy* 37: 171–183.

Brooks-Gunn, Jeanne, Wen-Jui Han, and Jane Waldfogel. 2002. "Maternal Employment and Child Cognitive Outcomes in the First Three Years of Life: The NICHD Study of Early Child Care." *Child Development* 73 (July-August): 1052–1072.

Brown, Clare. 2000. "'Judge Me All You Want': Cigarette Smoking and the Stigmatization of Smoking." Paper presented at the Annual Meeting of the Society for the Study of Social Problems.

Brown, Cynthia. 1993. "The Vanished Native Americans." *The Nation* 257 (October 11): 384–389.

Brown, Elaine. 1992. *A Taste of Power: A Black Women's Story.* Pantheon: New York.

Brown, Lester R., Christopher Flavin, and Hilary French. 2000. *State of the World 2000: World-Watch Institute Report on Progress Toward A Sustainable Society.* New York: W. W. Norton and Co.

Brown, Michael K., Martin Carnoy, Troy Duster, Elliott Currie, David B. Oppenheimer, Marjorie Schulz, and David Wellman. 2005. *White-washing Race: The Myth of a Color-Blind Society.* Berkeley, CA: University of California Press.

Brown, Roger. 1986. *Social Psychology,* 2nd ed. New York: Free Press.

Brown, Ryan P., and Robert A. Josephs. 1999. "A Burden of Proof: Stereotype Relevance and Gender Differences in Math Performance." *Journal of Personality and Social Psychology* 76 (No. 2): 246–257.

Browne, Irene (ed.). 1999. *Latinas and African American Women at Work: Race, Gender, and Economic Inequality.* New York: Russell Sage Foundation.

Bryant, Alyssa N. 2003. "Changes in Attitudes toward Women's Roles: Predicting Gender-Role Traditionalism among College Students." *Sex Roles* 48 (February): 131–142.

Bryant, Susan L., and Lillian M. Range. 1997. "Type and Severity of Child Abuse and College Students' Lifetime Suicidality." *Child Abuse and Neglect* 21 (December): 1169–1176.

Buchanan, Christy M., Eleanor Maccoby, and Sanford M. Dornsbusch. 1996. *Adolescents After Divorce.* Cambridge, MA: Harvard University Press.

Budig, Michelle J. 2002. "Male Advantage and the Gender Composition of Jobs: Who Rides the Glass Escalator?" *Social Problems* 49 (May): 257–277.

Bullard, Robert D. 1994a. *Dumping in Dixie: Race, Class, and Environmental Quality.* Boulder, CO: Westview.

Bullard, Robert D. 1994b. *Unequal Protection: Environmental Justice and Communities of Color.* San Francisco, CA: Sierra Club Books.

Bureau of Justice Statistics. 2005. *Prisoners in 2004.* Washington, DC: U.S. Department of Justice.

Bureau of Labor Statistics. 2005. *A Profile of the Working Poor: 2003.* Washington, DC: U.S. Department of Labor. www.bls.gov/cps/cpswp2003.pdf

Burris, Val. 2000. "The Myth of Old Money Liberalism: The Politics of the Forbes 400 Richest Americans." *Social Problems* 47 (Summer): 360–378.

Burt, Cyril. 1966. "The Genetic Determination of Differences in Intelligence: A Study of Monozygotic Twins Reared Together and Apart." *British Journal of Psychology* 57: 137–153.

Butsch, Richard. 1992. "Class and Gender in Four Decades of Television Situation Comedy: Plus ça Change . . ." *Critical Studies in Mass Communication* 9: 387–399.

Butterfield, Fox. 2000. "Racial Disparities Seen as Pervasive in Juvenile Justice." *The New York Times* (April 26): 1.

Buunk, Bram and Ralph B. Hupka. 1987. "Cross-Cultural Differences in the Elicitation of Sexual Jealousy." *Journal of Sex Research* 23 (February): 12–22.

Calhoun, Craig. 1994. *Social Theory and the Politics of Identity.* Cambridge, MA: Blackwell.

Cameron, Deborah. 1998. "Gender, Language, and Discourse: A Review Essay." *Signs* 23 (Summer): 945–974.

Campbell, Anne. 1987. "Self-Definition by Rejection: The Case of Gang Girls." *Social Problems* 34 (December): 451–466.

Campenni, C. Estelle. 1999. "Gender Stereotyping of Children's Toys: A Comparison of Parents and Nonparents." *Sex Roles* 40 (January): 121–138.

Cantor, Joanne. 2000. "Media Violence." *Journal of Adolescent Health* 27 (August): 30–34.

Caplow, Theodore. 1991. *American Social Trends.* New York: Harcourt Brace Jovanovich.

Cardenas, Jose A. 1996. "Ending the Crisis in the K-12 System." Pp. 51–70 in *Educating a New Majority: Transforming America's Educational System for Diversity,* edited by Laura I. Rendon and Richard O. Hope. San Francisco, CA: Jossey-Bass.

Carmichael, Stokely, and Charles V. Hamilton. 1967. *Black Power: The Politics of Liberation in America.* New York: Vintage.

Carnoy, Martin, Manuel Castells, Stephen S. Cohen, and Fernando Henrique Cardoso. 1993. *The New Global Economy in the Information Age.* University Park, PA: The Pennsylvania State University Press.

Carr, C. Lynn. 1998. "Tomboy Resistance and Conformity: Agency in Social Psychological Gender Theory." *Gender & Society* 12 (October): 528–553.

Carroll, J. B., 1956. *Language, Thought, and Reality: Selected Writings of Benjamin Lee Whorf.* Cambridge, MA: MIT Press.

Carson, Rachel. 1962. *The Silent Spring.* New York: Knopf.

Carter, Timothy S. 1999. "Ascent of the Corporate Model in Environmental-Organized Crime." *Crime, Law and Social Change* 21: 1–30.

Cashmore, Ellis. 1991. "Black Cops Inc." Pp. 87–108 in *Out of Order: Policing Black People,* edited by Ellis Cashmore and Eugene McLaughlin. New York: Routledge.

Cassen, Robert. 1994. "Population and Development: Old Debates, New Conclusions." *U.S.-Third World Policy Perspectives* 19: 282.

Cassidy, J., and P. R. Shaver, eds. 1999. *Handbook of Attachment: Theory, Research, and Critical Applications.* New York: Guilford.

Cassidy, Linda, and Rose Marie Hurrell. 1995. "The Influence of Victim's Attire on Adolescents' Judgments of Date Rape." *Adolescence* 30 (Summer): 319–323.

Catanzarite, Lisa, and Vilma Ortiz. 1996. "Family Matters, Work Matters? Poverty among Women of Color and White Women." Pp. 121–140 in *For Crying Out Loud: Women's Poverty in the United States,* edited by Diane Dujon and Ann Withorn. Boston, MA: South End Press.

Centeno, Miguel A. and Eszter Hargittai. 2003. "Defining a

Global Geography." *The American Behavioral Scientist* 44 (no. 10).

Centers, Richard. 1949. *The Psychology of Social Classes.* Princeton, NJ: Princeton University Press.

Chafetz, Janet. 1984. *Sex and Advantage.* Totowa, NJ: Rowman and Allanheld.

Chalfant, H. Paul, Robert E. Beckley, and C. Eddie Palmer. 1987. *Religion in Contemporary Society,* 2nd ed. Palo Alto, CA: Mayfield.

Chambliss, William J., and Howard F. Taylor. 1989. *Bias in the New Jersey Courts.* Trenton, NJ: Administrative Office of the Courts.

Chang, Jung. 1991. *Wild Swans: Three Daughters of China.* New York: Simon & Schuster.

Chen, Elsa, Y. F. 1991. "Conflict Between Korean Greengrocers and Black Americans." Unpublished senior thesis, Princeton University.

Chernin, Kim. 1991. *The Obsession.* New York: Harper Colophon.

Chesney-Lind, Meda. 1992. "Women's Prisons: Putting the Brakes on the Building Binge." *Corrections Today* 54 (August): 30–34.

Chia, Rosina C., Jamie L. Moore, Ka Nei Lam, C. J. Chuang, and B. S. Cheng. 1994. "Cultural Differences in Gender Role Attitudes between Chinese and American Students." *Sex Roles* 31 (July): 23–30.

Child Welfare Information Gateway. 2007. www.childwelfare.gov

Chipman, Susan. 1991. "Word Problems: Where Bias Creeps In." Unpublished Research Report, Office of Naval Research, Arlington, VA.

Chipuer, H. M., M. J. Rovine, and R. Plomin. 1990. "LISREL Modeling: Genetic and Environmental Influences on IQ Revisited." *Intelligence* 14: 11–29.

Chiricos, Ted, Sarah Eschholz, and Marc Gertz. 1997. "Crime, News and Fear of Crime: Toward an Identification of Audience Effects." *Social Problems* 44 (August): 342–357.

Chodorow, Nancy. 1978. *The Reproduction of Mothering: Psychoanalysis and the Study of Gender.* Berkeley, CA: University of California Press.

Chodorow, Nancy. 1994. *Femininities, Masculinities, Sexualities: Freud and Beyond.* Lexington, KY: University of Kentucky Press.

Chow, Esther Ngan-Ling, Doris Wilkinson, and Maxine Baca Zinn (eds.) 1996. *Race, Class, & Gender: Common Bonds, Different Voices.* Thousand Oaks, CA: Sage.

Churchill, Ward. 1992. *Struggle for the Land.* Toronto: Between the Lines.

Cialdini, R. B. 1993. *Influence: Science and Practice,* 3rd ed. New York: Harper Collins.

Cicourel, Aaron V. 1968. *The Social Organization of Juvenile Justice.* New York: Wiley.

Clark, Kenneth B., and Mamie P. Clark. 1947. "Racial Identification and Preference in Negro Children." Pp. 602–611 in *Readings in Social Psychology,* edited by T. M. Newcomb and E. L. Hartley. New York: Holt.

Clawson, Dan, and Naomi Gerstl. 2002. "Caring for Our Young: Child Care in Europe and the United States." *Contexts* 1 (Fall-Winter): 28–35.

Clawson, Rosalee A., and Rakuya Trice. 2000. "Poverty As We Know It: Media Portrayals of the Poor." *The Public Opinion Quarterly* 64 (Spring): 53–64.

Clifford, N. M., and Elaine Walster. 1973. "Research Note: The Effects of Physical Attractiveness on Teacher Expectations." *Sociology of Education* 46: 248–258.

Coale, Ansley. 1974. "The History of the Human Population." *Scientific American* 231 (3): 40–51.

Coale, Ansley. 1986. "Population Trends and Economic Development." Pp. 96–104 in *World Population and the U.S. Population Policy: The Choice Ahead,* edited by J. Menken. New York: W.W. Norton.

Cohen, Philip N. 1999. "Racial-Ethnic and Gender Differences in Returns to Cohabitation and Marriage: Evidence from the Current Population Survey." Paper presented at Annual Meeting of the Population Association of America, www.census.gov

Cole, David. 1999. "The Color of Justice." *The Nation* (October 11): 12–15.

College Board. 2006. *College Bound Seniors 2006: A Profile of SAT Program Test Takers.* New York: College Board.

Colley, Ann, and Zazie Todd. 2002. "Gender-Linked Differences in the Style and Content of E-mails to Friends." *Journal of Language and Social Psychology* 21 (December): 380–393.

Collins, Patricia Hill. 1990. *Black Feminist Theory: Knowledge, Consciousness and the Politics of Empowerment.* Cambridge, MA: Unwin Hyman.

Collins, Patricia Hill. 1998. *Fighting Words: Black Women and the Search for Justice.* Minneapolis, MN: University of Minnesota Press.

Collins, Patricia Hill. 2004. *Black Sexual Politics: African Americans, Gender, and the New Racism.* New York: Routledge.

Collins, Patricia Hill, Lionel A. Maldonado, Dana Y. Takagi, Barrie Thorne, Lynn Weber, and Howard Winant. 1995. "On West and Fenstermaker's 'Doing Difference.'" *Gender & Society* 9 (August): 491–505.

Collins, Randall. 1979. *The Credential Society.* New York: Academic Press.

Collins, Randall. 1988. *Theoretical Sociology.* San Diego, CA: Harcourt Brace Jovanovich.

Collins, Randall. 1994. *Four Sociological Traditions.* New York: Oxford.

Collins, Randall, and Michael Makowsky. 1972. *The Discovery of Society.* New York: Random House.

Collins, Sharon M. 1989. "The Marginalization of Black Executives." *Social Problems* 36: 317–331.

Collins-Lowry, Sharon M. 1997. *Black Corporate Executives: The Making and Breaking of a Black Middle Class.* Philadelphia, PA: Temple University Press.

Coltrane, Scott, and Melinda Messineo. 2000. "The Perpetuation of Subtle Prejudice: Race and Gender Imagery in 1990s Television Advertising." *Sex Roles* 42 (March): 363–389.

Conley, Dalton. 1999. *Being Black, Living in the Red: Race, Wealth, and Social Policy in America.* Berkeley, CA: University of California Press.

Connell, R. W. 1992. "A Very Straight Gay: Masculinity, Homosexual Experience, and the Dynamics of Gender." *American Sociological Review* 57 (December): 735–751.

Conrad, Peter, and Joseph W. Schneider. 1992. *Deviance and Medicalization: From Badness to Sickness,* expanded edition. Philadelphia, PA: Temple University Press.

Constable, Nicole. 1997. *Maid to Order in Hong Kong: Stories of Filipina Workers.* Ithaca, NY: Cornell University Press.

Cook, Christopher D. 2000. "Temps Demand a New Deal." *The Nation* (March 27): 13–19.

Cooksey, Elizabeth C., and Ronald R. Rindfuss. 2001. "Patterns of Work and Schooling in Young Adulthood." *Sociological Forum* 16 (December): 731–755.

Cookson, Peter W., Jr., and Caroline Hodges Persell. 1985. *Preparing for Power: America's Elite Boarding Schools.* New York: Basic Books.

Cooley, Charles Horton. 1902. *Human Nature and Social Order.* New York: Scribner's.

Cooley, Charles Horton. 1967 [1909]. *Social Organization.* New York: Schocken Books.

Coontz, Stephanie. 1992. *The Way We Never Were.* New York: Basic Books.

Cooper, Alvin, and Coralie R. Scherer. 1999. "Sexuality on the Internet: From Sexual Exploration to Pathological Expression." *Professional Psychology Research and Practice* 30, www.sex-centre.com/Internetsex_Folder/MSNBC_Study_pp.htm

Cortina, Lilia M., Suzanna Swan, Louise F. Fitzgerald, and Craig Walo. 1998. "Sexual Harassment and Assault: Chilling the Climate for Women in Academia." *Psychology of Women Quarterly* 22 (September): 419–441.

Coser, Lewis. 1977. *Masters of Sociological Thought.* New York: Harcourt Brace Jovanovich.

Costello, Mark. 2004. "Throwing Away The Key." *The New York Times Magazine* (June 6): 41–43.

Craig, Maxine Leeds. 2002. *Ain't I a Beauty Queen? Black Women, Beauty and the Politics of Race.* New York: Oxford University Press.

Crane, Diana (ed.). 1994. *The Sociology of Culture: Emerging Theoretical Perspectives.* Cambridge, England: Blackwell.

Crawford, Mary. 1995. *Talking Difference: On Gender and Language.* Thousand Oaks, CA: Sage.

Croll, Elisabeth. 1995. *Changing Identities of Chinese Women.* London: Zed Books.

Crouse, James, and Dale Trusheim. 1988. *The Case Against the SAT.* Chicago, IL: University of Chicago Press.

Crowley, Sue M. 1998. "Men's Self-Perceived Adequacy as the Family Breadwinner: Implications for Their Psychological, Marital, and Family Well-Being." *Journal of Family and Economic Issues* 19 (Spring): 7–23.

Csikszentmihalyi, Mihaly, and Barbara Schneider. 2000. *Becoming Adult: How Teenagers Prepare for the World of Work.* New York: Basic Books.

Cunningham, Mick. 2001. "The Influence of Parental Attitudes and Behaviors on Children's Attitudes toward Gender and Household Labor in Early Adulthood." *Journal of Marriage and Family* 63 (February): 111–122.

Currie, Dawn. 1997. "Decoding Femininity: Advertisements and Their Teenage Readers." *Gender & Society* 11 (August): 453–477.

Curtiss, Susan. 1977. *Genie: A Psycholinguistic Study of a Modern-Day "Wild Child."* New York: Academic Press.

Czerniawski, Amanda. 2007. "From Average to Ideal: The Evolution of the Height and Weight Table in the United States." *Social Science History* 31 (2): 273–296.

Dahlhamer, James, and Joanne M. Nigg. 1993. "An Empirical Investigation of Rumoring: Anticipating Disaster under Conditions of Uncertainty." Paper presented at the Annual Meetings of the Southern Sociological Society, April, Chattanooga, TN.

Dahrendorf, Rolf. 1959. *Class and Class Conflict in Industrial Society.* Stanford, CA: Stanford University Press.

D'Alessio, Stewart J., and Lisa Stolzenberg. 2003. "Race and the Probability of Arrest." *Social Forces* 81 (June): 1381–1397.

Dalton, Susan E., and Denise D. Bielby. 2000. "'That's Our Kind of Constellation': Lesbian Mothers Negotiate Institutionalized Understandings of Gender Within the Family." *Gender & Society* 14 (February): 36–61.

Das Gupta, Monisha. 1997. "'A Chambered Nautilus': The Contradictory Nature of Puerto Rican Women's Role in the Construction of a Transnational Community." *Gender & Society* 11 (October): 627–655.

Davis, Angela. 1981. *Women, Race, and Class.* New York: Random House.

Davis, James A., and Tom Smith. 1984. *General Social Survey Cumulative File, 1972–1982.* Ann Arbor, MI: Inter-University Consortium for Political and Social Research.

Davis, Kingsley. 1945. "The World Demographic Transition." *Annals of the American Academy of Political and Social Sciences* 237: 1–11.

Davis, Kingsley, and Wilbert E. Moore. 1945. "Some Principles of Stratification." *American Sociological Review* 10 (April): 242–247.

DeCarlo, Scott. 2006. "CEO Compensation." *Forbes,* April 20. www.forbes.com

Deegan, Mary Jo. 1988. "W. E. B. Du Bois and the Women of Hull-House, 1895–1899." *The American Sociologist* 19 (Winter): 301–311.

Deegan, Mary Jo. 1990. *Jane Addams and the Men of the Chicago School, 1892–1918.* New Brunswick, NJ: Transaction Books.

DeFleur, Lois. 1975. "Biasing Influences on Drug Arrest Records: Implications for Deviance Research." *American Sociological Review* 40 (March): 88–103.

Dellinger, Kristen, and Christine L. Williams. 1997. "Makeup at Work: Negotiating Appearance Rules in the Workplace." *Gender & Society* 11 (April): 151–177.

D'Emilio, John. 1998. *Sexual Politics, Sexual Communities: The Making of a Homosexual Minority in the United States, 1940–1970,* rev. edition. Chicago, IL: University of Chicago Press.

Demeny, Paul. 1991. "Tradeoffs between Human Numbers and Material Standards of Living." Pp. 408–421 in *Resources, Environment, and Population: Present Knowledge, Future Options,* edited by Kingsley Davis and Michail S. Bernstam. New York: Population Council.

DeNavas-Walt, Carmen, and Robert W. Cleveland. 2002. *Money Income in the United States 2002.*

Washington, DC: U.S. Census Bureau.

DeNavas-Walt, Carmen, Bernadette Proctor, and Cheryl Lee. 2006. *Income, Poverty, and Health Insurance Coverage in the Untied States: 2005.* Washington, DC: U.S. Census Bureau, #P60–231.

DeWitt, Karen. 1995. "Blacks Prone to Job Dismissal in Organizations." *The New York Times* (April 20): A19.

DiAcosta, Diego. 1998. "A Sociolinguistic Study of Two Baptist Churches." Senior thesis, Princeton University, Department of Sociology, Princeton, NJ.

Diaz-Duque, Ozzie F. 1989. "Communication Barriers in Medical Settings: Hispanics in the United States." *International Journal of the Sociology of Language* 79: 93–102.

Dill, Bonnie Thornton. 1988. "Our Mothers' Grief: Racial Ethnic Women and the Maintenance of Families." *Journal of Family History* 13 (October): 415–431.

Dillard, J.L. 1972. *Black English: Its Historical Usage in the United States.* New York: Random House.

DiMaggio, Paul, and Francie Ostrower. 1990. "Participation in the Arts by Black and White Americans." *Social Forces* 68 (March): 753–778.

DiMaggio, Paul J., and Walter W. Powell. 1991. "Introduction." Pp. 1–38 in *The New Institutionalism in Organizational Analysis,* edited by W. W. Powell and P. J. DiMaggio. Chicago, IL: University of Chicago Press.

Dines, Gail, and Jean M. Humez. 2002. *Gender, Race, and Class in Media,* 2nd ed. Thousand Oaks, CA: Sage.

Dion, K. K. 1972. "Physical Attractiveness and Evaluating Children's Transgressions." *Journal of Personality and Social Psychology* 24: 285–290.

Dixit, Avinash K. and Susan Sneath. 1997. *Games of Strategy.* New York: W. W. Norton.

Dollard, John, Neal E. Miller, Leonard W. Doob, O. H. Mowrer, and Robert R. Sears. 1939. *Frustration and Aggression.* New Haven, CT: Yale University Press.

Domhoff, G. William. 2002. *Who Rules America?* New York: McGraw-Hill.

Dominguez, Silvia, and Celeste Watkins. 2003. "Creating Networks for Survival and Mobility: Social Capital Among African-American and Latin-American Low-Income Mothers." *Social Problems* 50 (1): 111–135.

Donnerstein, Edward, Daniel Linz, and Steven Penrod. 1987. *The Question of Pornography: Research Findings and Policy Implications.* New York: Free Press.

Douglass, Jack D. 1967. *The Social Meanings of Suicide.* Princeton, NJ: Princeton University Press.

Dovidio, J. F. and S. L. Gaertner (eds.). 1986. *Prejudice, Discrimination, and Racism.* New York: Academic Press.

Dowd, James J., and Laura A. Dowd. 2003. "The Center Holds: From Subcultures to Social Worlds." *Teaching Sociology* 31 (January): 20–37.

Downey, Liam. 2005. "The Unintended Significance of Race: Environmental Racial Inequality in Detroit." *Social Forces* 83 (March): 971–1008.

Doyal, I. 1990. "Hazards of Hearth and Home." *Women's Studies International Forum* 13: 501–517.

Driggs, Ken. 1990. "After The Manifesto: Modern Polygamy and Fundamentalist Mormons." *Journal of Church and State* 32 (Spring): 367–389.

Du Bois, W.E.B. 1901. "The Freedmen's Bureau." *Atlantic Monthly* 86: 354–365.

DuBois, Cathy L., Deborah E. Knapp, Robert H. Faley, and Gary A. Kustis. 1998. "An Empirical Examination of Same and Other Gender Sexual Harassment in the Workplace." *Sex Roles* 9–10 (November): 731–749.

Dudley, Carl S., and David A. Roozen. 2001. *Faith Communities Today: A Report on Religion in the United States Today.* Hartford, CT: Hartford Institute for Religion Research, Hartford Seminary.

Due, Linnea. 1995. *Joining the Tribe: Growing Up Gay & Lesbian in the '90s.* New York: Doubleday.

Duneier, Mitchell. 1999. *Sidewalk.* New York: Farrar, Strauss and Giroux.

Dunlop, Riley E., and Angela G. Mertig. 1992. *American Environmentalism: The U.S. Environmental Movement, 1970–1990.* Philadelphia, PA: Taylor and Francis.

Dunne, Gillian A. 2000. "Opting into Motherhood: Lesbians Blurring the Boundaries and Transforming the Meaning of Parenthood and Kinship." *Gender & Society* 14 (February): 11–35.

Durkheim, Emile. 1897 [1951]. *Suicide.* Glencoe, IL: Free Press.

Durkheim, Emile. 1912 [1947]. *Elementary Forms of Religious Life.* Glencoe, IL: Free Press.

Durkheim, Emile. 1938 [1950]. *The Rules of Sociological Method.* Glencoe, IL: Free Press.

Durkheim, Emile. 1964 [1895]. *The Division of Labor in Society.* New York: Free Press.

Dworkin, Shari L., and Michael A. Messner. 1999. "Just Do . . . What?" Pp. 341–361 in *Revisioning Gender,* edited by Myra Marx Ferree, Judith Lorber, and Beth B. Hess. Thousand Oaks, CA: Sage.

Dyson, Michael Eric. 2006. *Come Hell or High Water: Hurricane Katrina and the Color of Disaster.* New York: Basic Books.

Eckberg, Douglass Lee. 1992. "Social Influences on Belief in Creationism." *Sociological Spectrum* 12 (April-June): 145–165.

Edin, Kathryn. 1991. "Surviving the Welfare System: How AFDC Recipients Make Ends Meet in Chicago." *Social Problems* 38 (November): 462–472.

Edin, Kathryn. 2000. "What Do Low-Income Single Mothers Say About Marriage?" *Social Problems* 47 (February): 112–133.

Edin, Kathryn, and Laura Lein. 1997. *Making Ends Meet: How Single Mothers Survive Welfare and Low-Wage Work.* New York: Russell Sage Foundation.

Edin, Kathyrn, and Maria Kefalas. 2005. *Promises I Can Keep: Why Poor Women Put Motherhood before Marriage.* Berkeley, CA: University of California Press.

Edison/Mikofsky. 2004. National Survey, *New York Times* (Nov. 4), p. 4. www.cnn.com

Edmondson, Brad. 2000. "Immigration Nation." *Preservation* (January-February): 31–38.

Ehrlich, Paul. 1968. *The Population Bomb.* New York: Ballantine Books.

Ehrlich, P. 1990. *The Population Explosion.* New York: Ballantine Books.

Ehrlich, Paul R., and Anne H. Ehrlich. 1991. *The Population Explosion.* New York: Touchstone/Simon & Schuster.

Ehrlich, Paul R., and Jianguo Liu. 2002. "Some Roots of Terrorism." *Population and Environment,* vol. 24, no. 2, Human Sciences Press.

Ehrlich, Paul R. Anne H. Ehrlich, and John P. Holdren. 1977. *Ecoscience: Population, Resources, Environment.* San Franciso, CA: Freeman.

Eichenwald, Kurt. 1996. "The Two Faces of Texaco." *The New York Times* (November 10): Section 3: 1ff.

Eichenwald, Kurt. 2002. "Two Ex-Officials at WorldCom Are Charged in Huge Fraud." *The New York Times* (August 2): A1 and A11.

Eisenberg, Daniel, and Maggie Sieger. 2003. "The Doctor Won't See You Now: The Soaring Cost of Malpractice." *Time* (June): 46–60.

Eitzen, D. Stanley, and Maxine Baca Zinn. 2006. *In Conflict and Order,* 11th ed. Boston, MA: Allyn and Bacon.

Eitzen, D. Stanley, and Maxine Baca Zinn. 2007. *In Conflict and Order,* 11th ed., Boston: Allyn and Bacon.

Eliason, Michele J., and Salome Raheim. 1996. "Categorical Measurement of Attitudes About Lesbian, Gay, and Bisexual People." *Journal of Gay and Lesbian Social Services* 4: 51–65.

Elizabeth, Vivienne. 2000. "Cohabitation, Marriage, and the Unruly Consequences of Difference." *Gender & Society* 14 (February): 87–110.

Elliott, Marta. 2001. "Gender Differences in the Causes of Depression." *Women and Health* 22: 163–177.

Ellison, Christopher G., and John P. Bartkowski. 2002. "Conservative Protestantism and the Division of Household Labor among Married Couples." *Journal of Family Issues* 23 (November): 950–985.

Ellison, Christopher G., and David A. Gay. 1989. "Black Political Participation Revisited: A Test of Compensatory, Ethnic Community, and Public Arena Models." *Social Science Quarterly* 70 (March): 101–119.

Emerson, Joan P. 1970. "Behaviors in Private Places: Sustaining Definitions of Reality in Gynecological Examinations." Pp. 74–97 in *Recent Sociology,* vol. 2, edited by H. P. Dreitzel. New York: Collier.

Emerson, Rana A. 2002. "Where My Girls At?: Negotiating Black Womanhood in Music Videos." *Gender & Society* 16 (February): 115–135.

Enloe, Cynthia. 1989. *Bananas, Beaches, and Bases: Making Feminist Sense of International Politics.* Berkeley, CA: University of California Press.

Enloe, Cynthia. 1993. *The Morning After: Sexual Politics at the End of the Cold War.* Berkeley, CA: University of California Press.

Enloe, Cynthia. 2001. *Bananas, Beaches, and Bases: Making Feminist Sense of International Politics,* updated edition. Berkeley, CA: University of California Press.

Epps, Edgar T. 2002. "Race, Class and Educational Opportunity: Trends in the Sociology of Education." In *2001 Race Odyssey: African Americans and Sociology,* edited by Bruce R. Hare. Syracuse, NY: Syracuse University Press.

Epstein, Steven. 1996. *Impure Science: AIDS, Activism, and the Politics of Knowledge.* Berkeley, CA: University of California Press.

Erikson, Eric. 1980. *Identity and the Life Cycle.* New York: W.W. Norton.

Erikson, Kai. 1966. *Wayward Puritans: A Study in the Sociology of Deviance.* New York: Wiley.

Ermann, M. David, and Richard J. Lundman. 1992. *Corporate and Governmental Deviance.* New York: Oxford University Press.

Esbensen-Finn, Aage, Elizabeth Piper Deschenes, and Thomas L. Winfree, Jr. 1999. "Differences between Gang Girls and Gang Boys: Results from a Multisite Survey." *Youth and Society* 31 (September): 27–53.

Eschholz, Sarah, Jana Bufkin, and Jenny Long. 2002. "Symbolic Reality Bites: Women and Racial/Ethnic Minorities in Modern Film." *Sociological Spectrum* 22 (July–August): 299–334.

Espenshade, Thomas J. 1995. "Unauthorized Immigration to the United States." *Annual Review of Sociology* 21: 195–216.

Espenshade, Thomas J., Lauren E. Hale, and Chang Y. Chung. 2005. "The Frog Pond Revisited: High School Academic Context, Class Rank, and Elite College Admission." *Sociology of Education* 78 (October): 269–293.

Essed, Philomena. 1991. *Understanding Everyday Racism.* Newbury Park, CA: Sage.

Esterberg, Kristin G. 2003. "New Right." *Encyclopedia of Lesbian, Gay, Bisexual and Transgendered History in America.* New York: Charles Scribner's Sons.

Etzioni, Amatai. 1975. *A Comparative Analysis of Complex Organization: On Power, Involvement, and Their Correlates,* rev. ed. New York: Free Press.

Etzioni, Amatai, John Wilson, Bob Edwards, and Michael W. Foley. 2001. "A Symposium on Robert D. Putnam's *Bowling Alone: The Collapse and Revival of American Community.*" *Contemporary Sociology* 30 (May): 223–230.

Evans, Peter B., Dietrich Rueschemeyer, and Theda Skocpol. 1985. *Bringing the State Back In.* Cambridge, MA: Cambridge University Press.

Farber, Naomi. 1990. "The Significance of Race and Class in Marital Decisions Among Unmarried Adolescent Mothers." *Social Problems* 37 (February): 51–63.

Fattah, Ezzat A., 1994. "The Interchangeable Roles of Victim and Victimizer." *HEUNI* Papers 3: 1–26.

Fausto-Sterling, Anne. 1992. *Myths of Gender: Biological Theories About Women and Men.* New York: Basic Books.

Fausto-Sterling, Anne. 2000. *Sexing the Body: Gender Politics and the Construction of Sexuality.* New York: Basic Books.

Feagin, Joe R. 2000. *Racist America: Roots, Future Realities, and Racial Reparations.* New York: Routledge.

Feagin, Joe R., and Clairece B. Feagin. 1993. *Racial and Ethnic Relations,* 4th ed. Englewood Cliffs, NJ: Prentice Hall.

Feagin, Joe R., and Vera Hernán. 1995. *White Racism.* New York: Routledge.

Federal Bureau of Investigation. 2000. *Uniform Crime Reports.* Washington, DC: U.S. Department of Justice.

Federal Bureau of Investigation. 2002. *Uniform Crime Reports.* Washington, DC: U.S. Department of Justice

Federal Bureau of Investigation. 2005. *Uniform Crime Reports.* Washington, DC: U.S. Department of Justice.

Federal Election Commission. 2001. "PAC Financial Activity through December 31, 2000," www.fec.gov

Fein, Melvyn L. 1988. "Resocialization: A Neglected Paradigm." *Clinical Sociology.*6: 88–100.

Ferber, Abby. 1998. *White Man Falling: Race, Gender, and White Supremacy.* Lanham, MD: Rowman and Littlefield.

Ferber, Abby. 1999. "What White Supremacists Taught a Jewish Scholar about Identity." *The Chronicle of Higher Education* (May 7): 86–87.

Fernald, Anne, and Hiromi Morikawa. 1993. "Common Themes and Cultural Variations in Japanese and American Mothers' Speech to Infants." *Child Development* 64 (June): 637–656.

Festinger, Leon, Stanley Schachter, and Kurt Back. 1950. *Social Pressures in Informal Groups: A Study of Human Factors in Housing.* Stanford, CA: Stanford University Press.

Fields, Jason. 2003. *Children's Living Arrangements and Characteristics, Detailed Tables, Current Population Report P20–547.* Washington, DC: U.S. Census Bureau, www.census.gov

Fields, Jason. 2004. *America's Families and Living Arrangements: 2003.* Current Population Reports, P20–553. Washington, DC: U.S. Bureau of the Census.

Fine, Michelle, and Lois Weis. 1998. *The Unknown City: The Lives of Poor and Working-Class Young Adults.* Boston, MA: Beacon Press.

Finer, Lawrence B., and Stanley K. Henshaw. 2003. "Abortion Incidence and Services in the United States in 2000." *Perspectives on Sexual and Reproductive Health* 35: 6–15.

Fischer, Claude. 1981. *To Dwell among Friends: Personal Networks in Town and City.* Chicago, IL: University of Chicago Press.

Fischer, Claude S., Michael Hout, Mártin Sánchez Jankowski, Samuel R. Lucas, Ann Swidler, and Kim Voss. 1996. *Inequality by Design: Cracking the Bell Curve Myth.* Princeton, NJ: Princeton University Press.

Fisher, Bonnie S., Francis T. Cullen, and Michael G. Turner. 2000. *Sexual Victimization of College Women.* Washington, DC: Bureau of Justice Statistics.

Fitzgibbon, Marian, and Melinda Stolley. 2002. "Minority Women: The Untold Story," NOVA Online. http://www.pbs.org/wgbh/nova/thin/minorities.html

Flanagan, William G. 1995. *Urban Sociology: Images and Structure.* Boston, MA: Allyn and Bacon.

Fleisher, Mark. 2000. *Dead End Kids: Gang Girls and the Boys They Know.* Madison, WI: University of Wisconsin Press.

Fleming, Jacqueline, and Nancy Garcia. 1998. "Are Standardized Tests Fair to African Americans? Predictive Validity of the SAT in Black and White Institutions." *Journal of Higher Education* 69 (September-October): 471–495.

Fletcher, Michael. 2000. "In The Targeted States, a Striking Turnout of Black Voters." *The Washington Post* (November 17): A29.

Flowers, M.L. 1977. "A Laboratory Test of Some Implications of Janis' Groupthink Hypothesis." *Journal of Personality and Social Psychology* 35 (December): 888–896.

Folbre, Nancy. 2001. *The Invisible Heart: Economics and Family Values.* New York: New Press.

Frankbgerg, Erika, and Chungmei Lee. 2002. *Race in American Public Schools: Rapidly Resegregating School Districts.* Cambridge, MA: The Civil Rights Project, Harvard University. Website: www.civilrightsproject.harvard.edu

Frankenberg, Erica, and Chungmei Lee. 2006. "Rapidly Resegregating School Districts." Pp. 247–353 in Elizabeth Higginbotham and Margaret L. Andersen, eds., *Race and Ethnicity in Society: The Changing Landscape.* Belmont, CA: Wadsworth.

Frazier, E. Franklin. 1957. *The Black Bourgeoisie.* New York: Collier Books.

Fredrickson, George M. 2003. *Racism: A Short History.* Princeton, NJ: Princeton University Press.

Freedle, Roy O. 2003. "Correcting the SATs Ethnic and Social Class Bias: A Method of Re-Estimating SAT Scores." *Harvard Educational Review* 73 (Spring): 1–43.

Freedman, Allan. 1997. "Lawyers Take a Back Seat in the 105th Congress." *Congressional Quarterly* 55 (January 4): 27–30.

Freedman, Estelle B., and John D'Emilio. 1988. *Intimate Matters: A History of Sexuality in America.* New York: Harper & Row.

Freeman, Jo. 1983a. "A Model for Analyzing the Strategic Options of Social Movement Organizations." Pp. 193–210 in *Social Movements of the Sixties and Seventies,* edited by Jo Freeman. New York: Longman.

Freud, Sigmund. 1901 [1965]. *The Psychopathology of Everyday Life,* translated by Alan Tyson and edited by James Strachey. New York: W.W. Norton.

Freud, Sigmund. 1923 [1960]. *The Ego and the Id,* translated by Joan Riviere. New York: W.W. Norton.

Freud, Sigmund. 1930 [1961]. *Civilization and Its Discontents,* translated by James Strachey. New York: W.W. Norton,

Fried, Amy. 1994. "'It's Hard to Change What We Want to Change': Rape Crisis Centers as Organizations." *Gender & Society* 4 (December): 562–583.

Friedman, Thomas L. 1999. *The Lexus and the Olive Tree.* New York: Farrar, Strauss, and Giraux.

Fryberg, Stephanie. 2003. "Really: You Don't Look Like an American Indian: Social Representations and Social Group Identities." Ph.D. Dissertation, Department of Psychology, Stanford University.

Frye, Marilyn. 1983. *The Politics of Reality.* Trumansburg, NY: The Crossing Press.

Fukuda, Mari. 1994. "Nonverbal Communication within Japanese and American Corporations." Unpublished manuscript, Princeton University.

Fullerton, Howard F. 1999. "Labor Force Projections to 2008: Steady Growth and Changing." *Monthly Labor Review* 122: 19–32.

Furstenberg, Frank. 1998. "Relative Risk: What Is the Family Doing to Our Children?" *Contemporary Sociology* 27 (May): 223–225.

Furstenberg, Frank F., Jr., and Christine Winquist Nord. 1985. "Parenting Apart: Patterns of Childrearing After Marital Disruption." *Journal of Marriage and the Family* 47 (November): 898–904.

Gagné, Patricia, and Richard Tewksbury. 1998. "Conformity Pressures and Gender Resistance among Transgendered Individuals." *Social Problems* 45 (February): 81–101.

Gallagher, Sally K. 2003. *Evangelical Identity and Gendered Family Life.* New Brunswick, NJ: Rutgers University Press.

Gallup Organization. 2002. "Confidence in Institutions." June. Princeton, NJ: Gallup Organization, www.gallup.com

Gallup Organization. 2003. "Homosexual Relations." Princeton, NJ: The Gallup Organization.

Gallup Organization. 2006. "Tobacco and Smoking." The Gallup Poll. Princeton, NJ: The Gallup Organization. www.gallup.com

Gallup, George, Jr. 2003. "Current Views on Premarital, Extramarital Sex." Princeton, NJ: The Gallup Poll.

Gamoran, Adam. 1972. "The Variable Effects of High School Tracking." *American Sociological Review* 57: 812–828.

Gamoran, Adam, and Robert D. Mare. 1989. "Secondary School Tracking and Educational Inequality: Compensation, Reinforcement, or Neutrality?" *American Journal of Sociology* 94: 1146–1183.

Gamson, Joshua. 1995. "Featured Essay." *Contemporary Sociology* 24 (May): 294–298.

Gamson, Joshua. 1998. *Freaks Talk Back: Tabloid Talk Shows and Sexual Nonconformity.* Chicago, IL: University of Chicago Press.

Gamson, Joshua. 1998. "Publicity Traps: Television Talk Shows and Lesbian, Gay, Bisexual, and Transgender Visibility." *Sexualities* 1: 11–41.

Gamson, William A. 1992. "The Social Psychology of Collective Action." Pp. 53–76 in *Frontiers in Social Movement Theory,* edited by Aldon D. Morris and Carol McClurg Mueller. New Haven, CT: Yale University Press.

Gamson, William, and Andre Modigliani. 1974. *Conceptions of Social Life: A Text-Reader for Social Psychology.* Boston, MA: Little, Brown.

Gans, Herbert. 1979. *Deciding What's News: A Study of the CBS Evening News, NBC Nightly News, Newsweek and Time.* New York: Pantheon.

Gans, Herbert J. 1999. *Popular Culture and High Culture: An Analysis and Evaluation of Taste.* New York: Basic Books.

Gardner, Howard. 1999 [1993]. *Frames of Mind: The Theory of Multiple Intelligences.* New York: Basic Books.

Garfinkel, Harold. 1967. *Studies in Ethnomethodology.* Englewood Cliffs, NJ: Prentice Hall.

Garrow, David J. 1981. *The FBI and Martin Luther King.* New York: Norton.

Gastil, John. 1990. "Generic Pronouns and Sexist Language: The Oxymoronic Character of Masculine Generics." *Sex Roles* 23 (December): 629–643.

Gates, Henry Louis, Jr. 1988. *The Signifying Monkey: A Theory of African-American Literary Criticism.* New York: Oxford University Press.

Gates, Henry Louis, Jr. 1992. "Integrating the American Mind." Pp. 105–120 in *Loose Canons: Notes on the Culture Wars,* edited by Henry Louis Gates, Jr. New York: Oxford University Press.

Gelles, Richard J. 1999. "Family Violence." Pp. 1–24 in *Family Violence: Prevention and Treatment,* edited by Robert J. Hampton, Thomas P. Gallota, Gerald R. Adams, Earl H. Potter III, and Roger P. Weissberg. Newbury Park, CA: Sage.

Genovese, Eugene. 1972. *Roll, Jordan, Roll: The World the Slaves Made.* New York: Pantheon.

Gerami, Shahin, and Melodye Lehnerer. 2001. "Women's Agency and Houshold Diplomacy: Negotiating Fundamentalism." *Gender & Society* 15 (August): 556–573.

Gersch, Beate. 1999. "Class in Daytime Talk Television." *Peace Review* 11 (June): 275–281.

Gerson, Kathleen. 1993. *No Man's Land: Men's Changing Commitments to Family and Work.* New York: Basic Books.

Gerstel, Naomi, and Sally Gallagher. 2001. "Men's Caregiving: Gender and the Contingent Character of Care." *Gender & Society* 15 (April): 197–217.

Gerth, Hans, and C. Wright Mills (eds.). 1946. *From Max Weber: Essays in Sociology.* New York: Oxford University Press.

Giannarelli, Linda, and James Barsimantov. 2000. *Child Care Expenses of America's Families.* Washington. DC: Urban Institute.

Gilbert, Daniel R., Susan T. Fiske, and Gardner Lindzey (eds.). 1998. *The Handbook of Social Psychology,* 4th ed. New York: Oxford University and McGraw-Hill.

Gilbert, D. T. and P. S. Malone. 1995. "The Correspondence Bias." *Psychological Bulletin* 117: 21–38.

Gilens, Martin. 1996. "Race and Poverty in America: Public Misperceptions and the American News Media." *The Public Opinion Quarterly* 60 (Winter): 515–541.

Gilkes, Cheryl Townsend. 2000. *"If It Wasn't For the Women . . ." Black Women's Experience and Womanist Culture in Church and Community."* Maryknoll, NY: Orbis Books.

Gilligan, Carol. 1982. *In a Different Voice: Psychological Theory and Women's Development.* Cambridge, MA: Harvard University Press.

Gimlin, Debra. 1996. "Pamela's Place: Power and Negotiation in the Hair Salon." *Gender & Society* 10 (October): 505–526.

Gimlin, Debra. 2002. *Body Work: Beauty and Self-Image in American Culture.* Berkeley, CA: University of California Press.

Gitlin, Todd. 2002. M*edia Unlimited: How the Torrent of Images and Sounds Overwhelms Our Lives.* New York: Metropolitan Books.

Glassner, Barry. 1999. *Culture of Fear: Why American Are Afraid of the Wrong Things.* New York: Basic Books.

Glazer, Nathan. 1970. *Beyond the Melting Pot: The Negroes, Puerto Ricans, Jews, Italians, and Irish of New York City.* Cambridge, MA: MIT Press.

Glazer, Nona. 1990. "The Home as Workshop: Women as Amateur Nurses and Medical Care

Providers." *Gender & Society* 4: 479–499.

Glenn, Evelyn Nakano. 1986. *Issei, Nisei, War Bride: Three Generations of Japanese American Women in Domestic Service.* Philadelphia, PA: Temple University Press.

Glenn, Evelyn Nakano. 2002. *Unequal Freedom: How Race and Gender Shaped American Citizenship and Labor.* Cambridge, MA: Harvard University Press.

Glenn, Norval, and Elizabeth Marquardt. 2001. "Hooking Up, Hanging Out, and Hoping for Mr. Right." New York: Institute for American Values.

Glock, Charles, and Rodney Stark. 1965. *Religion and Society in Tension.* Chicago, IL: Rand McNally.

Gluckman, Amy, and Betsy Reed (eds.). 1997. *Homo Economics: Capitalism, Community, and Lesbian and Gay Life.* New York: Routledge.

Goffman, Erving. 1961. *Asylums: Essays on the Social Situation of Mental Patients and Other Inmates.* Garden City, NY: Anchor.

Goffman, Erving. 1963. *Stigma: Notes on the Management of Spoiled Identity.* Englewood Cliffs, NJ: Prentice Hall.

Goldberger, Arthur S. 1979. "Heritability." *Econometrica* 46: 327–347.

Goldman, Noreen, Yorenzo Moreno, and Charles F. Westoff. 1989. *Peru Experimental Study: An Evaluation of Fertility and Child Health Information.* Princeton, NJ: Office of Population Research.

Goldstein, A.P. 1994. *The Ecology of Aggression.* New York: Plenum.

Gonzales, P., H. Blanton, and K. Williams. 2002. "The Effect of Stereotype Threat and Double-Minority Status on Test Performance in Latino Women." *Personality and Social Psychology Bulletin* 28: 659–670.

González, Tina Esther. 1996. *Social Control of Medical Professionals, Cultural Notions of Pain, and Doctors as Social Scientists.* Unpublished junior thesis, Princeton University.

Gordon, Linda. 1977. *Woman's Body / Woman's Right.* New York: Penguin.

Gordon, Margaret T., and Stephanie Riger. 1989. *The Female Fear.* New York: Free Press.

Gore, Albert. 2006. *An Inconvenient Truth: The Planetary Emergency of Global Warming and What We Can Do About It.* Emmaus, PA: Rodale.

Gottfredson, Michael R., and Travis Hirschi. 1990. *A General Theory of Crime.* Stanford, CA: Stanford University Press.

Gottfredson, Michael R., and Travis Hirschi. 1995. "National Crime Control Policies." *Society* 32 (January-February): 30–36.

Gough, Kathleen. 1984. "The Origin of the Family." Pp. 83–99 in *Women: A Feminist Perspective,* 3rd ed., edited by Jo Freeman. Palo Alto, CA: Mayfield.

Gould, Stephen Jay. 1981. *The Mismeasure of Man.* New York: W.W. Norton.

Gould, Stephen Jay. 1999. "The Human Difference." *The New York Times* (July 2).

Graefe, Deborah Roempke, and Daniel T. Lichtor. 1999. "Life Course Transitions of American Children: Parental Cohabitation, Marriage, and Single Motherhood." *Demography* 36 (May): 205–217.

Graham, Lawrence Otis. 1999. *Our Kind of People: Inside America's Black Upper Class.* New York: Harper Collins.

Gramsci, Antonio. 1971. *Selections from the Prison Notebooks of Antonio Gramsci,* edited by Quintin Hoare and Geoffrey Nowell. London: Lawrence and Wishart.

Granovetter, Mark. 1973. "The Strength of Weak Ties." *American Journal of Sociology* 78 (May): 1360–1380.

Granovetter, Mark. 1974. *Getting a Job: A Study of Contacts and Careers.* Cambridge, MA: Harvard University Press.

Granovetter, Mark S. 1995. "Afterward 1994: Reconsiderations and a New Agenda." Pp. 139–182 in *Getting A Job,* 2nd ed. by Mark S. Granovetter. Chicago, IL: University of Chicago Press.

Grant, Don Sherman, II, and Ramiro Martinez, Jr. 1997. "Crime and the Restructuring of the U.S. Economy: A Reconsideration of the Class Linkages." *Social Forces* 75 (March): 769–799.

Green, Gary Paul, Leam M. Tigges, and Daniel Diaz. 1999. "Racial and Ethnic Differences in Job Search Strategies In Atlanta, Boston, and Los Angeles." *Social Science Quarterly* 80 (June): 263–290.

Greenfeld, Lawrence A., and Tracy L. Snell. 1999. *Women Offenders.* Washington, DC: U.S. Bureau of Justice Statistics.

Greenhouse, Linda. 1996. "Gay Rights Laws Can't Be Banned, High Court Rules." *The New York Times* (May 21): A1ff.

Greenhouse, Steven. 2000. "Poll of Working Women Finds Them Stressed." *The New York Times,* April 3.

Grimal, Pierre (ed.). 1963. *Larousse World Mythology.* New York: Putnam.

Grindstaff, Laura. 2002. *The Money Shot: Trash Class, and the Making of TV Talk Shows.* Chicago, IL: University of Chicago Press.

Gruber, James E. 1982. "Blue-Collar Blues: The Sexual Harassment of Women Autoworkers." *Work and Occupations* 3 (August): 271–298.

Guinther, J. 1988. *The Jury in America.* New York: Facts on File Publications.

Guttmacher Institute. 2006a. "Contraceptive Use." New York: Guttmacher Institute. www.guttmacher.org

Guttmacher Institute. 2006b. "Facts on American Teens' Sexual and Reproductive Health." New York: Guttmacher Institute. www.guttmacher.org

Habermas, Jürgen. 1970. *Toward a Rational Society: Student Protest, Science, and Politics.* Boston, MA: Beacon Press.

Hadaway, C. Kirk, and Penny Long Marler. 1993. "All in the Family: Religious Mobility in America." *Review of Religious Research* 35 (December): 97–116.

Haddad, Yvonne Yazbeck, Jane I. Smith, and John L. Esposito (eds.). 2003. *Religion and Immigration: Christian, Jewish, and Muslim Experiences in the United States.* Walnut Creek, CA: AltaMira Press.

Hagan, John. 1993. "The Social Embeddedness of Crime and Unemployment." *Criminology* 31 (November): 465–491.

Hall, Edward T. 1966. *The Hidden Dimension.* New York: Doubleday.

Hall, Edward T., and Mildred Hall. 1987. *Hidden Differences: Doing Business with the Japanese.* New York: Anchor Press/Doubleday.

Hall, Elaine J. 1993. "Waitering/ Waitressing: Engendering in the Work of Table Servers." *Gender & Society* (September): 329–346.

Hallinan, Maureen T. 2003. *Ability Grouping and Student Learning.* Washington, DC: Brookings Papers on Educational Policy.

Hamer, Dean H., Stella Hu, Victoria L. Magnuson, Nan Hu, and Angela M. L. Pattatucci. 1993. "A Linkage Between DNA Markers on the X Chromosome and Male Sexual Identification." *Science* 261 (July): 321–327.

Hamer, Jennifer. 2001. *What It Means to Be Daddy.* New York: Columbia University Press.

Hamilton, Mykol C. 1988. "Using Masculine Generics: Does Generic He Increase Male Bias in the User's Imagery?" *Sex Roles* 19 (December): 785–799.

Handlin, Oscar. 1951. *The Uprooted.* Boston, MA: Little, Brown.

Haney, C., C. Banks, and P. G. Zimbardo. 1973. "Interpersonal Dynamics in a Simulated Prison." *International Journal of Criminology and Penology* 1: 69–97.

Haney, Lynne. 1996. "Homeboys, Babies, Men in Suits: The State and the Reproduction of Male Dominance." *American Sociological Review* 61 (October): 759–778.

Hans, Valerie, and Ramiro Martinez. 1994. "Intersections of Race, Ethnicity, and the Law." *Law and Human Behavior* 18 (June): 211–221.

Harris, Marvin. 1974. *Cows, Pigs, Wars, and Witches: The Riddles of Culture.* New York: Vintage.

Harrison, Bennett, and Harry Bluestone. 1982. *The Deindustrialization of America: Plant Closings, Community Abandonment, and the Dismantling of Basic Industry.* New York: Basic Books.

Harrison, Roderick. 2000. "Inadequacies of Multiple Response Race Data in the Federal Statistical System." Manuscript. Joint Center for Political and Economic Studies and Howard University, Department of Sociology, Washington, DC.

Harry, Joseph. 1979. "The Marital 'Liaisons' of Gay Men." *The Family Coordinator* 28 (October): 622–629.

Hartman, Chester, and Gregory D. Squires, eds. 2006. *There Is No Such Thing as a Natural Disaster: Race, Class, and Hurricane Katrina.* New York: Routledge.

Hastorf, Albert, and Hadley Cantril. 1954. "They Saw a Game: A Case Study." *Journal of Abnormal and Social Psychology* 40 (2): 129–134.

Hauan, Susan M., Nancy S. Landale, and Kevin T. Leicht. 2000. "Poverty and Work Effort among Urban Latino Men." *Work and Occupations* 27 (May): 188–222.

Hauser, Robert M., Howard F. Taylor, and Troy Duster. 1995. "The Bell Curve." *Contemporary Sociology* 24 (March): 149–161.

Hawley, Amos H. 1986. *Human Ecology: A Theoretical Essay.* Chicago, IL: University of Chicago Press.

Hays, Sharon. 2003. *Flat Broke with Children: Women in the Age of Welfare Reform.* New York: Oxford University Press.

Hearnshaw, Leslie. 1979. *Cyril Burt: Psychologist.* Ithaca, NY: Cornell University Press.

Heilman, Brice E., and Paul J. Kaiser. 2002. "Religion, Identity, and Politics in Tanzania." *Third World Quarterly* 23 (August): 691–709.

Heimer, Karen. 1997. "Socioeconomic Status, Subcultural Definitions and Violent Delinquency." *Social Forces* 75 (March): 799–833.

Hendy, Helen M., Cheryl Gustitus, and Jamie Leitzel-Schwalm. 2001. "Social Cognitive Predictors of Body Images in Preschool Children." *Sex Roles* 44 (May): 557–569.

Henslin, James M. 1993. "Doing the Unthinkable." Pp. 253–262 in *Down to Earth Sociology,* 7th ed., edited by James M. Henslin. New York: Free Press.

Herman, Judith. 1981. *Father-Daughter Incest.* Cambridge, MA: Harvard University Press.

Herrnstein, Richard J., and Charles Murray. 1994. *The Bell Curve: Intelligence and Class Structure in American Life.* New York: Free Press.

Herskovits, Melville J. 1941. *The Myth of the Negro Past.* New York: Harper and Brothers.

Hesse-Biber, Sharlene Hagy. 2007. *The Cult of Thinness,* 2nd ed. New York: Oxford University Press.

Higginbotham, A. Leon. 1978. *In the Matter of Color: Race and the American Legal Process.* New York: Oxford University Press.

Higginbotham, Elizabeth. 2001. *Too Much to Ask: Black Women in the Era of Integration.* Chapel Hill, NC: University of North Carolina Press.

Hill, C.T., Z. Rubin, and L.A. Peplau. 1976. "Breakups Before Marriage: The End of 103 Affairs." *Journal of Social Issues* 32: 147–168.

Hill, Lori Diane. 2001. *Conceptualizing Educational Attainment Opportunities of Urban Youth: The Effects of School Capacity, Community Context and Social Capital.* Ph.D. dissertation, University of Chicago.

Hirschi, Travis. 1969. *Causes of Delinquency.* Berkeley: University of California Press.

Hirschman, Charles. 1994. "Why Fertility Changes." *Annual Review of Sociology* 20: 203–223.

Hochschild, Arlie. 1983. *The Managed Heart: Commercialization of Human Feelings.* Berkeley, CA: University of California Press.

Hochschild, Arlie [Russell]. 1997. *The Time Bind: When Work Becomes Home and Home Becomes Work.* New York: Metropolitan Books.

Hochschild, Arlie Russell, with Anne Machung. 1989. *The Second Shift: Working Parents and the Revolution at Home.* New York: Viking.

Hoffman, Saul D., and Greg J. Duncan. 1988. "What *Are* the Economic Consequences of Divorce?" *Demography* 25 (November): 641–645.

Hoffnung, Michele. 2004. "Wanting It All: Career, Marriage, and Motherhood During College-Educated Women's 20s." *Sex Roles* 50(May): 711–723.

Hofstadter, Richard. 1944. *Social Darwinism in American Thought.* Philadelphia, PA: University of Pennsylvania Press.

Hollander, Jocelyn A. 2002. "Resisting Vulnerability: The Social Reconstruction of Gender in Interaction." *Social Problems* 49 (November): 474–496.

Hollingshead, August B., and Frederick C. Redlich. 1958. *Social Class and Mental Illness: A Community Study.* New York: Wiley.

Holm, Maj. Gen. Jeanne. 1992. *Women in the Military.* Novato, CA: Presidio Press.

Holmes, Schuyler. 2000. "Environmental Racism and Classism in Toxic Waste Dumping in Cleveland, Ohio." Junior thesis. Princeton University.

Homans, George. 1974. *Social Behavior: Its Elementary Forms,* revised ed. New York: Harcourt Brace Jovanovich.

Hondagneu-Sotelo, Pierrette. 2001. *Doméstica: Immigrant Workers Cleaning and Caring in the Shadows of Affluence.* Berkeley, CA: University of California Press.

Hondagneu-Sotelo, Pierrette, and Ernestine Avila. 1997. "'I'm Here, but I'm There': The Meanings of Latina Transnational Motherhood." *Gender & Society* 11(5): 548–571.

Hong, Laurence K. 1978. "Risky Shift and Cautious Shift: Some Direct Evidence on the Culture-Value Theory." *Social Psychology* 41 (December): 342–346.

Horowitz, Ruth. 1995. *Teen Mothers: Citizens or Dependents?* Chicago, IL: University of Chicago Press.

Horton, Hayward Derrick, Beverlyn Lundy Allen, Cedric Herring, and Melvin E. Thomas. 2000. "Lost in the Storm: The Sociology of the Black Working Class, 1850 to 1990." *American Sociological Review* 65 (February): 128–137.

Horton, Jacqueline A., ed. 1995. *The Women's Health Data Book,* 2nd ed. Washington, DC: Jacobs Institute of Women's Health.

House, James S. 1980. *Occupational Stress and the Mental and Physical Health of Factory Workers.* Ann Arbor, MI: Survey Research Center.

Hoyt, Wendy, and Lori R. Kogan. 2001. "Satisfaction with Body Image and Peer Relationships for Males and Females in a College Environment." *Sex Roles* 45 (August): 199–215.

Hudson, Ken. 1999. "No Shortage of 'Nonstandard' Jobs." *Briefing Paper.* Washington, DC: Economic Policy Institute, December.

Hughes, Langston. 1967. *The Big Sea.* New York: Knopf.

Humphries, Drew. 1999. *Crack Mothers: Pregnancy, Drugs, and the Media.* Columbus, OH: Ohio State University Press.

Hunter, Andrea, and Sherrill L. Sellers. 1998. "Feminist Attitudes Among African American Women and Men." *Gender & Society* 12 (February): 81–99.

Hunter, Margaret. 2002. "Rethinking Epistemology, Methodology, and Racism: or, Is White Sociology Really Dead?" *Race and Society* 5 (No. 2): 119–138.

Inciardi, James A. 2001. *The War on Drugs: The Continuing Saga of the Mysteries and Miseries of Intoxication, Addiction, Crime, and Public Policy,* 3rd ed. Boston, MA: Allyn and Bacon.

Inglehart, Ronald, and Wayne E. Baker. 2000. "Modernization, Cultural Change, and the Persistence of Traditional Values." *American Sociological Review* 65 (February): 19–51.

International Labour Organization. 2002. *Every Child Counts: Estimates on Child Labour,* www.ilo.org

Irwin, Katherine. 2001. "Legitimating the First Tattoo: Moral Passage through Informal Interaction." *Symbolic Interaction* 24 (March): 49–73.

Jackson, Linda A., Ruth E. Fleury, and Donna M. Lewandowski. 1996. "Feminism: Definitions, Support, and Correlates of Support Among Male and Female College Students." *Sex Roles* 34 (May): 687–693.

Jackson, Pamela Brayboy. 1992. "Specifying the Buffering Hypothesis: Support, Strain, and Depression." *Social Psychology Quarterly* 55: 363–378.

Jackson, Pamela Braboy. 2000. "Stress and Coping among Black Elites in Organizational Settings." Manuscript.

Jackson, Pamela B., Peggy A. Thoits, and Howard F. Taylor. 1994. "The Effects of Tokenism on America's Black Elite." Paper read before the American Sociological Association. August, Los Angeles, CA.

Jackson, Pamela B., Peggy A. Thoits, and Howard F. Taylor. 1995. "Composition of the Workplace and Psychological Well-Being: The Effects of Tokenism on America's Black Elite." *Social Forces* 74 (December): 543–557.

Jacobs, Jerry A., and Ronnie J. Steinberg. 1990. "Compensating Differentials and the Male-Female Wage Gap: Evidence from the New York State Comparable Worth Study." *Social Forces* 69 (December): 430–469.

Jacobs, Jerry A., and Kathleen Gerson. 2004. *The Time Divide: Work, Family, and Gender Inequality.* New York: Oxford University Press.

Jacobs, Lawrence R., and James A. Morone. 2004. "Health and Wealth." *The American Prospect* (June): A20–21.

Jang, Kerry L., W. J. Livesly, and Philip A. Vernon. 1996. "Heritability of the Big Five Personality Dimensions and Their Facets: A Twin Study." *Journal of Personality* 64: 577–589.

Janis, Irving L. 1982. *Groupthink: Psychological Studies of Policy Decisions and Fiascos,* 2nd ed. Boston, MA: Houghton Mifflin.

Janus, Samuel S., and Cynthia L. Janus. 1993. *The Janus Report on Sexual Behavior.* New York: Wiley.

Jencks, Christopher. 1993. *Rethinking Social Policy: Race, Poverty, and the Underclass.* New York: Harper Collins.

Jencks, Christopher, and Meredith Phillips (eds.). 1998. *The Black-White Test Score Gap.* Washington, DC: Brookings Institution Press.

Jencks, Christopher, Susan Bartlett, Mary Corcoran, James Crouse, David Eaglesfield, Gregory Jackson, Kent McClelland, Peter Mueser, Michael Olneck, Joseph Schwartz, Sherry Ward, and Jill Williams. 1979. *Who Gets Ahead? The Determinants of Economic Success in America.* New York: Basic Books.

Jencks, Christopher, Marshall Smith, Henry Ackland, Mary Jo Bane, David Cohen, Herbert Gintis, Barbara Heyns, and Stephan Michelson. 1972. *Inequality: A Reassessment of the Effect of Family and Schooling in America.* New York: Basic Books.

Jenness, Valerie. 1995. "Social Movement Growth, Domain Expansion, and Framing Processes: The Gay/Lesbian Movement and Violence Against Gays and Lesbians as a Social Problem." *Social Problems* 42 (February): 145–170.

Jenness, Valerie, and Kendal Broad. 1997. *Hate Crimes: New Social*

Movements and the Politics of Violence. New York: Aldine de Gruyter.

Jenness, Valerie, and Kendal Broad. 2002. *Hate Crimes: New Social Movements and the Politics of Violence.* New York: Aldine de Gruyter.

Jennings, M. K., and R. G. Niemi. 1974. *The Political Character of Adolescence.* Princeton, NJ: Princeton University Press.

Johnson, Kim K. P. 1995. "Attributions About Date Rape: Impact of Clothing, Sex, Money Spent, Date Type, and Perceived Similarity." *Family and Consumer Sciences Research Journal* 23 (March): 292–310.

Johnson, Norris R., James G. Stember, and Deborah Hunter. 1977. "Crowd Behavior as Risky Shift: A Laboratory Experiment." *Sociometry* 40 (2): 183–187.

Johnston, David Cay. 2000. "Corporations' Taxes are Falling Even as Individual Burden Rises." *The New York Times* (February 20): A1ff.

Jones, Diane Carlson. 2001. "Social Comparison and Body Image: Attractiveness Comparison to Models and Peers among Adolescent Girls and Boys." *Sex Roles* 45 (November): 645–664.

Jones, Edward E., Amerigo Farina, Albert H. Hastorf, Hazel Markus, Dale T. Miller, and Robert A. Scott. 1986. *Social Stigma: The Psychology of Marked Relationships.* New York: W. H. Freeman.

Jones, James M. 1997. *Prejudice and Racism,* 2nd ed. New York: McGraw-Hill.

Jordan, Winthrop D. 1969. *The White Man's Burden: Historical Origins of Racism in the United States.* New York: Oxford University Press.

Jordon, Winthrop D. 1968. *White Over Black: American Attitudes Toward the Negro 1550–1812.* Chapel Hill, NC: University of North Carolina Press.

Joseph, Janice. 1997. "Fear of Crime among Black Elderly." *Journal of Black Studies* 27 (May): 698–717.

Jucha, Robert. 2002. *Terrorism.* Belmont, CA: Wadsworth.

Kadushin, Charles. 1974. *The American Intellectual Elite.* Boston, MA: Little, Brown.

Kall, Denise. 2002. *Smoking Gun or Organizational Haze? The Tobacco Companies' Response to Cancer Research.* Unpublished senior thesis, University of Delaware.

Kalleberg, Arne L., Barbara F. Reskin, and Ken Hudson. 2000. "Bad Jobs in America: Standard and Nonstandard Employment Relations and Job Quality in the United States." *American Sociological Review* 65 (April): 256–278.

Kalmijn, Matthijs. 1991. "Status Homogamy in the United States." *American Journal of Sociology* 97 (September): 496–523.

Kalmijn, Matthijs. 1999. "Father Involvement in Childrearing and the Perceived Stability of Marriage." *Journal of Marriage and the Family* 61 (May): 409–421.

Kalof, Linda. 1999. "The Effects of Gender and Music Video Imagery on Sexual Attitudes." *Journal of Social Psychology* 139 (June): 378–385.

Kamin, Leon J. 1974. *The Science and Politics of IQ.* Potomac, MD: Lawrence Erlbaum.

Kamin, Leon J. 1995. "Behind the Curve." *Scientific American* 272 (February): 99–103.

Kamin, Leon J., and Arthur S. Goldberger. 2002. "Twin Studies in Behavioral Research: A Skeptical View." *Theoretical Population Biology* 61: 83–95.

Kane, Emily W. 2006. "'No Way My Boys are Going to be Like That!'": Parents' Responses to Children's Gender Nonconformity." *Gender & Society* 20(2):149–176.

Kanter, Rosabeth Moss. 1977. *Men and Women of the Corporation.* New York: Basic Books.

Kaplan, Elaine Bell. 1996. *Not Our Kind of Girl: Unraveling the Myths of Black Teenage Motherhood.* Berkeley, CA: University of California Press.

Kasarda, Jack. 1999. "Industrial Restructuring and the Changing Location of Jobs." In *State of the Union: America in the 1990s: Economic Trends,* edited by Reynolds Farley. New York: Russell Sage Foundation.

Katz, I., J. Wackenhut, and R. G. Hass. 1986. "Racial Ambivalence, Value Duality, and Behavior." Pp. 35–60 in *Prejudice, Discrimination, and Racism,* edited by J. F. Dovidio and S. L. Gaertner. New York: Academic Press.

Keil, Thomas J., and Gennaro F. Vito. 1995. "Factors Influencing the Use of 'Truth in Sentencing' Law in Kentucky Murder Cases: A Research Note." *American Journal of Criminal Justice* 20 (Fall): 105–111.

Keith, Verna M. 1997. "Life Stress and Psychological Well-Being among Married and Unmarried Blacks." Pp. 95–116 in *Family Life in Black America.* Thousand Oaks, CA: Sage.

Kelley, J. R., J. W. Jackson, and S. L. Huston-Comeaux. 1999. "The Effects of Time Pressure and Task Differences on Influence Models and Accuracy in Decision-Making Groups." *Personality and Social Psychology Bulletin* 23: 10–22.

Kendall, Lori. 2002. "'Oh No! I'm a Nerd!': Hegemonic Masculinity on an Online Forum." *Gender & Society* 14 (April): 256–274.

Kennedy, Randall. 2003. *Interracial Intimacies: Sex, Marriage, Identity and Adoption.* New York: Pantheon.

Kennedy, Robert E. 1989. *Life Choices: Applying Sociology,* 2nd ed. New York: Holt, Rinehart and Winston.

Kephart, W. H. 1993. *Extraordinary Groups: An Examination of Unconventional Life,* rev. ed. New York: St. Martin's Press.

Kerr, N. L. 1992. "Issue Importance and Group Decision Making." Pp. 68–88 in *Group Process and Productivity,* edited by S. Worchel, W. Wood, and J. A. Simpson. Newbury Park, CA: Sage.

Kessler, Suzanne J. 1990. "The Medical Construction of Gender: Case Management of Intersexed Infants." *Signs* 16 (Autumn): 3–26.

Khashan, Hilal, and Lina Kreidie. 2001. "The Social And Economic Correlates of Islamic Religiosity." *World Affairs* 1654 (Fall): 83–96.

Kilborn, Peter T., and Lynette Clemetson. 2002. "Gains of 90s Did Not Lift All, Census Shows." *The New York Times,* June 5: A1.

Kim, Elaine H. 1993. "Home is Where the *Han* is: A Korean American Perspective on the Los Angeles Upheavals." Pp. 215–235 in *Reading Rodney King/*

Reading Urban Uprising, edited by Robert Gooding-Williams. New York: Routledge.

Kimmel, Michael. 2000. "Saving the Males: The Sociological Implications of Virginia Military Institute and the Citadel." *Gender & Society* 14 (August): 494–516.

Kimmel, Michael S. 2001. "Masculinity as Homophobia: Fear, Shame, and Silence in the Construction of Gender Identity." Pp. 266–287 in *The Masculinities Reader,* edited by Stephen M. Whitehead and Frank J. Barrett. Cambridge, England: Polity.

Kimmel, Michael S., and Michael A. Messner. 2004. *Men's Lives,* 6th ed. Boston, MA: Allyn and Bacon.

Kitano, Harry. 1976. *Japanese Americans: The Evolution of a Subculture,* 2nd ed. New York: Prentice Hall.

Kitsuse, John I. 1980. "Coming Out All Over: Deviants and the Politics of Social Problems." *Social Problems* 28 (October): 1–13.

Kitsuse, John I., and Aaron V. Cicourel. 1963. "A Note on the Uses of Official Statistics." *Social Problems* 11 (Fall): 131–139.

Klapp, Orin. 1972. *Currents of Unrest: Introduction to Collective Behavior.* New York: Holt, Rinehart, and Winston.

Klausner, Samuel Z., and Edward F. Foulks. 1982. *Eskimo Capitalists: Oil, Alcohol, and Politics.* Totowa, NJ: Allanheld.

Kleinfeld, Judith S. 2002. "The Small World Problem." *Society* 39:61–66.

Klinger, Lori J., James A. Hamilton, and Peggy J. Cantrell. 2001. "Children's Perceptions of Aggression and Gender-Specific Content in Toy Commercials." *Social Behavior and Personality* 29: 11–20.

Kluegel, J. R., and Lawrence Bobo. 1993. "Dimensions of Whites' Beliefs About the Black-White Socioeconomic Gap." Pp. 127–147 in *Race and Politics in American Society,* edited by P. Sniderman, P. Tetlock, and E. Carmines. Stanford, CA: Stanford University Press.

Kniss, Fred. 2003. "Church and State." *Contexts* 2 (Spring) 62–63.

Knoke, David. 1992. *Political Networks: The Structural Perspective.* New York: Cambridge University Press.

Kochen, M. (ed.). 1989. *The Small World.* Norwood, NJ: Ablex Press.

Kocieniewski, David, and Robert Hanley. 2000. "Racial Profiling Was Routine, New Jersey Says." *The New York Times* (November 28): 1.

Kohlberg, Lawrence. 1969. "'Stage and Sequence': The Cognitive-Developmental Approach to Socialization." Pp. 347–480 in *Handbook of Socialization and Research,* edited by D.A. Goslin. Chicago, IL: Rand McNally.

Kohut, Andrew. 2007. "Muslim Americans: Middle Class and Mostly Mainstream." New York, NY: Pew Research Center.

Kovel, Johnathan. 1970. *White Racism: A Psychohistory.* New York: Pantheon.

Krasnodemski, Memory. 1996. "Justified Suffering: Attribution Theory Applied to Perceptions of the Poor in America." Unpublished senior thesis, Princeton University.

Kray, Susan. 1993. "Orientalization of an 'Almost White' Woman: The Interlocking Effects of Race, Class, Gender and Ethnicity in American Mass Media." *Critical Studies in Mass Communication* 10 (December): 349–366.

Kuhn, Harold W. and Sylvia Nasar. 2002. *The Essential John Nash.* Princeton, NJ: Princeton University Press.

Kulis, Stephen, Flavio Francisco Marsiglia, and Michael L. Hecht. 2002. "Gender Labels and Gender Identity as Predictors of Drug Use among Ethnically Diverse Middle School Students." *Youth and Society* 33 (March): 442–475.

Kurz, Demi. 1989. "Social Science Perspectives on Wife Abuse: Current Debates and Future Directions." *Gender & Society* 3 (December): 489–505.

Kurz, Demie. 1995. *For Richer for Poorer: Mothers Confront Divorce.* New York: Routledge.

Ladner, Joyce A. 1986. "Teenage Pregnancy: Implications for Black Americans." Pp. 65–84 in *The State of Black America 1986,* edited by James D. Williams. New York: National Urban League.

LaFlamme, Darquise, Andree Pomerrleau, and Gerard Malcuit. 2002. "A Comparison of Fathers' and Mothers' Involvement in Childcare and Stimulation Behaviors during Free-Play with Their Infants at 9 and 15 Months." *Sex Roles* 11–12 (December): 507–518.

LaFrance, Marianne, 2002. "Smile Boycotts and Other Body Politics." *Feminism & Psychology* 12 (August): 319–323.

Lamanna, Mary Ann, and Agnes Riedman. 2003. *Marriage and Families: Making Choices in a Diverse Society.* Belmont, CA: Wadsworth.

Lamb, Michael. 1998. "Cybersex: Research Notes on the Characteristics of the Visitors to Online Chat Rooms." *Deviant Behavior* 19 (April–June): 121–135.

Lamont, Michèle. 1992. *Money, Morals, and Manners: The Culture of the French and the American Upper-Middle Class.* Chicago, IL: University of Chicago Press.

Lamphere, Louise, Alex Stepick, and Guillermo Grenier (eds.). 1994. *Newcomers in the Workplace: Immigrants and the Restructuring of the U.S. Economy.* Philadelphia, PA: Temple University Press.

Landry, Bart. 2000. *Black Working Wives: Pioneers of the Americana Family Revolution.* Berkeley, CA: University of California Press.

Langhinrichsen-Rohling, Jennifer. Peter Lewinsohn, Paul Rohde, John Seeley, Candice M. Monson, Kathryn A. Meyer, and Richard Langford. 1998. "Gender Differences in the Suicide-Related Behaviors of Adolescents and Young Adults." *Sex Roles* 39 (December): 839–854.

Langman, Lauren, and Douglas Morris. 2002. "Internetworked Social Movements: The Promises and Prospects for Global Justice." Paper presented at the International Sociological Association, Brisbane, Australia.

Larana, Enrique, Hank Johnston, and Joseph R. Gusfield (eds.). 1994. *New Social Movements: From Ideology to Identity.* Philadelphia, PA: Temple University Press.

Lau, Bonnie. 2002. *Stereotype Threat: Minority Education, Intelligence Theory, Affirmative Action, and an Empirical Study of Women in the Quantitative Domain.* Senior thesis, Princeton University.

Lau, Bonnie, and Howard F. Taylor. 2003. "Ethnic and Gender Stereotype Threat Among Asian Women in Quantitative Test Performance." Manuscript, Department of Sociology, Princeton University.

Laumann, Edward O., John H. Gagnon, Robert T. Michael, and Stuart Michaels. 1994. *The Social Organization of Sexuality: Sexual Practices in the United States.* Chicago, IL: University of Chicago Press.

Laveist, Thomas A., and Amani Nuru-Jeter. 2002. "Is Doctor-Patient Race Concordance Associated with Greater Satisfaction with Care?" *Journal of Health and Social Behavior* 43 (September): 296–306.

Lavelle, Louis. 2001. "Executive Pay." *Business Week,* April 16, www.businessweek.com

Lawson, Helene M., and Kira Leck. 2006. "Dynamics of Internet Dating." *Social Science Computer Review* 24 (Summer): 189–208.

Lederman, Douglas. 1992. "Men Outnumber Women and Get Most of the Money in Big-Time Sports Programs." *The Chronicle of Higher Education* 38 (April): A1ff.

Lee, Gary R., Karen Seccombe, and Constance L. Shehan. 1991. "Marital Status and Personal Happiness: An Analysis of Trend Data." *Journal of Marriage and the Family* 53(4): 839–844.

Lee, Matthew T., and M. David Ermann. 1999. "Pinto 'Madness' as a Flawed Landmark Narrative: An Organizational and Network Analysis." *Social Problems* 46 (February): 30–47.

Lee, Sharon M. 1993. "Racial Classification in the U.S. Census: 1890–1990." *Ethnic and Racial Studies* 16(1): 75–94.

Lee, Sharon M. 1994. "Poverty and the U.S. Asian Population." *Social Science Quarterly* 75 (September): 541–559.

Lee, Stacey J. 1996. *Unraveling the "Model Minority" Stereotype: Listening to Asian American Youth.* New York: Teacher's College Press.

Legge, Jerome S. Jr., 1993. "The Persistence of Ethnic Voting: African Americans and Jews in the 1989 New York Mayoral Campaign." *Contemporary Jewry* 14: 133–146.

Lehmann, Nicholas. 1999. *The Big Test: The Secret History of the American Meritocracy.* New York: Farrar, Straus, and Giroux.

Leidner, Robin. 1993. *Fast Food, Fast Talk: Service Work and the Routinization of Everyday Life.* Berkeley, CA: University of California Press.

Lemert, Edwin M. 1972. *Human Deviance, Social Problems, and Social Control.* Englewood Cliffs, NJ: Prentice Hall.

Lempert, Lora Bex, and Marjorie L. DeVault. 2000. "Guest Editors' Introduction: Special Issue on Emergent and Reconfigured Forms of Family Life." *Gender & Society* 14 (February): 6–10.

Lenski, Gerhard, Jean Lenski, and Patrick Nolan. 1998. *Human Societies: An Introduction to Macro-Sociology,* 8th ed. New York: McGraw-Hill.

Leonhardt, David. 2006. "Gender Pay Gap, Once Narrowing, Is Stuck in Place." *The New York Times,* December 24: A1.

Lever, Janet. 1978. "Sex Differences in the Complexity of Children's Play and Games." *American Sociological Review* 43 (August): 471–483.

Levine, Adeline. 1982. *Love Canal: Science, Politics, and People.* Lexington, MA: Lexington Books.

Levine, Felice J., and Katherine J. Rosich. 1996. *Social Causes of Violence: Crafting a Science Agenda.* Washington, DC: American Sociological Association.

Levine, Lawrence. 1984. "William Shakespeare and the American People: A Study in Cultural Transformation." *American Historical Review* 89 (1): 34–66.

Levy, Ariel. 2005. *Female Chauvinist Pigs: Women and the Rise of Raunch Culture.* New York: Free Press.

Levy, Marion J. 1949. *The Structure of Society.* Princeton, NJ: Princeton University Press.

Lewin, Tamar. 2002. "Study Links Working Mothers to Slower Learning." *The New York Times.* July 17: A14.

Lewis, Dan A., Amy Bush Stevens, and Kristen Shook Slack. 2002. *Illinois Families Study, Welfare Reform in Illinois: Is the Moderate Approach Working?* Evanston, IL: Institute for Poverty Research, Northwestern University.

Lewis, Oscar. 1960. *Five Families: Mexican Case Studies in the Culture of Poverty.* New York: Basic Books.

Lewis, Oscar. 1966. "The Culture of Poverty." *Scientific American* 215 (October): 19–25.

Lewontin, Richard. 1996. *Human Diversity.* New York: W. H. Freeman.

Lieberson, Stanley. 1980. *A Piece of the Pie: Black and White Immigrants Since 1880.* Berkeley, CA: University of California Press.

Lin, Nan. 1989. "The Small World Technique as a Theory Construction Tool." Pp. 231–238 in *The Small World,* edited by M. Kochen. Norwood, NJ: Ablex Press.

Linton, Ralph. 1937. *The Study of Man.* New York: Appleton-Century.

Livingston, I. L. 1994. *Handbook of Black American Health: The Mosaic of Conditions, Issues and Prospects.* Westport, CT: Greenwood.

Lo, Clarence. 1990. *Small Property Versus Big Government: Social Origins of the Property Tax Revolt.* Berkeley, CA: University of California Press.

Locklear, Erin M. 1999. *Where Race and Politics Collide: The Federal Acknowledgement Process and Its Effects on Lumsee and Pequot Indians.* Unpublished senior thesis, Princeton University.

Logan, John, and Harvey L. Molotch. 1987. *Urban Fortunes: The Political Economy of Place* Berkeley, CA: University of California Press.

Logio-Rau, Kim. 1998. "Here's Looking at You, Kid: Race, Gender and Health Behaviors among Adolescents." Unpublished doctoral dissertation, University of Delaware.

Lombardo, William K., Gary A. Cretser, and Scott C. Roesch. 2001. "For Crying Out Loud-The Differences Persist into the '90s." *Sex Roles* 45 (December): 529–547.

Lorber, Judith. 1994. *Paradoxes of Gender.* New Haven, CT: Yale University Press.

Louie, Miriam Ching Yoon. 2001. *Sweatshop Warriors: Immigrant Women Workers Take on the Global Factory.* Cambridge, MA: South End Press.

Lovejoy, Meg. 2001. "Disturbances in the Social Body: Differences in

Body Image and Eating Problems Among African American and White Women." *Gender & Society* 15 (April): 239–261.

Lucal, Betsy. 1994. "Class Stratification in Introductory Textbooks: Relational or Distributional Models?" *Teaching Sociology* 22 (April): 139–150.

Lucas, Samuel R. 1999. *Tracking Inequality: Stratification and Mobility in American High Schools.* New York: Teachers College Press.

Luckenbill, David F. 1986. "Deviant Career Mobility: The Case of Male Prostitutes." *Social Problems* 33 (April): 283–296.

Luker, Kristin. 1975. *Taking Chances: Abortion and the Decision Not to Contracept.* Berkeley, CA: University of California Press.

Luker, Kristin. 1984. *Abortion and the Politics of Motherhood.* Berkeley, CA: University of California Press.

Luker, Kristin. 1996. *Dubious Conceptions: The Politics of Teenage Pregnancy.* Cambridge, MA: Harvard University Press.

Lupton, Deborah. 2002. "Road Rage: Drivers' Understandings and Experiences." *Journal of Sociology* 36 (September): 275–290.

Lyons, Linda. 2002. "Teen Attitudes Contradict Sex-Crazed Stereotype." *The Gallup Poll.* January 29, www.gallup.com

Machel, Graca. 1996. *Impact of Armed Conflict on Children.* New York: UNICEF/United Nations.

MacKinnon, Catherine. 1983. "Feminism, Marxism, Method, and the State: An Agenda for Theory." *Signs* 7 (Spring): 635–658.

MacKinnon, Catherine A. 2006. "Feminism, Marxism, Method, and the State: An Agenda for Theory." Pp. 829–868 in D. Kennedy and W.F. Fisher, *The Canon of American Legal Thought.* Princeton: Princeton University Press.

MacKinnon, Neil J., and Tom Langford. 1994. "The Meaning of Occupational Prestige Scores: A Social Psychological Analysis and Interpretation." *Sociological Quarterly* 35 (May): 215–145.

Mackintosh, N. J. 1995. *Cyril Burt: Fraud or Framed?* Oxford, England: Oxford University Press.

Madriz, Esther. 1997. *Nothing Bad Happens to Good Girls: Fear of Crime in Women's Lives.* Berkeley, CA: University of California Press.

Malat, Jennifer. 2001. "Social Distance and Patients' Rating of Healthcare Providers." *Journal of Health and Social Behavior* 42 (December): 36–372.

Malcomson, Scott L. 2000. *The American Misadventure of Race.* New York: Farrar, Strauss, and Giroux.

Maldonado, Lionel, A. 1997. "Mexicans in the American System: A Common Destiny." In *Ethnicity in the United States: An Institutional Approach,* edited by William Velez. Bayside, NY: General Hall.

Maldonado, Lionel A., and Charles V. Willie. 1996. "Developing a 'Pipeline' Recruitment Program for Minority Faculty." Pp. 330–371 in *Educating a New Majority: Transforming America's Educational System for Diversity,* edited by Laura I. Rendon and Richard O. Hope. San Francisco, CA: Jossey-Bass.

Malthus, Thomas Robert. 1798 [1926]. *First Essay on Population 1798.* London: Macmillan.

Mandel, Daniel. 2001. "Muslims on the Silver Screen." *Middle East Quarterly* 8 (Spring): 19–30.

Manning, Winton H., and Rex Jackson. 1984. "College Entrance Examinations: Objective Selection or Gatekeeping for the Economically Privileged," In *Perspectives on Bias in Mental Testing,* ed. by Cecil R. Reynolds and Robert T. Brown. Edited by New York: Plenum.

Marciniak, Liz-Marie. 1998. "Adolescent Attitudes Toward Victim Precipitation of Rape." *Violence and Victims* 12 (Fall): 287–300.

Marcuse, Herbert. 1964. *One-Dimensional Man.* Boston, MA: Beacon Press.

Margolin, Leslie. 1992. "Deviance on Record: Techniques for Labeling Child Abusers in Official Documents." *Social Problems* 39 (February): 58–70.

Marks, Carole. 1989. *Farewell, We're Good and Gone: The Great Black Migration.* Bloomington, IN: Indiana University Press.

Marks, Carole, and Deana Edkins. 1999. *The Power of Pride: Stylemakers and Rulebreakers of the Harlem Renaissance.* New York: Crown.

Marquis, Christopher. 2003. "Total of Unmarried Couples Surged in 2000 Census." *The New York Times* (March 13): A32.

Marshall, Susan. 1997. *Splintered Sisterhood: Gender & Class in the Campaign against Woman Suffrage.* Madison, WI: University of Wisconsin Press.

Martin, Patricia Yancey, and Robert Hummer. 1989. "Fraternities and Rape on Campus." *Gender & Society* 3 (December): 457–473.

Martin, Susan Ehrlich. 1989. "Sexual Harassment: The Link between Gender Stratification, Sexuality, and Women's Economic Status." Pp. 57–75 in *Women: A Feminist Perspective,* 4th ed., edited by Jo Freeman. Mountain View, CA: Mayfield.

Martinez, Ramiro. 1996. "Latinos and Lethal Violence: The Impact of Poverty and Inequality." *Social Problems* 43 (May): 131–146.

Martinez, Ramiro. 2002. *Latino Homicide: Immigration, Violence and Community.* New York: Routledge.

Marx, Anthony. 1997. *Making Race and Nation: A Comparison of the United States, South Africa, and Brazil.* New York: Cambridge University Press.

Marx, Gary T. 1967. "Religion: Opiate or Inspiration of Civil Rights Militancy among Negroes." *American Sociological Review* 32 (February): 64–72.

Marx, Gary T., and Douglas McAdam. 1994. *Collective Behavior and Social Movements: Process and Structure.* Englewood Cliffs, NJ: Prentice Hall.

Marx, Karl. 1967 [1867]. *Capital.* F. Engels (ed.). New York: International Publishers.

Marx, Karl. 1972 [1843]. "Contribution to the Critique of Hegel's *Philosophy of Right.*" Pp. 11–23 in *The Marx-Engels Reader,* edited by Robert C. Tucker. New York: W. W. Norton.

Massey, D., and M. Fischer. 2007. "Stereotype Threat and Academic Performance: New Findings from a Racially Diverse Sample of College Freshmen." *Du Bois Review* 2 (1): 45–67.

Massey, Douglas S. 1993. "Latino Poverty Research: An Agenda for the 1990s." *Social Science*

Research Council Newsletter 47 (March): 7–11.

Massey, Douglas S. 2005. *Strangers in a Strange Land: Humans in an Urbanizing World.* New York: Norton.

Massey, Douglas S., and Nancy A. Denton. 1993. *American Apartheid: Segregation and the Making of the Underclass.* Cambridge, MA: Harvard University Press.

Mathews, Linda. 1996. "More Than Identity Rests on a New Racial Category." *The New York Times* (July 6): 1–7.

Mauer, Marc. 1999. *Race to Incarcerate.* New York: The New Press.

Mazzuca, Josephine. 2002. "More Accepting of Homosexuals-Canada or U.S.?" *The Gallup Poll.* Princeton, NJ: Gallup Organization, www.gallup.com

McAdam, Doug. 1982. *Political Process and the Development of Black Insurgency 1930–1970.* Chicago, IL: University of Chicago Press.

McAdam, Doug. 1999. *Political Process and the Development of Black Insurgency. 1930–1970,* 2nd ed. Chicago, IL: University of Chicago Press.

McCabe, Janice. 2005. "What's in a Label? The Relationship between Feminist Self-Identification and 'Feminist' Attitudes among U.S. Women and Men." *Gender & Society* 19(4): 480–505.

McCall, Leslie. 2001. *Complex Inequality: Gender, Class, and Race in the New Economy.* New York: Routledge.

McCarthy, John, and Mayer Zald. 1973. *The Trend of Social Movements in America: Professionalism and Resource Mobilization.* Morristown, NJ: General Learning Press.

McCauley, C. 1989. "The Nature of Social Influence in Groupthink: Compliance and Internalization." *Journal of Personality and Social Psychology* 57 (August): 250–260.

McClelland, Susan. 2003. "A Grim Toll on the Innocent." *Maclean's* (May 12): 20.

McCloskey, Laura Ann. 1996. "Socioeconomic and Coercive Power Within the Family." *Gender & Society* 10 (August): 449–463.

McCollum, Chris. 2002. "Relatedness and Self-Definition: Two Dominant Themes in Middle-Class Americans' Life Stories." *Ethos* 30: 113–139.

McComb, Chris. 2001 (May 4). "Few Say It's Ideal for Both Parents to Work Full Time Outside of Home." Princeton, NJ: The Gallup Poll.

McCord, William, and Arline McCord. 1986. *Paths to Progress: Bread and Freedom in Developing Societies.* New York: W.W. Norton.

McGuire, Gail M., and Barbara F. Reskin. 1993. "Authority Hierarchies at Work: The Impacts of Race and Sex." *Gender & Society* 7 (December): 487–506.

McGuire, Meredith. 2001. *Religion: The Social Context,* 5th ed. Belmont, CA: Wadsworth.

McIntyre, Rusty B., Rene M. Paulson, and Charles G. Lord. 2003. "Alleviating Women's Mathematics Stereotype Threat Through Salience of Group Achievements." *Journal of Experimental Social Psychology* 39: 83–90.

McLoyd, Vonnie C., Ana Mari Cauce, David Takeuchi, and Leon Wilson. 2000. "Marital Processes and Parental Socialization in Families of Color: A Decade Review of Research." *Journal of Marriage and the Family* 62 (November): 1070–1093.

McManus, Patricia A., and Thomas A. DiPrete. 2001. "'Losers and Winners': The Financial Consequences of Separation and Divorce for Men." *American Sociological Review* 66 (April): 246–268.

Mead, George Herbert. 1934. *Mind, Self, and Society.* Chicago. IL: University of Chicago Press.

Meng, Susan. 2001. "Pet Sanctuary." *Forbes* (October 8): 236.

Meredith, Martin. 2003. *Elephant Destiny: Biography of an Endangered Species in Africa.* New York: HarperCollins.

Mernissi, Fataima. 1987. *Beyond the Veil: Male-Female Dynamics in Modern Muslim Society.* Bloomington, IN: Indiana University Press.

Merton, Robert K. 1957. *Social Theory and Social Structure.* New York: Free Press.

Merton, Robert K. 1968. "Social Structure and Anomie." *American Sociological Review* 3: 672–682.

Merton, Robert, and Alice K. Rossi. 1950. "Contributions to the Theory of Reference Group Behavior." Pp. 279–334 in *Continuities in Social Research Studies, Scope and Method of "The American Soldier,* edited by Robert K. Merton and Paul F. Lazarsfeld. New York: Free Press.

Messerschmidt, James W. 1997. *Crime as Structured Action: Gender, Race, Class and Crime in the Making.* Thousand Oaks, CA: Sage.

Messner, Michael A. 2002. *Taking the Field: Women, Men, and Sports.* Minneapolis, MN: University of Minnesota Press.

Metz, Michael A., B. R. Rosser-Simon, and Nancy Strapko. 1994. "Differences in Conflict-Resolution Styles Among Heterosexual, Gay, and Lesbian Couples." *Journal of Sex Research* 31: 293–308.

Meyer, David S., and Joshua Gamson. 1995. "The Challenge of Cultural Elites: Celebrities and Social Movements." *Sociological Inquiry* 65 (Spring): 181–206.

Meyer, Madonna Harrington. 1994. "Gender, Race, and the Distribution of Social Assistance: Medicaid Use Among the Elderly." *Gender & Society* 8 (March): 8–28.

Mickelson, Roslyn Arlin (ed.) 2000. *Children on the Streets of the Americas: Globalization, Homelessness, and Education in the United States, Brazil, and Cuba.* New York: Routledge.

Milgram, Stanley. 1974. *Obedience to Authority: An Experimental View.* New York: Harper & Row.

Milkie, Melissa A. 1999. "Social Comparisons, Reflected Appraisals, and Mass Media: The Impact of Pervasive Beauty Images on Black and White Girls' Self-Concepts." *Social Psychology Quarterly* 62 (June): 190–210.

Milkie, Melissa A. 2002. "Gendered Division of Childrearing: Ideals, Realities, and the Relationship to Parental Well-being." *Sex Roles* 47 (July): 21–38.

Milkie, Melissa, and Pia Peltola. 1999. "Playing All the Roles: Gender and the Work—Family Balancing Act." *Journal of Marriage and the Family* 61 (May): 476–490.

Miller, Eleanor. 1986. *Street Women.* Philadelphia, PA: Temple University Press.

Miller, Eleanor. 1991. "Jeffrey Dahmer, Racism, Homophobia, and Feminism." *SWS Network News* 8 (December): 2.

Miller, Matthew, and Tatiana Seraphin. 2006. "America's 400 Richest." *Forbes,* September 21. http://www.forbes.com

Miller, Susan L. 1997. "The Unintended Consequences of Current Criminal Justice Policy." Talk presented at Research on Women Series, University of Delaware, Newark, DE.

Mills, C. Wright. 1956. *The Power Elite.* New York: Oxford University Press.

Mills, C. Wright. 1959. *The Sociological Imagination.* New York: Oxford University Press.

Milner, Murray, Jr. 2004. *Freaks, Geeks, and Cool Kids: American Teenagers, Schools, and the Culture of Consumption.* New York: Routledge.

Miner, Horace. 1956. "Body Ritual Among the Nacirema." *American Anthropologist* 58: 503–507.

Mintz, Beth, and Michael Schwartz. 1985. *The Power Structure of American Business.* Chicago, IL: University of Chicago Press.

Mirandé, Alfredo. 1979. "Machismo: A Reinterpretation of Male Dominance in the Chicano Family." *The Family Coordinator* 28: 447–449.

Mirandé, Alfredo. 1985. *The Chicano Experience.* Notre Dame, IN: Notre Dame University Press.

Mishel, Lawrence, Jared Bernstein, and Sylvia Allegretto. 2005. *The State of Working America 2004/2005.* Washington, DC: Economic Policy Institute.

Misra, Joy, Stephanie Moller, and Maria Karides. 2003. "Envisioning Dependency: Changing Media Depictions of Welfare in the 20th Century." *Social Problems* 50 (November): 482–504.

Mitchell, G., Stephanie Obradovich, Fred Herring, Chris Tromberg, and Alysson L. Burns. 1992. "Reproducing Gender in Public Places: Adults' Attention to Toddlers in Three Public Locales." *Sex Roles* 26 (September): 323–330.

Mizruchi, Mark S. 1992. *The Structure of Corporate Political Action: Interfirm Relations and Their Consequences.* Cambridge, MA: Harvard University Press.

Mizutami, Osamu. 1990. *Situational Japanese.* Tokyo: The Japan Times.

Moen, Phyllis, Jungmeen E. Kim, and Heather Hofmeister. 2001. "Couples' Work/Retirement Transitions, Gender, and Marital Quality." *Social Psychology Quarterly* 64 (March): 55–71.

Moen, Phyllis. 2003. *It's About Time: Couples and Careers.* Ithaca, NY: Cornell University Press.

Moen, Phyllis with Donna Dempster-McClain, Joyce Altobelli, Wipas Wimonsate, Lisa Dahl, Patricia Roehling, and Stephen Sweet. 2004. *The New "Middle" Workforce.* The Bronfenbrenner Life Course Center and Cornell Careers Institute, www.lifecourse.cornell

Moller, Laura C., Shelley Hymel, and Kenneth H. Rubin. 1992. "Sex Typing in Play and Popularity in Middle Childhood." *Sex Roles* 26 (April): 331–353.

Montada, L., and M. Lerner, Jr. (eds.). 1998. *Responses to Victimization and Beliefs in a Just World.* New York: Plenum.

Montagne, Peter. 1995. "Review of Julian L. Simon, ed., *The State of Humanity.*" Cambridge, MA: Blackwell.

Montgomery, James D. 1992. "Job Search and Network Composition: Implications of the Strength of Work Ties Hypothesis." *American Sociological Review* 57 (October): 586–596.

Moore, David W., and Joseph Carroll. 2004. "Support for Gay Marriage/Civil Unions Edges Upward." *The Gallup Poll.* Princeton, NJ: Gallup Organization. Website: www.gallup.com

Moore, David W. 2005. (August 15). "Gender Stereotypes Prevail on Working outside the Home." Princeton, NJ: The Gallup Poll.

Moore, Gwen. 1979. "The Structure of a National Elite Network." *American Sociological Review* 44 (October): 673–692.

Moore, Joan. 1976. *Hispanics in the United States.* Englewood Cliffs, NJ: Prentice Hall.

Moore, Joan W., and John M. Hagedorn. 1996. "What Happens to Girls in Gangs?" Pp. 205–218 in *Gangs in America,* edited by C. Ronald Huff. Thousand Oaks, CA: Sage.

Moore, Robert B. 1992. "Racist Stereotyping in the English Language." Pp. 317–328 in *Race, Class, and Gender: An Anthology,* 2nd ed., edited by Margaret L. Andersen and Patricia Hill Collins. Belmont, CA: Wadsworth.

Moore, Valerie A. 2001. "'Doing' Racialized and Gendered Age to Organize Peer Relations: Observing Kids in Summer Camp." *Gender & Society* 15 (December): 835–858.

Moreland, Richard L., and Scott R. Beach. 1992. "Exposure Effects in the Classroom: The Development of Affinity among Students." *Journal of Experimental Social Psychology* 28: 255–276.

Morgan, Lewis H. 1877. *Ancient Society, or Researches in the Lines of Human Progress, from Savagery Through Barbarism to Civilization.* Cambridge, MA: Harvard University Press.

Morlan, Patricia A. 2005. *Are We Taught to Blame the Poor? Attributions for Poverty: Sociodemographic Predictors and the Effects of Higher Education.* Senior thesis, Princeton University.

Morning, Ann. 2005. "Race." *Contexts* 4 (Fall): 44–46.

Morris, Aldon. 1984. *The Origins of the Civil Rights Movement: Black Communities Organizing for Change.* New York: Free Press.

Morris, Aldon D. 1999. "A Retrospective on the Civil Rights Movement: Political and Intellectual Landmarks." *Annual Review of Sociology* 25: 517–539.

Morris, Aldon, and Carol McClurg Mueller (eds.). 1992. *Frontiers in Social Movement Theory.* New Haven, CT: Yale University Press.

Moskos, Charles. 1988. *Soldiers and Sociology.* Alexandria, VA: U.S. Army Research Institute for the Social and Behavioral Sciences.

Moskos, Charles C., and John Sibley Butler. 1996. *All That We Can Be: Black Leadership and Racial Integration the Army Way.* New York: Basic Books.

Moynihan, Daniel P. 1965. *The Negro Family: The Case for National Action.* Washington, DC: Office of Policy Planning and Research, U.S. Department of Labor.

Muller, Thomas, and Thomas J. Espenshade. 1985. *The Fourth Wave: California's Newest Immigrants.* Washington, DC: The Urban Institute Press.

Myers, Steven Lee. 2000. "Survey of Troops Finds Antigay Bias Common in Service." The *New York Times* (March 24): 1.

Myers, Walter D. 1998. *Amistad Affair.* New York: NAL/Dutton.

Myerson, Allen R. 1998. "Rating the Bigshots: Gates vs. Rockefeller." *The New York Times* (May 24): 4.

Myrdal, Gunnar. 1944. *An American Dilemma: The Negro Problem and Modern Democracy.* 2 Vols. New York: Harper and Row.

NAACP Legal Defense and Education Fund. 2006. *Annual Report.* Washington, DC: NAACP Legal Defense Fund.

Nack, Adina. 2000. "Damaged Goods: Women Managing the Stigma of AIDS." *Deviant Behavior* 21: 95–121.

Nagel, Joane. 1996. *American Indian Ethnic Renewal: Red Power and the Resurgence of Identity and Culture.* New York: Oxford University Press.

Nagel, Joane. 2003. *Race, Ethnicity, and Sexuality: Intimate Intersections, Forbidden Frontiers.* New York: Oxford University Press.

Namboodiri, Krishnan. 1988. "Ecological Demography: Its Place in Sociology." *American Sociological Review* 53 (August): 619–633.

Nanda, Serena. 1998. *Neither Man Nor Woman: The Hijras of India.* Belmont, CA: Wadsworth.

Nash, John F. 1951. "Non-Cooperative Games." *Annals of Mathematics* 54: 286–295.

National Cancer Institute. 2002. *Cancer among Blacks and Other Minorities: Statistical Profiles.* Washington, DC: U. S. Department of Health and Human Services, National Institutes of Health.

National Center for Health Statistics. 2002. *National Vital Statistics Report* 50 (No. 6, March 21).

National Center for Health Statistics. 2004. *Health United States.* Rockville, MD: U.S. Department of Health and Human Services.

National Center for Health Statistics. 2006. National Health United States. Hyattsville, MD: U.S. Department of Health and Human Services.

National Center on Elder Abuse. 2007. NCEA Fact Sheet. Elder Abuse Prevention and Incidence. Washington, DC: National Center on Elder Abuse. www.elderabusecenter.org

National Coalition against Domestic Violence. 2001. www.ncadv.org

National Coalition for the Homeless. 2006. "Fact Sheets." Washington, DC www.nationalhomeless.org

National Law Center on Homelessness and Poverty. 2004. *Homelessness in the United States and the Human Right to Housing.* www.nlchp.org

National Opinion Research Center. 2004. *General Social Survey,* www.norc.uchicago.edu

National Television Violence Study. 1997. *National Television Violence Study.* Thousand Oaks, CA: Sage.

National Urban League. 2006. *The State of Black America.* New York: National Urban League.

Nee, Victor. 1973. *Longtime Californ': A Documentary Study of an American Chinatown.* New York: Pantheon Books.

Neppl, Tricia K., and Ann D. Murray. 1997. "Social Dominance and Play Patterns among Preschoolers: Gender Comparisons." *Sex Roles* 36 (March): 381–394.

Newman, Katherine S. 2006. *Rampage: The Social Roots of School Shootings.* New York: Basic Books.

Newman, Katherine. 1999. *No Shame in My Game: The Working Poor in the Inner City.* New York: Russell Sage Foundation/Vintage Books.

Newport, Frank. 2000. "Women's Most Pressing Concerns Today are Money, Family, Health, and Stress." *The Gallup Poll Monthly* (March): 40–41.

Newport, Frank. 2001. "American Attitudes Toward Homosexuality Continue to Be More Tolerant." *The Gallup Poll* (June 4): Princeton, NJ: Gallup Organization.

Niebuhr, Gustav. 1998. "Makeup of American Religion Is Looking More Like Mosaic, Data Say." *The New York Times* (April 12): 14.

Nolan, Patrick, and Gerhard Lenski. 2005. *Human Societies.* New York: Paradigm Publishers.

Norris, Pippa, and Ronald Inglehart. 2002. "Islamic Culture and Democracy: Testing the 'Clash of Civilizations' Thesis." *Comparative Sociology* 1: 235–263.

Oakes, Jeannie. 1985. *Keeping Track: How Schools Structure Inequality.* New Haven, CT: Yale University Press.

Oakes, Jeannie. 1990. "Multiplying Inequalities: The Effects of Race, Social Class and Tracking on Opportunities to Learn Mathematics and Science." RAND for National Science Foundation.

Oakes, Jeannie, and Martin Lipton. 1996. "Developing Alternatives to Tracking and Grading." Pp. 168–200 in *Educating A New Majority: Transforming America's Educational System for Diversity,* edited by Laura I. Rendon and Richard O. Hope. San Francisco, CA: Jossey-Bass.

Ogburn, William F. 1922. *Social Change with Respect to Cultural and Original Nature.* New York: B. W. Huebsch.

O'Leary, Carol. 2002. "The Kurds of Iraq: Recent History, Future Prospects." *Middle East Review of International Affairs* 6 (December 2002).

Oliver, Melvin, and Thomas M. Shapiro. 1995. *Black Wealth / White Wealth: A New Perspective on Racial Inequality.* New York: Routledge.

Oliver, Melvin L., and Thomas M. Shapiro. 2001. "Wealth and Racial Stratification." Pp. 222–240 in *America Becoming: Racial Trends and Their Consequences,* edited by Neil Smelser, William Julius Wilson, and Faith Mitchell. Washington, DC: National Academies Press.

Ollivier, Michele. 2000. "'Too Much Money off Other People's Backs': Status in Late Modern Societies." *Canadian Journal of Sociology* 25 (Fall): 441–470.

Omi, Michael, and Howard Winant. 1994. *Racial Formation in the United States,* 2nd ed. New York: Routledge.

O'Neil, John. 2002. "Parent Smoking and Teenage Sex." *The New York Times,* September 3: F7.

Ornstein, Norman J., Thomas E. Mann, and Michael J. Malbin. 1996. *Vital Statistics on Congress, 1995–1996.* Washington, DC: Congressional Quarterly, Inc.

Orzechowski, Shawna, and Peter Sepielli. 2003. *Net Worth and Asset Ownership of Households: 1998 and 2000.* Washington, DC: U.S. Census Bureau, P70–88.

Owens, Sarah E. 1998. "The Effects of Race, Gender, Their Interactions, and Selected School Variables upon Educational Aspirations and Achievements." Senior thesis, Department of Sociology, Princeton University.

Padavic, Irene, and Barbara Reskin. 2002. *Women and Men at Work,* 2nd ed. Thousand Oaks, CA: Sage.

Page, Charles H. 1946. "Bureaucracy's Other Face." *Social Forces* 25 (October): 89–94.

Pager, Devah. 2005. "Double Jeopardy: Race, Crime, and Getting a Job." *Wisconsin Law Review* 2: 617–660.

Pager, Devah. 2003. "The Mark of a Criminal Record." *American Journal of Sociology* 108 (5): 937–975.

Pain, Emil. 2002. "The Social Nature of Extremism and Terrorism." *Social Sciences* 33: 55–68.

Pantoja, Adrian D., and Gary M. Segura. 2003. "Does Ethnicity Matter? Descriptive Representation in Legislatures and Political Alienation among Latinos." *Social Science Quarterly* 84 (June): 441–460.

Park, Robert E., and Ernest W. Burgess. 1921. *Introduction to the Science of Society.* Chicago, IL: University of Chicago Press.

Parreñas, Rhacel Salazar. 2001. *Servants of Globalization: Women, Migration, and Domestic Work.* Stanford, CA: Stanford University Press.

Parsons, Talcott (ed.). 1947. *Max Weber: The Theory of Social and Economic Organization.* New York: Free Press.

Parsons, Talcott. 1951a. *The Social System.* Glencoe, IL: Free Press.

Parsons, Talcott. 1951b. *Toward a General Theory of Action.* Cambridge, MA: Harvard University Press.

Parsons, Talcott. 1966. *Societies: Evolutionary and Comparative Perspectives.* Englewood Cliffs, NJ: Prentice Hall.

Paternoster, Raymond, and Robert Brame. 2003. "An Empirical Analysis of Maryland's Death Sentencing System with Respect to the Influence of Race and Legal Jurisdiction." www.urhome.umd.edu/newsdesk

Pattillo-McCoy, Mary. 1999. *Black Picket Fences: Privilege and Peril among the Black Middle Class.* Chicago, IL: University of Chicago Press.

Paulus, P. B., T. S. Larey, and M. T. Dzindolet. 2001. "Creativity in Groups and Teams." Pp. 319–338 in M. E. Turner, ed., *Groups At Work: Theory and Research.* Mahwah, NJ: Earlbaum.

Pavalko, Eliza K., Krysia N. Mossakowski, and Vanessa J. Hamilton. 2003. "Does Perceived Discrimination Affect Health? Longitudinal Relationships between Work Discrimination and Women's Physical and Emotional Health." *Journal of Health and Social Behavior* 43 (March): 18–33.

Payne, Deborah M., John T. Warner, and Roger D. Little. 1992. "Tied Migration and Returns to Human Capital: The Case of Military Wives." *Social Science Quarterly* 73 (June): 324–339.

Pedraza, Silvia. 1996a. "Cuba's Refugees: Manifold Migrations." Pp. 263–279 in *Origins and Destinies: Immigration, Race, and Ethnicity in America,* edited by Silvia Pedraza and Rubén Rumbaut. Belmont, CA: Wadsworth.

Pedraza, Silvia, and Rubén G. Rumbaut. (eds.). 1996. *Origins and Destinies: Immigration, Race, and Ethnicity in America.* Belmont, CA: Wadsworth.

Pellow, David N. 2003. *Garbage Wars: The Struggle for Environmental Justice in Chicago.* Cambridge, MA: MIT Press.

Pennock-Roman, Maria. 1994. "College Major and Gender Differences in the Prediction of College Grades." College Board Research Report No. 94-2, ETS Research Reports No. 94-24.

Peoples, James, and Garrick Bailey. 2003. *Humanity: An Introduction to Cultural Anthropology,* 6th ed. Belmont, CA: Wadsworth.

Peralta, Robert L. 2002. *Getting Trashed in College: Doing Alcohol, Doing Gender, Doing Violence,* Ph.D. dissertation, University of Delaware.

Perez, Denise N. 1996. *A Case for Derailment: Tracking in the American Public School System as an Obstacle to the Improvement of Minority Education.* Senior thesis, Department of Sociology, Princeton University.

Perrow, Charles. 1986. *Complex Organization: A Critical Essay,* 3rd ed. New York: Random House.

Perrow, Charles. 1994. "The Limit of Safety: The Enhancement of a Theory of Accidents." *Journal of Contingencies and Crisis Management* 22: 212–220.

Perrow, Charles, David M. Mednicoff, and Kathleen Tierney. 2005. "A Symposium on the 9/11 Commission Report." *Contemporary Sociology* 34 (March), pp. 99–120.

Perry-Jenkins, Maureen, Rena L. Repetti, and Anne C. Crouter. 2000. "Work and Family in the 1990s." *Journal of Marriage and the Family* 62 (November): 981–998.

Persell, Caroline Hodges. 1990. *Understanding Society: An Introduction to Sociology.* New York: Harper & Row.

Pescosolido, Bernice A., Steven A. Tuch, and Jack K. Martin. 2001. "The Profession of Medicine and the Public: Examining Americans' Changing Confidence in Physician Authority from the Beginning of the 'Health Care Crisis' to the Era of Health Care Reform." *Journal of Heath and Social Behavior* 42 (March): 1–16.

Petersen, Trond, Ishak Saporta, and Mark-David L. Seidel. 2000. "Offering A Job: Meritocracy and Social Networks." *American Journal of Sociology* 106 (November): 763–816.

Peterson, Richard R. 1996a. "A Re-Evaluation of the Economic Consequences of Divorce." *American Sociological Review* 61 (June): 528–536.

Peterson, Richard R. 1996b. "A Re-Evaluation of the Economic Consequences of Divorce: Reply to Weitzman." *American Sociological Review* 61 (June): 539–540.

Pettigrew, Thomas F. 1992. "The Ultimate Attribution Error: Extending Allport's Cognitive Analysis of Prejudice." Pp. 401–419 in *Readings About the Social Animal,* edited by Elliott Aronson. New York: Freeman.

Pew Internet and American Life Project. 2006. Teens and Parents Survey. www.pewinternet.org

Piaget, Jean. 1926. *The Language and Thought of the Child.* New York: Harcourt.

Polce-Lynch, Mary, Barbara J. Myers, Wendy Kliewer, and Christopher Kilmartin. 2001. "Adolescent Self-Esteem and Gender: Exploring Relations to Sexual Harassment, Body Image, Media Influence, and Emotional Expression." *Journal of Youth and Adolescence* 30 (April): 225–244.

Pollock, Philip H., and M. Elliot Vittas. 1995. "Who Bears the Burdens of Environmental Pollution: Race, Ethnicity, and Environmental Equity in Florida." *Social Science Quarterly* 76 (June): 294–310.

Portes, Alejandro, and Rubén G. Rumbaut. 1996. *Immigrant America: A Portrait,* 2nd ed. Berkeley, CA: University of California Press.

Portes, Alejandro, and Rubén G. Rumbaut. 2001. *Legacies: The Story of the Immigrant Second Generation.* Berkeley, CA: University of California Press.

Portes, Alejandro. 2002. "English-Only Triumphs, But the Costs Are High." *Contexts* 1 (February): 10–15.

Press, Andrea. 2002. "The Paradox of Talk." *Contexts* 1 (Fall–Winter): 69–70.

Press, Eyal. 1996. "Barbie's Betrayal." *The Nation* (December 30): 11–16.

Press, Julie E., and Eleanor Townsley. 1998. "Wives and Husbands' Reporting: Gender, Class, and Social Desirability." *Gender & Society* 12 (April): 188–218.

Presser, Harriet B. 2004. "The Economy that Never Sleeps." *Contexts* 3 (Spring): 42–49.

Preves, Sharon E. 2003. *Interzex and Identity: The Contested Self.* New Brunswick, NJ: Rutgers University Press.

Pruitt, D. G. 1971. "Choice Shifts in Group Discussion." *Journal of Personality and Social Psychology* 20 (December): 339–360.

Prus, Robert, and C. R. D. Sharper. 1991. *Road Hustler.* New York: Kauffman and Greenbery.

Punch, Maurice. 1996. *Dirty Business: Exploring Corporate Misconduct; Analysis and Cases.* Thousand Oaks, CA: Sage.

Putnam, Robert D. 2000. *Bowling Alone: The Collapse and Revival of American Community.* New York: Simon and Schuster.

Raboteau, Albert J. 1978. *Slave Religion: The "Invisible Institution" in the Antebellum South.* New York: Oxford University Press.

Raley, Sara B., Marybeth J. Mattingly, and Suzanne M. Bianchi. 2006. "How Dual Are Dual-Income Couples? Documenting Change from 1970 to 2001." *Journal of Marriage and Family* 68(1): 11–28.

Rampersad, Arnold. 1986. *The Life of Langston Hughes: Vol. I: 1902–1941. I, Too, Sing America.* New York: Oxford University Press.

Rampersad, Arnold. 1988. *The Life of Langston Hughes: Vol. II: 1941–1967. I Dream A World.* New York: Oxford University Press.

Rashid, Ahmed. 2000. *Taliban: Militant Islam, Oil, and Fundamentalism in Central Asia.* New Haven, CT: Yale University Press.

Read, Jen'nan Ghazal. 2003. "The Sources of Gender Role Attitudes among Christian and Muslim Arab-American Women." *Sociology of Religion* 64 (Summer): 207–222.

Read, Piers Paul. 1974. *Alive: The Story of the Andes Survivors.* Philadelphia, PA: Lippincott.

Reich, Robert. 1991. *The Work of Nations: Preparing Ourselves for 21st Century Capitalism.* New York: Knopf.

Reiman, Jeffrey. 2004. Executive Pay." *BusinessWeek* (April 17). www.businessweek.com

Reiman, Jeffrey. 2007. *The Rich Get Richer and the Poor Get Prison.* 8th edition. Boston, MA: Allyn and Bacon.

Reingold, Jennifer. 2000. "Executive Pay." *Business Week,* April 17, www.businessweek.com

Rendon, Laura I., and Richard O. Hope (eds.). 1996. *Educating a New Majority: Transforming America's Educational System for Diversity.* San Francisco, CA: Jossey-Bass.

Renzetti, Claire. 1992. *Violent Betrayal: Partner Abuse in Lesbian Relationships.* Newbury Park, CA: Sage.

Reskin, Barbara. 1988. "Bringing the Men Back In: Sex Differentiation and the Devaluation of Women's Work." *Gender & Society* 2 (March): 58–81.

Reskin, Barbara F., and Irene Padavic. 1988. "Supervisors as Gatekeepers: Male Supervisors' Responses to Women's Integration in Plant Jobs." *Social Problems* 35 (December): 536–550.

Rich, Adrienne. 1980. "Compulsory Heterosexuality and Lesbian Existence." *Signs* 5 (Summer): 631–660.

Ridley, Matt. 2003. *Nature via Nurture: Genes, Experience and What Makes Us Human.* New York: Harper Collins.

Rieff, David. 1993. "A Global Culture?" *World Policy Journal* 10 (Winter): 73–81.

Riesman, David. 1970[1950]. *The Lonely Crowd: A Study of the Changing American Character.* New Haven, CT: Yale University Press.

Rifkin, Jeremy. 1989. *Entropy: Into the Greenhouse World,* rev. ed. New York: Bantam Books.

Rifkin, Jeremy. 1998. "The Biotech Century: Human Life as Intellectual Property." *The Nation* 266 (April): 11–89.

Rindfuss, Ronald R., Elizabeth C. Cooksey, and Rebecca L. Sutterlin. 1999. "Young Adult Occupational Achievement: Early Expectations versus Behavioral Reality." *Work and Occupations* 26 (May): 220–263.

Risman, Barbara, and Pepper Schwartz. 2002. "After the Sexual Revolution: Gender Politics in Teen Dating." *Contexts* 1 (Spring): 16–24.

Ritzer, George. 2000. *The McDonaldization of Society: An Investigation into the Changing Character of Contemporary Social Life,* 3rd ed. Thousand Oaks, CA: Pine Forge Press.

Robbins, Thomas. 1988. *Cults, Convents, and Charisma: The Sociology of New Religious Movements* Beverly Hills, CA: Sage.

Roberts, Dorothy. 1997. *Killing the Black Body: Race, Reproduction and the Meaning of Liberty.* New York: Vintage Books.

Robertson, Tatsha, and Garrance Burke. 2001. "Fighting Terror: Concerned Family;" "Shock, Worry for Family of U.S. Man Captured with Taliban." *The Boston Globe,* (December 4): A1.

Robey, Bryant. 1982. "A Guide to the Baby Boom." *American Demographics* 4 (September): 16–21.

Robinson, Dawn T., and Lynn Smith-Lovin. 2001. "Getting a Laugh: Gender, Status, and Humor in Task Discussions." *Social Forces* 80 (September): 123–158.

Robinson, Jo Ann Gibson. 1987. *The Montgomery Bus Boycott and the Women Who Started It.* Knoxville, TN: The University of Tennessee Press.

Rodney, Walter. 1974. *How Europe Underdeveloped Africa.* Washington, DC: Howard University Press.

Rodney, Walter. 1981. *How Europe Underdeveloped Africa,* revised edition. Washington, DC: Howard University Press.

Rodriguez, Clara E. 1989. *Puerto Ricans: Born in the U.S.A.* Boston, MA: Unwin Hyman.

Rodriguez, Clara E. 2006. "Changing Race." Pp. 22–25 in Elizabeth Higginbotham and Margaret L. Andersen, eds., *Race and Ethnicity in Society: The Changing Landscape.* Belmont, CA: Wadsworth.

Roethlisberger, Fritz J., and William J. Dickson. 1939. *Management and the Worker.* Cambridge, MA: Harvard Univesity Press.

Romain, Suzanne. 1999. *Communicating Gender.* Mahwah, NJ: Erlbaum.

Roof, Wade Clark. 1999. *Spiritual Marketplace.* Princeton, NJ: Princeton University Press.

Root, Maria P. P. (ed.). 1996. *The Multiracial Experience: Racial Borders as the New Frontier.* Thousand Oaks, CA: Sage.

Root, Maria. P. P. 2001. *Love's Revolution: Interracial Marriage.* Philadelphia, PA: Temple University Press.

Rose, Tricia. 1994. *Black Noise: Rap Music and Black Culture in Contemporary America.* Middletown, CT: Wesleyan University Press.

Rosenau, Pauline Marie. 1992. *Post-Modernism and the Social Sciences: Insights, Inroads, and Intrusions.* Princeton, NJ: Princeton University Press.

Rosenberg, Yuvai. 2004. "Lost Youth." *American Demographics* (March): 17–19.

Rosengarten, Danielle. 2000. "Modern Times." *Dollars & Sense* (September): 4.

Rosenhan, David L. 1973. "On Being Sane in Insane Places." *Science* 179 (January 19): 250–258.

Rosenthal, Robert, and Lenore Jacobson. 1968. *Pygmalian in the Classroom: Teacher Expectations and Pupils' Intellectual Development.* New York: Holt, Rinehart and Winston.

Rostow, W.W. 1978. *The World Economy: History and Prospect.* Austin, TX: University of Texas Press.

Rubin, Gayle. 1975. "The Traffic in Women." Pp. 157–211 in *Toward an Anthropology of Women,* edited by Rayna Reiter. New York: Monthly Review Press.

Ruef, Martin, Howard Aldrich, and N. Carter. 2003. "The Structure of Foundation Teams: Homophily, Strong Ties, and Isolation Among U.S. Entrepreneurs." *American Sociological Review* 68: 195–222.

Rueschmeyer, Dietrich, and Theda Skocpol. 1996. *States, Social Knowledge, and the Origins of Modern Social Policies.* Princeton, NJ: Princeton University Press.

Rumbaut, Rubén G. 1996. "Origins and Destinies: Immigration, Race, and Ethnicity in Contemporary America." Pp. 21–42 in *Origins and Destinies: Immigration, Race, and Ethnicity in America,* edited by Sylvia Pedraza and Rubén G. Rumbaut. Belmont, CA: Wadsworth.

Rupp, Leila J., and Verta Taylor. 2003. *Drag Queens at the 801 Cabaret.* Chicago, IL: University of Chicago Press.

Rust, Paula C. 1993. "'Coming Out' in the Age of Social Constructionism: Sexual Identity Formation among Lesbian and Bisexual Women." *Gender & Society* 7 (March): 50–77.

Rust, Paula. 1995. *Bisexuality and the Challenge to Lesbian Politics.* New York: New York University Press.

Ryan, Charlotte. 1996. "Battered in the Media: Mainstream News Coverage of Welfare Reform." *Radical America* 26 (August): 29–41.

Ryan, William. 1971. *Blaming the Victim.* New York: Pantheon.

Rymer, Russ. 1993. *Genie: A Scientific Study.* New York: HarperCollins.

Saad, Lydia. 2002. "There's No Place Like Home to Spend an Evening, Say Most Americans" *The Gallup Poll.* Princeton, NJ: Gallup Organization, www.gallup.com

Saad, Lydia. 2007. "Tolerance for Gay Rights at High-Water Mark." *The Gallup Poll.* Princeton, NJ: The Gallup Organization.

Sadker, Myra, and David Sadker. 1994. *Failing at Fairness: How America's Schools Cheat Girls.* New York: Scribner's.

Saguy, Abigail. 2003. *What Is Sexual Harassment? From Capitol Hill to the Sorbonne.* Berkeley, CA: University of California Press.

Saks, M. J., and M. W. Marti. 1997. "A Meta-Analysis of the Effects of Jury Size." *Law and Human Behavior* 21: 451–467.

Sampson, Robert J. 1987. "Urban Black Violence: The Effect of Male Joblessness and Family Disruption." *American Journal of Sociology* 93: 348–382.

Sanchez, Laura, Wendy D. Manning, and Pamela J. Smock. 1998. "Sex-Specialized or Collaborative Mate Selection? Union Transitions among Cohabitors." *Social Science Research* 27 (September): 280–304.

Sanchez, Lisa Gonzalez. 1999. "Reclaiming Salsa." *Cultural Studies* 13 (April): 237–250.

Sanday, Peggy. 2002. *Women at the Center: Life in a Modern Matriarchy.* Ithaca, NY: Cornell University Press.

Sandnabba, N. Kenneth, and Christian Ahlberg. 1999. "Parents' Attitudes and Expectations About Children's Cross-Gender Behavior." *Sex Roles* 40 (February): 249–263.

Santelli, John et al. 2007. "Explaining Recent Declines in Adolescent Pregnancy in the United States: The Contribution of Abstinence and Increased Contraceptive Use." *American Journal of Public Health American Journal of Public Health* 97(1): 150–156.

Sapir, Edward. 1921. *Language: An Introduction to the Study of Speech.* New York: Harcourt Brace.

Sassen, Saskia. 1991. *The Global City: New York, London, Tokyo.* Princeton, NJ: Princeton University Press.

Saunders, Laura. 1990. "America's Richest Congressmen." *Forbes* 145 (February 19): 44–45.

Sawhill, Isabel, and Sara McClanahan (eds.). 2006. "Introducing the Issue." *The Future of Children* 16 (Fall): 3–17.

Scarpitti, Frank R., Margaret L. Andersen, and Laura O'Toole. 1997. *Social Problems,* 3rd ed. New York: Addison Wesley Longman.

Schacht, Steven P. 1996. "Misogyny on and off the 'Pitch': The Gendered World of Male Rugby

Players." *Gender & Society* 10 (October): 550–565.

Scheff, Thomas J. 1966. *Being Mentally Ill: A Sociological Theory.* Chicago, IL: Aldine.

Scheff, Thomas. 1984. *Being Mentally Ill: A Sociological Theory,* 2nd ed. New York: Aldine de Gruyter.

Scheid, Teresa L. 2003. "Managed Care and the Rationalization of Mental Health Services." *Journal of Health and Social Behavior* 44 (June): 142–161.

Schlosser, Eric. 2001. *Fast Food Nation: The Dark Side of the All-American Meal.* New York: Houghton Mifflin.

Schmitt, Eric. 2001. "Segregation Growing among U.S. Children." *The New York Times* (May 6): 28.

Schmitt, Frederika E., and Patricia Yancey Martin. 1999. "Unobtrusive Mobilization by an Institutionalized Rape Crisis Center: 'It Comes From the Victims.'" *Gender & Society* 13: 364–384.

Schneider, Barbara, and David Stevenson. 1999. *The Ambitious Generation: Motivated but Directionless.* New Haven, CT: Yale University Press.

Schneider, Beth. 1984. "Peril and Promise: Lesbians' Workplace Participation." Pp. 211–230 in *Women Identified Women,* edited by Trudy Darty and Sandee Potter. Mountain View, CA: Mayfield.

Schumann, Howard, Charlotte Steeh, Lawrence Bobo, and Maria Krysan. 1997. *Racial Attitudes in America: Trends and Interpretations.* Cambridge, MA: Harvard University Press.

Schur, Edwin M. 1984. *Labeling Women Deviant.* New York: Random House.

Schwartz, John, and Matthew L. Wald. 2003. "NASA's Failings Go Far Beyond Foam Hitting Shuttle, Panel Says." *The New York Times* (June 7): 1 and 12.

Schwartz, Pepper, and Virginia Rutter. 1998. *The Gender of Sexuality.* Thousand Oaks, CA: Sage.

Scully, Diana. 1990. *Understanding Sexual Violence: A Study of Convicted Rapists.* Boston, MA: Unwin Hyman.

Seidman, Steven. 2003. *The Social Construction of Sexuality.* New York: Oxford University Press.

Sen, Amartya. 2000. "Population and Gender Equity." *The Nation* (July 24–31): 16–18.

Sen, Amartya. 2002. "How to Judge Globalism." *The American Prospect,* Special Supplement (Winter): A2–A6.

Settersten, Richard A., Jr., and Loren D. Lovegreen. 1998. "Educational Experiences throughout Adult Life: New Hopes or No Hope for Life-Course Flexibility?" *Research on Aging* 20 (July): 506–538.

Shaberoff, Philip. 1988. "Water Supplies in Ground Held Generally Safe." *The New York Times* (October 8): 1, 6.

Shapiro, Thomas M. 2005. *The Hidden Cost of Being African American: How Wealth Perpetuates Inequality.* New York: Oxford University Press.

Sharp, Susan F., Toni L. Terling-Watt, Leslie A. Atkins, Jay Trace Gilliam, and Anna Sanders. 2000. "Purging Behavior in a Sample of College Females: A Research Note on General Strain Theory and Female Deviance." *Deviant Behavior* 22: 171–188.

Shibutani, Tomatsu. 1961. *Society and Personality: An Interactionist Approach to Social Psychology.* Englewood Cliffs, NJ: Prentice Hall.

Shibutani, Tomatsu. 1966. *Improvised News.* Indianapolis, IN: Bobbs-Merrill.

Shih, Margaret, Todd L. Pittinsky, and Nalini Ambady. 1999. "Stereotype Susceptibility: Identiy Salience and Shifts in Quantitative Performance." *Psychological Science* 10 (January): 80–83.

Shilts, Randy. 1988. *And the Band Played On.* New York: Penguin.

Sigall, H., and N. Ostrove. 1975. "Beautiful but Dangerous: Effects of Offender Attractiveness and Nature of Crime on Judicial Judgment." *Journal of Personality and Social Psychology* 31: 410–414.

Sigelman, Lee, and Paul J. Wahlbeck. 1999. "Gender Proportionality in Intercollegiate Athletics: The Mathematics of Title IX Compliance." *Social Science Quarterly* 80 (September): 518–538.

Signorielli, Nancy. 1991. *A Sourcebook on Children and Television.* New York: Greenwood Press.

Signorielli, Nancy, and Aaron Bacue. 1999. "Recognition and Respect: A Content Analysis of Prime-Time Television Characters across Three Decades." *Sex Roles* 40 (April): 527–544.

Signorielli, Nancy, Douglas McLeod, and Elaine Healy. 1994. "Gender Stereotypes in MTV Commercials: The Beat Goes On." *Journal of Broadcasting and Electronic Media* 38 (Winter): 91–101.

Silverthorne, Zebulon A. and Vernon L Quinsey. 2000. "Sexual Partner Age Preferences of Homosexual and Heterosexual Men and Women." *Archives of Sexual Behavior* 29 (No. 1): 67–76.

Simile, Catherine. 1995. *Disaster Settings and Mobilization for Contentious Collective Action: Case Studies of Hurricane Hugo and the Loma Prieta Earthquake.* Unpublished dissertation, University of Delaware.

Simmel, Georg. 1902. "The Number of Members as Determining the Sociological Form of the Group." *The American Journal of Sociology* 8 (July): 1–46.

Simmons, Wendy W. 2001. "Majority of Americans Say More Women in Political Office Would Be Positive for the Country." The Gallup Poll. Princeton, NJ.: Gallup Organization. www.gallup.com

Simon, David R. 2007. *Elite Deviance,* 8th ed. Boston, MA: Allyn and Bacon.

Simons, Marlise. 1989. "The Amazon's Savvy Indians." *The New York Times* Magazine (February): 36ff.

Simons, Ronald L. 1996. "The Effect of Divorce on Adult and Child Adjustment." Pp. 3–20 in *Understanding Differences between Divorced and Intact Families.* Thousand Oaks, CA: Sage.

Simpson, George Eaton, and J. Milton Yinger. 1985. *Racial and Cultural Minorities: An Analysis of Prejudice and Discrimination,* 5th ed. New York: Plenum.

Sirvananadan, A. 1995. "La trahision des clercs. (Racism)." *New Statesman and Society* 8: 20–22.

Sjøberg, Gideon. 1965. *The Preindustrial City: Past and Present.* New York: Free Press.

Skocpol, Theda. 1979. *States and Social Revolutions: A Comparative Analyses of France, Russia, and China.* New York: Cambridge University Press.

Skocpol, Theda. 1992. *Protecting Soldiers and Mothers: The Origins of Social Policy in the United States.* Cambridge, MA: Belknap Press.

Slavin, Robert E. 1993. "Ability Grouping in the Middle Grades: Achievement Effects and Alternatives." *The Elementary School Journal* 93 (No. 5): 535–552.

Smelser, Neil J. 1992a. "Culture: Coherent or Incoherent." Pp. 3–28 in *Theory of Culture,* edited by R. Münch and N. J. Smelser. Berkeley, CA: University of California Press.

Smith, A. Wade. 1989. "Educational Attainment as a Determinant of Social Class Among Black Americans." *Journal of Negro Education* 58 (Summer): 416–429.

Smith, Barbara. 1983. "Homophobia: Why Bring It Up?" *Interracial Books for Children Bulletin* 14: 112–113.

Smith, Jesse L. 2006. "The Interplay Among Stereotypes, Performance-Avoidance Goals, and Women's Math Performance Expectations." *Sex Roles* 54 (3/4): 287–295.

Smith, M. Dwayne, Joel A. Devine, and Joseph F. Sheley. 1992. "Crime and Unemployment: Effects Across Age and Race Categories." *Sociological Perspectives* 35 (Winter): 551–572.

Smith, Richard B., and Robert A. Brown. 1997. "The Impact of Social Support on Gay Male Couples." *Journal of Homosexuality* 33: 39–61.

Smith, Tom W. 1992. "Changing Racial Labels: From 'Colored' to 'Negro' to 'Black' to 'African American.'" *Public Opinion Quarterly* 56: 496–512.

Smith, Tom W. 1999. "The Religious Right and Anti-Semitism." *Review of Religious Research* 40 (March): 244–258.

Snipp, C. Matthew. 1989. *American Indians: The First of This Land.* New York: Russell Sage Foundation.

Snipp, C. Matthew. 1996. "The First Americans: American Indians." Pp. 390–404 in *Origins and Destinies: Immigration, Race, and Ethnicity in America,* edited by Sylvia Pedraza and Rubén G. Rumbaut. Belmont, CA: Wadsworth.

Snow, David A., and Robert D. Benford. 1988. "Ideology, Frame Resonance, and Participant Mobilization." *International Social Movements Research* 1: 197–217.

Snow, David A., Susan G. Baker, Leon Anderson, and Michael Martin. 1986. "The Myth of Pervasive Mental Illness among the Homeless." *Social Problems* 33 (June): 407–423.

Sorokin, Pitrim. 1941. *The Crisis of Our Age.* New York: Dutton.

Sotirovic, Mira. 2000. "Effects of Media Use on Audience Framing and Support for Welfare." *Mass Communication & Society* 2–3 (Spring–Summer): 269–296.

Sotirovic, Mira. 2001. "Media Use and Perceptions of Welfare." *Journal of Communication* 51 (December): 750–774.

Spangler, Eve, Marsha A. Gordon, and Ronald Pipkin. 1978. "Token Women: An Empirical Test of Kanter's Hypothesis." *American Journal of Sociology* 84: 160–170.

Speed, A., and S.W. Gangestad. 1997. "Romantic Popularity and Mate Preferences: A Peer-Nomination Study." *Personality and Social Psychology Bulletin* 23: 928–936.

Spencer, Herbert. 1882. *The Study of Sociology.* London: Routledge.

Spengler, Oswald. 1932. *The Decline of the West.* New York: Knopf.

Spitzer, Steven. 1975. "Toward a Marxian Theory of Deviance." *Social Problems* 22: 638–651.

Sprock, June, and Carol Y. Yoder. 1997. "Women and Depression: An Update on the Report of the APA Task Force." *Sex Roles* 36 (March): 269–303.

Stacey, Judith, and Timothy J. Bibliarz. 2001. (How) Does the Sexual Orientation of Parents Matter?" *American Sociological Review* 66 (April): 159–183.

Stack, Carol. 1974. *All Our Kin: Strategies for Survival in a Black Community.* New York: Harper Colophon Books.

Starr, Paul. 1982. *The Social Transformation of American Medicine.* New York: Basic Books.

Stearns, Linda Brewster, and Charlotte Wilkinson Coleman. 1990. "Industrial and Labor Market Structures and Black Male Employment in the Manufacturing Sector." *Social Science Quarterly* 71 (June): 285–298.

Steele, Claude M. 1992. "Race and the Schooling of Black Americans." *The Atlantic Monthly* 269 (April): 68–78.

Steele, Claude M. 1996. "A Burden of Suspicion: How Stereotypes Shape the Intellectual Identities and Performance of Women and African-Americans." Paper read before the Princeton Conference on Higher Education, Princeton, NJ, February.

Steele, Claude M. 1999. "Thin Ice: Stereotype Threat and Black College Students." *The Atlantic Monthly* (August): 44–54.

Steele, Claude M., and Joshua Aronson. 1995. "Stereotype Vulnerability and African American Intellectual Performance." Pp. 409–421 in *Readings About the Social Animal,* 7th ed, edited by Elliot Aronson. New York: W. H. Freeman.

Steffensmeier, Darrell, and Stephen Demuth. 2000. "Ethnicity and Sentencing Outcomes in U.S. Federal Courts: Who Is Punished More Harshly?" *American Sociological Review* 65 (October): 705–729.

Steffensmeier, Darrell, Jeffery Ulmer, and John Kramer. 1998. "The Interaction of Race, Gender, and Age in Criminal Sentencing: The Punishment Cost of Being Young, Black and Male." *Criminology* 36 (November): 763–797.

Steinberg, Ronnie. 1992. "Gendered Instructions: Cultural Lag and Gender Bias in the Hay System of Job Evaluation." *Work and Occupations* 19 (November): 387–424.

Stemja, Marie-Louise, Sonja Olin Lauritzen, and Per Tillgren. 2004. "Social Thinking and Cultural Images: Teenagers' Notions of Tobacco Use." *Social Science & Medicine* 59: 573–83.

Stern, Jessica. 2003. *Terror in the Name of God: Why Religious Militants Kill.* New York: Ecco.

Stern, Kenneth S. 1996. *A Force upon the Plain.* New York: Simon and Schuster.

Sternberg, Robert J. 1988. *The Triarchic Mind: A New Theory of Human Intelligence.* New York: Penguin.

Stevens, William K. 1994. "Poor Lands Success in Cutting Birthrate Upsets Old Theories." *The New York Times* (January 2): 1, 8.

Stewart, Abigail J., Anne P. Copeland, Nia Lane Chester, Janet E. Malley, and Nicole B. Barenbaum. 1997. *Separating Together: How Divorce Transforms Families.* New York: Guilford Press.

Stewart, Susan D. 2001. "Contemporary American Stepparenthood: Integrating Cohabiting and Nonresident Stepparents." *Population Research and Policy* 20 (August): 345–364.

Stoller, Eleanor Palo, and Rose Campbell Gibson. 2000. *Worlds of Difference: Inequality in the Aging Experience.* Thousand Oaks, CA: Pine Forge Press.

Stombler, Mindy, and Irene Padavic. 1997. "Sister Acts: Resisting Men's Domination in Black and White Fraternity Little Sister Programs." *Social Problems* 44 (May): 257–275.

Stoner, J.A.F. 1961. *A Comparison of Individual and Group Decisions Involving Risk.* Unpublished master's thesis, MIT.

Stotik, Jeffrey, Thomas E. Shriver, and Sherry Cable. 1994. "Social Control and Movement Outcome: The Case of AIM." *Sociological Focus* 27 (February): 53–66.

Stover, R. G., and C. A. Hope. 1993. *Marriage, Family and Intimate Relationships.* New York: Harcourt Brace Jovanovich.

Sudnow, David N. 1967. *Passing On: The Social Organization of Dying.* Englewood Cliffs, NJ: Prentice Hall.

Sullivan, Maureen. 1996. "Rozzie and Harriet? Gender and Family Patterns of Lesbian Coparents." *Gender & Society* 12 (December): 747–767.

Sullivan, Teresa A., Elizabeth Warren, and Jay Lawrence Westbrook. 2000. *The Fragile Middle Class: Americans in Debt.* New Haven, CT: Yale University Press.

Suro, Roberto. 2000. "Beyond Economics." *American Demographics* (February): 48–55.

Surratt, Hilary L., and James A. Inciardi. 1998. "Unraveling the Concept of Race in Brazil." *Journal of Psychoactive Drugs* 30 (3): 255–260.

Sussal, Carol M. 1994. "Empowering Gays and Lesbians in the Workplace." *Journal of Gay and Lesbian Social Services* 1: 89–103.

Sussman, N. M., and D. H. Tyson. 2000. "Sex and Power: Gender Differences in Computer-Mediated Interactions." *Computers in Human Behavior* 16 (July): 381–394.

Sutherland, Edwin H. 1940. "White Collar Criminality." *American Sociological Review* 5 (February): 1–12.

Sutherland, Edwin H., and Donald R. Cressey. 1978. *Criminology,* 10th ed. New York: Lippincott.

Swidler, Ann. 1986. "Culture in Action: Symbols and Strategies." *American Sociological Review* 51: 273–286.

Switzer, J.Y. 1990. "The Impact of Generic Word Choices: An Empirical Investigation of Age- and Sex-Related Differences." *Sex Roles* 22: 69–82.

Szasz, Thomas S. 1974. *The Myth of Mental Illness.* New York: Harper & Row.

Takaki, Ronald. 1989. *Strangers from a Different Shore: A History of Asian Americans.* New York: Penguin.

Tannen, Deborah. 1990. *You Just Don't Understand: Women and Men in Conversation.* New York: William Morrow.

Taylor, Howard F. 1973. "Linear Models of Consistency: Some Extensions of Blalock's Strategy." *American Journal of Sociology* 78 (March): 1192–1215.

Taylor, Howard F. 1980. *The IQ Game: A Methodological Inquiry into the Heredity–Environment Controversy.* New Brunswick, NJ: Rutgers University Press.

Taylor, Howard F. 1992a. "Intelligence." Pp. 941–949 in *Encyclopedia of Sociology,* edited by E. F. Borgatta and M. L. Borgatta. New York: Macmillan.

Taylor, Howard F. 1992b. "The Structure of a National Black Leadership Network: Preliminary Findings." Unpublished manuscript, Princeton University.

Taylor, Howard F. 1995. "Symposium on 'The Bell Curve'." *Contemporary Sociology* 24 (March): 153–158.

Taylor, Howard F. 2002. "Deconstructing the Bell Curve: Racism, Classism, and Intelligence Measurement in America. " In *2001 Race Odyssey.": African Americans and Sociology,* edited by Bruce R. Hare. Syracuse, NY: Syracuse University Press.

Taylor, Howard F. 2006. "Defining Race." *Race and Ethnicity in U.S. Society: The Changing Landscape,* edited by Elizabeth Higginbotham and Margaret L. Andersen. Belmont, CA: Wadsworth.

Taylor, Howard F. 2007. *Race and Class and the Bell Curve in America.* Manuscript, Princeton University, Princeton, NJ.

Taylor, Shelley E., Letitia Anne Peplau, and David O. Sears. 2006. *Social Psychology,* 12th Ed. Upper Saddle River, NJ: Prentice Hall.

Teaster, Pamela B. 2000. "A Response to the Abuse of Vulnerable Adults: The 200 Survey of State Adult Protective Services." Washington, DC: National Center on Elder Abuse, www.elderabusecenter.org

Telles, Edward E. 2004. *Race in Another America: The Significance of Skin Color in Brazil.* Princeton, NJ: Princeton University Press.

Thakkar, Reena R., Peter M. Gutierrez, Carly L. Kuczen, and Thomas R. McCanne. 2000. "History of Physical and/or Sexual Abuse and Current Suicidality in *College Women. Child Abuse and Neglect* 24 (October): 1345–1354.

The New York Times. 2005. *Class Matters.* New York: Times Books.

Thoits, Peggy A. 1991. "On Merging Identity Theory and Stress Research." *Social Psychology Quarterly* 54: 101–112.

Thomas, William I. 1931. *The Unadjusted Girl.* Boston, MA: Little, Brown.

Thomas, William I. 1966 [1931]. "The Relation of Research to the Social Process." Pp. 289–305 in *W. I. Thomas on Social Organization and Social Personality,* edited by Morris Janowitz. Chicago, IL: University of Chicago Press.

Thomas, William I. with Dorothy Swaine Thomas. 1928. *The Child in America.* New York: Knopf.

Thompson, Becky. 1994. *A Hunger So Wide and So Deep: American Women Speak Out on Eating Problems.* Minneapolis, MN: University of Minnesota Press.

Thorne, Barrie. 1993. *Gender Play: Girls and Boys in School.* New Brunswick, NJ: Rutgers University Press.

Thorne, Barrie, and Zella Luria. 1986. "Sexuality and Gender in Children's Daily Worlds." *Social Problems* 33 (February): 176–190.

Thorne, Barrie, with Marilyn Yalom. 1992. *Rethinking the Family: Some Feminist Questions.* Boston, MA: Northeastern University Press.

Thornton, Russell. 1987. *American Indian Holocaust and Survival: A Population History.* Norman, OK: University of Oklahoma Press.

Thornton, Russell. 2001. "Trends among American Indians in the United States." Pp. 135–169 in *America Becoming: Racial Trends and Their Consequences,* vol. I, edited by Neil J. Smelser, William Julius Wilson, and Faith Mitchell. Washington, DC: National Academy Press.

Tichenor, Veronica. 2005. "Maintaining Men's Dominance: Negotiating Identity and Power When She Earns More." *Sex Roles: A Journal of Research* 53(3–4): 191–205.

Tienda, Marta, and Haya Stier. 1996. "Generating Labor Market Inequality: Employment Opportunities and the Accumulation of Disadvantage." *Social Problems* 43 (May): 147–165.

Tierney, Patrick. 2000. *Darkness in El Dorado: How Scientists and Journalists Devastated the Amazon.* New York: W.W. Norton.

Tilly, Charles. 1986. *The Contentious French.* Cambridge, MA: Harvard University Press.

Tilly, Louise, and Joan Scott. 1978. *Women, Work, and Family.* New York: Holt, Rinehart, and Winston.

Tjaden, Patricia, and Nancy Thoennes. 2000. *Extent, Nature, and Consequences of Intimate Partner Violence.* Washington, DC: National Institute of Justice and the Centers for Disease Control and Prevention.

Tönnies, Ferdinand. 1963 [1887]. *Community and Society (Gemeinschaft and Gesellschaft).* New York: Harper & Row.

Toobin, Jeffrey. 1996. "The Marcia Clark Verdict." *The New Yorker* (September 9): 58–71.

Toynbee, Arnold J., and Jane Caplan. 1972. *A Study of History.* New York: Oxford University Press.

Travers, Jeffrey, and Stanley Milgram. 1969. "An Experimental Study of the Small World Problem." *Sociometry* 32: 425–443.

Treiman, Donald J. 2001. "Occupations, Stratification and Mobility." Pp. 297–313 in *The Blackwell Companion to Sociology,* edited by Judith R. Blau. Malden, MA: Blackwell.

Trent, Katherine, and Sharon L. Harlan. 1990. "Household Structure Among Teenage Mothers in the United States." *Social Science Quarterly* 71 (September): 439–457.

Tuan, Yi-Fu. 1984. *Dominance and Affection: The Making of Pets.* New Haven, CT: Yale University Press.

Tuchman, Gaye. 1979. "Women's Depiction by the Mass Media." *Signs* 4 (Spring): 528–542.

Tumin, Melvin M. 1953. "Some Principles of Stratification." *American Sociological Review* 18 (August): 387–393.

Turkle, Sherry. 1995. *Life on the Screen: Identity in the Age of the Internet.* New York: Simon & Schuster.

Turner, Jonathan. 1974. *The Structure of Sociological Theory.* Homewood, IL: Dorsey Press.

Turner, Ralph, and Lewis Killian. 1988. *Collective Behavior,* 3rd ed. Englewood Cliffs, NJ: Prentice Hall.

Turner, Ralph, Joanne M. Nigg, and Denise Paz. 1986. *Waiting for Disaster: Earthquake Watch in California.* Berkeley, CA: University of California Press.

Turner, Terence. 1969. "Tchikrin: A Central Brazilian Tribe and Its Symbolic Language of Body Adornment." *Natural History* 78 (October): 50–59.

Tyree, Andrea, and R. Hicks. 1988. "Sex and the Second Moment of Occupational Basic Distributions." *Social Forces* 66 (June): 1028–1037.

U.S. Bureau of Justice Statistics. 2005. *Criminal Victimization in the United States 2005, Statistical Tables.* Washington, DC: U.S. Bureau of Justice Statistics.

U.S. Bureau of Labor Statistics 2006. *Percent of All Workers with Access to Employer Asisstance for Child Care.* www.bls.gov

U.S. Bureau of Labor Statistics. 2006a. *Employment Characteristics of Families in 2005,* USDL 06-731. Washington, DC: U.S. Department of Labor.

U.S. Bureau of Labor Statistics. 2003. *Pilot Survey on the Incidence of Child Care Resource and Referral Services in June 2000.* Washington, DC: U.S. Department of Labor.

U.S. Census Bureau. 2004. *Detailed Income Tables, Table P2.* Washington, DC: U.S. Department of Commerce, www.census.gov

U.S. Census Bureau. 2004. *American Fact Finder* and *Fedstats.* www.census.gov and www.fedstats.gov

U.S. Census Bureau. 2006. *Statistical Abstract of the United States 2005.* Washington, DC: U.S. Department of Commerce.

U.S. Census Bureau. 2007. *Statistical Abstract of the United States: 2006.* Washington, DC: U.S. Department of Commerce.

U.S. Conference of Mayors. 2005. *A Status Report on Hunger and Homelessness in America's Cities.* www.usmayors.org

U.S. Department of Defense. 2003, www.dod.gov

U.S. Department of Labor. 2006. *Employment and Earnings.* Washington, DC: U.S. Department of Labor.

Uleman, J. S., L. S. Newman, and G. B. Moskowitz. 1996. "People as Flexible Interpreters: Evidence and Issues From Spontaneous Trait Inference." Pp. 211–279 in *Advances in Experimental Social Psychology,* vol. 28, edited by Mark Zanna. Boston, MA: Academic Press.

Ullman, Sarah E., George Karabatsos, and Mary P. Koss. 1999. "Alcohol and Sexual Assault in a National Sample of College Women." *Journal of Interpersonal Violence* 14 (June): 603–625.

UNICEF. 2000. *Domestic Violence against Women and Girls.* New York: United Nations.

United Nations. 2000. *Human Development Report.* New York: United Nations.

United Nations. 2005a. *The World's Women: Progress in Statistics.* New York: United Nations.

United Nations. 2005b. *Human Development Report.* New York: United Nations.

United Nations. 2006a. *In Depth Study on All Forms of Violence against Women.* New York: United Nations.

United Nations. 2006b (August). United Nations Statistics Division: *Statistics and Indicators on Women and Men.* www.unstats.un.org/demographic/products/default.htm

United Nations Development Fund for Women. 2005. *Progress of the World's Women 2005: Women, Work, and Poverty.* New York: United Nations.

United Nations News Center. 2007. "One Third of All Iraqis Live in Poverty, Study Says." February 13. New York: United Nations.

Urban Institute, The. 2000. *A New Look at Homelessness in America.* Washington, DC: The Urban Institute, www.urban.org

Uvin, Peter. 1998. *Aiding Violence.* West Hartford, CT: Kumarian Press.

Vail, D. Angus. 1999. "Tattoos Are Like Potato Chips . . .You Can't Have Just One: The Process of Becoming and Being a Collector." *Deviant Behavior* 20: 253–273.

Van Ausdale, Debra, and Joe R. Feagin. 1996. "The Use of Racial and Ethnic Concepts by Very Young Children." *American Sociological Review* 61 (October): 779–793.

Van Ausdale, Deborah, and Joe R. Feagin. 2000. *The First R: How Children Learn Race and Racism.* Lanham, MD: Rowan and Littlefield.

Vanlandingham, Mark. 1993. *Two Perspectives on Risky Sexual Practices among Northern Thai Males: The Health Belief Model and the Theory of Reasoned Action.* Unpublished doctoral dissertation, Princeton University.

Vanneman, Reeve, and Lynn Weber Cannon. 1987. *The American Perception of Class.* Philadelphia: Temple University Press.

VanRoosmalen, Erica, and Susan A. McDaniel. 1998. "Sexual Harassment in Academia: A Hazard to Women's Health." *Women and Health* 28: 33–54.

Varelas, Nicole, and Linda A. Foley. 1998. "Blacks' and Whites' Perceptions of Interracial and Intraracial Date Rape." *Journal of Social Psychology* 138 (June): 392–400.

Vaughan, Diane. 1996. *The Challenger Launch Decision: Risky Technology, Culture, and Deviance at NASA.* Chicago, IL: The University of Chicago Press.

Veblen, Thorstein. 1899 [1953]. *The Theory of the Leisure Class: An Economic Study of Institutions.* New York: The New American Library.

Voss, Laurie Scarborough. 1997. "Teasing, Disputing, and Playing: Cross-Gender Interactions and Space Utilization among First and Third Graders." *Gender & Society* 11 (April): 238–256.

Waite, Linda J. 2000. "Trends in Men's and Women's Well-Being in Marriage." Pp. 368–392 in *The Ties that Bind: Trends in Men's and Women's Well-Being in Marriage.* New York: Aldine de Gruyter.

Waldfogel, Jane, Wen-Jui Han, and Jeanne Brooks-Gunn. 2002. "The Effects of Early Maternal Employment in Child Cognitive Development." *Demography* 39: 369–392.

Walker-Barnes, Chanequa J., and Craig A. Mason. 2001. "Perceptions of Risk Factors for Female Gang Involvement among African American and Hispanic Women." *Youth and Society* 32 (March): 303–336.

Wallerstein, Immanuel M. 1974. *The Modern World System: Capitalist Agriculture and the Origins of the European World Economy in the Sixteenth Century.* New York: Academic Press.

Wallerstein, Immanuel M. 1979. *The Capitalist World-Economy.* New York: Cambridge University Press.

Wallerstein, Immanuel M. 1980. *The Modern World-System II.* New York: Academic Press.

Wallerstein, Immanuel M. 1989. *The Modern World System III: The Second Era of Great Expansion of the Capitalist World-Economy, 1730–1840.* New York: Academic Press.

Walsh, Anthony. 1994. "Homosexual and Heterosexual Child Molestation: Case Characteristics and Sentencing Differentials." *International Journal of Offender Therapy and Comparative Criminology* 38: 339–353.

Walters, Suzanna Danuta. 1999. "Sex, Text, and Context: (In) Between Feminism and Cultural Studies." Pp. 222–260 in *Revisioning Gender,* edited by Myra Marx Ferree, Judith Lorber, and Beth B. Hess. Thousand Oaks, CA: Sage.

Walters, Vivienne. 1993. "Stress, Anxiety, and Depression: Women's Accounts of Their Health Problems." *Social Science and Medicine* 36 (February): 393–402.

Warner, Rebecca L., and Brent S. Steel. 1999. "Child Rearing as a Mechanism for Social Change: The Relationship of Child Gender to Parents' Commitment to Gender Equity." *Gender & Society* 13 (August): 503–517.

Washington, Scott. 2006. "Racial Taxonomy." Unpublished manuscript, Princeton University, Princeton, NJ.

Wasserman, Stanley, and Katherine Faust (eds.). 1994. *Social Network Analysis: Methods and Applications.* Cambridge, MA: Cambridge University Press.

Waters, Mary C. 1990. *Ethnic Options: Choosing Identities in America.* Berkeley, CA: University of California Press.

Waters, Mary C., and Peggy Levitt, eds. 2002. *The Changing Face of Home: The Transnational Lives of the Second Generation.* New York: Russell Sage Foundation Press.

Watkins, Susan. 1987. "The Fertility Transition: Europe and the Third World Compared." *Sociological Forum* 2 (Fall): 645–673.

Watt, Toni Terling, and Susan F. Sharp. 2001. "Gender Differences in Strains Associated with Suicidal Behavior among Adolescents." *Journal of Youth and Adolescence* 30 (June): 333–348.

Watts, Duncan J. 1999. "Networks, Dynamics, and the Small-World Phenomenon." *American Journal of Sociology* 105 (September): 493–527.

Watts, Duncan J. and Stephen H. Strogatz. 1998. "Collective Dynamics of 'Small World' Networks." *Nature* 393: 440–442, www.sentencing project.org

Weber, Max. 1958 [1904]. *The Protestant Ethic and the Spirit of Capitalism.* New York: Scribner's.

Weber, Max. 1962 [1913]. *Basic Concepts in Sociology.* New York: Greenwood.

Weber, Max. 1978 [1921]. *Economy and Society: An Outline of Inter-*

pretive Sociology, Guenther Roth and Claus Wittich (eds.). Berkeley, CA: University of California Press.

Webster, Murray, Jr., and Stuart J. Hysom. 1998. "Creating Status Characteristics." *American Sociological Review* 63 (June): 351–378.

Weeks, John R. 2004. *Population: An Introduction to Concepts and Issues,* 9th ed. Belmont, CA: Wadsworth.

Weinrath, Michael, and John Gartrell. 1996. "Victimization and Fear of Crime." *Violence and Victims* 11 (Fall): 187–197.

Weisburd, David, Cynthia M. Lum, and Anthony Petrosino. 2001. "Does Research Design Affect Study Outcomes in Criminal Justice?" *The Annals of the American Academy of Political and Social Science* 578 (November): 50–70.

Weisburd, David, Stanton Wheeler, Elin Waring, and Nancy Bode. 1991. *Crimes of the Middle Class: White Collar Defenders in the Courts.* New Haven, CT: Yale University Press.

Weitz, Rose. 2001. *The Sociology of Health, Illness, and Health Care: A Critical Approach.* Belmont, CA: Wadsworth.

Weitz, Rose. 2004. *The Sociology of Health, Illness, and Health Care: A Critical Approach,* 3rd ed. Belmont, CA: Wadsworth.

Weitzman, Lenore J. 1985. *The Divorce Revolution: The Unexpected Consequences for Women and Children in America.* New York: Free Press.

Weitzman, Lenore J. 1996. "A Re-Evaluation of the Economic Consequences of Divorce: Reply to Peterson." *American Sociological Review* 61 (June): 537–538.

Welch, Susan, and Lee Sigelman. 1993. "The Politics of Hispanic Americans: Insights from the National Surveys, 1980–1988." *Social Science Quarterly* 74 (March) 76–94.

Wenk, Deeann, and Patricia Garrett. 1992. "Having a Baby: Some Predictions of Maternal Employment Around Childbirth." *Gender & Society* 6 (March): 49–65.

West, Candace, and Sarah Fenstermaker. 1995. "Doing Difference." *Gender & Society* 9 (February): 8–37.

West, Candace, and Don Zimmerman. 1987. "Doing Gender." *Gender & Society* 1 (June): 125–151.

West, Carolyn M. 1998. "Leaving a Second Closet: Outing Partner Violence in Same-Sex Couples." Pp. 163–183 in *Partner Violence: A Comprehensive Review of 20 Years of Research,* edited by Jana L. Jasinksi and Linda M. Williams. Thousand Oaks, CA: Sage.

Western, Bruce. 2006. *Punishment and Inequality in America.* New York: Russell Sage Foundation.

Wharton, Amy S., and Deborah K. Thorne. 1997. "When Mothers Matter: The Effects of Social Class and Family Arrangements on African American and White Women's Perceived Relations with Their Mothers." *Gender & Society* 11 (October): 656–681.

White, Deborah Gray. 1985. *Ar'n't I a Woman?: Female Slaves in the Plantation South.* New York: W. W. Norton.

White, Jonathan R. 2002. *Terrorism: An Introduction.* Belmont, CA: Wadsworth.

White, Paul. 1998. "The Settlement Patterns of Developed World Migrants in London." *Urban Studies* 35 (October): 1725–1744.

Whiten, A., J. Goodall, W. C. McGrew, T. Nishida, V. Reynolds, Y. Sugiyama, C. E. G. Tutin, R. W. Wrangham, and C. Boesch. 1999. "Cultures in Chimpanzees." *Nature* 399 (June 17): 682–684.

Whorf, Benjamin. 1956. *Language, Thought, and Reality: Selected Writings.* Cambridge, MA: Technology Press, MIT.

Whyte, William F. 1943. *Street Corner Society.* Chicago, IL: University of Chicago Press.

Wickelgren, Ingrid. 1999. "Discovery of 'Gay Gene' Questioned." *Science* 284 (April 23): 571.

Wilcox, Clyde, Lara Hewitt, and Dee Allsop. 1996. "The Gender Gap in Attitudes Toward the Gulf War: A Cross-National Perspective." *Journal of Peace Research* 33 (February): 67–82.

Wilder, Esther I., and Toni Terling Watt. 2002. "Risky Parental Behavior and Adolescent Sexual Activity at First Coitus," *The Milbank Quarterly* 80 (September): 481–524.

Wilkerson, Isabel. 2005. "In a Strange Land: From New Orleans to Sallisaw; Scattered in a Storm's Wake and Caught in a Clash of Cultures." *The New York Times,* October 9.

Wilkie, Jane. 1993. "Changes in U.S. Men's Attitudes Toward the Family Provider Role, 1972 to 1989." *Gender & Society* 7 (June): 261–279.

Williams, Christine L. 1989. *Gender Differences at Work: Women and Men in Nontraditional Occupations.* Berkeley, CA: University of California Press.

Williams, Christine L. 1992. "The Glass Escalator: Hidden Advantages for Men in the 'Female' Professions." *Social Problems* 39 (August): 253–267.

Williams, Christine L. 1995. *Still a Man's World: Men Who Do Women's Work.* Berkeley, CA: University of California Press.

Williams, David R., and Chiquita Collins, 1995. "U.S. Socioeconomic and Racial Differences in Health: Patterns and Explanations." *Annual Review of Sociology* 21: 349–386.

Willie, Charles Vert. 1979. *The Caste and Class Controversy.* Bayside, NY: General Hall.

Wilson, Barbara J., Stacy L. Smith, W. James Potter, Dale Kunkel, Daniel Linz, Carolyn M. Colvin, and Edward Donnerstein. 2002. "Violence in Children's Television Programming: Assessing the Risks." *Journal of Communication* 52 (March): 5–35.

Wilson, James Q., and Richard J. Herrnstein. 1985. *Crime and Human Nature.* New York: Simon and Schuster.

Wilson, John F. 1978. *Religion in American Society: The Effective Presence.* Englewood Cliffs, NJ: Prentice Hall.

Wilson, John. 1994. "Returning to the Fold." *Journal for the Scientific Study of Religion* 33 (June): 148–161.

Wilson, William Julius. 1978. *The Declining Significance of Race: Blacks and Changing American Institutions.* Chicago, IL: University of Chicago Press.

Wilson, William Julius. 1987. *The Truly Disadvantaged: The Inner City, the Underclass and Public Policy.* Chicago, IL: University of Chicago Press.

Wilson, William Julius. 1996. *When Work Disappears: The World of the New Urban Poor.* New York: Knopf.

Winnick, Louis. 1990. "America's 'Model Minority.'" *Commentary* 90 (August): 222–229.

Winseman, Albert L. 2004. "'Born Agains' Wield Political, Economic Influence." *The Gallup Poll.* Princeton, NJ: Gallup Organization, www.gallup.com

Wirth, Louis. 1938. "Urbanism as a Way of Life." *American Journal of Sociology* 40: 1–24.

Wolcott, Harry F. 1996. "Peripheral Participation and the Kwakiutl Potlatch." *Anthropology and Education Quarterly* 27 (December): 467–492.

Wolf, Naomi. 1991. *The Beauty Myth: How Images of Beauty Are Used Against Women.* New York: William Morrow.

Woo, Deborah. 1998. "The Gap Between Striving and Achieving." Pp. 247–256 in *Race, Class, and Gender: An Anthology,* 3rd ed., edited by Margaret L. Andersen and Patricia Hill Collins. Belmont, CA: Wadsworth.

Wood, Julia T. 1994. *Gendered Lives: Communication, Gender and Culture.* Belmont, CA: Wadsworth.

Wood, Richard L. 2002. *Faith in Action: Religion, Race, and Democratic Organizing in America.* Chicago, IL: University of Chicago Press.

Woods, T., B. Kurts-Costes, and S. Rowley. 2005. "The Development of Stereotypes About the Rich and Poor: Age, Race, and Family Income Differences in Beliefs." *Journal of Youth and Adolescence* 34 (5): 437–445.

Worchel, Stephen, Joel Cooper, George R. Goethals, and James L. Olsen. 2000. *Social Psychology.* Belmont, CA: Wadsworth.

Workman, Jane E., and Elizabeth W. Freeburg. 1999. "An Examination of Date Rape, Victim Dress, and Perceiver Variables within the Context of Attribution Theory." *Sex Roles* 41 (August): 261–277.

World Bank. 2004. *Water Resources Sector Strategy: Strategic Directions for World Bank Engagement.* www.worldbank.org

World Bank. 2007. "GNI per Capita 2005." *Quick References Tables,* www.worldbank.org

World Health Organization. 2000. *The State of Food Insecurity in the World: 2000.* Geneva: World Health Organization. www.who .int

World Health Organization. 2002, www.who.org

Wright, Erik Olin. 1979. *Class Structure and Income Determination.* New York: Academic Press.

Wright, Erik Olin. 1985. *Classes.* London: Verso.

Wright, Erik Olin, Karen Shire, Shu-Ling Hwang, Maureen Dolan, and Janeen Baxter. 1992. "The Non-Effects of Class on the Gender Division of Labor in the Home: A Comparative Study of Sweden and the United States." *Gender & Society* 6 (June): 252–282.

Wright, Laurence. 1994. "One Drop of Blood." *The New Yorker* 70 (July 25): 46–55.

Wrong, Dennis. 1961. "The Oversocialized Conception of Man in Modern Sociology." *American Sociological Review* 26 (April): 183–192.

Wuthnow, Robert. 1988. *After Heaven: Spirituality in America Since the 1950s.* Berkeley, CA: University of California Press.

Wuthnow, Robert (ed.). 1994. *I Come Away Stronger: How Small Groups Are Shaping American Religion.* Grand Rapids, MI: Eerdmans.

Wysocki, Diane Kholos. 1998. "Let Your Fingers Do the Talking: Sex on an Adult Chat-Line." *Sexualities* 1 (November): 425–452.

Xu, Wu, and Ann Leffler. 1992. "Gender and Race Effects on Occupational Prestige, Segregation, and Earnings." *Gender & Society* 6 (September): 376–392.

Xuewen, Sheng, Norman Stockman, and Norman Bonney. 1992. "The Dual Burden: East and West (Women's Working Lives in China, Japan, and Great Britain)." *International Sociology* 7 (June): 209–223.

Yoder, Janice. 1991. "Rethinking Tokenism: Looking Beyond Numbers." *Gender & Society* 5 (June): 178–192.

Young, J. W. 1994. "Differential Prediction of College Grades by Gender and by Ethnicity: A Replication Study." *Educational and Psychological Measurement* 54: 1022–1029.

Zajonc, Robert B. 1968. "Attitudinal Effects of Mere Exposure." *Journal of Personality and Social Psychology.* Monograph Supplement, Part 2: 1–29.

Zald, Mayer, and John McCarthy. 1975. "Organizational Intellectuals and the Criticism of Society." *Social Service Research* 49: 344–362.

Zhou, Min, and Carl L. Bankston. 2000. "Immigrant and Native Minority Groups, School Performance, and the Problem of Self-Esteem." Paper presented at the Southern Sociological Society, April, New Orleans, LA.

Zimbardo, Phillip G., Ebbe B. Ebbesen, and Christina Maslach. 1977. *Influencing Attitudes and Changing Behavior.* Reading, MA: Addison-Wesley.

Zola, Irving Kenneth. 1989. "Toward the Necessary Universalizing of a Disability Policy." *Milbank Quarterly* 67 (Suppl. 2): 401–428.

Zola, Irving Kenneth. 1993. "Self, Identity and the Naming Question: Reflections on the Language of Disability." *Social Science and Medicine* 36 (January): 167–173.

Zones, Jane Sprague. 1997. "Beauty Myths and Realities and Their Impact on Women's Health." Pp. 249–275 in *Women's Health: Complexities and Differences,* edited by Sheryl B. Ruzek, Virginia L. Olesen, and Adele E. Clarke. Columbus, OH: Ohio State University Press.

Zorza, J. 1991. "Woman Battering: A New Cause of Homelessness," *Clearinghouse Review* 25 (4).

Zuberi, Tukufu. 2001. *Thicker Than Blood: How Racial Statistics Lie.* Minneapolis, MN: University of Minneapolis Press.

Zuckerman, Phil. 2002. "The Sociology of Religion of W. E. B. Du Bois." *Sociology of Religion* 63: 239–253.

Zweigenhaft, Richard L., and G. William Domhoff. 2006. *Diversity in the Power Elite: How It Happened, Why It Matters.* Lanham, MD: Rowman and Littlefield.

Zwerling, Craig, and Hilary Silver. 1992. "Race and Job Dismissals in a Federal Bureaucracy." *American Sociological Review* 57 (October): 651–660.

Zwick, Rebecca, ed. 2004. *Rethinking the SAT: The Future of Standardized Testing in University Admissions.* New York: RoutledgeFalmer.

Photo Credits

Chapter 1

p.1 ©Gideon Mendel/CORBIS; p.2 ©Lara Jo Regan/Getty Images/Liaison; p.3 ©Spencer Platt/Getty Images; p.6L ©William Thomas Cain/Getty Images; p.6R ©Lindsay Hebberd/CORBIS; p.9 ©Jeff Greenberg /The Image Works; p.10 AP Images/Eugene Hoshiko; p.12T ©Culver Pictures; p.12B ©Bettmann/CORBIS; p.13L ©Bettmann/CORBIS; p.13R ©AKG/London; p.14 ©CORBIS; p.15 Special Collections and Archives, W. E. B. Du Bois Library, University of Massachusetts at Amherst; p.18 ©Nina Berman/Redux; p.19 ©Sonda Dawes/The Image Works; p.20 ©Bob Daemmrich/PhotoEdit; p.27 ©Dwayne Newton/PhotoEdit.

Chapter 2

p.33 ©Peter Turnley/CORBIS; p.34 ©Joan Bamburger/Anthro Photos; p.35 ©David R. Frazier Photolibrary, Inc./Alamy; p.36TL ©David Reed/CORBIS; p.36ML ©Lindsay Hebberd/CORBIS; p.36TR ©Paul Chesley/Stone/Getty Images; p.36MR © Jack Kurtz/The Image Works; p.36B ©Chris Ryan/Getty Images/OJO Images; p.37 ©Tami Chappell/Reuters/Landov; p.43 AP Images/Paul Sakuma; p.46L ©Robert Sullivan/AFP/Getty Images; p.46R AP Images/Chris O'Meara; p.49 AP Images/Amy Sancetta; p.50 ©Dennis MacDonald/PhotoEdit; p.51 ©Frank Micelotta/American Idol/Getty Images for Fox; p.59L The Granger Collection, New York; p.59R ©Michael Newman/PhotoEdit.

Chapter 3

p.63 ©Jim Erickson/CORBIS; p.64 ©Keith Meyers/The New York Times/Redux; p.68 ©Yellow Dog Productions/Getty Images/The Image Bank; p.69 Virginia Sherwood/© NBC/Courtesy: Everett Collection; p.70T AP Images; p.70B ©Amy Etra/PhotoEdit; p.74 ©Paul Chesley/Getty Images/Stone; p.78 ©Kate Powers/Getty Images/Taxi; p.79 ©Caroline Penn/CORBIS; p.83L AP Images/Rick Bowmer; p.83R ©CHRIS WILKINS/AFP/Getty Images; p.86 ©piluhin/Alamy; p.87TL ©Juliet Highet/The Hutchison Library; p.87TR ©Judy Griesedieck/CORBIS; p.87BL ©E. A. Heiniger/Photo Researchers, Inc.; p.87BR ©Sami Sallinen/Panos Pictures; p.88 ©CNN/Getty Images.

Chapter 4

p.93 © Darrin Klimek/Getty Images/Digital Vision; p.94 ©Bob Daemmrich/The Image Works; p.95L ©SHOUT/Alamy; p.95R ©MARK EDWARDS/Peter Arnold Inc.; p.98L ©Jeff Greenberg/PhotoEdit; p.98R AP Images/Sergei Grits; p.102 ©Jeff Greenberg /PhotoEdit; p.104 ©Lon C. Diehl/PhotoEdit; p.105 ©Denis Boissavy/Getty Images/Taxi; p.106 ©Michael Newman/PhotoEdit ; p.107 ©Nina Leen/Time Life Pictures/Getty Images; p.108 ©Alex Mares-Manton/Asia Images/Jupiter Images; p.111 AP Images/The Fort Collins Coloradoan, Rich Abrahamson; p.114T ©Bob Daemmrich/The Image Works ; p.114M AP Images/Matt Sayles; p.114B AP Images/Joanne Carole; p.115T AP Images/Kevork Djansezian; p.115M ©Jeff Greenberg/PhotoEdit.; p.115B ©Syracuse Newspapers/Michelle Gabel/The Image Works; p.117 ©Colin Young-Wolff/PhotoEdit.

Chapter 5

p.121 ©Scott Olson/Getty Images/Reportage; p.122 ©John Neubauer/PhotoEdit; p.123 ©Momatiuk Eastcott/The Image Works; p.125 ©DoD, Staff Sgt. Dawn Hinniss, U.S. Air Force; p.126 ©David R. Frazier/The Image Works; p.127 AP Images/Bebeto Matthews; p.130 From the film Obedience © 1965 by Stanley Milgram, © renewed 1993 by Alexandra Milgram, and distributed by Penn State Media Sales.; p.133 ©Sutcliffe SIPA Press; p.137L ©Michael Soforonski/SIPA Press; p.137TR AP Images/NASA; p.137BR AP Images/Tyler Morning Telegraph, Dr. Scott Lieberman; p.138 AP Images/Tony Dejak; p.139 ©Bananastock/Jupiter Images.

Chapter 6

p.145 ©Mario Tama/Getty Images; p.147 RAWA/WorldPictures News; p.148 ©Jeff Greenberg/PhotoEdit; p.151 ©Tery Eiler/Stock Boston, LLC; p.153 AP Images/Carolyn Kaster; p.155 ©David Young-Wolff/PhotoEdit; p.159 ©Mark Peterson/CORBIS; p.160 ©Keith Erskine /Alamy; p.165 Abbot Genser /©HBO/courtesy Everett Collection; p.167L AP Images/David J. Phillip; p.167R ©Daniel Acker/Bloomberg News/Landov; p.173 AP Images.

Chapter 7

p.178 ©Mitchell Funk/Getty Images/Photographer's Choice; p.179L ©Tony Freeman/PhotoEdit; p.179R ©Image Source Pink/Getty Images; p.180L ©Jeff Greenberg/Peter Arnold, Inc.; p.180M ©David R. Frazier/PhotoEdit; p.180R ©Stock Image/Alamy; p.181 ©Peter Kramer/Getty Images for CFDA; p.182TL ©Jeff Greenburg/PhotoEdit; p.182TR AP Images/Don Ryan; p.182BL AP Images/Richard Vogel; p.182BR ©Bob Collins/The Image Works; p.183TL ©Mark Wilson/Getty Images; p.183TR AP Images/Dave Martin; p.183 BL AP Images/Cheryl Gerber; p.183BR AP Images/Alex Brandon; p.187L ©Tony Freeman/PhotoEdit; p.187R ©Anne Dowie; p.188 ©Jim West/The Image Works; p.189 AP Images/Steve Helber/POOL; p.196 ©John Burke/Index Stock; p.200 AP Images/Bill Haber.

Chapter 8

p.209 ©Paula Bronstein/Getty Images; p.211L ©EPA/Wilson Wen/Landov; p.211R ©Peter Cade/Getty Images/Stone; p.212 ©DPA/DJD/The Image Works; p.213 ©Reuters/

CORBIS; p.217L © R. Azzi/Woodfin Camp & Associates; p.217R © JORGEN SCHYTTE/Peter Arnold, Inc.; p.218 ©Oliver Knight/Alamy; p.219 ©David Bacon/The Image Works; p.221 ©Rachel Epstein/ PhotoEdit; p.223 © Reuters/Damir Sagolj/Landov; p.227 AP Images/ Bassim Daham/FILE.

Chapter 9

p.233 ©Frans Lemmens/Getty Images/Photographer's Choice; p.234 ©Cameraman/The Image Works; p.235 ©David Maxwell/EPA/Landov; p.242 ©Tony Savino/Sygma/ CORBIS; p.243 ©Reuters/Jason Reed/Landov; p.244 ©Brown Brothers; p.247 ©Hazel Hankin; p.250 ©Bob Daemmrich/PhotoEdit; p.252 ©Bernard Boutrit/Woodfin Camp and Associates; p.254 AP Images/ Ben Sklar.

Chapter 10

p.261 ©David De Lossy/Getty Images/PhotoDisc; p.263 ©Serena Nanda; p.266 B © Markus Moellenberg/zefa/CORBIS; p.266TL ©Mike Watson Images/SuperStock; p.266TR ©Stockbyte/Getty Images; p.267TL AP Images/Stuart Ramson; p.267TR ©Karl Prouse/Catwalking/ Contributor/Getty Images; p.267BL ©ThinkStock/SuperStock; p.267BR ©Digital Vision Ltd./SuperStock; p.268T ©David Frazier/The Image Works; p.268B ©Rob Walls/Alamy; p.269L ©Tony Freeman/PhotoEdit; p.269R ©WoodyStock/Alamy; p.271 ©John Powell Photographer/Alamy; p.277T AP Images/Lincoln Journal Star, Robert Becker; p.277B ©David Bacon/The Image Works; p.278 ©Richard Hutchings/PhotoEdit; p.279 ©Bob Daemmrich/The Image Works; p.284 ©Rick Steele/UPI/ Landov; p.285 AP Images/Fred Beckham.

Chapter 11

p.289 ©Jonathan Nourok/Getty Images/Stone; p.290L ©Bill Aron/ PhotoEdit; p.290R ©John Burke/ Superstock; p.296 AP Images/Peter Lennihan; p.297 AP Images/Jennifer Graylock; p.299 ©Sam Yeh/ AFP Photo/Getty Images; p.301 ©Terry Schmitt/UP/Landov; p.305T ©Topham/The Image Works; p.305B © Bob Daemmrich/PhotoEdit; p.307 ©Graeme Robertson/Getty Images.

Chapter 12

p.311 ©Tracy Kahn/CORBIS; p.312L ©LWA/Dann Tardif/Getty Images/ Blend Images; p.312R ©Rachel Epstein/PhotoEdit; p.315 ©Robin Nelson/PhotoEdit; p.317 ©Ariel Skelley/Jupiter Images; p.323 AP Images/Danny Feld, Warner Bros; p.331 ©Bob Daemmrich/PhotoEdit; p.335T AP Images/Gregorio Borgia; p.335B ©Ethel Wolvovitz/The Image Works; p.338 ©JOHN GRESS/Reuters/Landov.

Chapter 13

p.343 ©Rod Morata/Getty Images/ Stone; p.344 ©Bob Daemmrich/The Image Works; p.350 ©Kaku Kurita; p.351 ©Lara Jo Regan/Getty Images; p.356L © Mystic Seaport, Rosenfeld Collection, Mystic, Connecticut; p.356R ©Rob Crandell/ The Image Works; p.357 ©Peter Hvizdak/The Image Works; p.359 ©Tracy Frankel/Getty Images/ Photographer's Choice; p.361 ©Chip Somodevilla/Getty Images; p.365 ©Mike Siluk/The Image Works; p.366 Artist Gilles Barbier's "Nursing Home" 2002.

Chapter 14

p.371 ©Andy Sacks/Getty Images/ Stone; p.374 AP Images; p.376L ©David Young Wolff/PhotoEdit; p.376R ©Bob Daemmrich/The Image Works; p.377 UN Photo/Mark Garten; p.382 AP Images; p.384 ©Bob Daemmrich/PhotoEdit; p.385 ©Mystic Seaport, Rosenfeld Collection, Mystic, Connecticut; p.387 ©AFP/Getty Images; p.388 ©Fritz Hoffmann/The Image Works; p.397 ©Jim Pickerell/Stock Boston, LLC.

Chapter 15

p.401 ©Christopher Pillitz/Getty Images/Reportage; p.404 Jacob Lawrence, The Migration of the Negro, The Philips Collection. ©2007 Gwendolyn Knight Lawrence/Artists Rights Society (ARS), New York; p.412 ©Dilip Mehta/Contact Press Images; p.415 ©Joel Sartore/National Geographic Image Collection; p.417 AP Images/EyePress; p.419 ©Jim West/PhotoEdit.

Chapter 16

p.425 ©Hisham Ibrahim /Getty Images/Photographer's Choice; p.426L ©Hulton-Deutsch Collection/ CORBIS; p.426R ©Mark Richards/ PhotoEdit; p.435 ©Terry Wild Studio Inc.; p.436 ©Explorer/Photo Researchers, Inc.; p.438 " Disappearing World" Granda TV/The Hutchison Library; p.439 ©Tricia Wachtendorf, courtesy of the Disaster Research Center, University of Delaware; p.443T AP Images/Susan Walsh; p.443B ©Thomas Frey/EPA/Landov; p.444L Mystic Seaport, Rosenfeld Collection, Mystic, Connecticut; p.444M ©Ezra Shaw/Getty Images; p.444R ©Nick Laham/Getty Images; p.445 AP Images/Eric Risberg.

Name Index

Subject Index

Page numbers in **bold** refer to definitions of vocabulary terms.
Page numbers in *italic* refer to boxed material or illustrations.

Dear Student,

I hope you enjoyed reading *Sociology: The Essentials,* Fifth Edition. With every book that I publish, my goal is to enhance your learning experience. If you have any suggestions that you feel would improve this book, I would be delighted to hear from you. All comments will be shared with the authors. My email address is Chris.Caldeira@Cengage.com, or you can mail this form (no postage required). Thank you.

School and address: ____

Department: ____

Instructor's name: ____

1. What I like most about this book is: ____

____

____

2. What I like least about this book is:

____

____

3. I would like to say to the authors of this book . . .

____

____

4. In the space below, or in an email to Chris.Caldeira@Cengage.com, please write specific suggestions for improving this book and anything else you'd care to share about your experience using this book.

____

____

____

____

____

____

DO NOT STAPLE. TAPE HERE.

DO NOT STAPLE. TAPE HERE.

FOLD HERE

NO POSTAGE NECESSARY IF MAILED IN THE UNITED STATES

BUSINESS REPLY MAIL

FIRST-CLASS MAIL PERMIT NO. 34 BELMONT CA

POSTAGE WILL BE PAID BY ADDRESSEE

Attn: Chris Caldeira, Sociology
Thomson Wadsworth
10 Davis Drive
Belmont, CA 94002-9801

FOLD HERE

OPTIONAL:

Your name: ________________________________ Date: ______________

May we quote you, either in promotion for *Sociology: The Essentials,* Fifth Edition, or in future publishing ventures?

Yes: __________ No: __________

Sincerely yours,

Margaret L. Andersen
Howard F. Taylor